Mathematical Modeling
—— and ——
Computation in Finance

Mathematical Modeling and Computation in Finance

with Exercises and Python and MATLAB computer codes

Cornelis W Oosterlee

*Centrum Wiskunde & Informatica (CWI) &
Delft University of Technology, The Netherlands*

Lech A Grzelak

Delft University of Technology, The Netherlands

NEW JERSEY · LONDON · SINGAPORE · BEIJING · SHANGHAI · HONG KONG · TAIPEI · CHENNAI · TOKYO

Published by

World Scientific Publishing Europe Ltd.
57 Shelton Street, Covent Garden, London WC2H 9HE
Head office: 5 Toh Tuck Link, Singapore 596224
USA office: 27 Warren Street, Suite 401-402, Hackensack, NJ 07601

Library of Congress Control Number: 2019950785

British Library Cataloguing-in-Publication Data
A catalogue record for this book is available from the British Library.

MATHEMATICAL MODELING AND COMPUTATION IN FINANCE
With Exercises and Python and MATLAB Computer Codes

Copyright © 2020 by Cornelis W. Oosterlee and Lech A. Grzelak

All rights reserved.

ISBN 978-1-78634-794-7
ISBN 978-1-78634-805-0 (pbk)

For any available supplementary material, please visit
https://www.worldscientific.com/worldscibooks/10.1142/Q0236#t=suppl

Desk Editor: Shreya Gopi

Typeset by Stallion Press
Email: enquiries@stallionpress.com

Dedicated to Anasja, Wim, Mathijs, Wim en Agnes (Kees)
Dedicated to my mum, brother, Anna and my whole family (Lech)

Preface

This book is discussing the interplay of stochastics (applied probability theory) and numerical analysis in the field of quantitative finance. The contents will be useful for people working in the financial industry, for those aiming to work there one day, and for anyone interested in quantitative finance.

Stochastic processes, and stochastic differential equations of increasing complexity, are discussed for the various asset classes, reaching to the models that are in use at financial institutions. Only in exceptional cases, solutions to these stochastic differential equations are available in closed form.

The typical models in use at financial institutions have changed over time. Basically, each time when the behavior of participants in financial markets changes, the corresponding stochastic mathematical models describing the prices may change as well. Also financial regulation will play its role in such changes. In the book we therefore discuss a variety of models for stock prices, interest rates as well as foreign-exchange rates. A basic notion in such a diverse and varying field is *"don't fall in love with your favorite model"*.

Financial derivatives are products that are based on the performance of another, uncertain, underlying asset, like on stock, interest rate or FX prices. Next to the modeling of these products, they also have to be priced, and the risk related to selling these products needs to be assessed. Option valuation is also encountered in the financial industry during the calibration of the stochastic models of the asset prices (fitting the model parameters of the governing SDEs so that model and market values of options match), and also in risk management when dealing with counterparty credit risk.

Advanced risk management consists nowadays of taking into account the risk that a counterparty of a financial contract may default (CCR, Counterparty Credit Risk). Because of this risk, fair values of option prices are adjusted, by means of the so-called Valuation Adjustments. We will also discuss the Credit Valuation Adjustment (CVA) in the context of risk management and derive the governing equations.

Option values are governed by partial differential equations, however, they can also be defined as expectations that need to be computed in an efficient, accurate and robust way. We are particularly interested in stochastic volatility based models, with the well-known Heston model serving as the point of reference.

As the computational methods to value these financial derivatives, we present a Fourier-based pricing technique as well as the Monte Carlo pricing method. Whereas Fourier techniques are useful when pricing basic option contracts, like European options, within the calibration procedure, Monte Carlo methods are often used when more involved option contracts, or more involved asset price dynamics are being considered.

By gradually increasing the complexity of the stochastic models in the different chapters of the book, we aim to present the mathematical tools for defining appropriate models, as well as for the efficient pricing of European options. From the equity models in the first 10 chapters, we move to short-rate and market interest rate models. We cast these models for the interest rate into the Heath-Jarrow-Morton framework, show relations between the different models, and we explain a few interest rate products and their pricing as well.

It is sometimes useful to combine SDEs from different asset classes, like stock and interest rate, into a correlated set of SDEs, or, in other words, into a system of SDEs. We discuss the hybrid asset price models with a stochastic equity model and a stochastic interest rate model.

Summarizing, the reader may encounter a variety of stochastic models, numerical valuation techniques, computational aspects, financial products and risk management applications while reading this book. The aim is to help readers progress in the challenging field of computational finance.

The topics that are discussed are relevant for MSc and PhD students, academic researchers as well as for quants in the financial industry. We expect knowledge of applied probability theory (Brownian motion, Poisson process, martingales, Girsanov theorem, ...), partial differential equations (heat equation, boundary conditions), familiarity with iterative solution methods, like the Newton-Raphson method, and a basic notion of finance, assets, prices, options.

Acknowledgment

Here, we would like to acknowledge different people for their help in bringing this book project to a successful end.

First of all, we would like to thank our employers for their support, our groups at CWI — Center for Mathematics & Computer Science, in Amsterdam and at the Delft Institute of Applied Mathematics (DIAM), from the Delft University of Technology in the Netherlands. We thank our colleagues at the CWI, at DIAM, and at Rabobank for their friendliness. Particularly, Nada Mitrovic is acknowledged for all the help. Vital for us were the many fruitful discussions, cooperations and input from our group members, like from our PhD students, the post-docs and also the several guests in our groups. In particular, we thank our dear colleagues Peter Forsyth, Luis Ortiz Gracia, Mike Staunton, Carlos Vazquez, Andrea Pascucci, Yuying Li, and Karel in't Hout for their insight and the discussions.

Proofreading with detailed pointers and suggestions for improvements has been very valuable for us and for this we would like to thank in particular Natalia Borovykh, Tim Dijkstra, Clarissa Elli, Irfan Ilgin, Marko Iskra, Fabien Le Floc'h, Patrik Karlsson, Erik van Raaij, Sacha van Weeren, Felix Wolf and Thomas van der Zwaard.

We got inspired by the group's PhD and post-doctoral students, in alphabetic order, Kristoffer Andersson, Anastasia Borovykh, Ki Wai Chau, Bin Chen, Fei Cong, Fang Fang, Qian Feng, Andrea Fontanari, Xinzheng Huang, Shashi Jain, Prashant Kumar, Coen Leentvaar, Shuaiqiang Liu, Peiyao Luo, Marta Pou, Marjon Ruijter, Beatriz Salvador Mancho, Luis Souto, Anton van der Stoep, Maria Suarez, Bowen Zhang, Jing Zhao, and Hisham bin Zubair. We would like to thank Shreya Gopi, her team and Jane Sayers at World Scientific Publishing for the great cooperation.

Our gratefulness to our families cannot be described in words.

Using this Book

The book can be used as a textbook for MSc and PhD students in applied mathematics, quantitative finance or similar studies. We use the contents of the book ourselves for two courses at the university. One is called "Computational Finance", which is an MSc course in applied mathematics, in a track called Financial Engineering, where we discuss most of the first 10 chapters, and the other course is "Special Topics in Financial Engineering", where interest rate models and products but also the risk management in the form of counterparty credit risk and credit valuation adjustment are treated. At other universities, these courses are also called "Financial Engineering" or "Quantitative Finance".

Exercises are attached to each chapter, and the software used to get the numbers in the tables and the curves in the figures is available.

Below most tables and figures in the book there are MATLAB and Python icons

indicating that the corresponding MATLAB and Python computer codes are available. In the e-book version, clicking on the icons will lead, via a hyperlink, to the corresponding codes. All codes are also available on a special webpage.

> On the webpage www.QuantFinanceBook.com the solutions to all odd-numbered exercises are available.
> Also the Python and MATLAB computer codes can be found on the webpage. Instructors can access the full set of solutions by registering at https://www.worldscientific.com/worldscibooks/10.1142/q0236

The computer codes come with a no warranty disclaimer.

No warranty disclaimer

The computer codes provided in this book are provided by the authors "as is", meant to support the theory and numerical results in this book. Any express or implied warranties are disclaimed. We do not guarantee that these codes are free of errors. They are not written according to some standards, and they are not intended for any general application.

Using the computer codes for other purposes than checking the solutions of the specific problems in the book is at the user's own risk and responsibility. The authors and publisher disclaim all liabilities for direct or consequential damages resulting from the use of the programs.

We wish you enjoyable reading!

Contents

Preface	vii
Acknowledgment	ix
Using this Book	xi

1 Basics about Stochastic Processes   **1**
   1.1 Stochastic variables 1
      1.1.1 Density function, expectation, variance 1
      1.1.2 Characteristic function 3
      1.1.3 Cumulants and moments 4
   1.2 Stochastic processes, martingale property 9
      1.2.1 Wiener process 11
      1.2.2 Martingales 12
      1.2.3 Iterated expectations (Tower property) 13
   1.3 Stochastic integration, Itô integral 14
      1.3.1 Elementary processes 14
      1.3.2 Itô isometry 17
      1.3.3 Martingale representation theorem 20
   1.4 Exercise set 25

2 Introduction to Financial Asset Dynamics   **27**
   2.1 Geometric Brownian motion asset price process 27
      2.1.1 Itô process 29
      2.1.2 Itô's lemma 30
      2.1.3 Distributions of $S(t)$ and $\log S(t)$ 34
   2.2 First generalizations 38
      2.2.1 Proportional dividend model 38
      2.2.2 Volatility variation 39

	2.2.3	Time-dependent volatility	39
2.3	Martingales and asset prices		40
	2.3.1	$\mathbb{P}$-measure prices	41
	2.3.2	$\mathbb{Q}$-measure prices	42
	2.3.3	Parameter estimation under real-world measure $\mathbb{P}$	44
2.4	Exercise set		49

3 The Black-Scholes Option Pricing Equation   51
- 3.1 Option contract definitions 51
  - 3.1.1 Option basics 52
  - 3.1.2 Derivation of the partial differential equation 56
  - 3.1.3 Martingale approach and option pricing 60
- 3.2 The Feynman-Kac theorem and the Black-Scholes model 61
  - 3.2.1 Closed-form option prices 63
  - 3.2.2 Green's functions and characteristic functions 66
  - 3.2.3 Volatility variations 71
- 3.3 Delta hedging under the Black-Scholes model 73
- 3.4 Exercise set 78

4 Local Volatility Models   81
- 4.1 Black-Scholes implied volatility 81
  - 4.1.1 The concept of implied volatility 82
  - 4.1.2 Implied volatility; implications 86
  - 4.1.3 Discussion on alternative asset price models 86
- 4.2 Option prices and densities 89
  - 4.2.1 Market implied volatility smile and the payoff 89
  - 4.2.2 Variance swaps 96
- 4.3 Non-parametric local volatility models 102
  - 4.3.1 Implied volatility representation of local volatility 105
  - 4.3.2 Arbitrage-free conditions for option prices 107
  - 4.3.3 Advanced implied volatility interpolation 111
  - 4.3.4 Simulation of local volatility model 114
- 4.4 Exercise set 117

5 Jump Processes   121
- 5.1 Jump diffusion processes 121
  - 5.1.1 Itô's lemma and jumps 124
  - 5.1.2 PIDE derivation for jump diffusion process 127
  - 5.1.3 Special cases for the jump distribution 128
- 5.2 Feynman-Kac theorem for jump diffusion process 130
  - 5.2.1 Analytic option prices 131
  - 5.2.2 Characteristic function for Merton's model 133
  - 5.2.3 Dynamic hedging of jumps with the Black-Scholes model 137
- 5.3 Exponential Lévy processes 139
  - 5.3.1 Finite activity exponential Lévy processes 142

		5.3.2 PIDE and the Lévy triplet 143

        5.3.2 PIDE and the Lévy triplet 143
        5.3.3 Equivalent martingale measure 145
  5.4 Infinite activity exponential Lévy processes 146
        5.4.1 Variance Gamma process 146
        5.4.2 CGMY process 151
        5.4.3 Normal inverse Gaussian process 155
  5.5 Discussion on jumps in asset dynamics 156
  5.6 Exercise set 159

6 The COS Method for European Option Valuation   163
  6.1 Introduction into numerical option valuation 164
        6.1.1 Integrals and Fourier cosine series 164
        6.1.2 Density approximation via Fourier cosine expansion 165
  6.2 Pricing European options by the COS method 169
        6.2.1 Payoff coefficients 172
        6.2.2 The option Greeks 173
        6.2.3 Error analysis COS method 174
        6.2.4 Choice of integration range 177
  6.3 Numerical COS method results 182
        6.3.1 Geometric Brownian Motion 183
        6.3.2 CGMY and VG processes 184
        6.3.3 Discussion about option pricing 187
  6.4 Exercise set 189

7 Multidimensionality, Change of Measure, Affine Processes   193
  7.1 Preliminaries for multi-D SDE systems 193
        7.1.1 The Cholesky decomposition 194
        7.1.2 Multi-D asset price processes 197
        7.1.3 Itô's lemma for vector processes 198
        7.1.4 Multi-dimensional Feynman-Kac theorem 200
  7.2 Changing measures and the Girsanov theorem 201
        7.2.1 The Radon-Nikodym derivative 202
        7.2.2 Change of numéraire examples 204
        7.2.3 From $\mathbb{P}$ to $\mathbb{Q}$ in the Black-Scholes model 206
  7.3 Affine processes 211
        7.3.1 Affine diffusion processes 211
        7.3.2 Affine jump diffusion processes 216
        7.3.3 Affine jump diffusion process and PIDE 217
  7.4 Exercise set 219

8 Stochastic Volatility Models   223
  8.1 Introduction into stochastic volatility models 224
        8.1.1 The Schöbel-Zhu stochastic volatility model 224
        8.1.2 The CIR process for the variance 225
  8.2 The Heston stochastic volatility model 231
        8.2.1 The Heston option pricing partial differential equation . . . 233

       8.2.2   Parameter study for implied volatility skew and smile . . . 236
       8.2.3   Heston model calibration 238
  8.3 The Heston SV discounted characteristic function 242
       8.3.1   Stochastic volatility as an affine diffusion process 242
       8.3.2   Derivation of Heston SV characteristic function 244
  8.4 Numerical solution of Heston PDE 247
       8.4.1   The COS method for the Heston model 248
       8.4.2   The Heston model with piecewise constant parameters . . . 250
       8.4.3   The Bates model 251
  8.5 Exercise set 255

9 Monte Carlo Simulation   257
  9.1 Monte Carlo basics 257
       9.1.1   Monte Carlo integration 260
       9.1.2   Path simulation of stochastic differential equations 265
  9.2 Stochastic Euler and Milstein schemes 266
       9.2.1   Euler scheme 266
       9.2.2   Milstein scheme: detailed derivation 269
  9.3 Simulation of the CIR process 274
       9.3.1   Challenges with standard discretization schemes 274
       9.3.2   Taylor-based simulation of the CIR process 276
       9.3.3   Exact simulation of the CIR model 278
       9.3.4   The Quadratic Exponential scheme 279
  9.4 Monte Carlo scheme for the Heston model 283
       9.4.1   Example of conditional sampling and integrated
              variance 283
       9.4.2   The integrated CIR process and conditional sampling . . . 285
       9.4.3   Almost exact simulation of the Heston model 288
       9.4.4   Improvements of Monte Carlo simulation 292
  9.5 Computation of Monte Carlo Greeks 294
       9.5.1   Finite differences 295
       9.5.2   Pathwise sensitivities 297
       9.5.3   Likelihood ratio method 301
  9.6 Exercise set 306

10 Forward Start Options; Stochastic Local Volatility Model   309
  10.1 Forward start options 309
       10.1.1 Introduction into forward start options 310
       10.1.2 Pricing under the Black-Scholes model 311
       10.1.3 Pricing under the Heston model 314
       10.1.4 Local versus stochastic volatility model 316
  10.2 Introduction into stochastic-local volatility model 319
       10.2.1 Specifying the local volatility 320
       10.2.2 Monte Carlo approximation of SLV expectation 327
       10.2.3 Monte Carlo AES scheme for SLV model 330
  10.3 Exercise set 336

11 Short-Rate Models — 339
- 11.1 Introduction to interest rates 339
  - 11.1.1 Bond securities, notional 340
  - 11.1.2 Fixed-rate bond 341
- 11.2 Interest rates in the Heath-Jarrow-Morton framework 343
  - 11.2.1 The HJM framework 343
  - 11.2.2 Short-rate dynamics under the HJM framework 347
  - 11.2.3 The Hull-White dynamics in the HJM framework 349
- 11.3 The Hull-White model 352
  - 11.3.1 The solution of the Hull-White SDE 352
  - 11.3.2 The HW model characteristic function 353
  - 11.3.3 The CIR model under the HJM framework 356
- 11.4 The HJM model under the T-forward measure 359
  - 11.4.1 The Hull-White dynamics under the T-forward measure . . 360
  - 11.4.2 Options on zero-coupon bonds under Hull-White model . . 362
- 11.5 Exercise set 365

12 Interest Rate Derivatives and Valuation Adjustments — 367
- 12.1 Basic interest rate derivatives and the Libor rate 368
  - 12.1.1 Libor rate 368
  - 12.1.2 Forward rate agreement 370
  - 12.1.3 Floating rate note 371
  - 12.1.4 Swaps 371
  - 12.1.5 How to construct a yield curve 375
- 12.2 More interest rate derivatives 378
  - 12.2.1 Caps and floors 378
  - 12.2.2 European swaptions 383
- 12.3 Credit Valuation Adjustment and Risk Management 386
  - 12.3.1 Unilateral Credit Value Adjustment 392
  - 12.3.2 Approximations in the calculation of CVA 395
  - 12.3.3 Bilateral Credit Value Adjustment (BCVA) 396
  - 12.3.4 Exposure reduction by netting 397
- 12.4 Exercise set 400

13 Hybrid Asset Models, Credit Valuation Adjustment — 405
- 13.1 Introduction to affine hybrid asset models 406
  - 13.1.1 Black-Scholes Hull-White (BSHW) model 406
  - 13.1.2 BSHW model and change of measure 408
  - 13.1.3 Schöbel-Zhu Hull-White (SZHW) model 413
  - 13.1.4 Hybrid derivative product 416
- 13.2 Hybrid Heston model 417
  - 13.2.1 Details of Heston Hull-White hybrid model 418
  - 13.2.2 Approximation for Heston hybrid models 420
  - 13.2.3 Monte Carlo simulation of hybrid Heston SDEs 428
  - 13.2.4 Numerical experiment, HHW versus SZHW model 431
- 13.3 CVA exposure profiles and hybrid models 433

   13.3.1 CVA and exposure 434
   13.3.2 European and Bermudan options example 434
 13.4 Exercise set 439

14 Advanced Interest Rate Models and Generalizations   445
 14.1 Libor market model 446
   14.1.1 General Libor market model specifications 446
   14.1.2 Libor market model under the HJM framework 449
 14.2 Lognormal Libor market model 451
   14.2.1 Change of measure in the LMM 452
   14.2.2 The LMM under the terminal measure 453
   14.2.3 The LMM under the spot measure 454
   14.2.4 Convexity correction 457
 14.3 Parametric local volatility models 460
   14.3.1 Background, motivation 460
   14.3.2 Constant Elasticity of Variance model (CEV) 461
   14.3.3 Displaced diffusion model 467
   14.3.4 Stochastic volatility LMM 470
 14.4 Risk management: The impact of a financial crisis 475
   14.4.1 Valuation in a negative interest rates environment 476
   14.4.2 Multiple curves and the Libor rate 479
   14.4.3 Valuation in a multiple curves setting 484
 14.5 Exercise set 486

15 Cross-Currency Models   489
 15.1 Introduction into the FX world and trading 490
   15.1.1 FX markets 490
   15.1.2 Forward FX contract 491
   15.1.3 Pricing of FX options, the Black-Scholes case 493
 15.2 Multi-currency FX model with short-rate interest rates 495
   15.2.1 The model with correlated, Gaussian interest rates 496
   15.2.2 Pricing of FX options 498
   15.2.3 Numerical experiment for the FX-HHW model 505
   15.2.4 CVA for FX swaps 508
 15.3 Multi-currency FX model with interest rate smile 510
   15.3.1 Linearization and forward characteristic function 514
   15.3.2 Numerical experiments with the FX-HLMM model 517
 15.4 Exercise set 522

References   525

Index   541

CHAPTER 1

Basics about Stochastic Processes

In this chapter:

We introduce some basics about stochastic variables and stochastic processes, like probability density functions, expectations and variances. The *basics about stochastic processes* are presented in **Section 1.1**. Martingales and the *martingale property* are explained in **Section 1.2**. The stochastic *Itô integral* is discussed in quite some detail in **Section 1.3**.
These basic entities from probability theory are fundamental in financial mathematics.

Keywords: stochastic processes, stochastic integral Itô integral, martingales.

1.1 Stochastic variables

We first discuss some known results from probability theory, and start with some facts about stochastic variables and stochastic processes.

1.1.1 Density function, expectation, variance

A real-valued random variable X is often described by means of its cumulative distribution function (CDF),

$$F_X(x) := \mathbb{P}[X \leq x],$$

and its probability density function (PDF),

$$f_X(x) := \mathrm{d}F_X(x)/\mathrm{d}x.$$

Let X be a continuous real-valued random variable with PDF $f_X(x)$. The expected value of X, $\mathbb{E}[X]$, is defined as:

$$\mathbb{E}[X] = \int_{-\infty}^{+\infty} x f_X(x) \mathrm{d}x = \int_{-\infty}^{+\infty} x \frac{\mathrm{d}F_X(x)}{\mathrm{d}x} \mathrm{d}x = \int_{-\infty}^{+\infty} x \mathrm{d}F_X(x),$$

provided the integral $\int_{-\infty}^{+\infty} |x| f_X(x) \mathrm{d}x$ is finite.

The variance of X, $\mathrm{Var}[X]$ is defined as:

$$\mathrm{Var}[X] = \int_{-\infty}^{+\infty} (x - \mathbb{E}[X])^2 f_X(x) \mathrm{d}x,$$

provided the integral exists.

For a continuous random variable X and some constant $a \in \mathbb{R}$ the expectation of an indicator function is related to the CDF of X, as follows,

$$\mathbb{E}[\mathbb{1}_{X \leq a}] = \int_{\mathbb{R}} \mathbb{1}_{x \leq a} f_X(x) \mathrm{d}x = \int_{-\infty}^{a} f_X(x) \mathrm{d}x =: F_X(a),$$

where $F_X(\cdot)$ is the CDF of X, and where the notation $\mathbb{1}_{X \in \Omega}$ stands for the indicator function of the set Ω, defined as follows:

$$\mathbb{1}_{X \in \Omega} = \begin{cases} 1 & X \in \Omega, \\ 0 & X \notin \Omega. \end{cases} \quad (1.1)$$

Definition 1.1.1 (Survival probability) *The survival probability is directly linked to the CDF. If X is a random variable which denotes the lifetime, for example, in a population, then $\mathbb{P}[X \leq x]$ indicates the probability of not reaching the age x. The survival probability, defined by*

$$\mathbb{P}[X > x] = 1 - \mathbb{P}[X \leq x] = 1 - F_X(x),$$

then indicates the probability of surviving a lifetime of length x. ◀

A basic and well-known example of a random variable is the normally distributed random variable. A normally distributed stochastic variable X, with expectation μ and variance σ^2, is governed by the following probability distribution function:

$$F_{\mathcal{N}(\mu,\sigma^2)}(x) = \mathbb{P}[X \leq x] = \frac{1}{\sigma\sqrt{2\pi}} \int_{-\infty}^{x} \exp\left(\frac{-(z-\mu)^2}{2\sigma^2}\right) \mathrm{d}z. \quad (1.2)$$

Variable X then is said to have an $\mathcal{N}(\mu, \sigma^2)$-normal distribution. The corresponding probability density function reads:

$$f_{\mathcal{N}(\mu,\sigma^2)}(x) = \frac{\mathrm{d}}{\mathrm{d}x} F_{\mathcal{N}(\mu,\sigma^2)}(x) = \frac{1}{\sigma\sqrt{2\pi}} \exp\left(\frac{-(x-\mu)^2}{2\sigma^2}\right). \quad (1.3)$$

Example 1.1.1 (Expectation, normally distributed random variable)
Let us consider $X \sim \mathcal{N}(\mu, 1)$. By the definition of the expectation, we find:

$$\mathbb{E}[X] = \int_{\mathbb{R}} x f_X(x) \mathrm{d}x = \frac{1}{\sqrt{2\pi}} \int_{\mathbb{R}} x \exp\left(-\frac{(x-\mu)^2}{2}\right) \mathrm{d}x$$

$$= \frac{1}{\sqrt{2\pi}} \int_{\mathbb{R}} (z+\mu) \exp\left(-\frac{z^2}{2}\right) \mathrm{d}z$$

$$= \frac{1}{\sqrt{2\pi}} \int_{\mathbb{R}} z \exp\left(-\frac{z^2}{2}\right) \mathrm{d}z + \frac{\mu}{\sqrt{2\pi}} \int_{\mathbb{R}} \exp\left(-\frac{z^2}{2}\right) \mathrm{d}z,$$

where we have used $z = x - \mu$. The function in the first integral is an odd function, i.e., $g(-x) = -g(x)$, so that the integral over $\mathbb{R}$ is equal to zero. The second integral is recognized as an integral over the PDF of $\mathcal{N}(0,1)$, so it equals $\sqrt{2\pi}$. Therefore, the expectation of $\mathcal{N}(\mu, 1)$ is equal to μ, i.e., indeed $\mathbb{E}[X] = \mu$. ♦

1.1.2 Characteristic function

In computational finance, we typically work with the density function of certain stochastic variables, like for stock prices or interest rates, as it forms the basis for the computation of the variable's expectation or variance. For some basic stochastic variables the density function is known in closed-form (which is desirable), however, for quite a few relevant stochastic processes in finance we do not know the corresponding density. Interestingly, we will see, for example, in Chapters 5, 7 and 8 in this book, that for some of these processes we can derive expressions for other important functions that contain important information regarding expectations and other quantities. One of these functions is the so-called *characteristic function*, another is the *moment-generating function*. So, we will also introduce these functions here.

The *characteristic function* (ChF), $\phi_X(u)$ for $u \in \mathbb{R}$ of the random variable X, is the Fourier-Stieltjes transform of the cumulative distribution function $F_X(x)$, i.e., with i the imaginary unit,

$$\boxed{\phi_X(u) := \mathbb{E}\left[e^{iuX}\right] = \int_{-\infty}^{+\infty} e^{iux} \mathrm{d}F_X(x) = \int_{-\infty}^{+\infty} e^{iux} f_X(x) \mathrm{d}x.} \quad (1.4)$$

A useful fact regarding $\phi_X(u)$ is that it uniquely determines the distribution function of X. Moreover, the moments of random variable X can also be derived by $\phi_X(u)$, as

$$\mathbb{E}\left[X^k\right] = \frac{1}{i^k} \frac{\mathrm{d}^k}{\mathrm{d}u^k} \phi_X(u)\Big|_{u=0},$$

with i again the imaginary unit, for $k \in \{0, 1, ...\}$, assuming $\mathbb{E}[|X|^k] < \infty$.

A relation between the characteristic function and the *moment generating function*, $\mathcal{M}_X(u)$, exists:

$$\mathbb{E}[X^k] = \frac{1}{i^k}\frac{\mathrm{d}^k}{\mathrm{d}u^k}\phi_X(u)\Big|_{u=0} \stackrel{\text{def}}{=} \frac{1}{i^k}\frac{\mathrm{d}^k}{\mathrm{d}u^k}\int_{-\infty}^{+\infty}\mathrm{e}^{iux}\mathrm{d}F_X(x)\Big|_{u=0} = \frac{\mathrm{d}^k}{\mathrm{d}u^k}\phi_X(-iu)\Big|_{u=0},$$

where the equation's right-hand side represents the moment-generating function, defined as

$$\mathcal{M}_X(u) := \phi_X(-iu) = \mathbb{E}\left[\mathrm{e}^{uX}\right]. \tag{1.5}$$

A relation exists between the moments of a positive random variable Y and the characteristic function for the log transformation $\phi_{\log Y}(u)$. For $X = \log Y$, the corresponding characteristic function reads:

$$\phi_{\log Y}(u) = \mathbb{E}\left[\mathrm{e}^{iu\log Y}\right] = \int_0^\infty \mathrm{e}^{iu\log y}f_Y(y)\mathrm{d}y$$

$$= \int_0^\infty y^{iu} f_Y(y)\mathrm{d}y. \tag{1.6}$$

Note that we use $\log Y \equiv \log_e Y \equiv \ln Y$. By setting $u = -ik$, we have:

$$\phi_{\log Y}(-ik) = \int_0^\infty y^k f_Y(y)\mathrm{d}y \stackrel{\text{def}}{=} \mathbb{E}[Y^k]. \tag{1.7}$$

The derivations above hold for those variables for which the characteristic function for the log-transformed variable is available.

Example 1.1.2 (Density, characteristic function of normal distribution) In Figure 1.1 the CDF, PDF (left side picture) and the characteristic function (right side picture) of the normal distribution, $\mathcal{N}(10,1)$, are displayed. The PDF and CDF are very smooth functions, whereas the characteristic function ("the Fourier transform of the density function") is an oscillatory function in the complex plane. ♦

Another useful function is the *cumulant characteristic function* $\zeta_X(u)$, defined as the logarithm of the characteristic function, $\phi_X(u)$:

$$\zeta_X(u) = \log \mathbb{E}\left[\mathrm{e}^{iuX}\right] = \log \phi_X(u).$$

The kth moment, $m_k(\cdot)$, and the kth cumulant, $\zeta_k(\cdot)$, can be determined by:

$$m_k(\cdot) = (-i)^k \frac{\mathrm{d}^k}{\mathrm{d}u^k}\phi_X(u)\Big|_{u=0}, \quad \zeta_k(\cdot) = (-i)^k \frac{\mathrm{d}^k}{\mathrm{d}u^k}\log\phi_X(u)\Big|_{u=0}, \tag{1.8}$$

where $(-i)^k \equiv i^{-k}$ for $k \in \mathbb{N}$, and with $\phi_X(u)$ defined in (1.4).

1.1.3 Cumulants and moments

Some properties follow directly from the definition of the characteristic function, such as

$$\phi_X(0) = 1, \quad \phi_X(-i) = \mathbb{E}\left[\mathrm{e}^X\right].$$

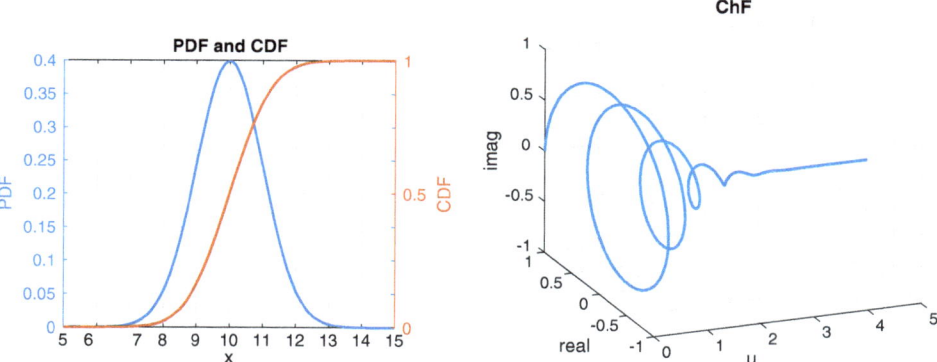

Figure 1.1: *The CDF, PDF and characteristic function for an $\mathcal{N}(10,1)$ random variable.*

By the definition of the moment-generating function, we also find

$$\mathcal{M}_X(u) \equiv \int_{\mathbb{R}} e^{ux} f_X(x) \mathrm{d}x. \tag{1.9}$$

Assuming that all moments of X are finite, the moment-generating function (1.9) admits a MacLaurin series expansion:

$$\begin{aligned}\mathcal{M}_X(u) &= \int_{\mathbb{R}} e^{ux} f_X(x) \mathrm{d}x \\ &= \sum_{k=0}^{\infty} \frac{u^k}{k!} \int_{\mathbb{R}} x^k f_X(x) \mathrm{d}x =: \sum_{k=0}^{\infty} m_k \frac{u^k}{k!},\end{aligned} \tag{1.10}$$

with the *raw moments*, $m_k = \int_{\mathbb{R}} x^k f_X(x) \mathrm{d}x$, for $k = 0, 1, \ldots$.
On the other hand, the cumulant-generating function is defined as:

$$\zeta_X(u) \equiv \log \mathcal{M}_X(u). \tag{1.11}$$

Using again the MacLaurin expansion gives us:

$$\zeta_X(u) = \sum_{k=0}^{\infty} \frac{\partial^k \zeta_X(u)}{\partial u^k}\bigg|_{u=0} \frac{u^k}{k!} =: \sum_{k=0}^{\infty} \zeta_k \frac{u^k}{k!}, \tag{1.12}$$

where $\zeta_k \equiv \frac{\partial^k \zeta_X(u)}{\partial u^k}\bigg|_{u=0}$ is the k-th cumulant.

By (1.11) a recurrence relation between the raw moments and the cumulants is obtained, i.e.,

$$\mathcal{M}_X(u) = \sum_{k=0}^{\infty} m_k \frac{u^k}{k!} = \exp\left(\sum_{k=0}^{\infty} \zeta_k \frac{u^k}{k!}\right).$$

Using the results from (1.11), the first four derivatives of the cumulant-generating function can be expressed as,

$$\frac{d\zeta_X(t)}{dt} = \frac{1}{\phi_X(-it)}\frac{d\phi_X(-it)}{dt},$$

$$\frac{d^2\zeta_X(t)}{dt^2} = -\frac{1}{(\phi_X(-it))^2}\left(\frac{d\phi_X(-it)}{dt}\right)^2 + \frac{1}{\phi_X(-it)}\frac{d^2\phi_X(-it)}{dt^2},$$

$$\frac{d^3\zeta_X(t)}{dt^3} = \frac{2}{(\phi_X(-it))^3}\left(\frac{d\phi_X(-it)}{dt}\right)^3 - \frac{3}{(\phi_X(-it))^2}\frac{d\phi_X(-it)}{dt}\frac{d^2\phi_X(-it)}{dt^2}$$
$$+ \frac{1}{\phi_X(-it)}\frac{d^3\phi_X(-it)}{dt^3},$$

$$\frac{d^4\zeta_X(t)}{dt^4} = -\frac{6}{(\phi_X(-it))^4}\left(\frac{d\phi_X(-it)}{dt}\right)^4 + \frac{12}{(\phi_X(-it))^3}\left(\frac{d\phi_X(-it)}{dt}\right)^2\frac{d^2\phi_X(-it)}{dt^2}$$
$$- \frac{3}{(\phi_X(-it))^2}\left(\frac{d^2\phi_X(-it)}{dt^2}\right)^2 - \frac{4}{(\phi_X(-it))^2}\frac{d\phi_X(-it)}{dt}\frac{d^3\phi_X(-it)}{dt^3}$$
$$+ \frac{1}{\phi_X(-it)}\frac{d^4\phi_X(-it)}{dt^4}.$$

With these derivatives of the cumulant-generating function, and the identity $\phi_X(0) = 1$, the first four cumulants are found as,

$$\zeta_1 = \frac{d\zeta_X(t)}{dt}\Big|_{t=0}, \quad \zeta_2 = \frac{d^2\zeta_X(t)}{dt^2}\Big|_{t=0},$$
$$\zeta_3 = \frac{d^3\zeta_X(t)}{dt^3}\Big|_{t=0}, \quad \zeta_4 = \frac{d^4\zeta_X(t)}{dt^4}\Big|_{t=0}.$$

For a random variable X, with μ its mean X, σ^2 its variance, γ_3 the skewness and γ_4 the kurtosis by the relation between the cumulants and moments, the following equalities can be found,

$$\zeta_1 = m_1 = \mu,$$
$$\zeta_2 = m_2 - m_1^2 = \sigma^2,$$
$$\zeta_3 = 2m_1^3 - 3m_1 m_2 + m_3 = \gamma_3 \sigma^3,$$
$$\zeta_4 = -6m_1^4 + 12m_1^2 m_2 - 3m_2^2 - 4m_1 m_3 + m_4 = \gamma_4 \sigma^4, \tag{1.13}$$

Recall that skewness is a measure of asymmetry around the mean of a distribution, whereas kurtosis is a measure for the tailedness of a distribution.

When the moments or the cumulants are available, the associated density can, at least formally, be recovered. Based on Equation (1.4), the probability density

Basics about Stochastic Processes

function can be written as the inverse Fourier transform of the characteristic function:

$$\boxed{f_X(x) = \frac{1}{2\pi}\int_{\mathbb{R}} \phi_X(u) e^{-iux} du.} \qquad (1.14)$$

By the definition of the characteristic function and the MacLaurin series expansion of the exponent around zero, we find:

$$\phi_X(u) = \mathbb{E}\left[e^{iuX}\right] = \sum_{k=0}^{\infty} \frac{(iu)^k}{k!} \mathbb{E}[X^k] = \sum_{k=0}^{\infty} \frac{(iu)^k}{k!} m_k, \qquad (1.15)$$

with m_k as in (1.10). Thus, Equation (1.14) equals:

$$f_X(x) = \frac{1}{2\pi} \sum_{k=0}^{\infty} \frac{m_k}{k!} \int_{\mathbb{R}} (iu)^k e^{-iux} du. \qquad (1.16)$$

Recall the definition of the *Dirac delta function and its k-th derivative*, as follows,

$$\delta(x) = \begin{cases} +\infty, & x = 0, \\ 0, & x \neq 0. \end{cases} \qquad (1.17)$$

and

$$\delta(x) = \frac{1}{2\pi}\int_{\mathbb{R}} e^{-iux} du, \quad \delta^{(k)}(x) = \frac{d^k}{dx^k}\delta(x) = \frac{1}{2\pi}\int_{\mathbb{R}} (-iu)^k e^{-iux} du.$$

Now, that Equation (1.16) is also written as:

$$f_X(x) = \sum_{k=0}^{\infty} (-1)^k \frac{m_k}{k!} \delta^{(k)}(x). \qquad (1.18)$$

Two-dimensional densities

The joint CDF of two random variables, X and Y, is the function $F_{X,Y}(\cdot,\cdot): \mathbb{R}^2 \to [0,1]$, which is defined by:

$$F_{X,Y}(x,y) = \mathbb{P}[X \leq x, Y \leq y].$$

If X and Y are continuous variables, then the *joint PDF of X and Y* is a function $f_{X,Y}(\cdot,\cdot): \mathbb{R}^2 \to \mathbb{R}^+ \cup \{0\}$, such that:

$$f_{X,Y}(x,y) = \frac{\partial^2 F_{X,Y}(x,y)}{\partial x \partial y},$$

> For any event A, it follows that
> $$\mathbb{P}[(X,Y) \in A] = \iint_A f_{X,Y}(x,y)\mathrm{d}x\mathrm{d}y.$$
> We will also use the vector notation in this book, particularly from Chapter 7 on, where we will then write $\mathbf{X} = [X,Y]^\mathrm{T}$ and $F_\mathbf{X}$, $f_\mathbf{X}$, respectively.

As the joint PDF is a true probability function, we have $f_{X,Y}(x,y) \geq 0$, for any $x, y \in \mathbb{R}$ and
$$\int_{-\infty}^{+\infty}\int_{-\infty}^{+\infty} f_{X,Y}(x,y)\mathrm{d}x\mathrm{d}y = 1.$$

Given the joint distribution of (X,Y), the expectation of a function $h(X,Y)$ is calculated as:
$$\mathbb{E}[h(X,Y)] = \int_{-\infty}^{+\infty}\int_{-\infty}^{+\infty} h(x,y)f_{X,Y}(x,y)\mathrm{d}x\mathrm{d}y.$$

The *conditional PDF of Y, given $X = x$*, is defined as:
$$f_{Y|X}(y|x) = \frac{f_{X,Y}(x,y)}{f_X(x)}, \quad -\infty < y < \infty.$$

Moreover, the *conditional expectation of X, given $Y = y$*, is defined as the mean of the conditional PDF of X, given $Y = y$, i.e.,
$$\mathbb{E}[X|Y = y] = \int_{-\infty}^{+\infty} x f_{X|Y}(x|y)\mathrm{d}x = \int_{-\infty}^{+\infty} x \frac{f_{X,Y}(x,y)}{f_Y(y)}\mathrm{d}y.$$

Based on the joint PDF, we can also determine the *marginal densities*, as follows,
$$f_X(x) = \frac{\mathrm{d}}{\mathrm{d}x}F_X(x) = \frac{\mathrm{d}}{\mathrm{d}x}\mathbb{P}[X \leq x]$$
$$= \frac{\mathrm{d}}{\mathrm{d}x}\int_{-\infty}^{x}\left(\int_{-\infty}^{+\infty} f_{X,Y}(u,y)\mathrm{d}y\right)\mathrm{d}u$$
$$= \int_{-\infty}^{+\infty} f_{X,Y}(x,y)\mathrm{d}y,$$

and similarly for $f_Y(y)$.

Example 1.1.3 (Bivariate normal density functions) In this example, we show three bivariate normal density and distribution functions, with $\mathbf{X} = [X,Y]^\mathrm{T}$, and
$$\mathbf{X} \sim \mathcal{N}\left(\begin{bmatrix}0\\0\end{bmatrix}, \begin{bmatrix}1, \rho\\\rho, 1\end{bmatrix}\right),$$

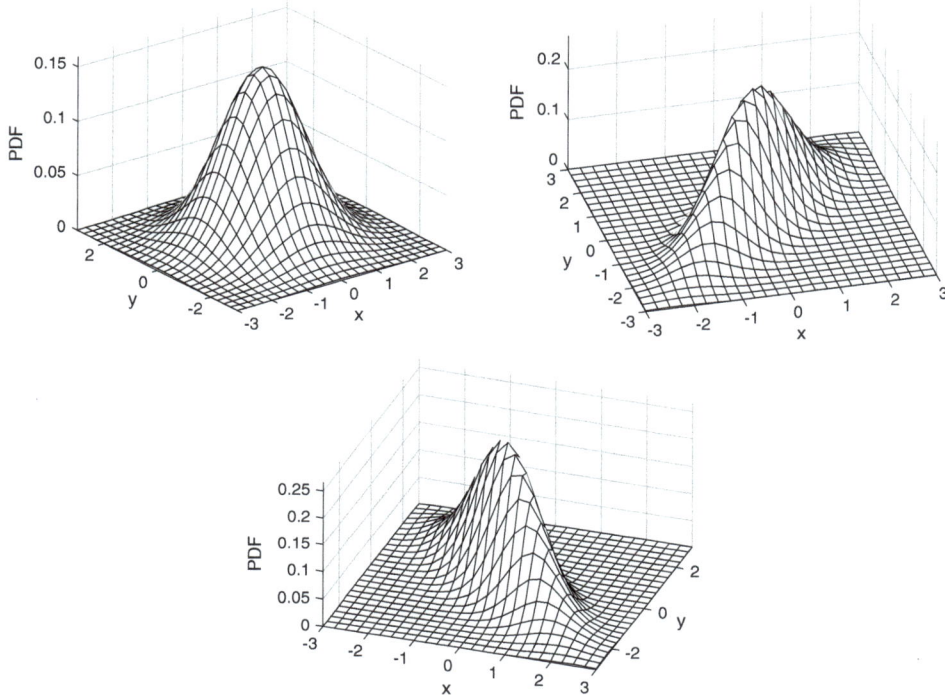

Figure 1.2: *Examples of two-dimensional normal probability density functions, with $\rho = 0$ (first), $\rho = 0.8$ (second) and $\rho = -0.8$ (third).*

in which the correlation coefficient is varied. Figure 1.2 displays the functions for $\rho = 0$ (first row), $\rho = 0.8$ (second row) and $\rho = -0.8$ (third row). Clearly, the correlation coefficient has an impact of the *direction* in these functions. ♦

1.2 Stochastic processes, martingale property

We will often work with stochastic processes for the financial asset prices, and give some basic definitions for them here.

A stochastic process, $X(t)$, is a collection of random variables indexed by a *time* variable t.

Suppose we have a set of calendar dates/days, $T_1, T_2, \ldots, T_m$. Up to *today*, we have observed certain state values of the stochastic process $X(t)$, see Figure 1.3. The past is known, and we therefore "see" the historical asset path. For the future we do not know the precise path but we may simulate the future according to some asset price distribution.

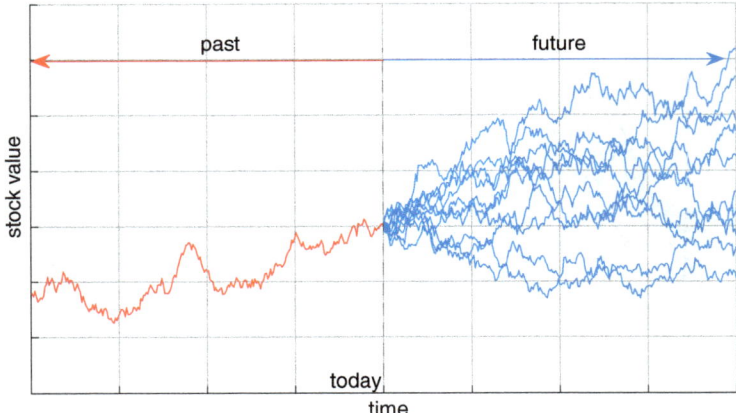

Figure 1.3: *Past and present in an asset price setting. We do not know the precise future asset path but we may simulate it according to some price distribution.*

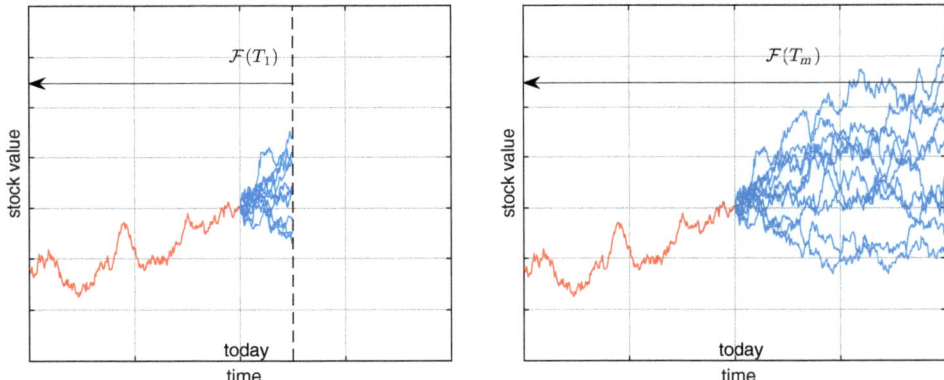

Figure 1.4: *Filtration figure, with $\mathcal{F}(t_0) \subseteq \mathcal{F}(T_1) \subseteq \mathcal{F}(T_2) \ldots \subseteq \mathcal{F}(T_m)$. When $X(t)$ is $\mathcal{F}(t_0)$ measurable this implies that at time t_0 the value of $X(t)$ is known. $X(T_1)$ is $\mathcal{F}(T_1)$ measurable, but $X(T_1)$ is a "future realization" which is not yet known at time t_0 ("today") and thus not $\mathcal{F}(t_0)$ measurable.*

The mathematical tool which helps us describe the knowledge of a stochastic process up-to a certain time T_i is the *sigma-field*, also known as sigma-algebra. The ordered sequence of sigma-fields is called a *filtration*, $\mathcal{F}(T_i) := \sigma(X(T_j) : 1 \leq j \leq i)$, generated by the sequence $X(T_j)$ for $1 \leq j \leq i$. The information available at time T_i is thus described by a filtration, see also Figure 1.4. As we consider a sequence of observation dates, $T_1, \ldots, T_i$, we deal in fact with a sequence of filtrations, $\mathcal{F}(T_1) \subseteq \cdots \subseteq \mathcal{F}(T_i)$.

If we write that a process is $\mathcal{F}(T)$-measurable, we mean that at any time $t \leq T$, the realizations of this process are known. A simple example for this may be the market price of a stock and its historical values, i.e., we know the stock values up to today exactly, but we do not know any future values. We then say "the stock is today measurable". However, when we deal with an SDE model for the stock price, the value may be T measurable, as we know the distribution for the period T of a financial contract.

A stochastic process $X(t)$, $t \geq 0$, is said to be adapted to the filtration $\mathcal{F}(t)$, if

$$\sigma(X(t)) \subseteq \mathcal{F}(t).$$

By the term "adapted process" we mean that a stochastic process "cannot look into the future". In other words, for a stochastic process $X(t)$ its realizations (paths), $X(s)$ for $0 \leq s < t$, are known at time s *but not yet at time t*.

1.2.1 Wiener process

Definition 1.2.1 (Wiener process) *A fundamental stochastic process, which is also commonly used in the construction of stochastic differential equations (SDEs) to describe asset price movements, is the Wiener process, also called Brownian motion. Mathematically, a Wiener process, $W(t)$, is characterized by the following properties:*

a. $W(t_0) = 0$, *(technically: $\mathbb{P}[W(t_0) = 0] = 1$)*,

b. $W(t)$ *is almost surely[a] continuous*,

c. $W(t)$ *has independent increments, i.e.* $\forall\, t_1 \leq t_2 \leq t_3 \leq t_4$, $W(t_2) - W(t_1) \perp\!\!\!\perp W(t_4) - W(t_3)$, *with distribution* $W(t) - W(s) \sim \mathcal{N}(0, t-s)$ *for* $0 = t_0 \leq s < t$, *i.e. the normal distribution with mean 0 and variance $t - s$.* ◀

---

[a]Almost surely convergence means that for a sequence of random variables X_m, the following holds: $\mathbb{P}\left[\lim_{m \to \infty} X_m = X\right] = 1$.

Example 1.2.1 Examples of processes that are adapted to the filtration $\mathcal{F}(t)$ are:

- $W(t)$ and $W^2(t) - t$, with $W(t)$ a Wiener process.

- $\max_{0 \leq s \leq t} W(s)$ and $\max_{0 \leq s \leq t} W^2(s)$.

Examples of processes that are not adapted to the filtration $\mathcal{F}(t)$ are:

- $W(t+1)$,

- $W(t) + W(T)$ for some $T > t$. ◆

1.2.2 Martingales

An important notion when dealing with stochastic processes is the martingale property.

> **Definition 1.2.2 (Martingale)** *Consider a probability space $(\Omega, \mathcal{F}, \mathbb{Q})$, where Ω is the set of all possible outcomes, $\mathcal{F}(t)$ is the sigma-field, and $\mathbb{Q}$ is a probability measure. A right continuous process $X(t)$ with left limits (so-called cádlág[a] process) for $t \in [0, T]$, is said to be a martingale with respect to the filtration $\mathcal{F}(t)$ under measure $\mathbb{Q}$, if for all $t < \infty$, the following holds:*
>
> $$\mathbb{E}[|X(t)|] < \infty,$$
>
> *and*
>
> $$\mathbb{E}[X(t)|\mathcal{F}(s)] = X(s), \quad \text{with} \quad s < t,$$
>
> *where $\mathbb{E}[\cdot|\mathcal{F}]$ is the conditional expectation operator under measure $\mathbb{Q}$.* ◀
>
> ---
> [a]"continue à droite, limite à gauche" (French for right continuous with left limits).

The definition implies that the best prediction of the expectation of a martingale's future value is its present value, and

$$\begin{aligned}\mathbb{E}\left[X(t+\Delta t) - X(t)|\mathcal{F}(t)\right] &= \mathbb{E}[X(t+\Delta t)|\mathcal{F}(t)] - \mathbb{E}[X(t)|\mathcal{F}(t)] \\ &= X(t) - X(t) = 0,\end{aligned} \quad (1.19)$$

for some time interval $\Delta t > 0$.

Proposition 1.2.1 *The Wiener process $W(t), t \in [0, T]$ is a martingale.* ◀

Proof We check the martingale properties. First of all, $\mathbb{E}[|W(t)|] < \infty$, since

$$\mathbb{E}\left[|W(t)|\right] = \int_{-\infty}^{+\infty} |x| \frac{1}{\sqrt{2\pi t}} e^{-\frac{x^2}{2t}} dx = \frac{2}{\sqrt{2\pi t}} \int_0^{+\infty} x e^{-\frac{x^2}{2t}} dx.$$

By setting $z = \frac{x^2}{2t}$, so that $t dz = x dx$, we find

$$\mathbb{E}\left[|W(t)|\right] = \frac{2t}{\sqrt{2\pi t}} \int_0^{+\infty} e^{-z} dz = \frac{2t}{\sqrt{2\pi t}} < \infty \text{ for finite } t.$$

For Wiener process $W(t), t \in [0, T]$, we also find, using (1.19), that

$$\begin{aligned}\mathbb{E}[W(t)|\mathcal{F}(s)] &= \mathbb{E}[W(s) + [W(t) - W(s)]|\mathcal{F}(s)] \\ &= \mathbb{E}[W(s)|\mathcal{F}(s)] + \mathbb{E}[W(t) - W(s)|\mathcal{F}(s)] \\ &= W(s) + 0 = W(s), \; \forall \, s, t > 0.\end{aligned}$$ ∎

1.2.3 Iterated expectations (Tower property)

Another important and useful concept is the concept of iterated expectations. The law of iterated expectations, also called the *tower property*, states that for any given random variable $X \in L^2$ (where L^2 indicates a so-called Hilbert space for which $\mathbb{E}[X^2(t)] < \infty$), which is defined on a probability space $(\Omega, \mathcal{F}, \mathbb{Q})$, and for any sigma-field $\mathcal{G} \subseteq \mathcal{F}$, the following equality holds:

$$\mathbb{E}[X|\mathcal{G}] = \mathbb{E}[\mathbb{E}[X|\mathcal{F}]|\mathcal{G}], \quad \text{for} \quad \mathcal{G} \subseteq \mathcal{F}.$$

If we consider another random variable Y, which is defined on the sigma-field $\mathcal{G}$, so that $\mathcal{G} \subseteq \mathcal{F}$, then the above equality can be written as

$$\mathbb{E}[Y] = \mathbb{E}[\mathbb{E}[Y|X]], \quad \text{for} \quad \sigma(Y) \subseteq \sigma(X).$$

Assuming that both random variables, X and Y, are continuous on $\mathbb{R}$ and are defined on the same sigma-field, we can prove the equality given above, as follows

$$\mathbb{E}[\mathbb{E}[Y|X]] = \int_{\mathbb{R}} \mathbb{E}[Y|X=x] f_X(x) \mathrm{d}x$$
$$= \int_{\mathbb{R}} \left(\int_{\mathbb{R}} y f_{Y|X}(y|x) \mathrm{d}y \right) f_X(x) \mathrm{d}x.$$

By the definition of the conditional density, i.e. $f_{Y|X}(y|x) = f_{Y,X}(y,x)/f_X(x)$, we have:

$$\mathbb{E}[\mathbb{E}[Y|X]] = \int_{\mathbb{R}} \left(\int_{\mathbb{R}} y \frac{f_{Y,X}(y,x)}{f_X(x)} \mathrm{d}y \right) f_X(x) \mathrm{d}x$$
$$= \int_{\mathbb{R}} y \left(\int_{\mathbb{R}} f_{Y,X}(y,x) \mathrm{d}x \right) \mathrm{d}y \quad (1.20)$$
$$= \int_{\mathbb{R}} y f_Y(y) \mathrm{d}y \stackrel{\text{def}}{=} \mathbb{E}[Y].$$

The conditional expectation will be convenient when dealing with continuous distributions. Let us take two independent random variables X and Y. Using the conditional expectation we can show the following equality:

$$\mathbb{P}[X < Y] = \int_{\mathbb{R}} \mathbb{P}[X < y] f_Y(y) \mathrm{d}y = \int_{\mathbb{R}} F_X(y) f_Y(y) \mathrm{d}y,$$

which can be proven as follows,

$$\mathbb{P}[X < Y] = \mathbb{E}[\mathbf{1}_{X<Y}]$$
$$= \mathbb{E}[\mathbb{E}[\mathbf{1}_{X<Y}|Y=y]] = \int_{\mathbb{R}} \mathbb{P}[X < y|Y=y] f_Y(y) \mathrm{d}y.$$

Since X and Y are independent variables, we have:

$$\mathbb{P}[X < Y] = \int_{\mathbb{R}} \mathbb{P}[X < y] f_Y(y) \mathrm{d}y = \int_{\mathbb{R}} F_X(y) f_Y(y) \mathrm{d}y.$$

The result given above can be used to show an example of a so-called *convolution*, which, for constant $c \in \mathbb{R}$ and two independent random variables X and Y, is defined as:

$$\mathbb{P}[X+Y<c] = \int_{\mathbb{R}} F_Y(c-x)f_X(x)\mathrm{d}x = \int_{\mathbb{R}} F_X(c-y)f_Y(y)\mathrm{d}y.$$

These integrals can be recognized as two expectations, $\mathbb{E}[F_Y(c-X)]$ and $\mathbb{E}[F_X(c-Y)]$.

1.3 Stochastic integration, Itô integral

For any differentiable function $\xi(t)$, one can use the following relation:

$$\int_0^T g(t)\mathrm{d}\xi(t) = \int_0^T g(t)\left(\frac{\mathrm{d}\xi(t)}{\mathrm{d}t}\right)\mathrm{d}t. \tag{1.21}$$

However, when $\xi(t)$ is a Wiener process, i.e. $\xi(t) \equiv W(t)$, Equality (1.21) is not valid, as the Brownian motion is nowhere differentiable.

Riemann-Stieltjes integration cannot be used when the integrand is based on a Wiener process. *Stochastic integration* can however be applied with the calculus, which is developed by the Japanese mathematician Kiyoshi Itô (1915–2008).

We consider the following stochastic differential equation,

$$\mathrm{d}I(t) = g(t)\mathrm{d}W(t), \quad \text{for } t \geq 0. \tag{1.22}$$

which is equivalent to the following *Itô integral*:

$$I(T) = \int_0^T g(t)\mathrm{d}W(t), \quad \text{for } T \geq 0, \tag{1.23}$$

where, in a certain interval $[0,T]$, the function $g(t)$ may represent a stochastic process, $g(t) := g(t,\omega)$. The variable ω then represents *randomness*, i.e. $\omega \in \Omega$ given the filtration $\mathcal{F}(t)$. The function $g(t)$ needs to satisfy the following two conditions:

1. $g(t)$ is $\mathcal{F}(t)$-measurable for any time t (in other words, the "process" $g(t)$ is an adapted process).

2. $g(t)$ is square-integrable, i.e.: $\mathbb{E}\left[\int_0^T g^2(t)\mathrm{d}t\right] < \infty$, $\forall\, T \geq 0$.

1.3.1 Elementary processes

For a given partition, $0 = t_0 < t_1 < \cdots < t_m = T$, of the time interval $[0,T]$, we make use of *elementary processes*, $\{g_m(t)\}_{m=0}^{\infty}$, with $g_m(t)$ a piecewise constant function. With the help of these elementary processes, we can formally define the Itô integral, as follows.

Basics about Stochastic Processes

> **Definition 1.3.1** *For any square-integrable adapted process $g(t) = g(t,\omega)$, with continuous sample paths, the Itô integral is given by:*
>
> $$I(T) \stackrel{\text{def}}{=} \int_0^T g(t)\mathrm{d}W(t) = \lim_{m\to\infty} I_m(T), \quad \text{in} \quad L^2. \tag{1.24}$$
>
> *Here, $I_m(T) = \int_0^T g_m(t)\mathrm{d}W(t)$ for some elementary process $\{g_m(t)\}_{m=0}^\infty$, satisfying:*
>
> $$\lim_{m\to\infty} \mathbb{E}\left[\int_0^T (g_m(t) - g(t))^2 \mathrm{d}t\right] = 0. \tag{1.25}$$
> ◂

The existence of a sequence of elementary processes is presented with the help of the following theorems.

> **Theorem 1.3.1 (Dominated Convergence Theorem in L^p)**
> Let $\{\xi_m\}_{m\in\mathbb{N}}$ be a sequence of functions in L^p, $p > 0$, such that there exists a real-valued function $\bar{\xi} \in L^p$ with $|\xi_m| < \bar{\xi}$ for all $m \in \mathbb{N}$. Assume that $\{\xi_m\}_{m\in\mathbb{N}} \to \xi$, in a pointwise fashion. Then
>
> $$\|\xi_m - \xi\|_{L^p} = \lim_{m\to\infty} \left(\mathbb{E}\left[|\xi_m - \xi|^p\right]\right)^{\frac{1}{p}} = 0.$$
>
> The proof of this theorem can be found in stochastic calculus textbooks. We will use $p = 1$ and $p = 2$ below.

> **Theorem 1.3.2** *A sequence of elementary processes $\{g_m(t)\}_{m=0}^\infty$ exists, such that:*
>
> $$\lim_{m\to\infty} \mathbb{E}\left[\int_0^T |g_m(t) - g(t)|^2 \mathrm{d}t\right] = 0. \tag{1.26}$$

Let us, for simplicity, assume here $T \in \mathbb{N}$. The objective of a proof of Theorem 1.3.2 is to find a sequence of elementary processes, $g_1(t), g_2(t), \ldots$, such that Equation (1.26) holds. To achieve this, we define the following elementary processes:

$$g_m(t) = \begin{cases} m \int_{\frac{k-1}{m}}^{\frac{k}{m}} g(s)\mathrm{d}s & \text{if } t \in \left[\frac{k-1}{m}, \frac{k}{m}\right) \text{ for } k = 1, 2, \ldots, mT, \\ 0 & \text{otherwise,} \end{cases} \tag{1.27}$$

The construction implies that $g_m(t)$ is in essence a step function, i.e. it is constant on each interval $t \in \left[\frac{k-1}{m}, \frac{k}{m}\right)$. However, because $g(t)$ is stochastic, each path of $g(t)$ yields a different constant realization of $g_m(t)$. Commonly, the function

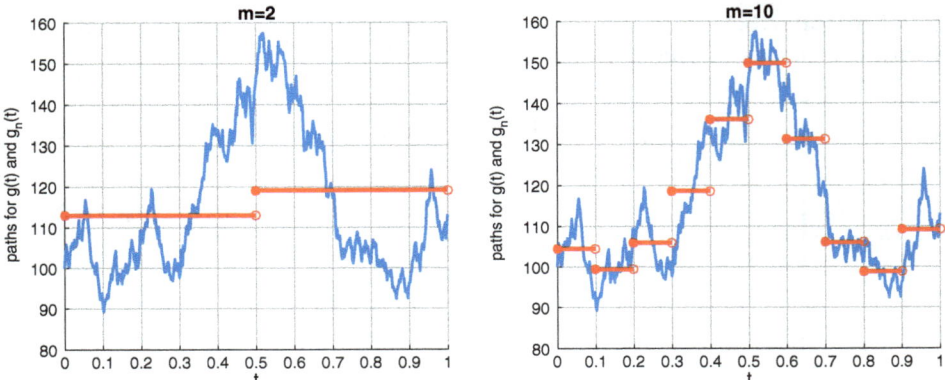

Figure 1.5: *Random step functions, approximating a stochastic function $g(t)$, with $m = 2$ and $m = 10$, respectively.*

in (1.27) is therefore called a *random step function*. We present an example in Figure 1.5.

By the Cauchy-Schwartz inequality, we obtain the following relation:

$$\int_{\frac{k-1}{m}}^{\frac{k}{m}} |g_m(t)|^2 dt = \int_{\frac{k-1}{m}}^{\frac{k}{m}} \left| m \int_{\frac{k-1}{m}}^{\frac{k}{m}} g(z) dz \right|^2 dt = \left(\frac{k}{m} - \frac{k-1}{m} \right) \left| m \int_{\frac{k-1}{m}}^{\frac{k}{m}} g(z) dz \right|^2$$

$$\leq m \left(\frac{k}{m} - \frac{k-1}{m} \right) \int_{\frac{k-1}{m}}^{\frac{k}{m}} g^2(z) dz = \int_{\frac{k-1}{m}}^{\frac{k}{m}} g^2(z) dz \quad a.s. \quad (1.28)$$

Where the abbreviation "a.s." stands for the *almost surely*, which essentially means that this inequality needs to hold for any realization $g(t)$.

We now move on to the main part of the proof, i.e., showing the equality in (1.26).

By the assumption of the a.s. continuity sample paths of $g(t)$, we have:

$$\lim_{m \to \infty} \int_0^T |g_m(t) - g(t)|^2 dt = 0 \quad a.s. \quad (1.29)$$

Defining $\xi_m := \int_0^T |g_m(t) - g(t)|^2 dt$, we thus have $\lim_{m \to \infty} \xi_m = 0$.

Based on the inequality $(a + b)^2 \leq 2(a^2 + b^2)$, the following holds true:

$$\xi_m \equiv \int_0^T |g_m(t) - g(t)|^2 dt \leq 2 \int_0^T |g_m(t)|^2 dt + 2 \int_0^T |g(t)|^2 dt.$$

We then find, using the inequality in (1.28),

$$\xi_m \leq 4 \int_0^T |g(t)|^2 dt =: \bar{\xi}. \quad (1.30)$$

Basics about Stochastic Processes

Since $\xi_m \to 0$ a.s. when $m \to \infty$ (1.29), $|\xi_m| < \bar{\xi}$ a.s. (1.30), and $\mathbb{E}[\bar{\xi}] < \infty$ by Theorem 1.3.1, we have:

$$||\xi_m - \xi||_{L^1} = \lim_{m \to \infty} \mathbb{E}\left[(\xi_m - 0)^1\right] = \lim_{m \to \infty} \mathbb{E}\left[\int_0^T |g_m(t) - g(t)|^2 \mathrm{d}t\right] = 0.$$

1.3.2 Itô isometry

Let's look at the discrete version of the Itô integral in some more detail:

$$I(T) := \int_0^T g(t)\mathrm{d}W(t) = \lim_{m \to \infty} \sum_{i=0}^{m-1} g(t_i)(W(t_{i+1}) - W(t_i)), \quad (1.31)$$

with $t_i = i\frac{T}{m}$. A proof of existence of the limit at the right-hand-side of (1.31) can be found in Theorem 1.3.3; more details and information regarding uniqueness can be found in the standard literature, see, for example, [Shreve, 2004].

The particular choice for the evaluation of $g(t)$ at the left-hand point of $[t_i, t_{i+1})$ is specific to Itô's calculus. If one evaluates the function at the mid-point, i.e., using $g((t_{i+1} + t_i)/2)$, integration is according to Stratonovich. Itô's integration has a preference in finance, as the left-hand time point indicates present time, whereas a stock price at the mid-point would rely on time points *in the future*.

We can make use of the property $\mathbb{E}[I(T)] \equiv \mathbb{E}[I(T)|\mathcal{F}(t_0)] = 0$ (a property we will encounter frequently in this chapter), which can be seen from the following derivation,

$$\mathbb{E}[I(T)] = \mathbb{E}\left[\lim_{m \to \infty} \sum_{i=0}^{m-1} g(t_i)(W(t_{i+1}) - W(t_i))\right]$$

$$= \lim_{m \to \infty} \sum_{i=0}^{m-1} \mathbb{E}[g(t_i)(W(t_{i+1}) - W(t_i))]$$

$$= \lim_{m \to \infty} \sum_{i=0}^{m-1} \mathbb{E}[g(t_i)|\mathcal{F}(t_0)]\,\mathbb{E}[W(t_{i+1}) - W(t_i)]. \quad (1.32)$$

Since increments of a Brownian motion are independent with respect to stochastic variables and functions up to time t_i, and since the increments $W(t_{i+1}) - W(t_i)$ are normally distributed with zero mean, the second expectation in (1.32) equals zero, $\mathbb{E}[W(t_{i+1}) - W(t_i)] = 0$, for any index i, and thus $\mathbb{E}[I(T)] = 0$.

This will be different in the case of Stratonovich' calculus, where a term $\mathbb{E}\left[g(t_{i+\frac{1}{2}})(W(t_{i+1}) - W(t_i))\right]$ would not be equal to 0.

The Itô isometry property states that for any stochastic process $g(t)$, satisfying the usual regularity conditions, the following equality holds,

$$\boxed{\mathbb{E}\left[\left(\int_0^T g(t)\mathrm{d}W(t)\right)^2\right] = \int_0^T \mathbb{E}[g^2(t)]\mathrm{d}t.} \quad (1.33)$$

To prove this equality, we make again use of an equally spaced partitioning, $0 = t_0 < t_1 < \cdots < T = t_m$, and write

$$\int_0^T g(t) dW(t) = \lim_{m \to \infty} \sum_{i=0}^{m-1} g(t_i) \left(W(t_{i+1}) - W(t_i) \right)$$

$$= \lim_{m \to \infty} \sum_{i=0}^{m-1} g(t_i) \Delta W_i, \quad (1.34)$$

with $\Delta W_i := W(t_{i+1}) - W(t_i)$. The square of the integral reads:

$$\left[\int_0^T g(t) dW(t) \right]^2 = \lim_{m \to \infty} \left[\sum_{i=0}^{m-1} g(t_i) \Delta W_i \right]^2$$

$$= \lim_{m \to \infty} \left[\sum_{i=0}^{m-1} g^2(t_i) \Delta W_i^2 + 2 \sum_{i=0}^{m-1} \sum_{j=i+1}^{m-1} g(t_i) g(t_j) \Delta W_i \Delta W_j \right].$$

We take the expectations at both sides of the equality above, and, since for $i \neq j$ ΔW_i is independent of ΔW_j (independent increments), the expectation of the double sum is equal to zero, i.e.

$$\mathbb{E} \left[\int_0^T g(t) dW(t) \right]^2 = \lim_{m \to \infty} \sum_{i=0}^{m-1} \mathbb{E} \left[g^2(t_i) \Delta W_i^2 \right].$$

We now use the tower property of expectations, i.e.,

$$\mathbb{E}[X | \mathcal{F}(s)] = \mathbb{E} \left[\mathbb{E}[X | \mathcal{F}(t)] | \mathcal{F}(s) \right], \quad s < t.$$

By setting $s = t_0 \equiv 0$ and $t = t_i$ in the expression above, we have:

$$\mathbb{E} \left[\int_0^T g(t) dW(t) \right]^2 = \lim_{m \to \infty} \sum_{i=0}^{m-1} \mathbb{E} \left[\mathbb{E} \left[g^2(t_i) \Delta W_i^2 | \mathcal{F}(t_i) \right] | \mathcal{F}(0) \right]$$

$$= \lim_{m \to \infty} \sum_{i=0}^{m-1} \mathbb{E} \left[g^2(t_i) \mathbb{E} \left[\Delta W_i^2 | \mathcal{F}(t_i) \right] | \mathcal{F}(0) \right]. \quad (1.35)$$

Since increments of Brownian motion are independent and the variance of $W(t_{i+1}) - W(t_i)$ equals $t_{i+1} - t_i$, we find

$$\mathbb{E} \left[(W(t_{i+1}) - W(t_i))^2 | \mathcal{F}(t_i) \right] = \mathbb{E} \left[(W(t_{i+1}) - W(t_i))^2 \right] = t_{i+1} - t_i, \quad (1.36)$$

so that Equation (1.35) reads:

$$\mathbb{E}\left[\int_0^T g(t)\mathrm{d}W(t)\right]^2 = \lim_{m\to\infty} \sum_{i=0}^{m-1} \mathbb{E}\left[g^2(t_i)|\mathcal{F}(0)\right](t_{i+1}-t_i) = \int_0^T \mathbb{E}[g^2(t)]\mathrm{d}t.$$

which defines the Itô isometry.

The following theorem, where the proof is based on the Itô isometry, confirms the existence of the Itô integral.[1]

Theorem 1.3.3 (Existence of Itô integral) *The Itô integral, as defined in (1.24), exists.*

Proof To show the existence of the integral, it is sufficient to show that

$$I_m(T) = \int_0^T g_m(t)\mathrm{d}W(t),$$

converges to some element in L^2. With the elementary process $g_m(t)$ a (random) step function on each interval of the time partition, we can write:

$$I_m(T) = \sum_{i=0}^{m-1} g_m(t_i)(W(t_{i+1}) - W(t_i)).$$

To show that the limit in Equation (1.24) exists, we consider the following limit:

$$\lim_{n,m\to\infty} (I_n(T) - I_m(T)) = 0, \text{ in } L^2. \tag{1.37}$$

If we can show that the limit (1.37) equals 0 in L^2, this implies the existence of the limit of $I_m(T)$.

For any $n > 0$ and $m > 0$ we have:

$$\mathbb{E}\left[I_n(T) - I_m(T)\right]^2 = \mathbb{E}\left[\int_0^T (g_n(t) - g_m(t))\,\mathrm{d}W(t)\right]^2$$

$$= \int_0^T \mathbb{E}\left[g_n(t) - g_m(t)\right]^2 \mathrm{d}t, \tag{1.38}$$

where the second step comes from Itô's isometry. Equation (1.38) can be rewritten as:

$$\int_0^T \mathbb{E}\left[g_n(t) - g_m(t)\right]^2 \mathrm{d}t = \int_0^T \mathbb{E}\left[(g_n(t) - g(t)) + (g(t) - g_m(t))\right]^2 \mathrm{d}t. \tag{1.39}$$

---

[1] For uniqueness, we refer again to standard literature.

Again using the inequality $(a+b)^2 \leq 2a^2 + 2b^2$, we have:

$$\int_0^T \mathbb{E}\left[g_n(t) - g_m(t)\right]^2 \mathrm{d}t \leq 2\int_0^T \mathbb{E}\left[g_n(t) - g(t)\right]^2 \mathrm{d}t + 2\int_0^T \mathbb{E}\left[g_m(t) - g(t)\right]^2 \mathrm{d}t.$$

Using the results from Theorem 1.3.2 we find:

$$\lim_{n \to \infty} \mathbb{E}\left[g_n(t) - g(t)\right]^2 \mathrm{d}t = 0, \text{ and } \lim_{m \to \infty} \mathbb{E}\left[g_m(t) - g(t)\right]^2 \mathrm{d}t = 0,$$

and therefore:

$$0 \leq \lim_{n,m \to \infty} \mathbb{E}\left[I_n(T) - I_m(T)\right]^2 \leq 0, \tag{1.40}$$

which implies, by the squeeze theorem of sequences, that the limit of $I_m(T)$ exists. This concludes the proof. ∎

Theorem 1.3.4 (Itô integral is a martingale) *For any* $g(t) \in L^2$, *the stochastic integral* $I(T) := \int_0^T g(t)\mathrm{d}W(t)$ *is a martingale with respect to the filtration* $\mathcal{F}(T), T \geq 0$.

Proof As shown in Theorem 1.3.3, $I(T)$ exists, so $\mathbb{E}[I(T)] < \infty$. For any random process $g(t)$ in L^2, we have:

$$\mathbb{E}[I(t + \Delta t)|\mathcal{F}(t)] = \mathbb{E}[I(t + \Delta t) - I(t) + I(t)|\mathcal{F}(t)]$$
$$= \mathbb{E}[I(t + \Delta t) - I(t)|\mathcal{F}(t)] + \mathbb{E}[I(t)|\mathcal{F}(t)] = 0 + I(t),$$

using the property of independent increments and the property of measurability. ∎

1.3.3 Martingale representation theorem

Theorem 1.3.5 (Martingale Representation Theorem) *With* $t_0 \leq t \leq T$, *let* $W(t)$, *with* $W(t_0) = W_0$, *be a Brownian motion on* $(\Omega, \mathcal{F}, \mathbb{P})$, *and* $\mathcal{F}(t)$, *a filtration generated by the Brownian motion. With* $X(t)$, *a martingale relative to this filtration, there is an adapted process* $g(t)$, *such that:*

$$\mathrm{d}X(t) = g(t)\mathrm{d}W(t), \quad \text{or} \quad X(t) = X_0 + \int_0^t g(z)\mathrm{d}W(z).$$

For a proof, see Øksendal [2000].

The theorem above states that if the process $X(t)$ is a martingale, adapted to the filtration generated by the Brownian motion $W^{\mathbb{P}}(t)$, then process $X(t)$ needs to be of the following form

$$\mathrm{d}X(t) = g(t)\mathrm{d}W^{\mathbb{P}}(t),$$

for some process $g(t)$.

As the integral formulation, in (1.23), is equivalent to the SDE in (1.22), we can conclude that an SDE without any drift term is a martingale.

Example 1.3.1 (Solution of Itô integral) In this example, we solve the following stochastic integral,

$$I(T) = \int_0^T W(t)\mathrm{d}W(t). \tag{1.41}$$

Following Definition 1.3.1, we have $g(t) := W(t)$, consider an equally spaced partition $0 = t_0 < t_1 < \cdots < t_m = T$, with $t_i = i\frac{T}{m}$, with an equidistant time increment $\Delta t = (t_{i+1} - t_i)$ and define a sequence of some elementary functions satisfying (1.25), as follows:

$$g_m(t) = \begin{cases} W(0), & \text{for } 0 \le t < t_1, \\ W(t_1), & \text{for } t_1 \le t < t_2, \\ \vdots \\ W(t_m), & \text{for } t_{m-1} \le t < t_m. \end{cases} \tag{1.42}$$

Let us verify for this sequence of elementary functions that the condition in (1.25) is satisfied, i.e.

$$\lim_{m \to \infty} \mathbb{E}\left[\int_0^T (g_m(t) - g(t))^2 \mathrm{d}t\right] = \lim_{m \to \infty} \mathbb{E}\left[\int_0^T (g_m(t) - W(t))^2 \mathrm{d}t\right]$$

$$= \lim_{m \to \infty} \sum_{i=0}^{m-1} \int_{t_i}^{t_{i+1}} \mathbb{E}\left[W(t_i) - W(t)\right]^2 \mathrm{d}t.$$

As $t_i < t$, the last term can be written as

$$\lim_{m \to \infty} \sum_{i=0}^{m-1} \int_{t_i}^{t_{i+1}} \mathbb{E}\left[W(t_i) - W(t)\right]^2 \mathrm{d}t = \lim_{m \to \infty} \sum_{i=0}^{m-1} \int_{t_i}^{t_{i+1}} (t - t_i)\mathrm{d}t$$

$$= \lim_{m \to \infty} \sum_{i=0}^{m-1} \frac{1}{2}(t_{i+1} - t_i)^2.$$

Using the fact that $\Delta t = t_{i+1} - t_i$, we have:

$$\lim_{m \to \infty} \sum_{i=0}^{m-1} \frac{1}{2}(\Delta t)^2 = 0. \tag{1.43}$$

This is because

$$\lim_{m\to\infty}\sum_{i=0}^{m-1}\frac{1}{2}(\Delta t)^2 = \lim_{m\to\infty}\sum_{i=0}^{m-1}\frac{1}{2}\left[(i+1)\frac{T}{m}-i\frac{T}{m}\right]^2 = \lim_{m\to\infty}\frac{1}{2}\sum_{i=0}^{m-1}\frac{T^2}{m^2},$$

which converges to 0 for $m \to \infty$.

So, the condition in (1.25) holds, and we continue with the discrete version of the integral in (1.41):

$$\int_0^T W(t)\mathrm{d}W(t) = \lim_{m\to\infty}\sum_{i=0}^{m-1} W(t_i)\left(W(t_{i+1}) - W(t_i)\right). \qquad (1.44)$$

By basic algebra[2] the right-hand side of (1.44) can be simplified to:

$$\sum_{i=0}^{m-1} W(t_i)(W(t_{i+1}) - W(t_i)) = \frac{1}{2}\sum_{i=0}^{m-1}\left(W^2(t_{i+1}) - W^2(t_i)\right)$$

$$-\frac{1}{2}\sum_{i=0}^{m-1}(W(t_{i+1}) - W(t_i))^2.$$

The first element at the right-hand side of the expression above can be recognized as a *telescopic sum*, i.e.

$$\sum_{i=0}^{m-1}\left(W^2(t_{i+1}) - W^2(t_i)\right) = \cancel{W^2(t_1)} - W^2(t_0) + \cancel{W^2(t_2)} - \cancel{W^2(t_1)} + \cdots$$

$$= W^2(t_m) - W^2(t_0),$$

as $t_m = T$, $t_0 = 0$ and $W^2(t_0) \equiv 0$, giving the following simplification:

$$\lim_{m\to\infty}\sum_{i=0}^{m-1} W(t_i)(W(t_{i+1}) - W(t_i)) = \frac{1}{2}W^2(T) - \frac{1}{2}\lim_{m\to\infty}\sum_{i=0}^{m-1}(W(t_{i+1}) - W(t_i))^2.$$

$$(1.45)$$

To finalize the task of calculating the integral in (1.41), we calculate the sum at the right-hand side of (1.45) for which we take the expectation:

$$\lim_{m\to\infty}\mathbb{E}\left[\sum_{i=0}^{m-1}(W(t_{i+1}) - W(t_i))^2\right] = \lim_{m\to\infty}\sum_{i=0}^{m-1}\mathbb{E}\left[W(t_{i+1}) - W(t_i)\right]^2$$

$$= \lim_{m\to\infty}\sum_{i=0}^{m-1}(t_{i+1} - t_i) = T. \qquad (1.46)$$

---

[2] $x(y-x) = \frac{1}{2}(y^2 - x^2) - \frac{1}{2}(y-x)^2.$

We need to verify whether T is the limit of the summation in L^2. This, by definition, can be confirmed by the following expectation:

$$\lim_{m\to\infty} \mathbb{E}\left[\sum_{i=0}^{m-1} (W(t_{i+1}) - W(t_i))^2 - T\right]^2$$

$$= \lim_{m\to\infty} \mathbb{E}\left[\sum_{i=0}^{m-1} (W(t_{i+1}) - W(t_i))^2\right]^2$$

$$- 2T \lim_{m\to\infty} \mathbb{E}\left[\sum_{i=0}^{m-1} (W(t_{i+1}) - W(t_i))^2\right] + T^2.$$

Using the results from (1.46), we find:

$$\lim_{m\to\infty} \mathbb{E}\left[\sum_{i=0}^{m-1} (W(t_{i+1}) - W(t_i))^2 - T\right]^2 = \lim_{m\to\infty} \mathbb{E}\left[\sum_{i=0}^{m-1} (W(t_{i+1}) - W(t_i))^2\right]^2 - T^2.$$

The expectation at the right-hand side can be further simplified, as follows,

$$\mathbb{E}\left[\sum_{i=0}^{m-1} (W(t_{i+1}) - W(t_i))^2\right]^2 = \mathbb{E}\left[\sum_{i=0}^{m-1} (\Delta t)^2 Z_i^4 + 2\sum_{i=0}^{m-1}\sum_{j=i+1}^{m-1} (\Delta t)^2 Z_i^2 Z_j^2\right].$$

with $Z_i = \mathcal{N}(0,1)$. Because the fourth moment of a standard normal random variable equals 3, i.e. $\mathbb{E}[Z_i^4] = 3$, and any Z_i is mutually independent of Z_j, for $i \neq j$, we get

$$\mathbb{E}\left[\sum_{i=0}^{m-1} (\Delta t)^2 Z_i^4 + 2\sum_{i=0}^{m-1}\sum_{j=i+1}^{m-1} (\Delta t)^2 Z_i^2 Z_j^2\right] = 3\sum_{i=0}^{m-1} (\Delta t)^2 + 2\sum_{i=0}^{m-1}\sum_{j=i+1}^{m-1} (\Delta t)^2$$

$$= 3(\Delta t)^2 m + (\Delta t)^2 (m^2 - m)$$

$$= T^2 + 2T\frac{1}{m}.$$

In the limit case $m \to \infty$, we find:

$$\lim_{m\to\infty} \mathbb{E}\left[\sum_{i=0}^{m-1} (W(t_{i+1}) - W(t_i))^2 - T\right]^2 = \lim_{m\to\infty}\left[T^2 + 2T\frac{1}{m} - T^2\right] = 0,$$

implying convergence of the sum in (1.46) to T, in L^2.

This result implies the following solution for the summation in (1.45):

$$\lim_{m\to\infty} \sum_{i=0}^{m-1} W(t_i)(W(t_{i+1}) - W(t_i)) = \frac{1}{2} W^2(T) - \frac{1}{2} T. \tag{1.47}$$

and the solution of the integral in (1.41) is therefore given by:

$$\boxed{\int_0^T W(t)\mathrm{d}W(t) = \frac{1}{2}W^2(T) - \frac{1}{2}T.} \qquad (1.48)$$

♦

We conclude this introductory part with the following short summary.

The Itô integral defined by (1.23) has a number of important properties, like

a. For every time $t \geq 0$, $I(t)$ is $\mathcal{F}(t)$-measurable,

b. $\mathbb{E}[I(t)|\mathcal{F}(0)] = 0$,

c. $\mathbb{E}[I(t)|\mathcal{F}(s)] = I(s)$, for $s < t$, which is the martingale property, and the dynamics $\mathrm{d}I(t)$ do not contain a drift term,

d. Itô isometry:
$$\mathbb{E}\left[\int_0^T g(t)\mathrm{d}W(t)\right]^2 = \int_0^T \mathbb{E}[g^2(t)]\mathrm{d}t,$$

e. For $0 \leq a < b < c$, it follows that
$$\int_a^c g(t)\mathrm{d}W(t) = \int_a^b g(t)\mathrm{d}W(t) + \int_b^c g(t)\mathrm{d}W(t),$$

f. Another equality which holds true is:
$$\int_a^c (\alpha \cdot g(t) + h(t))\mathrm{d}W(t) = \alpha \int_a^c g(t)\mathrm{d}W(t) + \int_a^c h(t)\mathrm{d}W(t),$$

with $a < c$, $\alpha \in \mathbb{R}$, and $h(t) \in \mathcal{F}(t)$.

Proofs for these statements are given in stochastic calculus textbooks, such as [Shreve, 2004].

Basics about Stochastic Processes

1.4 Exercise set

Exercise 1.1 With
$$F_{\mathcal{N}(0,1)}(x) = \frac{1}{\sqrt{2\pi}} \int_{-\infty}^{x} e^{-\frac{z^2}{2}} dz,$$
show that
$$F_{\mathcal{N}(0,1)}(x) + F_{\mathcal{N}(0,1)}(-x) = 1.$$

Exercise 1.2 Use $\mathbb{E}[X+Y] = \mathbb{E}[X] + \mathbb{E}[Y]$, and $\mathbb{E}[\alpha X] = \alpha \mathbb{E}[X]$ ($\alpha \in \mathbb{R}$), to show,
$$\mathbb{E}[(X - \mathbb{E}[X])^2] = \mathbb{E}[X^2] - (\mathbb{E}[X])^2;$$
and show with this, that,
$$\mathrm{Var}[\alpha X] = \alpha^2 \mathrm{Var}[X], \text{ with } \alpha \in \mathbb{R}.$$

Exercise 1.3 Suppose $X \sim \mathcal{N}(\mu, \sigma^2)$ and $Y = a + bX$ ($b \neq 0$). Determine $\mathbb{E}[Y]$, $\mathrm{Var}[Y]$ and the distribution function of Y. Determine $\mathbb{E}[e^X]$.

Exercise 1.4 Show that the standard normal distribution function, $F_{\mathcal{N}(0,1)}(x)$, can be calculated with the so-called *error function*,
$$\mathrm{erf}(x) = \frac{2}{\sqrt{\pi}} \int_0^x e^{-s^2} ds,$$
with the help of the formula,
$$F_{\mathcal{N}(0,1)}(x) = \frac{1 + \mathrm{erf}\left(\frac{x}{\sqrt{2}}\right)}{2}.$$

Exercise 1.5 With $X_1, \ldots, X_n$ i.i.d. random variables with the same distribution, expectation μ and variance σ^2. Random variable $\bar{X}$ is defined as the arithmetic average of these variables, which is the *sample mean*,
$$\bar{X} = \frac{1}{n} \sum_{k=1}^{n} X_k.$$

a. Show that $\mathbb{E}[\bar{X}] = \mu$.
b. Show that $\mathrm{Var}[\bar{X}] = \sigma^2/n$. The random variable $\bar{v}_N^2$, which is defined as:
$$\bar{v}_N^2 := \frac{\sum_{k=1}^{N}(X_k - \bar{X})^2}{N-1},$$
is the *sample variance*.
c. Show that $\sum_{k=1}^{N}(X_k - \bar{X})^2 = \sum_{k=1}^{N} X_k^2 - N\bar{X}^2$.
d. Show that $\mathbb{E}[\bar{v}_N^2] = \sigma^2$.

Exercise 1.6 For a given Brownian motion $W(t)$:

a. Solve analytically the expectation,
$$\mathbb{E}\left[W^4(t) - \frac{1}{2}W^3(t)\right].$$

b. Find analytically $\mathrm{Var}[Z(t)]$, with
$$Z(t) = W(t) - \frac{t}{T}W(T-t), \quad \text{for } 0 \leq t \leq T.$$

Exercise 1.7 Show theoretically that
$$\int_0^t W(z)\mathrm{d}z = \int_0^t (t-z)\mathrm{d}W(z).$$

Exercise 1.8 With standard Brownian motion $W(t)$, $t \geq 0$, calculate the integral
$$\int_{z=0}^T \int_{s=0}^z \mathrm{d}W(s)\mathrm{d}W(z).$$

Exercise 1.9 Show that, for a continuously, differentiable function $g(t)$, the process
$$X(t) = g(t)W(t) - \int_0^t \frac{\mathrm{d}g(z)}{\mathrm{d}z}W(z)\mathrm{d}z,$$
is a martingale, and subsequently show that
$$\mathbb{E}[\mathrm{e}^{2t}W(t)] = \mathbb{E}\left[\int_0^t 2\mathrm{e}^{2z}W(z)\mathrm{d}z\right].$$

Exercise 1.10 A time continuous stochastic process $\{X(t); t \in \mathcal{T}\}$ is called a Gaussian process, if for any set of time indices, $t_1,\ldots,t_m$, all linear combinations of $(X(t_1),\ldots,X(t_m))$ are governed by a univariate normal distribution.

Given a Gaussian process $X(t)$, $t > 0$ with $X(0) = 0$. Determine the following covariance,
$$2\mathbb{C}\mathrm{ov}[X(s), X(t)] = \mathbb{E}[X^2(s)] + \mathbb{E}[X^2(t)] - \mathbb{E}[(X(t) - X(s))^2], \; 0 < s < t.$$

Exercise 1.11 Consider the following SDE,
$$\mathrm{d}X(t) = \mu\mathrm{d}t + \sigma\mathrm{d}W(t), \quad X(t_0) = x_0, \tag{1.49}$$
with some constants μ and σ. Show that, by choosing $t_0 = 0$, the integrated process $X(t)$ follows the following distribution:
$$\int_0^T X(t)\mathrm{d}t \sim \mathcal{N}\left(x_0 T + \frac{1}{2}\mu T^2, \frac{1}{3}\sigma^2 T^3\right).$$

CHAPTER 2

Introduction to Financial Asset Dynamics

In this chapter:

In **Section 2.1**, we present the mathematical basis of stochastic models for financial asset prices. In particular, we focus on the *Geometric Brownian Motion stock price model*. *Itô's lemma* is discussed in this chapter, as it plays an important role for many derivations.
In **Section 2.2** some first variations to the basic geometric Brownian motion process are presented. The *martingale property of financial asset prices* is explained in **Section 2.3**.

Keywords: model for asset prices, geometric Brownian motion, martingales.

2.1 Geometric Brownian motion asset price process

A stock or share is a financial asset, which represents ownership of a tiny piece of a company, and is traded on financial markets. According to the efficient market hypothesis, the stock price is determined by the present value of the company plus the expectations of the companies' future performance. These expectations give rise to uncertainty in the asset price, as seen in the form of bid and ask prices offered by the participants in the financial markets. Asset prices thus have an element of randomness, commonly modeled by *stochastic differential equations* (SDEs). Closed-form solutions to these SDEs are available only in exceptional cases. They can serve as a validation for numerical techniques, or as building blocks for SDE asset price models of increasing complexity.

The most commonly used asset price process in finance is the geometric Brownian Motion (GBM) model, where the *logarithm of the asset price* follows an arithmetic Brownian motion, driven by a Wiener process $W(t)$.

> The asset price $S(t)$ is said to follow a GBM process, when it satisfies the following SDE:
>
> $$dS(t) = \mu S(t)dt + \sigma S(t) dW^{\mathbb{P}}(t), \quad \text{with} \quad S(t_0) = S_0, \qquad (2.1)$$
>
> where the Brownian motion $W^{\mathbb{P}}(t)$ is under the so-called real-world measure $\mathbb{P}$, $\mu = \mu^{\mathbb{P}}$ denotes the *drift parameter*, i.e. a constant deterministic growth rate of the stock, and σ is the (constant) percentage volatility parameter. Model (2.1) is also referred to as the *Samuelson model*. This is a short-hand notation for the integral formulation,
>
> $$S(t) = S_0 + \int_{t_0}^{t} \mu S(z) dz + \int_{t_0}^{t} \sigma S(z) dW^{\mathbb{P}}(z). \qquad (2.2)$$

The amount by which an asset price differs from its expected value is determined by the volatility parameter σ. Volatility is thus a statistical measure of the tendency of an asset to rise or fall sharply within a period of time. It can be calculated, for example, by the variance of the asset prices, measured within a certain time period. A high volatile market implies that prices have large deviations from their mean value in short periods of time. Although both μ and σ in (2.1) are assumed to be constant, the extension to time-dependent functions is also possible. It should however be noticed that these parameter values are estimates for the growth rate and the volatility *in the future*, i.e., for $t > t_0$.

Before we dive deeper into our first financial asset stochastic process, we have a look at the (deterministic) money savings account, which we denote by $M(t)$.

Definition 2.1.1 (Markov process) *A stochastic process is a Markov process, if the conditional probability distribution of future states depends only on the present state, and not on the history. In a financial setting, this implies that we assume that the current stock price contains all information of the past asset prices. The adapted stock price process $S(t)$ on a filtered probability space has the Markov property, if for each bounded and measurable function $g : \mathbb{R}^N \to \mathbb{R}$,*

$$\mathbb{E}[g(S(t))|\mathcal{F}(s)] = \mathbb{E}[g(S(t))|S(s)], \quad s \leq t. \qquad (2.3)$$

◀

Definition 2.1.2 (Money-savings account) *The simplest concept in finance is the time value of money. One unit of currency today is worth more than one unit in a year's time, when interest rates are positive, and less than one unit in the case of negative interest rates.. Particularly, we will focus our attention on compounded interest, which is defined as interest on earlier interest payments (on an initial notional amount). This interest can be either discretely or continuously compounded. Receiving m discrete interest payments at a rate of r/m per year, on an initial notional $M(0) = 1$, gives, after one year $T = 1$, the* amount,

$$M(T) = \left(1 + \frac{r}{m}\right)^m.$$

When the interest payments come in increasingly smaller time intervals, at a proportionally smaller interest rate (take the limit $m \to \infty$), this defines the continuous interest rate. It can be shown, by basic calculus, that

$$\lim_{m \to \infty} \left(1 + \frac{r}{m}\right)^m = e^{m \log(1 + \frac{r}{m})} = e^r.$$

At a time point t, we will have the amount e^{rt} on the bank account. This is the so-called money-savings account.

With an amount $M(t)$ at time t on the money-savings account, the Taylor expansion indicates that at time point $M(t + \Delta t)$, the amount will have increased by,

$$M(t + \Delta t) - M(t) \approx \frac{dM}{dt}\Delta t + \dots.$$

The change in money is proportional to the initial amount, the interest rate r and the time period the money is on the account, giving,

$$\frac{dM(t)}{dt} = rM(t). \qquad (2.4)$$

Starting with $M(0) = 1$ at $t_0 = 0$, we will have $M(t) = e^{rt}$, at time point t. On the other hand, if we wish to receive $M(T) = 1$ at a future time point $t = T$, we need to put the amount $e^{-r(T-t)}$ on our money-savings account at time point $t_0 = 0$. Clearly, money grows exponentially in continuous time, however, with a very small (positive) interest rate r, the growth is still rather limited in the long run.

Compared to the stochastic GBM asset dynamics in Equation (2.1), we find a drift term in the deterministic interest rate dynamics in (2.4) with $\mu = r$, and $\sigma = 0$. Stochastic interest rates, in the form of so-called short-rates and Libor rates, will also be discussed in detail in this book, starting from Chapter 11. These are important stochastic models for example for options on interest rates. ◀

2.1.1 Itô process

Itô's lemma is fundamental for stochastic processes, as it enables us to handle the Wiener increment $dW(t)$ as in (2.1), when $dt \to 0$ (similar to a Taylor expansion for deterministic variables and functions). By Itô's lemma we can derive solutions to SDEs, and we can derive pricing partial differential equations (PDEs) for financial derivative products in the subsequent chapter. We first discuss some issues related to the so-called *Itô processes*.

> **Definition 2.1.3 (Itô process)** Let us consider the following SDE, corresponding to the Itô process $X(t)$,
>
> $$\mathrm{d}X(t) = \bar{\mu}(t,X(t))\mathrm{d}t + \bar{\sigma}(t,X(t))\mathrm{d}W(t), \quad \text{with} \quad X(t_0) = X_0, \qquad (2.5)$$
>
> with two general functions for the drift $\bar{\mu}(t,x)$ and the volatility $\bar{\sigma}(t,x)$. These two functions cannot be "just any" functions; they need to satisfy the following two Lipschitz conditions:
>
> $$|\bar{\mu}(t,x) - \bar{\mu}(t,y)|^2 + |\bar{\sigma}(t,x) - \bar{\sigma}(t,y)|^2 \leq K_1 |x-y|^2,$$
> $$|\bar{\mu}(t,x)|^2 + |\bar{\sigma}(t,x)|^2 \leq K_2(1 + |x|^2),$$
>
> for some constants $K_1, K_2 \in \mathbb{R}^+$ and x and y in $\mathbb{R}$. The two conditions above state that the drift and volatility terms should not increase too rapidly. When these conditions hold, then, with probability one, a continuous, adapted solution of (2.5) exists, and the solution satisfies $\sup_{0 \leq t \leq T} \mathbb{E}[X^2(t)] < \infty$. ◀

2.1.2 Itô's lemma

With stochastic process $X(t)$ determined by (2.5), another process $Y(t)$ can be defined as a function of t and $X(t)$, i.e., $Y(t) := g(t,X)$.[1] $Y(t)$ is a stochastic process and its SDE can also be determined. The procedure to derive the SDE for process $Y(t)$ is given by Itô's lemma below.

Before we present Itô's lemma however, we give some heuristics. To derive the dynamics $\mathrm{d}Y(t)$ for $Y(t) = g(t,X)$, we may take a look at the 2D Taylor series expansion around some point (t_0, X_0), i.e.,

$$\begin{aligned}
g(t,X) = g(t_0, X_0) &+ \left.\frac{\partial g(t,X)}{\partial t}\right|_{t=t_0} \Delta t + \frac{1}{2}\left.\frac{\partial^2 g(t,X)}{\partial t^2}\right|_{t=t_0} (\Delta t)^2 \\
&+ \left.\frac{\partial g(t,X)}{\partial X}\right|_{X=X_0} \Delta X + \frac{1}{2}\left.\frac{\partial^2 g(t,X)}{\partial X^2}\right|_{X=X_0} (\Delta X)^2 \\
&+ \left.\frac{\partial^2 g(t,X)}{\partial t \partial X}\right|_{X=X_0, t=t_0} \Delta X \Delta t + \cdots,
\end{aligned} \qquad (2.6)$$

with $\Delta t = t - t_0$, and $\Delta X = X - X_0$. For $t \to t_0$ and $X \to X_0$, and $\mathrm{d}t = \lim_{t \to t_0} t - t_0$, $\mathrm{d}X = \lim_{X \to X_0} X - X_0$, we may write (2.6) as follows:

$$\begin{aligned}
\mathrm{d}g(t,X) = \frac{\partial g}{\partial t}\mathrm{d}t &+ \frac{1}{2}\frac{\partial^2 g}{\partial t^2}(\mathrm{d}t)^2 + \frac{\partial g}{\partial X}\mathrm{d}X + \frac{1}{2}\frac{\partial^2 g}{\partial X^2}\mathrm{d}X^2 \\
&+ \frac{\partial^2 g}{\partial t \partial X}\mathrm{d}X\mathrm{d}t + \cdots.
\end{aligned} \qquad (2.7)$$

---

[1] Here $X = X(t)$ serves as an independent variable.

In (2.7) we encounter infinitely many terms. Many of those terms however can be neglected in the limit $dt \to 0$. When the time increment dt goes to 0, terms $(dt)^2$ tend to 0 much faster than the terms with dt. This holds true for any term $(dt)^n$, with $n > 1$. Conventionally, this convergence behavior is described by the little-o notation, i.e. $(dt)^2 = o(dt)$.

It is common to denote $(dt)^2 = 0$, but we need to keep in mind that this equality actually means "order $dt \to 0$".

Remark 2.1.1 (Little-o and big-O notation) *By definition,*

$$g(x) = O(h(x)), \quad \text{if } |g(x)| < c \cdot h(x),$$

for some constant "c" and for sufficiently large x. Little-o describes the following asymptotic limit,

$$g(x) = o(h(x)), \quad \text{if } \lim_{x \to \infty} \frac{g(x)}{h(x)} = 0.$$

Informally, big-O can be thought of as "$g(x)$ does not grow faster than $h(x)$" and of little-o as "$g(x)$ grows much slower than $h(x)$". ▲

Example 2.1.1 (Little-o and big-O) As an example, the function $g(x) = 5x^2 - 1x + 9$ is $O(x^2)$, but it is not $o(x^2)$, because $\lim_{x \to \infty} \frac{g(x)}{x^2} = 5$.

The following statements are true for little-o,

$$x^2 = o(x^3), \quad x^2 = o(x!), \quad \log(x^2) = o(x).$$

The following statements are true under big-O notation,

$$x^2 = O(x^2), \quad x^2 = O(x^2 + x), \quad x^2 = O(100x^2),$$

however, they are not true when little-o would be used. ◆

Itô's table

The equality in (2.7) can be simplified, by neglecting the higher-order dt-terms, by writing,

$$dg(t, X) = \frac{\partial g}{\partial t} dt + \frac{\partial g}{\partial X} dX + \frac{1}{2} \frac{\partial^2 g}{\partial X^2} (dX)^2. \tag{2.8}$$

We need to make statements about the term $dXdX$, which, in the case of Equation (2.5), reads:

$$(dX)^2 = \bar{\mu}^2(t, X)(dt)^2 + \bar{\sigma}^2(t, X)(dW)^2 + 2\bar{\mu}(t, X)\bar{\sigma}(t, X)dW dt. \tag{2.9}$$

Two specific terms $dtdW$ and $dWdW$ need to be determined. The expectation of $dtdW$ equals 0 (because an expectation of a Brownian increment, scaled by a constant, equals zero) and the standard deviation equals $dt^{\frac{3}{2}}$ (as the standard deviation of dW is equal to $\sqrt{dt}$). We have $(dt)^{\frac{3}{2}}$, implying that $dtdW$ goes to

Table 2.1: *Itô multiplication table for Wiener process.*

	dt	dW(t)
dt	0	0
dW(t)	0	dt

0 rapidly when $dt \to 0$. Regarding the other term, the expectation of $dW dW$ is equal to dt implying that $dW dW$ is of order dt when $dt \to 0$.

So, when deriving the Itô dynamics, we make use of the Itô multiplication table, where the cross terms involving the Wiener process are handled as in Table 2.1, see also the discussion in Privault [1998].

Remark 2.1.2 *The approximation $(dW)^2 = dt$.*
Let us consider the following expectation, $\mathbb{E}\left[(dW)^2\right]$. By the same steps as in the derivations of the Itô isometry, see (1.36), we have

$$\mathbb{E}\left[(dW)^2\right] = \lim_{\Delta t \to 0} \mathbb{E}\left[(W(t+\Delta t) - W(t))^2\right] = \lim_{\Delta t \to 0} \Delta t = dt, \quad (2.10)$$

and the variance is equal to:

$$\begin{aligned}
\mathbb{V}ar\left[(dW)^2\right] &= \lim_{\Delta t \to 0} \mathbb{V}ar\left[(W(t+\Delta t) - W(t))^2\right] \\
&= \lim_{\Delta t \to 0} \mathbb{E}\left[(W(t+\Delta t) - W(t))^4\right] \\
&\quad - \lim_{\Delta t \to 0} \left(\mathbb{E}\left[(W(t+\Delta t) - W(t))^2\right]\right)^2 \\
&= \lim_{\Delta t \to 0} 3(\Delta t)^2 - \lim_{\Delta t \to 0} (\Delta t)^2 = \lim_{\Delta t \to 0} 2(\Delta t)^2 \\
&= 2(dt)^2.
\end{aligned}$$

We conclude that the variance of $(dW)^2$ converges to zero much faster than the expectation, when $\Delta t \to 0$. Because of this, we have as a stochastic calculus rule,

$$(dW)^2 = dt,$$

as the variance approaches zero rapidly in the limit. ▲

We can therefore write (2.9) as:

$$\boxed{(dX)^2 \approx \bar{\sigma}^2(t, X) dt.}$$

By collecting all building blocks we will write the dynamics of $g(t, X)$ as follows:

$$dg(t, X) = \frac{\partial g}{\partial t} dt + \left(\bar{\mu}(t, X) \frac{\partial g}{\partial X} + \frac{1}{2} \bar{\sigma}^2(t, X) \frac{\partial^2 g}{\partial X^2}\right) dt + \frac{\partial g}{\partial X} \bar{\sigma}(t, X) dW(t).$$

We now have the following lemma:

> **Theorem 2.1.1 (Itô's lemma)** *Suppose a process $X(t)$ follows the Itô dynamics,*
>
> $$\mathrm{d}X(t) = \bar{\mu}(t, X(t))\mathrm{d}t + \bar{\sigma}(t, X(t))\mathrm{d}W(t), \text{ with } X(t_0) = X_0,$$
>
> *where drift $\bar{\mu}(t, X(t))$ and diffusion $\bar{\sigma}(t, X(t))$ satisfy the standard Lipschitz conditions on the growth of these functions (as in Definition 2.1.3).*
> *Let $g(t, X)$ be a function of $X = X(t)$ and time t, with continuous partial derivatives, $\partial g/\partial X$, $\partial g^2/\partial X^2$, $\partial g/\partial t$. A stochastic variable $Y(t) := g(t, X)$ then also follows an Itô process, governed by the same Wiener process $W(t)$, i.e.,*
>
> $$\mathrm{d}Y(t) = \left(\frac{\partial g}{\partial t} + \bar{\mu}(t, X)\frac{\partial g}{\partial X} + \frac{1}{2}\frac{\partial^2 g}{\partial X^2}\bar{\sigma}^2(t, X)\right)\mathrm{d}t + \frac{\partial g}{\partial X}\bar{\sigma}(t, X)\mathrm{d}W(t).$$
>
> *The formal proof of this theorem is somewhat involved, and can be found in several textbooks on stochastic processes [Shreve, 2004].*

Again, Itô's lemma above is a short-hand notation for the integral formulation,

$$Y(t) = Y_0 + \int_{t_0}^{t} \left(\frac{\partial g}{\partial z} + \bar{\mu}(z, X)\frac{\partial g}{\partial X} + \frac{1}{2}\frac{\partial^2 g}{\partial X^2}\bar{\sigma}^2(z, X)\right) \mathrm{d}z$$

$$+ \int_{t_0}^{t} \frac{\partial g}{\partial X}\bar{\sigma}(z, X)\mathrm{d}W(z). \quad (2.11)$$

Itô's lemma can be used to find the solutions of a number of interesting integrals and SDEs.

Example 2.1.2 (Solution of Itô integral by Itô calculus) Recall the stochastic integral in (1.41) in Example 1.3.1, $\int_0^T W(t)\mathrm{d}W(t)$. This integral can also be solved with the help of Itô's calculus. Consider the basic stochastic process, i.e., $X(t) = W(t)$, which does not have a drift term and its volatility coefficient equals 1, so $\mathrm{d}X(t) = 0 \cdot \mathrm{d}t + 1 \cdot \mathrm{d}W(t)$.

If we apply Itô's lemma to $g(X(t)) = X^2(t)$, we find:

$$\mathrm{d}g(X) = \frac{\partial g}{\partial X}\mathrm{d}X(t) + \frac{1}{2}\frac{\partial^2 g}{\partial X^2}(\mathrm{d}X(t))^2$$

$$= 2X(t)\mathrm{d}X(t) + (\mathrm{d}X(t))^2.$$

After substitution of $W(t) = X(t)$, we find for the dynamics of $g(X(t)) = W^2(t)$,

$$\mathrm{d}W^2(t) = 2W(t)\mathrm{d}W(t) + (\mathrm{d}W(t))^2.$$

With $\mathrm{d}W(t)\mathrm{d}W(t) = \mathrm{d}t$ and integration of both sides, we arrive at:

$$\int_0^T \mathrm{d}W^2(t) = 2\int_0^T W(t)\mathrm{d}W(t) + \int_0^T \mathrm{d}t,$$

which is equivalent to:

$$\int_0^T W(t)\mathrm{d}W(t) = \frac{1}{2}\int_0^T \mathrm{d}W^2(t) - \frac{1}{2}\int_0^T \mathrm{d}t = \frac{1}{2}W^2(T) - \frac{1}{2}T,$$

providing indeed the solution to the integral of interest.

Although the technique presented above is elegant and shorter than dealing with the partitioned domains, as in (1.42), it is not always easy to find a function $g(\cdot)$ so that Itô's lemma can be applied. ♦

Example 2.1.3 (Stochastic integral $\int_0^T W(t)\mathrm{d}t$) We show here that

$$I(T) = \int_0^T W(t)\mathrm{d}t = \int_0^T (T-t)\mathrm{d}W(t).$$

By using $X(t) = W(t)$ and applying Itô's lemma to $g(t, X(t)) = (T-t)X(t)$, we find:

$$\mathrm{d}g(X) = \frac{\partial g}{\partial t}\mathrm{d}t + \frac{\partial g}{\partial X}\mathrm{d}X(t)$$
$$= -X(t)\mathrm{d}t + (T-t)\mathrm{d}X(t).$$

Integrating both sides, with $X(t) = W(t)$ gives us:

$$\int_0^T \mathrm{d}((T-t)W(t)) = -\int_0^T W(t)\mathrm{d}t + \int_0^T (T-t)\mathrm{d}W(t).$$

Clearly, $\int_0^T \mathrm{d}((T-t)W(t)) = (T-T)W(T) - W(T-0)W(0) = 0$, so

$$\int_0^T W(t)\mathrm{d}t = \int_0^T (T-t)\mathrm{d}W(t). \qquad (2.12)$$
♦

In Figure 2.1 we present a path of the Brownian motion $W(t)$, as well as two integral values, one from Example 2.1.2 and another from the example above.

2.1.3 Distributions of $S(t)$ and $\log S(t)$

By Itô's lemma we can show that the random variable $S := S(t)$ in (2.1) is from a *lognormal distribution*, i.e. $\log S$ is normally distributed. Using $g(t, S) = X(t) = \log S$, we obtain $\mathrm{d}g/\mathrm{d}S = 1/S$ and $\mathrm{d}^2 g/\mathrm{d}S^2 = -1/S^2$, so that Itô's lemma gives us,

$$\mathrm{d}g(t, S) = \mathrm{d}X(t) = \left(\mu - \frac{1}{2}\sigma^2\right)\mathrm{d}t + \sigma \mathrm{d}W^{\mathbb{P}}(t), \quad g(t_0) = \log S_0. \qquad (2.13)$$

As the Wiener increment $\mathrm{d}W(t)$ is normally distributed, with expectation 0 and variance $\mathrm{d}t$, Equation (2.13) confirms that $\mathrm{d}g(t, S)$ is normally distributed, with expectation $(\mu - \frac{\sigma^2}{2})\mathrm{d}t$ and variance $\sigma^2 \mathrm{d}t$. The stochastic variable $Y(t) = g(t, S)$ represents a sum of the *increments* $\mathrm{d}g$ (in the limit an infinite sum can be represented by an integral), so that $Y(t) = g(t, S)$ is normally distributed, with expectation $\log S_0 + (\mu - \frac{\sigma^2}{2})(t - t_0)$ and variance $\sigma^2(t - t_0)$.

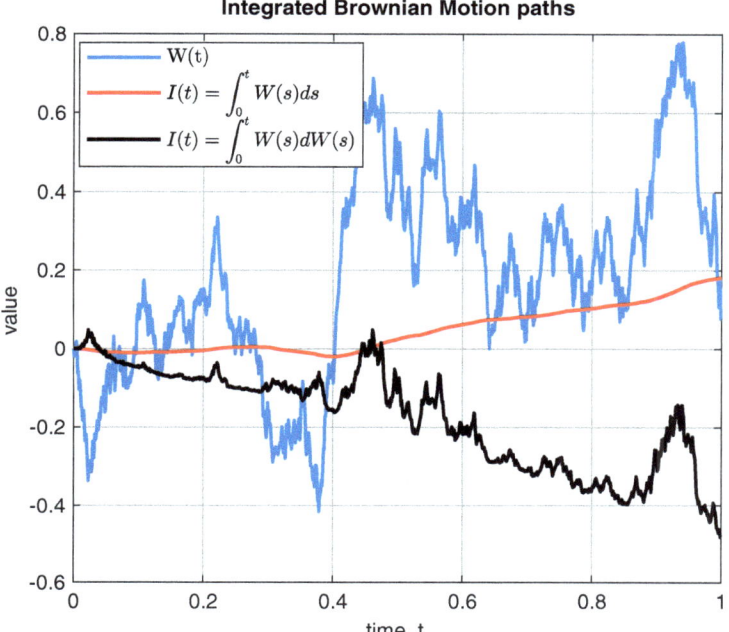

Figure 2.1: *Stochastic paths for $W(t)$, $\int_0^t W(s)\mathrm{d}s$ and $\int_0^t W(s)\mathrm{d}W(s)$, as a function of running time t.*

Example 2.1.4 (The BM and GBM distributions in time) Under the log-transformation, $X(t) = \log S(t)$, the right-hand side of the SDE (2.13) does not depend on $X(t)$, and therefore we can simply integrate both sides of the SDE,

$$\int_{t_0}^{T} \mathrm{d}X(t) = \int_{t_0}^{T} \left(\mu - \frac{1}{2}\sigma^2\right) \mathrm{d}t + \int_{t_0}^{T} \sigma \mathrm{d}W^{\mathbb{P}}(t), \qquad (2.14)$$

which yields the following solution,

$$X(T) = X(t_0) + \left(\mu - \frac{1}{2}\sigma^2\right)(T - t_0) + \sigma W^{\mathbb{P}}(T - t_0). \qquad (2.15)$$

Under the log-transformation, $X(T)$ is a normally distributed random variable with the following parameters:

$$X(T) \sim \mathcal{N}\left(X(t_0) + \left(\mu - \frac{1}{2}\sigma^2\right)(T - t_0), \sigma^2(T - t_0)\right).$$

With a back transformation, $S(T)$ is indeed the lognormally distributed random variable, given by:

$$S(T) \sim \exp(X(T)).$$

Furthermore, for $Z \sim \mathcal{N}(\mu, \sigma^2)$, the following equalities hold,

$$\mathbb{E}\left[e^Z\right] = e^{\mu + \frac{1}{2}\sigma^2}, \quad \operatorname{Var}\left[e^Z\right] = \left(e^{\sigma^2} - 1\right) e^{2\mu + \sigma^2}. \tag{2.16}$$

♦

Utilizing Itô's formula, we can thus determine the solution for $S(t)$ in (2.1), as

$$S(t) = S_0 \exp\left(\left(\mu - \frac{1}{2}\sigma^2\right)(t - t_0) + \sigma(W^{\mathbb{P}}(t) - W^{\mathbb{P}}(t_0))\right). \tag{2.17}$$

For $X(t) := \log S(t)$, we see, based on the findings in Example 2.1.4, that

$$F_{X(t)}(x) = \mathbb{P}[X(t) \leq x] = \tag{2.18}$$

$$:= \frac{1}{\sigma\sqrt{2\pi(t - t_0)}} \int_{-\infty}^{x} \exp\left(-\frac{(z - \log S_0 - (\mu - \frac{\sigma^2}{2})(t - t_0))^2}{2\sigma^2(t - t_0)}\right) dz,$$

and correspondingly the probability density function, $f_X(x) := f_{X(t)}(x)$, reads

$$f_X(x) = \frac{d}{dx} F_X(x) = \frac{1}{\sigma\sqrt{2\pi(t - t_0)}} \exp\left(-\frac{(x - \log S_0 - (\mu - \frac{\sigma^2}{2})(t - t_0))^2}{2\sigma^2(t - t_0)}\right). \tag{2.19}$$

The probability distribution function for $S(t)$ can now be obtained by the transformation $S(t) = \exp(X(t))$; for $x > 0$:

$$F_{S(t)}(x) = \mathbb{P}[S(t) \leq x] = \mathbb{P}[e^{X(t)} \leq x] = \mathbb{P}[X(t) \leq \log x] = F_{X(t)}(\log x)$$

$$= \frac{1}{\sigma\sqrt{2\pi(t - t_0)}} \int_{-\infty}^{\log x} \exp\left(-\frac{(z - \log S_0 - (\mu - \frac{\sigma^2}{2})(t - t_0))^2}{2\sigma^2(t - t_0)}\right) dz,$$

which is lognormal. The probability density function for $S(t)$, $f_S(x) := f_{S(t)}(x)$, is found to be

$$f_S(x) = \frac{1}{\sigma x \sqrt{2\pi(t - t_0)}} \exp\left(-\frac{(\log \frac{x}{S_0} - (\mu - \frac{\sigma^2}{2})(t - t_0))^2}{2\sigma^2(t - t_0)}\right), \quad x > 0. \tag{2.20}$$

In Figure 2.2 the time evolution of the probability density functions of the processes $X(t)$ and $S(t)$ are respectively presented. The parameter values are in the figure's caption. It can be seen from the graphs in Figure 2.2 that for a short time period the densities of $X(t)$ and $S(t)$ resemble each other well. This can be explained by the following relation between the normal and lognormal distributions. If $X \sim \mathcal{N}(0, 1)$, then $Y = \exp(X)$ is a lognormal distribution, and by the well-known MacLaurin expansion, see Equation (1.10), of the variable Y

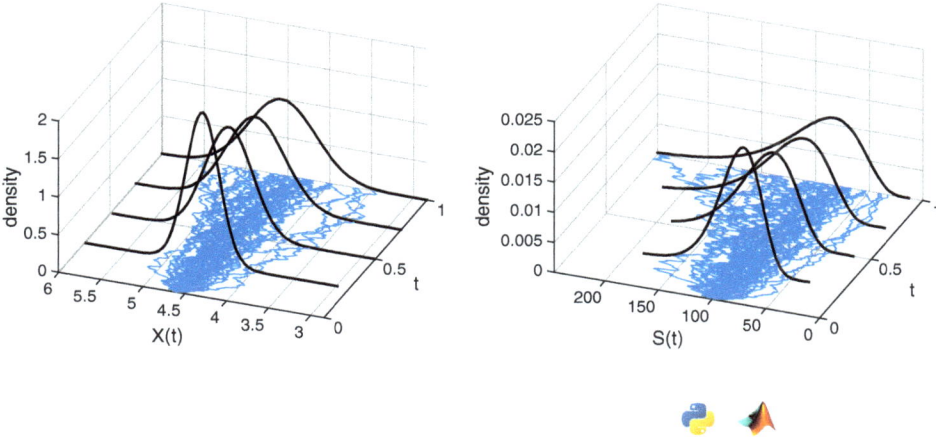

Figure 2.2: *Paths and the corresponding densities. Left: $X(t) = \log S(t)$ and Right: $S(t)$ with the following configuration: $S_0 = 100$, $\mu = 0.05$, $\sigma = 0.4$; $T = 1$.*

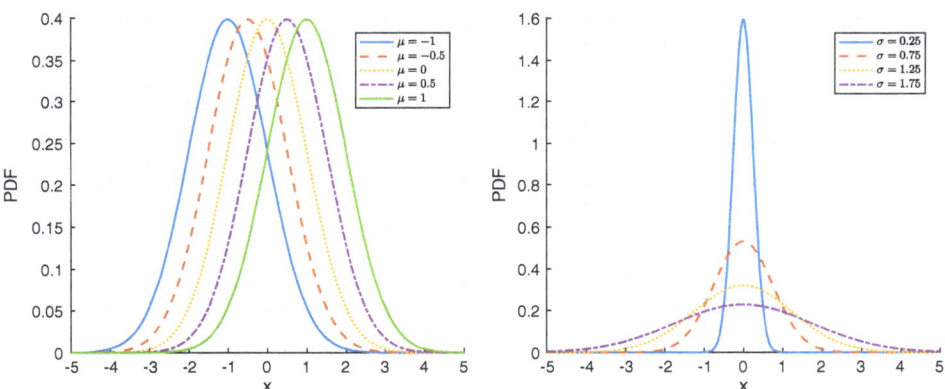

Figure 2.3: *Probability density function for the normal distribution. Left: the impact of parameter μ, with $\sigma = 1$; Right: the impact of σ, using $\mu = 0$.*

around 0, we find $Y \approx 1 + X \sim \mathcal{N}(1, 1)$. This approximation is insightful in the case when $T \to 0$.

Example 2.1.5 (Pictures of normal and lognormal distributions)
In Figures 2.3 and 2.4, graphical representations of the probability density functions, for different sets of parameters, are presented for respectively the normal and the lognormal distribution.

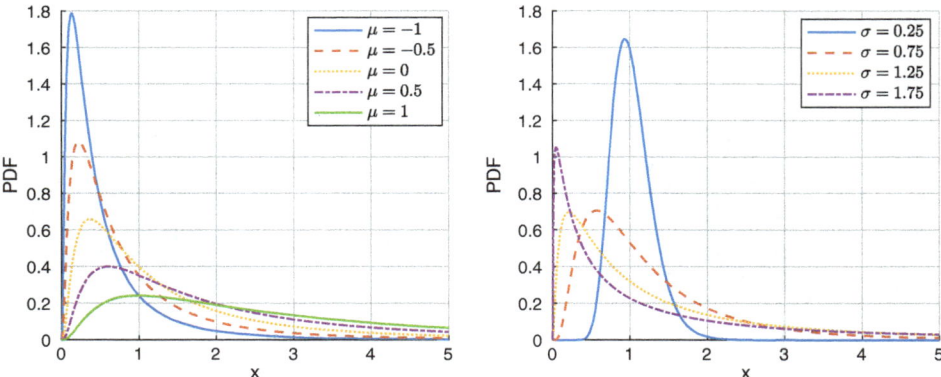

Figure 2.4: *Probability density function for the lognormal distribution. Left: the impact of parameter μ, with $\sigma = 1$; Right: impact of parameter σ, with $\mu = 0$.*

2.2 First generalizations

Here, we present some first variations of the basic GBM asset price process. We look into proportional dividends, which may be a first generalization of the GBM drift term, and into volatility generalizations, including a time-dependent volatility function.

2.2.1 Proportional dividend model

In the generic GBM stock price model in (2.1), a dividend payment has not been modeled. However, typically a company pays dividends once or twice a year. The exact amounts of dividend and the payment times may vary each year. At the time the dividend is paid, there will be a drop in the value of the stock.

Different mathematical models are available for the dividend payments, like deterministic or stochastic, continuous or discrete time models.

A continuous and constant dividend yield of size $q \cdot S(t)$, with a constant factor $q < 1$, is often used and it represents the fact that in a time instance dt the underlying asset pays out a *proportional dividend* of size $qS(t)dt$. This is considered a suitable model particularly for stock indices (in which multiple stocks are represented). Arbitrage considerations indicate that the asset price must fall by the amount of dividend payment, i.e., we have in the case of proportional dividend payment,

$$dS(t) = (\mu - q)S(t)dt + \sigma S(t)dW^{\mathbb{P}}(t), \qquad (2.21)$$

modeling a dividend paying stock with a continuous stream of dividends.

2.2.2 Volatility variation

In the GBM model, the stock dynamics include the volatility parameter σ. One may wonder why the volatility is often written as σ, and why it is not set equal to $\sigma S(t)$, as this term $\sigma S(t)$ is in front of the random driver $\mathrm{d}W(t)$. When we use the word volatility, we actually mean "the volatility of the stock *return*". The stock performance over a time period $[t, t+\Delta t]$ is determined by the ratio, $\frac{S(t+\Delta t)-S(t)}{S(t)}$. For an instantaneously small period $\Delta t \to 0$ this ratio is given by

$$\lim_{\Delta t \to 0} \frac{S(t+\Delta t) - S(t)}{S(t)} = \frac{\mathrm{d}S(t)}{S(t)} = \mu \mathrm{d}t + \boxed{\sigma} \mathrm{d}W^{\mathbb{P}}(t). \qquad (2.22)$$

From the stock return it is clear why σ is the volatility parameter of process $S(t)$.

Other asset models, in which the volatility in the $\mathrm{d}S(t)/S(t)$ term is written as a function of $S(t)$, are the so-called *parametric local volatility models*. Two well-known parametric local volatility models are the *quadratic model*, with its asset dynamics given by:

$$\frac{\mathrm{d}S(t)}{S(t)} = \mu \mathrm{d}t + \boxed{\sigma S(t)} \mathrm{d}W^{\mathbb{P}}(t),$$

and the *so-called 3/2 model*, with

$$\frac{\mathrm{d}S(t)}{S(t)} = \mu \mathrm{d}t + \boxed{\sigma \sqrt{S(t)}} \mathrm{d}W^{\mathbb{P}}(t).$$

These models form building blocks particularly for some stochastic interest rate processes. Some other parametric local volatility models will be discussed in follow-up sections.

2.2.3 Time-dependent volatility

Another generalization of the GBM asset price model, away from constant volatility, may be to use a time-dependent volatility coefficient, using $\sigma(t)$, i.e.

$$\mathrm{d}S(t) = \mu S(t)\mathrm{d}t + \sigma(t)S(t)\mathrm{d}W^{\mathbb{P}}(t), \quad S(t_0) = S_0, \qquad (2.23)$$

where $\sigma(t)$ is a deterministic function. In the example below, we will match the moments of a model with constant volatility and one with time-dependent volatility.

Example 2.2.1 (Time-dependent volatility and moment matching)
As an illustrative example of moment matching, we consider the following two stochastic processes, $X(t)$ and $Y(t)$, that are governed by the following SDEs:

$$\mathrm{d}X(t) = \left(\mu - \frac{1}{2}\sigma^2(t)\right)\mathrm{d}t + \sigma(t)\mathrm{d}W^{\mathbb{P}}(t),$$

$$\mathrm{d}Y(t) = \left(\mu - \frac{1}{2}\sigma_*^2\right)\mathrm{d}t + \sigma_*\mathrm{d}W^{\mathbb{P}}(t),$$

with a time-dependent volatility $\sigma(t)$ for process $X(t)$ and a constant volatility parameter σ_* for process $Y(t)$. The expectations of $X(T)$ and $Y(T)$ can be determined as

$$\mathbb{E}[X(T)] = X_0 + \int_0^T \left(\mu - \frac{1}{2}\sigma^2(t)\right) dt, \quad \mathbb{E}[Y(T)] = Y_0 + \left(\mu - \frac{1}{2}\sigma_*^2\right) T, \quad (2.24)$$

and the variance of process $X(t)$ reads:

$$\mathrm{Var}[X(T)] = \mathbb{E}[X^2(T)] - (\mathbb{E}[X(T)])^2 = \mathbb{E}\left[\int_0^T \sigma(t) dW^{\mathbb{P}}(t)\right]^2 = \int_0^T \sigma^2(t) dt,$$

where the last step is based on the Itô isometry (as in Section 1.3.2). The variance of process $Y(t)$ equals $\mathrm{Var}[Y(T)] = \sigma_*^2 T$. By equating the variances of the processes X and Y, i.e. $\mathrm{Var}[X(T)] = \mathrm{Var}[Y(T)]$, we find

$$\boxed{\sigma_* = \sqrt{\frac{1}{T} \int_0^T \sigma^2(t) dt}.} \quad (2.25)$$

Using this particular choice of σ_*, also the expectations of $X(T)$ and $Y(T)$ (with $X_0 = Y_0$) are equal, as in (2.24) we will get

$$\mathbb{E}[Y(T)] = Y_0 + \left(\mu - \frac{1}{2}(\sigma_*)^2\right) T$$

$$= X_0 + \left(\mu - \frac{1}{2}\frac{1}{T}\int_0^T \sigma^2(t) dt\right) T = \mathbb{E}[X(T)].$$

The processes X and Y are log-transformed GBM asset prices and these are normally distributed. In that case only the first two moments have to match in order to guarantee equality in distribution. ◆

With the moment matching technique, both processes will have exactly the same marginal distributions, however, their transitional distributions will differ.

2.3 Martingales and asset prices

The asset price models discussed will satisfy the condition $\mathbb{E}[|S(t)|] < \infty$. Pricing of financial derivatives, like financial options, heavily depends on calculating the expectation of some payoff with respect to an underlying process. Therefore, it is crucial that the moments exist.

The martingale property implies that a mathematical model of a financial product is free of arbitrage, i.e. there are no risk-free profits present in the mathematical model for the stochastic quantity under consideration. The probability measure under which the martingale property holds is often called the

Introduction to Financial Asset Dynamics 41

risk-neutral measure, and it is denoted by $\mathbb{Q}$. Particularly, discounted tradable assets will be martingales under the risk-neutral measure.

2.3.1 $\mathbb{P}$-measure prices

Financial stock price processes as observed in the stock exchanges are not martingales usually (i.e. they are not completely unpredictable).

Example 2.3.1 (Expectation of $S(t)$ under $\mathbb{P}$-measure) From Equation (2.23) (the time-dependent volatility case) we have,

$$S(t) = S_0 e^{\int_{t_0}^{t} (\mu - \frac{1}{2}\sigma^2(z))\mathrm{d}z + \int_{t_0}^{t} \sigma(z)\mathrm{d}W^{\mathbb{P}}(z)},$$

and the expectation is given by,

$$\mathbb{E}^{\mathbb{P}}\left[S(t)|\mathcal{F}(t_0)\right] = \mathbb{E}^{\mathbb{P}}\left[S_0 e^{\frac{1}{2}\int_{t_0}^{t}(\mu - \sigma^2(z))\mathrm{d}z + \int_{t_0}^{t}\sigma(z)\mathrm{d}W^{\mathbb{P}}(z)}\bigg|\mathcal{F}(t_0)\right]$$

$$= S_0 e^{\mu(t-t_0) - \frac{1}{2}\int_{t_0}^{t}\sigma^2(z)\mathrm{d}z} \mathbb{E}^{\mathbb{P}}\left[e^{\int_{t_0}^{t}\sigma(z)\mathrm{d}W^{\mathbb{P}}(z)}\bigg|\mathcal{F}(t_0)\right].$$

For a normally distributed $X \sim \mathcal{N}(0,1)$, it can be derived that $\mathbb{E}[e^{aX}] = e^{\frac{1}{2}a^2}$, and,

$$\int_{t_0}^{t} \sigma(z)\mathrm{d}W^{\mathbb{P}}(z) \sim \mathcal{N}\left(0, \int_{t_0}^{t} \sigma^2(z)\mathrm{d}z\right).$$

Therefore,

$$\mathbb{E}^{\mathbb{P}}\left[e^{\int_{t_0}^{t}\sigma(z)\mathrm{d}W^{\mathbb{P}}(z)}\bigg|\mathcal{F}(t_0)\right] = e^{\frac{1}{2}\int_{t_0}^{t}\sigma^2(z)\mathrm{d}z},$$

so that,

$$\mathbb{E}^{\mathbb{P}}\left[S(t)|\mathcal{F}(t_0)\right] = S_0 e^{\mu(t-t_0) - \frac{1}{2}\int_{t_0}^{t}\sigma^2(z)\mathrm{d}z} \mathbb{E}^{\mathbb{P}}\left[e^{\int_{t_0}^{t}\sigma(z)\mathrm{d}W^{\mathbb{P}}(z)}\bigg|\mathcal{F}(t_0)\right]$$

$$= S_0 e^{\mu(t-t_0) - \frac{1}{2}\int_{t_0}^{t}\sigma^2(z)\mathrm{d}z} e^{\frac{1}{2}\int_{t_0}^{t}\sigma^2(z)\mathrm{d}z}$$

$$= S_0 e^{\mu(t-t_0)}.$$ ♦

In a time interval Δt, we expect $S(t)$, as defined in Equation (2.1), to grow at some positive rate μ under the real-world probability measure $\mathbb{P}$. The stock price is governed by the lognormal distribution, with the expectation given by

$$\boxed{\mathbb{E}^{\mathbb{P}}[S(t)|\mathcal{F}(t_0)] = S_0 e^{\mu(t-t_0)},}$$

so that $\mathbb{E}^{\mathbb{P}}[S(t+\Delta t)|\mathcal{F}(t)] > S(t)$, i.e. $S(t)$ is not a martingale.[2] In particular, we may assume that $\mu > r$, where r is the risk-free interest rate (as otherwise one would not invest in the asset), so that also for the discounted asset price process, we find

$$\mathbb{E}^{\mathbb{P}}[e^{-r\Delta t}S(t+\Delta t)|\mathcal{F}(t)] > S(t).$$

---

[2] It is a so-called sub-martingale.

> Let us therefore consider the stock process, $S(t)$, under a different measure, $\mathbb{Q}$, as follows
> $$dS(t) = rS(t)dt + \sigma S(t)dW^{\mathbb{Q}}(t), \quad S(t_0) = S_0, \qquad (2.26)$$
> replacing the drift rate μ by the risk free interest rate r. The solution of Equation (2.26) reads:
> $$S(t) = S_0 \exp\left(\left(r - \frac{1}{2}\sigma^2\right)(t-t_0) + \sigma(W^{\mathbb{Q}}(t) - W^{\mathbb{Q}}(t_0))\right).$$

The concept of changing measures is very fundamental in computational finance. Chapter 7, particularly Section 7.2, is devoted to this concept, which is then used in almost all further chapters of this book.

2.3.2 $\mathbb{Q}$-measure prices

The stock price $S(t)$ is also governed by the lognormal distribution under this measure $\mathbb{Q}$, however, with the expectation given by:

$$\boxed{\mathbb{E}^{\mathbb{Q}}[S(t)|\mathcal{F}(t_0)] = S_0 e^{r(t-t_0)}.}$$

The expectation of $S(t)$, conditioned on time s with $s < t$, reads

$$\mathbb{E}^{\mathbb{Q}}[S(t)|\mathcal{F}(s)] = e^{\log S(s) + r(t-s)} = S(s)e^{r(t-s)} \neq S(s),$$

which implies that process $S(t)$ for $t \in [t_0, T]$ is also *not* a martingale under measure $\mathbb{Q}$.

With a *money-savings account*, i.e.,

$$M(t) = M(s)e^{r(t-s)}, \qquad (2.27)$$

we have

$$\mathbb{E}^{\mathbb{Q}}\left[\frac{S(t)}{M(t)}\bigg|\mathcal{F}(s)\right] = \frac{e^{-r(t-s)}}{M(s)}\mathbb{E}^{\mathbb{Q}}[S(t)|\mathcal{F}(s)]$$
$$= \frac{e^{-r(t-s)}}{M(s)}S(s)e^{r(t-s)} = \frac{S(s)}{M(s)}.$$

> The compensation term in the definition of the discounted process is related to the *money-savings account* $M(t)$, i.e.
> $$S(t)e^{-r(t-t_0)} \stackrel{\text{def}}{=} \frac{S(t)}{M(t)},$$
> where $dM(t) = rM(t)dt$, with $M(t_0) = 1$.

Introduction to Financial Asset Dynamics

The money-savings account $M(t)$ is the so-called *numéraire*, expressing the unit of measure in which all other prices are expressed. Numéraire is a French word, meaning the basic standard by which values are measured.

So, we have a probability measure, $\mathbb{Q}$ (the risk-neutral measure) under which the stock price, discounted by the risk-free interest rate r, becomes a martingale, i.e.

$$\mathbb{E}^{\mathbb{Q}}\left[e^{-r\Delta t}S(t+\Delta t)|\mathcal{F}(t)\right] = \mathbb{E}^{\mathbb{Q}}\left[\frac{M(t)}{M(t+\Delta t)}S(t+\Delta t)|\mathcal{F}(t)\right]$$

$$= M(t)\mathbb{E}^{\mathbb{Q}}\left[\frac{S(t+\Delta t)}{M(t+\Delta t)}|\mathcal{F}(t)\right]$$

$$= M(t)\frac{S(t)}{M(t)} = S(t). \qquad (2.28)$$

The asset price which satisfies relation (2.28) is thus a GBM asset price process, in which the drift parameter is set equal to the risk-free interest rate $\mu = r$, compare (2.1) and (2.26).

Example 2.3.2 (Asset paths) In Figure 2.5 we present some discrete paths that have been generated by a GBM process with $\mu = 0.15$, $r = 0.05$ and $\sigma = 0.1$. The left-hand picture in Figure 2.5 displays the paths of a discounted stock process, $S(t)/M(t)$, where $S(t)$ is defined under the real-world measure $\mathbb{P}$, where it has a drift parameter $\mu = 0.15$. The right-hand picture shows the paths for $S(t)/M(t)$, where $S(t)$ has drift parameter $r = 0.05$. Since $\mu > r$, the discounted stock price under the real-world measure will increase much faster than the stock process under the risk-neutral measure. In other words, under the real-world measure $S(t)/M(t)$ is a sub-martingale.

As we will see later, the valuation of financial options is related to the computation of expectations. The notations $\mathbb{E}^{\mathbb{Q}}[g(T,S)]$ and $\mathbb{E}^{\mathbb{P}}[g(T,S)]$ indicate that we take the expectation of a certain function, $g(T,S)$, which depends on the stochastic asset price process $S(t)$. Asset price $S(t)$ can be defined under a specific measure, like under $\mathbb{P}$ or $\mathbb{Q}$, where $\mathbb{P}[\cdot]$ and $\mathbb{Q}[\cdot]$ indicate the associated probabilities. If we take, for example, $g(T,S) = \mathbb{1}_{S(T)>0}$, then the following equalities hold:

$$\mathbb{E}^{\mathbb{P}}[\mathbb{1}_{S(T)>0}] = \int_\Omega \mathbb{1}_{S(T)>0} d\mathbb{P} = \mathbb{P}[S(T)>0], \qquad (2.29)$$

$$\mathbb{E}^{\mathbb{Q}}[\mathbb{1}_{S(T)>0}] = \int_\Omega \mathbb{1}_{S(T)>0} d\mathbb{Q} = \mathbb{Q}[S(T)>0], \qquad (2.30)$$

where the notation $\mathbb{1}_{X\in\Omega}$ stands for the indicator function of the set Ω, as in (1.1).

Superscripts $\mathbb{Q}$ and $\mathbb{P}$ indicate which process is used in the expectation operator to calculate the corresponding integrals. In other words, it indicates the underlying dynamics that have been used.

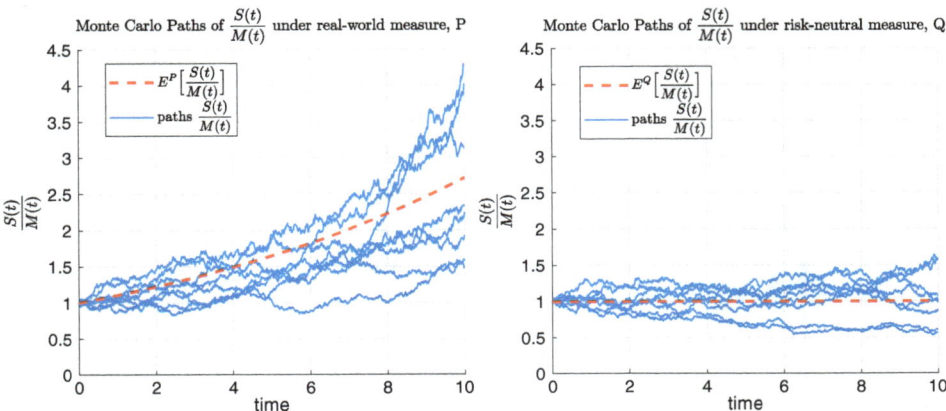

Figure 2.5: *Paths for the discounted stock process, $S(t)/M(t)$, with $S(t)$ under real-world measure $\mathbb{P}$ (left) and $S(t)$ under risk-neutral measure $\mathbb{Q}$ (right), with parameters $r = 0.05$, $\mu = 0.15$ and $\sigma = 0.1$.*

2.3.3 Parameter estimation under real-world measure $\mathbb{P}$

In this section we give insight in the parameter estimation for a stochastic process under the real-world measure $\mathbb{P}$. Estimation under the real-world measure means that the model parameters for a stochastic process are obtained by a *calibration to historical stock price values*. Calibration will be completely different under the risk-neutral measure $\mathbb{Q}$, as we will discuss in the chapter to follow. The parameters here will be estimated by using a very popular statistical estimation technique, which is called "maximum likelihood estimation method" (MLE) [Harris and Stocker, 1998]. The idea behind the method is to find the parameter estimates of the underlying probability distribution for a given data set.

As a first example of a stochastic process let us take the arithmetic Brownian motion process, which was introduced by Louis Bachelier [Bachelier, 1900] and is given by:

$$dX(t) = \mu dt + \sigma dW^{\mathbb{P}}(t), \tag{2.31}$$

with its solution:

$$X(t) = X(t_0) + \mu(t - t_0) + \sigma(W^{\mathbb{P}}(t) - W^{\mathbb{P}}(t_0)), \quad \text{or}$$

$$X(t) \sim \mathcal{N}\left(X(t_0) + \mu(t - t_0), \sigma^2(t - t_0)\right). \tag{2.32}$$

In order to forecast "tomorrow's" (at time $t + \Delta t$) value, given that we observed the values up to now (where "now" is time t), we can calculate the conditional expectation as follows,

$$\mathbb{E}^{\mathbb{P}}\left[X(t+\Delta t)|\mathcal{F}(t)\right] = \mathbb{E}^{\mathbb{P}}\left[X(t) + \mu(t + \Delta t - t) + \sigma(W^{\mathbb{P}}(t+\Delta t) - W^{\mathbb{P}}(t))|\mathcal{F}(t)\right]$$

$$= \mu \Delta t + \mathbb{E}^{\mathbb{P}}\left[X(t)|\mathcal{F}(t)\right] + \mathbb{E}^{\mathbb{P}}\left[\sigma(W^{\mathbb{P}}(t+\Delta t) - W^{\mathbb{P}}(t))|\mathcal{F}(t)\right].$$

Because of the measurability principle, $\mathbb{E}^{\mathbb{P}}[X(t)|\mathcal{F}(t)] = X(t)$ and by the property of independent increments the second expectation is equal to zero, i.e.,

$$\mathbb{E}^{\mathbb{P}}[X(t+\Delta t)|\mathcal{F}(t)] = X(t) + \mu \Delta t.$$

For the variance, we obtain,

$$\begin{aligned} &\text{Var}^{\mathbb{P}}[X(t+\Delta t)|\mathcal{F}(t)] \\ &= \text{Var}^{\mathbb{P}}[X(t) + \mu(t+\Delta t - t) + \sigma(W^{\mathbb{P}}(t+\Delta t) - W^{\mathbb{P}}(t))|\mathcal{F}(t)] \\ &= \text{Var}^{\mathbb{P}}[X(t)|\mathcal{F}(t)] + \text{Var}^{\mathbb{P}}\left[\sigma\sqrt{\Delta t}\widetilde{Z}|\mathcal{F}(t)\right], \end{aligned}$$

with $\widetilde{Z} \sim \mathcal{N}(0,1)$. By Equation (2.32) and the fact that[3] $\text{Var}^{\mathbb{P}}[X(t)|\mathcal{F}(t)] = 0$, we find:

$$\text{Var}^{\mathbb{P}}[X(t+\Delta t)|\mathcal{F}(t)] = \sigma^2 \Delta t.$$

So, we arrive at the prediction of the $t + \Delta t$ value, given the information[4] at time t, i.e. at the conditional random variable,

$$X(t+\Delta t)|X(t) \sim \mathcal{N}\left(X(t) + \mu \Delta t, \sigma^2 \Delta t\right), \qquad (2.33)$$

which is a normal random variable for which the parameters μ and σ have to be estimated.

Parameter estimation for arithmetic Brownian motion

Since the conditional distribution is normally distributed, the log-likelihood functional form is known. Assuming independence of the observations and by setting the observation values (typically historical values of some return or log-asset) $X(t_0), X(t_1), \ldots, X(t_m)$, the likelihood function $L(\hat{\mu}, \hat{\sigma}^2 | X(t_0), X(t_1), \ldots, X(t_m))$ is given by,

$$L(\hat{\mu}, \hat{\sigma}^2 | X(t_0), X(t_1), \ldots, X(t_m)) = \prod_{k=0}^{m-1} f_{X(t_{k+1})|X(t_k)}(X(t_{k+1})), \qquad (2.34)$$

with $\hat{\mu}$ and $\hat{\sigma}$ to be estimated, By Equation (2.33) we find:

$$f_{X(t_{k+1})|X(t_k)}(x) = f_{\mathcal{N}(X(t_k) + \hat{\mu}\Delta t, \hat{\sigma}^2 \Delta t)}(x), \qquad (2.35)$$

---

[3]Note that because of the measurability, we find:

$$\text{Var}^{\mathbb{P}}[X(t)|\mathcal{F}(t)] = \mathbb{E}^{\mathbb{P}}[X(t)^2|\mathcal{F}(t)] - \left(\mathbb{E}^{\mathbb{P}}[X(t)|\mathcal{F}(t)]\right)^2 = X^2(t) - X^2(t) = 0.$$

[4]By the notation $|X(t)$, we also indicate "the knowledge up-to time t", like when using $\mathcal{F}(t)$.

and Equation (2.34) reads,

$$L(\hat{\mu}, \hat{\sigma}^2 | X(t_0), X(t_1), \ldots, X(t_m))$$
$$= \prod_{k=0}^{m-1} f_{\mathcal{N}(X(t_k)+\hat{\mu}\Delta t, \hat{\sigma}^2 \Delta t)}(X(t_{k+1}))$$
$$= \prod_{k=0}^{m-1} \frac{1}{\sqrt{2\pi\hat{\sigma}^2 \Delta t}} \exp\left(-\frac{(X(t_{k+1}) - X(t_k) - \hat{\mu}\Delta t)^2}{2\hat{\sigma}^2 \Delta t}\right).$$

To simplify the maximization, we use the logarithmic transformation, i.e.,

$$\log L(\hat{\mu}, \hat{\sigma}^2 | \ldots) = \log\left(\frac{1}{\sqrt{2\pi\hat{\sigma}^2 \Delta t}}\right)^m - \sum_{k=0}^{m-1} \frac{(X(t_{k+1}) - X(t_k) - \hat{\mu}\Delta t)^2}{2\hat{\sigma}^2 \Delta t}.$$

Now,

$$\log\left(\frac{1}{\sqrt{2\pi\hat{\sigma}^2 \Delta t}}\right)^m = \log\left(2\pi\hat{\sigma}^2 \Delta t\right)^{-0.5m} = -0.5m \log\left(2\pi\hat{\sigma}^2 \Delta t\right),$$

by which we obtain,

$$\log L(\hat{\mu}, \hat{\sigma}^2 | \ldots) = -0.5m \log\left(2\pi\hat{\sigma}^2 \Delta t\right) - \sum_{k=0}^{m-1} \frac{(X(t_{k+1}) - X(t_k) - \hat{\mu}\Delta t)^2}{2\hat{\sigma}^2 \Delta t}.$$

To determine the maximum of the log-likelihood, the first-order conditions have to be satisfied for both parameters,

$$\frac{\partial}{\partial \hat{\mu}} \log L(\hat{\mu}, \hat{\sigma}^2 | \ldots) = 0, \quad \frac{\partial}{\partial \hat{\sigma}^2} \log L(\hat{\mu}, \hat{\sigma}^2 | \ldots) = 0.$$

This gives us the following estimators:

$$\hat{\mu} = \frac{1}{m\Delta t}(X(t_m) - X(t_0)), \quad \hat{\sigma}^2 = \frac{1}{m\Delta t} \sum_{k=0}^{m-1} (X(t_{k+1}) - X(t_k) - \hat{\mu}\Delta t)^2. \tag{2.36}$$

Based on these estimators $\hat{\mu}$, $\hat{\sigma}^2$, for given historical values $X(t_0), X(t_1), \ldots, X(t_m)$, we may determine the "historical" parameters for the process $X(t)$.

Log-normal distribution

Using the methodology above, in this numerical experiment we will forecast the stock value of an electric vehicle company. The ABM process we discussed earlier is not the most adequate process for this purpose, as it may give rise to negative

Introduction to Financial Asset Dynamics

stock values. An alternative is GBM, which may be closely related to ABM. Under the real-world measure $\mathbb{P}$ the following process for the stock is considered,

$$\mathrm{d}S(t) = \left(\mu + \frac{1}{2}\sigma^2\right)S(t)\mathrm{d}t + \sigma S(t)\mathrm{d}W^{\mathbb{P}}(t). \tag{2.37}$$

By the log-transformation and using Itô's lemma, we find,

$$\mathrm{d}X(t) := \mathrm{d}\log S(t) = \mu\mathrm{d}t + \sigma\mathrm{d}W^{\mathbb{P}}(t). \tag{2.38}$$

So, under the log-transformation of the process $S(t)$ in (2.37), we arrive at the estimates that were obtained from ABM in (2.36).

Now, given the observation of electric vehicle company stock prices $S(t_i)$ (in Figure 2.6), first we perform the log-transformation to obtain $X(t) := \log S(t)$ and subsequently estimate the parameters $\hat{\mu}$ and $\hat{\sigma}$ in (2.36). The data set contains closing prices between 2010 and 2018, giving us the following estimators $\hat{\mu} = 0.0014$ and $\hat{\sigma} = 0.0023$.

With $\hat{\mu}$ and $\hat{\sigma}$ for μ, σ in (2.37) determined, we may simulate "future" realizations for the stock prices. In Figure 2.6 (left) the historical stock prices are presented, whereas in the right-side figure the forecast for the future values is shown. In the experiments this forecast of the future prices is purely based on the historical stock realizations. This is the essence of working under the $\mathbb{P}$-real world measure. Note that by assuming a GBM process the stock prices are expected to grow in time. However, by changing the assumed stochastic process we may obtain a different forecast.

Estimating the parameter μ under measure $\mathbb{P}$ is thus related to fitting the process in (2.1) to *historical stock values*. Note that the paths in Figure 2.5 suggest that under the real-world measure the expected returns of the discounted

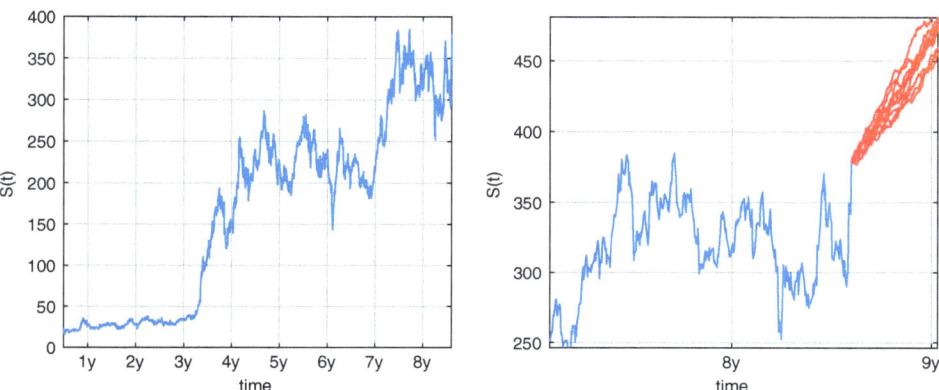

Figure 2.6: *Left: historical stock values of electric car company; Right: forecast for the future performance of the stock.*

stock are typically higher than one would expect under the risk-neutral measure. This is partially an argument why certain hedge funds work with processes under the ℙ-measure, where they aim to forecast the stock prices using Econometrics tools. Speculation is also typically done under this measure. Risk management, meaning analyzing the future behavior of assets in the real world, as it is often done by the regulator of financial institutions, is usually done under measure ℙ as well. In this case, specific asset scenarios (*back testing*) or even when *stress-testing*, extreme asset scenarios are provided under which the companies' balance sheets should be evaluated.

When we deal with financial option valuation in the chapters to follow, however, we need to fit the parameter values under the so-called *risk-neutral measure* ℚ. Then, we are mainly interested in the parameters *for a future time period*, and wish to extract the relevant information from financial products that may give us information about the expectations of the market participants about the future performance of the financial asset.

> This means that we aim to extract the relevant *implied asset information* from arbitrage-free option contracts and other financial derivatives. Financial institutions work under measure ℚ when pricing financial derivative products for their clients.

Pension funds should worry about ℙ measure valuation (liabilities and their performance and risk management in the real world), and also about ℚ measure calculations, when derivative contracts form a part of the pension investment portfolios. In essence, hedge funds may "bet" on the future, while banks and pension funds get a premium upfront and they hedge their position to keep the premium intact.

Introduction to Financial Asset Dynamics

2.4 Exercise set

Exercise 2.1 Apply Itô's lemma to find:

a. The dynamics of process $g(t) = S^2(t)$, where $S(t)$ follows a log-normal Brownian motion given by:
$$\frac{\mathrm{d}S(t)}{S(t)} = \mu \mathrm{d}t + \sigma \mathrm{d}W(t),$$
with constant parameters μ, σ and Wiener process $W(t)$.

b. The dynamics for $g(t) = 2^{W(t)}$, where $W(t)$ is a standard Brownian motion. Is this a martingale?

Exercise 2.2 Apply Itô's lemma to show that:

a. $X(t) = \exp(W(t) - \frac{1}{2}t)$ solves $\mathrm{d}X(t) = X(t)\mathrm{d}W(t)$,

b. $X(t) = \exp(2W(t) - t)$ solves $\mathrm{d}X(t) = X(t)\mathrm{d}t + 2X(t)\mathrm{d}W(t)$,

Exercise 2.3 Derive the Itô integration-by-parts rule, which reads,

$$\int_0^T \mathrm{d}X(t)\mathrm{d}Y(t) = (X(t)Y(t))|_{t=0}^{t=T} - \int_0^T X(t)\mathrm{d}Y(t) - \int_0^T Y(t)\mathrm{d}X(t). \qquad (2.39)$$

This can be written in differential form, as

$$\mathrm{d}(X(t) \cdot Y(t)) = Y(t) \cdot \mathrm{d}X(t) + X(t) \cdot \mathrm{d}Y(t) + \mathrm{d}X(t) \cdot \mathrm{d}Y(t). \qquad (2.40)$$

where the additional term, $\mathrm{d}X(t)\mathrm{d}Y(t)$ does not appear in the deterministic integration-by-parts rule.

Derive this rule by means of the following discrete sum,

$$\sum_{k=1}^N (X(t_{k+1}) - X(t_k))(Y(t_{i+k}) - Y(t_k)).$$

Exercise 2.4 For this exercise, it is necessary to download some freely available stock prices from the web.

a. Find two data sets with daily stock prices of two different stocks S_1 and S_2, that are "as independent as possible". Check the independence by means of a scatter plot of daily returns.

b. Find yourself two data sets with daily stock prices of two different stocks S_3 and S_4, that are "as dependent as possible". Check the dependence by means of a scatter plot of daily returns.

Exercise 2.5 Asset price process $S(t)$ is governed by the geometric Brownian motion,

$$S(t) = S_0 e^{(\mu - \frac{1}{2}\sigma^2)t + \sigma\sqrt{t}Z}, \text{ where } Z \sim \mathcal{N}(0,1)$$

with the lognormal density function,

$$f(x) = \frac{1}{x\sigma\sqrt{2\pi t}} \exp\left(\frac{-(\log(x/S_0) - (\mu - \sigma^2/2)t)^2}{2\sigma^2 t}\right), \text{ for } x > 0.$$

Determine $\mathbb{E}[S(t)]$ and $\mathbb{V}\mathrm{ar}[S(t)]$.

Exercise 2.6 Choose two realistic values: $0.1 \leq \sigma \leq 0.75$, $0.01 \leq \mu \leq 0.1$.

a. With $T = 3$, $S_0 = 0.7$, $\Delta t = 10^{-2}$ generate 10 asset paths that are driven by a geometric Brownian motion and the parameters above.

b. Plot for these paths the "running sum of square increments", i.e.
$$\sum_{k=1}^{m} (\Delta S(t_k))^2.$$

c. Use asset price market data (those from Exercise 2.4) and plot the asset return path and the running sum of square increments.

Exercise 2.7 Consider a stock price process, $S(t) = \exp(X(t))$, and determine the relation between the densities of $S(t)$ and $X(t)$. *Hint:* show that
$$f_{S(t)}(x) = \frac{1}{x} f_{X(t)}(\log(x)), \quad x > 0.$$

Exercise 2.8 Use Itô's lemma to prove that,
$$\int_0^T W^2(t) \mathrm{d}W(t) = \frac{1}{3} W^3(T) - \int_0^T W(t) \mathrm{d}t.$$

Exercise 2.9 Suppose that $X(t)$ satisfies the following SDE,
$$\mathrm{d}X(t) = 0.04 X(t) \mathrm{d}t + \sigma X(t) \mathrm{d}W^{\mathbb{P}}(t),$$
and $Y(t)$ satisfies:
$$\mathrm{d}Y(t) = \beta Y(t) \mathrm{d}t + 0.15 Y(t) \mathrm{d}W^{\mathbb{P}}(t).$$

Parameters β, σ are positive constants and both processes are driven by the same Wiener process $W^{\mathbb{P}}(t)$.
For a given process
$$Z(t) = 2 \frac{X(t)}{Y(t)} - \lambda t,$$
with $\lambda \in \mathbb{R}^+$.

a. Find the dynamics for $Z(t)$.

b. For which values of β and λ is process $Z(t)$ a martingale?

CHAPTER 3

The Black-Scholes Option Pricing Equation

In this chapter:

Financial derivatives are products that are based on the performance of some underlying asset, like a stock, an interest rate, an foreign-exchange rate, or a commodity price. They are introduced in this chapter, in **Section 3.1**. The fundamental pricing partial differential equation for the valuation of European options is derived in that section. For options on stocks, it is the famous *Black-Scholes equation*.

In this chapter, we also discuss a solution approach for the option pricing partial differential equation, which is based on the *Feynman-Kac theorem*, in **Section 3.2**. This theorem connects the solution of the Black-Scholes equation to the calculation of the discounted expected payoff function, under the risk-neutral measure. This formulation of the option price gives us a semi-analytic solution for the Black-Scholes equation. A hedging experiment is subsequently described in **Section 3.3**.

Keywords: Black-Scholes partial differential equation for option valuation, Itô's lemma, Feynman-Kac theorem, discounted expected payoff approach, martingale approach, hedging.

3.1 Option contract definitions

An option contract is a financial contract that gives the holder the right to trade (buy or sell) in an underlying asset in the future at a predetermined price. In fact, the option contract offers its holder the *optionality* to buy or sell the asset; there is no obligation.

If the underlying asset does not perform favorably, the holder of the option does not have to exercise the option and thus does not have to trade in the asset.

The counterparty of the option contract, the option seller (also called the *option writer*), is however obliged to trade in the asset when the holder makes use of the exercise right in the option contract. Options are also called *financial derivatives*, as their values are derived from the performance of another underlying asset. There are many different types of option contracts. Since 1973, standardized option contracts have been traded on regulated exchanges. Other option contracts are directly sold by financial companies to their customers.

3.1.1 Option basics

Regarding financial option contracts, there is the general distinction between call and put options.

A *call option* gives an option holder the right to *purchase an asset*, whereas a *put option* gives a holder the right to *sell an asset*, at some time in the future $t = T$, for a prescribed amount, *the strike price* denoted by K.

In the case of the so-called *European option*, there is one prescribed time point in the future in the option contract, which is called *the maturity time or expiry date* (denoted by $t = T$), at which the holder of the option may decide to trade in the asset for the strike price.

Example 3.1.1 (Call option example) Let's consider a specific scenario for a call option holder who bought, at $t = t_0 = 0$, a call option with expiry date $T = 1$ and strike price $K = 150$ and where the initial asset price $S_0 = 100$, see Figure 3.1 (left side). Suppose we are now at time point $t = t_0$. In the left picture of Figure 3.1 two possible future asset price path scenarios are drawn. When the blue asset path would occur, the holder of the call option would exercise the option at time $t = T$. At that time, the option holder would pay K to receive the asset worth $S(T) = 250$. Selling the asset immediately in the market would give the holder the positive amount $250 - 150 = 100$.

In the case the red colored asset path would happen, however, the call option would not be exercised by the holder as the asset would be worth less than K in the financial market. As the call option contract has the optionality to exercise, there is no need to exercise when $S(T) < K$ (why buy something for a price K when it is cheaper in the market?). ♦

Generally, the holder of a European call option will logically exercise the call option, when the asset price at maturity time $t = T$ is larger than the strike price, i.e. when $S(T) > K$. In the case of exercise, the option holder will thus pay an amount K to the writer, to obtain an asset worth $S(T)$. The profit in this case is $S(T) - K$, as the asset may immediately be sold in the financial market.

If $S(T) < K$, the holder will *not* exercise the European call option and the option is in that case valueless. In that case, the option holder would be able to buy the asset for less than the strike price K in the financial market, so it does not make any sense to make use of the right in the option contract.

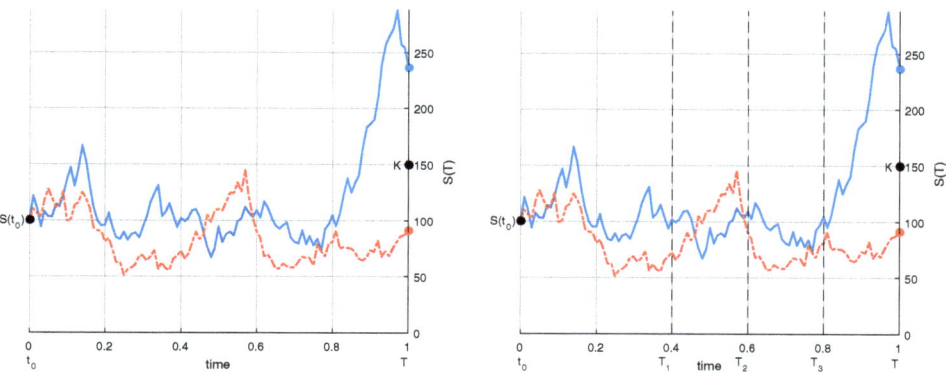

Figure 3.1: *Sketch of stochastic behavior of an asset; left: representation of a contract with only one decision point; right: a contract with more than one decision point, T_i, $i = 1, \ldots$.*

> With the payoff function denoted by $H(T, S)$, we thus find for a call option, see Figure 3.2 (left side picture),
>
> $$V_c(T, S) = H(T, S) := \max(S(T) - K, 0), \qquad (3.1)$$
>
> where the value of the call option at time t, for certain asset value $S = S(t)$ is denoted by $V_c(t, S)$, and $H(T, \cdot)$ is the option's payoff function.

The put option will be denoted by $V_p(t, S)$. Without any specification of the call or put features, the option is generally written as $V(t, S)$. However, we also sometimes use the notation $V(t, S; K, T)$, to emphasize the dependence on strike price K and maturity time T, when this appears useful. In the chapters dealing with options on interest rates we simply use the notation $V(t)$, because the option value may depend on many arguments in that case (too many to write down).

A European put option gives the holder the right to *sell* an asset at the maturity time T for a strike price K. The writer is then obliged to buy the asset, if the holder decides to sell.

> At the maturity time, $t = T$, a European put option has the value $K - S(T)$, if $S(T) < K$. The put option expires worthless if $S(T) > K$, see Figure 3.2 (right side), i.e.,
>
> $$V_p(T, S(T)) = H(T, S(T)) := \max(K - S(T), 0), \qquad (3.2)$$
>
> where a put option is denoted by $V_p(t, S(t))$.

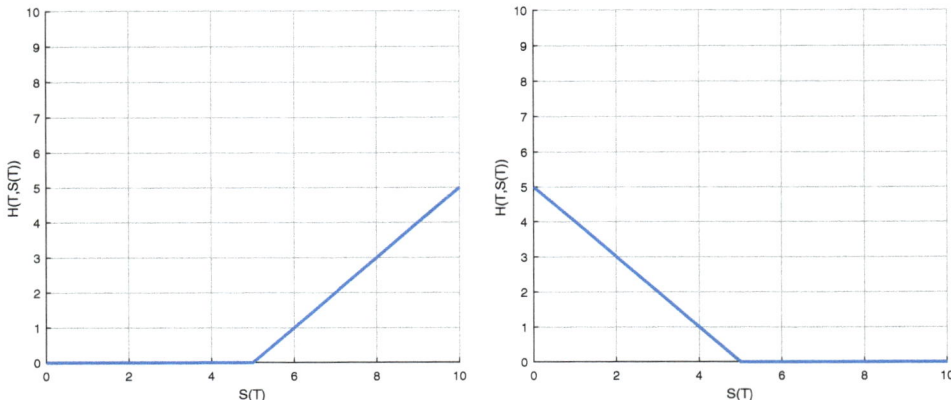

Figure 3.2: *The payoff $H(T, S(T))$ of European options with strike price K, at maturity time T; left: for a call option, right: for a put option.*

Definition 3.1.1 *The put and call payoff functions are convex functions.*
Remember that function $g(x)$ is called convex if,

$$g(\theta x_1 + (1-\theta)x_2) \leq \theta \cdot g(x_1) + (1-\theta) \cdot g(x_2), \ \forall x_1, x_2 \in \mathcal{C}, \forall \theta \in [0,1],$$

with $\mathcal{C}$ a convex set in $\mathbb{R}$. We have,

$$\max(\theta x_1 + (1-\theta)x_2, 0) \leq \theta \cdot \max(x_1, 0) + (1-\theta) \cdot \max(x_2, 0), \ \forall x_1, x_2 \in \mathcal{C}, \forall \theta \in [0,1].$$

The max functions in the payoffs, are convex operators, because the functions 0, $x - K$ and $K - x$ are all convex. ◀

Definition 3.1.2 (ITM, OTM, ATM) When for an underlying asset value at a time point $t < T$, i.e. $S(t)$, the corresponding option's payoff would be equal to zero (like $\max(K - S(t), 0) = 0$, for a put), we say that the option is "out-of-the money" (OTM). Its intrinsic option value ($K - S(t)$ for a put) then equals zero. When the option's intrinsic value is positive, the option is said to be in-the-money (ITM). When the intrinsic value is close to zero (like $K - S(t) \approx 0$, for a put), we call the option at-the-money (ATM).
The reasoning for puts and calls goes similarly, with the appropriate payoff functions. ◀

Some other financial derivative contracts include so-called *early exercise features*, meaning that there is *more than just one date*, $t = T$, to exercise these options. In the case of American call options, for example, the exercise of the option, i.e. paying K to purchase an asset worth $S(t)$, is permitted *at any time* during the life of the option, $t_0 \leq t \leq T$, whereas Bermudan options can be exercised *at certain dates*, T_i, $i = 1, \ldots m$, until the maturity time $T_m = T$ (see the right side picture of Figure 3.1).

An American option is thus a *continuous time equivalent* of the *discrete time* Bermudan option, assuming an increasing number of exercise dates, $m \to \infty$, with increasingly smaller time intervals, $\Delta t = \frac{T}{m} \to 0$. Options on individual stocks traded at regulated exchanges are typically American options.

Another class of options is defined as the *exotic options*, as they are governed by so-called exotic features in their payoff functions. These features can be path-dependency aspects of the stock prices, where the payoff does not only depend on the stock price $S(t)$ or $S(T)$, but also on certain functions of the stock price at different time points. These options are not traded on regulated exchanges but are sold *over-the-counter* (OTC), meaning that they are sold directly by banks and other financial companies to their counterparties. We will encounter some of these contracts, and the corresponding efficient pricing techniques, in the chapters to follow. Excellent overviews on the derivation of the governing PDEs in finance, for a variety of options, like exotic options and American options, include [Wilmott, 1998; Wilmott et al., 1995; Hull, 2012].

Put-call parity

The put-call parity for European options represents a fundamental result in quantitative finance. It is based on the following reasoning. Suppose we have two portfolios, one with a European put option and a stock, i.e., $\Pi_1(t,S) = V_p(t,S) + S(t)$, and a second portfolio with a European call option bought and the present value of cash amount K, i.e. $\Pi_2(t,S) = V_c(t,S) + Ke^{-r(T-t)}$. The term $e^{-r(T-t)}$ represents the discount factor, i.e., the value of €1 paying at time T as seen from today, with a continuously compounding interest rate r.

At maturity time T, we find

$$\Pi_1(T,S) = \max(K - S(T), 0) + S(T),$$

which equals K, if $S(T) < K$, and has value $S(T)$, if $S(T) > K$. In other words, $\Pi_1(T,S) = \max(K, S(T))$. The second portfolio has exactly the same value at $t = T$, i.e. $\Pi_2(T,S) = \max(K, S(T))$.

If $\Pi_1(T,S) = \Pi_2(T,S)$, then this equality should also be satisfied *at any time prior to the maturity time T*. Otherwise, an obvious arbitrage would occur, consisting of buying the cheaper portfolio and selling the more expensive one, so that at $t < T$ a positive cash amount results from this strategy. At $t = T$, both portfolios will have the same value so that buying one of the portfolios and selling the other at $t = T$ will result in neither profits nor losses at $t = T$, and the profit achieved at $t < T$ would remain. This is an arbitrage, i.e., a risk-free profit is obtained which is greater than just putting money on a bank account. This cannot be permitted. For any $t < T$, therefore the put-call parity

$$V_c(t,S) = V_p(t,S) + S(t) - Ke^{-r(T-t)}, \qquad (3.3)$$

should be satisfied. The interest rate r is supposed to be constant here.

In the case of a *dividend paying stock*, with a continuous stream of dividends, a put-call parity is also valid, but it should be modified. For the stock price $S(t)$

a dividend stream should be taken into account, i.e. $S(t)\mathrm{e}^{-q(T-t)}$, leading to the following adapted put-call parity relation,

$$V_c(t,S) = V_p(t,S) + S(t)\mathrm{e}^{-q(T-t)} - K\mathrm{e}^{-r(T-t)}. \qquad (3.4)$$

3.1.2 Derivation of the partial differential equation

An important question in quantitative finance is what is a fair value for a financial option at the time of selling the option, i.e. at $t = t_0$. In other words, how to determine $V(t_0, S_0)$, or, more generally, $V(t, S)$ for any $t \geq t_0$. Another question is how an option writer can reduce the risk of trading in the asset $S(T)$ at time $t = T$, for a fixed strike price K. In other words, how to manage the risk of the option writer?

Based on the assumption of a geometric Brownian Motion process for the asset prices $S(t)$, Fisher Black and Myron Scholes derived their famous partial differential equation (PDE) for the valuation of European options, published in 1973 in the Journal of Political Economy [Black and Scholes, 1973]. The Black-Scholes model is one of the most important models in financial derivative pricing.

The derivation to follow is based on the assumption that the interest rate r and volatility σ are constants or known functions of time. Further, a liquid financial market is assumed, meaning that assets can be bought and sold any time in arbitrary amounts. Short-selling is allowed, so that negative amounts of the asset can be traded as well, and transaction costs are not included (neither is a dividend payment during the lifetime of the option). In the model, there is no bid-ask spread for the stock prices nor for the option prices.

In this section, we will present the main ideas by Fisher Black and Myron Scholes. The derivation of the pricing PDE is based on the concept of a *replicating portfolio*, which is updated each time step. This is a portfolio, which is set up by the seller of the option and has essentially the same cash flows (payments) as the option contract which is sold by the option writer. A replicating portfolio can be either a static or a dynamic portfolio — it depends on whether one needs to update (re-balance) the financial position in time or not. A static portfolio is set up once, after selling the financial contract, and the portfolio is then not changed during the lifetime of the contract. A dynamic portfolio is regularly updated, based on the available newly updated market information. We follow the so-called dynamic delta hedge strategy with a dynamically changing (re-balanced) replicating portfolio.

Let's start with a stock price process, $S \equiv S(t)$, which is the underlying for the financial derivative contract, $V \equiv V(t, S)$, representing the value of a European option (sometimes also called "a *plain vanilla contingent claim*"). The underlying stock price process is assumed to be GBM, with the dynamics under the real-world measure $\mathbb{P}$ as in (2.1), i.e.,

$$\mathrm{d}S(t) = \mu S(t)\mathrm{d}t + \sigma S(t)\mathrm{d}W^{\mathbb{P}}(t).$$

As the price of the option $V(t, S)$ is a function of time t and the stochastic process $S(t)$, we will derive its dynamics, with the help of Itô's lemma, as follows:

$$\begin{aligned} dV(t,S) &= \frac{\partial V}{\partial t}dt + \frac{\partial V}{\partial S}dS + \frac{1}{2}\frac{\partial^2 V}{\partial S^2}(dS)^2 \\ &= \left(\frac{\partial V}{\partial t} + \mu S \frac{\partial V}{\partial S} + \frac{1}{2}\sigma^2 S^2 \frac{\partial^2 V}{\partial S^2}\right)dt + \sigma S \frac{\partial V}{\partial S}dW^{\mathbb{P}}. \end{aligned} \quad (3.5)$$

We construct a *portfolio* $\Pi(t, S)$, consisting of one option with value $V(t, S)$ and an amount, $-\Delta$, of stocks with value $S(t)$,

$$\boxed{\Pi(t, S) = V(t, S) - \Delta S(t).} \quad (3.6)$$

This portfolio thus consists of one long position in the option $V(t, S)$, and a short position of size Δ in the underlying $S(t)$.

Remark 3.1.1 (Short-selling of stocks) *When a trader holds a negative number of stocks, this implies that she/he has been short-selling stocks. Intuitively, it is quite confusing to own a negative amount of something. If you borrow money, then, in principle, you own a negative amount of money, because the money has to be paid back at some time in the future.*

Similarly, in financial practice, short-selling means that the trader borrows stocks from a broker, at some time t, for a fee. The stocks are subsequently sold, but the trader has to return the stocks to the broker at some later time. ▲

By Itô's lemma we derive the dynamics for an infinitesimal change in portfolio value $\Pi(t, S)$, based on the asset dynamics from Equation (2.1):

$$\begin{aligned} d\Pi &= dV - \Delta dS \\ &= \left(\frac{\partial V}{\partial t} + \mu S \frac{\partial V}{\partial S} + \frac{1}{2}\sigma^2 S^2 \frac{\partial^2 V}{\partial S^2}\right)dt + \sigma S \frac{\partial V}{\partial S}dW^{\mathbb{P}} - \Delta\left[\mu S dt + \sigma S dW^{\mathbb{P}}\right] \\ &= \left[\frac{\partial V}{\partial t} + \mu S \left(\frac{\partial V}{\partial S} - \Delta\right) + \frac{1}{2}\sigma^2 S^2 \frac{\partial^2 V}{\partial S^2}\right]dt + \sigma S \left(\frac{\partial V}{\partial S} - \Delta\right)dW^{\mathbb{P}}. \end{aligned} \quad (3.7)$$

The portfolio, although fully defined in terms of stocks and option, has random fluctuations, governed by Brownian motion $W^{\mathbb{P}}$. By choosing

$$\boxed{\Delta = \frac{\partial V}{\partial S},} \quad (3.8)$$

the infinitesimal change of portfolio $\Pi(t, S)$, in time instance dt, is given by:

$$d\Pi = \left(\frac{\partial V}{\partial t} + \frac{1}{2}\sigma^2 S^2 \frac{\partial^2 V}{\partial S^2}\right)dt, \quad (3.9)$$

which is now deterministic,[1] *as the* $dW^{\mathbb{P}}$-*terms cancel out.* Moreover, with Δ as defined by (3.8), the resulting dynamics of the portfolio do not contain the drift parameter μ, which drives the stock $S(t)$ under the real-world measure $\mathbb{P}$ in (2.1), anymore. The value of the portfolio still depends on volatility parameter σ, which is the representation of the uncertainty about the future behaviour of the stock prices.

The value of this portfolio should, on average, grow at the same *speed* (i.e., generate the same return) as money placed on a risk-free money-savings account. The bank account $M(t) = M(t_0)e^{r(t-t_0)}$, is modeled by means of $dM = rMdt$, which, for an amount $\Pi \equiv \Pi(t, S)$ can be expressed as,

$$d\Pi = r\Pi dt.$$

Here, r corresponds to the constant interest rate on a money-savings account. With the definitions in Equation (3.6) and the definition of Δ in Equation (3.8), the change in portfolio value is written as:

$$d\Pi = r\left(V - S\frac{\partial V}{\partial S}\right) dt. \quad (3.10)$$

Equating Equations (3.10) and (3.9), and dividing both sides by dt, gives us the Black-Scholes partial differential equation for the value of the option $V(t, S)$:

$$\boxed{\frac{\partial V}{\partial t} + rS\frac{\partial V}{\partial S} + \frac{1}{2}\sigma^2 S^2 \frac{\partial^2 V}{\partial S^2} - rV = 0.} \quad (3.11)$$

The Black-Scholes equation (3.11) is a parabolic PDE. With the "+"-sign in front of the diffusion term, i.e. $\frac{1}{2}\sigma^2 S^2 \frac{\partial^2 V}{\partial S^2}$, the parabolic PDE problem is *well-posed* when it is accompanied by a *final condition*. The natural condition for the Black-Scholes PDE is indeed a final condition, i.e. we know that,

$$V(T, S) = H(T, S),$$

where $H(T, S)$ is the payoff function, as in (3.1) for a call option, or (3.2) for a put option.

Notice that, except for this final condition, so far we did not specify the type of option in the derivation of the Black-Scholes equation. The equation holds for both calls, puts, and also even for some other option types.

PDE (3.11) is defined on a semi-infinite half space $(t, S) = [t_0 = 0, \ldots, T) \cup [0, S \to \infty)$, and there are no natural conditions at the outer boundaries. When we solve the PDE by means of a numerical method, we need to restrict our computations to a finite computational domain, and we thus need to define a finite domain $(t, S) = [t_0 = 0, \ldots, T) \cup [0, S_{max}]$. We then prescribe appropriate boundary conditions at $S = S_{max}$, as well as at $S = 0$, which is easily possible from an economic viewpoint (as we have information about the value of a call or put, when $S(t) = 0$ or when $S(t)$ is large).

---

[1] By "deterministic" we mean that the value is defined in terms of other variables, that nota bene, are stochastic, but without additional sources of randomness.

The Black-Scholes Option Pricing Equation

By solving the Black-Scholes equation, we thus know the fair value for the option price at any time $t \in [0, T]$, and at any future stock price $S(t) \in [0, S_{max}]$. The solution of the Black-Scholes PDE will be discussed in detail when the Feynman-Kac Theorem is introduced.

Hedge parameters

Next to these option values, other important information for an option writer is found in the so-called *hedge parameters* or the option Greeks. These are the *sensitivities of the option value* with respect to small changes in the problem parameters or in the independent variables, like the asset S, or the volatility σ.

We have already encountered the option delta, $\Delta = \partial V/\partial S$, in (3.8), representing the rate of change of the value of the option with respect to changes in S. In a replicating portfolio, with stocks to cover the option, Δ gives the number of shares that should be kept by the writer for each option issued, in order to cope with the possible exercise by the holder of the contract at time $t = T$. A negative number implies that short-selling of stocks should take place.

The *sensitivity of the option delta* is called the option *gamma*,

$$\boxed{\Gamma := \frac{\partial^2 V}{\partial S^2} = \frac{\partial \Delta}{\partial S}.}$$

A change in delta is an indication for the stability of a hedging portfolio. For small values of gamma, a writer may not yet need to update the portfolio frequently as the impact of changing the number of stocks (Δ) appears small. When the option gamma is large, however, the hedging portfolio appears only free of risk at a short time scale. There are several other important hedging parameters that we will encounter later in this book.

Dividends

The assumption that dividend payment on the stocks does not take place in the derivation of the Black-Scholes equation can be relaxed. As explained, at the time a dividend is paid, there will be a drop in the value of the stock, see also Equation (2.21). The value of an option on a dividend-paying asset is then also affected by these dividend payments, so that the Black-Scholes analysis must be modified to take dividend payments into account. The constant proportional dividend yield of size $qS(t)$ with $q < 1$, as in (2.21), is considered a satisfactory model for options on stock indices that are of European style. The dividend payment also has its effect on the hedging portfolio. Since we receive $qS(t)\mathrm{d}t$ for every asset held and we hold $-\Delta$ of the underlying, the portfolio changes by an amount $-qS(t)\Delta \mathrm{d}t$, i.e. we have to replace (3.7), by

$$\mathrm{d}\Pi = \mathrm{d}V - \Delta \mathrm{d}S - qS\Delta \mathrm{d}t.$$

Based on the same arguments as earlier, we can then derive the Black-Scholes PDE, where a *continuous stream of dividend payments* is modeled, as

$$\boxed{\frac{\partial V}{\partial t} + \frac{1}{2}\sigma^2 S^2 \frac{\partial^2 V}{\partial S^2} + (r-q)S\frac{\partial V}{\partial S} - rV = 0.} \qquad (3.12)$$

3.1.3 Martingale approach and option pricing

Based on the properties of martingales, we give an alternative derivation of the Black-Scholes option pricing PDE.

The following pricing problem, under the risk-neutral GBM model defined by (2.1), is connected to satisfying the martingale property, for a discounted option price:

$$\frac{V(t_0, S)}{M(t_0)} = \mathbb{E}^{\mathbb{Q}}\left[\frac{V(T,S)}{M(T)}\Big|\mathcal{F}(t_0)\right], \qquad (3.13)$$

with $M(t_0)$ the money-savings account at time t_0, where $M(t_0) = 1$, and $\mathcal{F}(t_0) = \sigma(S(s); s \leq t_0)$.

Since financial options are traded products, Equation (3.13) defines a *discounted option contract* to be a martingale, where the typical martingale properties should then be satisfied. These are properties like the expected value of a martingale should be equal to the present value, as this leads to no-arbitrage values when pricing these financial contracts.

We assume the existence of a differentiable function, $\frac{V(t,S)}{M(t)}$, so that:

$$\mathbb{E}^{\mathbb{Q}}\left[\frac{V(T,S)}{M(T)}\Big|\mathcal{F}(t)\right] = \frac{V(t,S)}{M(t)}. \qquad (3.14)$$

Using $M \equiv M(t)$ and $V \equiv V(t,S)$, the discounted option value V/M should be a martingale and its dynamics can be found, as follows,

$$d\left(\frac{V}{M}\right) = \frac{1}{M}dV - \frac{V}{M^2}dM = \frac{1}{M}dV - r\frac{V}{M}dt. \qquad (3.15)$$

Higher-order terms, like $(dM)^2 = O(dt^2)$, are omitted in (3.15), as they converge to zero rapidly with infinitesimally small time steps.

For the infinitesimal change dV of $V(t,S)$, we find under measure $\mathbb{Q}$ using Itô's lemma (3.5),

$$dV = \left(\frac{\partial V}{\partial t} + rS\frac{\partial V}{\partial S} + \frac{1}{2}\sigma^2 S^2 \frac{\partial^2 V}{\partial S^2}\right)dt + \sigma S\frac{\partial V}{\partial S}dW^{\mathbb{Q}}. \qquad (3.16)$$

As $\frac{V(t,S)}{M(t)}$ should be a martingale, Theorem 1.3.5 states that the dynamics of $\frac{V(t,S)}{M(t)}$ should not contain any dt-terms. Substituting Equation (3.16) into (3.15), setting the term in front of dt equal to zero, yields that in this case,

$$\frac{1}{M}\left(\frac{\partial V}{\partial t} + rS\frac{\partial V}{\partial S} + \frac{1}{2}\sigma^2 S^2 \frac{\partial^2 V}{\partial S^2}\right) - r\frac{V}{M} = 0, \qquad (3.17)$$

assuring that dt-terms are zero. Multiplying both sides of (3.17) by M, gives us the Black-Scholes pricing PDE (3.11), on the basis of the assumption that the martingale property should hold. More information on martingale methods can be found in [Pascucci, 2011].

3.2 The Feynman-Kac theorem and the Black-Scholes model

A number of different solution methods for solving the Black-Scholes PDE (3.11) are available. We are particularly interested in the solution via the Feynman-Kac formula, as given in the theorem below, which forms the basis for a closed-form expression for the option value. It can be generalized for pricing derivatives under other asset price dynamics as well, and forms the basis for option pricing by Fourier methods (in Chapter 6) as well as for pricing by means of Monte Carlo methods (in Chapter 9).

There are different versions of the Feynman-Kac theorem, as the theorem has originally been developed in the context of physics applications. We start with a version of the theorem which is related to option pricing for the case discussed so far, in which we deal with a deterministic interest rate r.

Theorem 3.2.1 (Feynman-Kac theorem) *Given the money-savings account, modeled by* $dM(t) = rM(t)dt$, *with constant interest rate* r, *let* $V(t,S)$ *be a sufficiently differentiable function of time* t *and stock price* $S = S(t)$. *Suppose that* $V(t,S)$ *satisfies the following partial differential equation, with general drift term,* $\bar{\mu}(t,S)$, *and volatility term,* $\bar{\sigma}(t,S)$:

$$\frac{\partial V}{\partial t} + \bar{\mu}(t,S)\frac{\partial V}{\partial S} + \frac{1}{2}\bar{\sigma}^2(t,S)\frac{\partial^2 V}{\partial S^2} - rV = 0, \qquad (3.18)$$

with a final condition given by $V(T,S) = H(T,S)$. *The solution* $V(t,S)$ *at any time* $t < T$ *is then given by:*

$$V(t,S) = e^{-r(T-t)}\mathbb{E}^{\mathbb{Q}}\left[H(T,S)|\mathcal{F}(t)\right] =: M(t)\mathbb{E}^{\mathbb{Q}}\left[\frac{H(T,S)}{M(T)}\bigg|\mathcal{F}(t)\right]$$

where the expectation is taken under the measure $\mathbb{Q}$, *with respect to a process* S, *which is defined by:*

$$dS(t) = \bar{\mu}(t,S)dt + \bar{\sigma}(t,S)dW^{\mathbb{Q}}(t), \quad t > t_0. \qquad (3.19)$$

Proof We present a short outline of the proof. Consider the term

$$\frac{V(t,S)}{M(t)} = e^{-r(t-t_0)}V(t,S),$$

for which we find the dynamics:

$$d(e^{-r(t-t_0)}V(t,S)) = V(t,S)d(e^{-r(t-t_0)}) + e^{-r(t-t_0)}dV(t,S). \quad (3.20)$$

Using $V := V(t,S)$, $S := S(t)$, $\bar{\mu} := \bar{\mu}(t,S)$, $\bar{\sigma} := \bar{\sigma}(t,S)$ and $W^{\mathbb{Q}} := W^{\mathbb{Q}}(t)$, and applying Itô's lemma to $V := V(t,S)$, we obtain:

$$dV = \left(\frac{\partial V}{\partial t} + \bar{\mu}\frac{\partial V}{\partial S} + \frac{1}{2}\bar{\sigma}^2\frac{\partial^2 V}{\partial S^2}\right)dt + \bar{\sigma}\frac{\partial V}{\partial S}dW^{\mathbb{Q}}.$$

By multiplying (3.20) by $e^{r(t-t_0)}$ and substituting the above expression in it, we find:

$$e^{r(t-t_0)}d(e^{-r(t-t_0)}V) = \underbrace{\left(\frac{\partial V}{\partial t} + \bar{\mu}\frac{\partial V}{\partial S} + \frac{1}{2}\bar{\sigma}^2\frac{\partial^2 V}{\partial S^2} - rV\right)}_{=0}dt + \bar{\sigma}\frac{\partial V}{\partial S}dW^{\mathbb{Q}},$$

in which the dt-term is equal to zero, because of the original PDE (3.18) which equals zero in the theorem. Integrating both sides gives us,

$$\int_{t_0}^{T} d(e^{-r(t-t_0)}V(t,S)) = \int_{t_0}^{T} e^{-r(t-t_0)}\bar{\sigma}\frac{\partial V}{\partial S}dW^{\mathbb{Q}}(t) \Leftrightarrow$$

$$e^{-r(T-t_0)}V(T,S) - V(t_0,S) = \int_{t_0}^{T} e^{-r(t-t_0)}\bar{\sigma}\frac{\partial V}{\partial S}dW^{\mathbb{Q}}(t).$$

We now take the expectation at both sides of this equation, with respect to the $\mathbb{Q}$-measure, and rewrite as follows:

$$V(t_0,S) = \mathbb{E}^{\mathbb{Q}}\left[e^{-r(T-t_0)}V(T,S)\Big|\mathcal{F}(t_0)\right] - \mathbb{E}^{\mathbb{Q}}\left[\int_{t_0}^{T} e^{-r(t-t_0)}\bar{\sigma}\frac{\partial V}{\partial S}dW^{\mathbb{Q}}(t)\Big|\mathcal{F}(t_0)\right].$$

According to the properties of an Itô integral, as defined in (1.23), $I(t_0) = 0$ and $I(t) = \int_0^t g(s)dW^{\mathbb{Q}}(s)$ is a martingale, so that we have that $\mathbb{E}^{\mathbb{Q}}[I(t)|\mathcal{F}(t_0)] = 0$ for all $t \geq t_0$.

The expectation of the Itô integral which is based on Wiener process $W^{\mathbb{Q}}(t)$ is therefore equal to 0, so it follows that

$$V(t_0,S) = \mathbb{E}^{\mathbb{Q}}\left[e^{-r(T-t_0)}V(T,S)\big|\mathcal{F}(t_0)\right] = \mathbb{E}^{\mathbb{Q}}\left[e^{-r(T-t_0)}H(T,S)\big|\mathcal{F}(t_0)\right].$$

This concludes this outline of the proof. ∎

By the Feynman-Kac theorem we can thus translate the problem of *solving the Black-Scholes PDE*, which is recovered by specific choices for drift term $\bar{\mu}(t,S) = rS$ and diffusion term $\bar{\sigma}(t,S) = \sigma S$, to the calculation of an expectation of a discounted payoff function under the $\mathbb{Q}$-measure.

A comprehensive discussion on the Feynman-Kac theorem and its various properties have been provided, for example, in [Pelsser, 2000].

> **Remark 3.2.1 (Log coordinates under the GBM model)**
> The Feynman-Kac theorem, as presented in Theorem 3.2.1, also holds true in logarithmic coordinates, i.e., when $X(t) = \log S(t)$. With this transformation of variables, the resulting log-transformed Black-Scholes PDE reads
> $$\frac{\partial V}{\partial t} + r\frac{\partial V}{\partial X} + \frac{1}{2}\sigma^2\left(-\frac{\partial V}{\partial X} + \frac{\partial^2 V}{\partial X^2}\right) - rV = 0, \quad (3.21)$$
> which comes with the transformed pay-off function, $V(T, X) = H(T, X)$. The Feynman-Kac theorem gives us, for the specific choice of $\bar{\mu} = (r - \frac{1}{2}\sigma^2)$ and $\bar{\sigma} = \sigma$, the following representation of the option value,
> $$V(t, X) = e^{-r(T-t)}\mathbb{E}^{\mathbb{Q}}\left[H(T, X)|\mathcal{F}(t)\right],$$
> where $H(T, X)$ represents the terminal condition in log-coordinates $X(t) = \log S(t)$, and
> $$dX(t) = \left(r - \frac{1}{2}\sigma^2\right)dt + \sigma dW^{\mathbb{Q}}(t). \quad (3.22)$$
> ▲

Note that we have used essentially the same *notation*, $V(\cdot, \cdot)$, for the option value when the asset price is an independent variable, i.e. $V(t, S)$, as when using the log-asset price as the independent variable, i.e. $V(t, X)$. From the context it will become clear how the option is defined.

Regarding the terminal condition, we use $H(\cdot, \cdot)$ also for the call and put payoff functions in log-coordinates, i.e. $H(T, X) = H_c(T, X) = \max(e^X - K, 0)$, $H(T, X) = H_p(T, X) = \max(K - e^X, 0)$, respectively.

Throughout the book, we will often use $\tau = T - t$, which is common practice, related to the fact that people are simply used to "time running forward" while we know the payoff function at $\tau = 0$ we do not know the value at $\tau = T - t_0$. Of course, with $d\tau = -dt$, a minus sign is added to a derivative with respect to τ.

3.2.1 Closed-form option prices

Some results will follow, where it is shown that for some payoff functions under the Black-Scholes dynamics an analytic solution, which is obtained via the Feynman-Kac theorem, is available for European options.

> **Theorem 3.2.2 (European call and put option)** A closed-form solution of the Black-Scholes PDE (3.11) for a European call option with a constant strike price K can be derived, with $H_c(T, S) = \max(S(T) - K, 0)$. The option value at any time t (in particular, also at time now, i.e. $t = t_0$), under the Black-Scholes model, of a call option can be written as:
> $$V_c(t, S) = e^{-r(T-t)}\mathbb{E}^{\mathbb{Q}}\left[\max(S(T) - K, 0)|\mathcal{F}(t)\right]$$
> $$= e^{-r(T-t)}\mathbb{E}^{\mathbb{Q}}\left[S(T)\mathbb{1}_{S(T)>K}|\mathcal{F}(t)\right] - e^{-r(T-t)}\mathbb{E}^{\mathbb{Q}}\left[K\mathbb{1}_{S(T)>K}|\mathcal{F}(t)\right],$$

The solution is then given by:

$$V_c(t, S) = S(t) F_{\mathcal{N}(0,1)}(d_1) - K e^{-r(T-t)} F_{\mathcal{N}(0,1)}(d_2), \qquad (3.23)$$

with

$$d_1 = \frac{\log \frac{S(t)}{K} + (r + \frac{1}{2}\sigma^2)(T-t)}{\sigma\sqrt{T-t}},$$

$$d_2 = \frac{\log \frac{S(t)}{K} + (r - \frac{1}{2}\sigma^2)(T-t)}{\sigma\sqrt{T-t}} = d_1 - \sigma\sqrt{T-t},$$

and $F_{\mathcal{N}(0,1)}(\cdot)$ the cumulative distribution function of a standard normal variable.

It is easy to show that this is indeed the solution by substitution of the solution is the Black-Scholes PDE.

Similarly, for a European put option, we find:

$$V_p(t, S) = K e^{-r(T-t)} F_{\mathcal{N}(0,1)}(-d_2) - S(t) F_{\mathcal{N}(0,1)}(-d_1),$$

with d_1 and d_2 as defined above. This solution can directly be found by writing out the integral form in (3.22), based on the log-transformed asset process.

Definition 3.2.1 (Closed-form expression for Δ) *Greek delta is given by $\Delta := \frac{\partial V}{\partial S}$, so by differentiation of the option price, provided in Theorem 3.2.2, we obtain,*

- *For call options,*

$$\Delta = \frac{\partial}{\partial S} V_c(t, S) = F_{\mathcal{N}(0,1)}(d_1). \qquad (3.24)$$

  with d_1 as in Theorem 3.2.2.

- *For put options,*

$$\Delta = \frac{\partial}{\partial S} V_p(t, S)$$
$$= -F_{\mathcal{N}(0,1)}(-d_1) = F_{\mathcal{N}(0,1)}(d_1) - 1, \qquad (3.25)$$

In particular, we are interested in the values of the Greeks, at $t = t_0$, i.e. for $S = S_0$. ◀

Example 3.2.1 (Black-Scholes solution) In this example, we will present a typical surface of option values. We will use a call option for this purpose, and the following parameter set:

$$S_0 = 10, r = 0.05, \sigma = 0.4, T = 1, K = 10.$$

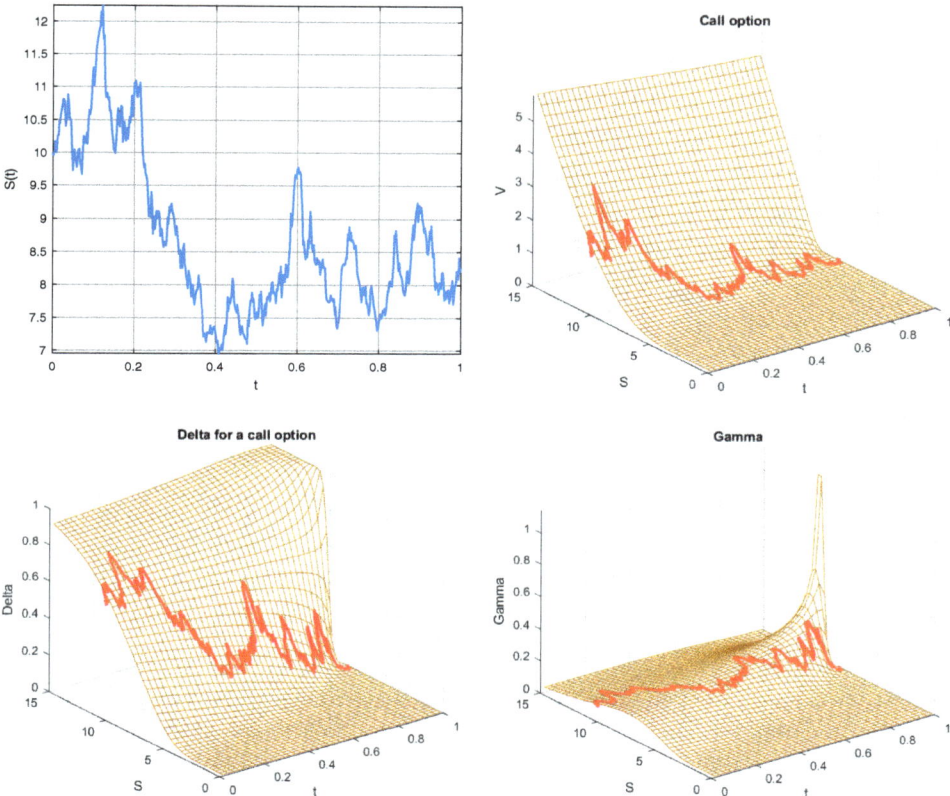

Figure 3.3: *Example of a call option with exercise date $T = 1$, the option surface $V_c(t, S)$ in the (t, S)-domain is computed. An asset path intersects the surface, so that at each point (t, S), the corresponding option value $V_c(t, S)$ can be read. The surfaces for the hedge parameters delta and gamma are also presented, with the projected stock price.*

Once the problem parameters have been defined, we can calculate the option values at any time t and stock price $S(t)$, by the solution in (3.23). Visualizing these call option values $V_c(t, S)$ in an (t, S)-plane, gives us the surface in Figure 3.3. From the closed-form solution, calculated by the Black-Scholes formula, we can also easily determine the first and second derivatives with respect to the option value, i.e., the option delta and gamma. They are presented in the same figure. When the stock price then moves at some time, $t > t_0$, we can "read" the corresponding option values, and also the option Greeks, see also Figure 3.3.

First of all, the option value surface is computed, and then the observed stock prices are projected on top of this surface. We also display the corresponding put option surface $V_p(t, S)$ in the (t, S)-plane in Figure 3.4.

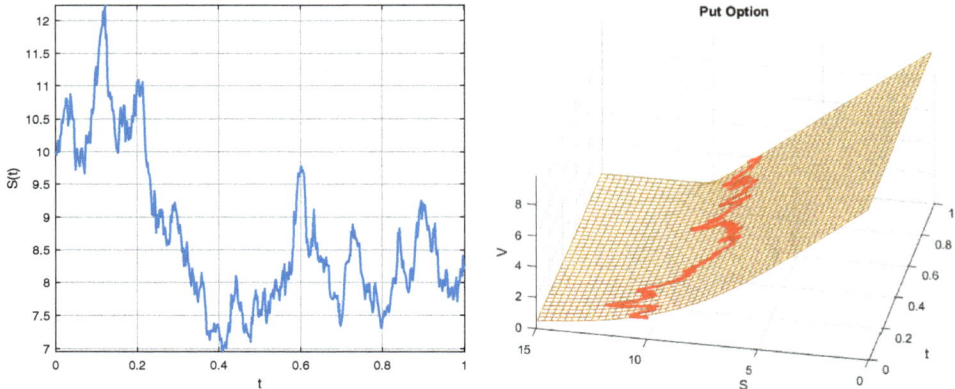

Figure 3.4: *Example of a put option, the option surface $V_p(t,S)$ in the (t,S)-domain is displayed with an asset path that intersects the surface.*

Digital option

As a first option contract away from the standard European put and call options, we discuss digital options, that are also called cash-or-nothing options. Digital options are popular for hedging and speculation. They are useful as building blocks for more complex option contracts. Consider the payoff of a cash-or-nothing digital call option, whose value is equal to 0 for $S(T) \leq K$ and its value equals K when $S(T) > K$, see Figure 3.5 (left side). The price at time t is then given by:

$$V(t,S) = e^{-r(T-t)}\mathbb{E}^{\mathbb{Q}}\left[K\mathbb{1}_{S(T)>K}|\mathcal{F}(t)\right]$$
$$= e^{-r(T-t)}K\mathbb{Q}\left[S(T) > K\right],$$

as $\mathbb{E}^{\mathbb{Q}}\left[\mathbb{1}_{S(T)>K}\right] := \mathbb{Q}\left[S(T) > K\right]$. Hence, the value of a cash-or-nothing call under the Black-Scholes dynamics, is given by

$$V(t,S) = Ke^{-r(T-t)}F_{\mathcal{N}(0,1)}(d_2), \qquad (3.26)$$

with d_2 defined in Theorem 3.2.2. Similarly, we can derive an expression for the solution of an *asset-or-nothing* option with payoff function in Figure 3.5 (right side).

3.2.2 Green's functions and characteristic functions

Based on the Feynman-Kac theorem we obtain interesting results and solutions by a clever choice of final condition $H(\cdot,\cdot)$ [Heston, 1993]. If, for example,[2]

$$H(T,X) := \mathbb{1}_{X(T) \geq \log K},$$

---

[2]Of course, it holds that $\mathbb{1}_{X(T) \geq \log K} = \mathbb{1}_{S(T) \geq K}$.

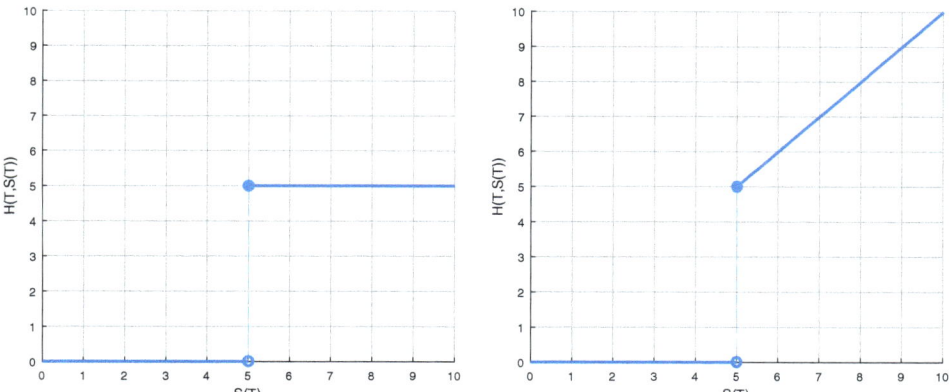

Figure 3.5: *Payoff functions for the cash-or-nothing (left), with $K\mathbb{1}_{S(T)\geq K}$ and $K=5$, and asset-or-nothing (right), with $S(T)\mathbb{1}_{S(T)\geq K}$ and $K=5$, call options.*

is chosen, the solution of the corresponding PDE is the *conditional probability* that $X(T)$ is greater than $\log K$.

As another example, we consider the following form of the final condition[a]:
$$\phi_X(u;T,T) = H(T,X) = e^{iuX(T)}.$$

By the Feynman-Kac theorem, the solution for $\phi_X := \phi_X(u;t,T)$ is then given by:
$$\phi_X(u;t,T) = e^{-r(T-t)}\mathbb{E}^{\mathbb{Q}}\left[e^{iuX(T)}\big|\mathcal{F}(t)\right], \tag{3.27}$$

which is by definition *the discounted characteristic function of $X(t)$*. We thus also know that $\phi_X(u;t,T)$ is a solution of the corresponding PDE:
$$\frac{\partial \phi_X}{\partial t} + \left(r - \frac{1}{2}\sigma^2\right)\frac{\partial \phi_X}{\partial X} + \frac{1}{2}\sigma^2 \frac{\partial^2 \phi_X}{\partial X^2} - r\phi_X = 0, \tag{3.28}$$

with terminal condition $\phi_X(u;T,T) = e^{iuX(T)}$.

---
[a]In order to emphasize that we will deal here with a characteristic function we use $\phi_X(u;t,T)$ instead of H.

In Chapter 6, we will work with the discounted characteristic function under the log-transformed GBM asset process, which is given by
$$\phi_X(u;t,T) = \exp\left[\left(r - \frac{\sigma^2}{2}\right)iu\tau - \frac{1}{2}\sigma^2 u^2 \tau - r\tau + iuX\right], \tag{3.29}$$

with $\tau = T - t$. For a stochastic process $X(t)$, $t > 0$, we denote its characteristic function by $\phi_X(u;t,T)$ or simply by $\phi_X(u,X,\tau)$ with $\tau = T - t$.

For a large number of stochastic underlying asset models for which the density is not known, the *characteristic function* (see the definition in Equation (1.4)), or the discounted characteristic function, is known, and this function also uniquely determines the transition density function.

Green's function

With the Feynman-Kac Theorem, we are able to make connections between a PDE and the integral formulation of a solution. By this, we can relate to well-known PDE terminology, for example, we can determine the Green's function which is related to the integral formulation of the solution of the PDE. We will see that it is, in fact, related to the asset price's density function. Moreover, with the Feynman-Kac Theorem we can derive the characteristic function, which we will need for option pricing with Fourier techniques in the chapters to come.

Based on the Feynman-Kac theorem, under the assumption of a constant interest rate, we can write the value of a European option as,

$$V(t_0, x) = e^{-r(T-t_0)} \int_{\mathbb{R}} H(T, y) f_X(T, y; t_0, x) dy, \qquad (3.30)$$

where $f_X(T, y; t_0, x)$ represents the *transition probability density function* in log-coordinates from state $\log S(t_0) = X(t_0) = x$ at time t_0, to state $\log S(T) = X(T) =: y$ at time T.

De facto, if we take $t_0 = 0$ and initial value x constant, we have here a marginal probability density function.

Taking a closer look at Equation (3.30), we recognize for the PDE (3.21) the *fundamental solution*, which is denoted here by $G_X(T, y; t_0, x) := e^{-r(T-t_0)} f_X(T, y; t_0, x)$, times the final condition $V(T, y) \equiv H(T, y)$, in the form of a convolution. The solution to the PDE can thus be found based on this fundamental solution by means of the Feynman-Kac theorem. The fundamental solution $e^{-r(T-t_0)} f_X(T, y; t_0, x)$ can be interpreted as the *Green's function*, which is connected to the following parabolic final value problem:

$$-\frac{\partial f_X}{\partial \tau} + r\frac{\partial f_X}{\partial x} + \frac{1}{2}\sigma^2 \left(-\frac{\partial f_X}{\partial x} + \frac{\partial^2 f_X}{\partial x^2}\right) - rf_X = 0, \qquad (3.31)$$

$$f_X(T, y; T, y) = \delta(y = \log K),$$

with $\delta(y = \log K)$ the Dirac delta function, which is nonzero for $y := X(T) = \log K$ and is 0 otherwise, and its integral equals one, see (1.17).

The Green's function, related to the log-space Black-Scholes PDE, is equal to a *discounted* normal probability density function, i.e.,

$$G_X(T, X; t_0, x) = \frac{e^{-r(T-t_0)}}{\sigma\sqrt{2\pi(T-t_0)}} \exp\left(-\frac{(X - x - (\mu - \frac{1}{2}\sigma^2)(T-t_0))^2}{2\sigma^2(T-t_0)}\right).$$

(3.32)

> The Green's function in derivative pricing, i.e. the discounted risk neutral probability density, has received the name *Arrow-Debreu security* in finance. We can write out the Green's function for the Black-Scholes operator *in the original (S,t) coordinates* explicitly, i.e. for $y = S(T), x = S(t_0)$, we find:
>
> $$G_S(T, y; t_0, x) = \frac{e^{-r(T-t_0)}}{\sigma y \sqrt{2\pi(T-t_0)}} \exp\left(-\frac{\left(\log\left(\frac{y}{x}\right) - (r - \frac{1}{2}\sigma^2)(T-t_0)\right)^2}{2\sigma^2(T-t_0)}\right), \quad (3.33)$$
>
> see also Equation (2.20). Density functions and Green's functions are known in closed-form only for certain stochastic models under basic underlying dynamics.

Arrow-Debreu security and the market implied density

A *butterfly spread option* is an option strategy which may give the holder a predetermined profit when the stock price movement stays within a certain range of stock prices. With $S(t_0) = K$, the butterfly spread option is defined by two long call options, one call with strike price $K_1 = K - \Delta K$, another call with strike price $K_3 = K + \Delta K$, $\Delta K > 0$, and by simultaneously two short call options with strike price $K_2 = K$. All individual options have the same maturity time $t = T$, with payoff,

$$V_B(T, S; K_2, T) = V_c(T, S; K_1, T) + V_c(T, S; K_3, T) - 2V_c(T, S; K_2, T), \quad (3.34)$$

see Figure 3.6. At the expiry date $t = T$, this construction will be profitable if the asset $S(T) \in [K - \Delta K, K + \Delta K]$; the butterfly spread has a zero value if $S(T) < K - \Delta K$ or $S(T) > K + \Delta K$. Since the butterfly option is a linear combination of individual option payoffs, it satisfies the Black-Scholes PDE, when we assume that stock $S(t)$ is modeled by a GBM.[3]

The butterfly spread option at time $t = t_0$ can be calculated as[4]

$$V_B(t_0, S; K, T) = e^{-r(T-t_0)} \mathbb{E}^{\mathbb{Q}}\left[V_B(T, S; K, T) | \mathcal{F}(t_0)\right], \quad (3.35)$$

based on the Feynman-Kac theorem.

This butterfly option is related to the so-called *Arrow-Debreu security* $V_{AD}(t, S)$, which is an imaginary security, that only pays out one unit at maturity date $t = T$, if $S(T) \equiv K$. The payoff is zero otherwise, i.e. $V_{AD}(T, S) = \delta(S(T) - K)$.

---

[3] Actually, the result here holds for any dynamics of S, it is not specific to GBM.
[4] We explicitly add the dependency of the option on the strike price K in the option argument list.

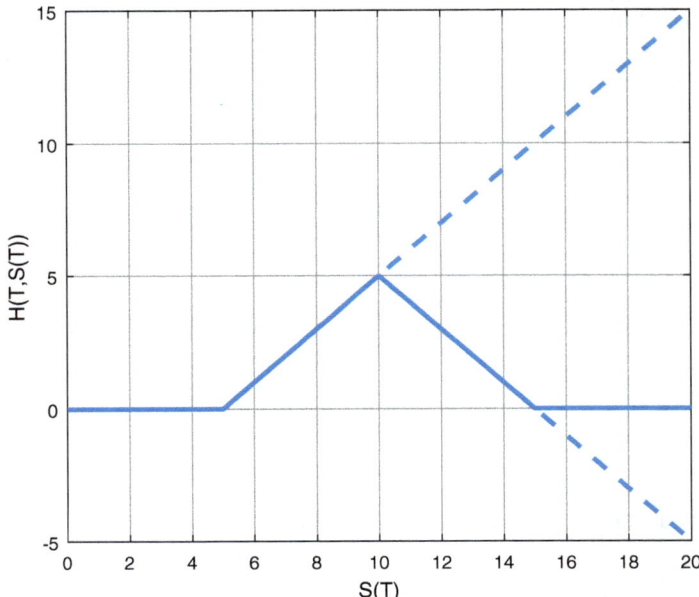

Figure 3.6: *Butterfly spread payoff function, with $K_1 = 5$, $K_2 = 10$ and $K_3 = 15$.*

The Arrow-Debreu security can be constructed by means of three European calls, like the butterfly spread option V_B:

$$V_{AD}(T, S; K, T) := \frac{1}{\Delta K^2}[V_c(T, S; K - \Delta K, T) + V_c(T, S; K + \Delta K, T)$$
$$- 2V_c(T, S; K, T)],$$

for small $\Delta K > 0$, with $V_c(t, S; K, T)$ representing the value of a European call at time t, with strike K, expiry date T.

> Note that for $\Delta K \to 0$, we have
> $$V_{AD}(T, S; K, T) = \frac{\partial^2 V_c(T, S; K, T)}{\partial K^2}.$$

On the other hand, in the limit $\Delta K \to 0$, we also find

$$\lim_{\Delta K \to 0} V_{AD}(t_0, S; K, T) = \lim_{\Delta K \to 0} e^{-r(T-t_0)} \mathbb{E}^{\mathbb{Q}}\left[V_{AD}(T, S; K, T) | \mathcal{F}(t_0)\right]$$
$$= e^{-r(T-t_0)} \mathbb{E}^{\mathbb{Q}}\left[\delta(S(T) - K) | \mathcal{F}(t_0)\right]$$
$$= e^{-r(T-t_0)} \int_0^\infty f_S(T, S(T); t_0, S_0) \cdot \delta(S(T) - K) \mathrm{d}S(T)$$
$$= e^{-r(T-t_0)} f_S(T, K; t_0, S_0), \qquad (3.36)$$

where $\delta(S(T) - K)$ is the Dirac delta function which is only nonzero when $S(T) = K$, $f_S(T, K; t_0, S_0)$ is the *transition risk-neutral density* with $S(T) \equiv K$. Transition densities model the evolution of a probability density through time, for example, between the time points s and t. Here we set time point $s = t_0$, which actually gives rise to the marginal distribution (the density of the marginal distribution is a special marginal density, for $s = t_0$).

The above expression shows a direct relation between the density and the butterfly spread option. With this connection, we can examine whether well-known properties of density functions (nonnegativity, integrating to one) are satisfied, for example for option prices observed in the financial markets. This property will be examined in Chapter 4 where local volatility asset models will be introduced.

3.2.3 Volatility variations

A generalization of the Black-Scholes model, which was already introduced in Equation (2.23) was to prescribe a time-dependent volatility coefficient $\sigma(t)$, instead of constant volatility.

Log-transformed GBM asset prices are normally distributed and in that case only the first two moments have to match to guarantee equality in distribution. By means of matching the moments of the GBM process and of a GBM with time-dependent volatility, it was shown in Example 2.2.1 under which conditions a model with time-dependent volatility and one with a constant volatility have identical first two moments. This implies, in fact, that European option values under a Black-Scholes model with a time-dependent volatility function are identical to the corresponding option values that are obtained by a log-transformed model based on the following *time-averaged, constant volatility function*,

$$\sigma_* = \sqrt{\frac{1}{T} \int_0^T \sigma^2(t) \mathrm{d}t}.$$

The valuation of European options under the GBM asset model with a time-dependent volatility can thus be performed also by using a constant volatility coefficient σ_*, as derived in (2.25), because the first two moments are identical. The difference between the two models is subtle and can only be observed when considering pricing exotic options (e.g. for path-dependent options).

Iterated expectations and stochastic volatility

We present another basic and often used application of the tower property of the expectation in finance. In this example we assume the following SDE for a stock price,

$$\mathrm{d}S(t) = rS(t)\mathrm{d}t + Y(t)S(t)\mathrm{d}W^{\mathbb{Q}}(t),$$

where $Y(t)$ represents a certain *stochastic volatility process* which has, for example, a lognormal distribution. After standard calculations, we obtain the following solution for $S(T)$, given by:

$$S(T) = S_0 \exp\left(\int_{t_0}^T \left(r - \frac{1}{2}Y^2(t)\right) dt + \int_{t_0}^T Y(t) dW^{\mathbb{Q}}(t)\right). \quad (3.37)$$

As Expression (3.37) contains integrals of the process $Y(t)$, it is nontrivial to determine a closed-form solution for the value of a European option.[5] A possible solution for the pricing problem is to use the *tower property of iterated expectations*, to determine the European option prices, conditioned on "realizations" of the volatility process $Y(t)$.

By the tower property of expectations, using $\mathbb{E} = \mathbb{E}^{\mathbb{Q}}$, the European call value can be reformulated as a discounted expectation with[6]:

$$\mathbb{E}\left[\max(S(T) - K, 0) \big| \mathcal{F}(t_0)\right]$$
$$= \mathbb{E}\left[\mathbb{E}\left[\max(S(T) - K, 0) \big| Y(t), t_0 \leq t \leq T\right] \big| \mathcal{F}(t_0)\right]. \quad (3.38)$$

Conditioned on the realizations of the variance process, the calculation of the inner expectation is equivalent to the Black-Scholes solution with a time-dependent volatility, i.e. for given realizations of $Y(t)$, $t_0 \leq t \leq T$, the asset value $S(T)$ in (3.37) is given by:

$$S(T) = S(t_0) \exp\left(\left(r - \frac{1}{2}\sigma_*^2\right)(T - t_0) + \sigma_*(W^{\mathbb{Q}}(T) - W^{\mathbb{Q}}(t_0))\right),$$

with

$$\sigma_*^2 = \frac{1}{T - t_0} \int_{t_0}^T Y^2(t) dt.$$

The solution of the inner expectation is then given by:

$$\mathbb{E}\left[\max(S(T) - K, 0) \big| \{Y(t)\}_{t_0}^T\right] = S(t_0) e^{r(T-t_0)} F_{\mathcal{N}(0,1)}(d_1) - K F_{\mathcal{N}(0,1)}(d_2),$$

with

$$d_1 = \frac{\log \frac{S(t_0)}{K} + (r + \frac{1}{2}\sigma_*^2)(T - t_0)}{\sigma_* \sqrt{T - t_0}}, \quad d_2 = d_1 - \sigma_* \sqrt{T - t_0},$$

$F_{\mathcal{N}(0,1)}$ being the standard normal cumulative distribution function.

We can substitute these results into Equation (3.38), giving:

$$\mathbb{E}[\max(S(T) - K, 0)] = \mathbb{E}\left[S(t_0) e^{r(T-t_0)} F_{\mathcal{N}(0,1)}(d_1) - K F_{\mathcal{N}(0,1)}(d_2)\right]$$
$$= S(t_0) e^{r(T-t_0)} \mathbb{E}\left[F_{\mathcal{N}(0,1)}(d_1)\right] - K \mathbb{E}\left[F_{\mathcal{N}(0,1)}(d_2)\right].$$

---

[5] If $Y(t)$ is normally distributed, the model is called the Schöbel-Zhu model, and if $Y(t)$ follows a Cox-Ingersol-Ross (CIR) square-root process, the system is called the Heston model. They are introduced in Chapter 8.

[6] The discount term $M(T)$ is omitted, only the expectation is displayed to save some space. The interest rates in this model are constant and do not influence the final result.

The option pricing problem under these nontrivial asset dynamics has been transformed into the calculation of an expectation of a normal CDF. The difficult part of this expectation is that both arguments of the CDF, d_1 and d_2, are functions of σ_*, which itself is a function of process $Y(t)$. One possibility to deal with the expectation is to use Monte-Carlo simulation, to be discussed in Chapter 9, another is to introduce an approximation, for example, $F_{\mathcal{N}(0,1)}(x) \approx g(x)$ (as proposed in [Piterbarg, 2005a], where $g(x)$ is a function, like $g(x) = a + be^{-cx}$).

3.3 Delta hedging under the Black-Scholes model

In this section we discuss some insightful details of the delta hedging strategy and of re-balancing a financial portfolio. The main concept of hedging is to eliminate risk or at least to reduce it when complete elimination is not possible.

The easiest way to eliminate the risk is to offset a trade. A so-called *back-to-back transaction* eliminates the risk associated with market movements. The main idea of such a transaction is as follows. Suppose a financial institution sells some financial derivative to a counterparty. The value of this derivative will vary, depending on market movements. One way to eliminate such movements is to buy "exactly the same", or a very similar, derivative from another counterparty. One may wonder about buying and selling the same derivatives at the same time, but a profit may be made from selling the derivative with some additional premium which is added to the price which is paid.

In Subsection 3.1.2 we discussed the concept of the replicating portfolio, under the Black-Scholes model, where the uncertainty of a European option was eliminated by the Δ hedge.

Based on the same strategy, consider the following portfolio:

$$\Pi(t, S) = V(t, S) - \Delta S. \tag{3.39}$$

The objective of delta hedging is that the value of the portfolio does not change when the underlying asset moves, so the derivative of portfolio $\Pi(t, S)$ w.r.t S needs to be equal to 0, i.e.,

$$\frac{\partial \Pi(t, S)}{\partial S} = \frac{\partial V(t, S)}{\partial S} - \Delta = 0 \Rightarrow \Delta = \frac{\partial V}{\partial S}, \tag{3.40}$$

with $V = V(t, S)$, and Δ given by Equation (3.24).

Suppose we sold a call $V_c(t_0, S)$ at time t_0, with maturity T and strike K. By selling, we obtained a cash amount equal to $V_c(t_0, S)$ and perform a *dynamic* hedging strategy until time T. Initially, at the inception time, we have

$$\Pi(t_0, S) := V_c(t_0, S) - \Delta(t_0)S_0.$$

This value may be negative when $\Delta(t_0)S_0 > V_c(t_0, S)$. If funds are needed for buying $\Delta(t)$ shares, we make use of a *funding account*, $\mathrm{PnL}(t) \equiv \mathrm{P\&L}(t)$. $\mathrm{PnL}(t)$ represents the total value of the option sold and the hedge, and it keeps track of the changes in the asset value $S(t)$.

Typically the funding amount $\Delta(t_0)S_0$ is then obtained from a trading desk of a treasury department.

Every day we may need to re-balance the position and hedge the portfolio. At some time $t_1 > t_0$, we then receive (or pay) interest over the time period $[t_0, t_1]$, which will amount to $\text{P\&L}(t_0)e^{r(t_1-t_0)}$. At t_1 we have $\Delta(t_0)S(t_1)$ which may be sold, and we will update the hedge portfolio. Particularly, we purchase $\Delta(t_1)$ stocks, costing $-\Delta(t_1)S(t_1)$. The overall $\text{P\&L}(t_1)$ account will become:

$$\text{P\&L}(t_1) = \underbrace{\text{P\&L}(t_0)e^{r(t_1-t_0)}}_{\text{interest}} - \underbrace{(\Delta(t_1) - \Delta(t_0))S(t_1)}_{\text{borrow}}. \tag{3.41}$$

Assuming a time grid with $t_i = i\frac{T}{m}$, the following recursive formula for the m time steps is obtained,

$$\text{P\&L}(t_0) = V_c(t_0, S) - \Delta(t_0)S(t_0), \tag{3.42}$$
$$\text{P\&L}(t_i) = \text{P\&L}(t_{i-1})e^{r(t_i-t_{i-1})} - (\Delta(t_i) - \Delta(t_{i-1}))S(t_i), \quad i = 1, \ldots, m-1.$$

At the option maturity time T, the option holder may exercise the option or the option will expire worthless. As the option writer, we will thus encounter a cost equal to the option's payoff at $t_m = T$, i.e. $\max(S(T) - K, 0)$. On the other hand, at maturity time we own $\Delta(t_m)$ stocks, that may be sold in the market. The value of the portfolio at maturity time $t_m = T$ is then given by:

$$\text{P\&L}(t_m) = \text{P\&L}(t_{m-1})e^{r(t_m - t_{m-1})}$$
$$- \underbrace{\max(S(t_m) - K, 0)}_{\text{option payoff}} + \underbrace{\Delta(t_{m-1})S(t_m)}_{\text{sell stocks}}. \tag{3.43}$$

In a perfect world, with continuous re-balancing, the $\text{P\&L}(T)$ would equal zero on average, i.e. $\mathbb{E}[\text{P\&L}(T)] = 0$. One may question the reasoning behind dynamic hedging if the profit made by the option writer on average equals zero. The profit in option trading, especially with OTC transactions, is to charge an additional fee (often called a "*spread*") at the start of the contract. At time t_0 the cost for the client is not $V_c(t_0, S)$ but $V_c(t_0, S) + spread$, where $spread > 0$ would be the profit for the writer of the option.

Example 3.3.1 (Expected P&L(T)) We will give an example where $\mathbb{E}[\text{P\&L}(T)|\mathcal{F}(t_0)] = 0$. We consider a three-period case, with $t_0, t_1, t_2 := T$, and $\Delta(t_i)$ is a deterministic function, as in the Black-Scholes hedging case. At t_0 an option is sold, with expiry date t_2 and strike price K. In a three-period setting, we have three equations for $\text{P\&L}(t)$ related to the initial hedging, the re-balancing and the final hedging, i.e.,

$$\text{P\&L}(t_0) = V_c(t_0, S) - \Delta(t_0)S(t_0),$$
$$\text{P\&L}(t_1) = \text{P\&L}(t_0)e^{r(t_1-t_0)} - (\Delta(t_1) - \Delta(t_0))S(t_1),$$
$$\text{P\&L}(t_2) = \text{P\&L}(t_1)e^{r(t_2-t_1)} - \max(S(t_2) - K, 0) + \Delta(t_1)S(t_2).$$

After collecting all terms, we find,

$$\text{P\&L}(t_2) = \Big[(V_c(t_0, S) - \Delta(t_0)S(t_0))\, e^{r(t_2-t_0)}$$
$$- (\Delta(t_1) - \Delta(t_0))\, S(t_1)e^{r(t_2-t_1)}\Big] - \max(S(t_2) - K, 0) + \Delta(t_1)S(t_2).$$

By the definition of a call option, we also have,

$$\mathbb{E}\left[\max(S(t_2) - K, 0)|\mathcal{F}(t_0)\right] = e^{r(t_2-t_0)}V_c(t_0, S), \qquad (3.44)$$

and because the discounted stock price, under the risk-neutral measure, is a martingale, $\mathbb{E}[S(t)|\mathcal{F}(s)] = S(s)e^{r(t-s)}$, the expectation of the P&L is given by:

$$\mathbb{E}[\text{P\&L}(t_2)|\mathcal{F}(t_0)] = \big(V_c(t_0, S) - \mathbb{E}[\max(S(t_2) - K, 0)|\mathcal{F}(t_0)]$$
$$+ \Delta(t_1)\mathbb{E}[S(t_2)|\mathcal{F}(t_0)] - \Delta(t_0)S(t_0)\big) \cdot e^{r(t_2-t_0)}$$
$$- (\Delta(t_1) - \Delta(t_0))\, \mathbb{E}[S(t_1)|\mathcal{F}(t_0)]e^{r(t_2-t_1)}.$$

Using the relation

$$\mathbb{E}[S(t_1)|\mathcal{F}(t_0)]e^{r(t_2-t_1)} = \mathbb{E}[S(t_2)|\mathcal{F}(t_0)] = S(t_0)e^{r(t_2-t_0)},$$

and by (3.44), the expression simplifies to,

$$\mathbb{E}[\text{P\&L}(t_2)|\mathcal{F}(t_0)] = -\Delta(t_1)S(t_0)e^{r(t_2-t_0)} + \Delta(t_1)\mathbb{E}[S(t_2)|\mathcal{F}(t_0)] = 0. \quad \blacklozenge$$

Example 3.3.2 (Dynamic hedge experiment, Black-Scholes model)
In this experiment we perform a dynamic hedge for a call option under the Black-Scholes model. For the asset price, the following model parameters are set, $S(t_0) = 1$, $r = 0.1$, $\sigma = 0.2$. The option's maturity is $T = 1$ and strike $K = 0.95$. On a time grid stock path $S(t_i)$ is simulated. Based on these paths, we perform the hedging strategy, according to Equations (3.42) and (3.43). In Figure 3.7 three stock paths $S(t)$ are presented, for one of them the option would be in-the-money at time T (upper left), one ends out-of-money (upper right) and is at-the-money (lower left) at time T. In the three graphs $\Delta(t_i)$ (green line) behaves like the stock process $S(t)$, however when the stock $S(t)$ give a call price (pink line) deep in or out of the money, $\Delta(t_i)$ is either 0 or 1. In Table 3.1 the results for these three paths are summarized. In all three cases, the initial hedge quantities are the same, however, they change over time with the stock paths. In all three cases the final P&L(t_m) is close to zero, see Table 3.1.

Example 3.3.3 (Re-balancing frequency) A crucial assumption in the Black-Scholes model is the omission of transaction costs. This basically implies that, free-of-charge, re-balancing can be performed at any time. In practice, however, this is unrealistic. Depending on the derivative details, re-balancing

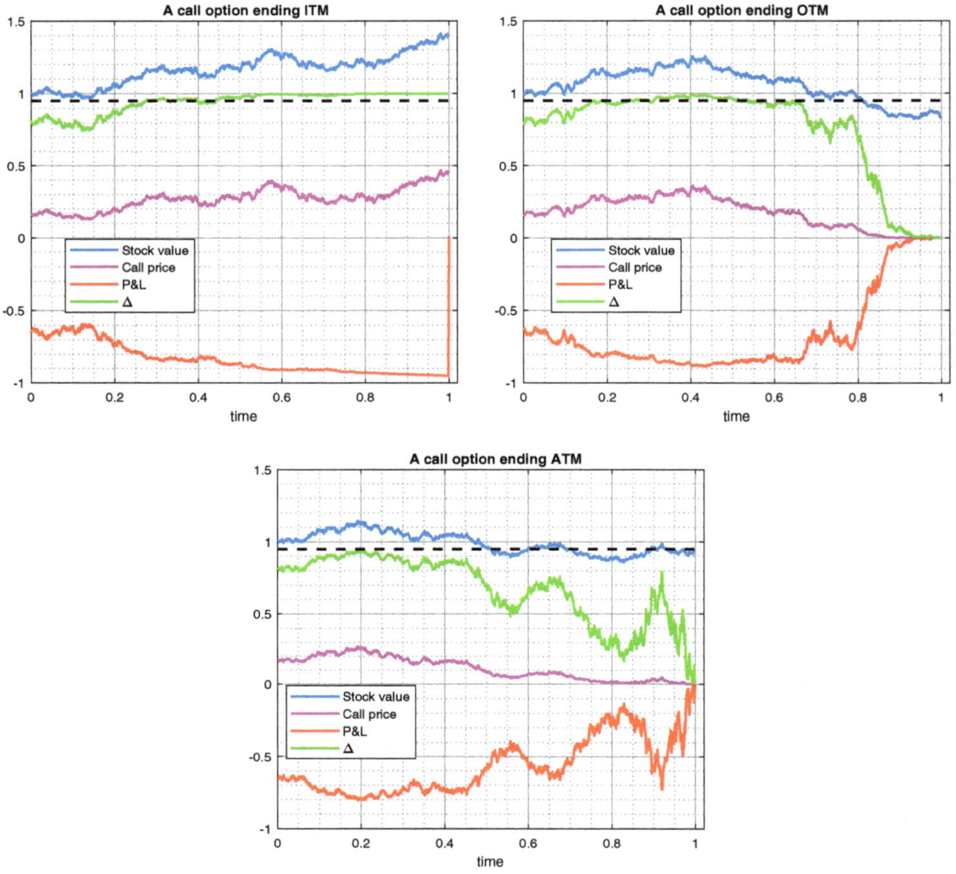

Figure 3.7: *Delta hedging a call option. Blue: the stock path, pink: the value of a call option, red: P&L(t) portfolio, and green: Δ.*

Table 3.1: *Hedge results for three stock paths.*

path no.	$S(t_0)$	P&L(t_0)	P&L(t_{m-1})	$S(t_m)$	$(S(t_m) - K)^+$	P&L(t_m)
path 1 (ITM)	1	−0.64	−0.95	1.40	0.45	$2.4 \cdot 10^{-4}$
path 2 (OTM)	1	−0.64	0.002	−0.08	0	$2.0 \cdot 10^{-4}$
path 3 (ATM)	1	−0.64	−0.010	0.96	0.01	$-2.0 \cdot 10^{-3}$

can take place daily, weekly or even on a monthly basis. When re-balancing takes place, transaction costs may need to be paid. So, for each derivative, the hedging frequency will be based on a balance between the cost and the impact of hedging.

The Black-Scholes Option Pricing Equation   77

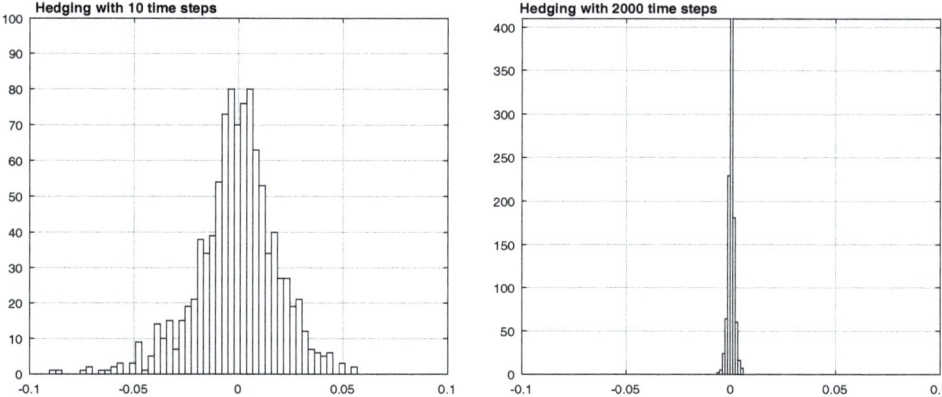

Figure 3.8: *The impact of the re-balancing frequency on the variance of the P&L(T) portfolio. Left: 10 re-balancing times, Right: 2000 re-balancing times.*

In Figure 3.8 the impact of the frequency of updating the hedge portfolio on the distribution of the P&L(T) is presented. Two simulations have been performed, one with 10 re-balancing actions during the option's life time and one with 2000 actions. It is clear that frequent re-balancing brings the variance of the portfolio P&L(T) down to almost 0.

3.4 Exercise set

Exercise 3.1 Today's asset price is $S_0 = €10$. A call on this stock with expiry in 60 days, and a strike price of $K = €10$ costs $V_c(t_0, S_0) = €1$. Suppose we bought either the stock, or we bought the call.

Fill the table below. The first row consists of possible asset prices $S(T)$, ranging from €8.5 to €11.5. Write in the second row the stock profit, in percentages, if we had bought the stock; in the third row the profit, in percentages, when having bought the option at t_0. Compare the profits and losses of having only the stock or having bought the option.

	stock price at $t = T$						
	€8.5	€9	€9.5	€10	€10.5	€11	€11.5
profit $S(T)$							
profit $V_c(T, S(T))$							

Exercise 3.2 Give at least six assumptions that form the foundation for the derivation of the Black-Scholes equation.

Exercise 3.3 Draw the payoff of a call option, as a function of strike price K, and draw in the same picture the call value at a certain time $t < T$ as a function of K.

Exercise 3.4 Assume we have a European call and put option (with the same exercise date $T = 1/4$, i.e., exercise in three months, and strike price $K = 10€$). The current stock price is 11€, and suppose a continuously compounded interest rate, $r = 6\%$. Define an arbitrage opportunity when both options are worth 2.5 €.

Exercise 3.5 Consider a portfolio Π consisting of one European option with value $V(S, t)$ and a negative amount, $-\Delta$, of the underlying stock, i.e.

$$\Pi(t) = V(S, t) - \Delta \cdot S(t).$$

With the choice $\Delta = \partial V / \partial S$, the change in this portfolio, in excess of the risk-free rate r, is given by,

$$d\Pi(t) - r\Pi(t)dt = \left(\frac{\partial V}{\partial t} + \frac{1}{2}\sigma^2 S^2 \frac{\partial^2 V}{\partial S^2} + rS\frac{\partial V}{\partial S} - rV \right) dt.$$

In the Black-Scholes argument for European options, this expression should be equal to zero, since this precludes arbitrage.

Give an example of an arbitrage trade, if the equality $d\Pi(t) = r\Pi(t)dt$ doesn't hold, for European options, by buying and selling the portfolio $\Pi(t)$ at the time points t and $t + \Delta t$, respectively.

Exercise 3.6 Derive the European option pricing equation assuming the underlying is governed by arithmetic Brownian motion,

$$dS(t) = \mu dt + \sigma dW^{\mathbb{P}}(t).$$

The Black-Scholes Option Pricing Equation

Exercise 3.7 Use a stochastic representation result (Feynman Kac theorem) to solve the following boundary value problem in the domain $[0, T] \times \mathbb{R}$.

$$\frac{\partial V}{\partial t} + \mu X \frac{\partial V}{\partial X} + \frac{1}{2}\sigma^2 X^2 \frac{\partial^2 V}{\partial X^2} = 0,$$
$$V(T, X) = \log(X^2),$$

where μ and σ are known constants.

Exercise 3.8 Consider the Black-Scholes option pricing equation,

$$\frac{\partial V}{\partial t} + rS\frac{\partial V}{\partial S} + \frac{1}{2}\sigma^2 S^2 \frac{\partial^2 V}{\partial S^2} - rV = 0.$$

a. Give boundary conditions (i.e., put values at $S = 0$ and for S "large") and the final condition (i.e. values at $t = T$) for a European put option.

b. Confirm that the expression

$$V_p(t, S) = -S F_{\mathcal{N}(0,1)}(-d_1) + K e^{-r(T-t)} F_{\mathcal{N}(0,1)}(-d_2),$$

with

$$d_{1,2} = \frac{\log(S/K) + (r \pm \frac{1}{2}\sigma^2)(T-t)}{\sigma\sqrt{T-t}},$$

(plus-sign for d_1, minus-sign for d_2), with $F_{\mathcal{N}(0,1)}(\cdot)$ the standard normal CDF, satisfies the Black-Scholes equation and is indeed the solution of a put option. (You may use the two identities $F_{\mathcal{N}(0,1)}(x) + F_{\mathcal{N}(0,1)}(-x) = 1$ and

$$S F'_{\mathcal{N}(0,1)}(d_1) - e^{-r(T-t)} K F'_{\mathcal{N}(0,1)}(d_2) = 0,$$

with $F'_{\mathcal{N}(0,1)}(\cdot)$ the derivative of $F_{\mathcal{N}(0,1)}(\cdot)$.)

c. Find the solution of the European call option with the help of the put-call parity relation.

Exercise 3.9 Confirm the analytic solution

a. of the plain vanilla call option, given in Theorem 3.2.2.

b. of the digital vanilla call option in Equation (3.26).

Exercise 3.10 Consider two portfolios: π_A, consisting of a call option plus $K e^{-rT}$ cash; π_B, consisting of one asset S_0.

a. Based on these two portfolios, determine the following bounds for the European call option value, at time $t = 0$:

$$\max(S_0 - K e^{-rT}, 0) \leq V_c(0, S_0) \leq S_0.$$

b. Derive the corresponding bounds for the time-zero value of the European put option $V_p(0, S_0)$, by the put-call parity relation.

c. Derive

$$\Delta = \frac{\partial V_c(t, S)}{\partial S},$$

and also the limiting behaviour of Δ, for $t \to T^-$.

c. Use again the put-call parity relation to derive the delta of a put option.

d. Give a financial argument which explains the value of $\partial V_p(t,S)/\partial S$ at expiry $t = T$, for an in-the-money option, as well as for an out-of-the-money option.

Exercise 3.11 A strangle is an option construct where an investor takes a long position in a call and in a put on the same share S with the same expiry date T, but with different strike prices (K_1 for the call, and K_2 for the put, with $K_1 > K_2$).

  a. Draw the payoff of the strangle. Give an overview of the different payments possible at the expiry date. Make a distinction between three different possibilities for stock price $S(T)$.

  b. Can the strange be valued with the Black-Scholes equation? Determine suitable boundary conditions, i.e., strangle values at $S = 0$ and for S "large", and final condition (value at $t = T$) for the strangle.

  c. When would an investor buy a strangle with $K_2 \ll K_1$, and $K_2 < S_0 < K_1$?

Exercise 3.12 Consider the Black-Scholes equation for a cash-or-nothing call option, with the "in-the-money" value, $V^{\text{cash}}(T, S(T)) = A$, for $t = T$.

  a. What are suitable boundary conditions, at $S(t) = 0$ and for $S(t)$ "large", for this option, where $0 \leq t \leq T$?

  b. The analytic solution is given by $V_c^{\text{cash}}(t, S) = Ae^{-r(T-t)} F_{\mathcal{N}(0,1)}(d_2)$. Derive a put-call parity relation for the cash-or-nothing option, for all $t \leq T$, and determine the value of a cash-or-nothing put option (same parameter settings).

  c. Give the value of the delta for the cash-or-nothing call, and draw in a picture the delta value just before expiry time, at $t = T^-$.

CHAPTER 4

Local Volatility Models

In this chapter:

The main problem with the Black-Scholes model is that it is not able to reproduce the *implied volatility smile*, which is present in many financial markets (see **Section 4.1**). This fact forms an important motivation to study and use *other asset models*. Alternative models are briefly discussed in Section 4.1.3. In this chapter, we give an introduction into asset prices based on *nonparametric local volatility* models. The basis for these models is a relation between (market) option prices and the implied probability density function, which we describe in **Section 4.2**. The derivation of the *nonparametric local volatility models* is found in **Section 4.3**. The pricing equation for the option values under these asset price processes appears to be a *PDE with time-dependent parameters*.

Keywords: implied volatility, smile and skew, alternative models for asset prices, implied asset density function, local volatility, arbitrage conditions, arbitrage-free interpolation.

4.1 Black-Scholes implied volatility

When the asset price is modeled by geometric Brownian motion (GBM) with constant volatility σ, the Black-Scholes model gives the unique fair value of an option contract on that underlying asset. This option price is a monotonically increasing function of the stock's volatility. High volatility means higher probability for an option to be in-the-money (ITM), see the definition in Definition 3.1.2, at the maturity date, so that the option becomes relatively more expensive.

4.1.1 The concept of implied volatility

European options in the financial markets are sometimes quoted in terms of their so-called *implied volatilities*, instead of the commonly known bid and ask option prices. In other words, the Black-Scholes implied volatility is then regarded as *a language* in which option prices can be expressed.

There are different ways to calculate the implied volatility, and we discuss a common computational technique.

For a given interest rate r, maturity T and strike price K, we have market put and call option prices on a stock S, that we will denote by $V_p^{mkt}(K,T)$, $V_c^{mkt}(K,T)$, respectively.

> Focussing on calls here, the Black-Scholes implied volatility σ_{imp} is the σ-value, for which
> $$V_c(t_0, S; K, T, \sigma_{\text{imp}}, r) = V_c^{mkt}(K,T), \tag{4.1}$$
> where $t_0 = 0$.
> The implied volatility σ_{imp} is defined as the volatility inserted as a parameter in the Black-Scholes solution that reproduces the market option price $V_c^{mkt}(K,T)$ at time $t_0 = 0$.
> Whenever we speak of implied volatility in this book, we actually mean "Black-Scholes implied volatility"!

Newton-Raphson iterative method

There is no general closed-form expression for the implied volatility in terms of the option prices, so a *numerical technique* needs to be employed to determine its value. A common method for this is the Newton-Raphson root-finding iteration. The problem stated in (4.1) can be reformulated as the following root-finding problem,

$$g(\sigma_{\text{imp}}) := V_c^{mkt}(K,T) - V_c(t_0 = 0, S_0; K, T, \sigma_{\text{imp}}, r) = 0. \tag{4.2}$$

Given an initial guess[1] for the implied volatility, i.e. $\sigma_{\text{imp}}^{(0)}$, and the derivative of $g(\sigma_{\text{imp}})$ w.r.t. σ_{imp}, we can find the next approximations, $\sigma_{\text{imp}}^{(k)}, k = 1, 2, \ldots$, by means of the *Newton-Raphson iterative process*, as follows

> $$\sigma_{\text{imp}}^{(k+1)} = \sigma_{\text{imp}}^{(k)} - \frac{g(\sigma_{\text{imp}}^{(k)})}{g'(\sigma_{\text{imp}}^{(k)})}, \quad \text{for } k \geq 0. \tag{4.3}$$

Within the context of the Black-Scholes model, European call and put prices, as well as their derivatives, are known in closed-form, so that the derivative in

---
[1] Iterants within an iterative process are here denoted by a superscript in brackets.

Local Volatility Models

Expression (4.3) can be found analytically as well, as

$$g'(\sigma) = -\frac{\partial V(t_0, S_0; K, T, \sigma, r)}{\partial \sigma} = -Ke^{-r(T-t_0)} f_{\mathcal{N}(0,1)}(d_2) \sqrt{T-t_0},$$

with $f_{\mathcal{N}(0,1)}(\cdot)$ the standard normal probability density function, $t_0 = 0$, and

$$d_2 = \frac{\log(\frac{S_0}{K}) + (r - \frac{1}{2}\sigma^2)(T-t_0)}{\sigma\sqrt{T-t_0}}.$$

Derivative $\frac{\partial V}{\partial \sigma}$ in $g'(\sigma)$ is, in fact, another option sensitivity, called the option's *vega*, the sensitivity with respect to volatility changes. It is another important hedge parameter. In the case where an analytic expression for the derivative is not known, additional approximations are required.

Example 4.1.1 We return to the example in Chapter 3, in Subsection 3.2.1, where the call option values and two hedge parameters, delta and gamma, were displayed for all time points and possible asset values. Here, in Figure 4.1, we will present the vega values in the (t, S)-surface, so that we see the corresponding values of the option vega at each point of the asset path followed. The following parameter set is used:

$$S_0 = 10, r = 0.05, \sigma = 0.4, T = 1, K = 10.$$

◆

Example 4.1.2 (Black-Scholes implied volatility) Consider a call option on a non-dividend paying stock $S(t)$. The current stock price is $S_0 = 100$, interest rate $r = 5\%$, the option has a maturity date in one year, struck at $K = 120$, and it is traded at $V_c^{mkt}(K,T) = 2$. The task is to find the option's implied volatility, σ_{imp}, i.e.

$$g(\sigma_{\text{imp}}) = 2 - V_c(0, 100; 120, 1y, \sigma_{\text{imp}}, 5\%) = 0.$$

The result is $\sigma_{\text{imp}} = 0.161482728841394....$ By inserting this value into the Black-Scholes equation, we obtain a call option value of 1.999999999999996.

Combined root-finding method

The Newton-Raphson algorithm converges quadratically, when the initial guess is *in the neighborhood* of a root. It is therefore important that the initial guess, $\sigma_{\text{imp}}^{(0)}$, is chosen *sufficiently close* to the root. A convergence issue is encountered with this method when the denominator in Expression (4.3) is very close to zero, or, in our context, when the option's vega is almost zero. The option vega is very small, for example, for deep ITM and OTM options, giving rise to convergence issues for the Newton-Raphson iterative method.

A basic root-finding procedure, like the bisection method may help to get a sufficiently close starting value for the Newton-Raphson iteration, or in the case

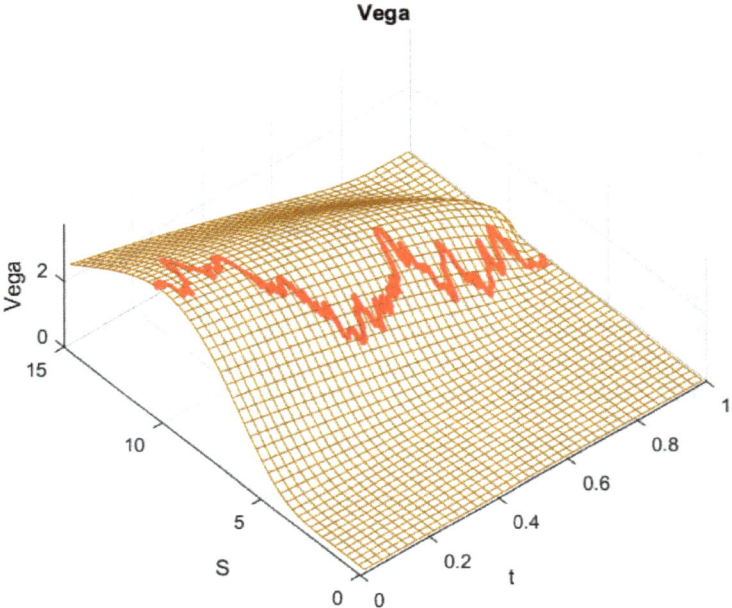

Figure 4.1: *Surface of vega values for a call with $T = 1$ in the (t, S) domain. An asset path intersects the surface so that at each $(t, S(t))$ the vega can be read.*

of serious convergence issues. Then, these two techniques form a *combined root-finding procedure*.

The method combines the efficiency of the Newton-Raphson method and the robustness of the bisection method. Similar to bisection, the combined root-finding method is based on the fact that the root should lie in an interval, here $[\sigma_l, \sigma_r]$. The method aims to find the root by means of the Newton-Raphson iteration. However, if the approximation falls outside $[\sigma_l, \sigma_r]$, the bisection method is employed to reduce the size of the interval. After some steps, the bisection method will have reduced the interval sufficiently, so that the Newton-Raphson stage will converge. The algorithm is presented below.

Combined root finding algorithm:

Given a function $g(\sigma) = 0$, find the root $\sigma = \sigma_{\text{imp}}$, with $g(\sigma_{\text{imp}}) = 0$.
Determine initial interval $[\sigma_{\text{imp}}^l, \sigma_{\text{imp}}^r]$
If $g(\sigma_{\text{imp}}^l) \cdot g(\sigma_{\text{imp}}^r) > 0$ *stop* (there is no zero in the interval)
If $g(\sigma_{\text{imp}}^l) \cdot g(\sigma_{\text{imp}}^r) < 0$ then
$\quad k = 1$
$\quad \sigma_{\text{imp}}^{(k)} = \frac{1}{2}(\sigma_{\text{imp}}^l + \sigma_{\text{imp}}^r)$

Local Volatility Models 85

$$\delta = -g(\sigma_{\text{imp}}^{(k)})/g'(\sigma_{\text{imp}}^{(k)})$$
while $\delta/\sigma_{\text{imp}}^{(k)} > \text{tol}$
  $\sigma_{\text{imp}}^{(k+1)} = \sigma_{\text{imp}}^{(k)} + \delta$
  if $\sigma_{\text{imp}}^{(k+1)} \notin [\sigma_{\text{imp}}^l, \sigma_{\text{imp}}^r]$, then
    if $g(\sigma_{\text{imp}}^l) \cdot g(\sigma_{\text{imp}}^{(k+1)}) > 0$, then $\sigma_{\text{imp}}^l = \sigma_{\text{imp}}^{(k)}$
    if $g(\sigma_{\text{imp}}^l) \cdot g(\sigma_{\text{imp}}^{(k+1)}) < 0$, then $\sigma_{\text{imp}}^r = \sigma_{\text{imp}}^{(k)}$
    $\sigma_{\text{imp}}^{(k+1)} = (\sigma_{\text{imp}}^l + \sigma_{\text{imp}}^r)/2$
  $\delta = -g(\sigma_{\text{imp}}^{(k+1)})/g'(\sigma_{\text{imp}}^{(k+1)})$
  $k = k+1$
continue

For the deep ITM and OTM options, for which the Newton-Raphson method does not converge, the combined method keeps converging, however, with a somewhat slower convergence rate. In the combined root-finding method above, derivative information is included, in the form of the option's vega in the Newton-Raphson step.

Remark 4.1.1 (Brent's method) *A derivative-free, black-box robust and efficient combined root-finding procedure was developed in the 1960s by van Wijngaarden, Dekker, and others [Dekker, 1969; Brent, 2013], which was subsequently improved by Brent [Brent, 1971]. The method, known as Brent's method is guaranteed to converge, so long as the function can be evaluated within an initial interval which contains a root. Brent's method combines the bisection technique with an inverse quadratic interpolation technique. The inverse quadratic interpolation technique is based on three previously computed iterates, $\sigma^{(k)}, \sigma^{(k-1)}, \sigma^{(k-2)}$ through which the inverse quadratic function is fitted. Iterate $\sigma^{(k+1)}$ is a quadratic function of y, and the $\sigma^{(k+1)}$-value at $y = 0$ is the next approximation for the root σ_{imp}.*

If the three point pairs are $[\sigma^{(k)}, g(\sigma^{(k)})], [\sigma^{(k-1)}, g(\sigma^{(k-1)})], [\sigma^{(k-2)}, g(\sigma^{(k-2)})]$, then the interpolation formula is given by,

$$\sigma^{(k+1)} = \frac{g(\sigma^{(k-1)})g(\sigma^{(k-2)})\sigma^{(k)}}{(g(\sigma^{(k)}) - g(\sigma^{(k-1)}))(g(\sigma^{(k)}) - g(\sigma^{(k-2)}))}$$
$$+ \frac{g(\sigma^{(k-2)})g(\sigma^{(k)})\sigma^{(k-1)}}{(g(\sigma^{(k-1)}) - g(\sigma^{(k-2)}))(g(\sigma^{(k-1)}) - g(\sigma^{(k)}))}$$
$$+ \frac{g(\sigma^{(k-1)})g(\sigma^{(k)})\sigma^{(k-2)}}{(g(\sigma^{(k-2)}) - g(\sigma^{(k-1)}))(g(\sigma^{(k-2)}) - g(\sigma^{(k)}))}. \quad (4.4)$$

When, however, that estimate lies outside the bisected interval, then it is rejected and a bisection step will take place instead.

If two consecutive approximations are identical, for example, $g(\sigma^{(k)}) = g(\sigma^{(k-1)})$, the quadratic interpolation is replaced by the secant method,

$$\sigma^{(k+1)} = \sigma^{(k-1)} - g(\sigma^{(k-1)}) \frac{\sigma^{(k-1)} - \sigma^{(k-2)}}{g(\sigma^{(k-1)}) - g(\sigma^{(k-2)})}. \tag{4.5}$$

▲

4.1.2 Implied volatility; implications

The volatility which is prescribed in the Black-Scholes solution is supposed to be a known constant or deterministic function of time. This is one of the basic assumptions in the Black-Scholes theory, leading to the Black-Scholes equation. This, however, is inconsistent with the observations in the financial market, because numerical inversion of the Black-Scholes equation based on market option prices, for different strikes and a fixed maturity time, exhibit a so-called *implied volatility skew or smile*. Figure 4.2 presents typical implied volatility shapes that we find in financial market data quotes. Here the implied volatility is shown against the strike dimension, varying K. Figure 4.3 shows two typical implied volatility surfaces in strike price and time dimension, where it is clear that the surface varies also in the time-wise direction. This is called the *term-structure* of the implied volatility surface.

It is without any doubt that the Black-Scholes model, and its notion of hedging option contracts by stocks and cash, forms the foundation of modern finance. Certain underlying assumptions are, however, questionable in practical market applications. To name a few, in the Black-Scholes world, delta hedging is supposed to be a continuous process, but in practice it is a discrete process (a hedged portfolio is typically updated once a week or so, depending on the type of financial derivative), and transaction costs for re-balancing of the portfolio are not taken into account. Empirical studies of financial time series have also revealed that the normality assumption for asset returns, $dS(t)/S(t)$, in the Black-Scholes theory cannot capture *heavy tails* and *asymmetries*, that are found to be present in log-asset returns in the financial market quotes [Rubinstein, 1994]. Empirical densities are usually peaked compared to the normal density; a phenomenon which is known by *excess of kurtosis*.

The main problem with the Black-Scholes model is however that it is *not able to reproduce the above mentioned implied volatility skew or smile*, that is commonly observed in many financial markets. This fact forms an important motivation to study and use alternative mathematical asset models.

In this chapter, the local volatility process is discussed in detail as a first alternative model for the stock price dynamics.

4.1.3 Discussion on alternative asset price models

To overcome the issues with the Black-Scholes dynamics, a number of alternative models have been suggested in the financial literature. These asset models include local volatility models [Dupire, 1994; Derman and Kani, 1998; Coleman *et al.*,

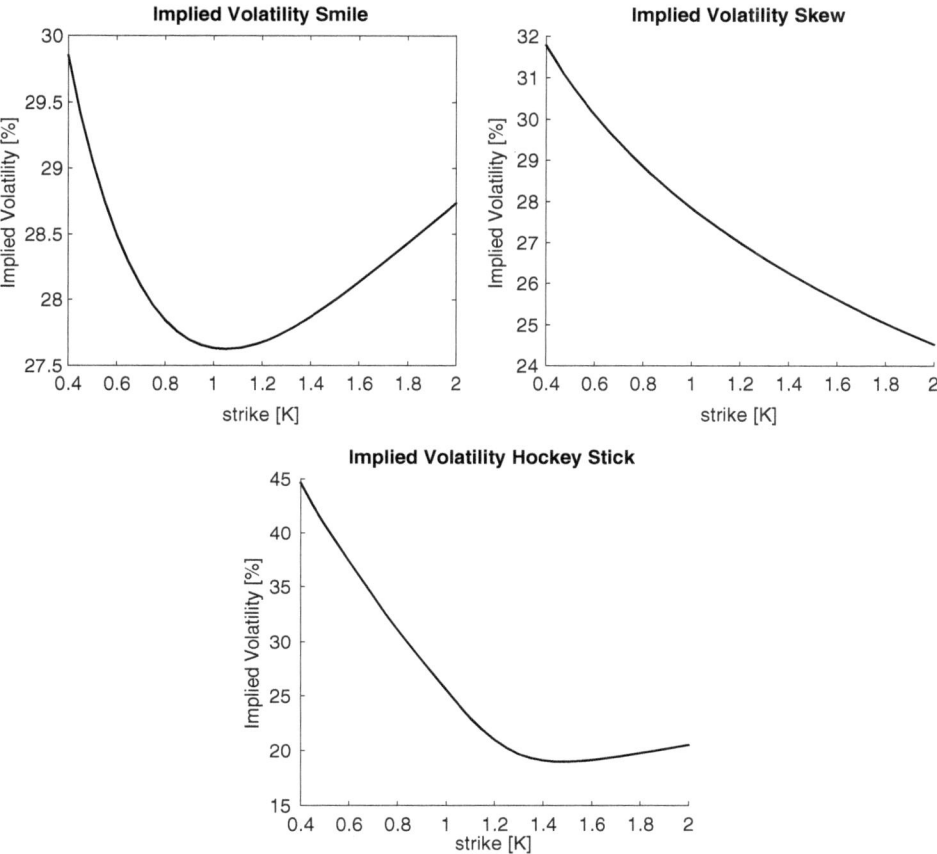

Figure 4.2: *Typical implied volatility shapes: a smile, a skew and the so-called hockey stick. The hockey stick can be seen as a combination of the implied volatility smile and the skew.*

1999; Dupire, 1994], stochastic volatility models [Hull and White, 1987; Heston, 1993], jump diffusion models [Kou, 2002; Kou and Wang, 2004; Merton, 1976] and, generally, Lévy models of finite and infinite activity [Barndorff, 1998; Carr et al., 2002; Eberlein, 2001; Matache et al., 2004; Raible, 2000]. It has been shown that several of these advanced asset models are, at least to some extent, able to generate the observed market volatility skew or smile.

Although many of the above mentioned models can be fitted to the option market data, a drawback is the need for the *calibration* to determine the open parameters of the underlying stock process, so that model and market option prices fit. An exception is formed by the so-called *local volatility* (LV) models. Since the input for the LV models are the market observed implied volatility values, the LV model can be calibrated *exactly* to any given set of arbitrage-free European vanilla option prices. Local volatility models therefore do not require

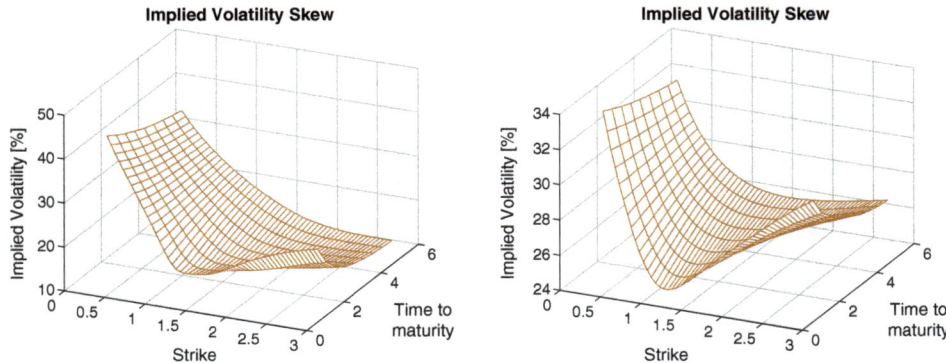

Figure 4.3: *Implied volatility surfaces. A pronounced smile for the short maturity and a pronounced skew for longer maturities T (left side figure), and a pronounced smile over all maturities (right side).*

calibration in the sense of finding the optimal model parameters, because the volatility function can be expressed in terms of the market quotes of call and/or put options. By the local volatility framework we can thus exactly reproduce market volatility smiles and skews.

For many years already, the LV model, as introduced by Dupire [1994] and Derman & Kani [1998], is considered a standard model for pricing and managing risks of financial derivative products.

Although well-accepted by practitioners for the accurate valuation of several option products, the LV model also has its limitations. For example, a *full matrix of option prices (meaning many strikes and maturity times)* is required to determine the implied volatilities. If a complete set of data is not available, as it often happens in practice, interpolation and extrapolation methods have to be applied, and they may give rise to inaccuracies in the option prices. There are other drawbacks, like an *inaccurate pricing of exotic options*, due to a local volatility model.

One may argue that an LV model does not reflect *"the dynamics of the market factors"*, because parameters are just fitted. The use of such a model outside the frame of its calibration, for different option products, different contract periods and thus volatilities, should be considered *with care*. This is an important drawback for example for the valuation of so-called forward starting options with longer maturities [Rebonato, 1999] to be covered in Chapter 10.

Example 4.1.3 Here we re-consider the two asset models from Example 2.2.1, one with a time-dependent volatility and another with a time-averaged constant volatility. In that example, we have shown that the same European option prices can be obtained by means of these two models with different asset dynamics.

Local Volatility Models

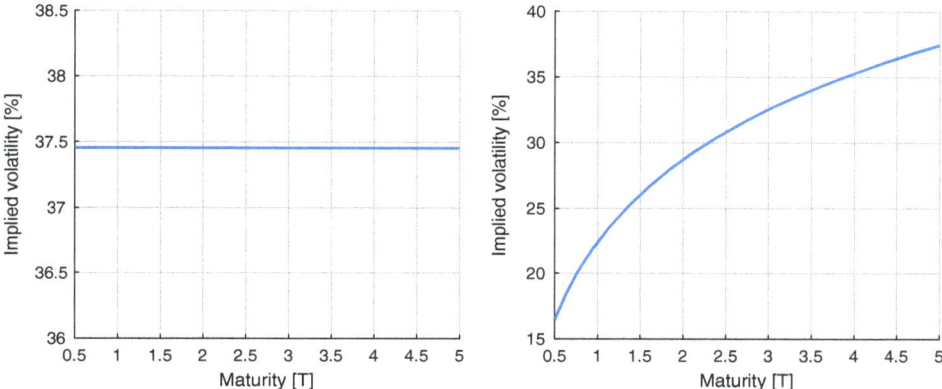

Figure 4.4: *Comparison of the volatility term structure (for ATM volatilities) for Black-Scholes model with constant volatility σ_* (left-side picture) versus a model with time-dependent volatility $\sigma(t)$ (right-side picture).*

In Figure 4.4, we now consider the implied volatility term-structure for these two models, one with a constant (*time-averaged*) volatility parameter σ_* and the other with a time-dependent volatility function $\sigma(t)$.

By means of the time-dependent volatility model, we can describe a so-called ATM volatility term structure (a time-dependent volatility structure based on at-the-money options), whereas for the model with the constant volatility model the volatility term structure is just a constant. Modeling the volatility as a time-dependent function may thus be desirable, especially in the context of hedging. Hedging costs are reduced typically by any model which explains the option pricing reality accurately. However, we typically do not know the specific time-dependent form of a volatility function beforehand. The *implied volatility* $\sigma_{\text{imp}}(T)$ may directly be obtained from a time-dependent volatility function $\sigma(t)$, via

$$\sigma_{\text{imp}}(T) = \sqrt{\frac{1}{T}\int_0^T \sigma^2(t)\mathrm{d}t},$$

by a property called *additivity of variance*. ♦

4.2 Option prices and densities

Local volatility models are based on the direct connection between option prices, implied density functions and Arrow-Debreu securities, as introduced in Section 3.2.2. The relation will become apparent in the present section.

4.2.1 Market implied volatility smile and the payoff

In this section we present a basic, but powerful, technique which was introduced by Breeden and Litzenberger [1978] for pricing European-style options with market implied volatility smile or skew patterns. The technique was actually developed 20 years before the local volatility model and it is based on the observation that the risk-neutral density for the underlying asset can be derived directly from market quotes of European call and put options. As discussed, see Equation (3.36), the calculation of the risk-neutral density from option prices requires second-order differentiation of the options with respect to the strike prices. This can be avoided with this approach as it is proposed to *differentiate the payoff function* instead of the option prices. Differentiation of the payoff is typically more stable as it does not involve any interpolation of market option prices.

Because the Breeden-Litzenberger framework is related to the marginal density of the stock S(T) at maturity time T, we cannot price any payoff function which depends on multiple maturity times. As a consequence, path-dependent option payoff functions cannot be priced in this framework.

In this section, we write in our notation the dependence on the strike price K and maturity time T explicitly in the arguments list of the option value. So, we use the notation $V(t_0, S_0; K, T)$.

Suppose we wish to price a basic European option with a well-known call or put payoff, which is also called *plain-vanilla payoff*, $H(T, S)$. As the payoff function depends only on $S(T)$, today's value is given by:

$$V(t_0, S_0; K, T) = e^{-r(T-t_0)} \int_0^\infty H(T,y) f_{S(T)}(y) \mathrm{d}y, \qquad (4.6)$$

with $f_{S(T)}(y) \equiv f_S(T, y; t_0, S_0)$ the risk-neutral density of the stock price process at time T; $H(T, y)$ is the payoff at time T.

The start of the derivation is to differentiate the call price $V_c(t_0, S_0; K, T)$, with payoff $\max(S - K, 0)$, with respect to strike K, as follows

$$\begin{aligned}\frac{\partial V_c(t_0, S_0; K, T)}{\partial K} &= e^{-r(T-t_0)} \frac{\partial}{\partial K} \int_K^{+\infty} (y - K) f_{S(T)}(y) \mathrm{d}y \\ &= -e^{-r(T-t_0)} \int_K^{+\infty} f_{S(T)}(y) \mathrm{d}y, \end{aligned} \qquad (4.7)$$

so that we find, for the second derivative with respect to K:

$$\boxed{\frac{\partial^2 V_c(t_0, S_0; K, T)}{\partial K^2} = e^{-r(T-t_0)} f_{S(T)}(K).} \qquad (4.8)$$

As presented earlier, the *implied stock density* can thus be obtained from (market) option prices via differentiation of the call prices (see Equation (3.36)), as

$$f_{S(T)}(y) = e^{r(T-t_0)} \frac{\partial^2}{\partial y^2} V_c(t_0, S_0; y, T). \qquad (4.9)$$

Local Volatility Models 91

With the help of the put-call parity (3.3), the density can also be obtained from market put option prices, i.e.,

$$f_{S(T)}(y) = e^{r(T-t_0)} \frac{\partial^2}{\partial y^2} \left(V_p(t_0, S_0; y, T) + S_0 - e^{-r(T-t_0)} y \right)$$

$$= e^{r(T-t_0)} \frac{\partial^2 V_p(t_0, S_0; y, T)}{\partial y^2}.$$

These expressions for $f_{S(T)}(y)$ can be inserted into Equation (4.6).

With two possibilities for implying the density for the stock, via calls or puts, one may wonder which options are the most appropriate. This question is related to the question which options are "liquidly available in the market". Liquidity means a lot of trading and therefore a small bid-ask spread leading to a better estimate of the stock's density. In practice, we cannot state that either calls or puts are most liquidly traded, but it is well-known that out-of-the-money (OTM) options, see the definition in Definition 3.1.2, are usually more liquid than in-the-money (ITM) options. To determine the market implied density for the stock price, one should preferably choose the OTM calls and puts.

Example 4.2.1 (Black-Scholes prices with derivative to the strike price)
The Black-Scholes solution that we have used in several earlier examples is considered, with parameter values chosen as:

$$S_0 = 10, r = 0.05, \sigma = 0.4, T = 1, K = 10.$$

In this example we present the call option values as a function of t and K, so we show the function V_c in the (t, K)-plane, see Figure 4.5. Note that the call option surface looks very differently when we plot it against varying strike prices K. In this section on the local volatility model, we also often encounter the derivatives of the option values with respect to strike price K. In Figure 4.5, we therefore also present these derivatives in the (t, K) plane. Notice that indeed the second derivative of the option with respect to the strike gives us the time-dependent evolution of a density function. ♦

We now start by splitting the integral in (4.6) into two parts:

$$V(t_0, S_0; K, T) = e^{-r(T-t_0)} \left(\int_0^{S_F} H(T, y) f_{S(T)}(y) \mathrm{d}y \right.$$

$$\left. + \int_{S_F}^{\infty} H(T, y) f_{S(T)}(y) \mathrm{d}y \right), \qquad (4.10)$$

with

$$S_F \equiv S_F(t_0, T) := e^{r(T-t_0)} S(t_0),$$

representing the *forward stock value*.

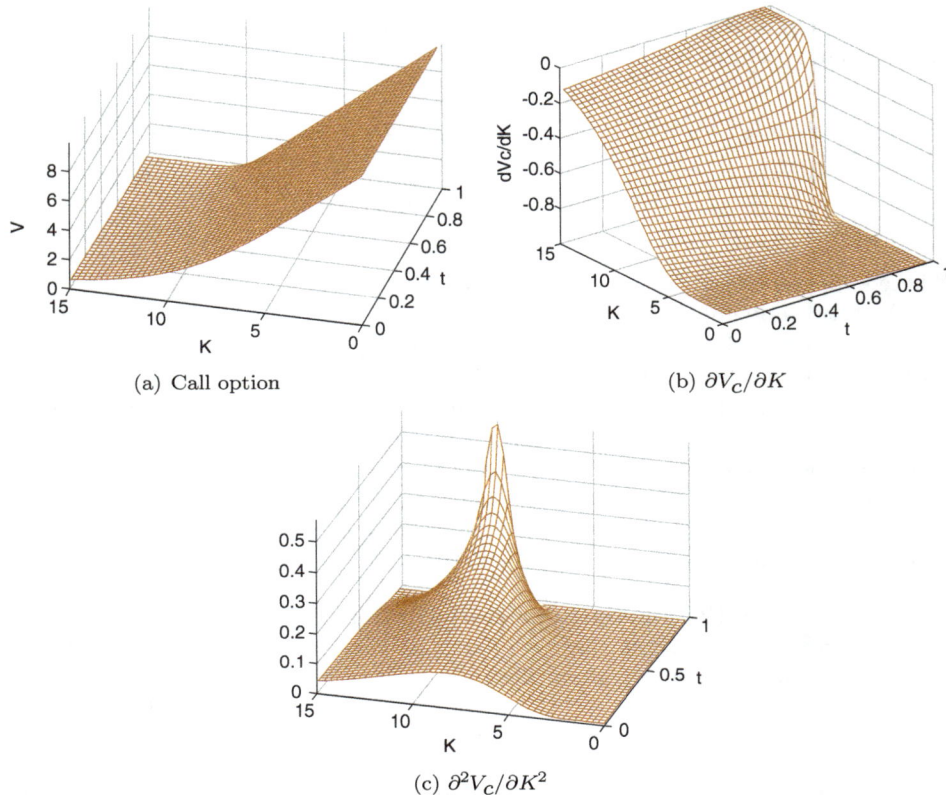

Figure 4.5: *Call option values as a function of running time t and strike K, plus the first derivative with respect to strike K, i.e. $\partial V_c/\partial K$ and the corresponding second derivative $\partial^2 V_c/\partial K^2$.*

For $S > K$ we have the OTM puts and for $S < K$ we have the OTM calls, so we calculate the first integral by put prices and the second by call prices, i.e.,

$$V(t_0, S_0; K, T) = \underbrace{\int_0^{S_F} H(T,y) \frac{\partial^2 V_p(t_0, S_0; y, T)}{\partial y^2} \, dy}_{I_1(y)}$$

$$+ \underbrace{\int_{S_F}^{\infty} H(T,y) \frac{\partial^2 V_c(t_0, S_0; y, T)}{\partial y^2} \, dy}_{I_2(y)}, \quad (4.11)$$

where $e^{r(T-t_0)}$ cancels out.

Local Volatility Models

To obtain an accurate estimate of $V(t_0, S_0; K, T)$, we need to compute second derivatives of puts and calls at time t_0. In practice, however, only a few market quotes for these options are available, and the computed results can be sensitive to the method used to calculate the derivatives. With the help of the integration-by-parts technique, we can *exchange differentiation of the options with differentiation of the payoff function*. Assume that the payoff $H(T, y)$ is twice differentiable with respect to y. By integration by parts of $I_1(y)$ in (4.11), we find:

$$\int_0^{S_F} I_1(y)\,dy = \left[H(T,y)\frac{\partial V_p(t_0, S_0; y, T)}{\partial y}\right]_{y=0}^{y=S_F}$$
$$- \int_0^{S_F} \frac{\partial H(T,y)}{\partial y}\frac{\partial V_p(t_0, S_0; y, T)}{\partial y}\,dy. \quad (4.12)$$

Applying again integration by parts to the last integral in (4.12), we find:

$$\int_0^{S_F} I_1(y)\,dy = \underbrace{\left[H(T,y)\frac{\partial V_p(t_0, S_0; y, T)}{\partial y}\right]_{y=0}^{y=S_F}}_{I_{1,1}(S_F)} - \underbrace{\left.\frac{\partial H(T,y)}{\partial y}V_p(t_0, S_0; y, T)\right|_{y=0}^{y=S_F}}_{I_{1,2}(S_F)}$$
$$+ \int_0^{S_F} \frac{\partial^2 H(T,y)}{\partial y^2}V_p(t_0, S_0; y, T)\,dy, \quad (4.13)$$

and for the second integral in (4.11), we find, similarly:

$$\int_{S_F}^{\infty} I_2(y)\,dy = \underbrace{\left[H(T,y)\frac{\partial V_c(t_0, S_0; y, T)}{\partial y}\right]_{y=S_F}^{y=\infty}}_{I_{2,1}(S_F)} - \underbrace{\left.\frac{\partial H(T,y)}{\partial y}V_c(t_0, S_0; y, T)\right|_{y=S_F}^{y=\infty}}_{I_{2,2}(S_F)}$$
$$+ \int_{S_F}^{\infty} \frac{\partial^2 H(T,y)}{\partial y^2}V_c(t_0, S_0; y, T)\,dy.$$

So, the option value $V(t_0, S_0; K, T)$ can be expressed as:

$$V(t_0, S_0; K, T) = I_{1,1}(S_F) - I_{1,2}(S_F) + I_{2,1}(S_F) - I_{2,2}(S_F)$$
$$+ \int_0^{S_F} \frac{\partial^2 H(T,y)}{\partial y^2}V_p(t_0, S_0; y, T)\,dy$$
$$+ \int_{S_F}^{\infty} \frac{\partial^2 H(T,y)}{\partial y^2}V_c(t_0, S_0; y, T)\,dy.$$

Let's have another look at the put-call parity. Differentiation of the put-call parity with respect to strike $y = K$, gives us:

$$\boxed{\frac{\partial V_c(t_0, S_0; y, T)}{\partial y} + e^{-r(T-t_0)} = \frac{\partial V_p(t_0, S_0; y, T)}{\partial y}.} \quad (4.14)$$

On the other hand, by evaluating the original put-call parity (3.3) at $K = S_F \equiv S_F(t_0, T)$, we find that $V_c(t_0, S_0; K, T) + e^{-r(T-t_0)} S_F = V_p(t_0, S_0; K, T) + S_0$, and therefore the ATM call and put options should satisfy the relation,

$$V_c(t_0, S_0; S_F, T) = V_p(t_0, S_0; S_F, T). \quad (4.15)$$

As a put option has value zero for strike $K = 0$, and a call option has value zero for "$K \to \infty$", we find, by (4.15),

$$I_{1,2} + I_{2,2} = \left.\frac{\partial H(T, y)}{\partial y}\right|_{y=S_F} V_p(t_0, S_0; y, T)$$

$$- \left.\frac{\partial H(T, y)}{\partial y}\right|_{y=S_F} V_c(t_0, S_0; y, T) = 0. \quad (4.16)$$

With the help of (4.14), we also find:

$$I_{1,1} + I_{2,1} = \left[H(T, y) \frac{\partial V_p(t_0, S_0; y, T)}{\partial y}\right]_{y=0}^{y=S_F} + \left[H(T, y) \frac{\partial V_c(t_0, S_0; y, T)}{\partial y}\right]_{y=S_F}^{y=\infty}$$

$$= \left[H(T, y) \frac{\partial V_c(t_0, S_0; y, T)}{\partial y}\right]_{y=0}^{y=\infty} + e^{-r(T-t_0)} \left(H(T, S_F) - H(T, 0)\right).$$

> The partial derivative of a call option with respect to the strike is given by the following expression:
>
> $$\frac{\partial V_c(t_0, S_0; K, T)}{\partial K} = -e^{-r(T-t_0)} \int_K^\infty f_{S(T)}(y) dy,$$
>
> $$= -e^{-r(T-t_0)} \left(1 - F_{S(T)}(K)\right), \quad (4.17)$$

where $F_{S(T)}(\cdot)$ is the CDF of stock $S(T)$ at time T. This implies,

$$\left[H(T, y) \frac{\partial V_c(t_0, S_0; y, T)}{\partial y}\right]_{y=0}^{y=\infty} = -e^{-r(T-t_0)} \left[H(T, \infty) \left(1 - F_{S(T)}(\infty)\right)\right.$$

$$\left. - H(T, 0) \left(1 - F_{S(T)}(0)\right)\right].$$

Since a stock price is nonnegative, $F_{S(T)}(0) = 0$. Assuming that for $y \to \infty$, CDF $F_{S(T)}(y)$ converges faster to 1 than $H(T, y)$ converges to infinity, gives,

$$\left[H(T, y) \frac{\partial V_c(t_0, S_0; K, T)}{\partial K}\right]_{y=0}^{y=\infty} = -e^{-r(T-t_0)} \left[0 - H(T, 0)(1 - 0)\right]$$

$$= e^{-r(T-t_0)} H(T, 0). \quad (4.18)$$

This means that the expressions for the terms $I_{1,1} + I_{2,1}$ are given by:

$$I_{1,1} + I_{2,1} = e^{-r(T-t_0)} H(T, 0) + e^{-r(T-t_0)} \left(H(T, S_F) - H(T, 0)\right)$$

$$= e^{-r(T-t_0)} H(T, S_F). \quad (4.19)$$

The pricing equation can thus be written as:

$$V(t_0, S_0; K, T) = e^{-r(T-t_0)} H(T, S_F) + \int_0^{S_F} V_p(t_0, S_0; y, T) \frac{\partial^2 H(T, y)}{\partial y^2} dy$$

$$+ \int_{S_F}^{\infty} V_c(t_0, S_0; y, T) \frac{\partial^2 H(T, y)}{\partial y^2} dy. \quad (4.20)$$

This equation consists of two parts and allows for an intuitive interpretation, [Carr and Madan, 1998]. We encounter the *forward value of the contract*, which is known today, and there is a *volatility smile correction*, which is given in terms of two integrals. As the final expression does not involve differentiation of the call prices to the strike, this representation is more stable than the version presented in Equation (4.11).

Example 4.2.2 (European option) The Breeden-Litzenberger technique is applied here to value European calls and puts, where we expect Equation (4.20) to collapse to the standard option pricing result. For $H(T, y) = \max(y - K, 0)$, Equation (4.20) reads:

$$V(t_0, S_0; K, T) = e^{-r(T-t_0)} \max(S_F - K, 0)$$

$$+ \int_0^{S_F} V_p(t_0, S_0; y, T) \frac{\partial^2 \max(y - K, 0)}{\partial y^2} dy$$

$$+ \int_{S_F}^{\infty} V_c(t_0, S_0; y, T) \frac{\partial^2 \max(y - K, 0)}{\partial y^2} dy, \quad (4.21)$$

with $S_F \equiv S_F(t_0, T) = e^{r(T-t_0)} S(t_0)$. The second derivative of the maximum operator is not differentiable everywhere. Using the indicator function,

$$\max(y - K, 0) = (y - K) \mathbb{1}_{y-K>0}(y),$$

we have

$$\frac{\partial \left((y - K) \mathbb{1}_{y-K>0}(y) \right)}{\partial y} = \mathbb{1}_{y-K>0}(y) + (y - K) \delta(y - K),$$

with $\delta(y)$ the Dirac delta function, see (1.17). Differentiating twice gives,

$$\frac{\partial^2 \left((y - K) \mathbb{1}_{y-K>0}(y) \right)}{\partial y^2} = \delta(y - K) + \delta(y - K) + (y - K) \delta'(y - K)$$

$$= \delta(y - K),$$

as $y\delta'(y) = -\delta(y)$. With these results, Equation (4.21) becomes, using the put-call parity from Equation (3.3),

$$V(t_0, S_0; K, T) = e^{-r(T-t_0)} \max(S_F - K, 0) + \int_0^{\infty} V_c(t_0, S_0; y, T) \delta(y - K) dy$$

$$+ \int_0^{S_F} \left(e^{-r(T-t_0)} y - S_0 \right) \delta(y - K) dy.$$

The last integral in the expression can be further reduced, as follows:

$$\int_0^{S_F} \left(e^{-r(T-t_0)}y - S_0\right) \delta(y - K) dy = e^{-r(T-t_0)} \int_0^{S_F} (y - S_F) \delta(y - K) dy,$$

where the right-hand side integral is only nonzero for $K < S_F(t_0, T)$ and because $\int_{-\infty}^{+\infty} g(y)\delta(y-a)dy = g(a)$ for any continuous function $g(y)$, we have

$$\int_0^{S_F} \left(e^{-r(T-t_0)}y - S_0\right) \delta(y - K) dy = -e^{-r(T-t_0)} (S_F - K) \mathbb{1}_{K<S_F}$$

$$= -e^{-r(T-t_0)} \max(S_F - K, 0).$$

So, the pricing equation yields:

$$V(t_0, S_0; K, T) = \int_0^\infty V_c(t_0, S_0; y, T)\delta(y - K)dy \equiv V_c(t_0, S_0; K, T),$$

which concludes this derivation. ◆

4.2.2 Variance swaps

So far we discussed models in which the underlying asset, i.e. the stock price, was specified. Volatility is considered to be a quantity, which is mainly used for improving the fit to the market implied volatility smile or skew. In this subsection we make a step towards volatility modeling and concentrate on the volatility itself. Via financial products called *variance swaps*, we can actually trade the volatility, just like any other stock or commodity.

> A variance swap is a forward contract which, at maturity T, pays the difference between the realized variance and a predefined strike price (multiplied by a certain notional amount). Generally, a forward contract is a contract which is not traded at a regulated exchange but is traded directly between two parties, over-the-counter.

The two trading parties agree to buy or to sell an asset at a pre-defined time in the future at an agreed price. The realized variance, which is then "swapped" with an amount K, can be measured in different ways, since there is no formally defined market convention.

Volatility can, for example, be measured indirectly, by the continuous observation of the stock performance. A model variance swap payoff is then defined as:

$$H(T, S) = \frac{252}{m} \sum_{i=1}^m \left(\log \frac{S(t_i)}{S(t_{i-1})}\right)^2 - K \qquad (4.22)$$

$$=: \sigma_v^2(T) - K,$$

for asset $S(t)$, with a given time-grid $t_0 < t_1 < \cdots < t_m = T$, the strike level K, and σ_v^2 is the realized variance of the stock over the life of the swap; 252 represents the number of business days in a given year. Typically, the strike K is set such that the value of the contract is initially equal to 0.

With a deterministic interest rate r, the contract value at time t_0 is given by,

$$V(t_0, S_0; K, T) = e^{-r(T-t_0)}\mathbb{E}^{\mathbb{Q}}\left[\sigma_v^2(T) - K|\mathcal{F}(t_0)\right], \tag{4.23}$$

and the strike value K at which the value of the contract initially equals zero is given by,

$$e^{-r(T-t_0)}\mathbb{E}^{\mathbb{Q}}\left[\sigma_v^2(T) - K|\mathcal{F}(t_0)\right] = 0, \tag{4.24}$$

so that $K = \mathbb{E}^{\mathbb{Q}}\left[\sigma_v^2(T)|\mathcal{F}(t_0)\right]$.

The limit of the log-term in (4.22), as the time grid gets finer, i.e., $\Delta t = t_i - t_{i-1} \to 0$, is written as,

$$\log\frac{S(t_i)}{S(t_{i-1})} = \log S(t_i) - \log S(t_{i-1}) \xrightarrow{\Delta t \to 0} \mathrm{d}\log S(t),$$

With the stock $S(t)$ governed by the following stochastic process:

$$\mathrm{d}S(t) = rS(t)\mathrm{d}t + \sigma(t)S(t)\mathrm{d}W(t), \tag{4.25}$$

with constant r and a stochastic process $\sigma(t)$ for the volatility, the dynamics under the log-transformation read,

$$\mathrm{d}\log S(t) = \left(r - \frac{1}{2}\sigma^2(t)\right)\mathrm{d}t + \sigma(t)\mathrm{d}W(t). \tag{4.26}$$

The Itô table gives, to leading order, $(\mathrm{d}\log S(t))^2 = \sigma^2(t)\mathrm{d}t$, and thus

$$\int_{t_0}^{T}(\mathrm{d}\log S(t))^2 = \int_{t_0}^{T}\sigma^2(t)\mathrm{d}t. \tag{4.27}$$

The rigorous proof of the corresponding derivations can be found in [O. Barndorff-Nielsen, 2006].

With (4.25) and (4.26), we have

$$\frac{\mathrm{d}S(t)}{S(t)} - \mathrm{d}\log S(t) = \frac{1}{2}\sigma^2(t)\mathrm{d}t. \tag{4.28}$$

The term $252/m$ in (4.22) annualizes the realized variance (it returns a year percentage), and in the continuous case this is modeled by $1/(T-t_0)$.

In the continuous case, the variance swap contract can thus be written as,

$$H(T, S) = \frac{1}{T-t_0}\int_{t_0}^{T}\sigma^2(t)\mathrm{d}t - K =: \sigma_v^2(T) - K. \tag{4.29}$$

The strike K at which the swap should be traded is then given by

$$K = \mathbb{E}^{\mathbb{Q}}\left[\frac{1}{T-t_0}\int_{t_0}^{T}\sigma^2(t)\mathrm{d}t\Big|\mathcal{F}(t_0)\right]$$

$$= \frac{2}{T-t_0}\mathbb{E}^{\mathbb{Q}}\left[\int_{t_0}^{T}\frac{\mathrm{d}S(t)}{S(t)} - \mathrm{d}\log S(t)\Big|\mathcal{F}(t_0)\right] \qquad (4.30)$$

$$= \frac{2}{T-t_0}\mathbb{E}^{\mathbb{Q}}\left[\int_{t_0}^{T}\frac{\mathrm{d}S(t)}{S(t)}\Big|\mathcal{F}(t_0)\right] - \frac{2}{T-t_0}\mathbb{E}^{\mathbb{Q}}\left[\log\frac{S(T)}{S(t_0)}\Big|\mathcal{F}(t_0)\right],$$

using (4.28). After simplifications and interchanging integration and taking the expectation, we find:

The *strike price* K for which the value of a variance swap equals zero at the inception time t_0, is given by

$$K = \frac{2}{T-t_0}\left(r(T-t_0) - \mathbb{E}^{\mathbb{Q}}\left[\log\frac{S(T)}{S(t_0)}\Big|\mathcal{F}(t_0)\right]\right), \qquad (4.31)$$

where $S(T)/S(t_0)$ represents the rate of return of the underlying stock.

Pricing equations for variance swaps

The VIX index is a known index which gives a measure for the implied volatility of the S&P 500 index. Its value is based on options on the S&P 500 index for which the time to maturity ranges from 23 to 37 days. An average of the implied volatility is calculated for 30 days options on the S&P500. The VIX index is also called the *fear index*, as it represents the market's expectation of the stock market volatility. A European counterpart to the VIX index is the VSTOXX Volatility index, where the underlying index is the Euro Stoxx 50 index.

We extract call and put market option prices on the S&P500. With the derivations from Section 4.2.1, we can derive pricing equations for the variance swaps. To determine strike K, we calculate the expectation in (4.31), with the help of these market option quotes.

To price the following contract

$$V(t_0, S_0; K, T) = e^{-r(T-t_0)}\mathbb{E}^{\mathbb{Q}}\left[H(T,S)\Big|\mathcal{F}(t_0)\right], \qquad (4.32)$$

with some payoff function $H(T,S)$, taking into account the volatility smile, the price can be determined via Equation (4.20), i.e.

$$V(t_0, S_0; K, T) = e^{-r(T-t_0)}H(T, S_F)$$
$$+ \int_{0}^{S_F} V_p(t_0, S_0; y, T)\frac{\partial^2 H(T,y)}{\partial y^2}\mathrm{d}y$$
$$+ \int_{S_F}^{\infty} V_c(t_0, S_0; y, T)\frac{\partial^2 H(T,y)}{\partial y^2}\mathrm{d}y, \qquad (4.33)$$

with the forward stock, $S_F \equiv S_F(t_0, T) := S(t_0)e^{r(T-t_0)}$, and where $V_c(t_0, S_0; y, T)$ and $V_p(t_0, S_0; y, T)$ are call and put option prices, respectively.

Strike price K at which a variance swap is traded, is given by:

$$K = \frac{2}{T - t_0}\left\{r(T - t_0) - \mathbb{E}^{\mathbb{Q}}\left[\log \frac{S(T)}{S(t_0)}\bigg|\mathcal{F}(t_0)\right]\right\}. \tag{4.34}$$

The expression in (4.34) can be modified by means of the forward stock,

$$K = -\frac{2}{T - t_0}\mathbb{E}^{\mathbb{Q}}\left[\log \frac{S(T)}{S(t_0)} - \log e^{r(T-t_0)}\bigg|\mathcal{F}(t_0)\right]$$

$$= -\frac{2}{T - t_0}\mathbb{E}^{\mathbb{Q}}\left[\log \frac{S(T)}{S_F}\bigg|\mathcal{F}(t_0)\right].$$

With $H(T, y) = \log \frac{y}{S_F}$, we find

$$\frac{\partial}{\partial y}H(T, y) = \frac{1}{y}, \qquad \frac{\partial^2}{\partial y^2}H(T, y) = -\frac{1}{y^2},$$

and the expectation in (4.34) can be calculated by Equations (4.32) and (4.33), as follows:

$$\mathbb{E}^{\mathbb{Q}}\left[\log \frac{S(T)}{S_F}\bigg|\mathcal{F}(t_0)\right] = e^{r(T-t_0)}V(t_0, S_0; K, T)$$

$$= \log \frac{S_F}{S_F} - e^{r(T-t_0)}\int_0^{S_F}\frac{1}{y^2}V_p(t_0, S_0; y, T)dy$$

$$- e^{r(T-t_0)}\int_{S_F}^{\infty}\frac{1}{y^2}V_c(t_0, S_0; y, T)dy$$

$$= -e^{r(T-t_0)}\left[\int_0^{S_F}\frac{1}{y^2}V_p(t_0, S_0; y, T)dy\right.$$

$$\left. + \int_{S_F}^{\infty}\frac{1}{y^2}V_c(t_0, S_0; y, T)dy\right],$$

so that,

$$K = -\frac{2}{T - t_0}\mathbb{E}^{\mathbb{Q}}\left[\log \frac{S(T)}{S_F}\bigg|\mathcal{F}(t_0)\right] \tag{4.35}$$

$$= \frac{2}{T - t_0}e^{r(T-t_0)}\left[\int_0^{S_F}\frac{1}{y^2}V_p(t_0, S_0; y, T)dy + \int_{S_F}^{\infty}\frac{1}{y^2}V_c(t_0, S_0; y, T)dy\right].$$

This expression can be efficiently calculated by, for example, a Gauss quadrature integration technique. The obtained results enable us to calculate the value of a variance swap for given market quotes of European call and put option prices. Once a market quote for a variance swap differs significantly from the value implied by the market call and put options, given in (4.35), we may find a volatility arbitrage possibility.

Remark 4.2.1 ($\mathbb{E}^{\mathbb{Q}}[\log S(T)|\mathcal{F}(t_0)]$ **in the Black-Scholes model)**
One may argue that $\mathbb{E}^{\mathbb{Q}}[\log S(T)|\mathcal{F}(t_0)]$ can be easily determined under the Black-Scholes model, as $\log S(T)$ is given by,

$$\log S(T) = \left(r - \frac{1}{2}\sigma^2\right)(T - t_0) + \sigma\left(W(T) - W(t_0)\right),$$

and the expectation is given by,

$$\mathbb{E}^{\mathbb{Q}}[\log S(T)|\mathcal{F}(t_0)] = \left(r - \frac{1}{2}\sigma^2\right)(T - t_0),$$

which is simply expressed in terms of the interest rate r and the volatility σ. However, it is unclear which volatility σ to use for the evaluation. A crucial difference with European option valuation is that these products (European options) are quoted in the financial market, so that the implied volatility can be determined. For $\mathbb{E}[\log S(T)]$ the situation is complicated as there are no liquid products that can be used to determine the volatility σ.

A solution to this problem is to use "all available market quotes" to derive the expectation, which is the idea behind the Breeden-Litzenberger method. ▲

Relation of variance swaps to VIX index

With the help of the Breeden-Litzenberger methodology, the equation in (4.35) provides a direct link between variance swaps and implied distributions. A key element of the equation is that the methodology is *essentially model-free*. The integrals are only based on market quotes for call and put options.

We will connect (4.35) to the VIX volatility index, and start with some strike,[2] $K_f < S_F(t_0, T)$, for which we decompose the integrals in (4.35), as follows:

$$K = \frac{2}{T - t_0} e^{r(T-t_0)} \left[\int_0^{K_f} \frac{1}{y^2} V_p(t_0, S_0; y, T) dy + \int_{K_f}^{\infty} \frac{1}{y^2} V_c(t_0, S_0; y, T) dy\right]$$

$$+ \frac{2}{T - t_0} e^{r(T-t_0)} \int_{K_f}^{S_F} \frac{1}{y^2} \left(V_p(t_0, S_0; y, T) - V_c(t_0, S_0; y, T)\right) dy. \quad (4.36)$$

By the put-call parity, for $y = K$, we have:

$$V_p(t_0, S_0; y, T) - V_c(t_0, S_0; y, T) = e^{-r(T-t_0)}y - S_0,$$

---

[2]Strike K_f should be the highest strike which is below the forward S_F.

so that for the last expression in (4.36), we write,

$$\frac{2}{T-t_0}e^{r(T-t_0)} \times \int_{K_f}^{S_F} \frac{1}{y^2}(e^{-r(T-t_0)}y - S_0)dy$$

$$= \frac{2}{T-t_0}e^{r(T-t_0)}\left(e^{-r(T-t_0)}(\log S_F - \log K_f) + S(t_0)\left(\frac{1}{S_F} - \frac{1}{K_f}\right)\right)$$

$$= \frac{2}{T-t_0}\left(\log \frac{S_F}{K_f} + \left(1 - \frac{S_F}{K_f}\right)\right). \qquad (4.37)$$

Application of a Taylor series expansion to the logarithm, ignoring terms higher than second-order, results in,

$$\log \frac{S_F}{K_f} \approx \left(\frac{S_F}{K_f} - 1\right) - \frac{1}{2}\left(\frac{S_F}{K_f} - 1\right)^2. \qquad (4.38)$$

With this, the expression in (4.37) becomes

$$\frac{2}{T-t_0}e^{r(T-t_0)}\int_{K_f}^{S_F} \frac{1}{y^2}\left(V_p(t_0, S_0; y, T) - V_c(t_0, S_0; y, T)\right)dy$$

$$\approx -\frac{1}{T-t_0}\left(\frac{S_F}{K_f} - 1\right)^2.$$

and thus (4.36) reads:

$$K \approx \frac{2}{T-t_0}e^{r(T-t_0)}\left[\int_0^{K_f} \frac{1}{y^2}V_p(t_0, S_0; y, T)dy + \int_{K_f}^{\infty} \frac{1}{y^2}V_c(t_0, S_0; y, T)dy\right]$$

$$- \frac{1}{T-t_0}\left(\frac{S_F}{K_f} - 1\right)^2.$$

Discretizing the integrals above, in the usual way discretizing the K dimension with N_K increments ΔK, gives,

$$K \approx \frac{2}{T-t_0}e^{r(T-t_0)}\left[\sum_{i=1}^{f-1}\frac{1}{K_i^2}V_p(t_0, S_0; K_i, T)\Delta K + \sum_{i=f}^{N_K}\frac{1}{K_i^2}V_c(t_0, S_0; K_i, T)\Delta K\right]$$

$$- \frac{1}{T-t_0}\left(\frac{S_F}{K_f} - 1\right)^2.$$

> This expression can be recognized as the square of the VIX index (denoted by VIX^2), which was defined by the Chicago Board Options Exchange (CBOE) in their white paper [CBOE White Paper], as follows,
>
> $$\text{VIX}^2 = \frac{2}{T-t_0} \sum_{i=1}^{N_K} \frac{\Delta K}{K_i^2} e^{r(T-t_0)} Q(K_i) - \frac{1}{T-t_0}\left[\frac{S_F}{K_f} - 1\right]^2. \quad (4.39)$$
>
> Here, $Q(K_i)$ represents the price of out-of-the money call and put options with strike prices K_i and K_f being the highest strike below the forward price $S_F = S_F(t_0, T) := S_0 e^{r(T-t_0)}$.

4.3 Non-parametric local volatility models

Although the Breeden-Litzenberger technique is useful, its applicability is essentially restricted to *plain vanilla*, European put and call options. Local volatility models, which have a wider range of applicability, are discussed next.

We consider the pricing of a call option under the *local volatility* model, based on a one-dimensional stochastic stock process. The term "local volatility" is used to indicate that the volatility of the process is a function of stock $S(t)$. The stock is not driven by an additional stochastic process, which is the case with the *stochastic volatility models*, where volatility is described by an additional stochastic process, as will be discussed in Chapter 8.

> The classical local volatility model is given by
>
> $$dS(t) = rS(t)dt + \sigma_{LV}(t,S)S(t)dW(t), \quad S(t_0) = S_0, \quad (4.40)$$
>
> with a constant (or deterministic) interest rate r.
> For a given option value $V(t, S)$, the no-arbitrage assumption gives rise to the following PDE, using Itô's lemma, as described in Chapter 2,
>
> $$\begin{cases} \dfrac{\partial V}{\partial t} + \dfrac{1}{2}\sigma_{LV}^2(t,S)S^2 \dfrac{\partial^2 V}{\partial S^2} + rS\dfrac{\partial V}{\partial S} - rV = 0, \\ V(T,S) = \max(S(T) - K, 0). \end{cases} \quad (4.41)$$
>
> PDE (4.41) is a *backward Kolmogorov partial differential equation.*

We may describe the evolution of a probability density function (PDF) by means of a Fokker-Planck PDE, which describes the *forward evolution of a PDF* in time. In that case, with an initial condition at t_0, the PDF is given by a Dirac delta function at time t_0.

Theorem 4.3.1 (Fokker-Planck PDE and SDEs) *The transition density $f_{S(t)}(y) \equiv f_S(t, y; t_0, S_0)$ associated to the general SDE for $S(t)$, $t_0 \leq t \leq T$,*

$$dS(t) = \bar{\mu}(t, S)dt + \bar{\sigma}(t, S)dW(t), \quad S(t_0) = S_0,$$

satisfies the Fokker-Planck (also called "forward Kolmogorov") PDE:

$$\begin{cases} \dfrac{\partial}{\partial t} f_{S(t)}(y) + \dfrac{\partial}{\partial y}[\bar{\mu}(t, y) f_{S(t)}(y)] - \dfrac{1}{2}\dfrac{\partial^2}{\partial y^2}[\bar{\sigma}^2(t, y) f_{S(t)}(y)] = 0, \\ f_{S(t_0)}(y) = \delta(y = S_0). \end{cases} \quad (4.42)$$

Local volatility modeling is based on the Fokker-Planck forward equation. In a local volatility framework the stock density is again directly connected to the market quotes of call and put options. After some calculations, the state-dependent volatility function, $\bar{\sigma}(t, S) = S \cdot \sigma_{LV}(t, S)$, can be described by means of financial market quotes.

Unfortunately, as only a limited number of options is quoted in the market for a particular stock, it is difficult to accurately recover the stock density from these few market quotes (one would require an infinite number of options in the "strike dimension" $K \in [0, \infty)$). We will discuss important conditions related to arbitrage-free option values and the related densities, to improve the accuracy of the approximation in Section 4.3.1.

The derivative of an option value with respect to maturity time T, is given by:

$$\frac{\partial V_c(t_0, S_0; K, T)}{\partial T} = \frac{\partial}{\partial T}\left(e^{-r(T-t_0)} \int_K^{+\infty} (y - K) f_{S(T)}(y) dy\right) \quad (4.43)$$

$$= -rV_c(t_0, S_0; K, T) + e^{-r(T-t_0)} \int_K^{+\infty} (y - K) \frac{\partial f_{S(T)}(y)}{\partial T} dy.$$

For the partial derivative in the last integral in (4.43), we employ the Fokker-Planck equation (4.42), with $\bar{\mu}(t, S(t)) = rS$ and $\bar{\sigma}(t, S) = \sigma_{LV}(t, S) \cdot S$, by which the integral can be written as

$$\int_K^{+\infty} (y - K) \frac{\partial f_{S(T)}(y)}{\partial T} dy = -r \int_K^{+\infty} (y - K) \frac{\partial (y f_{S(T)}(y))}{\partial y} dy \quad (4.44)$$

$$+ \frac{1}{2} \int_K^{+\infty} (y - K) \frac{\partial^2 (\sigma_{LV}^2(T, y) y^2 f_{S(T)}(y))}{\partial y^2} dy.$$

The first integral at the right-hand side is equal to[3]:

$$\int_K^{+\infty} (y - K) \frac{\partial (y f_{S(T)}(y))}{\partial y} dy = (y - K) y f_{S(T)}(y)\Big|_{y=K}^{+\infty} - \int_K^{+\infty} y f_{S(T)}(y) dy$$

$$= - \int_K^{+\infty} y f_{S(T)}(y) dy.$$

---

[3] Assuming that for $y \to +\infty$ the density $f_{S(T)}(y)$ decays *faster* to zero than y^2 grows to infinity.

By Equations (4.8) and (4.17), this integral can be expressed in terms of call option values, i.e.

$$
\begin{aligned}
-\int_K^{+\infty} y f_{S(T)}(y) \mathrm{d}y &= -\mathrm{e}^{r(T-t_0)} \int_K^{+\infty} y \frac{\partial^2 V_c(t_0, S_0; y, T)}{\partial y^2} \mathrm{d}y \\
&= -\mathrm{e}^{r(T-t_0)} \left[y \frac{\partial V_c(t_0, S_0; y, T)}{\partial y} \Big|_K^{+\infty} \right. \\
&\qquad\qquad \left. - \int_K^{+\infty} \frac{\partial V_c(t_0, S_0; y, T)}{\partial y} \mathrm{d}y \right] \\
&= -\mathrm{e}^{r(T-t_0)} \left[-y\mathrm{e}^{-r(T-t_0)} \left(1 - F_{S(T)}(y)\right) \Big|_K^{+\infty} \right. \\
&\qquad\qquad \left. - \int_K^{+\infty} \frac{\partial V_c(t_0, S_0; y, T)}{\partial y} \mathrm{d}y \right],
\end{aligned}
\qquad (4.45)
$$

since $F_{S(T)}(+\infty) = 1$ and under the assumption that $F_{S(T)}(y)$ converges to 1 faster than y to $+\infty$, we have

$$
y\mathrm{e}^{-r(T-t_0)} \left(1 - F_{S(T)}(y)\right) \Big|_K^{+\infty} = -K\mathrm{e}^{-r(T-t_0)} \left(1 - F_{S(T)}(K)\right). \qquad (4.46)
$$

Call option prices converge to 0 for $K \to \infty$, so that

$$
\int_K^{+\infty} \frac{\partial V_c(t_0, S_0; y, T)}{\partial y} \mathrm{d}y = -V_c(t_0, S_0; K, T). \qquad (4.47)
$$

By Equation (4.17), this gives,

$$
\begin{aligned}
-\int_K^{+\infty} y f_{S(T)}(y) \mathrm{d}y &= -\mathrm{e}^{r(T-t_0)} K \mathrm{e}^{-r(T-t_0)} \left(-\mathrm{e}^{r(T-t_0)} \frac{\partial V_c(t_0, S_0; K, T)}{\partial K} \right) \\
&\quad + \mathrm{e}^{r(T-t_0)} V_c(t_0, S_0; K, T) \\
&= \mathrm{e}^{r(T-t_0)} \left[K \frac{\partial V_c(t_0, S_0; K, T)}{\partial K} - V_c(t_0, S_0; K, T) \right].
\end{aligned}
$$

In a similar manner, the second integral in (4.44) can be determined, as

$$
\begin{aligned}
\int_K^{+\infty} (y - K) \frac{\partial^2 \left(\sigma_{LV}^2(T, y) y^2 f_{S(T)}(y)\right)}{\partial y^2} \mathrm{d}y &= \sigma_{LV}^2(T, K) K^2 f_{S(T)}(K) \\
&= \mathrm{e}^{r(T-t_0)} \sigma_{LV}^2(T, K) K^2 \frac{\partial^2 V_c(t_0, S_0; K, T)}{\partial K^2}.
\end{aligned}
$$

Collecting all terms obtained for (4.43), we find:

$$
\frac{\partial}{\partial T} V_c = -rV_c - rK \frac{\partial V_c}{\partial K} + rV_c + \frac{1}{2} \sigma_{LV}^2(T, K) K^2 \frac{\partial^2 V_c}{\partial K^2}. \qquad (4.48)
$$

Local Volatility Models

The expression for the local volatility function $\sigma_{LV}(T,K)$ can be found from,

$$\sigma_{LV}^2(T,K) = \frac{\dfrac{\partial V_c(t_0, S_0; K, T)}{\partial T} + rK\dfrac{\partial V_c(t_0, S_0; K, T)}{\partial K}}{\dfrac{1}{2}K^2\dfrac{\partial^2 V_c(t_0, S_0; K, T)}{\partial K^2}}. \quad (4.49)$$

Here $V_c(t_0, S_0; K, T)$ is the representation of the market call option prices at time t_0, initial stock value S_0, strike price K and maturity T.

In the local volatility model, the volatility $\sigma_{LV}(T,K)$ can thus be described by market quotes of option values, and the local volatility model may therefore *perfectly fit* to the market option quotes while the model stays parameter-free. In other words, there is no need for a calibration procedure.

4.3.1 Implied volatility representation of local volatility

From Equation (4.49) it is clear that the local volatility $\sigma_{LV}(T,K)$ can be expressed in terms of European option prices. An option price surface is thus needed to compute this so-called *Dupire's local volatility term*. Based the local volatility model, we may subsequently price options, for example, by means of the finite difference method for the resulting option valuation PDE.

In practice, however, not all derivatives in (4.49) can be obtained directly from the market option quotes and the derivatives may also need to be calculated from the market data by finite difference approximations. For each local volatility term, we need four option values for the finite difference approximations of the partial derivatives.

In the denominator of (4.49) we encounter an approximation of the density of the stock price, see Equation (4.9), for which finite difference approximations may lead to significant approximation errors for high strike prices. The errors may be particularly large when the stock density is close to 0. As a consequence, the local volatility function $\sigma_{LV}(T,K)$ will then also be unrealistically large.

Remark 4.3.1 *A number of improvements for the local volatility's accuracy have been presented in the literature. One approach is to parameterize the option surface by a certain two-dimensional (polynomial) function, $h(T,K)$, and fit it to market data $V_c^{mkt}(K,T)$. In this case, the partial derivatives with respect to time and strike price of $h(T,K)$ can be found analytically. The main problem of this approach is that it is difficult to determine a parametric function $h(T,K)$ which fits well to all market quotes while preserving the no-arbitrage assumptions.*

An issue when fitting a polynomial to option prices is that for OTM options small differences in prices may cause significant differences in the implied volatilities, due to the Black-Scholes inversion procedure. ▲

It is useful to *reformulate the LV model* in terms of the corresponding implied volatilities. However, the partial derivatives of the call prices with respect to the strike values are not equal to the partial derivatives of the implied volatilities.

We therefore derive the local volatility $\sigma_{LV}(T,K)$ in terms of the implied (Black-Scholes) volatilities, $\sigma_{\text{imp}}(T,K)$, which indicates[4] the implied volatility at time T and strike K. We know, from Equation (3.23), that arbitrage-free European call option prices are given by:

$$V_c(t_0, S_0; K, T) = S_0 F_{\mathcal{N}(0,1)}(d_1) - K e^{-r(T-t_0)} F_{\mathcal{N}(0,1)}(d_2),$$

with

$$d_1 = \frac{\log\left(\frac{S_0}{K}\right) + \left(r + \tfrac{1}{2}\sigma_{\text{imp}}^2(T,K)\right)(T-t_0)}{\sigma_{\text{imp}}(T,K)\sqrt{T-t_0}}, \quad d_2 = d_1 - \sigma_{\text{imp}}(T,K)\sqrt{T-t_0},$$

We will use,

$$y := \log\left(\frac{K}{S_F}\right) = \log\left(\frac{K}{S_0}\right) - r(T-t_0), \quad w := \sigma_{\text{imp}}^2(T,K)(T-t_0), \quad (4.50)$$

where S_F is again the forward price, defined as $S_F = S_F(t_0, T) := S_0 e^{r(T-t_0)}$. For the variables y and w, we define the call price, $c(y, w)$, by

$$\boxed{V_c(t_0, S_0; K, T) = S_0 \left[F_{\mathcal{N}(0,1)}(d_1) - e^y F_{\mathcal{N}(0,1)}(d_2) \right] =: c(y, w),} \quad (4.51)$$

where $d_1 = \tfrac{1}{2}\sqrt{w} - \frac{y}{\sqrt{w}}$, and $d_2 = d_1 - \sqrt{w}$.

Using the short-hand notation $V_c := V_c(t_0, S_0; K, T)$ and $c := c(y, w)$, and $\tau := T - t_0$, we obtain,

$$\frac{\partial V_c}{\partial K} = \frac{\partial c}{\partial y} \frac{1}{K} + \frac{\partial c}{\partial w} \frac{\partial w}{\partial K}, \quad (4.52)$$

thus

$$\frac{\partial^2 V_c}{\partial K^2} = \frac{1}{K^2}\left(\frac{\partial^2 c}{\partial y^2} - \frac{\partial c}{\partial y}\right) + \frac{2}{K}\frac{\partial w}{\partial K}\frac{\partial^2 c}{\partial w \partial y} + \frac{\partial^2 w}{\partial K^2}\frac{\partial c}{\partial w} + \left(\frac{\partial w}{\partial K}\right)^2 \frac{\partial^2 c}{\partial w^2}, \quad (4.53)$$

and,

$$\frac{\partial V_c}{\partial T} = -r\frac{\partial c}{\partial y} + \frac{\partial c}{\partial w}\frac{\partial w}{\partial T}. \quad (4.54)$$

After substituting these derivatives in the local volatility, Equation (4.49) is given by:

$$\sigma_{LV}^2(T,K) = \frac{\dfrac{\partial c}{\partial w}\dfrac{\partial w}{\partial T} + rK \dfrac{\partial c}{\partial w}\dfrac{\partial w}{\partial K}}{\dfrac{1}{2}\left(\dfrac{\partial^2 c}{\partial y^2} - \dfrac{\partial c}{\partial y}\right) + K\dfrac{\partial w}{\partial K}\dfrac{\partial^2 c}{\partial w \partial y} + \dfrac{1}{2}K^2\left[\dfrac{\partial^2 w}{\partial K^2}\dfrac{\partial c}{\partial w} + \left(\dfrac{\partial w}{\partial K}\right)^2 \dfrac{\partial^2 c}{\partial w^2}\right]}. \quad (4.55)$$

---

[4] We include here the additional arguments T and K to indicate the dependence of the implied volatility on the maturity and the strike price.

Local Volatility Models

The expression above can be simplified by the application of the following identities:

$$\frac{\partial^2 c}{\partial w^2} = \frac{\partial c}{\partial w}\left(-\frac{1}{8} - \frac{1}{2w} + \frac{y^2}{2w^2}\right), \quad \frac{\partial^2 c}{\partial w \partial y} = \frac{\partial c}{\partial w}\left(\frac{1}{2} - \frac{y}{w}\right), \quad \frac{\partial^2 c}{\partial y^2} = \frac{\partial c}{\partial y} + 2\frac{\partial c}{\partial w}.$$

The proofs for these equalities are based on standard derivations.

Local volatility expression (4.55) now becomes:

$$\sigma_{LV}^2(T,K) = \frac{\dfrac{\partial w}{\partial T} + rK\dfrac{\partial w}{\partial K}}{1 + K\dfrac{\partial w}{\partial K}\left(\dfrac{1}{2} - \dfrac{y}{w}\right) + \dfrac{1}{2}K^2\dfrac{\partial^2 w}{\partial K^2} + \dfrac{1}{2}K^2\left(\dfrac{\partial w}{\partial K}\right)^2\left(-\dfrac{1}{8} - \dfrac{1}{2w} + \dfrac{y^2}{2w^2}\right)}. \quad (4.56)$$

We may use Equation (4.50) to determine the remaining derivatives, which results in a relation between function w and implied volatility $\sigma_{\text{imp}} := \sigma_{\text{imp}}(T,k)$:

$$\frac{\partial w}{\partial T} = \sigma_{\text{imp}}^2 + 2(T - t_0)\sigma_{\text{imp}}\frac{\partial \sigma_{\text{imp}}}{\partial T},$$

$$\frac{\partial w}{\partial K} = 2(T - t_0)\sigma_{\text{imp}}\frac{\partial \sigma_{\text{imp}}}{\partial K},$$

$$\frac{\partial^2 w}{\partial K^2} = 2(T - t_0)\left(\frac{\partial \sigma_{\text{imp}}}{\partial K}\right)^2 + 2(T - t_0)\sigma_{\text{imp}}\frac{\partial^2 \sigma_{\text{imp}}}{\partial K^2}.$$

Local volatility function $\sigma_{LV}(T,K)$ can thus be expressed in terms of the implied volatilities σ_{imp}. In essence, the leading term in the denominator of Equation (4.56) is a first derivative, whereas in Equation (4.49) it is a second derivative term. Generally first derivatives can be computed numerically in a more stable way than second derivatives.

An example is presented in Section 4.3.4.

4.3.2 Arbitrage-free conditions for option prices

As already mentioned, in financial markets there are usually not enough market option quotes to approximate all sorts of financial derivatives, with different strike prices and maturity times, accurately, and interpolation and extrapolation of market option quotes often has to take place. In particular, an *arbitrage-free interpolation* between the available market data needs to be introduced. Arbitrage can occur in the time as well as in the strike direction, see an (t, S_t)-surface of option values in Figure 4.6.

An interpolation between different maturities and strike prices should satisfy the following conditions:

1. The so-called *calendar spread* condition:

$$V_c(t_0, S_0; K, T + \Delta T) - V_c(t_0, S_0; K, T) > 0.$$

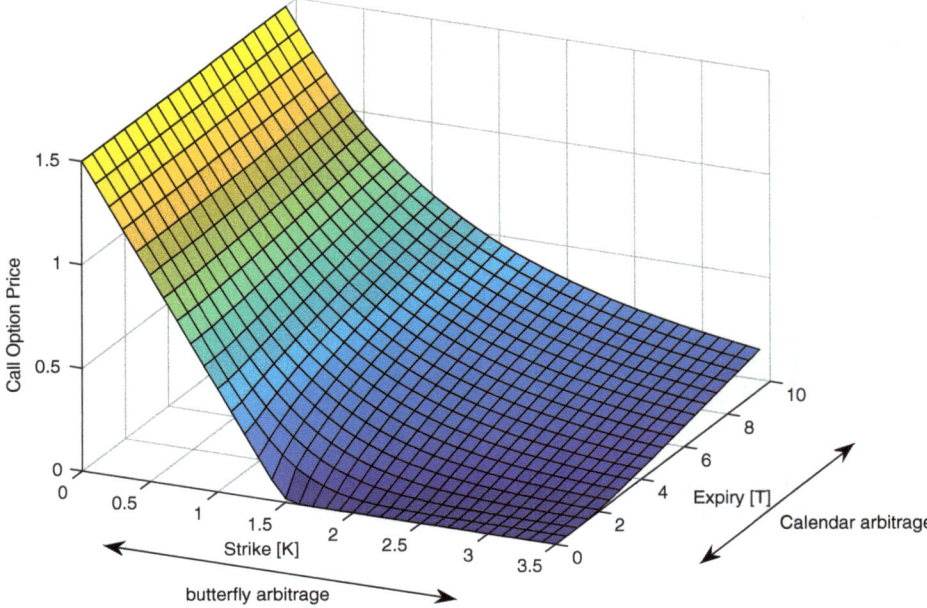

Figure 4.6: *Call prices and the two directions in which arbitrage may take place.*

If we divide both sides of the equation by ΔT and we let $\Delta T \to 0$, we find:

$$\lim_{\Delta T \to 0} \frac{1}{\Delta T} \left[V_c(t_0, S_0; K, T + \Delta T) - V_c(t_0, S_0; K, T) \right]$$
$$= \frac{\partial}{\partial T} V_c(t_0, S_0; K, T).$$

This condition can be interpreted as follows: for any two European-style options that have the same contract details, apart from the expiry date T, the contract which expires first has to be cheaper. This property is related to the fact that an option with a larger time to expiry, $T - t_0$, has a higher chance to get in-the-money.

2. *Monotonicity in the strike direction*:

$$V_c(t_0, S_0; K + \Delta K, T) - V_c(t_0, S_0; K, T) < 0, \quad \text{for calls},$$
$$V_p(t_0, S_0; K + \Delta K, T) - V_p(t_0, S_0; K, T) > 0, \quad \text{for puts}.$$

As before in the limit we have:

$$\lim_{\Delta K \to 0} \frac{1}{\Delta K} \left(V_c(t_0, S_0; K + \Delta K, T) - V_c(t_0, S_0; K, T) \right)$$
$$= \frac{\partial}{\partial K} V_c(t_0, S_0; K, T).$$

Local Volatility Models 109

In other words, *the most expensive* call option is the one with strike price $K = 0$. Since the call payoff value, $\max(S(T) - K, 0)$, is monotonically decreasing in strike K, any European call option with $K \neq 0$ has to be decreasing in price with decreasing K. The opposite can be concluded for put options, by using the call-put parity relation.

3. The so-called *butterfly condition* states,

$$V_c(t_0, S_0; K + \Delta K, T) - 2V_c(t_0, S_0; K, T) + V_c(t_0, S_0; K - \Delta K, T) \geq 0. \tag{4.57}$$

The nonnegativity of the expression above is understood from the following argument. If we buy one call option with strike $K + \Delta K$, sell two options with strike K and buy one call option with strike $K - \Delta K$, the resulting payoff cannot be negative (excluding transaction costs). Since any discounted nonnegative value stays nonnegative, Equation (4.57) needs to hold. As explained in Equation (3.36), the second derivative of the call option price can be related to the density of the stock, which obviously needs to be nonnegative for any strike K.

Example 4.3.1 (Exploiting calendar spread arbitrage) Consider the following European call options, $V_c(t_0, S_0; K, T_1)$ and $V_c(t_0, S_0; K, T_2)$, with $t_0 < T_1 < T_2$, zero interest rate and no dividend payment.

By Jensen's inequality, which states that for any convex function $g(\cdot)$ the following inequality holds

$$g(\mathbb{E}[X]) \leq \mathbb{E}[g(X)],$$

for any random variable X (see also Definition 3.1.1), we will show that $V_c(t_0, S_0; K, T_1) < V_c(t_0, S_0; K, T_2)$.

The value of a call, as seen from time $t = T_1$, with maturity time T_2, is given by:

$$V_c(T_1, S(T_1); K, T_2) = \mathbb{E}^{\mathbb{Q}}\left[\max(S(T_2) - K, 0) | \mathcal{F}(T_1)\right]$$
$$\geq \max(\mathbb{E}^{\mathbb{Q}}\left[S(T_2) | \mathcal{F}(T_1)\right] - K, 0). \tag{4.58}$$

With zero interest rate, $S(t)$ is a martingale and thus $\mathbb{E}^{\mathbb{Q}}\left[S(T_2) | \mathcal{F}(T_1)\right] = S(T_1)$. Therefore, (4.58) gives us:

$$V_c(T_1, S(T_1); K, T_2) \geq \max(S(T_1) - K, 0) = V_c(T_1, S(T_1); K, T_1). \quad \blacklozenge$$

A call option maturing "at a later time" should thus be more expensive than another one maturing "earlier", which can be generalized to any maturity time T.

Example 4.3.2 Examples for each of the three types of arbitrage are presented in Figure 4.7. The upper left figure illustrates the calendar spread arbitrage. For some strike prices the corresponding call option price with a longer maturity time is cheaper than the option with shorter maturity time. The upper right figure shows a case where the spread arbitrage is present and the lower figure shows a case with butterfly spread arbitrage. $\blacklozenge$

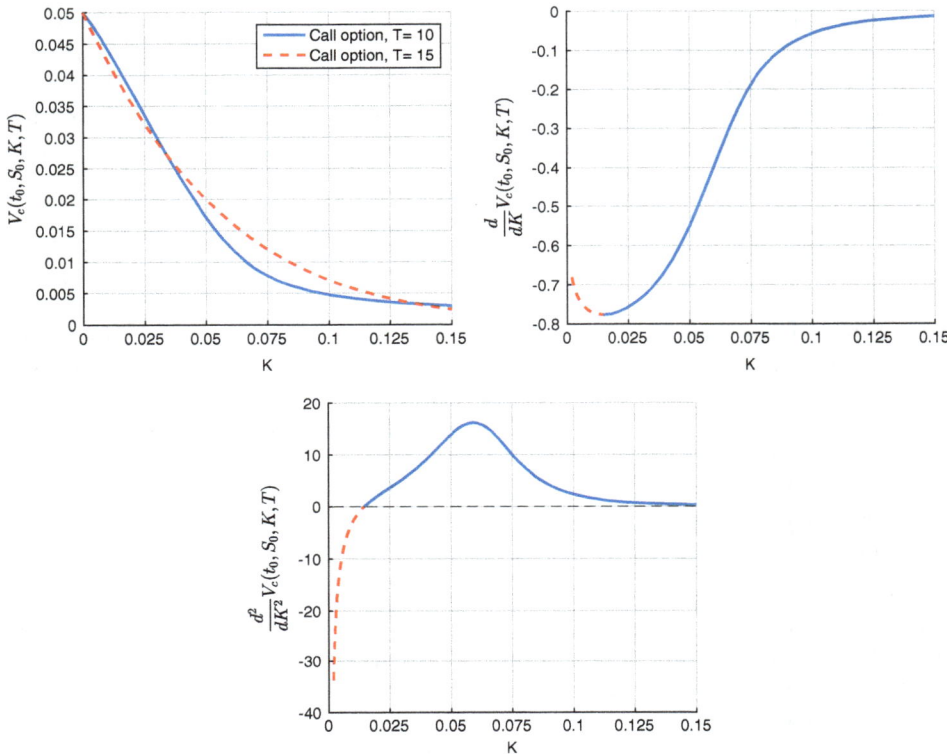

Figure 4.7: *Three types of arbitrage. Upper left: calendar spread arbitrage. Upper right: spread arbitrage. Lower: butterfly spread arbitrage.*

The spread and the butterfly arbitrage can be related to the PDF and CDF of the stock $S(T)$,

$$\frac{\partial}{\partial K} V_c(t_0, S_0; K, T) = e^{-r(T-t_0)} \left(F_{S(T)}(K) - 1 \right), \quad (4.59)$$

and,

$$\frac{\partial^2}{\partial K^2} V_c(t_0, S_0; K, T) = e^{-r(T-t_0)} f_{S(T)}(K). \quad (4.60)$$

As a consequence, by plotting the first and the second derivatives of the call option prices, arbitrage opportunities can immediately be recognized. The CDF of the stock price, $F_{S(T)}(K)$, needs to be monotone and the PDF of the stock price, $f_{S(T)}(K)$, needs to be nonnegative and should integrate to one.

4.3.3 Advanced implied volatility interpolation

The number of implied volatility market quotes for each maturity time T_i will vary for each market and for each asset class. For some stocks there are even hardly any implied volatility market quotes and for some other stocks there are only a few. The number of market quotes depends on the market's liquidity and on the particular market conventions. For example, in foreign-exchange (FX) markets (to be discussed in Chapter 15), one may observe from 3 to 7 implied volatility FX market quotes for each expiry date. The number of market quotes will depend on the particular currency pair. In Table 4.1, an example of FX market implied volatilities is presented. The ATM level (in bold letters) and four quotes for the ITM and OTM options are available.

Now, imagine that a financial institution should give an option price for a strike price which is not quoted in the FX market, e.g. for $K = 145$, which is not present in Table 4.1. For this purpose an interpolation technique is required, to interpolate between the available volatilities. The same holds true when pricing a nonstandard European-style derivative, for which the LV model or the Breeden-Litzenberger method is employed. These two models are based on a *continuum of traded strike prices*, and their implied volatilities.

There are different techniques to interpolate between the available market quotes, like linear, spline, or tangent spline, interpolation. One of the popular methods for the interpolation of implied volatilities is based on a paper by Hagan et al. [2002], as this interpolation is built on a specific implied volatility parametrization. This method is popular because of a direct relation between the interpolation and an arbitrage-free SDE model, the Stochastic Alpha Beta Rho (SABR) model.[5]

The parametrization gained in popularity because implied volatilities are analytically available, as an asymptotic expansion formula of an implied volatility was presented. The expression for the Black-Scholes implied volatility approximation

Table 4.1: *Implied volatilities for FX (USD/JPY) for $T = 1$ and forward $S_F(t_0, T) = 131.5$. The bold letters indicate the ATM level.*

$T = 1y$	$K = 110.0$	$K = 121.3$	**K = 131.5**	$K = 140.9$	$K = 151.4$
$\sigma_{imp}(T, K)$ [%]	14.2	11.8	**10.3**	10.0	10.7

---

[5]The SABR model is defined by the following system of SDEs:

$$dS_F(t,T) = \sigma(t)(S_F(t,T))^\beta dW_F(t), \quad S_F(t_0,T) = S_{F,0},$$
$$d\sigma(t) = \gamma\sigma(t)dW_\sigma(t), \quad \sigma(t_0) = \alpha.$$

where $dW_F(t)dW_\sigma(t) = \rho dt$ and the processes are defined under the T-forward measure $\mathbb{Q}^T$. The variance process $\sigma(t)$ is a lognormal distribution. Further, since for a constant $\sigma := \sigma(t)$ the forward $S_F(t,T)$ follows a CEV process, one can expect that the conditional SABR process, $S(T)$ given the paths of $\sigma(t)$ on the interval $0 \le t \le T$, is a CEV process as well. Systems of SDEs will be discussed in Chapter 7, and the CEV process in Chapter 14.

of the SABR model[6] reads:

$$\hat{\sigma}(T, K) = \frac{\hat{a}(K)\hat{c}(K)}{g(\hat{c}(K))}$$
$$\times \left[1 + \left(\frac{(1-\beta)^2}{24}\frac{\alpha^2}{(S_F(t_0)K)^{1-\beta}} + \frac{1}{4}\frac{\rho\beta\gamma\alpha}{(S_F(t_0)K)^{\frac{1-\beta}{2}}} + \frac{2-3\rho^2}{24}\gamma^2\right)T\right], \quad (4.61)$$

where

$$\hat{a}(K) = \frac{\alpha}{(S_F(t_0)\cdot K)^{\frac{1-\beta}{2}}\left(1 + \frac{(1-\beta)^2}{24}\log^2\left(\frac{S_F(t_0)}{K}\right) + \frac{(1-\beta)^4}{1920}\log^4\left(\frac{S_F(t_0)}{K}\right)\right)},$$

and

$$\hat{c}(K) = \frac{\gamma}{\alpha}(S_F(t_0)K)^{\frac{1-\beta}{2}}\log\frac{S_F(t_0)}{K}, \qquad g(x) = \log\left(\frac{\sqrt{1-2\rho x + x^2} + x - \rho}{1-\rho}\right).$$

In the case of *at-the-money* options, i.e. for $S_F(t_0) = K$, the formula reduces to,

$$\hat{\sigma}(T, K) \approx \frac{\alpha}{(S_F(t_0))^{1-\beta}}\left(1 + \left[\frac{(1-\beta)^2}{24}\frac{\alpha^2}{(S_F(t_0))^{2-2\beta}}\right.\right.$$
$$\left.\left. + \frac{1}{4}\frac{\rho\beta\alpha\gamma}{(S_F(t_0))^{1-\beta}} + \frac{2-3\rho^2}{24}\gamma^2\right]T\right).$$

Because the Hagan formula is derived from a model of an advanced 2D structure (the model falls into the class of the Stochastic Local Volatility models that will be discussed in Chapters 10), it can model a wide range of different implied volatility shapes. In Figure 4.8 the effect of different model parameters on the implied volatility shapes is shown. Notice that both parameters β and ρ have an effect on the implied volatility skew. In practice, β is often fixed, whereas ρ is used in a calibration. Parameter α controls the level of the implied volatility smile and γ the magnitude of the curvature of the smile.

Example 4.3.3 (Interpolation of implied volatilities) In this example, we will use the market implied volatility quotes from Table 4.1. For the given set of quotes, we compare two interpolation techniques for the implied volatilities, the linear interpolation with the Hagan interpolation. The linear interpolation is straightforward as it directly interpolates between any two market implied volatilities. For the Hagan interpolation the procedure is more involved. First, the model parameters $\alpha, \beta, \rho, \gamma$ need to be determined for which the parametrization fits best to the market quotes, i.e., minimize the difference $|\hat{\sigma}(T, K_i) - \sigma_{imp}(T, K_i)|$, for all K_i.

---

[6]Note that this is the *Black-Scholes* implied volatility and *not* the SABR implied volatility.

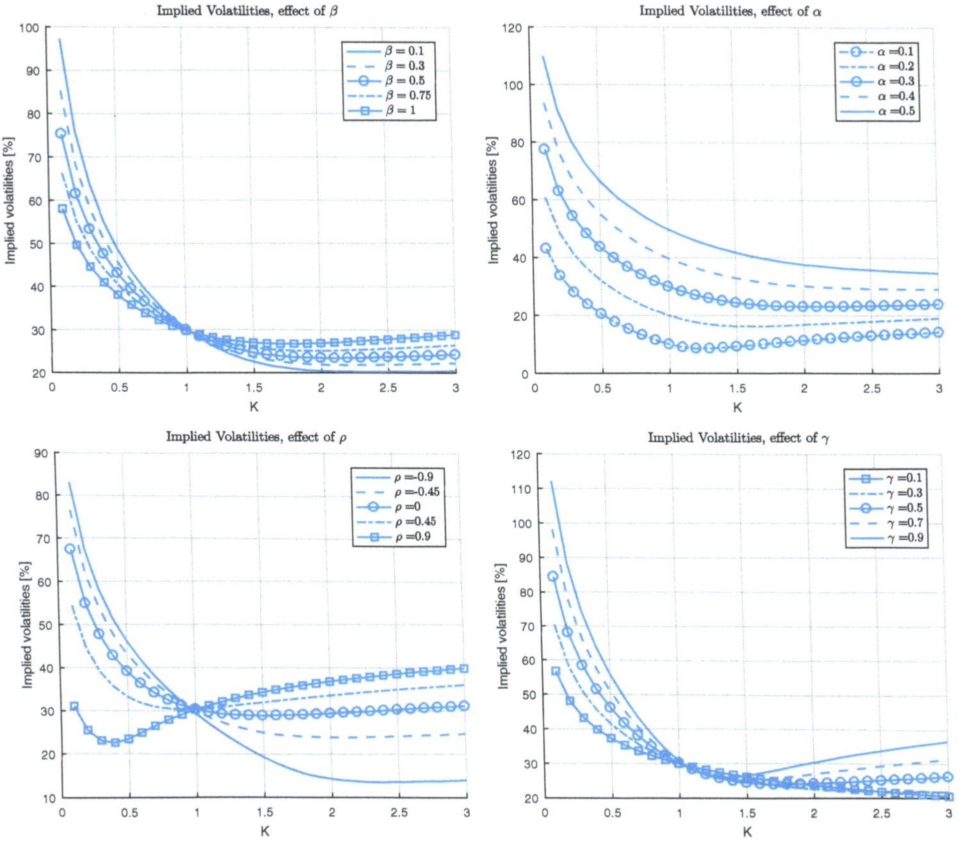

Local Volatility Models

Figure 4.8: *Different implied volatility shapes under Hagan's implied volatility parametrization, depending on different model parameters.*

In Figure 4.9 the results of the two interpolation methods are presented. The left side figure shows the market implied volatilities, with the linear interpolation and the Hagan interpolation. Both interpolation curves seem to fit very well to the market quotes. We also discussed that when market call and put option prices are available, it is easy to approximate the underlying density function, based on these quotes, as

$$f_{S(T)}(K) = e^{r(T-t_0)} \frac{\partial^2 V_c(t_0, S_0; K, T)}{\partial K^2}.$$

The right side figure shows the corresponding *"implied densities"* from the two interpolation techniques. Notice the nonsmooth nature of the density which is

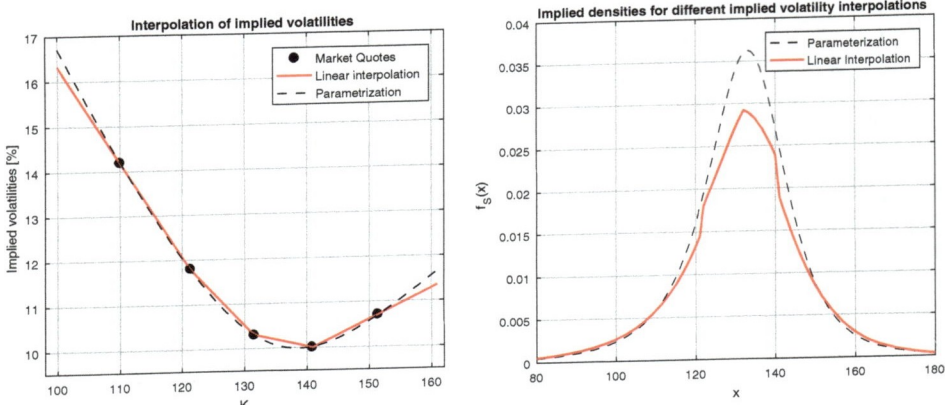

Figure 4.9: *Different interpolation techniques applied to FX market quotes. Left: interpolated implied volatilities. Right: the corresponding implied densities.*

based on linear interpolation. Integration of these density functions yields,

$$\text{Hagan}: \int_{\mathbb{R}} f_{S(T)}(z)\mathrm{d}z = 0.9969, \quad \text{Linear}: \int_{\mathbb{R}} f_{S(T)}(z)\mathrm{d}z = 0.8701.$$

In fact, the linear interpolation gives rise to a *"butterfly spread arbitrage"*, and is therefore unsatisfactory. The Hagan interpolation technique, on the other hand, yields much better results. However, this interpolation is also not perfect and it may give unsatisfactory results, in extreme market settings (for which, due to the parameter settings, the asymptotic volatility formula is not accurate anymore), like for example in the case of negative interest rates. The interpolation routine can however be improved, as discussed in [Grzelak and Oosterlee, 2016].

The SABR parametrization is often preferred to a standard linear or spline interpolation. Because the parametrization is derived from an arbitrage-free model, it fits to the purpose of interpolating market implied volatilities.

> When interpolating implied volatilities, it is important that *interpolation arbitrage* is avoided. An implied volatility parametrization which is derived from an arbitrage-free asset price model is often preferred to a standard interpolation technique which is available in the numerical software packages. ♦

4.3.4 Simulation of local volatility model

The expression for the local volatility is used within the SDE,

$$\mathrm{d}S(t) = rS(t)\mathrm{d}t + \sigma_{LV}(t, S(t))S(t)\mathrm{d}W(t).$$

Local Volatility Models 115

So, the local volatility term (4.49) at each time step $T = t$ in $K = S(t)$ is to be determined.

The local volatility framework relies on the available implied volatility surface, $\sigma_{\text{imp}}(T, K)$, for each expiry date T and strike price K. For each underlying only a few option market quotes are available, so that the industry standard is to parameterize the implied volatility surface by a 2D continuous function $\hat{\sigma}(T, K)$, which is calibrated to the market quotes, $\sigma_{\text{imp}}(T_i, K_j)$, at the set of available expiry dates and strike prices. The volatility parametrization $\hat{\sigma}(T, K)$ needs to satisfy the arbitrage-free conditions.

Let us consider the Hagan interpolation (4.61), using $\rho = 0, \beta = 1$, for which the implied volatilities are given in the following form:

$$\hat{\sigma}(T, K) = \frac{\gamma \log(S_F(t_0)/K)\left(1 + \frac{\gamma^2}{12}T\right)}{\log\left(\sqrt{1 + (\frac{\gamma}{\alpha}\log(S_F(t_0)/K))^2} + \frac{\gamma}{\alpha}\log(S_F(t_0)/K)\right)}, \quad (4.62)$$

and

$$\hat{\sigma}(T, S_F(t_0)) = \alpha\left(1 + \frac{\gamma^2}{12}T\right),$$

This functional form in (4.62) is commonly used in the industry [Grzelak and Oosterlee, 2016].

We start with the simulation of the local volatility model from (4.40), with $\sigma_{LV}(t, S)$ as defined in (4.56), and where $\sigma_{\text{imp}}(t, S(t)) = \hat{\sigma}(t, S(t))$ in (4.62).

The simulation of Equation (4.40) can be performed by means of an Euler discretization, which will be discussed in more detail in Chapter 9, i.e.,

$$s_{i+1} = s_i + rs_i\Delta t + \sigma_{LV}(t_i, s_i)s_i\big(W(t_{i+1}) - W(t_i)\big), \quad (4.63)$$

with $\sigma_{LV}(t_i, s_i)$ given by (4.56), i.e.

$$\sigma_{LV}^2(t_i, s_i) = \frac{\frac{\partial w}{\partial t_i} + rs_i\frac{\partial w}{\partial s_i}}{1 + s_i\frac{\partial w}{\partial s_i}\left(\frac{1}{2} - \frac{y}{w}\right) + \frac{1}{2}s_i^2\frac{\partial^2 w}{\partial s_i^2} + \frac{1}{2}s_i^2\left(\frac{\partial w}{\partial s_i}\right)^2\left(-\frac{1}{8} - \frac{1}{2w} + \frac{y^2}{2w^2}\right)}, \quad (4.64)$$

for $t_0 = 0$ and $w := t_i\hat{\sigma}^2(t_i, s_i)$,

$$\frac{\partial w}{\partial t_i} = \hat{\sigma}(t_i, s_i)^2 + 2t_i\hat{\sigma}(t_i, s_i)\frac{\partial \hat{\sigma}(t_i, s_i)}{\partial t_i}, \quad \frac{\partial w}{\partial s_i} = 2t_i\hat{\sigma}(t_i, s_i)\frac{\partial \hat{\sigma}(t_i, s_i)}{\partial s_i},$$

$$\frac{\partial^2 w}{\partial s_i^2} = 2t_i\left(\frac{\partial \hat{\sigma}(t_i, s_i)}{\partial s_i}\right)^2 + 2t_i\hat{\sigma}(t_i, s_i)\frac{\partial^2 \hat{\sigma}(t_i, s_i)}{\partial s_i^2}.$$

The local volatility component $\sigma_{LV}(t_i, s_i)$ in (4.64) needs to be evaluated for each realization s_i. In the case of a Monte Carlo simulation when many thousands asset paths are generated, the evaluation of (4.64) can become computationally intensive.

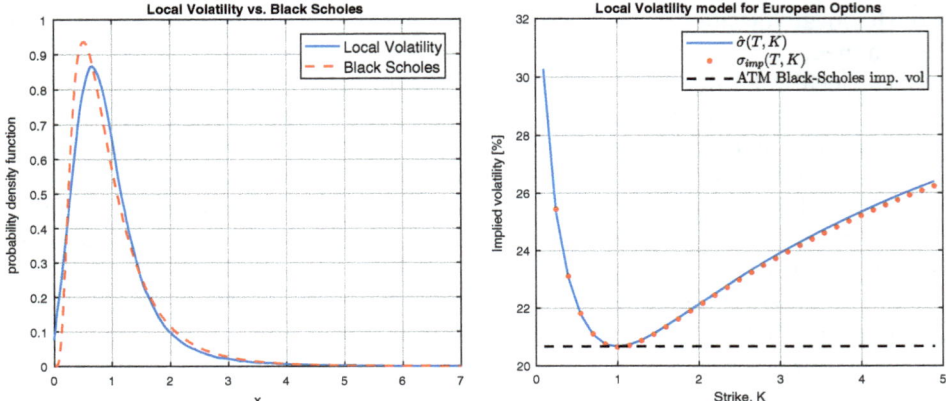

Figure 4.10: *Local volatility versus Black Scholes model. Left: the stock densities of the Black-Scholes and the local volatility model. Right: volatilities $\hat{\sigma}(T, K)$ and $\sigma_{imp}(T, K)$ obtained by the local volatility model, with $T = 10$, $\gamma = 0.2$, $S_0 = 1$ and $\alpha = 0.2$.*

As a sanity check, when the model has been implemented correctly the input volatilities $\hat{\sigma}(T, K)$ should resemble $\sigma_{\text{imp}}(T, K)$.

In the left side picture in Figure 4.10, we compare the local volatility density with the Black-Scholes log-normal density function in which the ATM volatility is substituted. In the right side figure the input volatilities $\hat{\sigma}(T, K)$ and the output implied volatilities $\sigma_{\text{imp}}(T, K)$, as obtained from the Monte Carlo simulation are shown. The local volatility model generates implied volatilities that resemble the input in the parametrization of $\sigma_{LV}(T, K)$ very well.

> We conclude with the following general statements regarding the LV model. Local volatility models resemble very well the implied volatilities that we observe in market option quotes, and they are used as an input to the LV model. Calibration to these input European options is thus highly accurate, and the model is "tailored" to these available option quotes. However, LV models may suffer from significant mispricing inaccuracy when dealing with financial derivatives products that depend on the volatility paths and, generally, on transition density functions. This information is simply not "coded" in the LV model framework. In Chapter 10 this problem will be discussed in further detail.

4.4 Exercise set

Exercise 4.1 Employ the Newton-Raphson iteration, the combined root-finding method, as well as Brent's method to find the roots of the following problems,

a. Compute the two solutions of the nonlinear equation,
$$g(x) = \frac{e^x + e^{-x}}{2} - 2x = 0.$$
Present the respective approximations for each iteration (so, the convergence) to the two solutions.

b. Acquire some option market data, either electronically or via a newspaper, of a company whose name starts with your initial or the first letter of your surname. Confirm that the option data contain an implied volatility skew or smile. Plot and discuss the implied volatility as a function of the strike price.

Exercise 4.2 Choose a set of option parameters for a deep in the money option. Confirm that the Newton-Raphson iteration has convergence issues, whereas the combined root-finding method converges in a robust way.

Exercise 4.3 With two new independent variables,
$$X_F := \log\left(\frac{Se^{r(T-t)}}{K}\right), \text{ and } t_* := \sigma\sqrt{T-t},$$

a. Rewrite the arguments $d_{1,2}$ in the Black-Scholes solution, in two variables that only depend on X_F and t_*.

b. Give an interpretation of the inequality $X_F \leq 0$.

c. Show that for the scaled put option value, $p = V_p(t, S(t))/S(t)$, we have,
$$p(X_F, t_*) = e^{-X_F} F_{\mathcal{N}(0,1)}(-d_2) - F_{\mathcal{N}(0,1)}(-d_1).$$

d. Determine
$$\frac{\partial p(X_F, t_*)}{\partial t_*}$$
and give an interpretation of this derivative.

Exercise 4.4 For $c(y, w) := S_0 \left[F_{\mathcal{N}(0,1)}(d_1) - e^y F_{\mathcal{N}(0,1)}(d_2) \right]$, with
$$d_1 = -\frac{y}{\sqrt{w}} + \frac{1}{2}\sqrt{w},$$
and $d_2 = d_1 - \sqrt{w}$, and y and w defined in (4.50) show the following equality:
$$\frac{\partial^2 c}{\partial w^2} = \frac{\partial c}{\partial w}\left(-\frac{1}{8} - \frac{1}{2w} + \frac{y^2}{2w^2}\right).$$

Exercise 4.5 Use the results from Exercise 4.4 to show that:
$$\frac{\partial^2 c}{\partial w \partial y} = \frac{\partial c}{\partial w}\left(\frac{1}{2} - \frac{y}{w}\right), \text{ and } \frac{\partial^2 c}{\partial y^2} = \frac{\partial c}{\partial y} + 2\frac{\partial c}{\partial w}.$$

Exercise 4.6 For which of the following payoff functions can option pricing be done in the Breeden-Litzenberger framework?

a. A digital option, where the payoff function is based on $\mathbb{1}_{S(T)>K}$,

b. A spread option based on the performance of two stocks, where the payoff reads $\max(S_1(T) - S_2(T), K)$,

c. A performance option, with as the payoff, $\max(S(T_2)/S(T_1) - K, 0)$,

d. A cliquet-type option, with payoff function, $\min(\max(\frac{S(T)}{S(t_0)} - K_1, 0), K_2)$.

Exercise 4.7 Let us assume a market in which $r = 0$, $S(t_0) = 1$ and the implied volatilities for $T = 4$ are given by the following formula,

$$\hat{\sigma}(K) = 0.510 - 0.591K + 0.376K^2 - 0.105K^3 + 0.011K^4, \qquad (4.65)$$

with an upper limit, which is given by $\hat{\sigma}(K) = \hat{\sigma}(3)$ for $K > 3$.

In the Breeden-Litzenberger framework any European-type payoff function can be priced with the following formula, in the case $r = 0$,

$$V(t_0, S_0) = V(T, S_0) + \int_0^{S_0} V_p(t_0, x) \frac{\partial^2 V(T, x)}{\partial x^2} dx + \int_{S_0}^{\infty} V_c(t_0, x) \frac{\partial^2 V(T, x)}{\partial x^2} dx, \qquad (4.66)$$

where $V(T, x)$ is one of the payoff functions at time T, that are given below. $V_p(t_0, x)$ and $V_c(t_0, x)$ are the put and call prices with strike price x. These prices can be obtained by evaluating the Black-Scholes formula for the volatilities in Equation (4.65).

Using Equation (4.66) compute numerically the option prices at $t = 0$, for the following payoff functions,

a. $V(T, S(T)) = \max(S^2(T) - 1.2S(T), 0)$

b. $V(T, S(T)) = \max(S(T) - 1.5, 0)$

c. $V(T, S(T)) = \max(1.7 - S(T), 0) + \max(S(T) - 1.4, 0)$

d. $V(T, S(T)) = \max(4 - S^3(T), 0) + \max(S(T) - 2, 0)$

Hint: Approximate the derivatives of the payoff $V(T, x)$, by finite differences. Note that increment δ, in $V(T, x + \delta)$, cannot be too small for accuracy reasons.

For which of the payoff functions above can the option price at $t = 0$ be determined without using the Breeden-Litzenberger model?

Exercise 4.8 Consider the SABR formula for the Black-Scholes implied volatility, i.e., Equation (4.61) with the following set of parameters, using $S_F(t_0) = S(t_0)e^{r(T-t_0)}$, $T = 2.5$, $\beta = 0.5$, $S(t_0) = 5.6$, $r = 0.015$, $\alpha = 0.2$, $\rho = -0.7$, $\gamma = 0.35$. Based on the relation between option prices and the stock's probability density function,

$$f_{S(T)}(x) = e^{r(T-t_0)} \frac{\partial^2 V_c(t_0, S_0; x, T)}{\partial x^2} \qquad (4.67)$$

$$\approx e^{r(T-t_0)} \frac{V_c(t_0, S_0; x + \Delta x, T) - 2V_c(t_0, S_0; x, T) + V_c(t_0, S_0; x - \Delta x, T)}{\Delta x^2},$$

compute the probability density function $f_{S(T)}(x)$. Is this probability density function free of arbitrage? Explain the impact of Δx on the quality of the PDF.

Exercise 4.9 With the parameters, $T = 1$, $S_0 = 10.5$, $r = 0.04$, we observe the following set of implied volatilities in the market,

Table 4.2: *Implied volatilities and the corresponding strikes.*

K	3.28	5.46	8.2	10.93	13.66	16.39	19.12	21.86
$\sigma_{imp}(T,K)[\%]$	31.37	22.49	14.91	9.09	6.85	8.09	9.45	10.63

Consider the following interpolation routines,

a) linear,

b) cubic spline,

c) nearest neighbor.

Which of these interpolations will give rise to the smallest values for the butterfly arbitrage? Explain how to reduce the arbitrage values. *Hint:* Assume a flat extrapolation prior to the first and behind the last strike price.

Exercise 4.10 Consider two expiry dates, $T_1 = 1y$ and $T_2 = 2y$, with as the parameter values, $r = 0.1$, $S_0 = 10.93$ and the following implied volatilities,

Table 4.3: *Implied volatilities and the corresponding strikes.*

K	3.28	5.46	8.2	10.93	13.66	16.39	19.12	21.86
$\sigma_{imp}(T_1,K)[\%]$	31.37	22.49	14.91	9.09	6.85	8.09	9.45	10.63
$\sigma_{imp}(T_2,K)[\%]$	15.68	11.25	7.45	4.54	3.42	4.04	4.72	5.31

Check for a calendar arbitrage based on these values.

Exercise 4.11 Consider two independent stock prices, $S_1(t)$ and $S_2(t)$, determine the following expectations using the Breeden-Litzenberger model,

$$\mathbb{E}^\mathbb{Q}\left[\log\left(S_1(T)^{S_2(T)}\right)\Big|\mathcal{F}(t_0)\right], \quad \mathbb{E}^\mathbb{Q}\left[\log\prod_{i=1}^{N}\frac{S_1(t_{i+1})}{S_1(t_i)}\frac{S_2(t_{i+1})}{S_2(t_i)}\Big|\mathcal{F}(t_0)\right].$$

Exercise 4.12 When pricing financial options based on the market probability density function, it is possible to use either Equation (4.11) or Equation (4.20). With the following payoff function, $V(T,S) = \max(S^2 - K, 0)$, and the settings as in Exercise 4.9, give arguments which of the two representations is more stable numerically, and present some corresponding numerical results.

CHAPTER 5

Jump Processes

In this chapter:

One way to deal with the market observed *implied volatility smile* is to consider asset prices based on jump processes. We first explain the *jump diffusion processes* in **Section 5.1** and detail the generalization of Itô's lemma to jump processes. The option valuation equations under these asset price processes with jumps turn out to be *partial integro-differential equations*. An analytic solution can be derived, based on the Feynman-Kac Theorem (see **Section 5.2**). The jump diffusion models can be placed in the class of exponential Lévy asset price processes. They are discussed in **Section 5.3**. We discuss Lévy jump models with a *finite* (**Section 5.3.1**) or an *infinite number of jumps* (**Section 5.4**). For certain specific Lévy asset models, numerical examples give insight in the impact of the different model parameters on the asset price density and the implied volatility. We end this chapter, in Section 5.5, with a general discussion on the use of jumps for asset price modeling in **Section 5.5**.

Important is that for the asset models in this chapter, the *characteristic function* can be derived. The characteristic function will form the basis of highly efficient Fourier option pricing techniques, in Chapter 6.

Keywords: alternative model for asset prices, jump diffusion process, Poisson process, Lévy jump models, impact on option pricing PDE, characteristic function, Itô's lemma.

5.1 Jump diffusion processes

We analyze the Black-Scholes model extended by independent jumps, that are driven by a Poisson process, the so-called *jump diffusion process*, and consider the following dynamics for the log-stock process, $X(t) = \log S(t)$, under the real-world measure $\mathbb{P}$:

$$dX(t) = \mu dt + \sigma dW^{\mathbb{P}}(t) + JdX_{\mathcal{P}}(t), \tag{5.1}$$

where μ is the drift, σ is the volatility, $X_{\mathcal{P}}(t)$ is a Poisson process and variable J gives the jump magnitude; J is governed by a distribution, F_J, of magnitudes. Processes $W^{\mathbb{P}}(t)$ and $X_{\mathcal{P}}(t)$ are assumed to be *independent*.

Before we start working with these dynamics, we first need to discuss some details of jump processes, in particular of the Poisson process, and also the relevant version of Itô's lemma.

Definition 5.1.1 (Poisson random variable) *A Poisson random variable, which is denoted by $X_{\mathcal{P}}$, counts the number of occurrences of an event during a given time period. The probability of observing $k \geq 0$ occurrences in the time period is given by*

$$\mathbb{P}[X_{\mathcal{P}} = k] = \frac{\xi_p^k e^{-\xi_p}}{k!}.$$

The mean of $X_{\mathcal{P}}$, $\mathbb{E}[X_{\mathcal{P}}] = \xi_p$, gives the average number of occurrences of the event, while for the variance, $\mathbb{V}\mathrm{ar}[X_{\mathcal{P}}] = \xi_p$. ◂

Definition 5.1.2 (Poisson process) *A Poisson process, $\{X_{\mathcal{P}}(t), t \geq t_0 = 0\}$, with parameter $\xi_p > 0$ is an integer-valued stochastic process, with the following properties*

- $X_{\mathcal{P}}(0) = 0$;

- $\forall \, t_0 = 0 < t_1 < \ldots < t_n$, *the increments* $X_{\mathcal{P}}(t_1) - X_{\mathcal{P}}(t_0)$, $X_{\mathcal{P}}(t_2) - X_{\mathcal{P}}(t_1), \ldots, X_{\mathcal{P}}(t_n) - X_{\mathcal{P}}(t_{n-1})$ *are independent random variables;*

- *for $s \geq 0, t > 0$ and integers $k \geq 0$, the increments have the Poisson distribution:*

$$\mathbb{P}[X_{\mathcal{P}}(s+t) - X_{\mathcal{P}}(s) = k] = \frac{(\xi_p t)^k e^{-\xi_p t}}{k!}. \tag{5.2}$$

◂

The Poisson process $X_{\mathcal{P}}(t)$ is a counting process, with the number of *jumps* in any time period of length t specified via (5.2). Equation (5.2) confirms the stationary increments, since the increments only depend on the length of the interval and not on initial time s.

Parameter ξ_p is the *rate* of the Poisson process, i.e. it indicates the number of jumps in a time period. The probability that exactly one event occurs in a small time interval, dt, follows from (5.2) as

$$\mathbb{P}[X_{\mathcal{P}}(s+dt) - X_{\mathcal{P}}(s) = 1] = \frac{(\xi_p dt) e^{-\xi_p dt}}{1!} = \xi_p dt + o(dt),$$

Jump Processes

and the probability that no event occurs in dt is

$$\mathbb{P}[X_{\mathcal{P}}(s+\mathrm{d}t) - X_{\mathcal{P}}(s) = 0] = \mathrm{e}^{-\xi_p \mathrm{d}t} = 1 - \xi_p \mathrm{d}t + o(\mathrm{d}t).$$

In a time interval dt, a jump will arrive with probability $\xi_p \mathrm{d}t$, resulting in:

$$\mathbb{E}[\mathrm{d}X_{\mathcal{P}}(t)] = 1 \cdot \xi_p \mathrm{d}t + 0 \cdot (1 - \xi_p \mathrm{d}t) = \xi_p \mathrm{d}t,$$

where $\mathrm{d}X_{\mathcal{P}}(t) = X_{\mathcal{P}}(s+\mathrm{d}t) - X_{\mathcal{P}}(s)$. The expectation is thus given by

$$\mathbb{E}[X_{\mathcal{P}}(s+t) - X_{\mathcal{P}}(s)] = \xi_p t.$$

With $X_{\mathcal{P}}(0) = 0$, the expected number of events in a time interval with length t, setting $s = 0$, equals

$$\mathbb{E}[X_{\mathcal{P}}(t)] = \xi_p t. \tag{5.3}$$

If we define another process, $\bar{X}_{\mathcal{P}}(t) := X_{\mathcal{P}}(t) - \xi_p t$, then $\mathbb{E}[\mathrm{d}\bar{X}_{\mathcal{P}}(t)] = 0$, so that process $\bar{X}_{\mathcal{P}}(t)$, which is referred to as the *compensated* Poisson process, is a martingale.

Example 5.1.1 (Paths of the Poisson process) We present some discrete paths that have been generated by a Poisson process, with $\xi_p = 1$. The left-hand picture in Figure 5.1 displays the paths by $\mathrm{d}X_{\mathcal{P}}(t)$, whereas the right-hand picture shows the same paths for the compensated Poisson process, $-\xi_p \mathrm{d}t + \mathrm{d}X_{\mathcal{P}}(t)$.

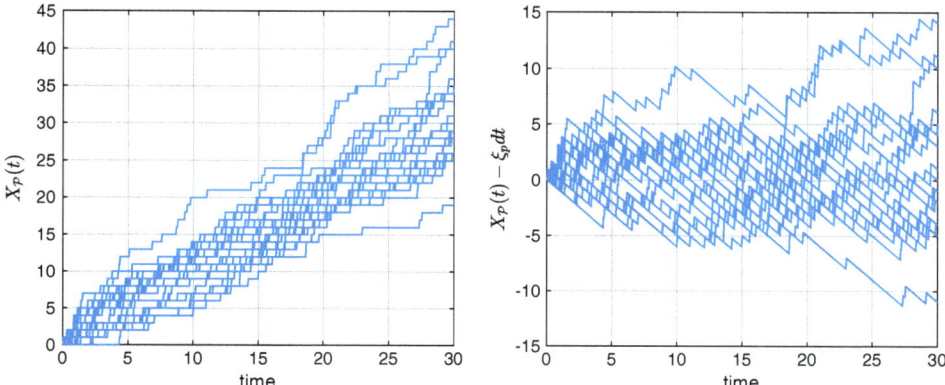

Figure 5.1: *Monte Carlo paths for the Poisson (left) and the compensated Poisson process (right), $\xi_p = 1$.*

> Given the following SDE:
>
> $$dX(t) = J(t)dX_{\mathcal{P}}(t), \quad (5.4)$$
>
> we may define the stochastic integral with respect to the Poisson process $X_{\mathcal{P}}(t)$, by
>
> $$X(T) - X(t_0) = \int_{t_0}^T J(t)dX_{\mathcal{P}}(t) := \sum_{k=1}^{X_{\mathcal{P}}(T)} J_k. \quad (5.5)$$
>
> Variable J_k for $k \geq 1$ is an i.i.d. sequence of random variables with a jump-size probability distribution F_J, so that $\mathbb{E}[J_k] = \mu_J < \infty$.

5.1.1 Itô's lemma and jumps

In Chapter 2, we have discussed that for traded assets the martingale property should be satisfied for the discounted asset price $S(t)/M(t)$, under the risk neutral pricing measure, see Section 2.3. Just substituting $\mu = r$ in (5.1) is however not sufficient to achieve this, as the jumps have an impact on the drift term too. A *drift adjustment* needs to compensate for average jump sizes.

In order to derive the dynamics for $S(t) = \exp(X(t))$, with $X(t)$ in (5.1), a variant of Itô's lemma which is related to the Poisson process, needs to be employed. We present a general result below for models with independent Poisson jumps.

> **Result 5.1.1 (Itô's lemma for Poisson process)** *We consider a càdlàg process, $X(t)$, defined as:*
>
> $$dX(t) = \bar{\mu}(t, X(t))dt + \bar{J}(t, X(t_-))dX_{\mathcal{P}}(t), \quad \text{with} \quad X(t_0) \in \mathbb{R}, \quad (5.6)$$
>
> *where $\bar{\mu}, \bar{J} : [0, \infty) \times \mathbb{R} \to \mathbb{R}$ are deterministic, continuous functions and $X_{\mathcal{P}}(t)$ is a Poisson process, starting at $t_0 = 0$.*
> *For a differentiable function, $g : [0, \infty) \times \mathbb{R} \to \mathbb{R}$, the Itô differential is given by:*
>
> $$dg(t, X(t)) = \left[\frac{\partial g(t, X(t))}{\partial t} + \bar{\mu}(t, X(t))\frac{\partial g(t, X(t))}{\partial X}\right]dt \quad (5.7)$$
> $$+ \left[g\left(t, X(t_-) + \bar{J}(t, X(t_-))\right) - g(t, X(t_-))\right]dX_{\mathcal{P}}(t),$$
>
> *where the left limit is denoted by $X(t_-) := \lim_{s \to t} X(s), s < t$, so that, by the continuity of $\bar{J}(\cdot)$, its left limit equals $\bar{J}(t, X(t_-))$.*

An intuitive explanation for Itô's formula in the case of jumps is that when a jump takes place, i.e. $dX_{\mathcal{P}}(t) = 1$, the process "jumps" from $X(t_-)$ to $X(t)$, with the jump size determined by function $\bar{J}(t, X(t))$ [Sennewald and Wälde, 2006],

resulting in the following relation:
$$g(t, X(t)) = g\left(t, X(t_-) + \bar{J}(t, X(t_-))\right).$$
After the jump at time t, the function $g(\cdot)$ is adjusted with the jump size, which was determined at time t_-.

In practical applications, stochastic processes that include both a Brownian motion and a Poisson process are encountered, as in Equation (5.1). The general dynamics of this combined process are given by:
$$dX(t) = \bar{\mu}(t, X(t))dt + \bar{J}(t, X(t_-))dX_{\mathcal{P}}(t) + \bar{\sigma}(t, X(t))dW(t), \quad \text{with} \quad X(t_0) \in \mathbb{R}, \tag{5.8}$$
with $\bar{\mu}(\cdot, \cdot)$ the drift, $\bar{\sigma}(\cdot, \cdot)$ the diffusion term, and $\bar{J}(\cdot, \cdot)$ the jump magnitude function. For a function $g(t, X(t))$ in (5.7) an extension of Result 5.1.1 is then required. Assuming that the Poisson process $X_{\mathcal{P}}(t)$ is *independent* of the Brownian motion $W(t)$, the dynamics of $g(t, X(t))$ are given by:

$$\begin{aligned}dg(t, X(t)) = &\left[\frac{\partial g(t, X(t))}{\partial t} + \bar{\mu}(t, X(t))\frac{\partial g(t, X(t))}{\partial X}\right.\\ &\left.+ \frac{1}{2}\bar{\sigma}^2(t, X(t))\frac{\partial^2 g(t, X(t))}{\partial X^2}\right]dt \\ &+ \left[g\left(t, X(t_-) + \bar{J}(t, X(t_-))\right) - g(t, X(t_-))\right]dX_{\mathcal{P}}(t) \\ &+ \bar{\sigma}(t, X(t))\frac{\partial g(t, X(t))}{\partial X}dW(t),\end{aligned} \tag{5.9}$$

Here we made use of the Itô multiplication Table 5.1, where the cross terms involving the Poisson process are also handled. An intuitive way to understand the Poisson process rule in Table 5.1 is found in the notion that the term $dX_{\mathcal{P}} = 1$ with probability $\xi_p dt$, and $dX_{\mathcal{P}} = 0$ with probability $(1 - \xi_p dt)$, which implies that
$$(dX_{\mathcal{P}})^2 = \begin{cases} 1^2 & \text{with probability } \xi_p dt, \\ 0^2 & \text{with probability } (1 - \xi_p dt) \end{cases}$$
$$= dX_{\mathcal{P}}.$$

To apply Itô's lemma to the function $S(t) = e^{X(t)}$, with $X(t)$ in (5.1), we substitute $\bar{\mu}(t, X(t)) = \mu$, $\bar{\sigma}(t, X(t)) = \sigma$ and $\bar{J}(t, X(t_-)) = J$ in (5.9), giving,
$$de^{X(t)} = \left(\mu e^{X(t)} + \frac{1}{2}\sigma^2 e^{X(t)}\right)dt + \sigma e^{X(t)}dW(t) + \left(e^{X(t)+J} - e^{X(t)}\right)dX_{\mathcal{P}}(t),$$

Table 5.1: *Itô multiplication table for Poisson process.*

	dt	dW(t)	$dX_{\mathcal{P}}(t)$
dt	0	0	0
dW(t)	0	dt	0
$dX_{\mathcal{P}}(t)$	0	0	$dX_{\mathcal{P}}(t)$

so that we obtain:

$$\frac{dS(t)}{S(t)} = \left(\mu + \frac{1}{2}\sigma^2\right)dt + \sigma dW(t) + \left(e^J - 1\right)dX_\mathcal{P}(t).$$

Until now, we have derived the dynamics for the stock $S(t)$ under the real-world measure $\mathbb{P}$. The next step is to derive the dynamics of the stock under the risk-neutral measure $\mathbb{Q}$.

For this, we check under which conditions the process $Y(t) := S(t)/M(t)$ is a martingale, or, in other words, the dynamics $dY(t) = \frac{dS(t)}{M(t)} - \frac{rS(t)dt}{M(t)}$, with $dY(t) = Y(t+dt) - Y(t)$, should have an expected value equal to zero:

$$\mathbb{E}\left[dY(t)\right] = \mathbb{E}\left[\mu S(t) + \frac{1}{2}\sigma^2 S(t) - rS(t)\right]dt + \mathbb{E}\left[\sigma S(t)dW(t)\right]$$
$$+ \mathbb{E}\left[\left(e^J - 1\right)S(t)dX_\mathcal{P}(t)\right].$$

From the properties of the Brownian motion and Poisson process, and the fact that all random components in the expression above are mutually independent,[1] we get:

$$\mathbb{E}\left[dY(t)\right] = \mathbb{E}\left[\mu S(t) + \frac{1}{2}\sigma^2 S(t) - rS(t)\right]dt + \mathbb{E}\left[\left(e^J - 1\right)S(t)\right]\xi_p dt$$
$$= \left(\mu - r + \frac{1}{2}\sigma^2 + \mathbb{E}\left[\xi_p(e^J - 1)\right]\right)\mathbb{E}[S(t)]dt.$$

By substituting $\mu = r - \frac{1}{2}\sigma^2 - \xi_p\mathbb{E}\left[e^J - 1\right]$, we have $\mathbb{E}[dY(t)] = 0$.

The term $\bar{\omega} := \xi_p\mathbb{E}\left[e^J - 1\right]$ is the so-called *drift correction term*, which makes the process a martingale.

The dynamics for stock $S(t)$ under the risk-neutral measure $\mathbb{Q}$ are therefore given by:

$$\boxed{\frac{dS(t)}{S(t)} = \left(r - \xi_p\mathbb{E}\left[e^J - 1\right]\right)dt + \sigma dW^\mathbb{Q}(t) + \left(e^J - 1\right)dX_\mathcal{P}^\mathbb{Q}(t).} \quad (5.10)$$

The process in (5.10) is often presented in the literature as the *standard jump diffusion model*.

The standard jump diffusion model is directly connected to the following $dX(t)$ dynamics:

$$\boxed{dX(t) = \left(r - \xi_p\mathbb{E}\left[e^J - 1\right] - \frac{1}{2}\sigma^2\right)dt + \sigma dW^\mathbb{Q}(t) + JdX_\mathcal{P}^\mathbb{Q}(t).} \quad (5.11)$$

---

[1] It can be shown that if $W(t)$ is a Brownian motion and $X_\mathcal{P}(t)$ a Poisson process with intensity ξ_p, and both processes are defined on the same probability space $(\Omega, \mathcal{F}(t), \mathbb{P})$, then the processes $W(t)$ and $X_\mathcal{P}(t)$ are independent.

5.1.2 PIDE derivation for jump diffusion process

We will determine the pricing PDE in the case the underlying dynamics are driven by the jump diffusion process. We depart from the following SDE:

$$dS(t) = \bar{\mu}(t, S(t))dt + \bar{\sigma}(t, S(t))dW^{\mathbb{Q}}(t) + \bar{J}(t, S(t))dX_{\mathcal{P}}^{\mathbb{Q}}(t), \qquad (5.12)$$

where, in the context of Equation (5.10), the functions $\bar{\mu}(t, S(t))$, $\bar{J}(t, S(t))$ and $\bar{\sigma}(t, S(t))$ are equal to

$$\bar{\mu}(t, S(t)) := \left(r - \xi_p \mathbb{E}\left[e^J - 1\right]\right) S(t), \quad \bar{\sigma}(t, S(t)) := \sigma S(t),$$
$$\bar{J}(t, S(t)) := (e^J - 1)S(t). \qquad (5.13)$$

The dynamics of process (5.12) are under the risk-neutral measure $\mathbb{Q}$, so that we may apply the martingale approach to derive the option pricing equation under the jump diffusion asset price dynamics. This means that, for a certain payoff $V(T, S)$, the following equality has to hold:

$$\frac{V(t, S)}{M(t)} = \mathbb{E}^{\mathbb{Q}}\left[\frac{V(T, S)}{M(T)}\bigg|\mathcal{F}(t)\right]. \qquad (5.14)$$

$V(t, S)/M(t)$ in (5.14) can be recognized as a martingale under the $\mathbb{Q}$-measure, so that the governing dynamics should not contain any drift terms. With Itô's lemma, the dynamics of V/M, using $M \equiv M(t), V \equiv V(t, S)$, are given by

$$d\frac{V}{M} = \frac{1}{M}dV - r\frac{V}{M}dt.$$

The dynamics of V are obtained by using Itô's lemma for the Poisson process, as presented in Equation (5.9). There, we set $g(t, S(t)) := V(t, S)$ and $\bar{J}(t, S(t)) := (e^J - 1)S$, which implies the following dynamics:

$$dV = \left(\frac{\partial V}{\partial t} + \bar{\mu}(t, S)\frac{\partial V}{\partial S} + \frac{1}{2}\bar{\sigma}^2(t, S)\frac{\partial^2 V}{\partial S^2}\right)dt + \bar{\sigma}(t, S)\frac{\partial V}{\partial S}dW^{\mathbb{Q}}(t)$$
$$+ \left(V(t, Se^J) - V(t, S)\right)dX_{\mathcal{P}}^{\mathbb{Q}}(t).$$

After substitution, the dynamics of V/M read:

$$d\frac{V}{M} = \frac{1}{M}\left(\frac{\partial V}{\partial t} + \bar{\mu}(t, S)\frac{\partial V}{\partial S} + \frac{1}{2}\bar{\sigma}^2(t, S)\frac{\partial^2 V}{\partial S^2}\right)dt + \frac{\bar{\sigma}(t, S)}{M}\frac{\partial V}{\partial S}dW^{\mathbb{Q}}$$
$$+ \frac{1}{M}\left(V(t, Se^J) - V(t, S)\right)dX_{\mathcal{P}}^{\mathbb{Q}} - r\frac{V(t, S)}{M}dt.$$

The jumps are independent of Poisson process $X_{\mathcal{P}}^{\mathbb{Q}}$ and Brownian motion $W^{\mathbb{Q}}$. Because V/M is a martingale, it follows that,

$$\left(\frac{\partial V}{\partial t} + \bar{\mu}(t, S)\frac{\partial V}{\partial S} + \frac{1}{2}\bar{\sigma}^2(t, S)\frac{\partial^2 V}{\partial S^2} - rV\right)dt$$
$$+ \mathbb{E}\left[(V(t, Se^J) - V(t, S))\right]\mathbb{E}\left[dX_{\mathcal{P}}^{\mathbb{Q}}\right] = 0. \qquad (5.15)$$

Based on Equation (5.3) and substitution of the expressions (5.13) in (5.15), the following pricing equation results:

$$\frac{\partial V}{\partial t} + \left(r - \xi_p \mathbb{E}\left[e^J - 1\right]\right) S \frac{\partial V}{\partial S} + \frac{1}{2}\sigma^2 S^2 \frac{\partial^2 V}{\partial S^2}$$
$$- (r + \xi_p)V + \xi_p \mathbb{E}\left[V(t, Se^J)\right] = 0, \qquad (5.16)$$

which is a *partial integro-differential equation* (PIDE), as we deal with partial derivatives, but the expectation gives rise to an integral term.

PIDEs are typically more difficult to be solved than PDEs, due to the presence of the additional integral term. For only a few models analytic solutions exist. An analytic expression has been found for the solution of (5.16) for Merton's model and Kou's model, the solution is given in the form of an infinite series [Kou, 2002; Merton, 1976], see Section 5.2.1 to follow. However, in the literature a number of numerical techniques for solving PIDEs are available (see for e.g. [He et al., 2006; Kennedy et al., 2009]).

Example 5.1.2 For the jump diffusion process under measure $\mathbb{Q}$, we arrive at the following option valuation PIDE, in terms of the prices S,

$$\begin{cases} -\dfrac{\partial V}{\partial t} = \dfrac{1}{2}\sigma^2 S^2 \dfrac{\partial^2 V}{\partial S^2} + (r - \xi_p \mathbb{E}[e^J - 1])S\dfrac{\partial V}{\partial S} - (r + \xi_p)V \\ \qquad\qquad + \xi_p \displaystyle\int_0^\infty V(t, Se^y) \mathrm{d}F_J(y), \quad \forall (t,S) \in [0,T] \times \mathbb{R}_+, \\ V(T,S) = \max(\bar{\alpha}(S(T) - K), 0), \quad \forall S \in \mathbb{R}_+, \end{cases}$$

with $\bar{\alpha} = \pm 1$ (call or put, respectively), and where $\mathrm{d}F_J(y) = f_J(y)\mathrm{d}y$.

In log-coordinates $X(t) = \log(S(t))$, the corresponding PIDE for $V(t, X)$ is given by

$$\begin{cases} -\dfrac{\partial V}{\partial t} = \dfrac{1}{2}\sigma^2 \dfrac{\partial^2 V}{\partial X^2} + (r - \dfrac{1}{2}\sigma^2 - \xi_p \mathbb{E}[e^J - 1])\dfrac{\partial V}{\partial X} - (r + \xi_p)V \\ \qquad\qquad + \xi_p \displaystyle\int_\mathbb{R} V(t, X + y) \mathrm{d}F_J(y), \quad \forall (t,X) \in [0,T] \times \mathbb{R}, \\ V(T,X) = \max\left(\bar{\alpha}(\exp(X(T)) - K), 0\right), \quad \forall X \in \mathbb{R}. \end{cases}$$

Notice the differences in the integral terms of the two PIDEs above. ◆

5.1.3 Special cases for the jump distribution

Depending on the cumulative distribution function for the jump magnitude $F_J(x)$, a number of jump models can be defined.

Two popular choices are:

⇒ *Classical Merton's model* [Merton, 1976]: The jump magnitude J is normally distributed, with mean μ_J and standard deviation σ_J. So, $dF_J(x) = f_J(x)dx$, where

$$f_J(x) = \frac{1}{\sigma_J\sqrt{2\pi}} \exp\left(-\frac{(x-\mu_J)^2}{2\sigma_J^2}\right). \tag{5.17}$$

⇒ *Non-symmetric double exponential* (Kou's model [Kou, 2002; Kou and Wang, 2004])

$$f_J(x) = p_1\alpha_1 e^{-\alpha_1 x}\mathbb{1}_{\{x\geq 0\}} + p_2\alpha_2 e^{\alpha_2 x}\mathbb{1}_{\{x<0\}}, \tag{5.18}$$

where p_1, p_2 are positive real numbers so that $p_1 + p_2 = 1$. To be able to integrate e^x over the real line it is required to have $\alpha_1 > 1$ and $\alpha_2 > 0$, and we obtain the expression

$$\mathbb{E}[e^{J_k}] = p_1\frac{\alpha_1}{\alpha_1 - 1} + p_2\frac{\alpha_2}{\alpha_2 + 1}. \tag{5.19}$$

Example 5.1.3 (Paths of jump diffusion process) In Figure 5.2 examples of paths for $X(t)$ in (5.11) and $S(t) = e^{X(t)}$, as in (5.10), are presented. Here, the classical Merton model is used, $J \sim \mathcal{N}(\mu_J, \sigma_J^2)$, where the jumps are symmetric, as described in (5.17).

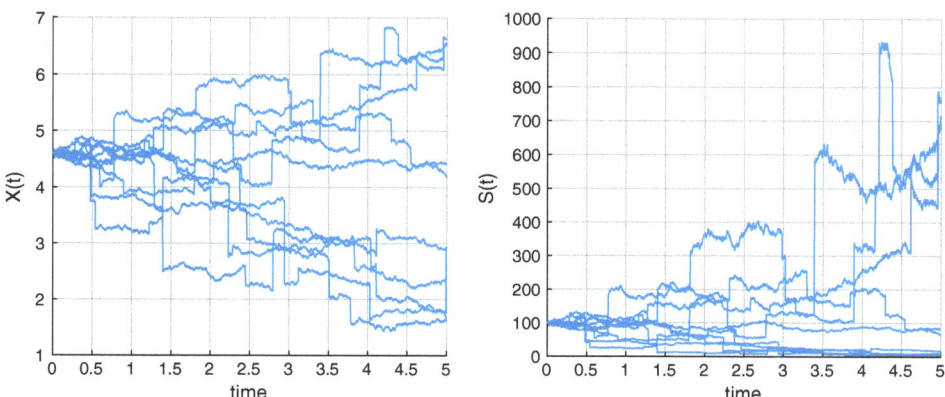

Figure 5.2: *Left side: Paths of process $X(t)$ (5.11); Right side: $S(t)$ in (5.10) with $S(t_0) = 100$, $r = 0.05$, $\sigma = 0.2$, $\sigma_J = 0.5$, $\mu_J = 0$, $\xi_p = 1$ and $T = 5$.*

5.2 Feynman-Kac theorem for jump diffusion process

The relation between the solution of a PDE and the computation of the discounted expected value of the payoff function, via the Feynman-Kac theorem (in Section 3.2), can be generalized to solving PIDEs that originate from the asset price processes $S(t)$ with jumps. As an example, we detail this for Merton's jump diffusion model.

With r constant, $S(t)$ is governed by the following SDE

$$\frac{\mathrm{d}S(t)}{S(t)} = \left(r - \xi_p \mathbb{E}\left[\mathrm{e}^J - 1\right]\right) \mathrm{d}t + \sigma \mathrm{d}W^{\mathbb{Q}}(t) + \left(\mathrm{e}^J - 1\right) \mathrm{d}X_{\mathcal{P}}^{\mathbb{Q}}(t),$$

and option value $V(t, S)$ satisfies the following PIDE,

$$\frac{\partial V}{\partial t} + \left(r - \xi_p \mathbb{E}\left[\mathrm{e}^J - 1\right]\right) S \frac{\partial V}{\partial S} + \frac{1}{2}\sigma^2 S^2 \frac{\partial^2 V}{\partial S^2} - (r + \xi_p)V$$
$$+ \xi_p \mathbb{E}\left[V(t, S\mathrm{e}^J)\right] = 0,$$

with $V(T, S) = H(T, S)$, and the term $(\mathrm{e}^J - 1)$ representing the size of a proportional jump, the risk-neutral valuation formula determines the option value, i.e.

$$V(t, S) = M(t)\mathbb{E}^{\mathbb{Q}}\left[\frac{1}{M(T)}H(T, S)|\mathcal{F}(t)\right] = \mathbb{E}^{\mathbb{Q}}\left[\mathrm{e}^{-r(T-t)}H(T, S)|\mathcal{F}(t)\right].$$

The discounted expected payoff formula resembles the well-known martingale property if we consider the quantity $\frac{V(t,S)}{M(t)}$, which should not contain a drift term, i.e. $\mathbb{E}^{\mathbb{Q}}\left[\mathrm{d}(V(t,S)/M(t))|\mathcal{F}(t)\right] = 0$. Equation (5.15) details this condition in the case of jump diffusion, confirming that then PIDE (5.16) should be satisfied. So, also in the case of this type of jump processes, we can solve the pricing PIDE in the form of the calculation of a discounted expected payoff, by means of the well-known Feynman-Kac theorem.

We will derive analytic option prices as well as the characteristic function for the jump diffusion process. Before we derive these, we present an application of the tower property of expectations for discrete random variables, as we will need this in the derivations to follow.

Example 5.2.1 (Tower property, discrete random variables) As an example of how convenient the conditional expectation can be used, we consider the following problem. Suppose $X_1, X_2, \ldots$ are independent random variables with the same mean μ and N_J is a nonnegative, integer-valued random variable

Jump Processes

which is independent of all X_i's. Then, the following equality holds:

$$\mathbb{E}\left[\sum_{i=1}^{N_J} X_i\right] = \mu \mathbb{E}[N_J]. \tag{5.20}$$

This equality is also known as *Wald's equation*.

Using the tower property (1.20) for discrete random variables z_1 and z_2, we find,

$$\mathbb{E}[\mathbb{E}[z_1|z_2]] = \sum_z \mathbb{E}[z_1|z_2=z]\,\mathbb{P}[z_2=z],$$

so that,

$$\mathbb{E}\left[\sum_{k=1}^{N_J} X_k\right] = \mathbb{E}\left[\mathbb{E}\left[\sum_{k=1}^{N_J} X_k \Big| N_J\right]\right]$$

$$= \sum_{n=1}^{\infty} \mathbb{E}\left[\sum_{k=1}^{n} X_k \Big| N_J = n\right] \mathbb{P}[N_J = n]$$

$$= \sum_{n=1}^{\infty} \mathbb{P}[N_J = n] \sum_{k=1}^{n} \mathbb{E}[X_k]. \tag{5.21}$$

Since the expectation for each X_k equals μ, we have:

$$\mathbb{E}\left[\sum_{k=1}^{N_J} X_k\right] = \sum_{n=1}^{\infty} \mathbb{P}[N_J = n] \sum_{k=1}^{n} \mu$$

$$= \mu \sum_{n=1}^{\infty} n\mathbb{P}[N_J = n] \stackrel{\text{def}}{=} \mu \mathbb{E}[N_J], \tag{5.22}$$

which confirms (5.20). ♦

5.2.1 Analytic option prices

In this section we derive the solution to call option prices for an asset under the risk-neutral *standard jump diffusion model*, as defined in Equation (5.10). Integrating the log-transformed asset price process in Equation (5.11), we find:

$$X(T) = X(t_0) + \int_{t_0}^{T} \left(r - \xi_p \mathbb{E}[e^J - 1] - \frac{1}{2}\sigma^2\right) dt$$

$$+ \int_{t_0}^{T} \sigma dW(t) + \int_{t_0}^{T} J(t) dX_{\mathcal{P}}(t)$$

$$= X(t_0) + \left(r - \xi_p \mathbb{E}[e^J - 1] - \frac{1}{2}\sigma^2\right)(T - t_0)$$

$$+ \sigma(W(T) - W(t_0)) + \sum_{k=1}^{X_{\mathcal{P}}(T)} J_k,$$

where $S(t) = e^{X(t)}$.

The price of a European call option under the jump diffusion process is given by,

$$V(t_0, S_0) = e^{-r(T-t_0)} \mathbb{E}^{\mathbb{Q}}\left[\max(S(T) - K, 0)\big|\mathcal{F}(t_0)\right], \qquad (5.23)$$

where

$$S(T) = \exp\left(\mu_X + \sigma\left(W(T) - W(t_0)\right) + \sum_{k=1}^{X_\mathcal{P}(T)} J_k\right), \qquad (5.24)$$

with $\mu_X := X(t_0) + \left(r - \xi_p \mathbb{E}[e^J - 1] - \frac{1}{2}\sigma^2\right)(T - t_0)$.

By conditioning the expectation in (5.23) on $X_\mathcal{P}(t)$, the number of jumps at time t, we find:

$$V(t_0, S_0) = e^{-r(T-t_0)}$$
$$\times \sum_{n \geq 0} (\mathbb{E}^{\mathbb{Q}}[\max(\exp\left(\mu_X + \sigma(W(T) - W(t_0)) + \sum_{k=1}^{n} J_k\right) - K, 0)\big|\mathcal{F}(t_0)]$$
$$\times \mathbb{P}[X_\mathcal{P}(t) = n]). \qquad (5.25)$$

Since the jump magnitudes J_k are assumed to be i.i.d. normally distributed, $J_k \sim \mathcal{N}(\mu_J, \sigma_J^2)$, it follows that,

$$\mu_X + \sigma\left(W(T) - W(t_0)\right) + \sum_{k=1}^{n} J_k \stackrel{d}{=} \mu_X + n\mu_J + \sqrt{\sigma^2 + \frac{n\sigma_J^2}{T-t_0}}\left(W(T) - W(t_0)\right),$$

so that the pricing formula is given by:

$$V(t_0, S_0) = e^{-r(T-t_0)} \sum_{n \geq 0} \mathbb{E}^{\mathbb{Q}}\big[\max\left(\exp\left(\hat{\mu}_X(n)\right.\right.$$
$$\left.\left. + \hat{\sigma}_X(n)\left(W(T) - W(t_0)\right)\right) - K, 0\right)\big|\mathcal{F}(t_0)\big]$$
$$\times \mathbb{P}\left[X_\mathcal{P}(t) = n\right], \qquad (5.26)$$

with

$$\hat{\mu}_X(n) := X(t_0) + \left(r - \xi_p \mathbb{E}[e^J - 1] - \frac{1}{2}\sigma^2\right)(T - t_0) + n\mu_J,$$

$$\hat{\sigma}_X(n) := \sqrt{\sigma^2 + \frac{n\sigma_J^2}{T-t_0}}.$$

Notice that the expectation in (5.26) resembles a call option price under a lognormally distributed stock process. However, the drift and the volatility terms depend on the number of jumps n, which is deterministic under conditioning.

Jump Processes

After further simplifications, we find:

$$\bar{V}(n) := \mathbb{E}^{\mathbb{Q}}\left[\max\left(\exp\left(\hat{\mu}_X(n) + \hat{\sigma}_X(n)\left(W(T) - W(t_0)\right)\right) - K, 0\right) \big| \mathcal{F}(t_0)\right]$$

$$= \exp\left(\hat{\mu}_X(n) + \frac{1}{2}\hat{\sigma}_X^2(n)(T - t_0)\right) \cdot F_{\mathcal{N}(0,1)}(d_1) - K \cdot F_{\mathcal{N}(0,1)}(d_2), \tag{5.27}$$

with

$$d_1 = \frac{\log \frac{S(t_0)}{K} + \left[r - \xi_p \mathbb{E}[e^J - 1] - \frac{1}{2}\sigma^2 + \hat{\sigma}_X^2(n)\right](T - t_0) + n\mu_J}{\hat{\sigma}_X(n)\sqrt{T - t_0}},$$

$$d_2 = \frac{\log \frac{S(t_0)}{K} + \left[r - \xi_p \mathbb{E}[e^J - 1] - \frac{1}{2}\sigma^2\right](T - t_0) + n\mu_J}{\hat{\sigma}_X(n)\sqrt{T - t_0}} = d_1 - \hat{\sigma}_X(n)\sqrt{T - t_0},$$

and $\mathbb{E}[e^J - 1] = e^{\mu_J + \frac{1}{2}\sigma_J^2} - 1$, when $J \sim \mathcal{N}(\mu_J, \sigma_J^2)$.

> Using Definition 5.1.2 for the Poisson distribution, the call option price is thus given by:
>
> $$V(t_0, S_0) = e^{-r(T-t_0)} \sum_{k \geq 0} \frac{(\xi_p(T - t_0))^k e^{-\xi_p(T-t_0)}}{k!} \bar{V}(k), \tag{5.28}$$
>
> with $\bar{V}(k)$ in (5.27).

Equation (5.28) indicates that with Merton jumps in the dynamics of stock process $S(t)$, the value of the call option is given by an *infinite sum of Black-Scholes call option prices*, with adjusted parameters.

We will also need the tower property of expectations in the following derivation.

5.2.2 Characteristic function for Merton's model

Merton's jump diffusion model under measure $\mathbb{Q}$ consists of a Brownian motion and a compound Poisson process, which, with $t_0 = 0$, is defined by

$$X(t) = X(t_0) + \bar{\mu}t + \sigma W(t) + \sum_{k=1}^{X_{\mathcal{P}}(t)} J_k. \tag{5.29}$$

Here, $\bar{\mu} = r - \frac{1}{2}\sigma^2 - \xi_p \mathbb{E}\left[e^J - 1\right]$, $\sigma > 0$, Brownian motion $W(t)$, Poisson process $X_{\mathcal{P}}(t), t \geq 0$ with parameter ξ_p and $\mathbb{E}[X_{\mathcal{P}}(t)|\mathcal{F}(0)] = \xi_p t$. The equation above can also be expressed in differential form as,

$$dX(t) = \bar{\mu}dt + \sigma dW(t) + J(t)dX_{\mathcal{P}}(t), \text{ with } X(t_0) \in \mathbb{R}. \tag{5.30}$$

In the Poisson process setting, the arrival of a jump is independent of the arrival of previous jumps, and the probability of two simultaneous jumps is equal to zero. As mentioned, variable J_k, $k \geq 1$, is an i.i.d. sequence of random variables with a jump-size probability distribution F_J, so that $\mathbb{E}[J_k] = \mu_J < \infty$.

> With all the sources of randomness *mutually independent*, we can determine the *characteristic function* of $X(t)$, i.e.,
>
> $$\phi_X(u) := \mathbb{E}\left[e^{iuX(t)}\right] \tag{5.31}$$
>
> $$= e^{iuX(0)} e^{iu\bar{\mu}t} \mathbb{E}\left[e^{iu\sigma W(t)}\right] \cdot \mathbb{E}\left[\exp\left(iu \sum_{k=1}^{X_\mathcal{P}(t)} J_k\right)\right].$$

The two expectations in Equation (5.31) can easily be determined. As $W(t) \sim \mathcal{N}(0,t)$, it follows that $\mathbb{E}\left[e^{iu\sigma W(t)}\right] = e^{-\frac{1}{2}\sigma^2 u^2 t}$. For the second expectation in (5.31), we first consider the summation:

$$\mathbb{E}\left[\exp\left(iu \sum_{k=1}^{X_\mathcal{P}(t)} J_k\right)\right] = \sum_{n \geq 0} \mathbb{E}\left[\exp\left(iu \sum_{k=1}^{X_\mathcal{P}(t)} J_k\right) \bigg| X_\mathcal{P}(t) = n\right] \mathbb{P}[X_\mathcal{P}(t) = n],$$

which results from the tower property of expectations, i.e. $\mathbb{E}[\mathbb{E}[X|Y]] = \mathbb{E}[X]$. $X_\mathcal{P}(t)$ is a Poisson process and by using (5.2), we have

$$\mathbb{E}\left[\exp\left(iu \sum_{k=1}^{X_\mathcal{P}(t)} J_k\right)\right] = \sum_{n \geq 0} \mathbb{E}\left[\exp\left(iu \sum_{k=1}^{n} J_k\right)\right] \frac{e^{-\xi_p t}(\xi_p t)^n}{n!}$$

$$= \sum_{n \geq 0} \frac{e^{-\xi_p t}(\xi_p t)^n}{n!} \left(\int_\mathbb{R} e^{iux} f_J(x) dx\right)^n. \tag{5.32}$$

The two nth powers at the right-hand side of (5.32) can be recognized as a Taylor expansion of an exponential function, giving the resulting expression:

$$\mathbb{E}\left[\exp\left(iu \sum_{k=1}^{X_\mathcal{P}(t)} J_k\right)\right] = e^{-\xi_p t} \sum_{n \geq 0} \frac{1}{n!} \left(\xi_p t \int_\mathbb{R} e^{iux} f_J(x) dx\right)^n$$

$$= \exp\left(\xi_p t \int_\mathbb{R} \left(e^{iux} f_J(x) dx - 1\right)\right)$$

$$= \exp\left(\xi_p t \int_\mathbb{R} \left(e^{iux} - 1\right) f_J(x) dx\right)$$

$$= \exp\left(\xi_p t \mathbb{E}[e^{iuJ} - 1]\right),$$

where we used that $\int_\mathbb{R} f_J(x) dx = 1$, and $J = J_k$ is an i.i.d. sequence of random variables with CDF $F_J(x)$ and PDF $f_J(x)$. Equation (5.31) can thus be written as:

Jump Processes

$$\phi_X(u) = \mathbb{E}\left[e^{iuX(t)}\right] \qquad (5.33)$$

$$= \exp\left(iu(X(0) + \bar{\mu}t) - \frac{1}{2}\sigma^2 u^2 t\right) \exp\left(\xi_p t \left(\mathbb{E}[e^{iuJ}] - 1\right)\right)$$

$$= \exp\left(iu(X(0) + \bar{\mu}t) - \frac{1}{2}\sigma^2 u^2 t + \xi_p t \int_{\mathbb{R}} \left(e^{iux} - 1\right) f_J(x) \mathrm{d}x\right),$$

with $\bar{\mu} = r - \frac{1}{2}\sigma^2 - \xi_p \mathbb{E}\left[e^J - 1\right]$.
For the Merton model, we have,

$$\mathbb{E}[e^{iuJ} - 1] = e^{iu\mu_J - \frac{1}{2}u^2 \sigma_J^2} - 1, \ \mathbb{E}[e^J - 1] = e^{\mu_J + \frac{1}{2}\sigma_J^2} - 1.$$

Remark 5.2.1 (Characteristic function for the Kou model) *The generic characteristic function for the jump diffusion model is found in Equation (5.33). For Kou's jump diffusion model, we have,*

$$\mathbb{E}[e^J] = p_1 \frac{\alpha_1}{\alpha_1 - 1} + p_2 \frac{\alpha_2}{\alpha_2 + 1},$$

and

$$\mathbb{E}[e^{iuJ}] = \frac{p_1 \alpha_1}{\alpha_1 - iu} + \frac{p_2 \alpha_2}{iu + \alpha_2}.$$

Therefore the characteristic function for Kou's model is given by,

$$\phi_{X_\mathcal{P}}(u) = \exp\left(iu\bar{\mu}t - \frac{1}{2}\sigma^2 u^2 t + \xi_p t \left[\frac{p_1 \alpha_1}{\alpha_1 - iu} + \frac{p_2 \alpha_2}{iu + \alpha_2} - 1\right]\right), \qquad (5.34)$$

with

$$\bar{\mu} = r - \frac{1}{2}\sigma^2 - \xi_p \left[p_1 \frac{\alpha_1}{\alpha_1 - 1} + p_2 \frac{\alpha_2}{\alpha_2 + 1} - 1\right].$$

▲

Example 5.2.2 (Densities of jump diffusion process) In Figure 5.3 some paths for jump diffusion process $X(t)$ as in (5.11), and $S(t) = e^{X(t)}$ as in (5.10), are shown with the corresponding densities. Merton's jump diffusion model is employed. These densities have been calculated using the jump diffusion characteristic function and by the COS method, which will be presented in the next chapter, i.e. Chapter 6.

In Figure 5.4, for the jump diffusion dynamics in (5.10), we study the effect of parameter σ_J on the corresponding distribution functions. The curve in the figure shows that as σ_J increases, the distribution of stock process $S(t)$ gets fatter tails. The jump magnitude J is a symmetric random variable (Merton's jump diffusion), implying that the stock may decrease or increase drastically within a short time period. Note that for $\sigma_J = 0$ the model is a standard GBM.

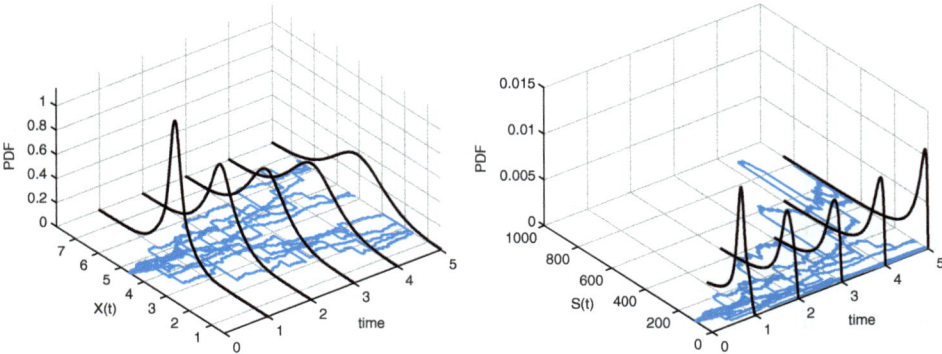

Figure 5.3: *Jump diffusion paths of $X(t)$ (left) and $S(t)$ (right), with $S(t_0) = 100$, $r = 0.05$, $\sigma = 0.2$, $\sigma_J = 0.5$, $\mu_J = 0$, $\xi_p = 1$ and $T = 5$.*

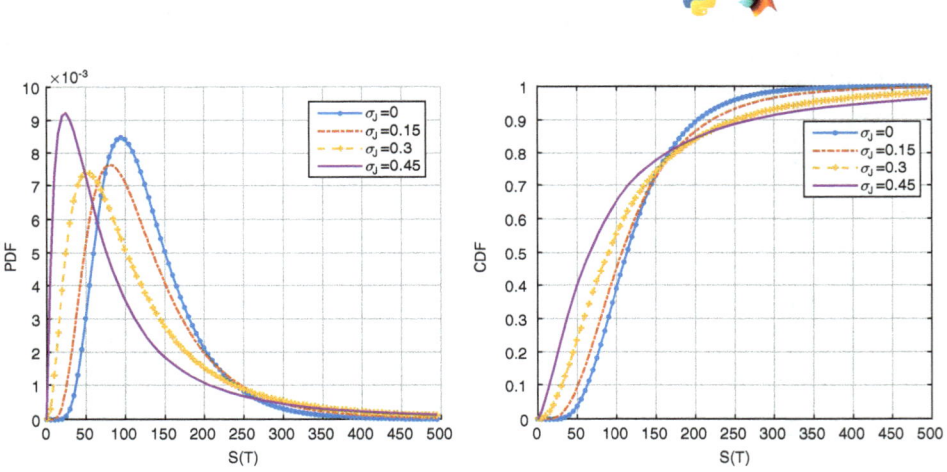

Figure 5.4: *PDF (left side figure) and CDF (right side) for $S(t)$ in (5.10) for varying σ_J, with $S(t_0) = 100$, $r = 0.05$, $\sigma = 0.2$, $\mu_J = 0$, $\xi_p = 1$ and $T = 5$.*

Example 5.2.3 (Implied volatilities for the jump diffusion model)

We illustrate the impact of the jump parameters in the Merton jump diffusion model on the implied volatilities. The jumps J are normally distributed, $J \sim \mathcal{N}(\mu_J, \sigma_J^2)$.

We analyse the influence of three parameters, ξ_p, σ_J and μ_J. In the experiment each parameter is varied individually, while the other parameters are fixed. Figure 5.5 shows the impact of the jump parameters in the jump diffusion process on the shape of the implied volatility.

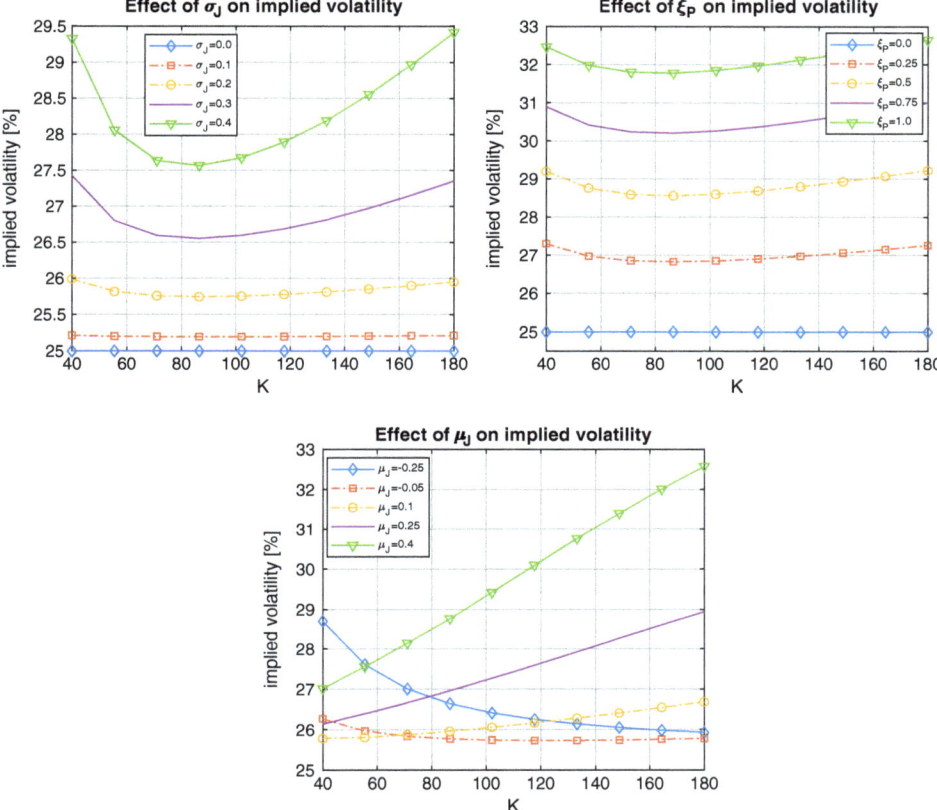

Figure 5.5: *Impact of different jump parameters on the shape of the implied volatility in Merton's jump diffusion model.*

Each of the jump parameters has a different effect on the shape of the implied volatility curve, i.e., σ_J has a significant impact on the curvature, ξ_p controls the overall level of the implied volatility, whereas μ_J influences the implied volatility slope (the skew).

5.2.3 Dynamic hedging of jumps with the Black-Scholes model

This experiment focuses on delta hedging, particularly when dealing with jumps in the asset dynamics. Recall the delta hedging experiment for the Black-Scholes dynamics in Section 3.3.

The stock process here is driven by geometric Brownian motion with a jump process, i.e. the jump diffusion process as defined in Equation (5.10).

The jumps are symmetric thus the probability of upward and downward jumps is equal. With the dynamics of the process specified, the option prices should be determined. Pricing of a call option can be done either by using Equation (5.28), or alternatively, one may use the COS method (to be presented in Chapter 6), which relies on the characteristic function of the jump diffusion process, see Equation (5.33). With the option prices, also the Greek $\Delta(t)$ can be computed, for example, by employing Equation (6.36) from the next chapter.

Some results when hedging a call option if the price process contains jumps are shown in Figure 5.6. It is clear from the figure that an occurring jump in the stock process has an immediate effect on the value of the option's delta Δ and on the corresponding P&L(t) portfolio as well. In Figure 5.7 the effect of different hedging frequencies on the variance of the P&L is presented. The figure at the left-hand side shows the P&L distribution after hedging 10 times and at the right-hand side we see the effect of 2000 hedging stages. Clearly, increasing the hedging frequency does not affect the distribution of the P&L. In both cases one may expect large losses due to the occurrence of the jump in the stock price process. In other words, delta-hedging alone is shown to be insufficient to reduce the (possibly large) losses.

In the literature, the hedging problem in the case of occurring jumps is sometimes reformulated as finding a strategy that minimizes *the expected variance* of a hedge portfolio minus the payoff of the option. In practice, one then hedges under jump diffusion by using *other options in the hedge portfolio*, to reduce the risk associated with a position in options.

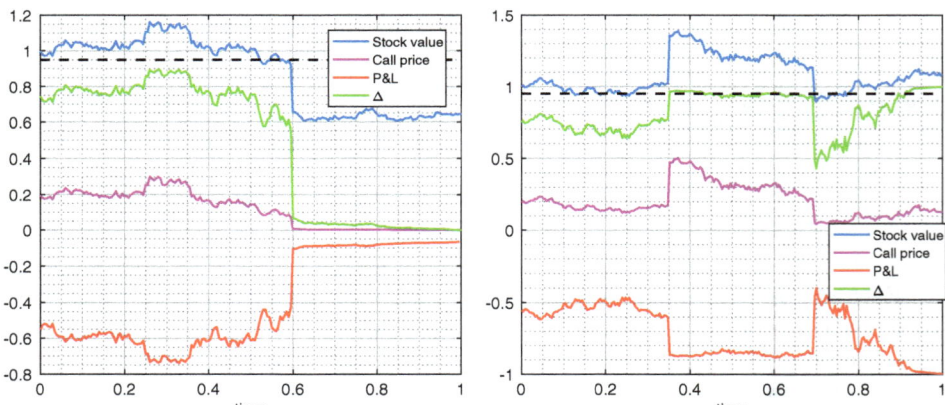

Figure 5.6: *Delta hedging a call option for a stock with jumps, where we however work with the Black-Scholes delta. Blue: the stock path, pink: value of the call option, red: P&L(t) portfolio, and green: the Δ. Left: a path with one jump time, right: two occurring jumps.*

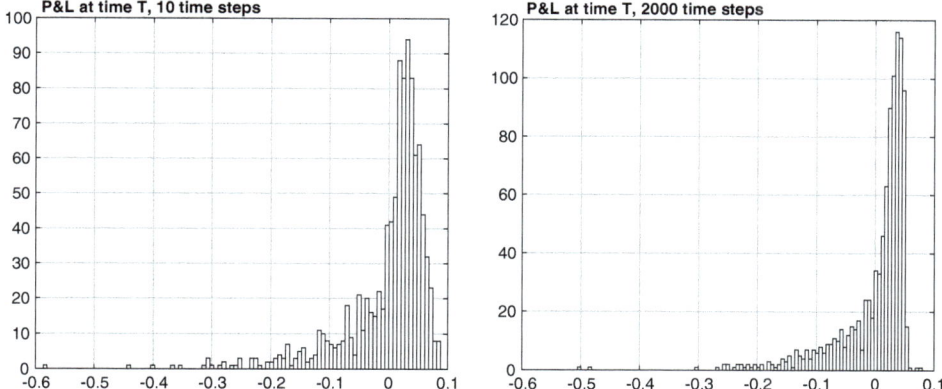

Figure 5.7: *The impact of the hedging frequency on the variance of the P&L(T) portfolio, with the stock following a jump diffusion process. Left: 10 times hedging during the option lifetime, Right: 2000 times hedging.*

Note that the minimization of the variance is done under a measure $\mathbb{Q}$, whereas the variance under the real-world measure is the target to minimize in practice. However, as long as the two measures have the same measure zero sets (which should be true), the real-word measure variance should also tend to zero, provided there are enough other options in the hedge portfolio.

The authors in [He *et al.*, 2006; Kennedy *et al.*, 2009] have shown however that a hedging portfolio which consists, next to $-\Delta$ stocks, *of a few options*, reduces the portfolio's variance under jump processes significantly.

5.3 Exponential Lévy processes

We can cast the processes discussed in the previous sections in the context of the more general class of *exponential Lévy processes*. Whereas we need to indicate the expected number of jumps in a time interval in jump diffusion processes, by means of parameter ξ_p, this is not required for certain other Lévy processes.

We consider again a stock, $S(t)$, and a bank account, $M(t)$, with deterministic interest rate r. Based on the definition of a *Lévy process*, $X_{\mathcal{L}}(t)$, we will generally model the asset process by means of an exponential Lévy process, i.e.

$$S(t) = S_0 e^{X_{\mathcal{L}}(t)}. \tag{5.35}$$

Definition 5.3.1 (Lévy process) *By a Lévy process $X_{\mathcal{L}}(t)$ for $t \geq t_0 = 0$, we mean any process starting at the origin, $X_{\mathcal{L}}(0) = 0$, so that on $(\Omega, \mathcal{F}(t), \mathbb{P})$, with $\mathcal{F}(t)$ the filtration generated by the Lévy process, the following conditions hold:*

- *Independent increments: For any $0 \leq s_1 < t_1 < s_2 < t_2$ the random variables $X_{\mathcal{L}}(t_1) - X_{\mathcal{L}}(s_1)$ and $X_{\mathcal{L}}(t_2) - X_{\mathcal{L}}(s_2)$ are independent,*

- *Stationary increments: For any $0 \leq s < t < \infty$, the law of $X_{\mathcal{L}}(t) - X_{\mathcal{L}}(s)$ only depends on the increments $t - s$,*

- *Stochastic continuity: For any $\epsilon > 0$ and $s, t > 0$ we have*

$$\lim_{s \to t} \mathbb{P}[|X_{\mathcal{L}}(t) - X_{\mathcal{L}}(s)| > \epsilon] = 0, \tag{5.36}$$

- *Sample paths are right-continuous with left limits, almost surely.* ◀

Each Lévy process can be characterized by a triplet $(\sigma_{\mathcal{L}}^2, F_{\mathcal{L}}, \mu_{\mathcal{L}})$. This set of three parameters is known as the *generating Lévy triplet*. The first parameter $\sigma_{\mathcal{L}}^2$ is the *Gaussian variance*, as it is associated with the Brownian part of a Lévy process; the second quantity, $F_{\mathcal{L}}$, is the *Lévy measure* and the third parameter, $\mu_{\mathcal{L}}$, corresponds to the *drift* parameter. If $F_{\mathcal{L}} \equiv 0$, $X_{\mathcal{L}}(t)$ represents a Brownian motion with drift. When $\sigma_{\mathcal{L}}^2 = 0$, $X_{\mathcal{L}}(t)$ is a purely non-Gaussian jump process. The Lévy measure is related to the expected number of jumps of a certain height in a time interval of length one [Papapantoleon, 2005].

The Lévy measure satisfies

$$F_{\mathcal{L}}(0) = 0, \quad \text{and} \quad \int_{\mathbb{R}} \min(1, x^2) F_{\mathcal{L}}(\mathrm{d}x) < \infty, \tag{5.37}$$

which means that the measure, $F_{\mathcal{L}}(\mathrm{d}x)$ does not have any *mass* at the origin, however, infinitely many (small) jumps may occur around zero.

Definition 5.3.2 *Let $X_{\mathcal{L}}$ be a Lévy process with triplet $(\sigma_{\mathcal{L}}^2, F_{\mathcal{L}}, \mu_{\mathcal{L}})$. Then,*

- *$X_{\mathcal{L}}$ is a finite activity process, if $F_{\mathcal{L}}(\mathbb{R}) < \infty$, as in that case almost all paths consist of a finite number of jumps.*

- *$X_{\mathcal{L}}$ is an infinite activity process, if $F_{\mathcal{L}}(\mathbb{R}) = \infty$, as then almost all paths consist of an infinite number of jumps. (see also [Sato, 2001]).* ◀

Remark 5.3.1 (Notation $F(\mathrm{d}x)$) *Basically, a measure is a quantity to indicate the size of a certain set. We are familiar with the Lebesgue measure on $\mathbb{R}^d$, which is a known representation for size. For $d = 1$ it is length, for $d = 2$ area, volume for $d = 3$, etc.*

The notation $F_{\mathcal{L}}(\mathrm{d}x)$ for a measure needs some explanation. Here, $\mathrm{d}x$ stands for an infinitesimal small interval $\mathrm{d}x := [x, x + \mathrm{d}x)$.

In the case of a Lebesgue measure, we may simply write,

$$F(\mathrm{d}x) = f(x)\mathrm{d}x, \text{ or } \int g(x)F(\mathrm{d}x) = \int g(x)f(x)\mathrm{d}x.$$

However, in the case we deal with a countable set, N, the measure will be a countable measure, in this case, we write

$$g(x)F_{\mathcal{L}}(\mathrm{d}x) := g(x)\delta_N(\mathrm{d}x).$$

Because $\delta(x = p) = 1$ for $p \in [x, x+\mathrm{d}x)$, and $\delta(x = p) = 0$ otherwise, we have

$$\int g(x)\delta_p(\mathrm{d}x) = g(p).$$

For a countable set N, we define the counting measure on N, for any function $g(x)$, as,

$$\int g(x)\delta_N(\mathrm{d}x) = \sum_{k=1}^{N} g(k).$$

The notation for the measure thus represents a simple summation in this case. Notice that we encountered this case already in Equation (5.5).

The notation $F_{\mathcal{L}}(\mathrm{d}x)$ is thus meant to unify the notation for both Lebesgue and countable measures. ▲

Brownian motion, the Poisson, (compound) Poisson processes and the jump diffusion process all belong to the *finite activity* Lévy processes, where paths typically consist of a continuous Brownian motion component and often a jump component. The jumps occur a finite number of times on each finite interval.

Lévy processes with paths that exhibit jumps *infinitely many times on a finite interval* are the *infinite activity* Lévy jump processes, such the Variance Gamma (VG) model [Madan et al., 1998], the CGMY model [Carr et al., 2002], and the Normal Inverse Gaussian model [Barndorff-Nielsen, 1997].

The family of Lévy processes is further characterized by the following fundamental result [Sato, 2001]:

Theorem 5.3.1 (Lévy-Khintchine representation) For all $u \in \mathbb{R}$ and $t \geq 0$,

$$\phi_{X_{\mathcal{L}}}(u) = \mathbb{E}\left[e^{iuX_{\mathcal{L}}(t)}\big|\mathcal{F}(0)\right] = e^{t\Psi(u)}, \quad (5.38)$$

with $t_0 = 0$, and,

$$\Psi(u) = -\frac{\sigma_{\mathcal{L}}^2}{2}u^2 + i\mu_{\mathcal{L}}u + \int_{\mathbb{R}}(e^{iux} - 1 - iux\mathbb{1}_{|x|\leq 1}(x))F_{\mathcal{L}}(\mathrm{d}x),$$

where $\sigma_{\mathcal{L}}$ is a nonnegative real number, $\mu_{\mathcal{L}}$ is real and $F_{\mathcal{L}}$ is the measure on $\mathbb{R}$ satisfying $F_{\mathcal{L}}(0) = 0$ and $\int_{\mathbb{R}} \min(1, x^2)\mathrm{d}F_{\mathcal{L}}(x) < \infty$.

In other words, for Lévy processes, we can derive the characteristic function.

Moreover, when the Lévy measure satisfies the following additional condition,
$$\int_{|x|\geq 1} |x| F_{\mathcal{L}}(\mathrm{d}x) < \infty,$$
there is no need to truncate the large jumps, like in the statement above, and we have
$$\Psi(u) = -\frac{\sigma_{\mathcal{L}}^2}{2} u^2 + i\mu_{\mathcal{L}} u + \int_{\mathbb{R}} (\mathrm{e}^{iux} - 1 - iux) F_{\mathcal{L}}(\mathrm{d}x).$$

Another distinction which is made in the case of Lévy processes is based on the notion of finite or infinite variation:

Definition 5.3.3 $X_{\mathcal{L}}$ with triplet $(\sigma_{\mathcal{L}}^2, F_{\mathcal{L}}, \mu_{\mathcal{L}})$ is a *finite variation process [Sato, 2001]*, if $\sigma_{\mathcal{L}}^2 = 0$ and $\int_{|x|<1} |x| F_{\mathcal{L}}(\mathrm{d}x) < \infty$. It is called an *infinite variation process*, if $\sigma_{\mathcal{L}}^2 \neq 0$ or $\int_{|x|\leq 1} |x| F_{\mathcal{L}}(\mathrm{d}x) = \infty$. ◂

A Lévy process is connected to infinitely divisible distributions. If we consider a stochastic variable $X_{\mathcal{L}}$ with a cumulative distribution function $F_{\mathcal{L}}$, we call $X_{\mathcal{L}}$ infinitely divisible if and only if for every $i = 1, 2, \ldots$ $X_{\mathcal{L}}$ can be decomposed as

$$X_{\mathcal{L}} = X_1 + X_2 + \cdots$$

for independent and identically distributed random variables X_i. These increments of a Lévy process are also infinitely divisible, as an increment over some time interval can again be decomposed into a sum of smaller-sized increments. From the independence and stationarity properties of the Lévy process, it follows that these smaller-sized increments are also independent and identically distributed, and so on, see Sato [2001].

Further basic information on Lévy processes can be found in [Bertoin, 1996]. The reader is referred to [Cont and Tankov, 2004; Schoutens, 2003] for additional information regarding Lévy processes in a financial context.

5.3.1 Finite activity exponential Lévy processes

If $F_{\mathcal{L}}$ is a *finite measure*, i.e. if

$$\int_{\mathbb{R}} F_{\mathcal{L}}(\mathrm{d}x) < \infty, \tag{5.39}$$

we can write

$$F_{\mathcal{L}}(\mathrm{d}x) =: \xi_p f_{X_{\mathcal{L}}}(x) \mathrm{d}x,$$

where $f_{X_\mathcal{L}}(x)$ is the probability density function and ξ_p the expected number of jumps (see Remark 5.3.1). The corresponding processes are called *finite activity processes*. Examples of finite activity Lévy processes are the Brownian motion discussed earlier, the Poisson process and also its extension the compounded Poisson process.

Example 5.3.1 (Brownian motion with drift) Drifted Brownian motion (the classical Samuelson model [Samuelson, 1965]) is also based on a Lévy process:

$$X(t) = \left(r - \frac{\sigma^2}{2}\right)t + \sigma W^\mathbb{Q}(t),$$

It is well-known [Karatzas and Shreve, 1998] that in this case there is only *one martingale measure*, and the process $\exp(-\frac{\sigma^2}{2}t + \sigma W^\mathbb{Q}(t))$, for $t \geq 0$, is a martingale. The Lévy-Khintchine exponent in (5.38) then reads

$$\psi(u) = -\frac{\sigma^2}{2}u^2 + \left(r - \frac{\sigma^2}{2}\right)iu,$$

so that the Lévy triplet equals $(\sigma^2, 0, r - \frac{\sigma^2}{2})$.

Notice that in the case of the jump diffusion process, the Lévy triplet is given by $(\sigma^2, \xi_p f_J, \mu)$. ♦

5.3.2 PIDE and the Lévy triplet

The underlying stochastic processes are defined under a risk-neutral measure $\mathbb{Q}$ when pricing options. In the case of finite activity jump processes for which $F_\mathcal{L}$ satisfies $\int_\mathbb{R} |x| dF_\mathcal{L}(x) < \infty$, Equation (5.38) may be written as

$$\mathbb{E}\left[e^{iuX_\mathcal{L}(t)}\right] = \exp\left[t\left(-\frac{\sigma_\mathcal{L}^2}{2}u^2 + i\mu_\mathcal{L} u + \int_\mathbb{R}(e^{iux} - 1 - iux)F_\mathcal{L}(dx)\right)\right], \quad (5.40)$$

where $(\sigma_\mathcal{L}^2, F_\mathcal{L}, \mu_\mathcal{L})$ is the *reduced Lévy triplet*, and the function in (5.40)

$$\psi_{X_\mathcal{L}}(u) := -\frac{\sigma_\mathcal{L}^2}{2}u^2 + i\mu_\mathcal{L} u + \int_\mathbb{R}(e^{iux} - 1 - iux)F_\mathcal{L}(dx), \quad (5.41)$$

is the *reduced Lévy-Khintchine exponent*. This is a convenient reformulation because it gives a direct link between the coefficients in the option pricing PIDE and the reduced triplet.

Let $V(t, S)$ be the value of a plain vanilla option on asset $S(t)$. If we assume that the discounted process, $e^{-rT}V(T, S)$, is a martingale, then we should have that $e^{-rt}V(t, S) = \mathbb{E}^\mathbb{Q}[e^{-rT}V(T, S)|\mathcal{F}(t)]$, which yields the well-known representation for the option,

$$V(t, S) = e^{-r(T-t)}\mathbb{E}^\mathbb{Q}[V(T, S(T))|\mathcal{F}(t)]. \quad (5.42)$$

To find a full characterization of the option pricing equation, we change variables, i.e. $X_\mathcal{L}(t) := \log S(t)$, and consider option value $V(t, X)$, so that

$$V(t, X) = e^{-r(T-t)} \mathbb{E}^\mathbb{Q}[V(T, X_\mathcal{L}(T))]. \tag{5.43}$$

Then, we rely on the following theorem (see e.g. [Raible, 2000]):

Theorem 5.3.2 Let $V(t, X) \in C^{1,2}([0, T) \times \mathbb{R}) \cap C^0([0, T] \times \mathbb{R})$ and assume (5.39). Then $V(t, X)$ satisfies the following PIDE:

$$\frac{\partial V}{\partial t} + \frac{1}{2}\sigma^2 \frac{\partial^2 V}{\partial X^2} + \left(r - \frac{1}{2}\sigma^2\right) \frac{\partial V}{\partial X} - rV$$
$$+ \int_\mathbb{R} \left(V(t, X + y) - V(t, X) - (e^y - 1)\frac{\partial V(t, X)}{\partial X}\right) F_\mathcal{L}(\mathrm{d}y) = 0,$$

$\forall (t, X) \in [0, T) \times \mathbb{R}$ and final condition $V(T, X)$. The reduced Lévy triplet is given by $(\sigma^2, F_\mathcal{L}, r - \frac{1}{2}\sigma^2)$, under risk-neutral measure $\mathbb{Q}$.

Incomplete markets and equivalent martingale measure

In the Black-Scholes world, we have a unique martingale measure, which transforms a geometric Brownian motion under measure $\mathbb{P}$ into another under measure $\mathbb{Q}$, with a different drift parameter (r instead of μ). The uniqueness of the risk-neutral measure $\mathbb{Q}$ guarantees a unique arbitrage-free pricing rule, and thus unique option prices. We call this market *a complete market*. The risk in the Black-Scholes model is in the Brownian motion driving the lognormal stock dynamics. As shown in Example 3.3, a dynamic, continuous hedge strategy, based on delta hedging with stocks and cash in infinitesimal time steps, may eliminate this risk completely, and also hedging with somewhat bigger time steps works reasonably well.

Lévy markets are generally *incomplete*, which means that not every financial option contract can be replicated by means of stocks and cash, or in other words, a perfect hedging strategy is usually not available. An example of this has been presented in Section 5.2.3. In contrast to the classical Black-Scholes model, option prices under general exponential Lévy processes usually cannot be obtained by replication arguments, based on delta hedging. However, an option price can still be determined, based on the no-arbitrage assumption.

In so-called *incomplete markets*, such as the market which consists of assets modeled by exponential Lévy jump processes, the hedging strategy should thus be different from the Black-Scholes set up. The occurrence of sudden jumps in the stock prices prohibits a perfect replication of the option value with stocks and cash. If a jump occurs, by the convexity of the option, a delta hedging writer will most certainly *lose money*, independent of the direction of the movement (Merton, 1976). The risk generated by the Brownian motion may still be hedged away by means of dynamically delta hedging.

Equation (5.28) confirms that the price of an option when jumps are present is given by an infinite sum of European options. The jump process is governed by a *possibly infinite number of different jump magnitudes*, and the so-called *jump risk* can only be completely hedged away by an *infinite number of hedge instruments*, see, e.g. [Naik and Lee, 1990]. However, this may often be well approximated by a finite number of jump magnitudes [He et al., 2006; Kennedy et al., 2009].

There is no perfect hedge in the case that jumps occur with very many different possible magnitudes, not even with continuous rebalancing, see the example in Section 5.2.3.

5.3.3 Equivalent martingale measure

In the Lévy framework the risk-neutral measure is therefore *not* unique, i.e. there exist infinitely many of them. The reason for this is the occurrence of the jumps in the Lévy framework. Due to this, we cannot perfectly replicate the option price, as we face the risk of a jump occurring in the stock price, which implies that there are multiple ways to price options in an arbitrage-free way.

A common approach to assigning values to options under Lévy processes is based on a change to a *convenient probability measure* and taking the expectation of the discounted prices.

This relates to switching measures from the real-world measure $\mathbb{P}$, to a risk-neutral measure $\mathbb{Q}$. Probability measure $\mathbb{Q}$ is known in the Lévy literature as the *Equivalent Martingale Measure* (EMM). It has the same null set as the market probability, and the discounted process, $e^{-rt}S(t)$, for $t \geq 0$, should be a martingale. The existence of such a measure is in some sense equivalent to the no-arbitrage assumption, see [Delbaen and Schachermayer, 1994], and for earlier related work, see [Harrison and Kreps, 1979; Harrison and Pliska, 1981].

Hence, one of the main issues in the context of Lévy asset price processes is to find a suitable EMM. Several techniques to find an EMM have been proposed in the literature. We outline a method that has been established in actuarial sciences, based on the Esscher transform [Gerber and Shiu, 1995; Raible, 2000].

The *EMM-condition* is given in this context by:

$$e^{-rt}\mathbb{E}^{\mathbb{Q}}[S(t)|\mathcal{F}(0)] = S_0. \tag{5.44}$$

We depart from Equation (5.35), and take the conditional expectation, setting $t_0 = 0$, i.e.,

$$\mathbb{E}^{\mathbb{Q}}[S(t)|\mathcal{F}(0)] = S_0 \mathbb{E}^{\mathbb{Q}}\left[e^{X_{\mathcal{L}}(t)}|\mathcal{F}(0)\right]. \tag{5.45}$$

Connecting this result to (5.44), gives us the following condition which needs to be satisfied:

$$\mathbb{E}^{\mathbb{Q}}\left[e^{X_{\mathcal{L}}(t)}|\mathcal{F}(0)\right] = e^{rt}. \tag{5.46}$$

By the Lévy-Khintchine representation, the characteristic function of $X_\mathcal{L}(t)$ in (5.35) is given by [Cont and Tankov, 2004], using $t_0 = 0$,

$$\phi_{X_\mathcal{L}}(u) = \mathbb{E}^\mathbb{Q}\left[e^{iuX_\mathcal{L}(t)}|\mathcal{F}(0)\right]$$

$$= \exp\left[t\left(-\frac{\sigma_\mathcal{L}^2}{2}u^2 + i\mu_\mathcal{L} u + \int_\mathbb{R}(e^{iux} - 1 - iux\mathbb{1}_{|x|\leq 1})F_\mathcal{L}(dx)\right)\right].$$

By substitution of $u = -i$, we arrive at:

$$\phi_{X_\mathcal{L}}(-i) = \mathbb{E}^\mathbb{Q}\left[e^{X_\mathcal{L}(t)}|\mathcal{F}(t_0)\right]$$

$$= \exp\left[t\left(\frac{\sigma_\mathcal{L}^2}{2} + \mu_\mathcal{L} + \int_\mathbb{R}(e^x - 1 - x\mathbb{1}_{|x|\leq 1})F_\mathcal{L}(dx)\right)\right].$$

It is common to assume that (5.35) is already formulated under the risk-neutral measure. To ensure that the relative price, $S(t)/M(t)$, is a martingale under the risk-neutral measure, we thus need to ensure that (see Equation (5.46)):

$$\phi_{X_\mathcal{L}}(-i) = e^{rt}, \qquad (5.47)$$

which is satisfied if we choose the drift $\mu_\mathcal{L} = r$ and

$$\frac{\sigma_\mathcal{L}^2}{2} + \int_\mathbb{R}(e^x - 1 - x\mathbb{1}_{|x|\leq 1})F_\mathcal{L}(dx) = 0. \qquad (5.48)$$

5.4 Infinite activity exponential Lévy processes

We here discuss some of the *infinite activity* Lévy jump processes. Infinite activity means that the asset paths may jump infinitely many times, for each finite interval. Moreover, jumps that are larger than a given magnitude occur only a finite number of times. Unlike the classical Samuelson model, or any jump diffusion model, infinite activity models may not have a continuous component.

From the discussion of the finite activity case, we benefit here regarding the relation between the PIDE and the Lévy triplet. Therefore we will shorten the discussion.

5.4.1 Variance Gamma process

The Variance Gamma (VG) process was introduced in financial modeling by Madan and Seneta [1990] as an alternative model to cope with the shortcomings of the Black-Scholes model. The VG process, which is a pure jump process, was derived by evaluating a Brownian motion with drift *at random times* that are driven by a Gamma process. Before we give details of the VG process, first the Gamma process is briefly explained.

Definition 5.4.1 (Gamma process) *A Gamma process $\Gamma(t; a, b)$ is a stochastic process whose increments, $\Gamma(t+\Delta t, a, b) - \Gamma(t, a, b)$, are independent Gamma distributed random variables with mean $a\Delta t$ and variance $b\Delta t$, on each interval of length Δt. More precisely,*

$$\Gamma(t + \Delta t, a, b) - \Gamma(t, a, b) \sim f_{\Gamma(a^2 \Delta t / b, b/a)}(x).$$

The corresponding density function, $f_{\Gamma(c,d)}(x)$, with scale parameter c and shape parameter d, is given by

$$f_{\Gamma(c,d)}(x) = \frac{1}{d^c \Gamma(c)} x^{c-1} e^{-x/d}, \ x > 0; \tag{5.49}$$

with $\Gamma(c)$ the Gamma function, which is defined by

$$\Gamma(c) = \int_0^\infty t^{c-1} e^{-t} dt. \tag{5.50}$$

Moreover, the corresponding characteristic function is given by,

$$\phi_{\Gamma(c,d)}(u) = (1 - idu)^{-c},$$

with the scale and shape parameter. ◀

Definition 5.4.2 *The VG process is defined by substituting a Gamma process $\Gamma(t, 1, \beta)$ for the time variable t in a Brownian motion with drift, with drift parameter $\theta \in \mathbb{R}$, volatility $\sigma_{VG} > 0$ and Wiener process $W(t)$, $t \geq 0$. This gives rise to a so-called subordinated process:*

$$\bar{X}_{VG}(t; \sigma_{VG}, \beta, \theta) := \theta \Gamma(t; 1, \beta) + \sigma_{VG} W(\Gamma(t; 1, \beta)), \tag{5.51}$$

where independence between the Gamma and Wiener processes is assumed.

Process $\Gamma(t; 1, \beta)$ corresponds to a unit mean rate time change, with its probability density function and characteristic function given by:

$$f_{\Gamma(t/\beta, \beta)}(x) = \frac{1}{\beta^{\frac{t}{\beta}} \Gamma(\frac{t}{\beta})} x^{\frac{t}{\beta} - 1} \exp\left(-\frac{x}{\beta}\right), \quad \phi_{\Gamma(t/\beta, \beta)}(u) = (1 - i\beta u)^{-\frac{t}{\beta}}. \tag{5.52}$$
◀

The three parameters defining the VG process $\bar{X}_{VG}$ are the volatility σ_{VG}, the variance β of the Gamma distributed time and the drift θ. Parameter θ measures the degree of *skewness* of the distribution, whereas β controls the *excess of kurtosis* with respect to the normal distribution [Carr et al., 1998]. In the symmetric case, $\theta = 0$, and for $t = 1$, the value of kurtosis is given by $3(1 + \beta)$, see [Madan and Seneta, 1990]. Since the kurtosis of a normal distribution is equal to 3, this indicates that β thus measures the *degree of "peakedness"* with respect to the normal distribution. A large value of β gives rise to fat tails of the probability density function, that are indeed observed in the empirical log-returns. Heuristically, for $\beta \to 0$, the time change is close to a linear time change, so that the VG process approximates a drifted Brownian motion.

VG asset price process

Definition 5.4.3 (Exponential VG asset dynamics) *In a market consisting of a bank account $M(t)$ with risk-free interest rate r, an asset $S(t)$ is governed by the exponential VG dynamics, when*

$$S(t) = S(t_0)e^{X_{\mathrm{VG}}(t)}, \tag{5.53}$$

where
$$X_{\mathrm{VG}}(t) = \mu_{\mathrm{VG}} t + \bar{X}_{\mathrm{VG}}(t; \sigma_{\mathrm{VG}}, \beta, \theta),$$

where μ_{VG} is the drift of the logarithmic asset price. The process $X_{\mathrm{VG}}, t \geq 0$ is a Lévy process, i.e. a process with stationary, independent increments. ◀

VG characteristic function

By the definition of the *characteristic function* and by using Equation (5.53), we have

$$\phi_{\log S(t)}(u) \stackrel{\text{def}}{=} \mathbb{E}\left[e^{iu \log S(t)} | \mathcal{F}(0)\right] = e^{iu \log S(0)} \mathbb{E}\left[e^{iu X_{\mathrm{VG}}(t)} | \mathcal{F}(0)\right].$$

Using the notation,

$$\varphi_{\mathrm{VG}}(u, t) := \mathbb{E}\left[e^{iu X_{\mathrm{VG}}(t)} | \mathcal{F}(0)\right], \tag{5.54}$$

we will derive the above expectation.

As $X_{\mathrm{VG}}(t)$ depends on the process $\bar{X}_{\mathrm{VG}}(t; \sigma_{\mathrm{VG}}, \beta, \theta)$, the following relation holds:

$$\mathbb{E}\left[e^{iu X_{\mathrm{VG}}(t)}\right] = e^{iu \mu_{\mathrm{VG}} t} \mathbb{E}\left[e^{\bar{X}_{\mathrm{VG}}(t; \sigma_{\mathrm{VG}}, \beta, \theta)}\right]. \tag{5.55}$$

By the tower property of expectations, we can write[2]:

$$\mathbb{E}\left[e^{iu X_{\mathrm{VG}}(t)}\right] = \mathbb{E}\left[\mathbb{E}\left[e^{iu X_{\mathrm{VG}}(t)} | \Gamma(t; 1, \beta)\right]\right]. \tag{5.56}$$

The Brownian motion is independent of Gamma process $\Gamma(t; 1, \beta)$, so that we find:

$$\mathbb{E}\left[e^{iu X_{\mathrm{VG}}(t)}\right] = \mathbb{E}\left[\mathbb{E}\left[e^{iu(\mu_{\mathrm{VG}} t + \theta \Gamma(t; 1, \beta) + \sigma_{\mathrm{VG}} W(\Gamma(t; 1, \beta)))} | \Gamma(t; 1, \beta)\right]\right]$$
$$= \mathbb{E}\left[e^{iu(\mu_{\mathrm{VG}} t + \theta \Gamma(t; 1, \beta))} \mathbb{E}\left[e^{iu \sigma_{\mathrm{VG}} W(\Gamma(t; 1, \beta))} | \Gamma(t; 1, \beta)\right]\right].$$

---

[2]For the sake of readability, the conditioning on filtration $\mathcal{F}(0)$ is not included in the notation.

Conditioning on $\Gamma(t;1,\beta)$ means that the expectation above is of the form, $\mathbb{E}[XY|X] = X\mathbb{E}[Y|X]$. The conditioning on $\Gamma(t;1,\beta)$ thus makes the quantity under the expectation operator *"locally constant"*, i.e.,

$$\mathbb{E}\left[e^{iu\sigma_{\text{VG}}W(\Gamma(t;1,\beta))}|\Gamma(t;1,\beta)\right] = \mathbb{E}\left[e^{iu\sigma_{\text{VG}}W(z)}|\Gamma(t;1,\beta) = z\right],$$

with z in the expectation at the right-hand side considered constant. The inner conditional expectation is, in essence, an expectation of a normally distributed random variable, so that we can write

$$\mathbb{E}\left[e^{iu\sigma_{\text{VG}}W(\Gamma(t;1,\beta))}|\Gamma(t;1,\beta)\right] = \mathbb{E}\left[e^{W((iu)^2\sigma_{\text{VG}}^2\Gamma(t;1,\beta))}|\Gamma(t;1,\beta)\right]$$

$$= e^{-\frac{1}{2}u^2\sigma_{\text{VG}}^2\Gamma(t;1,\beta)},$$

where we have used the calculation of the mean and variance of a lognormal random variable, as in (2.16). Thus, we find,

$$\mathbb{E}\left[e^{iuX_{\text{VG}}(t)}\right] = \mathbb{E}\left[e^{iu(\mu_{\text{VG}}t+\theta\Gamma(t;1,\beta))}e^{-\frac{1}{2}u^2\sigma_{\text{VG}}^2\Gamma(t;1,\beta)}\right]$$

$$= e^{iu\mu_{\text{VG}}t}\mathbb{E}\left[e^{i\left(u\theta+i\frac{1}{2}u^2\sigma_{\text{VG}}^2\right)\Gamma(t;1,\beta)}\right]. \quad (5.57)$$

With the characteristic function for the Gamma process in (5.52), the expression simplifies further:

$$\varphi_{\text{VG}}(u,t) = \mathbb{E}\left[e^{iuX_{\text{VG}}(t)}\right]$$

$$= e^{iu\mu_{\text{VG}}t}\phi_{\Gamma(\frac{t}{\beta},\beta)}\left(u\theta + i\frac{1}{2}u^2\sigma_{\text{VG}}^2\right)$$

$$\stackrel{\text{def}}{=} e^{iu\mu_{\text{VG}}t}\left(1 - i\beta\theta u + \frac{1}{2}\beta\sigma_{\text{VG}}^2 u^2\right)^{-\frac{t}{\beta}}. \quad (5.58)$$

Variance Gamma drift adjustment

As explained in Section 5.3.3, contrary to the classical framework of complete markets a rational option price cannot be obtained by replication of the option. We must therefore rely on the no-arbitrage assumption, and assume the existence of a risk-neutral probability measure, commonly known as the *Equivalent Martingale Measure (EMM)* $\mathbb{Q}$, so that the discounted process, $e^{-rt}S(t)$ for $t \geq t_0 = 0$, becomes a martingale. We assume that the parameters $\sigma_{\text{VG}}, \beta$ and θ are set to prescribe these risk-neutral asset dynamics, see, for example, [Cont and Tankov, 2004].

In a risk-neutral world, it is possible to determine the *adjusted drift term*, μ_{VG}, by substituting $u = -i$ in (5.58):

$$\varphi_{\text{VG}}(-i,t) = \mathbb{E}\left[e^{X_{\text{VG}}(t)}\right] = e^{\mu_{\text{VG}}t}\left(1 - \beta\theta - \frac{1}{2}\beta\sigma_{\text{VG}}^2\right)^{-\frac{t}{\beta}},$$

and comparing the result with the risk-neutrality condition, as explained in Equation (5.47),

$$\mathbb{E}^{\mathbb{Q}}\left[S(t)|\mathcal{F}(0)\right] = S_0 \mathbb{E}^{\mathbb{Q}}\left[e^{X_{\text{VG}}(t)}|\mathcal{F}(0)\right] = S_0 e^{rt}.$$

One arrives at

$$\mu_{\text{VG}} t - \frac{t}{\beta} \log\left(1 - \beta\theta - \frac{1}{2}\beta\sigma_{\text{VG}}^2\right) = rt,$$

which can be written as $\mu_{\text{VG}} = r + \bar{\omega}$, where the *drift correction term* reads

$$\boxed{\bar{\omega} = \frac{1}{\beta} \log\left(1 - \beta\theta - \frac{1}{2}\beta\sigma_{\text{VG}}^2\right),} \tag{5.59}$$

with the obvious condition $\theta + \sigma_{\text{VG}}^2/2 < 1/\beta$.

Remark 5.4.1 *Another way of expressing this "convexity coefficient" $\bar{\omega}$, which is commonly used in the literature, for example [Carr et al., 2002], is to use the fact that*

$$\varphi_{VG}(u,t) = e^{iu\mu_{VG}t} \phi_{\bar{X}_{VG}}(u),$$

see Equation (5.55), with $\bar{X}_{\text{VG}}$ in (5.51), which gives us the following expression for $\bar{\omega}$:

$$\bar{\omega} = -\frac{1}{t} \log\left(\phi_{\bar{X}_{VG}}(-i)\right).$$
▲

An important property of the VG process is that it can be decomposed as the difference of two Gamma processes, where the first one represents the gains and the second one the losses. This property may be readily derived by factoring the quadratic expression in u in Equation (5.58). Each factor can be identified as the characteristic function of a scaled Gamma process. From this factorization, the following Lévy density function $f_{\text{VG}}(y)$ follows:

$$f_{\text{VG}}(y) = \begin{cases} \dfrac{1}{\beta}\dfrac{\exp(-M|y|)}{|y|} & \text{if } y > 0, \\ \dfrac{1}{\beta}\dfrac{\exp(-G|y|)}{|y|} & \text{if } y < 0, \end{cases} \tag{5.60}$$

where the positive parameters M and G are given by

$$M = \left(\sqrt{\frac{\theta^2\beta^2}{4} + \frac{\beta\sigma_{\text{VG}}^2}{2}} + \frac{\theta\beta}{2}\right)^{-1}, \quad G = \left(\sqrt{\frac{\theta^2\beta^2}{4} + \frac{\beta\sigma_{\text{VG}}^2}{2}} - \frac{\theta\beta}{2}\right)^{-1}. \tag{5.61}$$

Here, the positive exponent M should be larger than 1 for the drift correction term $\bar{\omega}$ to be well-defined.

By Itô's lemma, the price of a European put option under the VG model, for $X(t) = \log(S(t))$, now satisfies the following PIDE [Cont and Tankov, 2004]:

$$\frac{\partial V}{\partial \tau} - (r + \bar{\omega})\frac{\partial V}{\partial X} + rV = \int_{\mathbb{R}} (V(t, X+y) - V(t, X)) f_{\text{VG}}(y) \mathrm{d}y,$$
$$H(\tau = 0, X) = \max(K - e^{X(T)}, 0), \qquad (5.62)$$

where τ denotes time to maturity, and $f_{\text{VG}}(y)$ is again the VG density function connected to the Lévy measure associated with the VG process.

5.4.2 CGMY process

Here, we will discuss the *CGMY model*, which is a generalization of the VG model. The underlying Lévy process is characterized in this case by the Lévy triplet $(0, F_{\text{CGMY}}, \mu_{\text{CGMY}})$, where the CGMY Lévy density, $f_{\text{CGMY}}(y)$ is specified as:

$$f_{\text{CGMY}}(y) = C\left[\frac{e^{-M|y|}}{|y|^{1+Y}}\mathbb{1}_{y>0} + \frac{e^{-G|y|}}{|y|^{1+Y}}\mathbb{1}_{y<0}\right]. \qquad (5.63)$$

The set of parameters (C, G, M, Y) has a direct interpretation for specific properties of the jump process. Constant C controls the *intensity* of the jumps, G and M model the *decay rate* of large positive and negative jumps, respectively, i.e. the larger the value of C, the greater the activity of the jumps. From the exponential decay of the Lévy density (5.63), we can deduce that for large values of G, large positive jumps become *less probable*, thereby increasing the occurrence of small positive jumps.

The parameters need to satisfy $C \geq 0$, $G \geq 0$, $M \geq 0$, and $Y < 2$. The condition $Y < 2$ is induced by the requirement that Lévy densities should integrate x^2 in the neighborhood of 0. With parameter $Y < 0$, we deal with a *finite activity process*; for $Y \in [0, 1]$, we have an *infinite activity process of finite variation*, because $\int_{|x|<1} x f_{\text{CGMY}}(x) \mathrm{d}x < \infty$, whereas for $Y \in (1, 2)$, the process is of *infinite activity and infinite variation*.

Sometimes, the CGMY model given above is augmented by a *diffusion* term. So, an additional stochastic term, i.e. independent Brownian motion, is added to the definition of the model. This extended version of the pure jump CGMY framework is then called the CGMYB (CGMY-Brownian motion) model. The corresponding Lévy triplet is given by $(\sigma^2_{\text{CGMYB}}, F_{\text{CGMYB}}, \mu_{\text{CGMYB}})$. This CGMYB model has five open parameters to be calibrated to option market quotes, whereas the CGMY model has four. The CGMY model encompasses several models.

- When $\sigma_{\text{CGMYB}} = 0$ and $Y = 0$, we obtain the VG model.

- When $C = 0$, the model reduces to the GBM model.

Characteristic function of CGMY process

As mentioned, when $\int_{\mathbb{R}} |x| \mathbb{1}_{|x| \leq 1}(x) f_{\mathcal{L}}(x) < \infty$, the finite variation representation for the Lévy-Khintchine theorem reads,

$$\psi_{X_{\mathcal{L}}}(u) = -\frac{\sigma_{\mathcal{L}}^2}{2} u^2 + iu\mu_{\mathcal{L}} + \int_{\mathbb{R}} (e^{iux} - 1) f_{\mathcal{L}}(x) dx.$$

We will provide the derivation of the characteristic function for the CGMY model in this case.

$$\psi_{X_{\text{CGMYB}}}(u) = iu\mu_{\text{CGMYB}} - \frac{\sigma_{\text{CGMYB}}^2}{2} u^2 + \int_{\mathbb{R}} (e^{iux} - 1) f_{\text{CGMY}}(x) dx.$$

First of all, we calculate the first term that comes from $f_{\text{CGMY}}(y)$,

$$C \int_0^\infty \frac{e^{-Mx}}{x^{Y+1}} \left(e^{iux} - 1\right) dx = C \int_0^\infty e^y y^{-Y-1} (M - iu)^Y dy$$
$$- C \int_0^\infty e^y y^{-Y-1} M^Y dy,$$

where in the first term on the right-hand side $y = x(M - iu)$ is used, and in the second term $y = Mx$. One can easily recognize the integral expressions as being Gamma functions,

$$C \int_0^\infty \frac{e^{-Mx}}{x^{Y+1}} \left(e^{iux} - 1\right) dx = C\Gamma(-Y)((M - iu)^Y - M^Y).$$

The other piece coming from the density is found as follows,

$$C \int_{-\infty}^0 \frac{e^{-G|x|}}{|x|^{Y+1}} \left(e^{iux} - 1\right) dx = -C \int_\infty^0 \frac{e^{-G|x|}}{|x|^{Y+1}} \left(e^{-iux} - 1\right) dx$$
$$= C \int_0^\infty \frac{e^{-G|x|}}{|x|^{Y+1}} \left(e^{-iux} - 1\right) dx,$$

which has the same form as the previous integral, apart from $u \to -u$. Therefore,

$$\psi_{X_{\text{CGMYB}}}(u) = iu\mu_{\text{CGMYB}} - \frac{\sigma_{\text{CGMYB}}^2}{2} u^2$$
$$+ C\Gamma(-Y)((M - iu)^Y - M^Y + (G + iu)^Y - G^Y).$$

Under the EMM, the process is a martingale,

$$\mathbb{E}\left[e^{X_{\text{CGMYB}}} | \mathcal{F}(0)\right] = e^{rt},$$

that is for $u = -i$, it follows that,

$$\log(\phi_{X_{\text{CGMYB}}}(-i)) = \mu_{\text{CGMYB}} t - \frac{\sigma_{\text{CGMYB}}^2}{2} t$$
$$+ Ct\Gamma(-Y)((M - iu)^Y - M^Y + (G + iu)^Y - G^Y)$$
$$= rt.$$

By defining $\mu_{\text{CGMYB}} - \frac{\sigma_{\text{CGMYB}}^2}{2} + \bar{\omega} = r$, one finds the *drift correction term*.

> Conveniently, the characteristic function of the CGMYB log-asset price can be found in closed-form, as:
>
> $$\phi_{\log S(t)}(u) = e^{iu \log S(0)} \mathbb{E}\left[e^{iu X_{\text{CGMY}}(t)} | \mathcal{F}(0)\right]$$
> $$= \exp\left[iu\left(\frac{1}{t}\log S(0) + r + \bar{\omega} - \frac{1}{2}\sigma^2_{\text{CGMYB}}\right)t - \frac{1}{2}\sigma^2_{\text{CGMYB}}u^2 t\right]$$
> $$\times \varphi_{\text{CGMY}}(u,t), \qquad (5.64)$$
>
> with
>
> $$\varphi_{\text{CGMY}}(u,t) := \exp\left(tC\Gamma(-Y)\left((M-iu)^Y - M^Y + (G+iu)^Y - G^Y\right)\right),$$
>
> where $\Gamma(x)$ is the Gamma function (5.50) and, as in the VG model,
>
> $$\bar{\omega} = -\frac{1}{t}\log\left(\phi_{X_{\text{CGMYB}}}(-i)\right).$$

One can verify that the parameters G and M represent, respectively, the smallest and largest finite moment in the model, as

$$\mathbb{E}[S^u(t)] := \phi_{\log S(t)}(-iu)$$

is infinite for $u < -G$ and for $u > M$.

Example 5.4.1 (CGMYB parameters and implied volatility smile)
For the CGMYB model, we also perform a set of numerical experiments to assess the impact of the different CGMYB parameters on the implied volatility surface. The base test parameter set is chosen here as $C = 1$, $G = 1$, $M = 5$, $Y = 0.5$, $\sigma_{\text{CGMYB}} = 0.2$ and $r = 0.1$. In Figure 5.8 a systematic variation of the C, G, M and Y-parameters is performed, and the corresponding implied volatility curves are displayed. We know that the parameter C is a measure of overall level of activity, G and M are measures of the skewness, and Y is a measure of the fine structure, but this is not clearly visible in the implied volatility curves. ♦

Connection of VG process to CGMY process

Returning to the VG process, we may write the Lévy-Khintchine representation of $X_{\text{VG}}(t)$ as follows[3]:

$$\mathbb{E}^{\mathbb{Q}}\left[e^{iu X_{\text{VG}}(t)}\right] = \exp\left[t\left(i\mu_{\text{VG}}u + \int_{\mathbb{R}}(e^{iux} - 1)f_{\text{VG}}(x)\mathrm{d}x\right)\right]. \qquad (5.65)$$

---

[3]In this representation the term $-iux\mathbb{1}_{x\leq 1}$ under the integration is missing. This is due to the finite variation of the VG process. This missing term is included in the drift μ_{VG} [Poirot and Tankov, 2006].

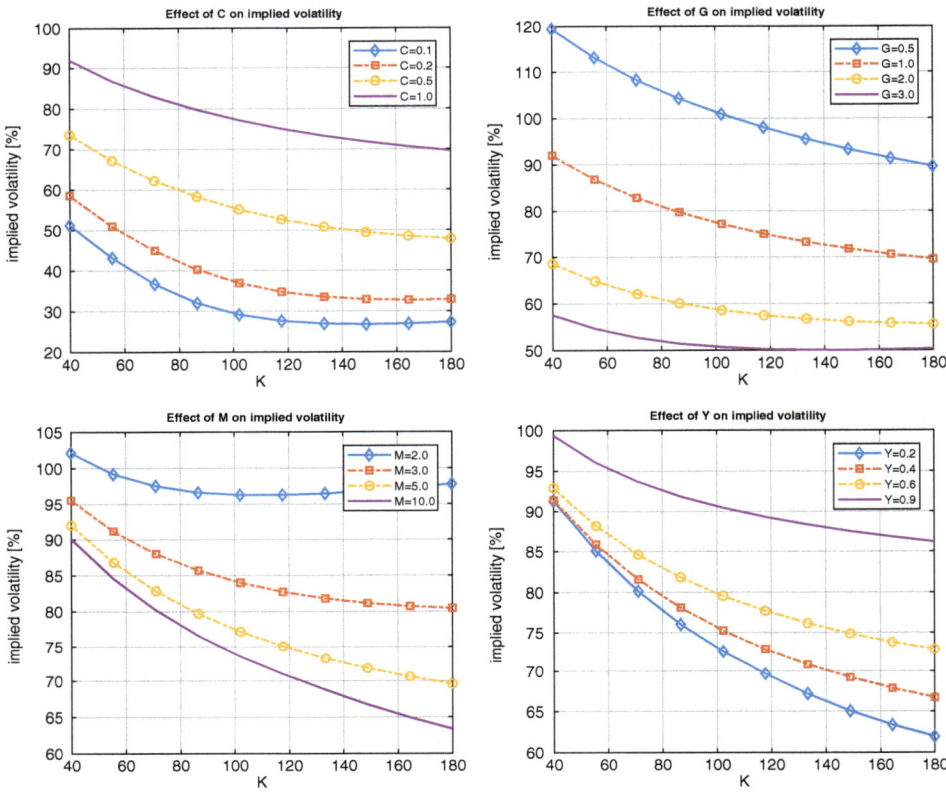

Figure 5.8: *Implied volatilities for the CGMYB model. The reference parameters are $C = 1$, $G = 1$, $M = 5$, $Y = 0.5$, $\sigma_{CGMYB} = 0.2$ and $r = 0.1$.*

We consider cases where the Lévy measure has a density, so that we have $F_{\mathrm{VG}}(\mathrm{d}x) = f_{\mathrm{VG}}(x)\mathrm{d}x$.

Similarly, in the risk-neutral world, $\mu_{\mathrm{VG}} = -r + \bar{\omega}$, where $\bar{\omega}$ is the drift correction term (5.59), see also [Hirsa and Madan, 2004], given by

$$\bar{\omega} = \int_{\mathbb{R}} (\mathrm{e}^y - 1) f_{\mathrm{VG}}(y) \mathrm{d}y. \tag{5.66}$$

This representation requires the specification of the Lévy measure $F_{\mathrm{VG}}(y)$. In [Carr et al., 2002] the Lévy density is given as

$$f_{\mathrm{VG}}(y) = C \left[\frac{\mathrm{e}^{-M|y|}}{|y|} \mathbb{1}_{y>0} + \frac{\mathrm{e}^{-G|y|}}{|y|} \mathbb{1}_{y<0} \right], \tag{5.67}$$

with constants C, G and M, defined as

$$C = \frac{1}{\beta}, \quad G = \left(\sqrt{\frac{1}{4}\theta^2\beta^2 + \frac{1}{2}\sigma_{\text{VG}}^2\beta} - \frac{1}{2}\theta\beta\right)^{-1},$$

$$M = \left(\sqrt{\frac{1}{4}\theta^2\beta^2 + \frac{1}{2}\sigma_{\text{VG}}^2\beta} + \frac{1}{2}\theta\beta\right)^{-1}. \quad (5.68)$$

By means of capital letters C, G and M, it is shown that the VG process is a particular case of the more general CGMY model [Carr et al., 2002].

With this parametrization, the drift correction term $\bar{\omega}$ has the following representation,

$$\bar{\omega} := \int_{\mathbb{R}} (1 - e^y) F_{\text{VG}}(\mathrm{d}y) = C \log\left[\left(1 + \frac{1}{G}\right)\left(1 - \frac{1}{M}\right)\right]. \quad (5.69)$$

To see this relation, observe that

$$\int_0^\infty \frac{e^{xy} - 1}{y} e^{-y} \mathrm{d}y = \sum_{k=1}^\infty \frac{x^k}{k!} \int_0^\infty y^{k-1} e^{-y} \mathrm{d}y = \sum_{k=1}^\infty \frac{x^k}{k} = -\log(1-x).$$

For $\bar{\omega}$ to be well-defined, $M > 1$ must be satisfied in (5.69).

After appropriate substitutions, the representations in (5.59) and (5.69) are indeed equivalent.

5.4.3 Normal inverse Gaussian process

This section discusses another well-known Lévy process, i.e. the Normal Inverse Gaussian process (NIG), which was introduced in [Barndorff-Nielsen, 1997], and is based on the inverse Gaussian process. This process belongs to a class of hyperbolic Lévy processes [Barndorff-Nielsen, 1978], and also gives rise to *fatter* tailed distributions than a standard normal process.

The Normal Inverse Gaussian (NIG) process is a variance-mean mixture of a Gaussian distribution with an inverse Gaussian.

The *density function* for the NIG process is available in closed form, and reads:

$$f_{\text{NIG}}(x) = \frac{K_{3,1}\left(\alpha\delta\sqrt{1 + (\frac{x}{\delta})^2}\right)}{\sqrt{1 + \left(\frac{x}{\delta}\right)^2}} \frac{\alpha}{\pi} \exp\left(\delta\sqrt{\alpha^2 - \beta^2} - \beta\right) \exp(\beta x), \quad (5.70)$$

with $x \in \mathbb{R}$, and parameters $\alpha > 0$, $0 < |\beta| < \alpha$, $\delta > 0$. The function $K_{3,\gamma}$ is the modified Bessel function of the third kind, with index γ, which is defined as,

$$K_{3,\gamma}(x) = \frac{1}{2}\int_0^\infty u^{(\gamma-1)} \exp\left(-\frac{1}{2}x\left(z + \frac{1}{z}\right)\right)\mathrm{d}z, \quad x > 0 \quad (5.71)$$

For the NIG process the moments can be found.

The α-parameter controls the *steepness of the density* in the sense that the steepness of the density increases monotonically with increasing α. This has implications also for the tail behavior: large values of α imply light tails, while smaller values of α imply heavier tails. The Gaussian distribution is obtained if $\alpha \to \infty$.

Parameter β is a *skewness or symmetry parameter*, and $\beta > 0$ gives rise to a density skew to the right, whereas $\beta < 0$ produces a density skew to the left. With $\beta = 0$ we encounter a density which is symmetric around 0. Parameter δ is a *scale parameter* in the sense that the rescaled parameters $\alpha \to \alpha\delta$ and $\beta \to \beta\delta$ are invariant under scale changes of x, as described in [Barndorff-Nielsen, 1997].

The *Lévy triplet* for the NIG process, $(\sigma_{\text{NIG}}^2, F_{\text{NIG}}, \mu_{\text{NIG}})$, is defined, as follows:

$$\sigma_{\text{NIG}}^2 = 0,$$

$$F_{\text{NIG}} = \frac{\alpha\delta}{\pi} \frac{\exp(\beta x) K_{3,1}(\alpha|x|)}{|x|} dx,$$

$$\mu_{\text{NIG}} = \frac{2\alpha\delta}{\pi} \int_0^1 \sinh(\beta x) K_{1,3}(\alpha x) dx.$$

The pure jump characteristic function of the NIG model reads

$$\varphi_{\text{NIG}}(u,t) = \exp\left(t\delta\left(\sqrt{\alpha^2 - \beta^2} - \sqrt{\alpha^2 - (\beta + iu)^2}\right)\right),$$

with parameters $\alpha, \delta > 0$ and $\beta \in (-\alpha, \alpha - 1)$.

The drift correction term, $\bar{\omega}$, which satisfies $\exp(-\bar{\omega}t) = \varphi_{\text{NIG}}(-i, t)$, making the process a martingale process, is given by

$$\bar{\omega} = \delta(\sqrt{\alpha^2 - (\beta + 1)^2} - \sqrt{\alpha^2 - \beta^2}).$$

Sometimes also a diffusion components is added to the NIG model, so that the process is governed by four parameters, similar as the CGMY process (without diffusion). We will denote the extended version of the NIG model by NIGB (Normal Inverse Gaussian process with Brownian motion). As a result, the dynamics of the NIGB model are driven by the parameters $(\sigma_{\text{NIGB}}, \alpha, \beta, \delta)$, where σ_{NIGB} is the volatility of the diffusion, and the characteristic function reads

$$\varphi_{\text{NIGB}}(u,t) = \exp\left(t\delta\left(\sqrt{\alpha^2 - \beta^2} - \sqrt{\alpha^2 - (\beta + iu)^2}\right) - \frac{\sigma_{\text{NIGB}}^2 u^2}{2} t\right).$$

5.5 Discussion on jumps in asset dynamics

Jump diffusion models and Lévy based models are attractive because they can, by definition, explain the jump patterns exhibited by some stocks [Sepp and Skachkov, 2003]. Figure 5.9 shows some historical stock values for the companies

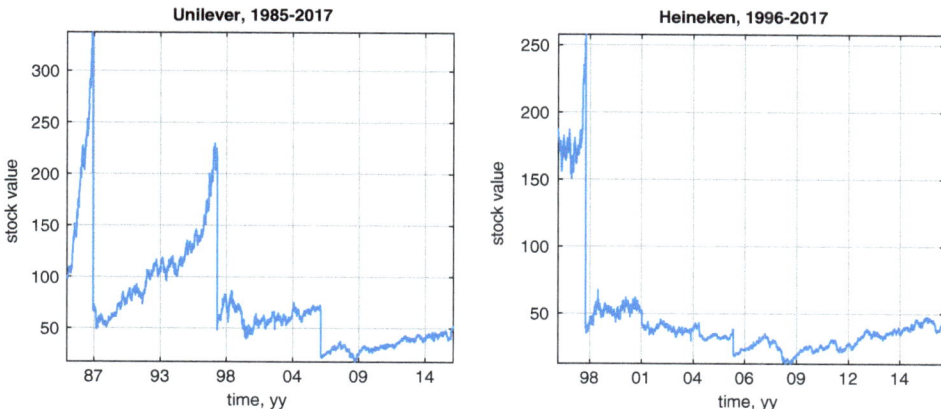

Figure 5.9: *Stock values for the two companies Unilever (left-hand picture) and Heineken (right-hand picture).*

Unilever and Heineken. Clearly the stocks exhibit some jump patterns. A closer inspection of the actual prices teaches us however that the stock price of Heineken is indeed related to a jump in value, whereas the jumps in Unilever's stock price were mainly due to so-called *stock splits* (when a company decides to increase the number of stocks, the individual stocks loose their value according to the decided split), that are connected to foreseen drops in the stock prices. The presence of jumps has been observed in the market, especially in times of financial turmoil, like in 1987, 2000 or 2008. Jump processes are superior to the Black-Scholes model in the sense that daily log-returns have heavy tails, and for longer time periods jump processes approach normality, which is consistent with empirical studies.

Models with jumps can capture the excess of kurtosis rather well. By introducing extra parameters, one can control the kurtosis and asymmetry of the log-return density, and one is also able to fit the smile in the implied volatility. Some studies also reveal that Lévy models are realistic for pricing options close to the maturity time [Das and Foresi, 1996].

> However, jump models have the undesired tendency to *increase the at-the-money volatility*. This is caused by a one-directional jump impact on the volatility, i.e. jumps can only increase volatility and not decrease it. Additionally, jumps are not *tradable quantities*, so that a portfolio that perfectly replicates any financial option cannot be constructed. In contrast to the Black-Scholes model, a hedging strategy for the writer of the option that will completely remove the risk of writing the option does thus not exist.

With jump processes we must rely on the no-arbitrage assumption, which implies the existence of an Equivalent Martingale Measure (EMM). Assuming that such a measure has been chosen, it is possible to derive a *partial integro-differential equation* satisfied by the option price. The corresponding numerical treatment to determine option prices may however be somewhat involved, due to the appearance of the integral term.

For all jump processes discussed in this chapter, however, we could determine the corresponding characteristic function. The characteristic function will be beneficial for the highly efficient valuation of European options by means of a Fourier pricing technique, in the forthcoming chapter.

5.6 Exercise set

Exercise 5.1 Jumps arrive in a stock index process due to two jumping assets, according to independent Poisson processes of, respectively, intensities $\xi_{p,1}$ and $\xi_{p,2}$.

  a. Determine the probability that asset S_1 jumps first.

  b. Determine the probability density function for the time until a first jump occurs in the stock index process.

  c. Choose $\xi_{p,1} = 0.3$ and $\xi_{p,2} = 0.5$ per year. Determine the probability of at least 2 jumps in the next 4 years.

  d. What is the probability of no jumps in the index in the next 2 years?

Exercise 5.2 Assuming major asset price jumps, leading to a financial crisis, are occurring in a particular stock with an intensity $\xi_p = 0.5$ per decade.

  a. From now on, what is the expected number of years and the standard deviation for the first major price jump to occur?

  b. What is the probability that the first fifteen years will be free of major jumps?

  c. What is the probability that the first asset price jump will occur in 4 years from now, and that it occurs 6 years from now?

Exercise 5.3 Increments, $X_\mathcal{P}(t_j) - X_\mathcal{P}(t_i)$, $t_i < t_j$ in different time intervals of the same length $t_j - t_i$ have the same distribution, regardless of the specific time point t_i.

$$\mathbb{P}[X_\mathcal{P}(t_j) - X_\mathcal{P}(t_i) = k] = \mathbb{P}[X_\mathcal{P}(t_j - t_i) - X_\mathcal{P}(0) = k]$$
$$= \mathbb{P}[X_\mathcal{P}(t_j - t_i)] = k]$$
$$= e^{-\xi_p(t_j - t_i)} \frac{(\xi_p(t_j - t_i))^k}{k!}, \quad k = 0, 1, \ldots.$$

where the second step is due to the fact that $X_\mathcal{P}(0) = 0$.

Given Poisson process $X_\mathcal{P}(t)$, $t \in [0, \infty)$ with intensity ξ_p. Determine the 2D joint distribution in the form of the so-called *probability mass function* of $(X_\mathcal{P}(t), X_\mathcal{P}(s))$, for any $0 < s < t$.

Exercise 5.4 Two stocks in a stock index exhibit jumps in their stock prices, S_1, S_2. These jumps occur as a Poisson process with a their means 1, and 3, every 5 years. The two stock price processes are independent. Determine the probability that 2 jumps occur in S_1 before 2 jumps are observed in S_2.

Exercise 5.5 Consider the event that exactly one asset jump arrives in a interval $[0, T]$. Determine the probability of this event happening.

This event is the same as the event that there is exactly one asset jump in the interval $[0, \frac{T}{2})$ and no jumps in the interval $[\frac{T}{2}, T]$, or no jumps in $[0, \frac{T}{2})$ and exactly one in $[\frac{T}{2}, T]$. Confirm that this calculation gives rise to exactly the same probability.

Exercise 5.6 Asset jumps arrive at the rate of 2 jumps per 5 years as a Poisson process. Find the probability that in a 15 years period, 1 jump arrives during the first 2 years and 1 jump arrive during the last 3 years.

With t_k the random time corresponding to the occurrence of the k-th event, $k = 1, 2, \ldots$. Let τ_k denote the i.i.d. exponential inter-arrival times, then $\tau_k = t_k - t_{k-1}$ and $t_0 = 0$. The time T_k, at which the k-th jump occurs in a Poisson process is given by $\tau_1 + \tau_2 + \cdots + \tau_k$. With $\xi_p = 1$ jump per 5 years, determine the mean and variance of the time until the arrival of the third jump.

Exercise 5.7 Under the Kou model the jump density is given by:
$$f_J(x) = p_1\alpha_1 e^{-\alpha_1 x} 1_{x \geq 0} + p_2\alpha_2 e^{\alpha_2 x} 1_{x < 0},$$
Determine $\mathbb{E}[e^{iuJ}]$ for $p_1 + p_2 = 1$ and $\alpha_1 > 1$ and $\alpha_2 > 0$.

Exercise 5.8 In Section 5.2.1 a closed-form expression for call option prices under the Merton's model has been derived. Derive the corresponding put option prices (without using the put-call parity) relation.

Exercise 5.9 Consider the problem of pricing a so-called Asian option with the payoff function given by,
$$V(T, S) = \max(A(T) - K, 0), \tag{5.72}$$
with the discrete average given by,
$$A(T) = \frac{1}{m} \sum_{i=1}^{m} S(T_i). \tag{5.73}$$

Define the quantity,
$$R(T_i) = \log\left(\frac{S(T_i)}{S(T_{i-1})}\right) = \log(S(T_i)) - \log(S(T_{i-1})),$$
and show that the following equality holds true,
$$A(T) = \frac{1}{m} S(T_0) e^{B_m}$$
$$= \frac{1}{m} S(T_0) e^{R(T_1) + \log(1 + \exp(B_{m-1}))},$$
with,
$$B_i = R(T_{m-i+1}) + \log(1 + \exp(B_{i-1})). \tag{5.74}$$

Exercise 5.10 For the jump diffusion process with the jump magnitude according to the Merton model (5.17), consider which model parameter ranges will give rise to an "*intensive and rare jump pattern*", and which ranges will give rise to *frequent and small-sized jumps*.

Make the same distinction for the Kou jump diffusion model (5.18), where a distinction between *small positive, and large negative jumps* and also *large positive and small negative jumps* can be made.

Exercise 5.11 For the Merton jump diffusion model (5.17) determine the European call and put option prices, as in Section 5.2.1, with Equation (5.28), with the following parameter set, $S_0 = 40, K = 40, r = 0.06, \sigma = 0.2$, and $T = 0.1, T = 1, T = 10$, and:

Set I: $\xi_p = 3, \mu_J = -0.2, \sigma_J = 0.2$;

Set II: $\xi_p = 8, \mu_J = -0.2, \sigma_J = 0.2$;

Set III: $\xi_p = 0.1, \mu_J = -0.9, \sigma_J = 0.45$.

Jump Processes

a. Determine the suitable number of terms in the summation in (5.28) to obtain stable option prices under this setting.

b. Compare your option value to the Black-Scholes option prices, in which the same problem parameters are used, but no jumps occur.

c. Moreover, choose $K = 50$ and all the other parameters as before (OTM calls) and compute the option prices.

Exercise 5.12 With the PDFs of the infinite activity Lévy processes available, for the VG, the CGMY and the NIG models, compute call and put option values by means of numerical integration with the composite Trapezoidal rule.

a. Set $S_0 = 40, K = 40, T = 0.5$, and vary in a very structured way the open asset model parameters in such a way that it is possible to evaluate the impact of skewness, kurtosis, fat tails on the option prices. Report your findings.

b. Compute the corresponding implied volatilities, and find a parameter set which generates an implied volatility smile and an implied volatility skew, respectively.

c. We wish to value options for many different strike prices, $K = \{0.25, \ldots, 0.75\}$, with steps of 0.05. Determine the prices under the NIG model for these different K-values, by numerical integration.

d. What is the complexity of this computation, if we have k different strike prices, and n integration points?

Exercise 5.13 Consider the following stochastic volatility jump model,

$$\mathrm{d}S(t) = rS(t)\mathrm{d}t + JS(t)\mathrm{d}W^{\mathbb{Q}}(t), \quad (5.75)$$

with

$$J = \begin{cases} 0.1 & \text{with} \quad \mathbb{Q}[J = 0.1] = \tfrac{1}{3}, \\ 0.2 & \text{with} \quad \mathbb{Q}[J = 0.2] = \tfrac{2}{3}, \end{cases} \quad (5.76)$$

with constant interest rate, $r = 0.05$, and initial stock value $S(t_0) = 1$.

a. By assuming independence between the jump J and the Brownian motion, use the concept of conditional expectations to derive the call option value, i.e.,

$$V(t_0, S_0) = \sum_{i=1}^{2} V_c(t_0, S_0; K, T, \sigma_i) \mathbb{Q}[J = \sigma_i],$$

with strike K and maturity T, where the notation $V_c(t_0, S_0; K, T, \sigma_i)$ indicates the Black-Scholes call option value and the implied volatility σ_i is given in (5.76).

b. For a maturity time $T = 2$ and a strip of strike prices $K = 0, 0.1, \ldots, 3$ perform a simulation of the model in (5.75) and compare the results to the solution obtained in the previous exercise.

CHAPTER 6

The
COS Method for European Option Valuation

In this chapter:

In this chapter we focus on an *efficient option pricing method*, which is based on the discounted expected payoff integral representation for *European options*. The method is called *the COS method*, as the key idea is to approximate the probability density function, which appears in the risk-neutral valuation formula, by a *Fourier cosine series expansion*. Fourier cosine series coefficients will have a closed-form relation with the *characteristic function*. As such, the COS method can be used for a quite general class of asset price processes, i.e. for those for which the characteristic function is available. These include the *exponential Lévy processes*, the *affine jump diffusion processes*, and thus also some *stochastic volatility models*, to be discussed in the chapters to follow. In **Section 6.1**, we introduce the Fourier cosine expansion for solving inverse Fourier integrals and for *density recovery from a characteristic function*. Based on this, we derive in **Section 6.2** the formulas for pricing European options and the corresponding hedge parameters. Numerical results are given in **Section 6.3**. For *smooth probability density functions*, the convergence is *exponential*. Since the computational complexity grows only linearly with the number of terms in the expansion, the COS method is *optimal in error convergence and in computational complexity* for European options.

Keywords: COS method, Fourier cosine expansions, characteristic function, density recovery, European option pricing, exponential error convergence.

6.1 Introduction into numerical option valuation

When valuing and risk-managing financial or insurance products, practitioners require fast, accurate and robust computations for option prices and for the sensitivities (delta, gamma, etc.). As we have seen in the previous chapters, the resulting pricing equations for the valuation of the corresponding financial derivatives products, set up with the help of Itô's lemma, are typically partial (-integro) differential equations, P(I)DEs. The Feynman-Kac theorem connects the solution of a P(I)DE to the conditional expectation of the value of a contract payoff function *under the risk-neutral measure*. Starting from these representations, we can apply several numerical techniques to calculate the option price itself. Broadly speaking, we can distinguish three types of computational methods:

1. Numerical solution of the partial-(integro) differential equation;
2. Numerical integration;
3. Monte Carlo simulation.

The distinction between the P(I)DE[1] and the integral representations is however subtle. Given the option pricing P(I)DE, one can formally write down the solution as a Green's function integral. This integral is the point of departure for numerical integration methods as well as for Monte Carlo simulation. In practice, all three computational techniques are employed, sometimes even at the same time, to provide a validation of numerical prices and sensitivities. Monte Carlo methods will be discussed in Chapter 9. Comprehensive overviews on the derivation of governing option valuation PDEs in finance include [Wilmott, 1998; Wilmott et al., 1995; Hull, 2012; Kwok, 2008]. Prominent work on P(I)DE discretization and solution techniques can be found in [d'Halluin et al., 2004; Forsyth and Vetzal, 2002; Zvan et al., 1998, 2001; In't Hout and Foulon, 2010; Haentjens and In't Hout, 2012; In 't Hout, 2017; Witte and Reisinger, 2011]. See also [Reisinger, 2012; Reisinger and Wittum, 2004, 2007; Suárez-Taboada and Vázquez, 2012; Pascucci et al., 2013; Bermúdez et al., 2006; Vázquez, 1998], among many others.

6.1.1 Integrals and Fourier cosine series

As we consider European options with initial date t_0 (at which the asset value $x = X(t_0)$ is known) and maturity date T, with $y = X(T)$, it is convenient to use the following simplified notation for the probability density function, $f_X(y) \equiv f_{X(T)}(y) := f_X(T, y; t_0, x)$.

---

[1] We write "P(I)DE" to discuss both the PDE as well as the PIDE (which appears in the case of asset processes with jumps).

> The density and its characteristic function, $f_X(y)$ and $\phi_X(u)$, form an example of a Fourier pair,[a]
> $$\phi_X(u) = \int_{\mathbb{R}} e^{iyu} f_X(y) dy, \qquad (6.1)$$
> and,
> $$f_X(y) = \frac{1}{2\pi} \int_{\mathbb{R}} e^{-iuy} \phi_X(u) du. \qquad (6.2)$$
>
> [a]Here we use the convention of the Fourier transform as often seen in the financial literature. Other conventions can also be used, and modifications to the methods are then straightforward.

6.1.2 Density approximation via Fourier cosine expansion

Fourier cosine series expansions give an optimal approximation of functions with a finite support [Boyd, 1989].

Remark 6.1.1 (Relation with Fourier series) *The general definition of the Fourier expansion of a function $g(x)$ on an interval $[-1, 1]$ is as follows,*

$$g(\theta) = {\sum_{k=0}^{\infty}}' \bar{A}_k \cos(k\pi\theta) + \sum_{k=1}^{\infty} \bar{B}_k \sin(k\pi\theta), \qquad (6.3)$$

where the prime at the sum, $\sum'$, indicates that the first term in the summation is weighted by one-half, and the coefficients are given by

$$\bar{A}_k = \int_{-1}^{1} g(\theta) \cos(k\pi\theta) d\theta, \quad \bar{B}_k = \int_{-1}^{1} g(\theta) \sin(k\pi\theta) d\theta. \qquad (6.4)$$

By setting $\bar{B}_k = 0$, we obtain the classical Fourier cosine expansion, by which we can represent even functions around $\theta = 0$ exactly. ▲

We can extend any function $g : [0, \pi] \to \mathbb{R}$ to become an *even function on* $[-\pi, \pi]$, as follows,

$$\bar{g}(\theta) = \begin{cases} g(\theta), & \theta \geq 0 \\ g(-\theta), & \theta < 0. \end{cases} \qquad (6.5)$$

Even functions can be expressed as Fourier cosine series. For a function $\bar{g}(\theta)$ supported on $[-\pi, \pi]$, the cosine expansion reads

$$\bar{g}(\theta) = {\sum_{k=0}^{\infty}}' \bar{A}_k \cdot \cos(k\theta), \qquad (6.6)$$

with

$$\bar{A}_k = \frac{1}{\pi} \int_{-\pi}^{\pi} \bar{g}(\theta) \cos(k\theta) d\theta = \frac{2}{\pi} \int_{0}^{\pi} g(\theta) \cos(k\theta) d\theta, \qquad (6.7)$$

> Because $g(\theta) = \bar{g}(\theta)$ on $[0, \pi]$, the Fourier cosine expansion of the (not necessarily even) function $g(\theta)$ on $[0, \pi]$ is also given by the Equations (6.6), (6.7).

The Fourier cosine series expansion, as used in the COS method, is based on a classical definition of the cosine series in the interval $[-\pi, \pi]$, with π being merely a scaling factor, and the function's maximum is attained at the domain boundary.

For functions supported on any other finite interval, say $[a, b] \in \mathbb{R}$, the Fourier cosine series expansion can be obtained via a *change of variables*:

$$\theta := \frac{y-a}{b-a}\pi, \quad y = \frac{b-a}{\pi}\theta + a.$$

It then reads

$$g(y) = \sum_{k=0}^{\infty}{}' \bar{A}_k \cdot \cos\left(k\pi \frac{y-a}{b-a}\right), \qquad (6.8)$$

with

$$\bar{A}_k = \frac{2}{b-a}\int_a^b g(y)\cos\left(k\pi\frac{y-a}{b-a}\right)\mathrm{d}y. \qquad (6.9)$$

Since any real function has a cosine expansion when it has finite support, the derivation of the approximation of a probability density function starts with a truncation of the infinite integration range in (6.2). Due to the conditions for the existence of a Fourier transform, the integrands in (6.2) have to decay to zero at $\pm\infty$ and we can truncate the integration range without losing significant accuracy.

Suppose $[a, b] \in \mathbb{R}$ is chosen such that the *truncated integral* approximates the infinite counterpart very well, i.e.,

$$\hat{\phi}_X(u) := \int_a^b e^{iuy}f_X(y)\mathrm{d}y \approx \int_{\mathbb{R}} e^{iuy}f_X(y)\mathrm{d}y = \phi_X(u). \qquad (6.10)$$

We relate equation (6.10) to (6.9), by recalling the well-known Euler formula:

$$e^{iu} = \cos(u) + i\sin(u),$$

which implies that $\mathrm{Re}\{e^{iu}\} = \cos(u)$, where $\mathrm{Re}\{\cdot\}$ denotes the real part of the argument. Based on this, for any random variable, X, and constant, $a \in \mathbb{R}$, the following equality holds:

$$\phi_X(u)e^{ia} = \mathbb{E}[e^{iuX+ia}] = \int_{-\infty}^{\infty} e^{i(uy+a)}f_X(y)\mathrm{d}y. \qquad (6.11)$$

By taking the real parts in (6.11), we find:

$$\mathrm{Re}\left\{\phi_X(u)e^{ia}\right\} = \mathrm{Re}\left\{\int_{-\infty}^{\infty} e^{i(uy+a)}f_X(y)\mathrm{d}y\right\} = \int_{-\infty}^{\infty} \cos(uy+a)f_X(y)\mathrm{d}y.$$

We substitute the Fourier argument $u = \frac{k\pi}{b-a}$, and multiply the characteristic function in (6.10), by $\exp\left(-i\frac{ka\pi}{b-a}\right)$, i.e.

$$\hat{\phi}_X\left(\frac{k\pi}{b-a}\right) \cdot \exp\left(-i\frac{ka\pi}{b-a}\right) = \int_a^b \exp\left(iy\frac{k\pi}{b-a} - i\frac{ka\pi}{b-a}\right) f_X(y) dy. \quad (6.12)$$

By taking the real part at both sides of the equation, we find:

$$\text{Re}\left\{\hat{\phi}_X\left(\frac{k\pi}{b-a}\right) \cdot \exp\left(-i\frac{ka\pi}{b-a}\right)\right\} = \int_a^b \cos\left(k\pi\frac{y-a}{b-a}\right) f_X(y) dy. \quad (6.13)$$

At the right-hand side of (6.13), we have the definition of $\bar{A}_k$ in (6.9), so

$$\bar{A}_k \equiv \frac{2}{b-a}\text{Re}\left\{\hat{\phi}_X\left(\frac{k\pi}{b-a}\right) \cdot \exp\left(-i\frac{ka\pi}{b-a}\right)\right\}. \quad (6.14)$$

It follows from (6.10) that $\bar{A}_k \approx \bar{F}_k$ with

$$\bar{F}_k := \frac{2}{b-a}\text{Re}\left\{\phi_X\left(\frac{k\pi}{b-a}\right) \cdot \exp\left(-i\frac{ka\pi}{b-a}\right)\right\}. \quad (6.15)$$

We now *replace $\bar{A}_k$ by $\bar{F}_k$* in the series expansion of $f_X(y)$ on $[a,b]$, i.e.,

$$\hat{f}_X(y) \approx \sum_{k=0}^{\infty}{}' \bar{F}_k \cos\left(k\pi\frac{y-a}{b-a}\right), \quad (6.16)$$

and *truncate the series summation*, so that

$$\hat{f}_X(y) \approx \sum_{k=0}^{N-1}{}' \bar{F}_k \cos\left(k\pi\frac{y-a}{b-a}\right). \quad (6.17)$$

Remember that the first term in the summation should be multiplied by one-half (indicated by the $\sum'$ symbol). *Forgetting to multiply the first term by one-half is an often made computer implementation mistake!*

The resulting error in the approximation $\hat{f}_X(y)$ consists of two parts: a series truncation error from (6.16) to (6.17) and an error originating from the approximation of $\bar{A}_k$ by $\bar{F}_k$.

Since the cosine series expansions of so-called *entire functions* (i.e. functions without any singularities[2] anywhere in the complex plane, except at ∞) exhibit an *exponential convergence* [Boyd, 1989], we can expect (6.17) to give highly accurate approximations, with a small value for N, to density functions that have no singularities on $[a,b]$.

---

[2] By "singularity" we mean, as in [Boyd, 1989], poles, fractional powers, logarithms, other branch points, and discontinuities in a function or in any of its derivatives.

Example 6.1.1 To demonstrate the highly efficient convergence, we use the approximation in (6.17) for the standard normal density function. The corresponding PDF and characteristic function are known in closed-form, and read, respectively,

$$f_{\mathcal{N}(0,1)}(y) = \frac{1}{\sqrt{2\pi}} e^{-\frac{1}{2}y^2}, \quad \phi_{\mathcal{N}(0,1)}(u) = e^{-\frac{1}{2}u^2}.$$

We determine the accuracy of the Fourier cosine expansion approximation of the PDF for different values of N in (6.17). Integration interval $[a, b] = [-10, 10]$, and the maximum absolute error is measured at $y = \{-5, -4, \ldots, 4, 5\}$.

Table 6.1 shows that a very small error is obtained with only a small number of terms N in the expansion. From the differences in the CPU times in the table, defined as "time(N)-time($N/2$)", we can observe a linear complexity. This technique is thus highly efficient for the recovery of the density function.

Figure 6.1 presents the convergence of the normal density function with different values of N. It is clear that for small values of N the normal density

Table 6.1: *Maximum error when recovering $f_X(y)$ from $\phi_X(u)$ by Fourier cosine expansion.*

N	4	8	16	32	64
Error	0.25	0.11	0.0072	4.04e-07	3.33e-17
CPU time (msec.)	0.046	0.061	0.088	0.16	0.29
Diff. in CPU (msec.)	–	0.015	0.027	0.072	0.13

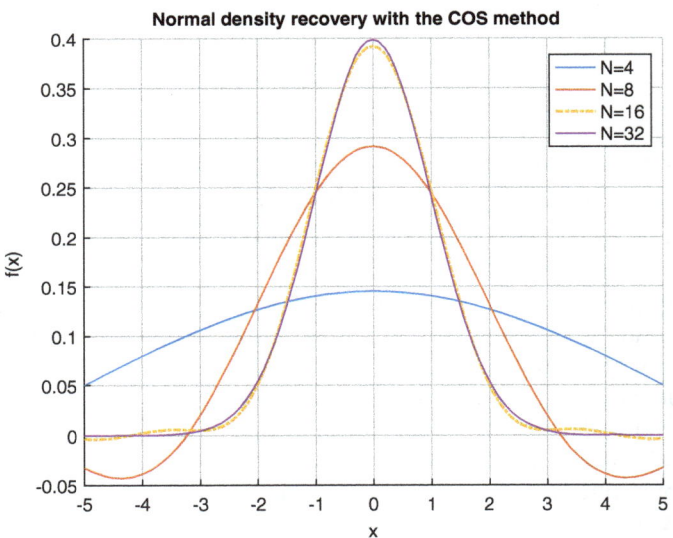

Figure 6.1: *Convergence of the normal density for different numbers of expansion terms.*

is not well approximated (even with negative values), however, the density converges exponentially for increasing N. ♦

Example 6.1.2 (Lognormal density recovery) In this example we will employ the COS method to approximate the probability density function of a *lognormal random variable*. Note that the characteristic function of lognormal random variables is *not known* in closed-form, and therefore we cannot simply apply the COS method to a lognormal characteristic function. To obtain the density of a lognormal random variable, we will employ the relation between a normal and lognormal random variable. With Y a lognormal random variable with parameters μ and σ^2, then $Y = e^X$, with $X \sim \mathcal{N}(\mu, \sigma^2)$, see also Equation (2.19). The CDF of Y in terms of the CDF of X can be calculated as,

$$F_Y(y) \stackrel{\text{def}}{=} \mathbb{P}[Y \leq y] = \mathbb{P}[e^X \leq y] = \mathbb{P}[X \leq \log(y)] = F_X(\log(y)). \quad (6.18)$$

By differentiation, we get:

$$f_Y(y) \stackrel{\text{def}}{=} \frac{\mathrm{d}F_Y(y)}{\mathrm{d}y} = \frac{\mathrm{d}F_X(\log(y))}{\mathrm{d}\log y} \frac{\mathrm{d}\log(y)}{\mathrm{d}y} = \frac{1}{y} f_X(\log(y)). \quad (6.19)$$

To numerically obtain the density function of a lognormal variable Y, we will thus work with the transformed density of X. Since X is normally distributed, and $\phi_X(u) = e^{i\mu u - \frac{1}{2}\sigma^2 u^2}$, by the COS method, we may recover the PDF of a lognormal random variable.

For this numerical experiment, we set $\mu = 0.5$ and $\sigma = 0.2$ and analyze the convergence of the COS method, in dependence of the number of expansion terms N. By comparing the convergence in Figures 6.1 and 6.2, we notice that the recovery of the lognormal PDF is computationally more intensive. The convergence is however still exponential. ♦

6.2 Pricing European options by the COS method

The COS formula for pricing European options is obtained by substituting for the density function its Fourier cosine series expansion. A probability density function tends to be smooth and therefore only a few terms in the expansion may already give a highly accurate approximation.

The point of departure for pricing European options with numerical integration techniques is the risk-neutral valuation formula. Following the notation from the previous chapters, we set $X(t) := \log S(t)$, with process $X(t)$ taking values $X(t) = x$ and $X(T) = y$. The value of a plain vanilla option is given by:

$$V(t_0, x) = e^{-r\tau}\mathbb{E}^{\mathbb{Q}}\left[V(T, y)|\mathcal{F}(t_0)\right] = e^{-r\tau}\int_{\mathbb{R}} V(T, y) f_X(T, y; t_0, x)\mathrm{d}y, \quad (6.20)$$

where $\tau = T - t_0$, $f_X(T, y; t_0, x)$ is the *transition probability density of $X(T)$*, and r is the interest rate. We will again use a *simpler notation* for the density, i.e. $f_X(y) \equiv f_{X(T)}(y) := f_X(T, y; t_0, x)$.

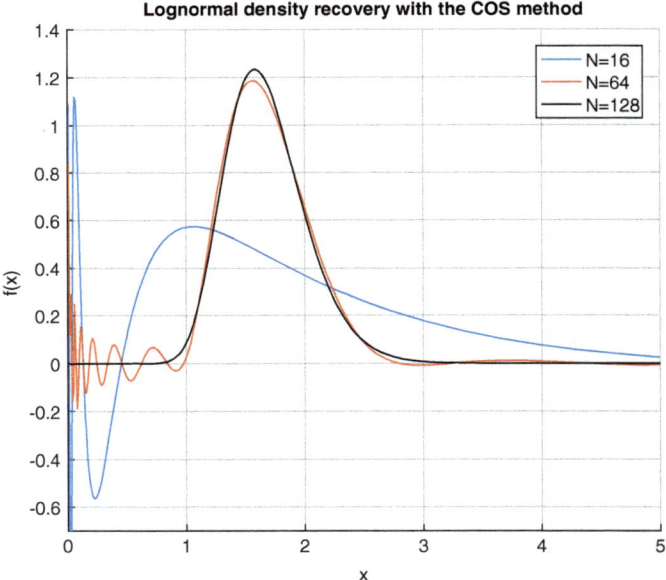

Figure 6.2: *Recovery of the lognormal PDF for different numbers of expansion terms.*

Since $f_X(y)$ rapidly decays to zero as $y \to \pm\infty$, we truncate the infinite integration range to $[a,b] \subset \mathbb{R}$, without losing significant accuracy, and for $S(t) = e^{X(t)}$, with $x := X(t_0)$ and $y := X(T)$, we obtain approximation V_I:

$$V(t_0, x) \approx V_\mathrm{I}(t_0, x) = e^{-r\tau} \int_a^b V(T,y) f_X(y) \mathrm{d}y, \qquad (6.21)$$

with $\tau = T - t_0$.

By the Roman subscript, as in variable V_I, subsequent numerical approximations are denoted, which turns out to be helpful in the error analysis in Section 6.2.3. We will give insight into the choice of $[a,b]$ in Section 6.2.4.

In the second step, since $f_X(y)$ is usually not known whereas the characteristic function is, the density is approximated by its Fourier cosine expansion in y, as in (6.8),

$$\hat{f}_X(y) = {\sum_{k=0}^{+\infty}}' \bar{A}_k(x) \cos\left(k\pi \frac{y-a}{b-a}\right), \qquad (6.22)$$

with

$$\bar{A}_k(x) := \frac{2}{b-a} \int_a^b \hat{f}_X(y) \cos\left(k\pi \frac{y-a}{b-a}\right) \mathrm{d}y, \qquad (6.23)$$

so that,

$$V_{\mathrm{I}}(t_0, x) = e^{-r\tau} \int_a^b V(T, y) \sum_{k=0}^{+\infty}{}' \bar{A}_k(x) \cos\left(k\pi \frac{y-a}{b-a}\right) dy. \qquad (6.24)$$

We interchange the summation and integration (application of Fubini's Theorem), and insert the following definition

$$H_k := \frac{2}{b-a} \int_a^b V(T, y) \cos\left(k\pi \frac{y-a}{b-a}\right) dy, \qquad (6.25)$$

resulting in

$$V_{\mathrm{I}}(t_0, x) = \frac{b-a}{2} e^{-r\tau} \cdot \sum_{k=0}^{+\infty}{}' \bar{A}_k(x) \cdot H_k. \qquad (6.26)$$

Here, the H_k are the *cosine series coefficients of the payoff function*, $V(T, y)$. Thus, from (6.21) to (6.26) we have transformed an integral of the product of two real functions, $f_X(y)$ and $V(T, y)$, into a product of their Fourier cosine series coefficients.

Due to the rapid decay rate of these coefficients, we may truncate the series summation to obtain approximation V_{II}:

$$V_{\mathrm{II}}(t_0, x) = \frac{b-a}{2} e^{-r\tau} \cdot \sum_{k=0}^{N-1}{}' \bar{A}_k(x) H_k. \qquad (6.27)$$

Similar to the derivation in Section 6.1.1, the coefficients $\bar{A}_k(x)$, as defined in (6.23), can be approximated by $\bar{F}_k(x)$, as defined in (6.15).

By replacing $\bar{A}_k(x)$ in (6.27) by $\bar{F}_k(x)$, we obtain

$$V(t_0, x) \approx V_{\mathrm{III}}(t_0, x) = e^{-r\tau} \sum_{k=0}^{N-1}{}' \mathrm{Re}\left\{\phi_X\left(\frac{k\pi}{b-a}\right) \exp\left(-ik\pi \frac{a}{b-a}\right)\right\} \cdot H_k, \qquad (6.28)$$

with $\tau = T - t_0$ and x a function of $S(t_0)$ (like $x = \log(S(t_0))$ or $x = \log(S(t_0)/K)$), and the characteristic function

$$\boxed{\phi_X(u) = \phi_X(u; t_0, T) \equiv \phi_X(u, x; t_0, T),} \qquad (6.29)$$

as the characteristic function also depends on variable x.

This is the *COS formula* for general underlying processes. The H_k-coefficients can be obtained analytically for plain vanilla and digital options, and (6.28) can be simplified for Lévy models, so that many strike prices can be calculated simultaneously.

Heuristically speaking, we decompose the probability density function into a weighted sum of many "density-like basis functions" by which option values can be obtained analytically. The payoff series coefficients, H_k, have to be recovered, for which analytic solutions are available for several option contracts.

6.2.1 Payoff coefficients

The payoff for European options can be conveniently written, in an *adjusted log-asset price*, i.e.
$$y(T) = \log \frac{S(T)}{K},$$
as
$$V(T, y) := [\bar{\alpha} \cdot K(e^y - 1)]^+ \quad \text{with} \quad \bar{\alpha} = \begin{cases} 1 & \text{for a call,} \\ -1 & \text{for a put,} \end{cases}$$
where the notation $[h(y)]^+$ denotes "$\max[h(y), 0)]$".

Before deriving the H_k-coefficients from the definition in (6.25), we need two basic results.

Result 6.2.1 *The cosine series coefficients, χ_k, of $g(y) = e^y$ on an integration interval $[c, d] \subset [a, b]$,*
$$\chi_k(c, d) := \int_c^d e^y \cos\left(k\pi \frac{y-a}{b-a}\right) dy, \tag{6.30}$$

and the cosine series coefficients, ψ_k, of $g(y) = 1$ on an integration interval $[c, d] \subset [a, b]$,
$$\psi_k(c, d) := \int_c^d \cos\left(k\pi \frac{y-a}{b-a}\right) dy, \tag{6.31}$$

are known analytically. Basic calculus shows that
$$\chi_k(c, d) := \frac{1}{1 + \left(\frac{k\pi}{b-a}\right)^2} \left[\cos\left(k\pi \frac{d-a}{b-a}\right) e^d - \cos\left(k\pi \frac{c-a}{b-a}\right) e^c \right.$$
$$\left. + \frac{k\pi}{b-a} \sin\left(k\pi \frac{d-a}{b-a}\right) e^d - \frac{k\pi}{b-a} \sin\left(k\pi \frac{c-a}{b-a}\right) e^c \right], \tag{6.32}$$

and
$$\psi_k(c, d) := \begin{cases} \left[\sin\left(k\pi \frac{d-a}{b-a}\right) - \sin\left(k\pi \frac{c-a}{b-a}\right)\right] \frac{b-a}{k\pi}, & k \neq 0, \\ (d - c), & k = 0. \end{cases} \tag{6.33}$$

> *Focusing on a call option, in the case that $a < 0 < b$, we obtain*
>
> $$H_k^{call} = \frac{2}{b-a} \int_0^b K(e^y - 1) \cos\left(k\pi \frac{y-a}{b-a}\right) dy$$
> $$= \frac{2}{b-a} K \left(\chi_k(0,b) - \psi_k(0,b)\right), \quad (6.34)$$
>
> *where χ_k and ψ_k are given by (6.32) and (6.33), respectively. Similarly, for a vanilla put, we find*
>
> $$H_k^{put} = \frac{2}{b-a} K \left(-\chi_k(a,0) + \psi_k(a,0)\right). \quad (6.35)$$
>
> *In the situation, where we deal with $a < b < 0$, we have, of course,*
>
> $$H_k^{call} = 0,$$
>
> *while for $0 < a < b$, the payoff coefficients H_k^{call} are defined by $c \equiv a$, $d \equiv b$ in (6.25) and in Result 6.2.1. For put options the relations are reversed.*

Coefficients for digital and gap options

The payoff coefficients H_k are different for different payoff functions. Cash-or-nothing options are often building blocks for constructing more complex option products. The payoff of the *cash-or-nothing call option* equals 0 if $S(T) \leq K$ and K if $S(T) > K$, see also Example 3.2.1. For this contract, the cash-or-nothing call coefficients, H_k^{cash}, can also be obtained analytically:

$$H_k^{cash} = \frac{2}{b-a} K \int_0^b \cos\left(k\pi \frac{y-a}{b-a}\right) dy = \frac{2}{b-a} K \psi_k(0,b).$$

with $\psi(a,b)$ as in (6.33).

For option contracts for which the H_k-coefficients can be obtained only numerically, the error convergence will be dominated by the numerical quadrature rule employed.

6.2.2 The option Greeks

Series expansions for the option Greeks Δ and Γ can be derived similarly. With $V \equiv V(t_0, S)$, $S \equiv S(t_0)$, $x := X(t_0)$, we find:

$$\Delta = \frac{\partial V}{\partial S} = \frac{\partial V}{\partial x} \frac{\partial x}{\partial S} = \frac{1}{S} \frac{\partial V}{\partial x}, \quad \Gamma = \frac{\partial^2 V}{\partial S^2} = \frac{1}{S^2} \left(-\frac{\partial V}{\partial x} + \frac{\partial^2 V}{\partial x^2}\right).$$

It follows that

$$\Delta \approx \frac{1}{S} e^{-r\tau} \sum_{k=0}^{N-1}{}' \operatorname{Re}\left\{\phi\left(\frac{k\pi}{b-a}\right) \exp\left(-ik\pi\frac{x-a}{b-a}\right) \frac{ik\pi}{b-a}\right\} \cdot H_k, \quad (6.36)$$

and

$$\Gamma \approx \frac{1}{S^2} e^{-r\tau} \sum_{k=0}^{N-1}{}' \operatorname{Re}\left\{\phi\left(\frac{k\pi}{b-a}\right) \exp\left(-ik\pi\frac{x-a}{b-a}\right)\right.$$

$$\left. \times \left[\left(\frac{ik\pi}{b-a}\right)^2 - \frac{ik\pi}{b-a}\right]\right\} \cdot H_k. \quad (6.37)$$

6.2.3 Error analysis COS method

In the derivation of the COS formula there were three stages in which approximation errors were introduced: the *truncation of the integration range* in the risk-neutral valuation formula (6.21), the substitution of the density by its *(truncated by N) cosine series expansion* (6.27), and the substitution of the series coefficients by the *characteristic function approximation* (6.28). Therefore, the overall error consists of *three parts*:

1. The *integration range truncation error*:

$$\epsilon_{\mathrm{I}} := V(t_0, x) - V_{\mathrm{I}}(t_0, x) = \int_{\mathbb{R}\setminus[a,b]} V(T, y) f_X(y) \mathrm{d}y. \quad (6.38)$$

2. The *series truncation error on $[a,b]$*:

$$\epsilon_{\mathrm{II}} := V_{\mathrm{I}}(t_0, x) - V_{\mathrm{II}}(t_0, x) = \frac{1}{2}(b-a)e^{-r\tau} \sum_{k=N}^{+\infty} \bar{A}_k(x) \cdot H_k, \quad (6.39)$$

where $\bar{A}_k(x)$ and H_k are defined by (6.23) and (6.25), respectively.

3. The *error related to approximating $\bar{A}_k(x)$ by $\bar{F}_k(x)$ in (6.15)*:

$$\epsilon_{\mathrm{III}} := V_{\mathrm{III}}(t_0, x) - V_{\mathrm{II}}(t_0, x) \quad (6.40)$$

$$= e^{-r\tau} \sum_{k=0}^{N-1}{}' \operatorname{Re}\left\{\int_{\mathbb{R}\setminus[a,b]} \exp\left(ik\pi\frac{y-a}{b-a}\right) f_X(y)\mathrm{d}y\right\} \cdot H_k.$$

We do not take any error in the coefficients H_k into account, as we have a closed-form solution, at least for plain vanilla European options.

The key to bound the errors lies in the decay rate of a cosine series coefficients. The convergence rate of the Fourier cosine series depends on the properties of the functions on the expansion interval. With the integration range $[a, b]$ chosen sufficiently large, for example, the overall error is dominated by ϵ_{II}.

Equation (6.39) indicates that ϵ_{II} depends on both $\bar{A}_k(x)$ and H_k, the series coefficients of the density and that of the payoff, respectively. The density is typically smoother than the payoff functions in finance and the coefficients $\bar{A}_k$ often decay faster than H_k. Consequently, the product of $\bar{A}_k$ and H_k converges faster than either $\bar{A}_k$ or H_k. Error ϵ_{II} is then dominated by the series truncation error of the density function, and *converges exponentially* in the case of density functions $f_X(x) \in \mathbb{C}^\infty([a, b])$ with nonzero derivatives.

Note that in the case of a *discontinuous* probability density function, we will encounter a slow algebraic convergence order, which can be related to the well-known *Gibbs phenomenon*, and is typically observed in Fourier series expansions of discontinuous functions.

With a properly chosen truncation of the integration range, the overall error converges either exponentially for density functions with nonzero derivatives, belonging to $\mathbb{C}^\infty([a, b] \subset \mathbb{R})$, or algebraically for density functions with a discontinuity in one of its derivatives. In essence, the COS method replaces the density in the risk-neutral valuation formula by a partial sum of the Fourier cosine series of the density. Therefore, when the integration range truncation error is not dominant, the overall error heavily depends on the properties of the density. For further details, we refer to [Fang, 2010].

In fact, the cosine expansion of a function $g(y)$ in y is equal to a Chebyshev series expansion of $g(\cos^{-1}(t))$ in t, see Remark 6.2.2.

Remark 6.2.1 (Cosine series) *We will give some intuition why the cosine expansion is used within the COS method, and not the full Fourier series. Basically, a cosine expansion requires fewer terms to approximate a linear function.*

The Fourier series representation of a function of a continuous variable x in an interval, defines a periodic extension of that function for all $x \in \mathbb{R}$. If the function values at the interval's end points are not equal, the periodic extension is not continuous, which will give rise to convergence problems near the interval's endpoints (known as the Gibb's phenomenon).

Essentially the same problem is observed in the discrete case, when using an N-point Discrete Fourier Transform (DFT), which implicitly assumes an N-point periodization of a data set. These convergence problems manifest themselves over the entire data set, slowing down the overall convergence of the DFT.

One way to overcome such problems is by considering the function to be an even function on an interval of twice the size. The periodic extension of this function will then be continuous at the interval's end points. With an even periodic extension of a data set, the sine functions, that are part of the complex exponential in a DFT, will have weight 0, so that a Discrete Cosine Transform (DCT) remains.

Convergence is one reason why the DCT is important in signal processing. A related important property is that the "energy of the signal", which is measured by the sums of the squares of its coefficients, is mainly contained in the low frequency coefficients. In other words, the DCT coefficients decay more rapidly than the corresponding DFT coefficients, so that a signal can often be approximated with fewer DCT coefficients. ▲

Remark 6.2.2 (Relation with Chebyshev polynomials) *The Fourier cosine expansion is equivalent to using Chebyshev polynomials to represent a function. Chebyshev polynomials form a set of orthogonal basis functions approximating a function $g(y)$ by a sum of $N+1$ orthogonal basis functions:*

$$g(y) \approx g_N(y) = \sum_{k=0}^{N} \bar{a}_k T_k(y). \tag{6.41}$$

The series coefficients in (6.41) are chosen in such a way that the residual, $|g(y) - g_N(y)|$, is minimized. The Chebyshev series expansion minimizes $\|g(y) - g_N(y)\|_{L^\infty}$. The Chebyshev approximation of a real-valued function is an optimal approximation of a function on a finite interval, in the min-max sense, i.e. the error of the Chebyshev approximation polynomial is minimal in the max-norm.[3] The set of orthogonal Chebyshev polynomials of the first kind, $T_k(y)$, is defined by polynomials of degree k, as

$$T_k(y) = \cos(k\theta), \ y = \cos\theta, \ \text{or} \ T_k(y) = \cos(k \arccos y).$$

The first few Chebyshev polynomials are given by:

$$T_0(y) = 1, \ T_1(y) = y, \ T_2(y) = 2y^2 - 1,$$
$$T_3(y) = 4y^3 - 3y, \ T_4(y) = 8y^4 - 8y^2 + 1, \ldots$$

following directly from the fact that $\cos(k\theta)$ is a polynomial of degree k in $\cos\theta$, and,

$$\cos 0 = 1, \ \cos(2\theta) = 2\cos^2\theta - 1,$$
$$\cos(3\theta) = 4\cos^3\theta - 3\cos\theta, \ \cos(4\theta) = 8\cos^4\theta - 8\cos^2\theta + 1.$$

A Chebyshev approximation is thus a cosine series under a variable transformation

$$g(\theta) = \sum_{k=0}^{\infty}{'} \bar{a}_k T_k(\theta), \ \text{where} \ T_k(\theta) = \cos\left(k \cos^{-1}(\theta)\right), \ \theta \in [-1,1], \tag{6.42}$$

---

[3] When a function is spatially periodic, the exponential Fourier series expansion is the optimal approximation.

with the coefficients

$$\bar{a}_k = \frac{2}{\pi} \int_{-1}^{1} \frac{1}{\sqrt{1-\theta^2}} g(\theta) T_k(\theta) \mathrm{d}\theta. \tag{6.43}$$

The identity $T_k(\cos y) = \cos(ky)$ *and the transformation,* $\theta = \cos y$, *give*

$$\bar{a}_k = \frac{2}{\pi} \int_0^{\pi} g(\cos y) \cos(ky) \mathrm{d}y.$$

so that the coefficients of the Chebyshev series of $g(\theta)$ *are identical to the Fourier cosine coefficients of* $g(\cos y)$ *on the interval* $[0, \pi]$, *and both expansions are equivalent under transformation. The expansion for functions supported on another finite interval* $[a, b]$ *can be found by the change of variables in (6.8).* ▲

6.2.4 Choice of integration range

The choice of integration range, $[a, b]$, is important for accurate option valuation by the COS method. An interval which is chosen too small will lead to a significant integration-range truncation error, whereas an interval which is set very large would require a large value for N (meaning many terms in the cosine expansion (6.28) to achieve a certain level of accuracy).

By the definition of the integration range, as given [Fang and Oosterlee, 2008; Fang, 2010; Fang and Oosterle, 2009], and centering the domain at $x + \zeta_1$, we use

$$[a, b] := \left[(x + \zeta_1) - L\sqrt{\zeta_2 + \sqrt{\zeta_4}}, \ (x + \zeta_1) + L\sqrt{\zeta_2 + \sqrt{\zeta_4}} \right], \tag{6.44}$$

with $L \in [6, 12]$ depending on a user-defined tolerance level and $\zeta_1, \ldots, \zeta_4$ being specific *cumulants* of the underlying stochastic process.

Notice the following difference: $x = \log S(t_0)$ when the COS formula is used for density recovery, whereas $x = \frac{\log S(t_0)}{K}$ when pricing options.

Remark 6.2.3 (Intuition integration range) *Integration range (6.44) is a rule-of-thumb, which is based on the cumulants of* $\log(S(T)/K)$.

Recall from Chapter 1, particularly in Equation (1.13), that the even k-th cumulant is proportional to the k-th power of the standard deviation. Taking square roots, as is done in (6.44), relates the terms appearing to the standard deviation. ▲

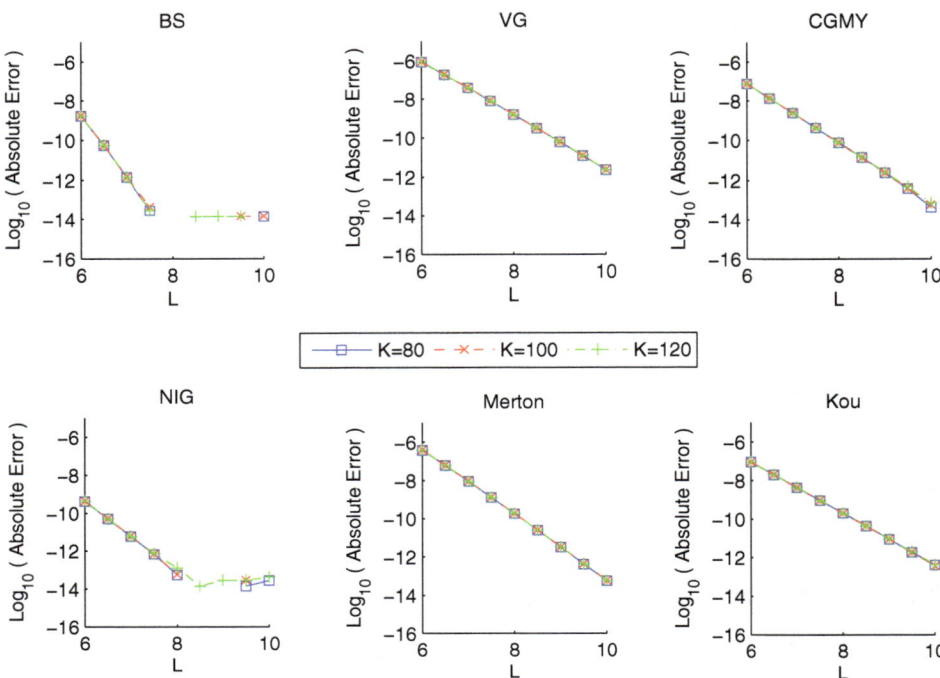

Figure 6.3: L versus the logarithm of the absolute errors for pricing call options by the COS method with $N = 2^{14}$, $T = 1$ and three different strike prices.

The relation between the user-defined tolerance level and the width of $[a, b]$, which is represented by L in (6.44) has been checked via numerical experiments, with the aim to determine one single value of L for different exponential Lévy asset price processes. With N chosen large, e.g. $N = 2^{14}$, the series truncation error is negligible and the integration range error, which has a direct relation to the user-defined tolerance level, dominates. The observed error for different values of L is presented in Figure 6.3. Again, *BS* denotes the Black-Scholes model (geometric Brownian Motion), *Merton* denotes the jump diffusion model developed in [Merton, 1976], and *Kou* is the jump diffusion model from [Kou, 2002]. *VG* stands for the Variance Gamma model [Madan et al., 1998], *CGMY* denotes the model from [Carr et al., 2002], *NIG* is short for the Normal Inverse Gaussian Lévy process [Barndorff-Nielsen, 1997]. Figure 6.3 shows that the integration range error decreases exponentially with L.

The use of $L = 8$ seems appropriate for all Lévy processes considered, and can be used as a reliable *rule of thumb*.

Cumulant ζ_4 is included in (6.44), because the density functions of many Lévy processes, for short maturity time T, will have sharp peaks and fat tails (accurately represented by ζ_4). Formula (6.44) is accurate in the range $T = 0.1$ to $T = 10$. It then defines an integration range which gives a truncation error around 10^{-12}. Larger values of parameter L would require larger N-values to reach the same level of accuracy.

Remark 6.2.4 (Simple integration range) *As an alternative integration range, which is much simpler to compute, we propose here the following,*

$$[a,b] = [-L\sqrt{T}, L\sqrt{T}]. \tag{6.45}$$

Clearly, when pricing options this integration range is not tailored to the density function for $\log\left(\frac{S(t)}{K}\right)$, *and it should therefore be considered with some care, but it has the advantages that the integration range depends neither on the cumulants nor on the strike K.*

This is particularly useful when we consider the computation of option values for multiple strike prices, $\mathbf{K}$, simultaneously and also when the cumulants are difficult to compute. We will use this simplified range for the multiple strike vector computation in Table 6.4, as well as for the computation of option prices under the Heston stochastic volatility in Table 8.2. ▲

Overview of characteristic functions

Here, we summarize some well-known characteristic functions for several exponential Lévy processes, that we already discussed in the previous chapters. The parameters used in the notation of these characteristic functions are the same we used in the respective chapters.

The characteristic function of the *log-asset price* is known, and the payoff was also represented as a function of the log-asset price, $x := \log(S_0/K)$ and $y := \log(S(T)/K)$.

The characteristic function of $X(t) = \log(S(t)/K)$ reads,

$$\phi_X(u; t_0, t) := e^{iuX(t_0)} \varphi_X(u, t) = e^{iux} \mathbb{E}[e^{iuX(t)}], \tag{6.46}$$

and is denoted by $\phi_X(u) := \phi_X(u; t_0, t)$.

The characteristic functions for several exponential Lévy processes are summarized in Table 6.2, see also [Cont and Tankov, 2004; Schoutens, 2003]; "GBM" stands for geometric Brownian Motion model; "Merton" and "Kou" denote the jump diffusion models, from Section 5.1, as developed in [Kou and Wang, 2004] and [Merton, 1976], "VG" and "CGMY" are the Variance Gamma and CGMY models from Sections 5.4.1 and 5.4.2, respectively; "NIGB" for Normal Inverse Gaussian with Brownian motion, see Section 5.4.3 and [Barndorff-Nielsen, 1997]. In the table, the *drift correction term*, $\bar{\omega}$, which is defined as

Table 6.2: *Characteristic functions for various models.*

GBM	$\varphi_X(u,t) = \exp\left(iu\mu t - \frac{1}{2}\sigma^2 u^2 t\right)$
	$\mu := r - \frac{1}{2}\sigma^2 - q$
Merton	$\varphi_X(u,t) = \exp\left(iu\mu t - \frac{1}{2}\sigma^2 u^2 t\right) \cdot \varphi_{\text{Merton}}(u,t)$
	$\varphi_{\text{Merton}}(u,t) = \exp\left[\xi_p t\left(\exp(i\mu_J u - \frac{1}{2}\sigma_J^2 u^2) - 1\right)\right]$
	$\mu := r - \frac{1}{2}\sigma^2 - q - \bar{\omega},$
	$\bar{\omega} = \xi_p(\exp(\frac{1}{2}\sigma_J^2 + \mu_J) - 1)$
Kou	$\varphi_X(u,t) = \exp\left(iu\mu t - \frac{1}{2}\sigma^2 u^2 t\right) \cdot \varphi_{\text{Kou}}(u,t)$
	$\varphi_{\text{Kou}}(u,t) = \exp\left[\xi_p t\left(\frac{p_1\alpha_1}{\alpha_1 - iu} + \frac{p_2\alpha_2}{\alpha_2 + iu} - 1\right)\right]$
	$\mu := r - \frac{1}{2}\sigma^2 - q + \bar{\omega},$
	$\bar{\omega} := \xi_p\left(1 - \frac{p_1\alpha_1}{\alpha_1 - 1} - \frac{p_2\alpha_2}{\alpha_2 + 1}\right)$
VG	$\varphi_X(u,t) = \exp\left(iu\mu t\right) \cdot \varphi_{\text{VG}}(u,t)$
	$\varphi_{\text{VG}}(u,t) = (1 - iu\theta\beta + \frac{1}{2}\sigma_{\text{VG}}^2 \beta u^2)^{-t/\beta}$
	$\mu := r - q + \bar{\omega},$
	$\bar{\omega} := (1/\beta) \cdot \log(1 - \theta\beta - \frac{1}{2}\sigma_{\text{VG}}^2 \beta)$
CGMYB	$\varphi_X(u,t) = \exp\left(iu\mu t - \frac{1}{2}\sigma_{\text{CGMYB}}^2 u^2 t\right) \cdot \varphi_{\text{CGMY}}(u,t)$
	$\varphi_{\text{CGMY}}(u,t) = \exp(Ct\Gamma(-Y)[(M - iu)^Y - M^Y + (G + iu)^Y - G^Y])$
	$\mu := r - \frac{1}{2}\sigma_{\text{CGMYB}}^2 - q + \bar{\omega},$
	$\bar{\omega} := -C \cdot \Gamma(-Y)[(M - 1)^Y - M^Y + (G + 1)^Y - G^Y]$
NIGB	$\varphi_X(u,t) = \exp\left(iu\mu t - \frac{1}{2}\sigma_{\text{NIGB}}^2 u^2 t\right) \cdot \varphi_{\text{NIG}}(u,t)$
	$\varphi_{\text{NIG}}(u,t) = \exp\left[\delta t \left(\sqrt{\alpha^2 - \beta^2} - \sqrt{\alpha^2 - (\beta + iu)^2}\right)\right]$
	$\mu := r - \frac{1}{2}\sigma_{\text{NIGB}}^2 - q + \bar{\omega},$
	$\bar{\omega} = \delta(\sqrt{\alpha^2 - (\beta + 1)^2} - \sqrt{\alpha^2 - \beta^2})$

$\bar{\omega} := -\frac{1}{t}\log\left(\varphi_X(-i,t)\right)$ (making these exponential processes martingales) is also included, in the general drift parameter μ. Also proportional dividend payment rate q, as in (2.21) is included in parameter μ.

Note that in the table a Brownian motion component has been added to the CGMY and NIG processes, so the CGMYB and NIGB processes are displayed. Except for the VG process, now all processes contain a diffusion term.

Overview of cumulants

Given the characteristic functions, in Table 6.2, the cumulants can be computed by means of the formula,

$$\zeta_n(X) = \frac{1}{i^n} \frac{\partial^n (t\Psi(u))}{\partial u^n}\bigg|_{u=0},$$

where the argument "$t\Psi(u)$" is the exponent of the characteristic function $\phi_X(u)$, i.e.

$$\phi_X(u) = e^{t\Psi(u)}, \quad t \geq 0.$$

The formulas for the specific cumulants are summarized in Table 6.3.

Table 6.3: *Cumulants of interest for the various models.*

GBM	$\zeta_1 = (r - q - \frac{1}{2}\sigma^2)t, \quad \zeta_2 = \sigma^2 t, \quad \zeta_4 = 0$
Merton	$\zeta_1 = t(r - q - \bar{\omega} - \frac{1}{2}\sigma^2 + \xi_p \mu_J) \qquad \zeta_2 = t\left(\sigma^2 + \xi_p \mu_J^2 + \sigma_J^2 \xi_p\right)$ $\zeta_4 = t\xi_p \left(\mu_J^4 + 6\sigma_J^2 \mu_J^2 + 3\sigma_J^4 \xi_p\right)$
Kou	$\zeta_1 = t\left(r - q + \bar{\omega} - \frac{1}{2}\sigma^2 + \frac{\xi_p p_1}{\alpha_1} - \frac{\xi_p p_2}{\alpha_2}\right) \quad \zeta_2 = t\left(\sigma^2 + 2\frac{\xi_p p_1}{\alpha_1^2} + 2\frac{\xi_p p_2}{\alpha_2^2}\right)$ $\zeta_4 = 24 t\xi_p \left(\frac{p_1}{\alpha_1^4} + \frac{p_2}{\alpha_2^4}\right)$
VG	$\zeta_1 = (r - q - \bar{\omega} + \theta)t \qquad\qquad \zeta_2 = (\sigma_{\text{VG}}^2 + \beta\theta^2)t$ $\zeta_4 = 3(\sigma_{\text{VG}}^4 \beta + 2\theta^4 \beta^3 + 4\sigma_{\text{VG}}^2 \theta^2 \beta^2)t$
CGMYB	$\zeta_1 = (r - q + \bar{\omega} - \frac{1}{2}\sigma_{\text{CGMYB}}^2)t + Ct\Gamma(1 - Y)\left(M^{Y-1} - G^{Y-1}\right)$ $\zeta_2 = \sigma_{\text{CGMYB}}^2 t + Ct\Gamma(2 - Y)\left(M^{Y-2} + G^{Y-2}\right)$ $\zeta_4 = C\Gamma(4 - Y)t\left(M^{Y-4} + G^{Y-4}\right)$
NIG	$\zeta_1 = (r - q + \bar{\omega} - \frac{1}{2}\sigma_{\text{NIGB}}^2 + \delta\beta/\sqrt{\alpha^2 - \beta^2})t$ $\zeta_2 = \delta\alpha^2 t(\alpha^2 - \beta^2)^{-3/2}$ $\zeta_4 = 3\delta\alpha^2(\alpha^2 + 4\beta^2)t(\alpha^2 - \beta^2)^{-7/2}$

Here, the drift correction $\bar{\omega}$ satisfies $\exp(-\bar{\omega}t) = \varphi_X(-i, t)$.

Efficient computation for Lévy models

Pricing formula (6.28) can be used for European options under any underlying process as long as the characteristic function is available. This is the case for exponential Lévy models and also for models from the class of regular affine processes of [Duffie et al., 2003], including the exponentially affine jump diffusion class of [Duffie et al., 2000]. The latter class will be discussed in some detail in the chapter to follow.

It is worth mentioning that the COS pricing formula (6.28) can be greatly simplified for exponential Lévy processes, as multiple options *for different strike prices* can be computed *simultaneously*.

For a given column vector of strikes, **K**, we consider the following transformation $\mathbf{x} = \log\left(\frac{S(t_0)}{\mathbf{K}}\right)$ and $\mathbf{y} = \log\left(\frac{S(T)}{\mathbf{K}}\right)$. For a Lévy process, the characteristic function can be represented by

$$\phi_\mathbf{x}(u; t_0, T) = \varphi_{X_\mathcal{L}}(u, T) \cdot e^{iu\mathbf{x}}. \tag{6.47}$$

In this case, the pricing formula simplifies, as we can write,

$$V(t_0, \mathbf{x}) \approx e^{-r\tau} \sum_{k=0}^{N-1}{}' \operatorname{Re}\left\{ \varphi_{X_\mathcal{L}}\left(\frac{k\pi}{b-a}, T\right) \exp\left(ik\pi\frac{\mathbf{x}-a}{b-a}\right) \right\} \mathbf{H}_k. \tag{6.48}$$

with $\tau = T - t_0$. Recalling the H_k-formulas for vanilla European options in (6.34) and (6.35), they can be presented as a *vector* multiplied by a scalar, i.e.

$$\mathbf{H}_k = U_k \mathbf{K},$$

where

$$U_k = \begin{cases} \dfrac{2}{b-a}\left(\chi_k(0, b) - \psi_k(0, b)\right) & \text{for a call,} \\ \dfrac{2}{b-a}\left(-\chi_k(a, 0) + \psi_k(a, 0)\right) & \text{for a put.} \end{cases} \tag{6.49}$$

As a result, the COS pricing formula reads[4]

$$\boxed{V(t_0, \mathbf{x}) \approx \mathbf{K} e^{-r\tau} \cdot \operatorname{Re}\left\{ \sum_{k=0}^{N-1}{}' \varphi_{X_\mathcal{L}}\left(\frac{k\pi}{b-a}, T\right) U_k \cdot \exp\left(ik\pi\frac{\mathbf{x}-a}{b-a}\right) \right\},} \tag{6.50}$$

where the summation can be written as a matrix-vector product if **K** (and therefore **x**) is a vector.

Equation (6.50) is an expression with independent variable **x**. It is therefore possible to obtain the option prices for different strikes in one single numerical experiment, by choosing a **K**-vector as the input vector.

6.3 Numerical COS method results

In this section, some numerical experiments are presented to confirm the efficiency and accuracy of the COS method. The focus is on the plain vanilla European options and different processes for the underlying asset are considered, like the

---

[4] Although the U_k values are real, we keep them in the curly brackets. This allows us to interchange $\operatorname{Re}\{\cdot\}$ and $\sum'$, and it simplifies the implementation in certain software packages.

Black-Scholes model and the infinite activity exponential Lévy VG and CGMY processes. These are numerical experiments with long and short maturity times.

The underlying density function for each individual experiment is also recovered with the help of the Fourier cosine series inversion, as presented in Section 6.1.1. This may help the reader to gain some insight in the relation between the error convergence and the properties of the densities.

All CPU times presented, in milliseconds, are determined after averaging the computing times obtained from 10^4 experiments, on a regular CPU.

6.3.1 Geometric Brownian Motion

The first set of call option experiments is performed under the GBM process with a short time to maturity. The parameters selected for this test are

$$S_0 = 100, \quad r = 0.1, \quad q = 0, \quad T = 0.1, \quad \sigma = 0.25. \tag{6.51}$$

The convergence behavior at three different strike prices, $K = 80, 100$, and 120, is computed and checked. For this computation, we have used the vector-valued version of the COS method, with the simplified integration range in Equation (6.45). The results for these strikes are thus obtained in one single computation, as explained in (6.50).

In Table 6.4, the CPU time and the error convergence information is displayed. The maximum error of the option values over the three strike prices is presented.

Remark 6.3.1 *In all numerical experiments a linear computational complexity is achieved by the COS method. By doubling N, performing the computations, and checking the differences between subsequent timings, we can distinguish the linear complexity from the computational overhead.* ▲

Example 6.3.1 (Cash-or-nothing option) We confirm that the convergence of the COS method does not depend on a discontinuity in the payoff function, provided an analytic expression for the coefficients H_k^{cash} is available, by pricing a cash-or-nothing call option. The underlying process is GBM, so that an analytic solution exists. Parameters selected for this test are

$$S_0 = 100, \quad K = 120, \quad r = 0.05, \quad q = 0, \quad T = 0.1, \quad \sigma = 0.2. \tag{6.52}$$

Table 6.4: *Error convergence and CPU time using the COS method for European calls under GBM, with parameters as in (6.51); $K = 80, 100, 120$; reference val. = $20.799226309\ldots$, $3.659968453\ldots$, and $0.044577814\ldots$, respectively.*

N	16	32	64	128	256
msec.	0.10	0.11	0.13	0.15	0.19
abs. err. $K = 80$	3.67	7.44e-01	1.92e-02	1.31e-07	5.68e-14
abs. err. $K = 100$	3.87	5.28e-01	1.52e-02	3.87e-07	1.44e-13
abs. err. $K = 120$	3.17	8.13e-01	2.14e-02	3.50e-07	1.26e-13

Table 6.5: *Error and CPU time for a cash-or-nothing call option with the COS method, with parameters as in (6.52); reference val. = 0.273306496....*

N	40	60	80	100	120	140
Error	-2.26e-01	2.60e-03	3.59e-05	-4.85e-07	1.29e-09	-9.82e-13
CPU time (msec.)	0.330	0.334	0.38	0.43	0.49	0.50

Table 6.6: *Maximum error when recovering $f_{CGMYB}(y)$ from $\phi_{CGMYB}(u)$ by the Fourier cosine expansion, with parameters as in (6.53). The grid for y is from $y_{\min} = -2$ to $y_{\max} = 12$ with 250 steps.*

	\multicolumn{6}{c}{$Y = 0.5$}					
N :	32	48	64	80	96	128
Error:	4.27e-04	1.34e-06	1.79e-09	1.01e-12	2.22e-16	< 1e-16
Time (msec.)	0.12	0.15	0.20	0.21	0.22	0.29
	\multicolumn{6}{c}{$Y = 1.5$}					
N :	8	16	24	32	40	48
Error:	6.14e-02	1.81e-032	8.25e-06	6.10e-09	8.13e-13	< 1e-16
Time (msec.)	0.09	0.10	0.11	0.12	0.13	0.14

Table 6.5 presents the exponential convergence of the COS method.   ◆

6.3.2 CGMY and VG processes

Especially when pricing call options, the solution's accuracy may exhibit *some sensitivity regarding the size of this truncated integration interval*. This holds specifically for call options under fat-tailed distributions, like under Lévy jump processes, or for options with an extremely long time to maturity.[5] A call payoff grows exponentially in log–stock price which may introduce cancellation errors in large domain sizes. Put options do not suffer from this, as their payoff value is bounded by strike value K. In [Fang and Oosterlee, 2008], European call options were priced with European put option values, in combination with the put-call parity (3.4).

We consider here the method's convergence for call options under the CGMY model in some more detail, *employing the put-call parity relation to price call options*. It has been reported in [Almendral and Oosterlee, 2007; Wang et al., 2007] that PIDE methods have difficulty solving for parameter $Y \in [1, 2]$. Here, we choose $Y = 0.5$ and $Y = 1.5$, respectively. Other parameters are selected as follows:

$$S_0 = 100, \quad K = 100, \quad r = 0.1, \quad \sigma_{\text{CGMYB}} = 0.2,$$
$$q = 0, \quad C = 1, \quad G = 5, \quad M = 5, \quad T = 1. \quad (6.53)$$

---
[5]This is mainly the case for insurance options with a life time of 30 years and higher.

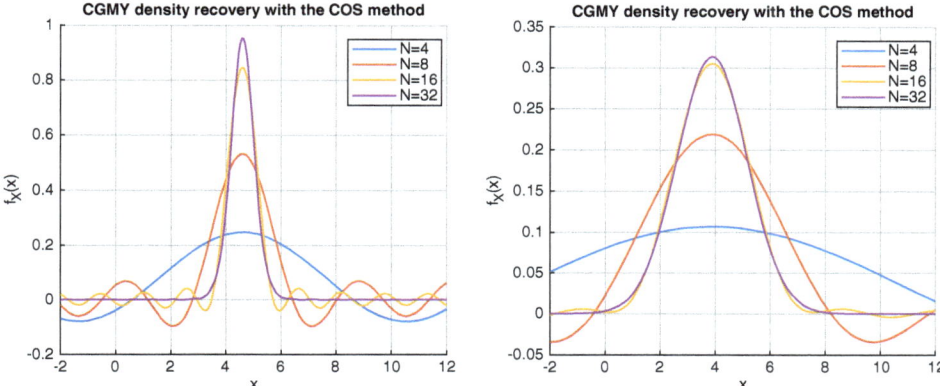

Figure 6.4: *Recovery of the CGMY density obtained for different numbers of expansion terms. Left: $Y = 0.5$ Right: $Y = 1.5$; other parameters are as in (6.53).*

Table 6.7: *Convergence of the COS method (accuracy and speed) for CGMY with $Y = 0.5$ and $Y = 1.5$; Other parameters as in (6.53); reference values: $(Y = 0.5) = 21.679593920471817\ldots$, $(Y = 1.5) = 50.27953397994453\ldots$.*

	$Y = 0.5$					
N :	32	48	64	80	96	128
Error:	3.794e-02	1.075e-03	2.597e-05	5.132e-07	8.023e-09	8.811e-13
Time (msec.)	0.61	0.69	0.78	0.89	0.95	1.11
	$Y = 1.5$					
N :	4	8	16	24	32	36
Error:	2.428e+00	6.175e-02	3.224e-04	1.565e-08	2.842e-14	<1e-16
Time (msec.)	0.53	0.56	0.58	0.63	0.68	0.72

The convergence of the COS method for the recovery of the CGMYB density function is presented in Table 6.6. In Figure 6.4, the recovered density functions for two Y-values are plotted. For large values of Y, the tails of the density function are fatter and the center of the distribution shifts.

Reference option values for the numerical experiments are computed by the COS method with $N = 2^{14}$. The numerical option results are presented in Table 6.7.

The COS method converges exponentially, and with a relatively small value of N, i.e., $N \leq 100$, the results are already accurate up to seven digits. The computational time spent is less than 0.1 millisecond. The convergence rate with $Y = 1.5$ in Table 6.7 is faster than that of $Y = 0.5$, because the density functions from a fat-tailed distribution can often be well represented by cosine basis functions.

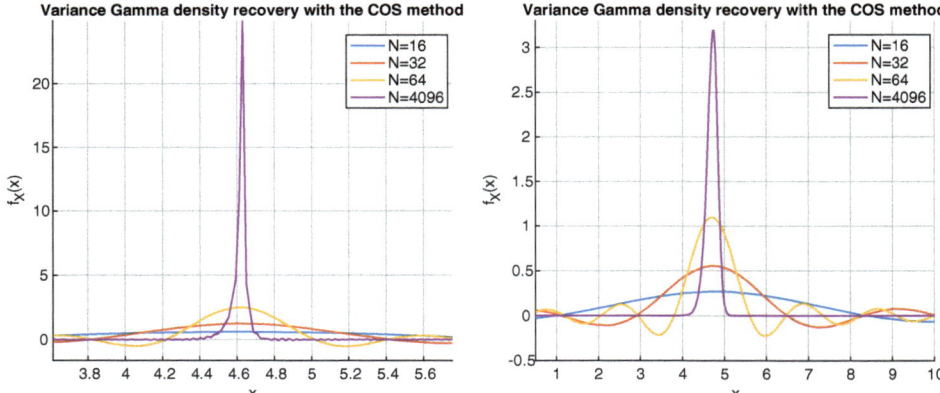

Figure 6.5: *Recovery of the VG density obtained for different numbers of expansion terms. Left: $T = 0.1$ Right: $T = 1$, with other parameters as in (6.54).*

Table 6.8: *Convergence of the COS method for a call under the VG model with $K = 90$ and other parameters as in (6.54).*

$T = 0.1$; Reference val. = 10.993703187...			$T = 1$; Reference val. = 19.099354724...		
N	Error	Time (msec.)	N	Error	Time (msec.)
64	2.05e-02	0.04	32	3.41e-01	0.05
128	1.99e-02	0.06	64	2.46e-01	0.06
256	8.30e-04	0.08	128	1.02e-02	0.09
512	3.06e-04	0.10	160	9.51e-04	0.07
1024	3.02e-05	0.14	512	3.40e-08	0.11

Example 6.3.2 (Variance Gamma) As another example, call options are priced under the VG process, which also belongs to the class of infinite activity Lévy processes, see Section 5.4.1. The VG process is usually parameterized with the parameters σ_{VG}, θ, and β.

The parameters selected in the numerical experiments are

$$S_0 = 100, \ r = 0.1, \ q = 0, \ \sigma_{VG} = 0.12, \ \theta = -0.14, \ \beta = 0.2. \tag{6.54}$$

Here we compare the convergence for $T = 1$ with $T = 0.1$.

Figure 6.5 presents the difference in shape of the two recovered VG density functions. For $T = 0.1$, the density is much more peaked. Results are summarized in Table 6.8. Note that for $T = 0.1$ the error convergence of the COS method is *algebraic instead of exponential*, in agreement with the recovered density function in Figure 6.5, which is clearly not in $C^\infty([a,b])$. In the extreme case, we would observe a delta function-like function for $T \to 0$. ◆

6.3.3 Discussion about option pricing

At a financial institution, one can distinguish a number of tasks that must be performed in order to price a financial derivative product. First of all, the financial product is defined, because specific customers may need it. Typically, there are underlying assets, like interest rate and stocks, that are modeled by means of SDEs. To determine values for the open parameters in the SDEs, one needs to calibrate, i.e. to fit typically mathematical European option prices, to the financial market option prices.

Plain vanilla European index options often form the basis for the calibration. European option prices are usually heavily traded, with small bid-ask spread, and they contain valuable market information about the future uncertainty for example, by means of the implied volatilities. These European option products may also be used for hedging strategies to reduce the risk associated with selling a specific derivatives product. As any pricing and risk management system has to be able to calibrate to plain vanilla European options, it is important to value these options *quickly and accurately*.

The financial derivative product of interest, possibly with complex contract features, is often priced by means of a *Monte Carlo simulation* or by numerical approximations of the governing P(I)DE.

> The choice of numerical method for option valuation is thus often based on whether the computation is meant for *model calibration*, for which the speed of a pricing method for European options is essential, or for the *pricing of a specific involved derivatives contract*, for which robustness of the numerical method is of importance.

The choice of stochastic model for the underlying assets also has a significant impact on the computational techniques. Ideally a (semi-)analytic option price is available for the European option prices, but this is only the case for the most basic asset price dynamics, like the Black-Scholes dynamics. Alternatively, accurate *approximate option values* (or approximate implied volatilities) by means of an asymptotic expansion may be used, see, for example, [Fouque et al., 2004; Hagan et al., 2002; Lorig et al., 2015], but the range of applicability may be limited. The class of Fourier-based methods represents highly efficient numerical pricing techniques for pricing European options under asset price dynamics for which the characteristic function is available, like the COS method from this chapter. Because of their speed in computation, they may thus fit well to the parameter calibration task.

Fourier methods are typically faster than P(I)DE and Monte Carlo methods for pricing single asset European options. A benchmark option pricing comparison has been presented in [von Sydow et al., 2015], in which the Fourier-based option pricing techniques showed superior performance for many basic options.

They are also referred to as transform methods, as the computation takes place on the basis of a Fourier transform, combined with numerical integration [Carr and Madan, 1999; O'Sullivan, 2005; Lee, 2004]. Fourier methods are computationally fast but it is a challenge to employ them for exotic payoff contracts.

Remark 6.3.2 (Fast Fourier Transform) *Numerical integration methods typically compute the Fourier integrals by applying equally spaced numerical integration rules and then employing the FFT algorithm by imposing the Nyquist relation to the grid sizes,* $\Delta y, \Delta u$, *in the y- and u-domains, respectively, i.e.*

$$\Delta y \cdot \Delta u \equiv \frac{2\pi}{N}, \qquad (6.55)$$

▲

where N represents the number of grid points. The grid values can then be obtained in $O(N \log_2 N)$ operations. However, the error convergence of equally spaced integration rules may not be very high; parameter N has to be a power of two, and the relation imposed on the grid sizes (6.55) prevents one from using coarse grids in both y and u computational domains.

Other highly efficient techniques for pricing plain vanilla options, under certain specified asset dynamics, include the Carr-Madan FFT-based method [Carr and Madan, 1999], the fast Gauss transform [Broadie and Yamamoto, 2003] and the double-exponential transformation [Mori and Sugihara, 2001; Yamamoto, 2005]. Also wavelet-based numerical techniques for option pricing and risk management, based on B-splines in [Ortiz-Gracia and Oosterlee, 2013; Kirkby, 2016], and on Shannon wavelets [Cattani, 2008] in the Shannon Wavelet Inverse Fourier Transform (SWIFT) method [Ortiz-Gracia and Oosterlee, 2016], have recently shown promising performance in robustness and efficiency.

We will not discuss these techniques here and refer the interested reader to the journal publications.

6.4 Exercise set

Exercise 6.1 Consider the following tasks.

a. Approximate the call and put *payoff functions*, with $K = 10$, by their Fourier cosine expansions on the domain $[0, 30]$. Plot the respective approximations of these functions with $N = 2, 5, 10$ and 40 cosine expansion terms.

b. Approximate the call payoff function, with $K = 10$, by its Fourier cosine expansion, but now on the domain $[10, 30]$. Plot the respective approximations of these functions with $N = 2, 5, 10$ and 40 cosine expansion terms. Describe the differences with the observations in item a.

c. Approximate the call and put payoff functions, with $K = 10$, by their Fourier *sine* expansions on the domain $[0, 30]$. Plot the respective approximations of these functions with $N = 2, 5, 10$ and 40 cosine expansion terms. Describe the differences with the observations in item a.

Exercise 6.2 Recover the PDFs of the VG, CGMY and NIG models, for specific choices of their model parameters so that the impact of the kurtosis, skewness and the fat-tailedness can be observed.

Exercise 6.3 Compare call option prices based on the characteristic function of Merton's jump diffusion model from the previous section, in (5.33), with the analytic option values based on the infinite sum given in Section (5.2.1).

Exercise 6.4 With the characteristic function of a random variable X, the density can be recovered by the COS method using the following formula,

$$\hat{f}_X(y) \approx \sum_{k=0}^{N-1} \bar{F}_k \cos\left(k\pi \frac{y-a}{b-a}\right), \tag{6.56}$$

with

$$\bar{F}_k = \frac{2}{b-a} \text{Re}\left\{\phi_X\left(\frac{k\pi}{b-a}\right) \cdot \exp\left(-i\frac{ka\pi}{b-a}\right)\right\}, \tag{6.57}$$

show that the CDF, $\hat{F}_X(y)$, can be approximated by means of the following formula,

$$\hat{F}_X(y) \approx \sum_{k=0}^{N-1} \bar{F}_k \psi(a, b, y), \tag{6.58}$$

with

$$\psi(a, b, y) = \begin{cases} \frac{b-a}{k\pi} \sin\left(\frac{k\pi(y-a)}{b-a}\right), & \text{for } k = 1, 2, \ldots, N-1, \\ (x-a), & \text{for } k = 0. \end{cases} \tag{6.59}$$

Exercise 6.5 The Gumbel distribution is an often encountered distribution in extreme value theory. This distribution may be used to model the distribution of the maximum value of a number of samples. A Gumbel distributed random variable, Y, with *scale parameter* $1/k_1 > 0$ and *location parameter* $k_2 \in \mathbb{R}$ is given by,

$$f_Y(x) = k_1 \exp\left(k_1(k_2 - y) - e^{k_1(k_2-y)}\right).$$

Show that the related characteristic function is given by,

$$\phi_Y(u) = \mathbb{E}\left[e^{iuY}\right] = \Gamma\left(1 - \frac{iu}{k_1}\right) e^{iuk_2},$$

where $\Gamma(\cdot)$ is the Gamma function, which is defined as $\Gamma(s) = \int_0^\infty z^{s-1} e^{-z} dz$.

Exercise 6.6 Compute the jump diffusion option prices that were previously computed in Exercise 5.11, by means of the COS method.

Use also Kou's jump diffusion model within the COS method, based on the three test parameter sets

Set I: $\xi_p = 8, p_1 = 0.4, p_2 = 0.6, \alpha_1 = 10, \alpha_2 = 5$;

Set II: $\xi_p = 8, p_1 = 0.2, p_2 = 0.8, \alpha_1 = 10, \alpha_2 = 5$;

Set III: $\xi_p = 0.1, p_1 = 0.2, p_2 = 0.8, \alpha_1 = 10, \alpha_2 = 50$;

Explain the observed densities, in dependence of the different parameter sets.

Exercise 6.7 Derive the first four cumulants for the CGMYB model, with the characteristic function defined in Equation (5.64).

Exercise 6.8 Repeat the computation of the option prices under the infinite Lévy models, as performed in Exercise 5.12, by means of the COS method. Can you confirm that identical option prices result? What is the complexity of computing the prices for a range of strike prices when using the COS method?

Exercise 6.9 When pricing call options, the solution's accuracy is sensitive to the width of the integration domain $[a, b]$. The payoff function of a call grows exponentially in log-stock price, which may result in larger round-off errors, under fat-tailed distributions. Often it is possible to avoid some of these errors by using the put-call parity, see Equation (3.3), which reads:

$$V_c(t_0, S) = V_p(t_0, S) + S(t_0) - K e^{-r(T-t_0)}.$$

Consider two models, Geometric Brownian Motion and the exponential Variance Gamma process and investigate, for different sets of self chosen parameters, the improvement of the COS method convergence when using the call-put parity when pricing a European call option with $T = 2$, $S_0 = 1$ and $K = 0.5, 1, 3$.

Exercise 6.10 The objective of this exercises is to implement the FFT algorithm for the recovery of a marginal density $f_X(x)$, of process $X(t)$.

    a. Use the definition of the characteristic function and show that for the density $f_X(x)$ the following equality holds:

$$f_X(x) = \frac{1}{2\pi} \left(\int_0^\infty e^{-iux} \phi_X(u) du + \overline{\int_0^\infty e^{-iux} \phi_X(u) du} \right).$$

    b. Use some properties of complex numbers to show that:

$$f_X(x) = \frac{1}{\pi} \text{Re}\left(\int_0^\infty e^{-iux} \phi_X(u) du \right).$$

c. Consider $\bar{\phi}_X(u, x) := e^{-iux}\phi_X(u)$ for which we apply the Trapezoidal integration rule on a domain $[0, u_{\max}]$, with $u_n = (n-1)\Delta u$. Define a grid x: $x_j = (j-1)\Delta x + x_{\min}$, for $j = 1, \ldots, N$, and show that

$$\bar{\phi}_X(u_n, x_j) = \phi_X(u_n)e^{-i(n-1)\Delta u[(j-1)\Delta x + x_{\min}]},$$

for any $n = 1, \ldots, N$ and $j = 1, \ldots, N$.

d. Use the discrete Fourier transform for function f_n given by:

$$\phi_j = \sum_{n=1}^{N} e^{-\frac{2\pi i}{N}(n-1)(j-1)} f_n. \quad (6.60)$$

Take $\Delta u \cdot \Delta x = \frac{2\pi}{N}$ and show that:

$$\sum_{n=1}^{N} \bar{\phi}_X(u_n, x_j) = \sum_{n=1}^{N} e^{-i\frac{2\pi}{N}(n-1)(j-1)} \left[e^{-iu_n x_{\min}} \phi_X(u_n) \right].$$

e. Derive the functions $\widehat{\phi}_X(u_n)$ and $\widehat{\phi}_1$, such that the following equality holds,

$$f_X(x_j) \approx \frac{1}{\pi} \operatorname{Re}\left[\Delta u \left(\sum_{n=1}^{N} e^{-i\frac{2\pi}{N}(n-1)(j-1)} \widehat{\phi}_X(u_n) - \widehat{\phi}_1 \right) \right]. \quad (6.61)$$

f. Implement all steps of the FFT density recovery as explained above and compare the result, in terms of computing time and accuracy, with a COS method density recovery, for a normal probability density function.

Exercise 6.11 Fourier-based techniques can be improved by addition of a so-called "spectral filter". This is particularly useful when a discontinuity is present in a characteristic function, since a discontinuity hampers the global convergence of the COS method. The idea of filtering is to alter the expansion coefficients in such a way that they decay faster. If the filter is well chosen, this will improve the COS method convergence rate away from a discontinuity in the characteristic function. The COS method for pricing of European-type payoff function, which is extended with an exponential filter reads,

$$\hat{f}_X(y) \approx \sum_{k=0}^{N-1} \hat{s}\left(\frac{k}{N}\right) \bar{F}_k \cos\left(k\pi \frac{y-a}{b-a}\right), \quad (6.62)$$

with $\bar{F}_k$ given in (6.57) and

$$\hat{s}(x) = \exp\left(-\alpha x^p\right),$$

where p must be an even number and $\alpha = -\log \epsilon_m$, and ϵ_m represents the machine epsilon, so that $\hat{s}(1) = \epsilon_m \approx 0$ within machine precision.

Consider the VG model with zero drift with the probability density function, which is given by,

$$f_{VG}(y) = \int_0^\infty \frac{1}{\sigma\sqrt{2\pi z}} \exp\left(-\frac{(y-\theta z)^2}{2\sigma^2 z}\right) \frac{z^{\frac{T}{\beta}-1} \exp\left(-\frac{z}{\beta}\right)}{\beta^{\frac{T}{\beta}} \Gamma\left(\frac{T}{\beta}\right)} dz \quad (6.63)$$

and the corresponding characteristic function is given by,

$$\phi_{VG}(u, T) = \left(1 - i\beta\theta u + \frac{1}{2}\beta\sigma_{VG}^2 u^2\right)^{-\frac{T}{\beta}}. \quad (6.64)$$

Using the "filtered COS algorithm" in (6.62) and the characteristic function in (6.64) recover the probability density function $\hat{f}_Y(y)$ and compare the results to those based on Equation (6.63). In the experiment consider the following set of parameters: $\theta = -0.14$, $\beta = 0.2$, $\sigma = 0.12$, $N = 500$, $p = 6$, $\epsilon = 1e - 16$ and for the following two maturities, $T = 0.1$, $T = 1$. Generate one figure with the obtained densities and a second figure presenting the log error, $\log(|f_Y(y) - \hat{f}_Y(y)|)$. In the graphs also include the density (and error) without inclusion of the filter.

CHAPTER 7

Multidimensionality, Change of Measure, Affine Processes

> **In this chapter:**
>
> To be equipped for more involved stochastic models for assets in the forthcoming chapters, we present in **Section 7.1** some mathematical tools and concepts for *multi-dimensional stochastic processes*. We explain the *Cholesky decomposition*, to change the formulation of the multi-dimensional processes from dependent to independent Brownian motions, the multi-dimensional Feynman-Kac theorem for option pricing, and, particularly, *the multi-dimensional Itô lemma* which forms the basis for the option valuation models. Another mathematical formalism that we will employ is related to *measure changes*, such as from the real-world to the risk-neutral measure. We will also discuss the change of numéraire in a more general setting in **Section 7.2**, and we give examples for it. *Affine processes* form an important class of asset price processes, and they will be described in **Section 7.3**. For these processes, the *characteristic function* is readily available.

Keywords: Multi-dimensional correlated stochastic processes, Cholesky decomposition, multi-dimensional Itô's lemma, Girsanov theorem for measure transformation, affine diffusion class of SDEs, characteristic function.

7.1 Preliminaries for multi-D SDE systems

We start with the definition of a system of SDEs in which the Brownian motions, $\mathbf{W}(t)$, are *correlated Brownian motions*, and work toward an SDE system with uncorrelated, or *independent*, Brownian motions, $\widetilde{\mathbf{W}}(t)$. Useful definitions and theorems for systems of SDEs, like the multi-dimensional Itô's lemma, or the class of affine diffusions, are often stated for systems with independent Brownian motions. A correlated SDE system is transformed to a form in which the volatilities and correlations are separated.

7.1.1 The Cholesky decomposition

The basis for the formulation of SDE systems based on correlated Brownian motions $\mathbf{W}(t)$ towards systems with independent Brownian motions $\widetilde{\mathbf{W}}(t)$, and vice versa, is found in the well-known Cholesky decomposition of a correlation matrix.

Definition 7.1.1 (Vector-valued Brownian motions) *In the case of correlated Brownian motions, it is well-known that $\mathbb{E}[W_i(t) \cdot W_j(t)] = \rho_{i,j} t$, if $i \neq j$, and $\mathbb{E}[W_i(t) \cdot W_i(t)] = t$, if $i = j$, for $i, j = 1, \ldots n$. Similarly, for the correlated Brownian increments, it follows that $\mathrm{d}W_i(t) \cdot \mathrm{d}W_j(t) = \rho_{i,j} \mathrm{d}t$, if $i \neq j$, and $\mathrm{d}W_i(t) \cdot \mathrm{d}W_i(t) = \mathrm{d}t$, if $i = j$.*

Two Brownian motions are said to be independent, if $\mathbb{E}[\widetilde{W}_i(t) \cdot \widetilde{W}_j(t)] = 0$, if $i \neq j$ and $\mathbb{E}[\widetilde{W}_i(t) \cdot \widetilde{W}_j(t)] = t$, if $i = j$, for $i, j = 1, \ldots n$. For the Brownian increments, we then have similarly, $\mathrm{d}\widetilde{W}_i(t) \cdot \mathrm{d}\widetilde{W}_j(t) = 0$, if $i \neq j$ and $\mathrm{d}\widetilde{W}_i(t) \cdot \mathrm{d}\widetilde{W}_j(t) = \mathrm{d}t$, if $i = j$. ◀

Example 7.1.1 (Examples of correlated Brownian motions)
In Figure 7.1, two Brownian motion paths are presented in each figure, where the correlation between the Brownian motions is varied. In the first figure, the two Brownian motions are governed by a negative correlation parameter $\rho_{1,2}$, in the second figure $\rho_{1,2} = 0$, while in the third figure a positive correlation $\rho_{1,2} > 0$ is used. The impact of correlation parameter $\rho_{1,2}$ on these paths becomes clear from the figures. ◆

A correlation structure can be cast into a correlation matrix $\mathbf{C}$.

Definition 7.1.2 (Cholesky decomposition) *Each symmetric positive definite matrix $\mathbf{C}$ has a unique factorization, the so-called Cholesky decomposition, of the form,*
$$\mathbf{C} = \mathbf{L}\mathbf{L}^{\mathrm{T}}, \tag{7.1}$$
where $\mathbf{L}$ is a lower triangular matrix with positive diagonal entries.
When a matrix is only symmetric positive semi-definite, this implies that at least one eigenvalue of the matrix equals zero. In that case, there is a dependency among the columns that make up the correlation matrix, which means that one process is a linear combination of the other processes, and can be eliminated. ◀

The following example shows the use of this decomposition for a correlation matrix.

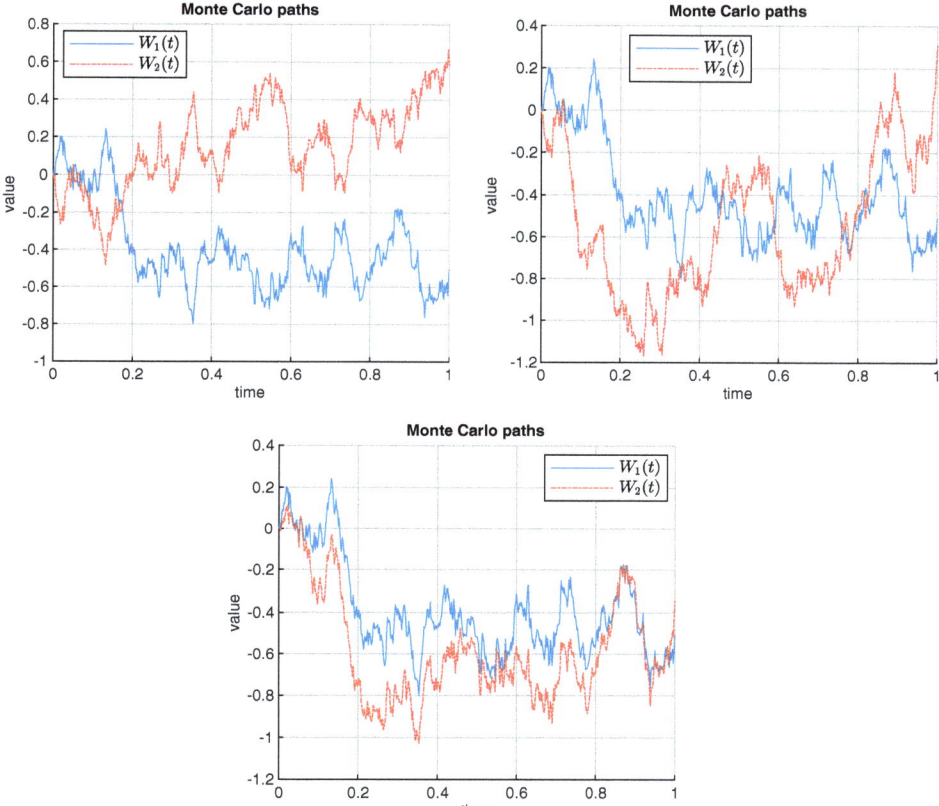

Figure 7.1: *Monte Carlo Brownian motion $W_i(t)$ paths with different correlations, $\mathbb{E}[W_1(t)W_2(t)] = \rho_{1,2}t$; middle: zero correlation; left: negative correlation ($\rho_{1,2} < 0$), right: positive correlation ($\rho_{1,2} > 0$).*

Example 7.1.2 (Example of Cholesky decomposition) For a given (2×2)-*correlation matrix* $\mathbf{C}$, we find the Cholesky decomposition (7.1), as

$$\mathbf{C} \stackrel{\text{def}}{=} \begin{bmatrix} 1 & \rho_{1,2} \\ \rho_{1,2} & 1 \end{bmatrix} = \begin{bmatrix} 1 & 0 \\ \rho_{1,2} & \sqrt{1-\rho_{1,2}^2} \end{bmatrix} \begin{bmatrix} 1 & \rho_{1,2} \\ 0 & \sqrt{1-\rho_{1,2}^2} \end{bmatrix}. \quad (7.2)$$

♦

It is a useful exercise to transform an SDE system with two independent Brownian motions, $\widetilde{\mathbf{W}}(t) = [\widetilde{W}_1(t), \widetilde{W}_2(t)]^{\mathrm{T}}$, into an SDE system with correlated Brownian motions with correlation coefficient $\rho_{1,2}$. This technique can directly be generalized to higher dimensions.

Example 7.1.3 (Correlating independent Brownian motions.) To correlate independent Brownian motions, we perform a matrix-vector multiplication, $\mathbf{L} \cdot \widetilde{\mathbf{W}}(t)$, with $\mathbf{L}$ a lower triangular matrix, as in (7.2), i.e.

$$\begin{bmatrix} 1 & 0 \\ \rho_{1,2} & \sqrt{1-\rho_{1,2}^2} \end{bmatrix} \begin{bmatrix} \widetilde{W}_1(t) \\ \widetilde{W}_2(t) \end{bmatrix} = \begin{bmatrix} \widetilde{W}_1(t) \\ \rho_{1,2}\widetilde{W}_1(t) + \sqrt{1-\rho_{1,2}^2}\widetilde{W}_2(t) \end{bmatrix}. \qquad (7.3)$$

Now, we define $W_1(t) := \widetilde{W}_1(t)$, $W_2(t) := \rho_{1,2}\widetilde{W}_1(t) + \sqrt{1-\rho_{1,2}^2}\widetilde{W}_2(t)$, and determine the covariance between the correlated Brownian motions $W_1(t)$ and $W_2(t)$ as

$$\begin{aligned} \mathbb{C}\mathrm{ov}[W_1(t)W_2(t)] &= \mathbb{E}[W_1(t)W_2(t)] - \mathbb{E}[W_1(t)]\mathbb{E}[W_2(t)] \\ &= \mathbb{E}\left[\widetilde{W}_1(t)\left(\rho_{1,2}\widetilde{W}_1(t) + \sqrt{1-\rho_{1,2}^2}\widetilde{W}_2(t)\right)\right] - 0 \\ &= \rho_{1,2}\mathbb{E}\left[(\widetilde{W}_1(t))^2\right] + \sqrt{1-\rho_{1,2}^2}\mathbb{E}[\widetilde{W}_1(t)]\mathbb{E}[\widetilde{W}_2(t)] \\ &= \rho_{1,2}\mathbb{E}[(\widetilde{W}_1(t))^2] = \rho_{1,2}\mathrm{Var}[\widetilde{W}_1(t)] = \rho_{1,2}t, \end{aligned}$$

using $\mathbb{E}[W_i(t)] = \mathbb{E}[\widetilde{W}_i(t)] = 0$.

The correlation between $W_1(t)$ and $W_2(t)$ then equals $\rho_{1,2}$, as desired.

Similarly, for increments of Brownian motion, the following expectation $\mathbb{E}[\Delta W_1(t)\Delta W_2(t)]$ can be determined, using the same heuristics as for the case $(\mathrm{d}W(t))^2$ in Remark 2.1.2,

$$\begin{aligned} \mathbb{E}[\Delta W_1(t)\Delta W_2(t)] &= \mathbb{E}\left[\Delta\widetilde{W}_1(t)\left(\rho_{1,2}\Delta\widetilde{W}_1(t) + \sqrt{1-\rho_{1,2}^2}\Delta\widetilde{W}_2(t)\right)\right] \\ &= \mathbb{E}\left[\rho_{1,2}\Delta\widetilde{W}_1(t)\Delta\widetilde{W}_1(t)\right] + \sqrt{1-\rho_{1,2}^2}\mathbb{E}\left[\Delta\widetilde{W}_1(t)\Delta\widetilde{W}_2(t)\right] \\ &= \rho_{1,2}\mathrm{d}t. \end{aligned}$$

due to the independence of $\widetilde{W}_1(t)$ and $\widetilde{W}_2(t)$.

By the definition of the variance of a sum of two random variables, we have $\mathrm{Var}[aX + bY] = a^2\mathrm{Var}[X] + b^2\mathrm{Var}[Y] + 2ab\mathbb{C}\mathrm{ov}[X,Y]$, which gives us,

$$\begin{aligned} \mathrm{Var}[(\Delta W_1(t))(\Delta W_2(t))] &= \mathrm{Var}[\sqrt{\Delta t}Z_1\sqrt{\Delta t}Z_2] \\ &= (\Delta t)^2\mathrm{Var}\left[\widetilde{Z}_1 \cdot \left(\rho_{1,2}\widetilde{Z}_1 + \sqrt{1-\rho_{1,2}^2}\widetilde{Z}_2\right)\right] \\ &= (\Delta t)^2\left(\rho_{1,2}^2\mathrm{Var}[\widetilde{Z}_1^2] + (1-\rho_{1,2}^2)\mathrm{Var}[\widetilde{Z}_1\widetilde{Z}_2]\right. \\ &\quad\left. + \rho_{1,2}\sqrt{1-\rho_{1,2}^2}\mathbb{C}\mathrm{ov}\left[\widetilde{Z}_1^2, \widetilde{Z}_1\widetilde{Z}_2\right]\right), \end{aligned}$$

where $\widetilde{Z}_1 \sim \mathcal{N}(0,1)$, $\widetilde{Z}_2 \sim \mathcal{N}(0,1)$.

Since $\widetilde{Z}_1$ and $\widetilde{Z}_2$ are independent standard normal variables, the last term is equal to zero. Therefore, by the definition of the variance, we get:

$$\text{Var}[(\Delta W_1(t))(\Delta W_2(t))] = (\Delta t)^2 \left(\rho_{1,2}^2 \text{Var}[\widetilde{Z}_1^2] + (1-\rho_{1,2}^2)\text{Var}[\widetilde{Z}_1\widetilde{Z}_2]\right)$$
$$= (\Delta t)^2 \left(\rho_{1,2}^2 \left(\mathbb{E}[\widetilde{Z}_1^4] - (\mathbb{E}[\widetilde{Z}_1^2])^2\right) + (1-\rho_{1,2}^2)\right).$$

Since the 4th moment of the standard normal distribution is equal to 3, i.e. $\mathbb{E}[\widetilde{Z}_1^4] = 3$, we find:

$$\text{Var}[(\Delta W_1(t))(\Delta W_2(t))] = (\Delta t)^2 \left(2\rho_{1,2}^2 + (1-\rho_{1,2}^2)\right)$$
$$= (\Delta t)^2 (\rho_{1,2}^2 + 1).$$

thus it converges to zero much faster than the expectation. ♦

7.1.2 Multi-D asset price processes

Let's first look at the following general system of *correlated SDEs*,

$$d\mathbf{X}(t) = \bar{\boldsymbol{\mu}}(t, \mathbf{X}(t))dt + \bar{\boldsymbol{\Sigma}}(t, \mathbf{X}(t))d\mathbf{W}(t), \quad 0 \leq t_0 < t, \tag{7.4}$$

where $\bar{\boldsymbol{\mu}}(t, \mathbf{X}(t)) : D \to \mathbb{R}^n$, $\bar{\boldsymbol{\Sigma}}(t, \mathbf{X}(t)) : D \to \mathbb{R}^{n \times n}$ and $\mathbf{W}(t)$ is a column vector of correlated Brownian motions in $\mathbb{R}^n$.

This SDE system can be written as:

$$\begin{bmatrix} dX_1 \\ \vdots \\ dX_n \end{bmatrix} = \begin{bmatrix} \bar{\mu}_1 \\ \vdots \\ \bar{\mu}_n \end{bmatrix} dt + \begin{bmatrix} \bar{\Sigma}_{1,1} & \cdots & \bar{\Sigma}_{1,n} \\ \vdots & \ddots & \vdots \\ \bar{\Sigma}_{n,1} & \cdots & \bar{\Sigma}_{n,n} \end{bmatrix} \begin{bmatrix} dW_1 \\ \vdots \\ dW_n \end{bmatrix} \Leftrightarrow$$

$$d\mathbf{X} = \bar{\boldsymbol{\mu}}dt + \bar{\boldsymbol{\Sigma}}d\mathbf{W}. \tag{7.5}$$

In the representation above, the number of Brownian motions is equal to the number of SDEs. This can be generalized, i.e. the number of underlying Brownian motions can be larger than the number of underlying processes.

The system in (7.5) is written down in a generic way, as each process $X_i(t)$ may be connected to the volatilities from the other processes $X_j(t)$. In practice, this is however unusual, as typically matrix $\bar{\boldsymbol{\Sigma}}$ is a diagonal matrix.

We are now ready to connect SDE system (7.4) to a stochastic process $\mathbf{X}(t)$ based on independent Brownian motions, i.e.,

$$d\mathbf{X}(t) = \bar{\boldsymbol{\mu}}(t, \mathbf{X}(t))dt + \bar{\boldsymbol{\sigma}}(t, \mathbf{X}(t))d\widetilde{\mathbf{W}}(t), \quad 0 \leq t_0 < t, \tag{7.6}$$

where $\widetilde{\mathbf{W}}(t)$ is a column vector of *independent* Brownian motions in $\mathbb{R}^n$.

Setting $\bar{\boldsymbol{\mu}} = \bar{\boldsymbol{\mu}}(t, \mathbf{X}(t))$, $\bar{\boldsymbol{\sigma}} = \bar{\boldsymbol{\sigma}}(t, \mathbf{X}(t))$ and $\widetilde{\mathbf{W}} = \widetilde{\mathbf{W}}(t)$, and Equation (7.6) can be written in the following matrix representation:

$$\begin{bmatrix} \mathrm{d}X_1 \\ \vdots \\ \mathrm{d}X_n \end{bmatrix} = \begin{bmatrix} \bar{\mu}_1 \\ \vdots \\ \bar{\mu}_n \end{bmatrix} \mathrm{d}t + \begin{bmatrix} \bar{\sigma}_{1,1} & \cdots & \bar{\sigma}_{1,n} \\ \vdots & \ddots & \vdots \\ \bar{\sigma}_{n,1} & \cdots & \bar{\sigma}_{n,n} \end{bmatrix} \begin{bmatrix} \mathrm{d}\widetilde{W}_1 \\ \vdots \\ \mathrm{d}\widetilde{W}_n \end{bmatrix}$$

$$= \bar{\boldsymbol{\mu}} \mathrm{d}t + \bar{\boldsymbol{\sigma}} \mathrm{d}\widetilde{\mathbf{W}}. \tag{7.7}$$

We first re-write system (7.5) as follows:

$$\begin{bmatrix} \mathrm{d}X_1 \\ \vdots \\ \mathrm{d}X_n \end{bmatrix} = \begin{bmatrix} \bar{\mu}_1 \\ \vdots \\ \bar{\mu}_n \end{bmatrix} \mathrm{d}t + \begin{bmatrix} \bar{\Sigma}_{1,1} & \cdots & \bar{\Sigma}_{1,n} \\ \vdots & \ddots & \vdots \\ \bar{\Sigma}_{n,1} & \cdots & \bar{\Sigma}_{n,n} \end{bmatrix} \begin{bmatrix} \mathrm{d}W_1 \\ \vdots \\ \mathrm{d}W_n \end{bmatrix}$$

$$= \begin{bmatrix} \bar{\mu}_1 \\ \vdots \\ \bar{\mu}_n \end{bmatrix} \mathrm{d}t + \begin{bmatrix} \bar{\Sigma}_{1,1} & \cdots & \bar{\Sigma}_{1,n} \\ \vdots & \ddots & \vdots \\ \bar{\Sigma}_{n,1} & \cdots & \bar{\Sigma}_{n,n} \end{bmatrix} \begin{bmatrix} 1 & 0 & \cdots \\ \rho_{1,2} & \sqrt{1 - \rho_{1,2}^2} & \cdots \\ \vdots & & \ddots & \vdots \\ \rho_{1,n} & \cdots & & \cdots \end{bmatrix} \begin{bmatrix} \mathrm{d}\widetilde{W}_1 \\ \vdots \\ \mathrm{d}\widetilde{W}_n \end{bmatrix}$$

$$= \bar{\boldsymbol{\mu}} \mathrm{d}t + \bar{\boldsymbol{\Sigma}} \mathbf{L} \mathrm{d}\widetilde{\mathbf{W}} =: \bar{\boldsymbol{\mu}} \mathrm{d}t + \bar{\boldsymbol{\sigma}} \mathrm{d}\widetilde{\mathbf{W}}. \tag{7.8}$$

With matrix $\bar{\boldsymbol{\Sigma}}$ the usual diagonal matrix, we have a similar expression. So, matrix $\bar{\boldsymbol{\sigma}} := \bar{\boldsymbol{\Sigma}} \mathbf{L}$, thus providing the connection between the two SDE formulations.

7.1.3 Itô's lemma for vector processes

Let us consider the process for $\mathbf{X}(t) = [X_1(t), X_2(t), \ldots, X_n(t)]^{\mathrm{T}}$ given by (7.6) and let a real-valued function $g \equiv g(t, \mathbf{X}(t))$ be sufficiently differentiable on $\mathbb{R} \times \mathbb{R}^n$. Increment $\mathrm{d}g(t, \mathbf{X}(t))$ is then governed by the following SDE:

$$\mathrm{d}g(t, \mathbf{X}(t)) = \frac{\partial g}{\partial t} \mathrm{d}t + \sum_{j=1}^{n} \frac{\partial g}{\partial X_j} \mathrm{d}X_j(t) + \frac{1}{2} \sum_{i,j=1}^{n} \frac{\partial^2 g}{\partial X_i \partial X_j} \mathrm{d}X_i(t) \mathrm{d}X_j(t). \tag{7.9}$$

Using the matrix notation from representation (7.7), we distinguish the drift and the volatility terms in $\mathrm{d}g(t, \mathbf{X}(\mathbf{t}))$, as follows:

$$dg(t, \mathbf{X}(t)) = \left(\frac{\partial g}{\partial t} + \sum_{i=1}^{n} \bar{\mu}_i(t, \mathbf{X}(t)) \frac{\partial g}{\partial X_i} \right.$$

$$\left. + \frac{1}{2} \sum_{i,j,k=1}^{n} \bar{\sigma}_{i,k}(t, \mathbf{X}(t)) \bar{\sigma}_{j,k}(t, \mathbf{X}(t)) \frac{\partial^2 g}{\partial X_i \partial X_j} \right) dt$$

$$+ \sum_{i,j=1}^{n} \bar{\sigma}_{i,j}(t, \mathbf{X}(t)) \frac{\partial g}{\partial X_i} d\widetilde{W}_j(t). \qquad (7.10)$$

This is *Itô's lemma for vector processes with independent increments*. It is derived by an application of the Taylor series expansion, and the Itô table.

Example 7.1.4 (2D correlated geometric Brownian motion) With two-dimensional Brownian motion $\mathbf{W}(t) = [W_1(t), W_2(t)]^T$, with correlation parameter ρ, we construct a portfolio consisting of two correlated stocks, $S_1(t)$ and $S_2(t)$, with the following dynamics:

$$\begin{aligned} dS_1(t) &= \mu_1 S_1(t) dt + \sigma_1 S_1(t) dW_1(t), \\ dS_2(t) &= \mu_2 S_2(t) dt + \sigma_2 S_2(t) dW_2(t), \end{aligned} \qquad (7.11)$$

with $\mu_1, \mu_2, \sigma_1, \sigma_2$ constants. By the Cholesky decomposition this system can be expressed in terms of independent Brownian motions as:

$$\begin{bmatrix} dS_1(t) \\ dS_2(t) \end{bmatrix} = \begin{bmatrix} \mu_1 S_1(t) \\ \mu_2 S_2(t) \end{bmatrix} dt + \begin{bmatrix} \sigma_1 S_1(t) & 0 \\ \rho \sigma_2 S_2(t) & \sqrt{1-\rho^2} \sigma_2 S_2(t) \end{bmatrix} \begin{bmatrix} d\widetilde{W}_1(t) \\ d\widetilde{W}_2(t) \end{bmatrix}. \qquad (7.12)$$

Application of the multi-dimensional Itô lemma (7.10) to a sufficiently smooth function, $g = g(t, S_1, S_2)$, with $S_i = S_i(t)$, $i = 1, 2$, gives us,

$$dg(t, S_1, S_2) = \left(\frac{\partial g}{\partial t} + \mu_1 S_1 \frac{\partial g}{\partial S_1} + \mu_2 S_2 \frac{\partial g}{\partial S_2} + \frac{1}{2} \sigma_1^2 S_1^2 \frac{\partial^2 g}{\partial S_1^2} + \frac{1}{2} \sigma_2^2 S_2^2 \frac{\partial^2 g}{\partial S_2^2} \right.$$
$$\left. + \rho \sigma_1 \sigma_2 S_1 S_2 \frac{\partial^2 g}{\partial S_1 \partial S_2} \right) dt + \sigma_1 S_1 \frac{\partial g}{\partial S_1} dW_1 + \sigma_2 S_2 \frac{\partial g}{\partial S_2} dW_2,$$

which is re-written in correlated Brownian motions form with $W_1(t)$ and $W_2(t)$. This result holds for any function $g(t, S_1, S_2)$ which satisfies the differentiability conditions. If we, for example, take $g(t, S_1, S_2) = \log S_1$ the result collapses to the well-known one-dimensional dynamics for the log-stock:

$$d \log S_1(t) = \left(\mu_1 - \frac{1}{2} \sigma_1^2 \right) dt + \sigma_1 dW_1(t).$$

♦

7.1.4 Multi-dimensional Feynman-Kac theorem

Also the Feynman-Kac theorem can be generalized to the multi-dimensional setting. Here we show that for the general two-dimensional PDE, with final condition $V(T, \mathbf{S})$,

$$\frac{\partial V}{\partial t} + \bar{\mu}_1(t, \mathbf{S})\frac{\partial V}{\partial S_1} + \bar{\mu}_2(t, \mathbf{S})\frac{\partial V}{\partial S_2} + \frac{1}{2}\bar{\sigma}_1^2(t, \mathbf{S})\frac{\partial^2 V}{\partial S_1^2} + \frac{1}{2}\bar{\sigma}_2^2(t, \mathbf{S})\frac{\partial^2 V}{\partial S_2^2}$$
$$+ \rho\bar{\sigma}_1(t, \mathbf{S})\bar{\sigma}_2(t, \mathbf{S})\frac{\partial^2 V}{\partial S_1 \partial S_2} - rV = 0.$$

$$V(T, \mathbf{S}) = H(\mathbf{S}),$$

the solution can be written as

$$V(t, \mathbf{S}) = e^{-r(T-t)}\mathbb{E}[H(\mathbf{S}(T))|\mathcal{F}(t)],$$

with the dynamics for $S_1(t)$ and $S_2(t)$, for $t > t_0$, given by,

$$dS_1(t) = \bar{\mu}_1(t, \mathbf{S})dt + \bar{\sigma}_1(t, \mathbf{S})dW_1^{\mathbb{Q}}(t),$$
$$dS_2(t) = \bar{\mu}_2(t, \mathbf{S})dt + \bar{\sigma}_2(t, \mathbf{S})dW_2^{\mathbb{Q}}(t),$$

with correlated Wiener processes, $dW_1(t)dW_2(t) = \rho dt$.

To prove the Feynman-Kac theorem for the two-dimensional case, consider,

$$\frac{V(t, \mathbf{S})}{M(t)} = e^{-r(t-t_0)}V(t, \mathbf{S}).$$

The corresponding dynamics, using $\bar{\mu}_i \equiv \bar{\mu}_i(t, \mathbf{S}), \bar{\sigma}_i \equiv \bar{\sigma}_i(t, \mathbf{S})$ are found to be,

$$d(e^{-r(t-t_0)}V(t, \mathbf{S})) = V(t, \mathbf{S})d(e^{-r(t-t_0)}) + e^{-r(t-t_0)}dV(t, \mathbf{S}). \quad (7.13)$$

Using the 2D version of Itô's lemma, gives us,

$$dV(t, \mathbf{S}) = \left(\frac{\partial V}{\partial t} + \bar{\mu}_1\frac{\partial V}{\partial S_1} + \bar{\mu}_2\frac{\partial V}{\partial S_2} + \frac{1}{2}\bar{\sigma}_1^2\frac{\partial^2 V}{\partial S_1^2} + \frac{1}{2}\bar{\sigma}_2^2\frac{\partial^2 V}{\partial S_2^2} + \rho\bar{\sigma}_1\bar{\sigma}_2\frac{\partial^2 V}{\partial S_1 \partial S_2}\right)dt$$
$$+ \bar{\sigma}_1\frac{\partial V}{\partial S_1}dW_1^{\mathbb{Q}}(t) + \bar{\sigma}_2\frac{\partial V}{\partial S_2}dW_2^{\mathbb{Q}}(t). \quad (7.14)$$

Multiplication by $e^{r(t-t_0)}$ and after inserting (7.14) into (7.13), we obtain,

$$e^{r(t-t_0)}d(e^{-r(t-t_0)}V(t, \mathbf{S})) = \left(\frac{\partial V}{\partial t} + \bar{\mu}_1\frac{\partial V}{\partial S_1} + \bar{\mu}_2\frac{\partial V}{\partial S_2} + \frac{1}{2}\bar{\sigma}_1^2\frac{\partial^2 V}{\partial S_1^2} + \frac{1}{2}\bar{\sigma}_2^2\frac{\partial^2 V}{\partial S_2^2}\right.$$
$$\left. + \rho\bar{\sigma}_1\bar{\sigma}_2\frac{\partial^2 V}{\partial S_1 \partial S_2} - rV\right)dt$$
$$+ \bar{\sigma}_1\frac{\partial V}{\partial S_1}dW_1^{\mathbb{Q}}(t) + \bar{\sigma}_2\frac{\partial V}{\partial S_2}dW_2^{\mathbb{Q}}(t)$$
$$= \bar{\sigma}_1\frac{\partial V}{\partial S_1}dW_1^{\mathbb{Q}}(t) + \bar{\sigma}_2\frac{\partial V}{\partial S_2}dW_2^{\mathbb{Q}}(t), \quad (7.15)$$

where, because V/M is a $\mathbb{Q}$-martingale, the drift term should be zero, and, as a consequence, all dt-terms cancel out. Integrating both sides, gives us

$$\int_{t_0}^{T} d(e^{-r(t-t_0)} V(t,\mathbf{S})) = \int_{t_0}^{T} e^{-r(t-t_0)} \bar{\sigma}_1 \frac{\partial V}{\partial S_1} dW_1^{\mathbb{Q}}(t)$$

$$+ \int_{t_0}^{T} e^{-r(t-t_0)} \bar{\sigma}_2 \frac{\partial V}{\partial S_2} dW_2^{\mathbb{Q}}(t),$$

or,

$$e^{-r(t-t_0)} V(t,\mathbf{S}) - V(t_0,\mathbf{S}) = \int_{t_0}^{T} e^{-r(t-t_0)} \bar{\sigma}_1 \frac{\partial V}{\partial S_1} dW_1^{\mathbb{Q}}(t)$$

$$+ \int_{t_0}^{T} e^{-r(t-t_0)} \bar{\sigma}_2 \frac{\partial V}{\partial S_2} dW_2^{\mathbb{Q}}(t).$$

Taking expectations with respect to the $\mathbb{Q}$-measure results in,

$$V(t_0,\mathbf{S}) = \mathbb{E}^{\mathbb{Q}}\left[e^{-r(t-t_0)} V(t,\mathbf{S}) | \mathcal{F}(t_0)\right] - \mathbb{E}^{\mathbb{Q}}\left[\int_{t_0}^{T} e^{-r(t-t_0)} \bar{\sigma}_1 \frac{\partial V}{\partial S_1} dW_1^{\mathbb{Q}}(t) | \mathcal{F}(t_0)\right]$$

$$- \mathbb{E}^{\mathbb{Q}}\left[\int_{t_0}^{T} e^{-r(t-t_0)} \bar{\sigma}_2 \frac{\partial V}{\partial S_2} dW_2^{\mathbb{Q}}(t) | \mathcal{F}(t_0)\right]. \tag{7.16}$$

Now, because of the properties of the Itô integral, we have that $I(t_0) = 0$ and $I(t) = \int_{t_0}^{t} g(z) dW^{\mathbb{Q}}(z)$ is a martingale, so that,

$$\mathbb{E}^{\mathbb{Q}}[I(t) | \mathcal{F}(t_0)] = 0,$$

for all $t > t_0$. Therefore, the integrals based on the Brownian motions in (7.16) are equal to zero, and thus,

$$V(t_0,\mathbf{S}) = \mathbb{E}^{\mathbb{Q}}[e^{-r(t-t_0)} V(t,\mathbf{S}) | \mathcal{F}(t_0)].$$

This concludes the derivation.

This derivation indicates how the Feynman-Kac theorem should be generalized to higher-dimensional problems.

7.2 Changing measures and the Girsanov theorem

When dealing with involved systems of SDEs, it is sometimes possible to reduce the complexity of the pricing problem by an appropriate *measure transformation*. Under the appropriate numéraire, processes may become martingales. Working with martingales is typically favorable as these processes are free of drift terms.[1]

---

[1] See again Theorem 1.3.5 the martingale representation theorem.

Although a process which is free of drift terms may still have an involved volatility structure, it is considered to be simpler to work with.

7.2.1 The Radon-Nikodym derivative

We first give some theoretical background. Given two absolutely continuous probability measures, $\mathbb{M} \sim \mathbb{N}$, on probability space $(\Omega, \mathcal{F})$ and a sigma-field, $\mathcal{G} \in \mathcal{F}$, the *Bayes theorem* [Bayes, 1763] for an $\mathcal{F}$-measurable random variable X states that

$$\mathbb{E}^{\mathbb{M}}[X|\mathcal{G}] = \frac{\mathbb{E}^{\mathbb{N}}\left[\frac{d\mathbb{M}}{d\mathbb{N}}X\Big|\mathcal{G}\right]}{\mathbb{E}^{\mathbb{N}}\left[\frac{d\mathbb{M}}{d\mathbb{N}}\Big|\mathcal{G}\right]}. \tag{7.17}$$

It implies that we can take expectations under a different measure, $\mathbb{N}$, once we know the measure transformation $d\mathbb{M}/d\mathbb{N}$.

Theorem 7.2.1 (Change of numéraire) *Assume that a numéraire $M(t)$ exists, as well as a probability measure $\mathbb{M}$, which is equivalent to a measure $\mathbb{Q}^0$, where the price of any traded asset X, relative to M, is a martingale under measure $\mathbb{M}$, i.e.*

$$\mathbb{E}^{\mathbb{M}}\left[\frac{X(T)}{M(T)}\Big|\mathcal{F}(t)\right] = \frac{X(t)}{M(t)}. \tag{7.18}$$

Let N be an arbitrary numéraire. A probability measure $\mathbb{N}$ exists, and is equivalent to $\mathbb{Q}^0$, so that the price of any attainable[a] derivative, i.e. $V(t) \equiv V(t,S)$, normalized by the quantity N, is a martingale under measure $\mathbb{N}$, i.e.

$$\mathbb{E}^{\mathbb{N}}\left[\frac{V(T)}{N(T)}\Big|\mathcal{F}(t)\right] = \frac{V(t)}{N(t)}. \tag{7.19}$$

The Radon-Nikodym derivative, which defines measure $\mathbb{N}$, is then given by:

$$\lambda_{\mathbb{M}}^{\mathbb{N}}(T) := \frac{d\mathbb{N}}{d\mathbb{M}}\Big|_{\mathcal{F}(T)} = \frac{N(T)M(t)}{N(t)M(T)}. \tag{7.20}$$

---

[a] A financial derivative is called attainable if it can be replicated by means of a self-financing strategy with almost surely the same payoff. Think of the European option under the Black-Scholes dynamics which can be replicated by stocks and cash.

Proof For both numéraires, $M(T)$ and $N(T)$, the following equalities hold:

$$\mathbb{E}^{\mathbb{M}}\left[\frac{X(T)}{M(T)}\Big|\mathcal{F}(t)\right] = \frac{X(t)}{M(t)}, \quad \mathbb{E}^{\mathbb{N}}\left[\frac{X(T)}{N(T)}\Big|\mathcal{F}(t)\right] = \frac{X(t)}{N(t)}. \tag{7.21}$$

We first re-write the first expectation in integral form:

$$\frac{X(t)}{M(t)} = \mathbb{E}^{\mathbb{M}}\left[\frac{X(T)}{M(T)}\Big|\mathcal{F}(t)\right] := \int_A \frac{X(T)}{M(T)} d\mathbb{M}. \tag{7.22}$$

From Equation (7.20) we know that

$$\mathrm{d}\mathbb{M} = \frac{N(t)}{N(T)}\frac{M(T)}{M(t)}\mathrm{d}\mathbb{N},$$

so that Equation (7.22) becomes:

$$\frac{X(t)}{M(t)} = \int_A \frac{X(T)}{M(T)}\mathrm{d}\mathbb{M} = \int_A \frac{X(T)}{M(T)}\left(\frac{N(t)}{N(T)}\frac{M(T)}{M(t)}\right)\mathrm{d}\mathbb{N}, \qquad (7.23)$$

from which, after simplification and using the fact that the processes are *known* at time t (measurability principle), we find:

$$\frac{X(t)}{M(t)} = \frac{N(t)}{M(t)}\mathbb{E}^{\mathbb{N}}\left[\frac{X(T)}{N(T)}\bigg|\mathcal{F}(t)\right]. \qquad (7.24)$$

Thus, for an arbitrary numéraire, a probability measure $\mathbb{N}$ exists, which is equivalent to the initial measure $\mathbb{Q}^0$, so that the price of an attainable derivative, $V(t)$, normalized by N, is a martingale under measure $\mathbb{N}$, and

$$\mathbb{E}^{\mathbb{N}}\left[\frac{V(T)}{N(T)}\bigg|\mathcal{F}(t)\right] = \frac{\mathbb{E}^{\mathbb{M}}\left[\frac{V(T)}{N(T)}\frac{\mathrm{d}\mathbb{N}}{\mathrm{d}\mathbb{M}}\bigg|\mathcal{F}(t)\right]}{\mathbb{E}^{\mathbb{M}}\left[\frac{\mathrm{d}\mathbb{N}}{\mathrm{d}\mathbb{M}}\bigg|\mathcal{F}(t)\right]} = \frac{\mathbb{E}^{\mathbb{M}}\left[\frac{V(T)}{N(T)}\frac{N(T)M(t)}{M(T)N(t)}\bigg|\mathcal{F}(t)\right]}{\mathbb{E}^{\mathbb{M}}\left[\frac{N(T)M(t)}{M(T)N(t)}\bigg|\mathcal{F}(t)\right]}.$$

After simplifications, the expectation above reads:

$$\mathbb{E}^{\mathbb{N}}\left[\frac{V(T)}{N(T)}\bigg|\mathcal{F}(t)\right] = \frac{\mathbb{E}^{\mathbb{M}}\left[\frac{V(T)}{M(T)}\bigg|\mathcal{F}(t)\right]}{\mathbb{E}^{\mathbb{M}}\left[\frac{N(T)}{M(T)}\bigg|\mathcal{F}(t)\right]}.$$

Now, using equalities from (7.21) gives us:

$$\frac{\mathbb{E}^{\mathbb{M}}\left[\frac{V(T)}{M(T)}\bigg|\mathcal{F}(t)\right]}{\mathbb{E}^{\mathbb{M}}\left[\frac{N(T)}{M(T)}\bigg|\mathcal{F}(t)\right]} = \frac{V(t)}{M(t)}\frac{M(t)}{N(t)} = \frac{V(t)}{N(t)}.$$

The last two equalities are due to the assumption that a numéraire M exists, so that the price of any traded asset relative to that numéraire is a martingale under measure $\mathbb{M}$. ∎

Theorem 7.2.1 holds for any $t_0 \leq t \leq T$, depending on the filtration. More generally, we present the following theorem with the time points t_0 and t.

> **Theorem 7.2.2 (The Girsanov Theorem)** The Girsanov theorem states the following: Suppose we have the following SDE for $S(t)$,
>
> $$\mathrm{d}S(t) = \bar{\mu}^{\mathbb{M}}(t, S(t))\mathrm{d}t + \bar{\sigma}(t, S(t))\mathrm{d}W^{\mathbb{M}}(t), \quad S(t_0) = S_0,$$
>
> where Brownian motion $W^{\mathbb{M}}(t)$ is defined under the measure $\mathbb{M}$, and $\bar{\mu}^{\mathbb{M}}(t, S(t))$ and $\bar{\sigma}(t, S(t))$ satisfy the usual Lipschitz conditions. $W^{\mathbb{N}}(t)$, which is defined by
>
> $$\mathrm{d}W^{\mathbb{N}}(t) = -\left(\frac{\bar{\mu}^{\mathbb{N}}(t, S(t)) - \bar{\mu}^{\mathbb{M}}(t, S(t))}{\bar{\sigma}(t, S(t))}\right)\mathrm{d}t + \mathrm{d}W^{\mathbb{M}}(t),$$
>
> is a Brownian motion under a measure $\mathbb{N}$, and the SDE for $S(t)$ under the measure $\mathbb{N}$ is given by
>
> $$\mathrm{d}S(t) = \bar{\mu}^{\mathbb{N}}(t, S(t))\mathrm{d}t + \bar{\sigma}(t, S(t))\mathrm{d}W^{\mathbb{N}}(t), \quad S(t_0) = S_0.$$

This holds for a drift $\bar{\mu}^{\mathbb{N}}(t, S(t))$, for which the ratio,

$$\frac{\bar{\mu}^{\mathbb{N}}(t, S(t)) - \bar{\mu}^{\mathbb{M}}(t, S(t))}{\bar{\sigma}(t, S(t))},$$

is bounded.

We may define the measure $\mathbb{N}$ by the following martingale,

$$\lambda_{\mathbb{M}}^{\mathbb{N}}(t) := \left.\frac{\mathrm{d}\mathbb{N}}{\mathrm{d}\mathbb{M}}\right|_{\mathcal{F}(t)} = \exp\left[-\frac{1}{2}\int_{t_0}^{t}\left(\frac{\bar{\mu}^{\mathbb{N}}(s, S(s)) - \bar{\mu}^{\mathbb{M}}(s, S(s))}{\bar{\sigma}(s, S(s))}\right)^2 \mathrm{d}s\right.$$
$$\left. + \int_{t_0}^{t}\frac{\bar{\mu}^{\mathbb{N}}(s, S(s)) - \bar{\mu}^{\mathbb{M}}(s, S(s))}{\bar{\sigma}(s, S(s))}\mathrm{d}W^{\mathbb{M}}(t)\right],$$

or, equivalently:

$$\mathrm{d}\lambda_{\mathbb{M}}^{\mathbb{N}}(t) = \lambda_{\mathbb{M}}^{\mathbb{N}}(t)\bar{\sigma}_\lambda(t)\mathrm{d}W^{\mathbb{M}}(t), \quad \lambda_{\mathbb{M}}^{\mathbb{N}}(t_0) = 1,$$

where

$$\bar{\sigma}_\lambda(t) = \frac{\bar{\mu}^{\mathbb{N}}(t, S(t)) - \bar{\mu}^{\mathbb{M}}(t, S(t))}{\bar{\sigma}(t, S(t))}. \tag{7.25}$$

Then, measure $\mathbb{N}$ is equivalent measure to $\mathbb{M}$. The proof can be found in [Girsanov, 1960].

> A numéraire needs to be a tradable asset. Processes like volatility or the short-rate are not observable and therefore they cannot be used as a numéraire.

7.2.2 Change of numéraire examples

Let us define two equivalent measures, $\mathbb{M}$ and $\mathbb{N}$, where the first measure, $\mathbb{M}$, is associated with numéraire $M(t)$, given by:

$$\mathrm{d}M(t) = \bar{\mu}_M(t, M(t))\mathrm{d}t + \bar{\Sigma}_M(t, M(t))\mathrm{d}W_M^{\mathbb{M}}(t), \quad M(t_0) = M_0,$$

and the second measure, ℕ, is generated by numéraire $N(t)$, whose dynamics read:

$$dN(t) = \bar{\mu}_N(t, N(t))dt + \bar{\Sigma}_N(t, N(t))dW_N^\mathbb{M}(t), \quad N(t_0) = N_0,$$

with a correlation between the Brownian motions, $dW_M^\mathbb{M}(t)dW_N^\mathbb{M}(t) = \rho dt$. Superscript "ℳ" in the underlying Brownian motions, $dW_M^\mathbb{M}(t)$ and $dW_N^\mathbb{M}(t)$, indicates that both processes are defined under the measure ℳ. Consider an asset $S(t)$, which fluctuates according to the following SDE:

$$dS(t) = \bar{\mu}_S^\mathbb{M}(t, S(t))dt + \bar{\Sigma}_S(t, S(t))dW_S^\mathbb{M}(t). \tag{7.26}$$

We discuss the derivation of the dynamics for process $S(t)$ under the measure ℕ here. As $S(t)$ may be correlated with both numéraires, we formulate all underlying processes in matrix notation, as follows

$$d\begin{bmatrix} M(t) \\ N(t) \\ S(t) \end{bmatrix} = \begin{bmatrix} \bar{\mu}_M \\ \bar{\mu}_N \\ \bar{\mu}_S^\mathbb{M} \end{bmatrix} dt + \begin{bmatrix} \bar{\Sigma}_M & 0 & 0 \\ 0 & \bar{\Sigma}_N & 0 \\ 0 & 0 & \bar{\Sigma}_S \end{bmatrix} \mathbf{L} d\widetilde{\mathbf{W}}^\mathbb{M}(t), \tag{7.27}$$

with $\mathbf{L}$ the Cholesky lower triangular matrix calculated from the correlation matrix (as in Section 7.1.3), $d\widetilde{\mathbf{W}}^\mathbb{M}(t)$ the column vector of independent Brownian increments, $\bar{\mu}$ and $\bar{\Sigma}$ represent the drift and volatility vectors.

Equivalently, we can present processes for $M(t)$, $N(t)$ and $S(t)$ as:

$$dM(t) = \bar{\mu}_M(t, M(t))dt + \bar{\sigma}_M(t, M(t))d\widetilde{\mathbf{W}}^\mathbb{M}(t),$$
$$dN(t) = \bar{\mu}_N(t, N(t))dt + \bar{\sigma}_N(t, N(t))d\widetilde{\mathbf{W}}^\mathbb{M}(t),$$
$$dS(t) = \bar{\mu}_S^\mathbb{M}(t, S(t))dt + \bar{\sigma}_S(t, S(t))d\widetilde{\mathbf{W}}^\mathbb{M}(t),$$

with $\bar{\sigma}_N(\cdot)$, $\bar{\sigma}_M(\cdot)$ and $\bar{\sigma}_S(\cdot)$ the row vectors resulting from multiplying the covariance matrix with matrix $\mathbf{L}$ in (7.27), i.e.

$$\bar{\sigma}_M(t, M(t)) = \begin{bmatrix} \bar{\Sigma}_M(t, M(t))\mathbf{L}_{1,1} & 0 & 0 \end{bmatrix},$$
$$\bar{\sigma}_N(t, N(t)) = \begin{bmatrix} \bar{\Sigma}_N(t, N(t))\mathbf{L}_{2,1} & \bar{\Sigma}_N(t, N(t))\mathbf{L}_{2,2} & 0 \end{bmatrix},$$
$$\bar{\sigma}_S(t, S(t)) = \begin{bmatrix} \bar{\Sigma}_S(t, S(t))\mathbf{L}_{3,1} & \bar{\Sigma}_S(t, S(t))\mathbf{L}_{3,2} & \bar{\Sigma}_S(t, S(t))\mathbf{L}_{3,3} \end{bmatrix},$$

with $\mathbf{L}_{i,j}$ the $(i,j)^{th}$-element of the Cholesky matrix.

Process $S(t)$ under the measure ℕ is of the following form,

$$dS(t) = \bar{\mu}_S^\mathbb{N}(t, S(t))dt + \bar{\sigma}_S(t, S(t))d\widetilde{\mathbf{W}}^\mathbb{N}(t). \tag{7.28}$$

We learn from the Girsanov theorem that a measure transformation does not affect the volatility terms, and therefore only a transformed drift $\bar{\mu}_S^\mathbb{N}(\cdot)$ has to be determined. To determine $\bar{\mu}_S^\mathbb{N}(t, S(t))$, we make again use of the Radon-Nikodym derivative,

$$\lambda_{\mathbb{M}}^{\mathbb{N}}(t) = \frac{d\mathbb{N}}{d\mathbb{M}}\bigg|_{\mathcal{F}(t)}. \tag{7.29}$$

By Girsanov's Theorem, i.e., Theorem 7.2.2, we know that the Radon-Nikodym derivative $\lambda_{\mathbb{M}}^{\mathbb{N}}(t)$ in (7.29) is a *martingale* under measure $\mathbb{M}$, driven by numéraire $M(t)$, and its dynamics are given by:

$$d\lambda_{\mathbb{M}}^{\mathbb{N}}(t) = \bar{\sigma}_\lambda(t)\lambda_{\mathbb{M}}^{\mathbb{N}}(t)d\widetilde{\mathbf{W}}^{M}(t), \tag{7.30}$$

where

$$\bar{\sigma}_\lambda(t) = \frac{\bar{\mu}_S^{\mathbb{N}}(t,S(t)) - \bar{\mu}_S^{\mathbb{M}}(t,S(t))}{\bar{\sigma}_S(t,S(t))}. \tag{7.31}$$

Moreover, the Radon-Nikodym derivative in (7.29) can be expressed, by Theorem 7.2.2, in terms of numéraires as:

$$\lambda_{\mathbb{M}}^{\mathbb{N}}(t) = \frac{N(t)}{N(t_0)}\frac{M(t_0)}{M(t)}.$$

By applying Itô's lemma and the fact that $\lambda_{\mathbb{M}}^{\mathbb{N}}(t)$ is a martingale under $\mathbb{M}$ (zero-drift dynamics), we find the following form of $d\lambda_{\mathbb{M}}^{\mathbb{N}}(t)$ under the measure $\mathbb{M}$:

$$d\lambda_{\mathbb{M}}^{\mathbb{N}}(t) = \frac{M(t_0)}{N(t_0)}\left(\frac{1}{M(t)}\bar{\sigma}_N(t,N(t))d\widetilde{\mathbf{W}}^{M}(t) - \frac{N(t)}{M^2(t)}\bar{\sigma}_M(t,M(t))d\widetilde{\mathbf{W}}^{M}(t)\right)$$
$$= \lambda_{\mathbb{M}}^{\mathbb{N}}(t)\left(\frac{\bar{\sigma}_N(t,N(t))}{N(t)} - \frac{\bar{\sigma}_M(t,M(t))}{M(t)}\right)d\widetilde{\mathbf{W}}^{M}(t). \tag{7.32}$$

By equating the volatility terms in the processes for $\lambda_{\mathbb{M}}^{\mathbb{N}}(t)$ in (7.30), (7.31) and (7.32), we find:

$$\frac{\bar{\mu}_S^{\mathbb{N}}(t,S(t)) - \bar{\mu}_S^{\mathbb{M}}(t,S(t))}{\bar{\sigma}_S(t,S(t))} = \frac{\bar{\sigma}_N(t,N(t))}{N(t)} - \frac{\bar{\sigma}_M(t,M(t))}{M(t)}. \tag{7.33}$$

Thus, the drift term $\bar{\mu}_S^{\mathbb{N}}(t,S(t))$ for process $S(t)$ under the measure $\mathbb{N}$ should equal:

$$\bar{\mu}_S^{\mathbb{N}}(t,S(t)) = \bar{\mu}_S^{\mathbb{M}}(t,S(t)) + \bar{\sigma}_S(t,S(t))\left(\frac{\bar{\sigma}_N(t,N(t))}{N(t)} - \frac{\bar{\sigma}_M(t,M(t))}{M(t)}\right). \tag{7.34}$$

With this result we can determine the drift transformation, when changing numéraires.

7.2.3 From $\mathbb{P}$ to $\mathbb{Q}$ in the Black-Scholes model

Recall that in the Black-Scholes model the stock is driven under the real-world $\mathbb{P}$ measure by $dS(t) = \mu S(t)dt + \sigma S(t)dW^{\mathbb{P}}(t)$, with constant parameters μ and σ. In Section 2.3, it was shown that under the risk-neutral $\mathbb{Q}$ measure the stock divided by the numéraire (which was the money-savings account, $dM(t) = rM(t)dt$) is a martingale. Here, we will relate these results to the concept of measure transformation. Applying Itô's lemma to $S(t)/M(t)$ with the stock defined by (2.1) gives us:

$$d\frac{S(t)}{M(t)} = \frac{1}{M(t)}dS(t) - \frac{S(t)}{M^2(t)}dM(t) + \frac{1}{M^3(t)}(dM(t))^2 - \frac{1}{M^2(t)}dS(t)dM(t).$$

Remember that by the Itô's table, see Table 2.1 in the second chapter, the third and the fourth terms will vanish, so that we should consider only

$$d\frac{S(t)}{M(t)} = \frac{1}{M(t)}dS(t) - \frac{S(t)}{M^2(t)}dM(t).$$

By substitution of the dynamics of $S(t)$ and $M(t)$, we further obtain:

$$d\frac{S(t)}{M(t)} = \frac{1}{M(t)}\left(\mu S(t)dt + \sigma S(t)dW^{\mathbb{P}}(t)\right) - rM(t)\frac{S(t)}{M^2(t)}dt$$

$$= \mu\frac{S(t)}{M(t)}dt + \sigma\frac{S(t)}{M(t)}dW^{\mathbb{P}}(t) - r\frac{S(t)}{M(t)}dt. \qquad (7.35)$$

Based on the arguments in Section 2.3, it is known that $S(t)/M(t)$ is a martingale and by Theorem 1.3.4 we know that Itô integrals are martingales when they do not contain any drift terms. Moreover, by Theorem 7.2.2, we also know that the volatility is not affected by a change of measure. Therefore, under the risk-neutral measure, the ratio $S(t)/M(t)$ should be of the following form:

$$d\frac{S(t)}{M(t)} = \sigma\frac{S(t)}{M(t)}dW^{\mathbb{Q}}(t), \qquad (7.36)$$

By equating (7.35) and (7.36), we find:

$$\mu\frac{S(t)}{M(t)}dt + \sigma\frac{S(t)}{M(t)}dW^{\mathbb{P}}(t) - r\frac{S(t)}{M(t)}dt = \sigma\frac{S(t)}{M(t)}dW^{\mathbb{Q}}(t),$$

which is equivalent (as both $S(t)$ and $M(t)$ are positive) to

$$dW^{\mathbb{Q}}(t) = \frac{\mu - r}{\sigma}dt + dW^{\mathbb{P}}(t). \qquad (7.37)$$

Equation (7.37) determines the measure transformation from the real-world $\mathbb{P}$ measure, to the risk-neutral $\mathbb{Q}$-measure, under the Black-Scholes model. So, we have,

$$dS(t) = \mu S(t)dt + \sigma S(t)dW^{\mathbb{P}}(t)$$

$$= \mu S(t)dt + \sigma S(t)\left(dW^{\mathbb{Q}}(t) - \frac{\mu - r}{\sigma}dt\right)$$

$$= rS(t)dt + \sigma S(t)dW^{\mathbb{Q}}(t), \qquad (7.38)$$

which is in accordance with the SDE (2.26).

Pricing under the stock measure

It is natural when dealing with financial derivatives to associate discounting to the money-savings account. However, a *change of measure* may be beneficial in several situations. Here, we will present some examples for which changing measures simplifies the pricing problem significantly.

Let us consider the Black-Scholes model under risk-neutral $\mathbb{Q}$-measure, where the dynamics for two stocks are driven by the following system of SDEs:

$$\begin{aligned} dS_1(t) &= rS_1(t)dt + \sigma_1 S_1(t)dW_1^{\mathbb{Q}}(t), \\ dS_2(t) &= rS_2(t)dt + \sigma_2 S_2(t)dW_2^{\mathbb{Q}}(t), \end{aligned} \qquad (7.39)$$

with r, σ_1, σ_2 constant, and correlation $dW_1^{\mathbb{Q}}(t)dW_2^{\mathbb{Q}}(t) = \rho dt$. The money-savings account is given by $dM(t) = rM(t)dt$, as usual.

The financial product under consideration is here defined by

$$V(t_0, S_1, S_2) = M(t_0)\mathbb{E}^{\mathbb{Q}}\left[\frac{S_1(T)}{M(T)}\mathbb{1}_{S_2(T)>K}\Big|\mathcal{F}(t_0)\right]. \qquad (7.40)$$

The payoff of this option product resembles a so-called *modified asset-or-nothing digital option*, which is based on $S_1(T)$ and $S_2(T)$. This product pays an amount $S_1(T)$, but only if stock $S_2(T)$ is above a certain pre-specified limit, K. The product holder is exposed to the risk connected to the correlation between the stocks S_1 and S_2.

We will now employ the change of measure technique, as presented above. Instead of the money-savings account $M(t)$ as the numéraire, we will consider stock $S_1(t)$ as the new numéraire. Note that if $S_1(t) = S_2(t)$, see Theorem 3.2.2 and Example 3.2.1, we can set $K = 0$ in the Black-Scholes solution to obtain the exact solution for the asset-or-nothing option price.

Application of the Radon-Nikodym technique gives us the following relation:

$$\lambda_{\mathbb{Q}}^{S_1}(T) = \frac{d\mathbb{Q}^{S_1}}{d\mathbb{Q}}\bigg|_{\mathcal{F}(T)} = \frac{S_1(T)M(t_0)}{S_1(t_0)M(T)}, \qquad (7.41)$$

where $\mathbb{Q}^{S_1}$ represents the measure for which stock $S_1(t)$ is the numéraire, and $\mathbb{Q}$ stands for the known risk-neutral measure. Under the new numéraire the value of the derivative transforms into:

$$\begin{aligned} V(t_0, S_1, S_2) &= M(t_0)\mathbb{E}^{S_1}\left[\frac{S_1(T)}{M(T)}\mathbb{1}_{S_2(T)>K}\frac{S_1(t_0)M(T)}{S_1(T)M(t_0)}\Big|\mathcal{F}(t_0)\right] \\ &= S_1(t_0)\mathbb{E}^{S_1}\left[\mathbb{1}_{S_2(T)>K}\Big|\mathcal{F}(t_0)\right] = S_1(t_0)\mathbb{Q}^{S_1}[S_2(T) > K]. \end{aligned}$$
$$(7.42)$$

To determine the probability of $S_2(T)$ being larger than K, let's consider the stock dynamics under measure $\mathbb{Q}^{S_1}$. Under this measure we have to ensure that all underlying processes are martingales. In the current model we have three assets, i.e. the money-savings account $M(t)$, stock $S_1(t)$ and stock $S_2(t)$. The drift terms for the stocks $S_1(t)$ and $S_2(t)$ need to be adjusted, so that the quantities $M(t)/S_1(t)$ and $S_2(t)/S_1(t)$ are martingales. By Itô's lemma, we find,

$$d\left(\frac{M(t)}{S_1(t)}\right) = \frac{1}{S_1(t)}dM(t) - \frac{M(t)}{S_1^2(t)}dS_1(t) + \frac{M(t)}{S_1^3(t)}(dS_1(t))^2$$

$$= \frac{M(t)}{S_1(t)}\left(\sigma_1^2 dt - \sigma_1 dW_1^{\mathbb{Q}}(t)\right).$$

This implies a measure transformation, as follows,

$$dW_1^{\mathbb{Q}}(t) := dW_1^{S_1}(t) + \sigma_1 dt, \tag{7.43}$$

and therefore the stock dynamics under the $\mathbb{Q}^{S_1}$-measure are given by:

$$\frac{dS_1(t)}{S_1(t)} = rdt + \sigma_1\left(dW_1^{S_1}(t) + \sigma_1 dt\right)$$

$$= \left(r + \sigma_1^2\right) dt + \sigma_1 dW_1^{S_1}(t).$$

For the second stock we have:

$$d\left(\frac{S_2(t)}{S_1(t)}\right) / \left(\frac{S_2(t)}{S_1(t)}\right) = (\sigma_1^2 - \rho\sigma_1\sigma_2)\, dt + \sigma_2 dW_2^{\mathbb{Q}}(t) - \sigma_1 dW_1^{\mathbb{Q}}(t).$$

By Equation (7.43), we now find:

$$d\left(\frac{S_2(t)}{S_1(t)}\right) / \left(\frac{S_2(t)}{S_1(t)}\right) = -\rho\sigma_1\sigma_2 dt + \sigma_2 dW_2^{\mathbb{Q}}(t) - \sigma_1 dW_1^{S_1}(t). \tag{7.44}$$

Here the term $\sigma_1 dW_1^{S_1}(t)$ is already under the required measure and we should bring the remaining term under the measure $\mathbb{Q}^{S_1}$. Expression (7.44) implies the following change of measure for the Brownian motion in the second stock process, $W_2^{\mathbb{Q}}(t)$,

$$dW_2^{S_1}(t) := dW_2^{\mathbb{Q}}(t) - \rho\sigma_1 dt.$$

Finally, the model under the stock-measure is given by:

$$dS_1(t) = (r + \sigma_1^2)S_1(t)dt + \sigma_1 S_1(t)dW_1^{S_1}(t),$$
$$dS_2(t) = (r + \rho\sigma_1\sigma_2)S_2(t)dt + \sigma_2 S_2(t)dW_2^{S_1}(t), \tag{7.45}$$
$$dM(t) = rM(t)dt.$$

The dynamics of $M(t)$ are not changed, as they do not involve any stochastic processes.

Returning to the question of calculating $\mathbb{Q}^{S_1}[S_2(T) > K]$ in Equation (7.42), we can now use the fact that stock process $S_2(T)$ under the $\mathbb{Q}^{S_1}$-measure has the following expression, see (7.45):

$$S_2(T) = S_2(t_0) \exp\left[\left(r + \rho\sigma_1\sigma_2 - \frac{1}{2}\sigma_2^2\right)(T - t_0) + \sigma_2\left(W_2^{S_1}(T) - W_2^{S_1}(t_0)\right)\right].$$

We recognize the lognormal distribution, so that the solution can be calculated based on the CDF of the lognormal distribution, like in the Black-Scholes model.

Example 7.2.1 We can also determine the dynamics for processes $S_1(t)$ and $S_2(t)$ under the $\mathbb{Q}^{S_1}$-measure by using the results in Section 7.2.2. Let us define vector $\mathbf{X}(t) := [S_1(t), S_2(t), M(t)]^T$, consisting of the stock processes and the money-savings account, and express System (7.39) in terms of independent Brownian motions:

$$d\mathbf{X}(t) = \bar{\boldsymbol{\mu}}^{\mathbb{Q}}(t, \mathbf{X}(t))dt + \bar{\boldsymbol{\sigma}}(t, \mathbf{X}(t))d\widetilde{\mathbf{W}}^{\mathbb{Q}}(t), \tag{7.46}$$

where:

$$\bar{\boldsymbol{\sigma}}(t, \mathbf{X}(t)) \equiv \begin{bmatrix} \sigma_1 S_1(t) & 0 & 0 \\ 0 & \sigma_2 S_2(t) & 0 \\ 0 & 0 & 0 \end{bmatrix} \begin{bmatrix} 1 & 0 & 0 \\ \rho & \sqrt{1-\rho^2} & 0 \\ 0 & 0 & 0 \end{bmatrix}$$

$$= \begin{bmatrix} \sigma_1 S_1(t) & 0 & 0 \\ \rho \sigma_2 S_2(t) & \sigma_2 \sqrt{1-\rho^2} S_2(t) & 0 \\ 0 & 0 & 0 \end{bmatrix}. \tag{7.47}$$

The model can thus be written as:

$$d\begin{bmatrix} S_1(t) \\ S_2(t) \\ M(t) \end{bmatrix} = \bar{\boldsymbol{\mu}}^{\mathbb{Q}}(t, \mathbf{X}(t))dt + \begin{bmatrix} \sigma_1 S_1(t) & 0 & 0 \\ \rho \sigma_2 S_2(t) & \sigma_2 \sqrt{1-\rho^2} S_2(t) & 0 \\ 0 & 0 & 0 \end{bmatrix} d\widetilde{\mathbf{W}}^{\mathbb{Q}}(t). \tag{7.48}$$

As we will change the numéraire to $S_1(t)$, we take the first row of the volatility matrix and apply Equation (7.34) for the transformation of the drift. However, since there is no volatility in the money-savings account process, i.e. $\bar{\sigma}_M = 0$ in (7.34), we obtain,

$$\bar{\boldsymbol{\mu}}^{S_1}(t, \mathbf{X}(t)) = \bar{\boldsymbol{\mu}}^{\mathbb{Q}}(t, \mathbf{X}(t)) + \bar{\boldsymbol{\sigma}}(t, \mathbf{X}(t)) \left[\frac{\sigma_1 S_1(t)}{S_1(t)}, \frac{0}{S_1(t)}, \frac{0}{S_1(t)}\right]^T. \tag{7.49}$$

After substitution and simplifications, we arrive at the following drift term:

$$\bar{\boldsymbol{\mu}}^{S_1}(t, \mathbf{X}(t)) = \begin{bmatrix} rS_1(t) \\ rS_2(t) \\ r \end{bmatrix} + \begin{bmatrix} \sigma_1 S_1(t) & 0 & 0 \\ \rho\sigma_2 S_2(t) & \sigma_2 S_2(t)\sqrt{1-\rho^2} & 0 \\ 0 & 0 & 0 \end{bmatrix} \begin{bmatrix} \sigma_1 \\ 0 \\ 0 \end{bmatrix}, \tag{7.50}$$

which can be written as:

$$\bar{\boldsymbol{\mu}}^{S_1}(t, \mathbf{X}(t)) = \begin{bmatrix} (r + \sigma_1^2) S_1(t) \\ (r + \rho\sigma_1\sigma_2) S_2(t) \\ r \end{bmatrix}. \tag{7.51}$$

It is easy to see that the drift term $\bar{\boldsymbol{\mu}}^{S_1}(t, \mathbf{X}(t))$ presented above equals the one derived in (7.45). ◆

Here, we briefly summarize the measure changes encountered in this book. Consider $X(t)$ to be a "tradable asset", three measures and the corresponding martingale property are as follows.

- Risk-neutral measure is associated with the money-savings account, $M(t)$, as the numéraire,

$$\mathrm{d}X(t) = \bar{\mu}^{\mathbb{Q}}(t)\mathrm{d}t + \bar{\sigma}(t)\mathrm{d}W^{\mathbb{Q}}(t) \implies \mathbb{E}^{\mathbb{Q}}\left[\frac{X(t)}{M(t)}\bigg|\mathcal{F}(t_0)\right] = \frac{X(t_0)}{M(t_0)}.$$

- Forward measure is associated with the ZCB, $P(t,T)$, as the numéraire, discussed in Chapter 11,

$$\mathrm{d}X(t) = \bar{\mu}^T(t)\mathrm{d}t + \bar{\sigma}(t)\mathrm{d}W^T(t) \implies \mathbb{E}^T\left[\frac{X(t)}{P(t,T)}\bigg|\mathcal{F}(t_0)\right] = \frac{X(t_0)}{P(t_0,T)}.$$

- Stock measure is associated with the stock, $S(t)$, as the numéraire,

$$\mathrm{d}X(t) = \bar{\mu}^S(t)\mathrm{d}t + \bar{\sigma}(t)\mathrm{d}W^S(t) \implies \mathbb{E}^S\left[\frac{X(t)}{S(t)}\bigg|\mathcal{F}(t_0)\right] = \frac{X(t_0)}{S(t_0)}.$$

7.3 Affine processes

In this section we describe a general class of vector-valued SDEs for which the characteristic function, $\phi_X(u;t,T)$, can be determined, often in closed form.

The class of affine diffusion (AD) stochastic processes is part of the more general affine jump diffusion (AJD) processes, which will be described in Section 7.3.2.

7.3.1 Affine diffusion processes

We start with a description of the AD class. For the asset dynamics, we refer to a probability space $(\Omega, \mathcal{F}(t), \mathbb{Q})$ and a Markovian n-dimensional affine process $\mathbf{X}(t)$, in some space $D \subset \mathbb{R}^n$.

Stochastic models in this class can be expressed by the following stochastic differential form, see also (7.6):

$$\mathrm{d}\mathbf{X}(t) = \bar{\boldsymbol{\mu}}(t,\mathbf{X}(t))\mathrm{d}t + \bar{\boldsymbol{\sigma}}(t,\mathbf{X}(t))\mathrm{d}\widetilde{\mathbf{W}}(t), \quad 0 \leq t_0 < t,$$

where $\bar{\boldsymbol{\mu}}(t,\mathbf{X}(t)) : D \to \mathbb{R}^n$, $\bar{\boldsymbol{\sigma}}(t,\mathbf{X}(t)) : D \to \mathbb{R}^{n \times n}$ and $\widetilde{\mathbf{W}}(t)$ is a column vector of *independent Brownian motions* in $\mathbb{R}^n$. The functions $\bar{\boldsymbol{\mu}}(t,\mathbf{X}(t))$ and $\bar{\boldsymbol{\sigma}}(t,\mathbf{X}(t))$ are assumed to satisfy certain conditions, like their derivatives of any order should exist and should be bounded (see [Milstein et al., 2004]).

For processes in the AD class, it is required that drift $\bar{\boldsymbol{\mu}}(t, \mathbf{X}(t))$, interest rate component $\bar{r}(t, \mathbf{X}(t))$, and the covariance matrix $\bar{\boldsymbol{\sigma}}(t, \mathbf{X}(t))\bar{\boldsymbol{\sigma}}(t, \mathbf{X}(t))^T$ are of *the affine form*, i.e.

$$\bar{\boldsymbol{\mu}}(t, \mathbf{X}(t)) = a_0 + a_1 \mathbf{X}(t), \text{ for any } (a_0, a_1) \in \mathbb{R}^n \times \mathbb{R}^{n \times n}, \quad (7.52)$$

and

$$\bar{r}(t, \mathbf{X}(t)) = r_0 + r_1^T \mathbf{X}(t), \text{ for } (r_0, r_1) \in \mathbb{R} \times \mathbb{R}^n, \quad (7.53)$$

and

$$(\bar{\boldsymbol{\sigma}}(t, \mathbf{X}(t))\bar{\boldsymbol{\sigma}}(t, \mathbf{X}(t))^T)_{i,j} = (c_0)_{ij} + (c_1)_{ij}^T \mathbf{X}_j(t), \quad (7.54)$$

with $(c_0, c_1) \in \mathbb{R}^{n \times n} \times \mathbb{R}^{n \times n \times n}$, and where in (7.54) it is meant that each element in the matrix $\bar{\boldsymbol{\sigma}}(t, \mathbf{X}(t))\bar{\boldsymbol{\sigma}}(t, \mathbf{X}(t))^T$ should be affine, as well as each vector element in the drift and interest rate vectors. For stochastic processes, it is relatively easy to satisfy the two first conditions (7.52) and (7.53). It is however not always trivial to verify the third condition (7.54), as it requires the analysis of the underlying covariance structure. A covariance structure is directly connected to the correlations between the driving Brownian motions, which makes it difficult to satisfy the condition (7.54) for several SDE systems.

The requirement on the interest rate process, $\bar{r}(t, \mathbf{X}(t))$ in (7.53) is related to the process used for *discounting* the asset value, for example by a stochastic interest rate process. In the Black-Scholes model, we simply have $\bar{r}(t, \mathbf{X}(t)) \equiv r$, so that in (7.53) $r_0 = r$.

When the dynamics of state vector $\mathbf{X}(t)$ are affine, it can be shown that the discounted characteristic function (ChF) is of the following form:

$$\phi_{\mathbf{X}}(\mathbf{u}; t, T) = \mathbb{E}^{\mathbb{Q}}\left[e^{-\int_t^T r(s)ds + i\mathbf{u}^T \mathbf{X}(T)} \Big| \mathcal{F}(t)\right] = e^{\bar{A}(\mathbf{u}, \tau) + \bar{\mathbf{B}}^T(\mathbf{u}, \tau)\mathbf{X}(t)}, \quad (7.55)$$

where the expectation is taken under the risk-neutral measure $\mathbb{Q}$, with time lag,[a] $\tau = T - t$, and $\bar{A}(\mathbf{u}, 0) = 0$ and $\bar{B}(\mathbf{u}, 0) = i\mathbf{u}^T$.

---

[a] The T (in brackets) indicates maturity time, whereas the superscript T in $\mathbf{u}^T$ denotes the vector's transpose.

The coefficients $\bar{A} := \bar{A}(\mathbf{u}, \tau)$ and $\bar{\mathbf{B}}^T := \bar{\mathbf{B}}^T(\mathbf{u}, \tau)$ in (7.55) satisfy the following complex-valued *Riccati* ordinary differential equations (ODEs),[2]

$$\boxed{\begin{aligned}\frac{d\bar{A}}{d\tau} &= -r_0 + \bar{\mathbf{B}}^T a_0 + \frac{1}{2}\bar{\mathbf{B}}^T c_0 \bar{\mathbf{B}}, \\ \frac{d\bar{\mathbf{B}}}{d\tau} &= -r_1 + a_1^T \bar{\mathbf{B}} + \frac{1}{2}\bar{\mathbf{B}}^T c_1 \bar{\mathbf{B}}.\end{aligned}} \quad (7.56)$$

---

[2] Although the functions have two arguments, it is only the τ to which the derivative is taken, so it is a set of ODEs, not PDEs.

These ODEs can be derived by inserting the general solution for the characteristic function (7.55), in the pricing PDE, which originates from the asset price process (7.6) with (7.52), (7.53), (7.54), see also Duffie-Pan-Singleton [Duffie et al., 2000].

The dimension of the ODEs for $\bar{\mathbf{B}}(\mathbf{u}, \tau)$ is directly related to the dimension of the state vector $\mathbf{X}(t)$.

Remark 7.3.1 *Multi-factor, vector-valued stochastic models for which the dimension $n > 1$ often provide an improved fit to the observed financial market data compared to the one-factor models. However, as the dimension of the SDE system increases, the ODEs to be solved to determine the characteristic function can become increasingly involved. If an analytic solution to the ODEs cannot be obtained, one can apply well-known numerical ODE techniques, like Runge-Kutta methods, instead. This requires some computational effort however, which may slow down the overall computations.*

An overview of topics regarding the characteristic function is found in [Ushakov, 1999]. ▲

So, by merely examining the system of SDEs, and checking the affinity conditions, we already know whether or not the characteristic function for the SDE system can be determined. This is beneficial. The statement may seem somewhat mysterious, but in fact it is a direct result of well-known PDE theory regarding solution for convection-diffusion reaction PDEs with constant coefficients. There is a direct connection to the derivations that were done in Section 3.2.2, in particular in Equation (3.27).

We will show an example of such a derivation for the Black-Scholes equation, in the example to follow.

Example 7.3.1 (Black-Scholes case and affine diffusion theory)
The GBM SDE (2.26) is a one-dimensional scalar equation, so the vectors and matrices in the affinity theory will collapse to scalars. It is clear from examing the GBM SDE that GBM is not in the class of affine diffusions. With the diffusion term being equal to σS, the term $\bar{\boldsymbol{\sigma}}(t, \mathbf{X}(t))\bar{\boldsymbol{\sigma}}(t, \mathbf{X}(t))^{\mathrm{T}}$ in the theory equals $\sigma^2 S^2$, which is clearly not linear in S.

We therefore first need to apply the log-transformation, $X(t) = \log S(t)$, because the use of Itô's lemma has shown that, due to the constant drift and diffusion terms after this transformation, we deal with affine log-asset dynamics. Under the risk-neutral dynamics, we have $a_0 = r - \frac{1}{2}\sigma^2$, $a_1 = 0$, $c_0 = \sigma^2$, $c_1 = 0$, $r_0 = r$, $r_1 = 0$ in Equations (7.52), (7.54) and (7.53), respectively.

In order to find the characteristic function, using $\tau = T - t$,

$$\phi_X(u; t, T) = e^{\bar{A}(u,\tau) + \bar{B}(u,\tau)X(t)},$$

we set up the following system of ODEs[3]

$$\begin{cases} \dfrac{\mathrm{d}\bar{B}}{\mathrm{d}\tau} = -r_1 + a_1\bar{B} + \dfrac{1}{2}\bar{B}c_1\bar{B}, \\ \dfrac{\mathrm{d}\bar{A}}{\mathrm{d}\tau} = -r_0 + a_0\bar{B} + \dfrac{1}{2}\bar{B}c_0\bar{B}, \end{cases}$$

---

[3]From now on, we will often use the notation $\bar{A} \equiv \bar{A}(u, \tau), \bar{B} \equiv \bar{B}(u, \tau)$ and so on.

which then reads,
$$\begin{cases} \dfrac{d\bar{B}}{d\tau} = 0, \\ \dfrac{d\bar{A}}{d\tau} = -r + \left(r - \tfrac{1}{2}\sigma^2\right)\bar{B} + \tfrac{1}{2}\sigma^2\bar{B}^2. \end{cases}$$

By the initial conditions, for $\tau = 0$, i.e. $\bar{B}(u,0) = iu, \bar{A}(u,0) = 0$, we obtain:
$$\begin{cases} \bar{B}(u,\tau) = iu, \\ \bar{A}(u,\tau) = \left[-r + iu\left(r - \tfrac{1}{2}\sigma^2\right) - \tfrac{1}{2}u^2\sigma^2\right]\tau. \end{cases}$$

The corresponding characteristic function is then given by:
$$\phi_X(u;t,T) = e^{iu\log S(t) + iu\left(r - \frac{1}{2}\sigma^2\right)\tau - \frac{1}{2}u^2\sigma^2\tau - r\tau},$$

which concludes this first example. ♦

Example 7.3.2 (Black-Scholes case, characteristic function) Based on classical PDE theory, and connected to the Feynman-Kac Theorem (3.2.1), we will assume that the discounted characteristic function of the following form,
$$\phi_X(u;t,T) = \exp\left(\bar{A}(u,\tau) + \bar{B}(u,\tau)X\right), \tag{7.57}$$
with the initial condition
$$\phi_X(u;T,T) = e^{iuX} \tag{7.58}$$
is the solution of the log-transformed Black-Scholes PDE (3.21).

Differentiation of this solution $\phi_X \equiv \phi_X(u;t,T)$ in (7.57), gives,
$$\frac{\partial \phi_X}{\partial \tau} = \phi_X\left(\frac{d\bar{A}}{d\tau} + X\frac{d\bar{B}}{d\tau}\right), \quad \frac{\partial \phi_X}{\partial X} = \phi_X \bar{B}, \quad \frac{\partial^2 \phi_X}{\partial X^2} = \phi_X \bar{B}^2.$$

Substituting these expressions in the log-transformed Black-Scholes PDE (3.21), leads to
$$-\left(\frac{d\bar{A}}{d\tau} + X\frac{d\bar{B}}{d\tau}\right) + \left(r - \frac{1}{2}\sigma^2\right)\bar{B} + \frac{1}{2}\sigma^2\bar{B}^2 - r = 0. \tag{7.59}$$

Now, the characteristic function (7.57) satisfies the log-transformed Black-Scholes PDE, if the terms in front of the X-variable in (7.59) will be equal to zero, and the remaining terms (the zero-th order terms) should also equal 0. To achieve this, the following set of ODEs should be satisfied, $\forall X$,
$$\frac{d\bar{B}}{d\tau} = 0, \quad \text{and} \quad \frac{d\bar{A}}{d\tau} = \left(r - \frac{1}{2}\sigma^2\right)\bar{B} + \frac{1}{2}\sigma^2\bar{B}^2 - r. \tag{7.60}$$

Incorporation of the initial condition in τ, i.e. Equation (7.58), to the solution of (7.60) gives us
$$\bar{B}(u,\tau) = iu, \quad \text{and} \quad \bar{A}(u,\tau) = \left(r - \frac{1}{2}\sigma^2\right)iu\tau - \frac{1}{2}\sigma^2 u^2 \tau - r\tau.$$

The discounted characteristic function under the log-transformed GBM asset process, is thus given by

$$\phi_X(u,t,T) = \exp\left(\left(r - \frac{\sigma^2}{2}\right)iu\tau - \frac{1}{2}\sigma^2 u^2 \tau - r\tau + iuX\right),$$

which is, of course, identical to the form obtained by application of the affine diffusion theory in Example 7.3.1.

This derivation is a specific example for the general theory regarding affine diffusion processes and the characteristic function in this section. ◆

Example 7.3.3 (2D GBM as an affine diffusion process) The 2D correlated GBM process, with $\mathbf{S}(t) = [S_1(t), S_2(t)]$, is given by the SDE system (7.11) in Example 7.1.4, with $\rho dt = dW_1(t)dW_2(t)$.

This 2D system is not affine, and therefore the logarithmic transformation, $\mathbf{X}(t) = [\log S_1(t), \log S_2(t)] =: [X_1(t), X_2(t)]$, should be performed, i.e.

$$\begin{cases} dX_1(t) = \left(\mu_1 - \frac{1}{2}\sigma_1^2\right)dt + \sigma_1 d\widetilde{W}_1(t), \\ dX_2(t) = \left(\mu_2 - \frac{1}{2}\sigma_2^2\right)dt + \sigma_2\left(\rho d\widetilde{W}_1(t) + \sqrt{1-\rho^2}d\widetilde{W}_2(t)\right), \end{cases} \quad (7.61)$$

with independent Brownian motions $\widetilde{W}_1(t)$, $\widetilde{W}_2(t)$. In matrix notation, this system reads,

$$\begin{bmatrix} dX_1(t) \\ dX_2(t) \end{bmatrix} = \begin{bmatrix} \mu_1 - \frac{1}{2}\sigma_1^2 \\ \mu_2 - \frac{1}{2}\sigma_2^2 \end{bmatrix} dt + \begin{bmatrix} \sigma_1 & 0 \\ \sigma_2\rho & \sigma_2\sqrt{1-\rho^2} \end{bmatrix} \begin{bmatrix} d\widetilde{W}_1(t) \\ d\widetilde{W}_2(t) \end{bmatrix} \Leftrightarrow$$

$$d\mathbf{X}(t) = \bar{\boldsymbol{\mu}}(t,\mathbf{X}(t))dt + \bar{\boldsymbol{\sigma}}(t,\mathbf{X}(t))d\widetilde{\mathbf{W}}(t). \quad (7.62)$$

Following the definitions for the AD processes in Section 7.3.1, we here find:

$$a_0 = \begin{bmatrix} \mu_1 - \frac{1}{2}\sigma_1^2 \\ \mu_2 - \frac{1}{2}\sigma_2^2 \end{bmatrix}, \quad r_0 = r, \text{ and } c_0 = \begin{bmatrix} \sigma_1^2 & \sigma_1\sigma_2\rho \\ \sigma_1\sigma_2\rho & \sigma_2^2 \end{bmatrix}.$$

Moreover, $a_1 = 0$, $c_1 = 0$ and $r_1 = 0$.

For the affine system of SDEs, we can now easily derive the 2D characteristic function (7.55), with $\mathbf{u} = [u_1, u_2]$, and $\tau = T - t$,

$$\phi_{\mathbf{X}}(\mathbf{u};t,T) = e^{\bar{A}(\mathbf{u},\tau) + \bar{B}_1(\mathbf{u},\tau)X_1(t) + \bar{B}_2(\mathbf{u},\tau)X_2(t)}, \quad (7.63)$$

with

$$\phi_{\mathbf{X}}(\mathbf{u};T,T) = e^{iu_1 X_1(T) + iu_2 X_2(T)}. \quad (7.64)$$

In the present setting, we thus have $\bar{A}(\mathbf{u},0) = 0$, $\bar{B}_1(\mathbf{u},0) = iu_1$ and $\bar{B}_2(\mathbf{u},0) = iu_2$.

Since a_1, c_1 and r_1 are zero vectors and matrices, the system of the ODEs, as in (7.56), is given by:

$$\frac{\mathrm{d}\bar{B}_1}{\mathrm{d}\tau} = 0, \quad \frac{\mathrm{d}\bar{B}_2}{\mathrm{d}\tau} = 0, \tag{7.65}$$

and

$$\frac{\mathrm{d}\bar{A}}{\mathrm{d}\tau} = -r + iu_1\left(\mu_1 - \frac{1}{2}\sigma_1^2\right) + iu_2\left(\mu_2 - \frac{1}{2}\sigma_2^2\right)$$
$$+ \frac{1}{2}(iu_1)^2\sigma_1^2 + \frac{1}{2}(iu_2)^2\sigma_2^2 + \rho\sigma_1\sigma_2(iu_1)(iu_2),$$

so that $\bar{B}_1(\mathbf{u},\tau) = iu_1$, $\bar{B}_2(\mathbf{u},\tau) = iu_2$, and the solution for $\bar{A}(\mathbf{u},\tau)$ reads:

$$\bar{A}(\mathbf{u},\tau) = -r\tau + iu_1\left(\mu_1 - \frac{1}{2}\sigma_1^2\right)\tau + iu_2\left(\mu_2 - \frac{1}{2}\sigma_2^2\right)\tau$$
$$- \frac{1}{2}u_1^2\sigma_1^2\tau - \frac{1}{2}u_2^2\sigma_2^2\tau - \rho\sigma_1\sigma_2 u_1 u_2 \tau. \tag{7.66}$$

The 2D characteristic function is therefore given by:

$$\phi_{\mathbf{X}}(\mathbf{u}; t, T) = \mathrm{e}^{iu_1 X_1(t) + iu_2 X_2(t) + \bar{A}(\mathbf{u},\tau)},$$

with the function $\bar{A}(\mathbf{u},\tau)$ from (7.66). ◆

7.3.2 Affine jump diffusion processes

In this section we provide a description of the price processes in the *affine jump diffusion (AJD) class*, which is more general than the AD class from Section 7.3.1. It gives us a more general framework to derive the characteristic function, for example, for an affine jump diffusion model. Basically all statements regarding the characteristic function from the AD class can be generalized towards the AJD class.

The class of AJD stochastic processes for the asset dynamics refers to a fixed probability space $(\Omega, \mathcal{F}(t), \mathbb{Q})$ and a Markovian n-dimensional affine process $\mathbf{X}(t) = [X_1(t), \ldots, X_n(t)]^{\mathrm{T}}$ in some space $\mathbb{R} \subset \mathbb{R}^n$.

The stochastic model of interest can now be expressed by the following stochastic differential form:

$$\mathrm{d}\mathbf{X}(t) = \bar{\mu}(t, \mathbf{X}(t))\mathrm{d}t + \bar{\sigma}(t, \mathbf{X}(t))\mathrm{d}\widetilde{\mathbf{W}}(t) + \mathbf{J}(t)^{\mathrm{T}}\mathrm{d}\mathbf{X}_{\mathcal{P}}(t), \tag{7.67}$$

where $\widetilde{\mathbf{W}}(t)$ is an $\mathcal{F}(t)$-standard column vector of *independent* Brownian motions in $\mathbb{R}^n$, $\bar{\mu}(t, \mathbf{X}(t)) : \mathbb{R} \to \mathbb{R}^n$, $\bar{\sigma}(t, \mathbf{X}(t)) : \mathbb{R} \to \mathbb{R}^{n \times n}$, and $\mathbf{X}_{\mathcal{P}}(t) \in \mathbb{R}^n$ is a vector of orthogonal Poisson processes, characterized by an intensity vector $\bar{\xi}(t, \mathbf{X}(t)) \in \mathbb{R}^n$.

$\mathbf{J}(t) \in \mathbb{R}^n$ is a vector governing the amplitudes of the jumps and is assumed to be a matrix of correlated random variables, that are independent of the state vector $\mathbf{X}(t)$ and of $\mathbf{X}_{\mathcal{P}}(t)$.

For processes in the AJD class, the drift term $\bar{\mu}(t,\mathbf{X}(t))$, covariance matrix $\bar{\sigma}(t,\mathbf{X}(t))\bar{\sigma}(t,\mathbf{X}(t))^{\mathrm{T}}$, and interest rate component $\bar{r}(t,\mathbf{X}(t))$ (as explained previously) should be affine, but also the jump intensity should be of *the affine form*, i.e.

$$\bar{\xi}(t,\mathbf{X}(t)) = l_0 + l_1 \mathbf{X}(t), \text{ with } (l_0, l_1) \in \mathbb{R}^n \times \mathbb{R}^n, \quad (7.68)$$

It can be shown that in this class, for a state vector $\mathbf{X}(t)$, the discounted characteristic function is also of the following form:

$$\phi_{\mathbf{X}}(\mathbf{u}; t, T) = \mathbb{E}^{\mathbb{Q}}\left[e^{-\int_t^T r(s)ds + i\mathbf{u}^{\mathrm{T}}\mathbf{X}(T)}\Big|\mathcal{F}(t)\right] = e^{\bar{A}(\mathbf{u},\tau) + \bar{\mathbf{B}}^{\mathrm{T}}(\mathbf{u},\tau)\mathbf{X}(t)},$$

with the expectation under risk-neutral measure $\mathbb{Q}$.

The coefficients $\bar{A}(\mathbf{u},\tau)$ and $\bar{\mathbf{B}}^{\mathrm{T}}(\mathbf{u},\tau)$ have to satisfy the following complex-valued *Riccati* ODEs, see the work by Duffie-Pan-Singleton [Duffie et al., 2000]:

$$\begin{aligned}\frac{\mathrm{d}\bar{A}}{\mathrm{d}\tau} &= -r_0 + \bar{\mathbf{B}}^{\mathrm{T}} a_0 + \frac{1}{2}\bar{\mathbf{B}}^{\mathrm{T}} c_0 \bar{\mathbf{B}} + l_0^{\mathrm{T}} \mathbb{E}\left[e^{\mathbf{J}(\tau)\bar{\mathbf{B}}} - 1\right], \\ \frac{\mathrm{d}\bar{\mathbf{B}}}{\mathrm{d}\tau} &= -r_1 + a_1^{\mathrm{T}} \bar{\mathbf{B}} + \frac{1}{2}\bar{\mathbf{B}}^{\mathrm{T}} c_1 \bar{\mathbf{B}} + l_1^{\mathrm{T}} \mathbb{E}\left[e^{\mathbf{J}(\tau)\bar{\mathbf{B}}} - 1\right],\end{aligned} \quad (7.69)$$

where the expectation, $\mathbb{E}[\cdot]$ in (7.69), is taken with respect to the jump amplitude $\mathbf{J}(t)$. The dimension of the (complex-valued) ODEs for $\bar{\mathbf{B}}(\mathbf{u},\tau)$ corresponds to the dimension of the state vector $\mathbf{X}(t)$. The interpretation of these ODEs for the AJD models remains the same as in the case of the affine diffusion models, presented in Section 7.3. We clarify the expression in (7.69) by means of the following example in Section 7.3.3.

7.3.3 Affine jump diffusion process and PIDE

Earlier, we have seen that for a stock price, driven by the following SDE,

$$\frac{\mathrm{d}S(t)}{S(t)} = \left(r - \xi_p \mathbb{E}\left[e^J - 1\right]\right)\mathrm{d}t + \sigma \mathrm{d}W^{\mathbb{Q}}(t) + \left(e^J - 1\right)\mathrm{d}X_{\mathcal{P}}^{\mathbb{Q}}(t), \quad (7.70)$$

the corresponding option pricing PIDE is given by:

$$\frac{\partial V}{\partial t} + \left(r - \xi_p \mathbb{E}\left[e^J - 1\right]\right) S \frac{\partial V}{\partial S} + \frac{1}{2}\sigma^2 S^2 \frac{\partial^2 V}{\partial S^2} - (r + \xi_p)V + \xi_p \mathbb{E}\left[V(t, Se^J)\right] = 0.$$

According to the affinity conditions in (7.67), this model (7.70) does not belong to the class of affine jump diffusions. We therefore consider the same model under log-asset transformation, $X(t) = \log S(t)$, for which the dynamics read,

$$\mathrm{d}X(t) = \left(r - \xi_p \mathbb{E}\left[e^J - 1\right] - \frac{1}{2}\sigma^2\right)\mathrm{d}t + \sigma \mathrm{d}W^{\mathbb{Q}}(t) + J\mathrm{d}X_{\mathcal{P}}^{\mathbb{Q}}(t).$$

For $V(\tau, X)$, and $\tau := T - t$, we find the following option pricing PIDE:

$$\frac{\partial V}{\partial \tau} = \left(r - \xi_p \mathbb{E}\left[e^J - 1\right] - \frac{1}{2}\sigma^2\right)\frac{\partial V}{\partial X} + \frac{1}{2}\sigma^2\frac{\partial^2 V}{\partial X^2}$$
$$- (r + \xi_p)V + \xi_p \mathbb{E}\left[V(T - \tau, X + J)\right]. \tag{7.71}$$

The discounted characteristic function is now of the following form,

$$\phi_X := \phi_X(u; t, T) = e^{\bar{A}(u,\tau) + \bar{B}(u,\tau)X(t)},$$

with initial condition $\phi_X(u; T, T) = e^{iuX(0)}$. Substitution of all derivatives results in,

$$\frac{\partial \phi_X}{\partial \tau} = \phi_X\left(\frac{d\bar{A}}{d\tau} + \frac{d\bar{B}}{d\tau}\right), \quad \frac{\partial \phi_X}{\partial X} = \phi_X \bar{B}, \quad \frac{\partial^2 \phi_X}{\partial x^2} = \phi_X \bar{B}^2,$$

and, because the expectation in (7.71) is taken only with respect to jump size $F_J(y)$,

$$\mathbb{E}[\phi_{X+J}] = \mathbb{E}\left[\exp\left(\bar{A}(u,\tau) + \bar{B}(u,\tau)(X + J)\right)\right]$$
$$= \phi_X \cdot \mathbb{E}\left[\exp\left(\bar{B}(u,\tau)J\right)\right]. \tag{7.72}$$

In the PIDE in (7.71), this gives us:

$$-\left(\frac{d\bar{A}}{d\tau} + X\frac{d\bar{B}}{d\tau}\right) + \left(r - \xi_p \mathbb{E}\left[e^J - 1\right] - \frac{1}{2}\sigma^2\right)\bar{B}$$
$$+ \frac{1}{2}\sigma^2 \bar{B}^2 - (r + \xi_p) + \xi_p \mathbb{E}\left[\exp\left(\bar{B} \cdot J\right)\right] = 0.$$

and the following system of ODEs needs to be solved,

$$\begin{cases} \frac{d\bar{B}}{d\tau} = 0, \\ \frac{d\bar{A}}{d\tau} = \left(r - \xi_p \mathbb{E}\left[e^J - 1\right] - \frac{1}{2}\sigma^2\right)\bar{B} + \frac{1}{2}\sigma^2 \bar{B}^2 - (r + \xi_p) + \xi_p \mathbb{E}\left[e^{\bar{B} \cdot J}\right]. \end{cases}$$

With the solution, $\bar{B}(u, \tau) = iu$, we find,

$$\bar{A}(u, \tau) = \left(r - \xi_p \mathbb{E}\left[e^J - 1\right] - \frac{1}{2}\sigma^2\right)iu\tau - \frac{1}{2}\sigma^2 u^2 \tau - (r + \xi_p)\tau + \xi_p \tau \mathbb{E}\left[e^{iuJ}\right].$$

The system of ODEs (7.73) forms an example of the ODE representation in (7.69), using $r_0 = r$, $r_1 = 0$, $a_0 = -\xi_p \mathbb{E}\left[e^J - 1\right] - \frac{1}{2}\sigma^2$, $a_1 = 0$, $c_0 = \sigma^2$, $c_1 = 0$, $l_0 = \xi_p$, $l_1 = 0$.

This concludes this chapter, in which we have presented some *mathematical equipment* to work with different vector-valued stochastic asset price processes in the chapters to follow.

7.4 Exercise set

Exercise 7.1 With $W_1(t), W_2(t)$ and $W_3(t)$ three Wiener processes, where $dW_1(t) \cdot dW_2(t) = dW_1(t) \cdot dW_3(t) = dW_2(t) \cdot dW_3(t) = \rho dt$, find the solution of the SDE

$$\frac{dS(t)}{S(t)} = \frac{3}{2}dt + dW_1(t) + dW_2(t) + dW_3(t), \quad S(0) = S_0.$$

Exercise 7.2 Consider a process governed by the Cox-Ingersoll-Ross (CIR) process, with dynamics given by,

$$dS(t) = \lambda(\theta - S(t))dt + \gamma\sqrt{S(t)}dW^{\mathbb{P}}(t), \tag{7.73}$$

with all parameters positive. Additionally, we have the usual money savings account, $dM(t) = rM(t)dt$.

a. Change the measure of the process (7.73) from $\mathbb{P}$ to $\mathbb{Q}$.

b. Derive the corresponding discounted characteristic function. Does the discounted characteristic function depend on the parameters λ and θ?

Exercise 7.3 Suppose we have the following SDE:

$$dS(t) = rS(t)dt + \sigma(t, \cdot)S(t)dW(t), \quad S(t_0) = S_0, \tag{7.74}$$

with constant interest rate r and a, possibly stochastic, volatility function $\sigma(t, \cdot)$.

a. Show that the model in (7.74) does not belong to the class of affine processes.

b. Under the log-transformation, $X(t) = \log S(t)$, the model is assumed to be affine, and its dynamics read:

$$dX(t) = \left(r - \frac{1}{2}\sigma^2(t, \cdot)\right)dt + \sigma(t, \cdot)dW(t), \quad X(t_0) = \log(S_0).$$

Since $X(t)$ is assumed to be affine, it is possible to derive its characteristic function $\phi_{X(t)}(u)$.

Determine the relation between the densities of $S(t)$ and $X(t)$, i.e., show that

$$f_{S(T)}(x) = \frac{1}{x}f_{X(T)}(\log(x)), \quad x > 0.$$

Exercise 7.4 The *displaced diffusion model* is described by the following SDE:

$$dS(t) = \sigma\left[\vartheta S(t) + (1-\vartheta)S(t_0)\right]dW^{\mathbb{P}}(t), \quad S(t_0) > 0, \tag{7.75}$$

with a constant volatility σ, and a so-called *displacement parameter* ϑ. Show that the process in (7.75) is a well-defined process under the risk-neutral measure $\mathbb{Q}$, by showing that $S(t)$ divided by the money-savings account $M(t)$ is a martingale.

Exercise 7.5 For a displaced diffusion model, as in (7.75), however with a nonzero interest rate r, the pricing of European options can be performed with the help of a measure change. In the case of a deterministic interest rate, the derivations can be obtained directly, however the presented methodology may be also used when the interest rates are assumed to be stochastic.

When the process follows the dynamics in (14.50), we may consider the forward rate $S^F(t, T) = \frac{S(t)}{P(t,T)}$.

Show that, with the zero-coupon bond as the numéraire, the forward stock $S^F(t,T)$ is a martingale under the corresponding measure $\mathbb{Q}^T$.

Exercise 7.6 Show that for two jointly normally distributed random variables $X_1 \stackrel{d}{=} \mathcal{N}(\mu_1, \sigma_1^2)$ and $X_2 \stackrel{d}{=} \mathcal{N}(\mu_2, \sigma_2^2)$, that are correlated with correlation coefficient ρ, the following result holds:

$$\mathbb{E}\left[X_1 \mathbb{1}_{X_2 > k}\right] = \left(\mu_1 + \rho \sigma_1 \frac{f_{\mathcal{N}(0,1)}\left(\frac{k-\mu_2}{\sigma_2}\right)}{1 - F_{\mathcal{N}(0,1)}\left(\frac{k-\mu_2}{\sigma_2}\right)}\right)(1 - F_{X_2}(k)),$$

with $f_{\mathcal{N}(0,1)}(\cdot)$ and $F_{\mathcal{N}(0,1)}(\cdot)$ the standard normal PDF and CDF, respectively, and $F_{X_2}(\cdot)$ is the CDF corresponding to random variable X_2.

Exercise 7.7 Consider the following system of SDEs:

$$dS(t) = \sigma(t)S(t)dW_1(t),$$
$$d\sigma(t) = \gamma \sigma(t) dW_2(t),$$

with initial values $S(t_0)$ and $\sigma(t_0)$, volatility parameter γ and correlation coefficient ρ. Show that, under the log-transformation for the stock, $X(t) = \log S(t)$, it follows that,

$$X(T) = X_0 - \frac{1}{2}\int_0^T \sigma^2(t)dt + \frac{1}{\gamma\rho}\left[\sigma(T) - \sigma(t_0) - \gamma\sqrt{1-\rho^2}\int_0^T \sigma(t)d\widetilde{W}_2(t)\right], \quad (7.76)$$

where $\widetilde{W}_2(t)$ is independent of the other Brownian motions.

Exercise 7.8 We use the Black-Scholes model with a stock and the money-savings account under the $\mathbb{Q}$ measure, given by,

$$dS(t) = rS(t)dt + \sigma S(t)dW(t),$$
$$dM(t) = rM(t)dt,$$

$r = 0.065$, $S_0 = 1$ and $\sigma = 0.4$, determine analytically, by means of measure change, the price of the following financial derivative,

$$V(t_0) = \mathbb{E}^{\mathbb{Q}}\left[\frac{1}{M(T)}\max\left(S^2(T) - S(T)K, 0\right)\bigg|\mathcal{F}(t_0)\right]. \quad (7.77)$$

Exercise 7.9 For a so-called exchange option, i.e. the option to exchange one asset for another one, an analytic solution is available after a change of measure. In this exercise we ask you to derive the analytic solution. The pay-off function is given by

$$V^{ex}(T, S_1(T), S_2(T)) = \max\left(S_1(T) - S_2(T), 0\right),$$

and the $\mathbb{Q}$-measure dynamics of $S_1(t), S_2(t)$ are governed by

$$dS_1(t) = rS_1(t)dt + \sigma_1 S_1(t)dW_1(t), \quad S_1(0) = S_{1,0},$$
$$dS_2(t) = rS_2(t)dt + \sigma_2 S_2(t)dW_2(t), \quad S_2(0) = S_{2,0},$$

with $dW_1(t)dW_2(t) = \rho dt$.

Derive the *Margrabe formula* [Margrabe, 1978], for the value of the exchange option at time $t_0 = 0$, which reads,

$$V^{ex}(t_0, S_1(t_0), S_2(t_0)) = S_{1,0} F_{\mathcal{N}(0,1)}(d_1) - S_{2,0} F_{\mathcal{N}(0,1)}(d_2),$$

$$\sigma = \sqrt{\sigma_1^2 - 2\rho\sigma_1\sigma_2 + \sigma_2^2},$$

$$d_1 = \frac{\log \frac{S_1(t_0)}{S_2(t_0)} + \frac{1}{2}\sigma^2(T - t_0)}{\sigma\sqrt{T - t_0}},$$

$$d_2 = \frac{\log \frac{S_1(t)}{S_2(t_0)} - \frac{1}{2}\sigma^2(T - t_0)}{\sigma\sqrt{T - t_0}} = d_1 - \sigma\sqrt{T - t_0},$$

with $F_{\mathcal{N}(0,1)}(\cdot)$ the cumulative distribution function of the standard normal variable. *Hint: Rewrite the equations in the variable $\frac{S_1(t)}{S_2(t)}$ and compare the resulting dynamics with those in Equation (7.44) in Section 7.2.3.*

Exercise 7.10 Let the following system of SDEs be given,

$$\mathrm{d}S(t)/S(t) = r(t)\mathrm{d}t + \sigma\mathrm{d}W_x(t), \quad S(t_0) = S_0 > 0, \tag{7.78}$$

$$\mathrm{d}r(t) = \lambda\left(\theta(t) - r(t)\right)\mathrm{d}t + \eta\mathrm{d}W_r(t), \quad r(t_0) = r_0, \tag{7.79}$$

where $W_x(t)$ and $W_r(t)$ are two *uncorrelated Brownian motions* with $\mathrm{d}W_x(t)\mathrm{d}W_r(t) = 0$.

Show that this system of SDEs is affine after the log-transformation for the $S(t)$ process.

Determine the Riccati ODEs governing the characteristic function of this system of SDEs. (This system is commonly known as the Black-Scholes Hull-White model, which will be discussed in Chapter 13).

CHAPTER 8

Stochastic Volatility Models

In this chapter:

The main problem with the Black-Scholes model is that it is not able to reproduce the *implied volatility smile*, which is present in many financial market option prices. We have already seen a few alternative SDE models in the previous chapters, but this implied volatility feature forms an important motivation to study and use *yet another asset model* in this chapter. We will here discuss an asset price SDE system based on *stochastic volatility* (**Section 8.1**). Connections to the theory presented in the previous chapter are provided. We use, for example, the *multi-d Itô's lemma* for stochastic volatility processes. Also the techniques to correlate independent Brownian motions will be useful when considering stochastic volatility models. We will particularly discuss the Heston stochastic volatility model, where the asset's variance follows a Cox-Ingersoll-Ross process (**Section 8.2**).
The pricing equations for the options under these asset price processes turn out to be involved. We deal with a *two-dimensional option pricing PDE*. The equations are derived based on the martingale approach, and also on a hedging argument in which an option enters the hedge portfolio, next to assets and bonds. Important is that for the stochastic volatility, the *discounted characteristic function* can be derived, which forms the basis of the Fourier option pricing techniques, see **Section 8.3**. For the asset models discussed, numerical examples give the reader insight in the *impact of the different model parameters*, see **Section 8.4**.

Keywords: stochastic volatility, Heston model, impact on option pricing PDE, implied volatility, discounted characteristic function, Itô's lemma, martingale approach, Bates stochastic volatility jump model.

8.1 Introduction into stochastic volatility models

In the previous chapters, we have discussed alternative asset price models, in order to deal with the nonconstant implied volatility, which is observed in the market. We have seen the local volatility models, in Chapter 4, as well as the asset models with jumps, in Chapter 5. Each of these models had their advantages and also disadvantages. The main disadvantage of the LV models is that the dynamics are very much tailored to a fixed observed market at a specific maturity time. This may be fine for the valuation of European options, but when dealing in particular with path-dependent options it is important that relevant properties of the underlying transition densities would be accurately modeled.

The asset models with jumps would be able to model the governing properties of transition densities, however, with these models it is not completely clear how an accurate replicating portfolio for hedging under jump dynamics would look. Jump models are often used in the commodity industry for valuation, where the underlying commodity prices may exhibit significant jumps. Hedging under jump processes would have to take place by choosing additional options in the hedge portfolio, but the number of option is not immediately clear. Also the calibration of the jump models may not be a trivial exercise, because of real-world versus risk-neutral dynamics, parameter modeling and other issues.

Another major step away from the assumption of constant volatility in asset pricing, was made by Hull and White [1987], Stein and Stein [1991], Heston [1993], Schöbel and Zhu [1999], modeling the *volatility as a diffusion process*. The idea to model volatility as a random variable is confirmed by practical financial data which indicates the highly variable and unpredictable nature of the stock price's volatility. The resulting models in this chapter are the so-called stochastic volatility (SV) models. The volatilities in these SV models "behave and move", to a certain extent, in a different way than the corresponding asset prices — a property which local volatility models do not have [Ren et al., 2007].

Asset return distributions under SV models also typically exhibit fatter tails than their lognormal counterparts, hereby being more realistic. However, the most significant argument to consider the volatility to be stochastic is the implied volatility smile/skew, which is present in the market, and can be accurately recovered by SV models, especially for options with a medium to long time to the maturity date.

With an additional stochastic process, which is correlated to the asset price process $S(t)$, we deal with a *system of SDEs* for which the theory from the previous chapter appears very useful.

8.1.1 The Schöbel-Zhu stochastic volatility model

A first obvious choice for a stochastic process for the volatility may be an Ornstein-Uhlenbeck (OU) mean reverting process, which reads as follows,

$$\mathrm{d}\sigma(t) = \kappa(\bar{\sigma} - \sigma(t))\mathrm{d}t + \gamma \mathrm{d}W_\sigma^{\mathbb{Q}}(t), \quad \sigma(t_0) = \sigma_0 > 0. \tag{8.1}$$

Here, the parameters $\kappa \geq 0$, $\bar{\sigma} \geq 0$ and $\gamma > 0$ are called the speed of mean reversion, the long-term mean of the volatility process and the volatility-of-volatility, respectively.

With a parameter, γ, in front of the Wiener increment in (8.1), the OU process for σ is just a Gaussian process, with as its mean
$$\lim_{t \to \infty} \mathbb{E}[\sigma(t)] = \bar{\sigma},$$
and variance
$$\lim_{t \to \infty} \text{Var}[\sigma(t)] = \gamma^2/(2\kappa).$$
An OU process models a mean reversion feature for the volatility. In other words, if the volatility exceeds its mean, it is driven back to the mean with the speed κ of mean reversion. The same is true if the volatility is below its mean.

The *Schöbel-Zhu asset price SV model* is connected to the OU process for volatility. It is defined by the following 2D system of SDEs:
$$\begin{cases} dS(t) = rS(t)dt + \sigma(t)S(t)dW_x(t) \\ d\sigma(t) = \kappa(\bar{\sigma} - \sigma(t))dt + \gamma dW_\sigma(t), \end{cases} \quad (8.2)$$
with all parameters positive and correlation $\rho_{x,\sigma}$ between the Brownian motions, $dW_x(t)dW_\sigma(t) = \rho_{x,\sigma}dt$.

This model has several favorable features. One of them is that this model fits, under the log-transformation, in the class of affine diffusions, see Section 7.3, which guarantees that the corresponding discounted characteristic function can be derived. This derivation is however not trivial, and requires the introduction of a new variable, $v(t) = \sigma^2(t)$, in an extended space.

The pricing of European options is however easily possible in the affine class, for example, by means of the COS method.

The OU process is a normally distributed process for which many properties are known. An *undesired property* of the OU process is that the variable may become **negative**, which is, of course, not a suitable property for a volatility.

8.1.2 The CIR process for the variance

To avoid negative volatilities, the asset's *variance* is modeled here by a so-called CIR process,
$$dv(t) = \kappa(\bar{v} - v(t))dt + \gamma\sqrt{v(t)}dW_v^{\mathbb{Q}}(t), \quad (8.3)$$
with κ the speed of mean reversion, $\bar{v}$ the long-term mean and where γ controls the volatility of the variance process. The CIR process was originally proposed by Cox et al. [1985] to model interest rates. The process, here defined under the risk-neutral $\mathbb{Q}$-measure, can be recognized as a mean reverting square-root process, which will be described in more detail in the following subsections. A square-root process precludes negative values for $v(t)$, and if $v(t)$ reaches zero it subsequently becomes positive. This is a useful property for a variance process.

Properties of the CIR process

Before we discuss the Heston SV asset price model in detail, we will first focus on the variance SDE (8.3) and its solution. SDE (8.3) models the asset's variance, not for the volatility itself (of course, $\sigma = \sqrt{v(t)}$). We will look, in detail, into the *near-singular behavior* of the CIR process near the origin, and discuss the relevant parameter sets giving rise to this phenomenon.

> In the CIR process (8.3), it is the *Feller condition*, i.e. $2\kappa\bar{v} \geq \gamma^2$, which guarantees that $v(t)$ stays positive; otherwise, if the Feller condition is not satisfied, $v(t)$ may reach zero, as indicated in [Feller, 1951; Cox et al., 1985]. When the Feller condition is not satisfied, the cumulative distribution exhibits a *near-singular behavior* near the origin, or, in other words, the left tail of the density may grow extremely fast in value. When the Feller condition is not satisfied, there is an accumulation of probability mass around 0. With more accumulation, the density needs to become extreme in order to fit the tail near the singularity around 0 with the continuous part of the density.

Definition 8.1.1 (CIR CDF and PDF) *Process $v(t)|v(s)$, $t > s > 0$, under the CIR dynamics, is known to be distributed as $\bar{c}(t,s)$ times a noncentral chi-squared random variable, $\chi^2(\delta, \bar{\kappa}(t,s))$, where δ is the "degrees of freedom" parameter and the noncentrality parameter is $\bar{\kappa}(t,s)$, i.e.,*

$$v(t)|v(s) \sim \bar{c}(t,s)\chi^2(\delta, \bar{\kappa}(t,s)), \quad t > s > 0, \tag{8.4}$$

with

$$\bar{c}(t,s) = \frac{1}{4\kappa}\gamma^2(1 - e^{-\kappa(t-s)}), \quad \delta = \frac{4\kappa\bar{v}}{\gamma^2}, \quad \bar{\kappa}(t,s) = \frac{4\kappa v(s)e^{-\kappa(t-s)}}{\gamma^2(1 - e^{-\kappa(t-s)})}. \tag{8.5}$$

The corresponding cumulative distribution function (CDF) is found to be:

$$F_{v(t)}(x) = \mathbb{Q}[v(t) \leq x] = \mathbb{Q}\left[\chi^2(\delta, \bar{\kappa}(t,s)) \leq \frac{x}{\bar{c}(t,s)}\right] = F_{\chi^2(\delta, \bar{\kappa}(t,s))}\left(\frac{x}{\bar{c}(t,s)}\right). \tag{8.6}$$

> Here,
>
> $$F_{\chi^2(\delta,\bar{\kappa}(t,s))}(y) = \sum_{k=0}^{\infty} \exp\left(-\frac{\bar{\kappa}(t,s)}{2}\right) \frac{\left(\frac{\bar{\kappa}(t,s)}{2}\right)^k}{k!} \frac{\gamma\left(k + \frac{\delta}{2}, \frac{y}{2}\right)}{\Gamma\left(k + \frac{\delta}{2}\right)}, \tag{8.7}$$
>
> with the lower incomplete Gamma function, $\gamma(a, z)$, and the Gamma function, $\Gamma(z)$,
>
> $$\gamma(a, z) = \int_0^z t^{a-1}e^{-t}dt, \quad \Gamma(z) = \int_0^{\infty} t^{z-1}e^{-t}dt, \tag{8.8}$$
>
> see also (5.50).

The corresponding density function (see, for example, [Johnson and Kotz, 1970], [Moser, Oct. 30–Nov. 2 2007]) reads:

$$f_{\chi^2(\delta,\bar{\kappa}(t,s))}(y) = \frac{1}{2}e^{-\frac{1}{2}(y+\bar{\kappa}(t,s))}\left(\frac{y}{\bar{\kappa}(t,s)}\right)^{\frac{1}{2}\left(\frac{\delta}{2}-1\right)} \mathcal{B}_{\frac{\delta}{2}-1}\left(\sqrt{\bar{\kappa}(t,s)y}\right), \qquad (8.9)$$

where

$$\mathcal{B}_a(z) = \left(\frac{z}{2}\right)^a \sum_{k=0}^{\infty} \frac{\left(\frac{1}{4}z^2\right)^k}{k!\Gamma(a+k+1)}, \qquad (8.10)$$

is a modified Bessel function of the first kind, see also [Abramowitz and Stegun, 1972; Gradshteyn and Ryzhik, 1996].

The density function for $v(t)$ can now be expressed as:

$$f_{v(t)}(x) := \frac{\mathrm{d}}{\mathrm{d}x}F_{v(t)}(x) = \frac{\mathrm{d}}{\mathrm{d}x}F_{\chi^2(\delta,\bar{\kappa}(t,s))}\left(\frac{x}{\bar{c}(t,s)}\right)$$

$$= \frac{1}{\bar{c}(t,s)}f_{\chi^2(\delta,\bar{\kappa}(t,s))}\left(\frac{x}{\bar{c}(t,s)}\right). \qquad (8.11)$$

By the properties of the noncentral chi-squared distribution the mean and variance of process $v(t)|v(0)$ are known explicitly:

$$\mathbb{E}[v(t)|\mathcal{F}(0)] = \bar{c}(t,0)(\delta + \bar{\kappa}(t,0)),$$
$$\mathbb{V}\mathrm{ar}[v(t)|\mathcal{F}(0)] = \bar{c}^2(t,0)(2\delta + 4\bar{\kappa}(t,0)). \qquad (8.12)$$

◀

As indicated in [Feller, 1951; Cox et al., 1985], with the parameter $\delta := 4\kappa\bar{v}/\gamma^2$, the Feller condition is equivalent to "$\delta \geq 2$". By defining another parameter, $q_F := (2\kappa\bar{v}/\gamma^2) - 1$, the Feller condition is satisfied, when

$$q_F := \frac{2\kappa\bar{v}}{\gamma^2} - 1 = \frac{\delta}{2} - 1 \geq 0.$$

Although each of the three parameters, κ, $\bar{v}$ and γ, in SDE (8.3) plays a unique role in the tuning of the shape and the magnitude of the variance density, the decay rate at the left tail can be well-characterized by the values of q_F, whose definition interval is $[-1,\infty)$. Based on the nonnegativeness of κ, $\bar{v}$ and γ, the near-singular problem occurs when $q_F \in [-1,0]$, which is thus directly related to the Feller condition.

Example 8.1.1 As an experiment which supports this insight, we will consider two parameter sets and analyze the corresponding PDFs and CDFs. There is one parameter set for which the Feller condition holds, i.e. $q_F = 2$, for $T = 5$, $\kappa = 0.5$, $v_0 = 0.2$, $\bar{v} = 0.05$, $\gamma = 0.129$ and one set for which the Feller condition is violated, $q_F = -0.5$, with $T = 5$, $\kappa = 0.5$, $v_0 = 0.2$, $\bar{v} = 0.05$, $\gamma = 0.316$.

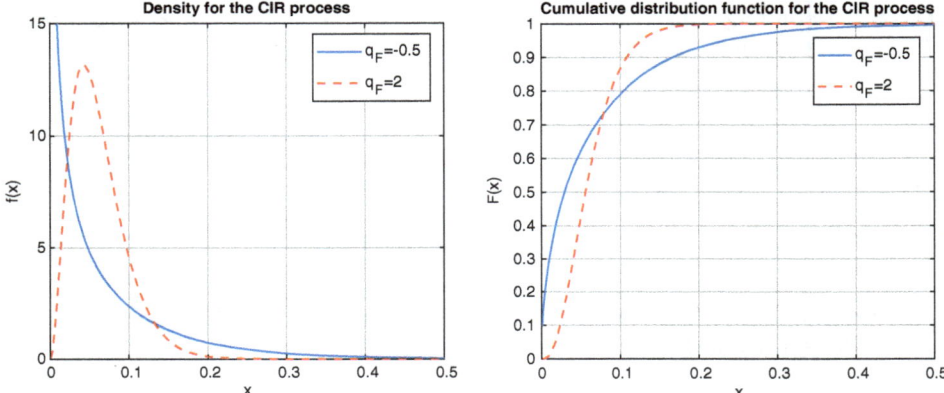

Figure 8.1: *Density and cumulative distribution functions for the CIR process. Blue straight lines: $q_F = -0.5$ with $T = 5$, $\kappa = 0.5$, $v_0 = 0.2$, $\bar{v} = 0.05$, $\gamma = 0.316$; Red dashed lines: $q_F = 2$ for $T = 5$, $\kappa = 0.5$, $v_0 = 0.2$, $\bar{v} = 0.05$, $\gamma = 0.129$.*

As is confirmed in Figure 8.1, the value of q_F determines the decay rate in the left-side tail of the variance density function, whereas the right-side tail always decays to zero rapidly. For $q_F > 0$, the density values tend towards zero in both tails. For q_F smaller and approaching 0, the decay of the left-side tail slows down. Near $q_F = 0$, the left tail stays almost constant. For $q_F \in [-1, 0]$, the left tail increases drastically in value.

The fact that q_F determines the decay rate of the PDF's left tail can be understood as follows. When q_F changes sign, the PDF's governing functions, $(\cdot)^{(\frac{1}{2}\delta - 1)} = (\cdot)^{q_F/2}$ and $\mathcal{B}_{q_F}(\cdot)$, change shape around the origin, i.e., from monotonically increasing functions they become monotonically decreasing.

Squared Bessel process

Some more background of this variance process is given, where we will see that after suitable transformations the variance process resembles the squared Bessel process. Properties of the squared Bessel process can thus be connected to the variance process.

For the parameter $\delta \geq 0$ and a stochastic process $Y(t) \geq 0$, the solution of the squared Bessel process of dimension δ, which is defined by the following SDE,

$$dY(t) = \delta dt + 2\sqrt{|Y(t)|}dW(t), \tag{8.13}$$

is unique and strong. Parameter δ is the degrees of freedom parameter in the squared Bessel process.

With,
$$\delta = \frac{4\kappa\bar{v}}{\gamma^2}, \quad \nu = \frac{\gamma^2}{4\kappa},$$
and the time-change,
$$\bar{Y}(t) = e^{-\kappa t} Y(\nu(e^{\kappa t} - 1)), \tag{8.14}$$
we can transform the squared Bessel process in (8.13) into the following process:
$$d\bar{Y}(t) = \kappa(\nu \cdot \delta - \bar{Y}(t))dt + 2\sqrt{\kappa\nu\bar{Y}(t)}dW.$$
By the definitions of ν and δ, it is easily seen that this $\bar{Y}$ process is equal to the variance process in the Heston model.

The squared Bessel process is a Markov process and the corresponding transition densities, for different parameter values, are known in closed-form.

First of all, in the case of $\delta = 0$, the solution of $Y(t)$ is identically zero, $Y(t) = 0$.

A few technical details regarding squared Bessel processes are well-known. They are presented in the form of results below.

Result 8.1.1 *For the standard squared Bessel process, as defined by SDE (8.13), the following statements hold true, see also [Andersen and Andreasen, 2010]:*

1. *All solutions to SDE (8.13) are nonexplosive.*

2. *For $\delta < 2$, $Y = 0$ is an attainable boundary for the solution of (8.13).*

3. *For $\delta \geq 2$, SDE (8.13) has a unique solution and zero is not attainable.*

4. *For $0 < \delta < 2$, SDE (8.13) does not have a unique solution, unless the boundary condition is specified for the solution to (8.13) at $Y = 0$.*

5. *For $\delta \leq 0$, there is a unique strong solution to SDE (8.13), and the boundary at zero is absorbing.*

These results have been proved in Appendix A of [Andersen and Andreasen, 2010], where the proof was based on the theory presented in [Borodin, 2002]. For the latter two cases in Result 8.1.1, the transition densities are known and are presented in the next result.

Result 8.1.2 (Transition density for squared Bessel process) *The transition density, $f_B(t, Y(t); s, Y(s))$, for the squared Bessel process is given by the following functions.*

1. For $\delta \leq 0$ and for $0 < \delta < 2$ in Equation (8.13), but only when the boundary is absorbing at $Y(t) = 0$, we have $Y(t) \geq 0, t > 0$, and

$$f_B(t, Y(t); s, Y(s)) = \frac{1}{2(t-s)} \left(\frac{Y(t)}{Y(s)}\right)^{\frac{\delta-2}{4}}$$

$$\times \exp\left(-\frac{Y(s) + Y(t)}{2(t-s)}\right) I_{|\frac{\delta-2}{2}|}\left(\frac{\sqrt{Y(s) \cdot Y(t)}}{t-s}\right),$$

(8.15)

2. For $0 < \delta < 2$, when $Y(t) = 0$ is a reflecting boundary, we find $Y(t) \geq 0, t > 0$, and

$$f_B(t, Y(t); s, Y(s)) = \frac{1}{2(t-s)} \left(\frac{Y(t)}{Y(s)}\right)^{\frac{\delta-2}{4}}$$

$$\times \exp\left(-\frac{Y(s) + Y(t)}{2(t-s)}\right) I_{\frac{\delta-2}{2}}\left(\frac{\sqrt{Y(s) \cdot Y(t)}}{t-s}\right).$$

(8.16)

By $I_a(x)$ in (8.15), (8.16), we denote the Bessel function, which is defined by

$$I_a(x) := \sum_{j=0}^{\infty} \frac{(x/2)^{2j+a}}{j!\Gamma(a+j+1)},$$

and $\Gamma(x)$ is again the Gamma function.

Proof See [Borodin, 2002] p.136 for the squared Bessel process transition densities. ∎

Remark 8.1.1 (Time-changed Brownian motion) *In (8.14) a time change was suggested to transform two processes. A time change is typically related to the "speed at which the paths of Brownian motion move in time". Brownian motion $W(t)$ has mean zero and a variance equal to time t. When looking at a different time νt, the increments $W(\nu \cdot t) - W(\nu \cdot s)$, $t \geq s \geq 0$, are normally distributed with mean zero and variance $\nu \cdot (t-s)$. Any scaling of the time parameter of a Brownian motion, $W(\nu \cdot t) \stackrel{d}{=} \sqrt{\nu} W(t)$, will thus result in another process, which is scaled by a factor $\sqrt{\nu}$.*

Consequently, by multiplication of $W(t)$ by a process $\bar{\nu}(t)$, which is independent of $W(t)$, we find equality in distribution,

$$\bar{\nu}(t) W(t) \stackrel{d}{=} W(\bar{\nu}^2(t) \cdot t).$$

(8.17)

From here on, we will not use the "d" for the equality in distribution, for notational simplicity.

The distribution at both sides of this equation remains the same, but the speed (i.e., time) at which paths of the Brownian motion $W(\bar{\nu}^2(t))$ evolve is controlled by process $\bar{\nu}(t)$. If we now consider the following time change, $\bar{\nu}(t) = (1-\rho^2)\int_0^t \sigma^2(z)\mathrm{d}z$, then the Brownian motion under this "clock" will have the same distribution as $\sqrt{1-\rho^2}\int_0^t \sigma(z)\mathrm{d}W(z)$, i.e.

$$\mathrm{d}W(\bar{\nu}(t)) = \sqrt{1-\rho^2}\sigma(t)\mathrm{d}W(t),$$

or, in integrated form,

$$W(\bar{\nu}(t)) = \int_0^{\bar{\nu}(t)} \mathrm{d}W(t) = \sqrt{1-\rho^2}\int_0^t \sigma(z)\mathrm{d}W(z).$$

A time-change does not only change the corresponding Brownian motion, but also the time in the drift term, since: $\mathrm{d}\bar{\nu}(t) = (1-\rho^2)\sigma^2(t)\mathrm{d}t$. ▲

8.2 The Heston stochastic volatility model

In this section we introduce the Heston SV model and the corresponding system of SDEs.

In the Heston SV model [Heston, 1993], we deal with two stochastic differential equations, one for the underlying asset price $S(t)$, and one for the variance process $v(t)$, that are described under the risk-neutral measure $\mathbb{Q}$, by

$$\begin{cases} \mathrm{d}S(t) = rS(t)\mathrm{d}t + \sqrt{v(t)}S(t)\mathrm{d}W_x^{\mathbb{Q}}(t), & S(t_0) = S_0 > 0 \\ \mathrm{d}v(t) = \kappa(\bar{v} - v(t))\mathrm{d}t + \gamma\sqrt{v(t)}\mathrm{d}W_v^{\mathbb{Q}}(t), & v(t_0) = v_0 > 0. \end{cases} \quad (8.18)$$

A correlation is defined between the underlying Brownian motions, i.e. $\mathrm{d}W_v^{\mathbb{Q}}(t)\mathrm{d}W_x^{\mathbb{Q}}(t) = \rho_{x,v}\mathrm{d}t$. Parameters $\kappa \geq 0$, $\bar{v} \geq 0$ and $\gamma > 0$ are called the speed of mean reversion, the long-term mean of the variance process and the volatility of the volatility, respectively.

Real-world measure and the Heston SV model

In the case of the Heston model it is convenient to depart from asset dynamics that are already under the risk-neutral measure (i.e. the measure under which the discounted asset price, $S(t)/M(t)$, is a martingale, with $M(t)$ as in (2.27)). The authors in [Wong and Heyde, 2006] provide a detailed discussion on measure transformations for SV models. In the case of the Heston SV model the measure transformation requires quite some effort, especially for parameter values for which the variance can become zero.

The model under the $\mathbb{Q}$-measure is governed by a mean reverting volatility CIR-type process. The impact of the measure change on the dynamics of the underlying Heston model processes is briefly explained based on the following Heston dynamics under the real-world $\mathbb{P}$-measure.

$$\begin{cases} dS(t) = \mu S(t)dt + \sqrt{v(t)}S(t)dW_x^{\mathbb{P}}(t), \\ dv(t) = \kappa(\bar{v}^{\mathbb{P}} - v(t))dt + \gamma\sqrt{v(t)}dW_v^{\mathbb{P}}(t), \end{cases} \quad (8.19)$$

with the correlation between the Brownian motions under the $\mathbb{P}$-measure given by $dW_x^{\mathbb{P}}(t)dW_v^{\mathbb{P}}(t) = \rho_{x,v}dt$. With help of the Cholesky decomposition, the model can be presented in terms of independent Brownian motions, as,

$$\begin{cases} dS(t) = \mu S(t)dt + \sqrt{v(t)}S(t)d\widetilde{W}_x^{\mathbb{P}}(t), \\ dv(t) = \kappa(\bar{v}^{\mathbb{P}} - v(t))dt + \gamma\sqrt{v(t)}\left(\rho_{x,v}d\widetilde{W}_x^{\mathbb{P}}(t) + \sqrt{1-\rho_{x,v}^2}d\widetilde{W}_v^{\mathbb{P}}(t)\right), \end{cases}$$

with the independent processes $\widetilde{W}_x^{\mathbb{P}}(t)$ and $\widetilde{W}_v^{\mathbb{P}}(t)$. We seek for a process, $\Phi(t)$, in

$$d\widetilde{W}_x^{\mathbb{Q}}(x) = \Phi(t)dt + d\widetilde{W}_x^{\mathbb{P}}(x),$$

so that $\frac{S(t)}{M(t)}$ becomes a martingale, with $M(t)$ the usual money-savings account. Application of Itô's lemma to $\bar{S}(t) = \frac{S(t)}{M(t)}$, gives,

$$\frac{d\bar{S}(t)}{\bar{S}(t)} = (\mu - r)dt + \sqrt{v(t)}d\widetilde{W}_x^{\mathbb{P}}(t). \quad (8.20)$$

We thus find the following measure transformation:

$$\begin{cases} d\widetilde{W}_x^{\mathbb{P}}(t) = d\widetilde{W}_x^{\mathbb{Q}}(t) - (\mu - r)/\sqrt{v(t)}dt, \\ d\widetilde{W}_v^{\mathbb{P}}(t) = d\widetilde{W}_v^{\mathbb{Q}}(t). \end{cases} \quad (8.21)$$

With this measure transformation, it is necessary to satisfy $v(t) > 0$.

Under the $\mathbb{Q}$-measure the stock dynamics of the Heston model are then given by

$$dS(t) = rS(t)dt + \sqrt{v(t)}S(t)d\widetilde{W}_x^{\mathbb{Q}}(t),$$

and the variance process by

$$dv(t) = \kappa(\bar{v}^{\mathbb{P}} - v(t))dt + \gamma\sqrt{v(t)}\left(\rho_{x,v}d\widetilde{W}_x^{\mathbb{P}}(t) + \sqrt{1-\rho_{x,v}^2}d\widetilde{W}_v^{\mathbb{P}}(t)\right)$$

$$= \kappa\left(\bar{v}^{\mathbb{P}} - \frac{\rho_{x,v}}{\kappa}\gamma(\mu - r) - v(t)\right)dt$$

$$+ \gamma\sqrt{v(t)}\left[\rho_{x,v}d\widetilde{W}_x^{\mathbb{Q}}(t) + \sqrt{1-\rho_{x,v}^2}d\widetilde{W}_v^{\mathbb{Q}}(t)\right].$$

By setting $\bar{v} := \bar{v}^{\mathbb{P}} - \rho_{x,v}\gamma(\mu - r)/\kappa$ and formulating by means of correlated Brownian motions, we obtain SDE (8.3) for the variance in the $\mathbb{Q}$-measure.

Although the measure change explained above relies on $v(t) > 0$, it is known that the CIR process can also reach the value zero, but subsequently returns to positive values.

In practice, it is difficult to satisfy the Feller condition (which guarantees positive variance), when working with financial market data, see, for example [Andersen, 2008], where it is stated that typically $2\kappa\bar{v} \ll \gamma^2$ in many practical cases. The case $v(t) = 0$ is more involved regarding a suitable measure change, however, the authors in [Desmettre, 2018] proposed a measure change also for this case. Essentially, they replace the constant drift term by a (time- and) volatility-dependent drift term, which attains the value r when the volatility equals zero.

8.2.1 The Heston option pricing partial differential equation

Here, we derive the Heston SV option pricing PDE basically in the two ways that were also used when deriving the Black-Scholes PDE. The Black-Scholes replicating portfolio argument can be used to determine the option pricing PDE under Heston dynamics. The construction of the replicating portfolio is, however, more involved when the value of the option depends on nonobservable quantities, like the stochastic volatility.

Therefore, we start the derivation of the Heston PDE by means of the martingale approach.

From this point on, we won't use the superscript $\mathbb{Q}$ anymore to indicate Brownian motion under the risk-neutral measure, unless explicitly required.

Martingale approach

For Heston's SV model (8.18), we consider the following pricing problem:

$$V(t, S, v) = M(t)\mathbb{E}^{\mathbb{Q}}\left[\frac{1}{M(T)}V(T, S, v)\Big|\mathcal{F}(t)\right], \tag{8.22}$$

with $\mathcal{F}(t) = \sigma(s, S, v; s \leq t)$ (the sigma-algebra).

Dividing (8.22) by $M(t)$ gives,

$$\mathbb{E}^{\mathbb{Q}}\left[\frac{1}{M(T)}V(T, S, v)\Big|\mathcal{F}(t)\right] = \frac{V(t, S, v)}{M(t)}. \tag{8.23}$$

Assume a differentiable function, the discounted option value, which should be a martingale. We can determine its dynamics using Itô's lemma,

$$d\left(\frac{V}{M}\right) = \frac{1}{M}dV - r\frac{V}{M}dt.$$

With the help of the two-dimensional Itô lemma, we find for an infinitesimal increment, dV, with the dynamics for $S(t)$ and $v(t)$ in the Heston model (8.18), the following dynamics,

$$dV = \left(\frac{\partial V}{\partial t} + rS\frac{\partial V}{\partial S} + \kappa(\bar{v}-v)\frac{\partial V}{\partial v} + \frac{1}{2}vS^2\frac{\partial^2 V}{\partial S^2} + \rho_{x,v}\gamma Sv\frac{\partial^2 V}{\partial S \partial v}\right.$$
$$\left. + \frac{1}{2}\gamma^2 v\frac{\partial^2 V}{\partial v^2}\right)dt + S\sqrt{v}\frac{\partial V}{\partial S}dW_x + \gamma\sqrt{v}\frac{\partial V}{\partial v}dW_v.$$

Note that the formulation above is already written in terms of correlated Brownian motions.

For the martingale property to hold, the dynamics should be free of dt-terms, which yields

$$\frac{1}{M}\left(\frac{\partial V}{\partial t} + rS\frac{\partial V}{\partial S} + \kappa(\bar{v}-v)\frac{\partial V}{\partial v} + \frac{1}{2}vS^2\frac{\partial^2 V}{\partial S^2}\right.$$
$$\left. + \rho_{x,v}\gamma Sv\frac{\partial^2 V}{\partial S \partial v} + \frac{1}{2}\gamma^2 v\frac{\partial^2 V}{\partial v^2}\right) - r\frac{V}{M} = 0, \qquad (8.24)$$

resulting in the *Heston stochastic volatility PDE*, for the valuation of options (after multiplication of (8.24) by M, of course).

Solutions to the Heston option pricing PDE can be found with the help of the *Feynman-Kac theorem*, by which we can describe the solution as the discounted expected payoff under the risk-neutral measure. For the Heston model, the conditional density function in the Feynman-Kac representation is not known, but we can determine its discounted characteristic function, *because the Heston model is within the affine diffusion framework*, see Section 8.3.

Replication and hedging approach

As the second approach to define the Heston option pricing PDE, we may construct a replicating portfolio with value $\Pi(t, S, v)$, again with the help of Itô's lemma.

The market completeness principle states however that the number of assets and the number of sources of randomness (i.e. Brownian motions) should be equal. Previously, in the Black-Scholes model, we had one stochastic differential equation, with one Brownian motion, so that the market was complete. In the case of jump diffusion dynamics, this was already not the case anymore, as we dealt, next to Brownian motion, with a Poisson jump process. The market was incomplete, and we had to resume to the concept of Equivalent Martingale Measure.

The situation is also nontrivial when considering the stochastic volatility processes. The difficulty is related to the fact that the volatility itself cannot easily be bought or sold on the market (to be precise, we can mimic derivatives to measure volatility, but volatility is not a unique measurable and traded quantity).

The hedging portfolio in the case of stochastic volatility consists of the option sold, with value $V(t, S, v)$, plus $-\Delta$ units of the underlying asset $S(t)$, and, in order to hedge the risk associated with the random volatility, also $-\Delta_1$ units of *another option*, which is bought with value $V_1(t, S, v; K_1, T)$.

This way we may deal with the two sources of randomness, i.e. in $S(t)$ and in $v(t)$. So,

$$\Pi(t, S, v) = V(t, S, v; K, T) - \Delta S - \Delta_1 V_1(t, S, v; K_1, T), \qquad (8.25)$$

where we explicitly included the dependence on K, T, K_1, T in the arguments of V and V_1, indicating that V_1 is an option with the same maturity time but with a different strike K_1.

Using the notation $V \equiv V(t, S, v; K, T)$ and $V_1 \equiv V_1(t, S, v; K_1, T)$, by Itô's lemma we derive the dynamics of an infinitesimal change of the portfolio $\Pi \equiv \Pi(t, S, v)$, as follows

$$d\Pi = \left(\frac{\partial V}{\partial t} + \frac{1}{2} v S^2 \frac{\partial^2 V}{\partial S^2} + \rho_{x,v} \gamma S v \frac{\partial^2 V}{\partial S \partial v} + \frac{1}{2} \gamma^2 v \frac{\partial^2 V}{\partial v^2} \right) dt$$

$$- \Delta_1 \left(\frac{\partial V_1}{\partial t} + \frac{1}{2} v S^2 \frac{\partial^2 V_1}{\partial S^2} + \rho_{x,v} \gamma S v \frac{\partial^2 V_1}{\partial S \partial v} + \frac{1}{2} \gamma^2 v \frac{\partial^2 V_1}{\partial v^2} \right) dt$$

$$+ \left(\frac{\partial V}{\partial S} - \Delta_1 \frac{\partial V_1}{\partial S} - \Delta \right) dS + \left(\frac{\partial V}{\partial v} - \Delta_1 \frac{\partial V_1}{\partial v} \right) dv.$$

By prescribing:

$$\frac{\partial V}{\partial S} - \Delta_1 \frac{\partial V_1}{\partial S} - \Delta = 0 \quad \text{and} \quad \frac{\partial V}{\partial v} - \Delta_1 \frac{\partial V_1}{\partial v} = 0, \qquad (8.26)$$

we eliminate the randomness from the portfolio dynamics.

As before, in order to avoid arbitrage opportunities, the return of this deterministic, risk-free portfolio must earn the risk-free interest rate, i.e., $d\Pi = r\Pi dt$ and by (8.25) we have $d\Pi = r(V - \Delta S - \Delta_1 V_1)dt$.

After application of (8.26) and re-ordering of some terms, we find the following equality:

$$\left(\frac{\partial V}{\partial t} + \frac{1}{2} v S^2 \frac{\partial^2 V}{\partial S^2} + \rho_{x,v} \gamma S v \frac{\partial^2 V}{\partial S \partial v} + \frac{1}{2} \gamma^2 v \frac{\partial^2 V}{\partial v^2} + rS \frac{\partial V}{\partial S} - rV \right) \Big/ \frac{\partial V}{\partial v}$$

$$= \left(\frac{\partial V_1}{\partial t} + \frac{1}{2} v S^2 \frac{\partial^2 V_1}{\partial S^2} + \rho_{x,v} \gamma S v \frac{\partial^2 V_1}{\partial S \partial v} + \frac{1}{2} \gamma^2 v \frac{\partial^2 V_1}{\partial v^2} + rS \frac{\partial V_1}{\partial S} - rV_1 \right) \Big/ \frac{\partial V_1}{\partial v},$$

where, at the left-hand side, option value V, is a function of the independent variables t, S, v, and, at the right-hand side, option value V_1 is a function of the same independent variables.

Both sides of this equation should, in such a setting, be equal to a specific function, $g(t, S, v)$, which *only depends on the independent variables S, v and time t*.

> The following choice, $g(t, S, v) = -\kappa(\bar{v} - v(t))$, like the term in (8.18), leads to the following option pricing PDE under the Heston SV dynamics:
>
> $$\frac{\partial V}{\partial t} + \frac{1}{2}vS^2\frac{\partial^2 V}{\partial S^2} + \rho_{x,v}\gamma Sv\frac{\partial^2 V}{\partial S\partial v} + \frac{1}{2}\gamma^2 v\frac{\partial^2 V}{\partial v^2}$$
> $$+ rS\frac{\partial V}{\partial S} + \kappa(\bar{v} - v)\frac{\partial V}{\partial v} - rV = 0. \qquad (8.27)$$
>
> Notice that this representation is identical to the PDE resulting from the martingale approach, in Equation (8.24).

In [Heath and Schweizer, 2000] however other functions $g(t, S, v)$ have been discussed.

8.2.2 Parameter study for implied volatility skew and smile

This section discusses the implied volatility patterns that can be generated by the Heston SV model. The variance, i.e. $v(t) = \sigma^2(t)$, in the Heston SV model is governed by the mean reverting CIR process, and each parameter has a specific effect on the implied volatility curve generated by the dynamics. We give some intuition about the parameter values and their impact on the implied volatility shape.

Insight into the effect of the model parameters may be useful, for example during the calibration, where an initial set of parameters needs to be prescribed, a-priori, before the calibration iteration starts, to fit the parameters of the Heston SV model, $\{\rho_{x,v}, v_0, \bar{v}, \kappa, \gamma\}$, to match the market data (see the calibration details in Section 8.2.3).

To analyze the parameter effects numerically, we use the following set of reference parameters,

$$T = 2, S_0 = 100, \kappa = 0.1, \gamma = 0.1, \bar{v} = 0.1, \rho_{x,v} = -0.75, v_0 = 0.05, r = 0.05.$$

A numerical study is performed by varying individual parameters while keeping the others fixed. For each parameter set, Heston SV option prices are computed (by means of the numerical solution of the Heston PDE) and they are inserted in a Newton-Raphson iteration to determine the Black-Scholes implied volatilities.

The first two parameters that are varied are the correlation parameter $\rho_{x,v}$ and the so-called vol-vol parameter γ. Figure 8.2 (left side) shows that, when $\rho_{x,v} = 0\%$, an increasing value of γ gives a more pronounced implied volatility *smile*. A higher volatility-of-volatility parameter thus increases the implied volatility *curvature*. We also see, in Figure 8.2 (right side), that when the correlation between stock and variance process, $\rho_{x,v}$, gets increasingly negative, the slope of the *skew* in the implied volatility curve increases.

Another parameter in the dynamics of the Heston model is parameter κ, the speed of mean reversion of $v(t)$. The left side graph in Figure 8.3 shows that κ has a limited effect on the implied volatility smile or skew, up to $1\% - 2\%$ only. However, κ determines the speed at which the volatility converges to the

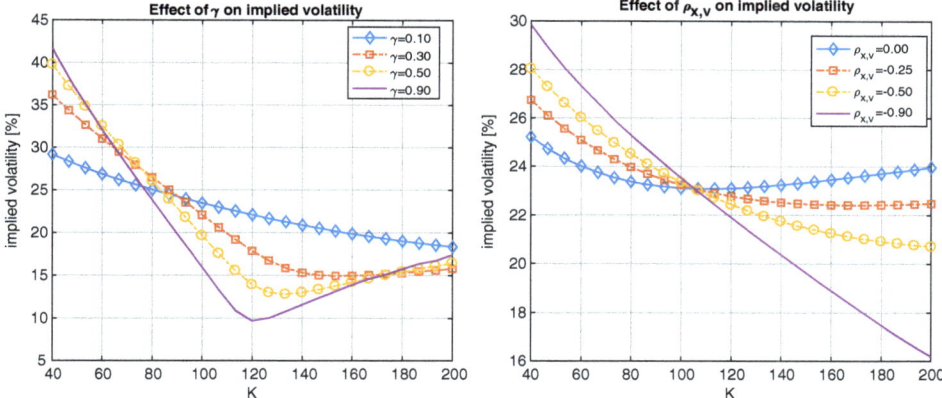

Figure 8.2: *Impact of variation of the Heston vol-vol parameter γ (left side), and correlation parameter $\rho_{x,v}$ (right side), on the implied volatility, as a function of strike price K.*

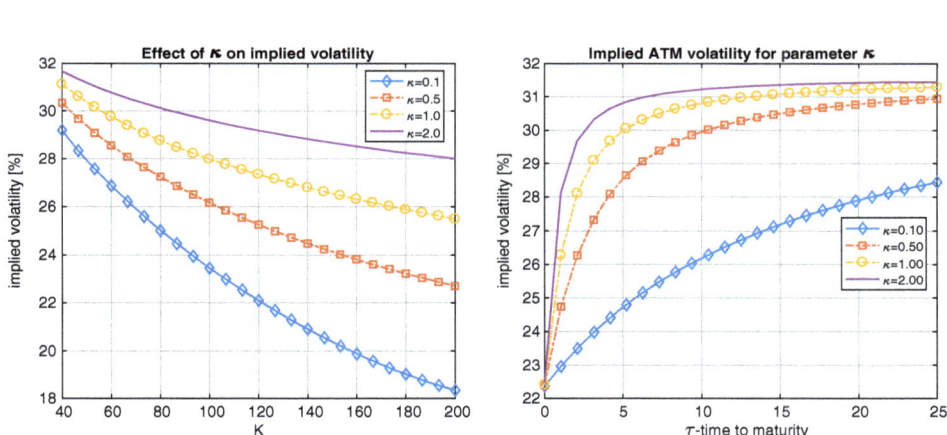

Figure 8.3: *Impact of variation of the Heston parameter κ on the implied volatility as a function of strike K (left side), impact of variation of κ on the ATM volatility, as a function of $\tau = T - t$ (right side).*

long-term volatility $\bar{v}$, see the right side graph in Figure 8.3, which shows the at-the-money (ATM) implied volatility for different κ-values. In this example, we have $\sqrt{\bar{v}} \approx 22.36\%$, and a large κ-value implies a faster convergence of the implied volatility to this $\sqrt{\bar{v}}$ value.

The remaining two parameters, the initial variance v_0, and the variance level $\bar{v}$, have a similar effect on the implied volatility curve, see Figure 8.4. The effect

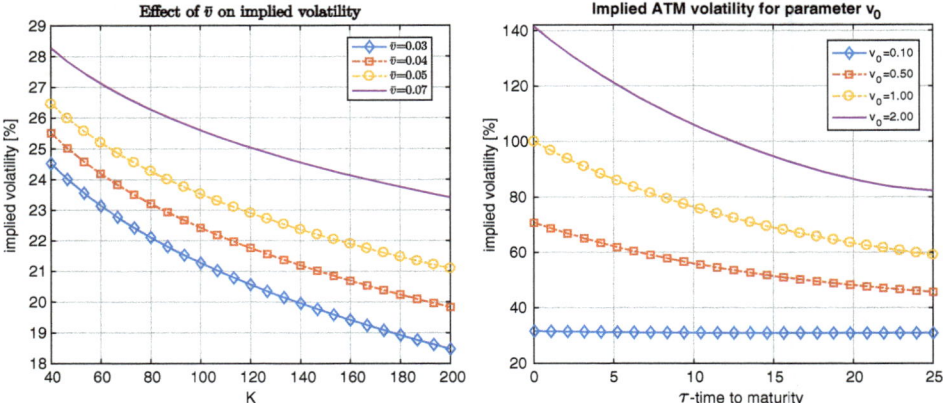

Figure 8.4: *Impact of changing v_0 and $\bar{v}$ on the Heston implied volatility; left side: $\bar{v}$ as a function of the strike K, right side: v_0 as a function of time to maturity $\tau = T - t$.*

of these two parameters also depends on the value of κ, which controls the speed at which the implied volatility converges from $\sqrt{v_0}$ to $\sqrt{\bar{v}}$ (or v_0 to $\bar{v}$).

8.2.3 Heston model calibration

During the calibration procedure of the Heston model we search for the model parameters such that the difference between market plain vanilla option prices, V^{mkt}, and the model prices, $V = V^H$, will be as small as possible. As mentioned, the parameters whose values we search for in this case are $\Omega = \{\rho_{x,v}, v_0, \bar{v}, \kappa, \gamma\}$. In order to determine these "optimal" parameters a target function is defined, which is based on the difference between the model and the market values. However, there is no unique target function.

The most common choices include,

$$\min_{\Omega} \sqrt{\sum_i \sum_j w_{i,j} \left(V_c^{mkt}(t_0, S_0; K_i, T_j) - V_c(t_0, S_0; K_i, T_j, \Omega)\right)^2}, \quad (8.28)$$

and

$$\min_{\Omega} \sqrt{\sum_i \sum_j w_{i,j} \left(\sigma_{imp}^{mkt}(t_0, S_0; K_i, T_j) - \sigma_{imp}(t_0, S_0; K_i, T_j, \Omega)\right)^2}, \quad (8.29)$$

where $V_c^{mkt}(t_0, S_0; K_i, T_j)$ is the call option price for strike K_i and maturity T_j in the market, $V_c(t_0, S_0; K_i, T_j, \Omega)$ is the Heston call option value; $\sigma_{imp}^{mkt}(\cdot)$, $\sigma_{imp}(\cdot)$ are the implied volatilities from the market and the Heston model, respectively, and $w_{i,j}$ is some weighting function.

Often, $w_{i,j} = 1$, however, the target function is often based only on the OTM call and put option prices, which implies a "natural" weighting.

The main difficulty when calibrating the Heston model is that the set Ω includes five parameters that need to be determined, and that the model parameters are not completely "independent". With this we mean that the effect of different parameters on the shape of the implied volatility smile may be quite similar. For this reason, one may encounter several "local minima" when searching for optimal parameter values. The insight in the effect of the different parameters on the implied volatility, like in the earlier example, may already give us a satisfactory initial parameter set.

The optimization can be accelerated by a reduction of the set of parameters to be optimized. By comparing the impact of the speed of mean reversion parameter κ and the curvature parameter γ (see Figures 8.2 and 8.3), it can be observed that these two parameters have a similar effect on the shape of the implied volatility. It is therefore common practice to prescribe (or fix) one of them. Practitioners often fix $\kappa = 0.5$ and optimize parameter γ. By this, the optimization reduces to four parameters.

Another parameter that may be determined in advance by using heuristics is the initial value of the variance process v_0. For maturity time T "close to today" (i.e. $T \to 0$), one would expect the stock price to behave very much *like in the Black-Scholes case*. The impact of a stochastic variance process should reduce to zero, in the limit $T \to 0$. For options with short maturities, the process may therefore be approximated by a process of the following form:

$$dS(t) = rS(t)dt + \sqrt{v_0}S(t)dW_x(t). \tag{8.30}$$

This suggests that for initial variance v_0 one may use the square of the ATM implied volatility of an option with the shortest maturity, $v_0 \approx \sigma_{imp}^2$, for $T \to 0$, as an accurate approximation.

In Table 8.1 some values of an observed implied volatility surface are presented. The strike values are presented as a percentage of the spot price S_0, so that the ATM strike level corresponds to $K = S_0 = 100$.

Initial value v_0 for the Heston model may thus be estimated by the square of the ATM volatility of an option with expiry $T = 1$ week, i.e., $v_0 = \sigma_{imp}^2(1w) = (0.05001)^2 = 0.0025$ (indicated by the boxed entry in Table 8.1). The value for the mean variance $\bar{v}$ may be connected to the 10y value, i.e. $\bar{v} = \sigma_{imp}^2(10y) = (0.07221)^2$.

As an accurate approximation for the initial guess for the parameters in the Heston model one may also use the connection of the Heston dynamics to a Black-Scholes dynamics with a time-dependent volatility function, as discussed in Section 3.2.3. In the Heston model we may, for example, *project* the variance process onto its expectation, i.e.,

$$dS(t) = rS(t)dt + \mathbb{E}[\sqrt{v(t)}]S(t)dW_x(t).$$

By this projection the parameters of the variance process $v(t)$ may be calibrated similar to the case of the time-dependent Black-Scholes model. A focus on a

Table 8.1: *Implied volatilities, in percentages, for different maturities and strikes. The strikes are percentages of S_0. "w" is week, "m" month and "y" is year.*

strike\\T	1w	1m	6m	1y	2y	5y	8y	10y
40	22.622	22.371	20.706	18.537	18.041	16.395	14.491	13.136
60	15.491	15.350	14.415	13.195	12.916	11.988	10.917	10.177
80	9.930	9.882	9.562	9.154	9.070	8.786	8.459	8.251
100	**5.001**	**5.070**	**5.515**	**6.041**	**6.196**	**6.639**	**7.015**	**7.221**
120	4.542	4.598	4.962	5.430	5.654	6.277	6.782	7.041
140	6.278	6.300	6.405	6.460	6.607	6.982	7.232	7.328
160	7.748	7.752	7.707	7.517	7.615	7.806	7.824	7.751
180	8.988	8.976	8.818	8.445	8.508	8.563	8.399	8.182

time-dependent volatility model would imply that a complete row of volatilities (bold faced in the table) in Table 8.1 will be considered during the calibration.

The Heston parameters are then determined, such that

$$\sigma^{ATM}(T_i) = \sqrt{\int_0^{T_i} \left(\mathbb{E}[\sqrt{v(t)}]\right)^2 dt},$$

where $\sigma^{ATM}(T_i)$ is the ATM implied volatility for maturity T_i in Table 8.1.

Another calibration technique for the Heston parameters is to use VIX index market quotes. As discussed in Remark 8.2.1 below, with different market quotes for different strike prices K_i and for different maturities T_j, we may also determine the optimal parameters by solving the following equalities, for all pairs (i,j),

$$K_{i,j} = \bar{v} + \frac{v_0 - \bar{v}}{\kappa(T_j - t_0)}\left(1 - e^{-\kappa(T_i - t_0)}\right). \tag{8.31}$$

When the initial values of the parameters have been determined, one can use the whole implied volatility surface in Table 8.1 to determine the optimal model parameters.

Remark 8.2.1 (Variance swap modeling under the Heston model)
Variance swap have been introduced in Section 4.2.2, where a first pricing methodology has been presented. Subsequently, a connection to the VIX index quotes was made.

We explained that variance swaps are quoted in terms of their strike prices, K, that are determined so that the value of the contract should be equal to 0 at the inception date t_0. When the volatility $\sigma(t)$ is modeled by a stochastic process, however, the variance swap pricing equation will be modified. If the stock prices are governed by the Heston SV model, then the expression in (4.30) will be modified and will read:

$$K = \frac{1}{T - t_0}\mathbb{E}^{\mathbb{Q}}\left[\int_{t_0}^{T} v(t)dt \Big| \mathcal{F}(t_0)\right] = \frac{1}{T - t_0}\int_{t_0}^{T}\mathbb{E}^{\mathbb{Q}}\left[v(t)\Big|\mathcal{F}(t_0)\right]dt. \tag{8.32}$$

Stochastic Volatility Models 241

The expectation of the variance in the Heston process is now given by,

$$\mathbb{E}^{\mathbb{Q}}\left[v(t)|\mathcal{F}(t_0)\right] = v_0 e^{-\kappa(t-t_0)} + \bar{v}\left(1 - e^{-\kappa(t-t_0)}\right), \qquad (8.33)$$

so that the strike K should be equal to

$$K = \bar{v} + \frac{v_0 - \bar{v}}{\kappa(T-t_0)}\left(1 - e^{-\kappa(T-t_0)}\right). \qquad (8.34)$$

▲

Example 8.2.1 (Calibration and volatility surface) In Figure 8.5 three examples of "the mode of calibration" are outlined and sketched. In the first figure, calibration will take place for one point on the implied volatility surface, say only to the ATM implied volatility, which resembles the Black-Scholes case. The second plot shows a calibration with a time-dependent volatility, where the term structure of volatility can thus be taken into account. The third figure

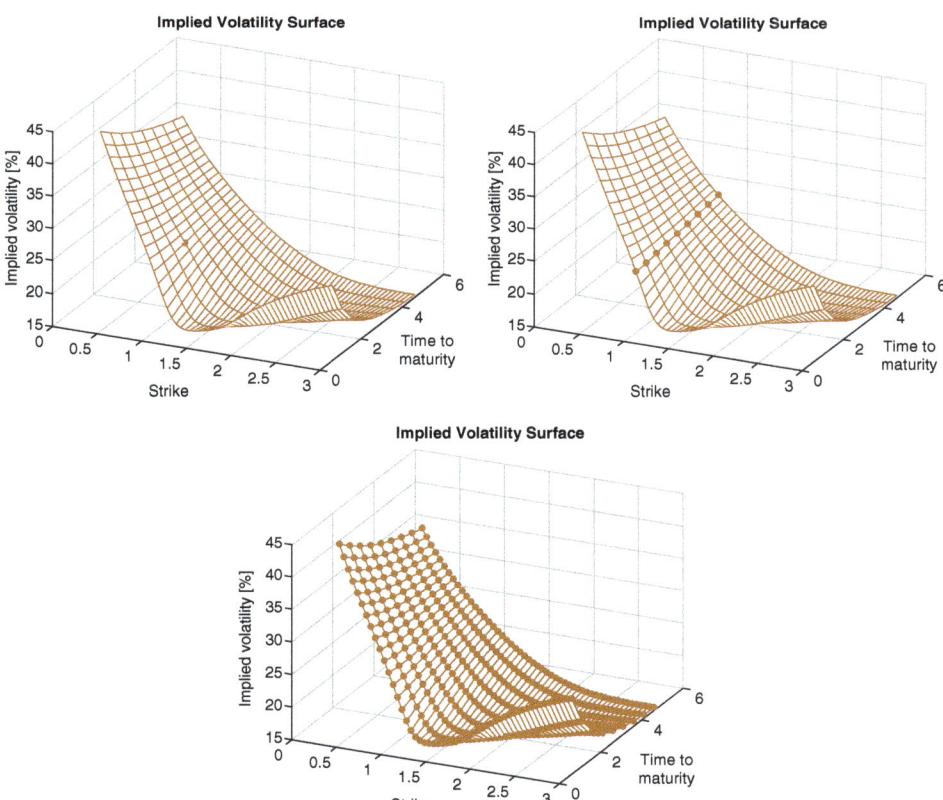

Figure 8.5: *Implied volatility surfaces and different modes of calibration: calibration to ATM (first picture), calibration to the term structure (second picture), calibration to the whole volatility surface (third picture).*

represents a calibration to all points of the implied volatility surface as may be established by a calibration of the Heston model. ♦

> Although the SV models have many desired features for pricing options, they cannot always be easily calibrated to a given set of arbitrage-free European vanilla market option prices. In particular, the accuracy of the Heston SV model for pricing short-maturity options in the equity market is typically not completely satisfactory [Engelmann et al., 2011].

8.3 The Heston SV discounted characteristic function

We will derive the discounted characteristic function, in detail in the present section, and start the discussion of casting the Heston SV model in the framework of AD processes.

8.3.1 Stochastic volatility as an affine diffusion process

Under risk-neutral measure $\mathbb{Q}$, a general set of stochastic models with a diffusive volatility structure can be written as:

$$\begin{cases} dS(t) = rS(t)dt + a(t,v)S(t)dW_x(t), \\ dv(t) = b(t,v)dt + c(t,v)dW_v(t), \end{cases} \quad (8.35)$$

with constant r, correlation $dW_x(t)dW_v(t) = \rho_{x,v}dt$, $|\rho_{x,v}| < 1$. Depending on the functions $a(t,v)$, $b(t,v)$ and $c(t,v)$, a number of different stochastic volatility models can be defined.

With $\mathbf{X}(t) = [X(t), v(t)]^{\mathrm{T}}$, under the log-transformation $X(t) = \log S(t)$, the model can be expressed in terms of the independent Brownian motions,

$$\begin{bmatrix} dX(t) \\ dv(t) \end{bmatrix} = \begin{bmatrix} \bar{\mu}_1(t,\mathbf{X}(t)) \\ \bar{\mu}_2(t,\mathbf{X}(t)) \end{bmatrix} dt + \begin{bmatrix} \bar{\sigma}_{1,1}(t,\mathbf{X}(t)) & \bar{\sigma}_{1,2}(t,\mathbf{X}(t)) \\ \bar{\sigma}_{2,1}(t,\mathbf{X}(t)) & \bar{\sigma}_{2,2}(t,\mathbf{X}(t)) \end{bmatrix} \begin{bmatrix} d\widetilde{W}_x(t) \\ d\widetilde{W}_v(t) \end{bmatrix},$$

with

$$\bar{\boldsymbol{\mu}}(t,\mathbf{X}(t)) = \begin{bmatrix} r - \frac{1}{2}a^2(t,v) \\ b(t,v) \end{bmatrix}, \quad \bar{\boldsymbol{\sigma}}(t,\mathbf{X}(t)) = \begin{bmatrix} a(t,v) & 0 \\ \rho_{x,v}c(t,v) & \sqrt{(1-\rho_{x,v}^2)}c(t,v) \end{bmatrix}.$$

We may determine under which conditions the system in (8.35) belongs to the *class of AD processes*. Obviously, the interest is constant here, $\bar{r}(t,\mathbf{X}(t)) = r$. For the drift terms of $\mathbf{X}(t)$ we have:

$$\bar{\boldsymbol{\mu}}(t,\mathbf{X}(t)) := \begin{bmatrix} r - \frac{1}{2}a^2(t,v) \\ b(t,v) \end{bmatrix} = \begin{bmatrix} a_0 + a_1 v \\ b_0 + b_1 v \end{bmatrix}, \quad \text{for } a_0, b_0, a_1, b_1 \in \mathbb{R}. \quad (8.36)$$

So, $a_0 = r$, $a_1 = -0.5$ and $a^2(t,v) = v$. The last equality implies that function $a(t,v)$ should be equal to $\sqrt{v(t)}$ (or equal to a constant or a deterministic function of time) to be in the AD class. In the second row, function $b(t,v)$ can be a linear function of time and of $v(t)$.

Regarding the covariance term,

$$\bar{\sigma}(t,\mathbf{X}(t))\bar{\sigma}(t,\mathbf{X}(t))^{\mathrm{T}} = \begin{bmatrix} a^2(t,v) & \rho_{x,v}a(t,v)c(t,v) \\ \rho_{x,v}a(t,v)c(t,v) & c^2(t,v) \end{bmatrix}$$

$$= \begin{bmatrix} c_{0,1,1}+c_{1,1,1}v & c_{0,1,2}+c_{1,1,2}v \\ c_{0,2,1}+c_{1,2,1}v & c_{0,2,2}+c_{1,2,2}v \end{bmatrix}. \quad (8.37)$$

With the previously determined $a(t,v) = \sqrt{v}$, it follows that,

$$a^2(t,v) = (\sqrt{v})^2 = c_{0,1,1} + c_{1,1,1}v,$$
$$\rho_{x,v}a(t,v)c(t,v) = \rho_{x,v}\sqrt{v}c(t,v) = c_{0,1,2} + c_{1,1,2}v,$$
$$c^2(t,v) = c_{0,2,2} + c_{1,2,2}v.$$

The three equations above will hold true, when $c(t,v) = \sqrt{v}$ or with $c(t,v)$ a deterministic function of time.

The *generator* of the SV model, for $\mathbf{X} = [X_1(t), X_2(t)]^{\mathrm{T}} =: [X(t), v(t)]^{\mathrm{T}}$, is given by,

$$\mathcal{A} = \sum_{i=1}^{2} \bar{\mu}_i(t,\mathbf{X})\frac{\partial}{\partial X_i} + \frac{1}{2}\sum_{i=1}^{2}\sum_{j=1}^{2}(\bar{\sigma}(t,\mathbf{X})\bar{\sigma}(t,\mathbf{X})^{\mathrm{T}})_{i,j}\frac{\partial^2}{\partial X_i \partial X_j}.$$

By the Feynman-Kac theorem, the corresponding PDE for $V(t,\mathbf{X})$ is then known to be:

$$\frac{\partial}{\partial t}V + \mathcal{A}V - rV = 0,$$

with the corresponding solution

$$V(t,\mathbf{X}) = e^{-r(T-t)}\mathbb{E}^{\mathbb{Q}}\left[H(T,\mathbf{X})|\mathcal{F}(t)\right],$$

and a terminal condition $V(T,\mathbf{X}) = H(T,\mathbf{X})$.

Heston dynamics in the AD class

The Heston dynamics from (8.18), with $X(t) = \log S(t)$, are thus given by the following system of SDEs:

$$\begin{cases} \mathrm{d}X(t) = \left(r - \frac{1}{2}v(t)\right)\mathrm{d}t + \sqrt{v(t)}\mathrm{d}W_x(t), \\ \mathrm{d}v(t) = \kappa(\bar{v} - v(t))\mathrm{d}t + \gamma\sqrt{v(t)}\mathrm{d}W_v(t), \end{cases} \quad (8.38)$$

with $\mathrm{d}W_x(t)\mathrm{d}W_v(t) = \rho_{x,v}\mathrm{d}t$. Model (8.38) is affine. To confirm the affinity of the covariance matrix, the SV Heston model can be expressed in terms of two independent Brownian motions, as in the representation (7.6),

$$\begin{bmatrix} \mathrm{d}X(t) \\ \mathrm{d}v(t) \end{bmatrix} = \begin{bmatrix} r - \frac{1}{2}v(t) \\ \kappa(\bar{v} - v(t)) \end{bmatrix} \mathrm{d}t + \begin{bmatrix} \sqrt{v(t)} & 0 \\ \rho_{x,v}\gamma\sqrt{v(t)} & \gamma\sqrt{(1-\rho_{x,v}^2)v(t)} \end{bmatrix} \begin{bmatrix} \mathrm{d}\widetilde{W}_x(t) \\ \mathrm{d}\widetilde{W}_v(t). \end{bmatrix}$$

In the notation from (7.54), we have:

$$\bar{\boldsymbol{\sigma}}(t,\mathbf{X}(t))\bar{\boldsymbol{\sigma}}(t,\mathbf{X}(t))^{\mathrm{T}} = \begin{bmatrix} v(t) & \rho_{x,v}\gamma v(t) \\ \rho_{x,v}\gamma v(t) & \gamma^2 v(t) \end{bmatrix} \quad (8.39)$$

which is affine in its state variables $\mathbf{X}(t) = [X(t), v(t)]^{\mathrm{T}}$ for constant γ and $\rho_{x,v}$.

8.3.2 Derivation of Heston SV characteristic function

The discounted characteristic function is of the following form:

$$\phi_{\mathbf{X}}(\mathbf{u}) := \phi_{\mathbf{X}}(\mathbf{u}; t, T) = \mathbb{E}^{\mathbb{Q}}\left[\exp\left(-r(T-t) + i\mathbf{u}^{\mathrm{T}}\mathbf{X}(T)\right)\big|\mathcal{F}(t)\right]$$
$$= \exp\left(\bar{A}(\mathbf{u},\tau) + \bar{\mathbf{B}}^{\mathrm{T}}(\mathbf{u},\tau)\mathbf{X}(t)\right), \quad (8.40)$$

with $\mathbf{u}^{\mathrm{T}} = [u, u_2]^{\mathrm{T}}$, $\tau = T - t$, $\bar{\mathbf{B}}(\mathbf{u},\tau) = [\bar{B}(\mathbf{u},\tau), \bar{C}(\mathbf{u},\tau)]^{\mathrm{T}}$. At time T, using $\mathbf{u} = [u, 0]^{\mathrm{T}}$, we have,

$$\phi_{\mathbf{X}}(u; T, T) = \mathbb{E}^{\mathbb{Q}}\left[e^{i\mathbf{u}^{\mathrm{T}}\mathbf{X}(T)}\big|\mathcal{F}(T)\right] = e^{i\mathbf{u}^{\mathrm{T}}\mathbf{X}(T)} = e^{iuX(T)}, \quad (8.41)$$

In variable τ, Equation (8.41) gives $\bar{A}(\mathbf{u},0) = 0$, $\bar{B}(\mathbf{u},0) = iu$ and $\bar{C}(u,0) = 0$. The following lemmas define the Riccati ODEs, from (7.56), and detail their solutions.

Lemma 8.3.1 (Heston ODEs) *The functions $\bar{A}(\mathbf{u},\tau)$, $\bar{B}(\mathbf{u},\tau)$ and $\bar{C}(\mathbf{u},\tau)$ in (8.40) satisfy the following system of ODEs:*

$$\frac{\mathrm{d}\bar{B}}{\mathrm{d}\tau} = 0, \quad \bar{B}(\mathbf{u},0) = iu,$$

$$\frac{\mathrm{d}\bar{C}}{\mathrm{d}\tau} = \bar{B}(\bar{B}-1)/2 - \left(\kappa - \gamma\rho_{x,v}\bar{B}\right)\bar{C} + \gamma^2\bar{C}^2/2, \quad \bar{C}(\mathbf{u},0) = 0,$$

$$\frac{\mathrm{d}\bar{A}}{\mathrm{d}\tau} = \kappa\bar{v}\bar{C} + r(\bar{B}-1), \quad \bar{A}(\mathbf{u},0) = 0,$$

with parameters κ, γ, $\bar{v}$, r and $\rho_{x,r}$ as in the Heston model (8.38). ◂

We give a sketch of the proof of Lemma 8.3.1. For $\mathbf{X}(t) = [X(t), v(t)]^T$, solution $\phi_{\mathbf{X}} := \phi_{\mathbf{X}}(u;t,T)$ satisfies the following pricing PDE:

$$0 = -\frac{\partial \phi_{\mathbf{X}}}{\partial \tau} + \left(r - \frac{1}{2}v\right)\frac{\partial \phi_{\mathbf{X}}}{\partial X} + \kappa(\bar{v} - v(t))\frac{\partial \phi_{\mathbf{X}}}{\partial v} + \frac{1}{2}\gamma^2 v \frac{\partial^2 \phi_{\mathbf{X}}}{\partial v^2} + \frac{1}{2}v\frac{\partial^2 \phi_{\mathbf{X}}}{\partial X^2}$$
$$+ \rho_{x,v}\gamma v \frac{\partial^2 \phi_{\mathbf{X}}}{\partial X \partial v} - r\phi_{\mathbf{X}}, \qquad (8.42)$$

subject to the condition $\phi_{\mathbf{X}}(u;T,T) = \exp(iuX(0))$, using $\tau = 0$. Since the PDE in (8.42) is affine, its solution is of the following form:

$$\phi_{\mathbf{X}}(u;t,T) = \exp\left(\bar{A}(\mathbf{u},\tau) + \bar{B}(\mathbf{u},\tau)X(t) + \bar{C}(\mathbf{u},\tau)v(t)\right). \qquad (8.43)$$

By substitution of this solution, and collecting the terms for X and v, we obtain the set of ODEs in the lemma.

Lemma 8.3.2 (The solution to the Heston model ODEs) *The solution to the system of ODEs specified in Lemma 8.3.1, with their initial conditions, is given by:*

$$\bar{B}(\mathbf{u},\tau) = iu,$$

$$\bar{C}(\mathbf{u},\tau) = \frac{1 - e^{-D_1\tau}}{\gamma^2(1 - ge^{-D_1\tau})}(\kappa - \gamma\rho_{x,v}iu - D_1),$$

$$\bar{A}(\mathbf{u},\tau) = r(iu - 1)\tau + \frac{\kappa\bar{v}\tau}{\gamma^2}(\kappa - \gamma\rho_{x,v}iu - D_1) - \frac{2\kappa\bar{v}}{\gamma^2}\log\left(\frac{1 - ge^{-D_1\tau}}{1 - g}\right),$$

for $D_1 = \sqrt{(\kappa - \gamma\rho_{x,v}iu)^2 + (u^2 + iu)\gamma^2}$ *and* $g = \dfrac{\kappa - \gamma\rho_{x,v}iu - D_1}{\kappa - \gamma\rho_{x,v}iu + D_1}.$ ◀

Proof We find $\bar{B}(\mathbf{u},\tau) = iu$, and thus the following ODEs have to be solved:

$$\begin{cases} \dfrac{d\bar{C}}{d\tau} = a_1 - a_2\bar{C} + a_3\bar{C}^2, & \bar{C}(\mathbf{u},0) = 0, \\ \dfrac{d\bar{A}}{d\tau} = b_1 + b_2\bar{C}, & \bar{A}(\mathbf{u},0) = 0. \end{cases} \qquad (8.44)$$

with $a_1 = -\frac{1}{2}(u^2 + iu)$, $a_2 = \kappa - \gamma\rho_{x,v}iu$ and $a_3 = \frac{1}{2}\gamma^2$, and $b_1 = r(iu - 1)$ and $b_2 = \kappa\bar{v}$.

First, we solve system (8.44) for $\bar{C}$, where

$$a_1 - a_2\bar{C} + a_3\bar{C}^2 = a_3(\bar{C} - r_+)(\bar{C} - r_-),$$

with:

$$r_\pm = \frac{1}{2a_3}(a_2 \pm D_1), \quad \text{and} \quad D_1 = \sqrt{a_2^2 - 4a_1a_3}.$$

By separation of variables, we have
$$\frac{1}{a_3(\bar{C}-r_+)(\bar{C}-r_-)}\mathrm{d}\bar{C} = \mathrm{d}\tau,$$
which is equivalent to:
$$\left(\frac{1/(r_+-r_-)}{a_3(\bar{C}-r_+)} - \frac{1/(r_+-r_-)}{a_3(\bar{C}-r_-)}\right)\mathrm{d}\bar{C} = \mathrm{d}\tau. \tag{8.45}$$
Integrating (8.45), gives us:
$$\frac{\log(\bar{C}-r_+)}{a_3(r_+-r_-)} - \frac{\log(\bar{C}-r_-)}{a_3(r_+-r_-)} = \tau + \widetilde{c}. \tag{8.46}$$
Since $a_3(r_+ - r_-) = D_1$ and $\bar{C}(\mathbf{u},0) = 0$, we find for the integration constant $\widetilde{c}$,
$$\widetilde{c} = \frac{1}{D_1}\log(-r_+) - \frac{1}{D_1}\log(-r_-),$$
and hence
$$\frac{r_-}{r_+} = \mathrm{e}^{-\widetilde{c}D_1} =: g.$$
Solving for $\bar{C}(\mathbf{u},\tau)$ in (8.46) yields:
$$\bar{C}(\mathbf{u},\tau) = \left(\frac{1-\mathrm{e}^{-D_1\tau}}{1-g\mathrm{e}^{-D_1\tau}}\right)r_-. \tag{8.47}$$
With $a_1 = -\frac{1}{2}(u^2+iu)$, $a_2 = \kappa - \gamma\rho_{x,v}iu$ and $a_3 = \frac{1}{2}\gamma^2$, function $\bar{C}(\mathbf{u},\tau)$ is determined.

We then solve the ODE for $\bar{A}(\mathbf{u},\tau)$:
$$\bar{A}(\mathbf{u},\tau) = b_1\tau + b_2\int_0^\tau \bar{C}(\mathbf{u},z)\mathrm{d}z = b_1\tau + b_2\left(\tau r_- - \frac{1}{a_3}\log\left(\frac{1-g\mathrm{e}^{-D_1\tau}}{1-g}\right)\right).$$
Herewith, the proof is finished, and the discounted characteristic function is thus determined. ∎

Correlation and moment explosion

We discuss the moments of the Heston model and the issue of the "*explosion of moments*", see also [Andersen and Piterbarg, 2007]. Moment explosion is undesirable in practice as it is a sign of an unstable model. As discussed in Section 1.1, from the function $\phi_{\log S(t)}(u)$ for the log-transformed asset process, $X(t) = \log S(t)$, we can can calculate the moments of $S(t)$, for $t > 0$, using $\mathbf{u} = [u,0]^\mathrm{T}$ in the Heston characteristic function. By the definition of the characteristic function for $\log S(t)$,
$$\phi_{\log S(t)}(u) = \mathbb{E}\left[\mathrm{e}^{iu\log S(t)}\right] = \int_0^\infty \mathrm{e}^{iu\log y}f_{S(t)}(y)\mathrm{d}y$$
$$= \int_0^\infty y^{iu}f_{S(t)}(y) = \mathbb{E}\left[(S(t))^{iu}\right], \tag{8.48}$$

substituting $u = -ik$, gives us,

$$\phi_{\log S(t)}(-ik) = \int_0^\infty y^k f_{S(t)}(y) \overset{\mathrm{d}}{=} \mathbb{E}\left[S^k(t)\right], \tag{8.49}$$

resulting in the kth moment.

Under the Heston model, the characteristic function reads,

$$\phi_{\log S(T)}(u) = e^{iu \log S(t_0) + \bar{C}(u,\tau) v(t_0) + \bar{A}(u,\tau)}, \tag{8.50}$$

and Lemma 8.3.1 states that three ODEs need to be solved to determine the functions $\bar{C}(u,\tau)$, $\bar{A}(u,\tau)$. In particular, with $u = -ik$, the ODE for $\bar{C}(u,\tau)$ is given by:

$$\frac{\mathrm{d}\bar{C}}{\mathrm{d}\tau} = \frac{1}{2}k(k-1) + (\gamma \rho_{x,v} k - \kappa)\bar{C} + \frac{1}{2}\gamma^2 \bar{C}^2. \tag{8.51}$$

It has been shown in [Andersen and Piterbarg, 2007] that, depending on the parameter set, the solution to Equation (8.51), and therefore $\mathbb{E}[S^k(t)]$, may become unstable, and it may even *explode* in finite time. The right-hand side of (8.51) is a quadratic polynomial,

$$\frac{\mathrm{d}\bar{C}}{\mathrm{d}\tau} = a_1 + a_2 \bar{C} + a_3 \bar{C}^2 =: h(\bar{C}), \tag{8.52}$$

with $a_1 = \frac{1}{2}k(k-1)$, $a_2 = (\gamma \rho_{x,v} k - \kappa)$ and $a_3 = \frac{1}{2}\gamma^2$. For the discriminant of $h(\bar{C})$, $D_h(\bar{C})$, we obtain,

$$D_h(\bar{C}) = a_2^2 - 4a_1 a_3 = (\gamma \rho_{x,v} k - \kappa)^2 - 4\frac{1}{2}k(k-1)\frac{1}{2}\gamma^2. \tag{8.53}$$

The existence of solutions is now related to the positions of the roots of this polynomial. We can examine for which values of parameter $\rho_{x,v}$, the discriminant stays positive, i.e.,

$$D_h(\bar{C}) > 0 \iff (\gamma \rho_{x,v} k - \kappa)^2 > k(k-1)\gamma^2.$$

After simplifications this gives us the following condition for the correlation coefficient $\rho_{x,v}$,

$$\boxed{\rho_{x,v} < \frac{\kappa}{\gamma k} - \sqrt{\frac{k-1}{k}}.}$$

If the above inequality does not hold, one may expect that $\mathbb{E}[S^k(t)] \to \infty$, $\forall k > 0$.

To guarantee higher-order moments of $S(t)$, the correlation $\rho_{x,v}$ even needs to converge to -1 to ensure a stable solution.

8.4 Numerical solution of Heston PDE

With the discounted characteristic function known in closed form, we can solve the risk-neutral valuation form, i.e. we can compute the discounted expectation, which results from the Feynman-Kac formula.

8.4.1 The COS method for the Heston model

For the Heston model, the COS pricing equation, see Equation (6.28), can thus directly be applied, see also Fang and Oosterlee [2011] for a more general Heston application. However, the COS method for the Heston model can be simplified. A vector of strike prices can be given as input to the Heston COS formula, and as a result a vector of European option values is computed.

Like for Lévy processes, as presented in Equation (6.48), this is possible since for $\mathbf{X}(t) = [X(t), v(t)]^T$, the characteristic function for $\mathbf{u} = [u, 0]^T$ can be expressed as:

$$\phi_{\mathbf{X}}(u; t_0, T) = \varphi_H(u, T; v(t_0)) \cdot e^{iuX(t_0)}, \tag{8.54}$$

with $v(t_0)$ the variance of the underlying at t_0.

The Heston part $\varphi_H(u, T; v(t_0))$ reads:

$$\varphi_H(u, T; v(t_0)) = \exp\left[\left(iur\tau + \frac{v(t_0)}{\gamma^2}\left(\frac{1 - e^{-D_1\tau}}{1 - ge^{-D_1\tau}}\right)(\kappa - i\rho_{x,v}\gamma u - D_1)\right)\right]$$

$$\times \exp\left[\frac{\kappa\bar{v}}{\gamma^2}\left(\tau(\kappa - i\rho_{x,v}\gamma u - D_1) - 2\log\left(\frac{1 - ge^{-D_1\tau}}{1 - g}\right)\right)\right],$$

with $\tau = T - t_0$, and,

$$D_1 = \sqrt{(\kappa - i\rho_{x,v}\gamma u)^2 + (u^2 + iu)\gamma^2} \quad \text{and} \quad g = \frac{\kappa - i\rho_{x,v}\gamma u - D_1}{\kappa - i\rho_{x,v}\gamma u + D_1}.$$

Remark 8.4.1 (Complex-valued characteristic function) *This discounted characteristic function is uniquely specified, since we evaluate the term $\sqrt{(x+yi)}$ in such a way that the real part is nonnegative, and we restrict the complex logarithm to its principal branch. In this case, the resulting characteristic function is the correct one for all complex-valued u parameters in the strip of analycity of the discounted characteristic function, as proven in [Lord and Kahl, 2010].* ▲

Using $\mathbf{X}(t) = \log\frac{S(t)}{\mathbf{K}}$, with a vector of strikes $\mathbf{K}$, we find the following COS formula:

$$\boxed{V(t_0, \mathbf{x}) \approx \mathbf{K}e^{-r\tau} \cdot \operatorname{Re}\left\{\sum_{k=0}^{N-1}{}' \varphi_H\left(\frac{k\pi}{b-a}, T; v(t_0)\right) U_k \cdot \exp\left(ik\pi\frac{\mathbf{X}(t_0) - a}{b-a}\right)\right\},}$$

(8.55)

with $\varphi_H(u, T; v(t_0))$ as in (8.54) and the remaining parameters described previously.

Stochastic Volatility Models

> The hedge parameters Δ and Γ are exactly as presented in Equations (6.36), (6.37).
>
> To obtain the COS formula for Vega, $\frac{\partial V}{\partial v}$ for the Heston model (8.55), for example,
>
> $$\frac{\partial V}{\partial v} \approx e^{-r\tau} \sum_{k=0}^{N-1}{'} \operatorname{Re}\left\{ \frac{\partial}{\partial v} \varphi_H\left(\frac{k\pi}{b-a}, T; v(t_0)\right) \exp\left(ik\pi \frac{x-a}{b-a}\right)\right\} \cdot H_k. \quad (8.56)$$
>
> With (8.54), we have
>
> $$\frac{\partial}{\partial v}\varphi_H\left(\frac{k\pi}{b-a}, T; v(t_0)\right) = \bar{C}(\mathbf{u}, \tau) \cdot \varphi_H\left(\frac{k\pi}{b-a}, T; v(t_0)\right),$$
>
> where $\bar{C}(\mathbf{u}, \tau)$ is from (8.47).

Example 8.4.1 (Numerical experiment for the Heston model) We use the Heston model and price call options by the COS method, with the following parameters:

$$S_0 = 100, \quad K = 100, \quad r = 0, \quad q = 0, \quad \kappa = 1.5768, \quad \gamma = 0.5751,$$
$$\bar{v} = 0.0398, \quad v_0 = 0.0175, \quad \rho_{x,v} = -0.5711, \quad \text{and} \quad T = 1. \quad (8.57)$$

We define the integration range, instead of (6.44), by Equation (6.45), which avoids the computation of the cumulants. Cumulant ζ_2 may become negative for sets of Heston parameters that do not satisfy the Feller condition $2\bar{v}\kappa > \gamma^2$, which is a reason for the use of the range in (6.45).

Here, we mimic the calibration situation, in which many option values are computed simultaneously. We compute the option values for 21 consecutive strike prices, ranging from $K = 50, 55, 60, \ldots, 150$; see the results in Table 8.2, where the maximum error over all strike prices is presented. With $N = 160$, the COS method can price the 21 options highly accurately, within 3 milliseconds. ◆

Table 8.2: *Error convergence and CPU time for call prices under the Heston model by the COS method, pricing 21 strikes, with $T = 1$, with parameters as in (8.57).*

N	32	64	96	128	160
CPU time (msec.)	0.85	1.45	2.04	2.64	3.22
max. abs. err.	1.43e-01	6.75e-03	4.52e-04	2.61e-05	4.40e-06

8.4.2 The Heston model with piecewise constant parameters

An often employed extension of the Heston model to improve the calibration fit to financial market data, is a Heston model with *time-dependent parameters*, i.e. for $X(t) := \log S(t)$, the system is governed by the following dynamics:

$$\begin{cases} dX(t) = \left(r(t) - \frac{1}{2}v(t)\right) dt + \sqrt{v(t)} dW_x(t), \\ dv(t) = \kappa(t)(\bar{v}(t) - v(t)) dt + \gamma(t)\sqrt{v(t)} dW_v(t), \end{cases} \quad (8.58)$$

with correlation $dW_x(t)dW_v(t) = \rho_{x,v}(t)dt$ and all remaining parameters strictly positive and time-dependent. When these parameters are deterministic functions of time, the model stays in the class of affine processes. The discounted characteristic function for $\mathbf{u} \in \mathbb{C}^2$ is then given by:

$$\phi_{\mathbf{X}}(\mathbf{u}; t_0, T) = \exp\left(\bar{A}(\mathbf{u}, \tau) + \bar{\boldsymbol{B}}^{\mathrm{T}}(\mathbf{u}, \tau) \mathbf{X}(t)\right),$$

with $\tau = T - t_0$, $\bar{\mathbf{B}}(\mathbf{u}, \tau) = [\bar{B}(\mathbf{u}, \tau), \bar{C}(\mathbf{u}, \tau)]^{\mathrm{T}}$. With $\mathbf{u} = [u, 0]^{\mathrm{T}}$, which gives $\bar{B}(\mathbf{u}, 0) = iu$ and $\bar{C}(\mathbf{u}, 0) = 0$, we can immediately determine the complex-valued ODEs,[1]

$$\frac{d\bar{C}}{d\tau} = \frac{1}{2}iu(iu - 1) - \left(\kappa(T-\tau) - \gamma(T-\tau)\rho_{x,v}(T-\tau)iu\right)\bar{C}$$
$$+ \frac{1}{2}\gamma^2(T-\tau)\bar{C}^2,$$
$$\frac{d\bar{A}}{d\tau} = \kappa(T-\tau)\bar{v}(T-\tau)\bar{C} + r(T-\tau)(iu - 1),$$

with the conditions $\bar{C}(\mathbf{u}, 0) = 0$ and $\bar{A}(\mathbf{u}, 0) = 0$.

Since in this generalized Heston model we deal with time-dependent parameters, the corresponding ODEs can not be solved analytically. One can apply one of the well-known numerical ODE methods, like fourth-order Runge-Kutta time-stepping, but such numerical techniques may increase computation time.

In the proof of Lemma 8.3.2, however, we derived the solution to general Riccati-type ODEs with constant parameters, i.e. to a general system of ODEs given by:

$$\begin{cases} \dfrac{d\bar{C}}{d\tau} = c_1 - c_2\bar{C} + c_3\bar{C}^2, & \bar{C}(\mathbf{u}, 0) = c_0, \\ \dfrac{d\bar{A}}{d\tau} = a_1 + a_2\bar{C}, & \bar{A}(\mathbf{u}, 0) = a_0, \end{cases} \quad (8.59)$$

---

[1] The ODE for $\bar{B}(\mathbf{u}, \tau)$ is trivial, with $\bar{B}(\mathbf{u}, \tau) = iu$.

with the solutions for $\bar{C}(\mathbf{u}, \tau)$ and $\bar{A}(\mathbf{u}, \tau)$ given by:

$$\bar{A}(\mathbf{u}, \tau) = a_0 + a_1\tau + a_2\left[\left(r_+ - \frac{D_1}{c_3}\right)\tau - \frac{1}{c_3}\log\left(\frac{ge^{-\tau D_1} - 1}{g - 1}\right)\right],$$

$$\bar{C}(\mathbf{u}, \tau) = \frac{r_+ge^{-\tau D_1} - r_-}{ge^{-\tau D_1} - 1},$$

and where

$$g = \frac{2c_3c_0 - (c_2 - D_1)}{2c_3c_0 - (c_2 + D_1)}, \quad r_\pm = \frac{1}{2c_3}(c_2 \pm D_1), \quad \text{and} \quad D_1 = \sqrt{c_2^2 - 4c_1c_3},$$

with $a_1 = r(T - \tau)(iu - 1)$, $a_2 = \kappa(T - \tau)\bar{v}(T - \tau)$, $c_1 = 1/2iu(iu - 1)$, $c_2 = \kappa(T - \tau) - \gamma(T - \tau)\rho_{x,v}(T - \tau)iu$ and $c_3 = 1/2\gamma^2(T - \tau)$.

Using the insight that the Riccati-type ODEs in (8.59) are valid for nonzero conditions a_0 and c_0, we will define a Heston model with *piece-wise constant parameters* r, κ, γ, $\rho_{x,v}$ and $\bar{v}$.

So, we define a time-grid $0 = \tau_0 \leq \tau_1 \cdots \leq \tau_{N-1} \leq \tau_N = \tau$. At each grid point τ_i, $i = 0, \ldots, N$, the model parameters will be evaluated at $T - \tau_i$, so that at the first interval $[0, \tau_1)$ we have the parameters of interval $[T - \tau_1, T)$, at the second interval $[\tau_1, \tau_2)$ we have parameters from $[T - \tau_2, T - \tau_1)$, etc.

Piece-wise constant parameters imply that the characteristic function can be evaluated recursively, i.e. at the first interval from $[0, \tau_1)$, we use the initial conditions $\bar{C}(\mathbf{u}, 0) = 0$ and $\bar{A}(\mathbf{u}, 0) = 0$. When the corresponding analytic solution is determined, as for Equation (8.59), we obtain two solutions, a_1 and c_1. For interval $[\tau_1, \tau_2)$, we then assign the initial conditions $\bar{C}(\mathbf{u}, \tau_1) = c_1$ and $\bar{A}(\mathbf{u}, \tau_1) = a_1$. This procedure will be repeated until the last time step, where the initial values c_{N-1} and a_{N-1} are used to evaluate $\bar{C}(\mathbf{u}, \tau_N)$ and $\bar{A}(\mathbf{u}, \tau_N)$.

8.4.3 The Bates model

The Bates model [Bates, 1996] also generalizes the Heston model by adding jumps to the Heston stock price process. The model is described by the following system of SDEs:

$$\begin{cases} \frac{dS(t)}{S(t)} = \left(r - \xi_p\mathbb{E}[e^J - 1]\right)dt + \sqrt{v(t)}dW_x(t) + \left(e^J - 1\right)dX_\mathcal{P}(t), \\ dv(t) = \kappa\left(\bar{v} - v(t)\right)dt + \gamma\sqrt{v(t)}dW_v(t), \end{cases}$$

with Poisson process $X_\mathcal{P}(t)$ with intensity ξ_p, and normally distributed jump sizes, J, with expectation μ_J and variance σ_J^2, i.e. $J \sim \mathcal{N}(\mu_j, \sigma_j^2)$. $X_\mathcal{P}(t)$ is assumed to be independent of the Brownian motions and of the jump sizes.

Under the log transformation, the Bates model reads

$$\begin{cases} dX(t) = \left(r - \frac{1}{2}v(t) - \xi_p\mathbb{E}[e^J - 1]\right)dt + \sqrt{v(t)}dW_x(t) + JdX_\mathcal{P}(t), \\ dv(t) = \kappa\left(\bar{v} - v(t)\right)dt + \gamma\sqrt{v(t)}dW_v(t). \end{cases}$$

The PIDE for the Bates model, under the risk-neutral measure, can now be derived and is given, for $V = V(t, X)$, by

$$\frac{\partial V}{\partial t} + \left(r - \frac{1}{2}v - \xi_p \mathbb{E}[e^J - 1]\right)\frac{\partial V}{\partial X} + \kappa(\bar{v} - v(t))\frac{\partial V}{\partial v} + \frac{1}{2}\gamma^2 v \frac{\partial^2 V}{\partial v^2} + \frac{1}{2}v\frac{\partial^2 V}{\partial X^2}$$
$$+ \rho_{x,v}\gamma v \frac{\partial^2 V}{\partial X \partial v} + \xi_p \mathbb{E}\left[V(t, X + J)\right] = (r + \xi_p)V.$$

The Bates model belongs to the class of affine jump diffusion processes, see Section 7.3.2. Therefore, we can derive the corresponding discounted characteristic function, for state-vector $\mathbf{X}(t) = [X(t), v(t)]^T$. The model is based on the same covariance matrix as the Heston model, see (8.39). The jump component, as in (7.68), is given by $\bar{\xi}_p(t, \mathbf{X}(t)) = \xi_{p,0} + \xi_{p,1}\mathbf{X}(t) = \xi_p$, and a_0 and a_1, are given by

$$a_0 = \begin{bmatrix} r - \xi_p \mathbb{E}[e^J - 1] \\ \kappa\bar{v} \end{bmatrix}, \quad a_1 = \begin{bmatrix} 0 & -\frac{1}{2} \\ 0 & -\kappa \end{bmatrix}. \tag{8.60}$$

With $\mathbf{J} = [J, 0]$ and $\bar{\mathbf{B}}(\mathbf{u}, \tau) = [\bar{B}(\mathbf{u}, \tau), \bar{C}(\mathbf{u}, \tau)]^T$, the affinity relations in the class of AJD processes provide the following system of ODEs:

Lemma 8.4.1 (Bates ODEs) *The functions $\bar{A}_{Bates}(\mathbf{u}, \tau)$, $\bar{B}(\mathbf{u}, \tau)$ and $\bar{C}(\mathbf{u}, \tau)$ in (7.69) satisfy the following system of ODEs:*

$$\frac{d\bar{B}}{d\tau} = 0,$$
$$\frac{d\bar{C}}{d\tau} = \frac{1}{2}\bar{B}(\bar{B} - 1) - \left(\kappa - \gamma\rho_{x,v}\bar{B}\right)\bar{C} + \frac{1}{2}\gamma^2\bar{C}^2,$$
$$\frac{d\bar{A}_{Bates}}{d\tau} = \kappa\bar{v}\bar{C} + r(\bar{B} - 1) - \xi_p \mathbb{E}[e^J - 1]\bar{B} + \xi_p \mathbb{E}\left[e^{J\bar{B}} - 1\right],$$

with initial conditions $\bar{B}(\mathbf{u}, 0) = iu$, $\bar{C}(\mathbf{u}, 0) = 0$ and $A_{Bates}(\mathbf{u}, 0) = 0$, and the parameters $\kappa, \gamma, \bar{v}, r$ and $\rho_{x,r}$ are as in the Heston model (8.38). ◂

The functions $\bar{B}(\mathbf{u}, \tau)$ and $\bar{C}(\mathbf{u}, \tau)$ are the same as for the standard Heston model. The Bates and the Heston model differ only in $\bar{A}_{Bates}(\mathbf{u}, \tau)$, which in the Bates model contains jump components. As the jumps J in the Bates model are normally distributed, with mean μ_J and variance σ_J^2, the two expectations in the ODEs for $\bar{A}_{Bates}(\mathbf{u}, \tau)$ are given by

$$\mathbb{E}\left[e^J - 1\right] = e^{\mu_J + \frac{1}{2}\sigma_J^2} - 1, \quad \mathbb{E}\left[e^{iuJ} - 1\right] = e^{iu\mu_J - \frac{1}{2}\sigma_J^2 u^2} - 1,$$

so that the ODE for $\bar{A}_{Bates}(\mathbf{u}, \tau)$ reads:

$$\frac{d\bar{A}_{Bates}}{d\tau} = \frac{d\bar{A}}{d\tau} - \xi_p iu\left(e^{\mu_J + \frac{1}{2}\sigma_J^2} - 1\right) + \xi_p\left(e^{iu\mu_J - \frac{1}{2}\sigma_J^2 u^2} - 1\right),$$

with $\frac{\mathrm{d}\bar{A}}{\mathrm{d}\tau}$ as derived for the Heston model, given in (8.3.2). The solution for $\bar{A}_{\text{Bates}}(\mathbf{u}, \tau)$ can also easily be found, as

$$\bar{A}_{\text{Bates}}(\mathbf{u}, \tau) = \bar{A}(\mathbf{u}, \tau) - \xi_p i u \tau \left(e^{\mu_J + \frac{1}{2}\sigma_J^2} - 1 \right) + \xi_p \tau \left(e^{i u \mu_J - \frac{1}{2}\sigma_J^2 u^2} - 1 \right).$$

Example 8.4.2 (Impact of Bates parameters on implied volatility)
Three plots in Figure 8.6 give an indication of the effect of the jump parameters in the Bates model on the implied volatilities. In the experiments we use $T = 1$, $S_0 = 100$; the Heston model parameters have been set to $\kappa = 1.2$, $\bar{v} = 0.05$, $\gamma = 0.05$, $\rho_{x,v} = -0.75$, $v_0 = 0.05$ and $r = 0r$. The three jump parameters, i.e. the intensity ξ_p, the jump mean μ_J and jump volatility σ_J, have been varied; the base case for these parameters is given by $\mu_J = 0, \sigma_J = 0.2, \xi_p = 0.1$. Figure 8.6 shows quite similar patterns for the implied volatilities regarding

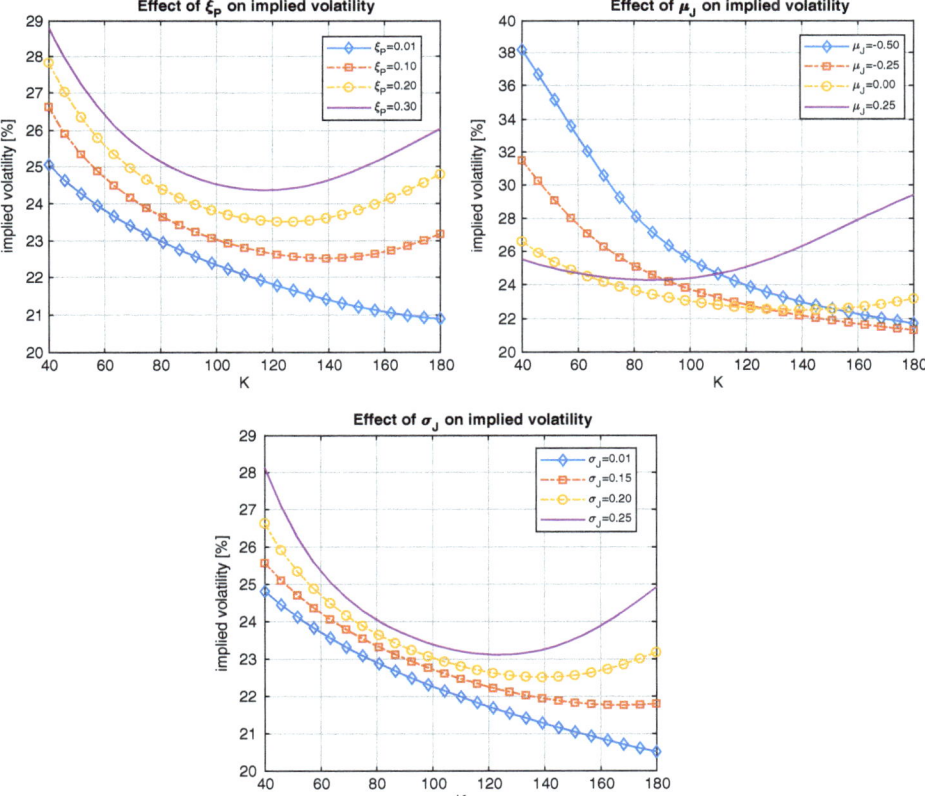

Figure 8.6: *Effect of jump parameters on the implied volatilities in the Bates model; effect of changes in ξ_p (first plot), μ_J (second plot), and σ_J (third plot).*

parameter changes in ξ_p and σ_J. An increase of the parameter values increases the curvature of the implied volatilities. The plots also indicate that the implied volatilities from the Bates model can be higher than those obtained by the Heston model. The explanation lies in the fact that the additional, uncorrelated jump component is present in the dynamics of the Bates model, which may increase the *realized* stock variance. An increase of volatility of the stock increases the option price as the option is more likely to end in the money.

In the case of parameter μ_J, which represents the *expected* location of the jumps, the implied volatilities reveal an irregular behavior. Parameter μ_J can take either positive or negative values, as it determines whether the stock paths will have upwards or downwards jumps. For negative μ_J-values the stock price is expected to have more downwards movements whereas for positive μ_J-values positive jumps in the stock values are expected.

Although the Bates model seems a very fine model from a theoretical point-of-view, we should keep in mind that it is generally difficult to get a good intuition for hedging under jump diffusion type models and that there are quite a few parameters to calibrate. Some of them may have a very similar impact, so that the risk of over-fitting and thus multiple parameter solutions giving very similar implied volatility curves is present. Too many parameters is not always preferred in the financial industry, and may also not be very elegant (with many parameters we may fit all sorts of features, but the predictive power of such a model may be limited).♦

Stochastic Volatility Models 255

8.5 Exercise set

Exercise 8.1 Let $X(t)$ be governed by the following dynamics:
$$\mathrm{d}X(t) = (2\mu X(t) + \sigma^2)\mathrm{d}t + \sigma\sqrt{X(t)}\mathrm{d}W(t),\ X(t_0) = X_0.$$
Derive which SDE is satisfied by the transformed process, $Y(t) = \sqrt{X(t)}$.

Exercise 8.2 Analyze the Schöbel-Zhu model in the context of the Cholesky decomposition and the instantaneous covariance matrix. Show that the Schöbel-Zhu system is an affine system.

Exercise 8.3 The Schöbel-Zhu model is governed by the volatility process $\sigma(t)$. Derive the distribution of the corresponding variance process $v(t) = \sigma^2(t)$.

Exercise 8.4 With $\gamma = 0$ the Heston model collapses to the Black-Scholes model with a time-dependent volatility parameter $\sigma(t)$. Derive the function $\sigma(t)$ in this case. Show the behaviour of the corresponding implied Black-Scholes volatility surface, for varying parameters $\kappa, \bar{v}$ and v_0.

Exercise 8.5 Show, with the help of the multi-dimensional Itô's lemma, that the martingale approach indeed results in the Heston partial differential equation.

Exercise 8.6 Confirm that the ODEs in Lemma 8.3.2 are correct.

Exercise 8.7 Compute the option Greeks delta, gamma, and vega for the Heston model by using the COS method and compare the results to the finite difference approximations. Employ the parameter values for the Heston model as presented in Example 8.4.1.

Exercise 8.8 Consider the Heston model with maturity time $T = 4$, constant interest rate $r = 0.05$, initial variance $v(t_0) = 0.02$ and initial stock price $S(t_0) = 100$, *with piece-wise constant parameters* that are given by,

$$\kappa(t) = 1.3\mathbb{1}_{0 \leq t < 1} + 0.9\mathbb{1}_{1 \leq t < 2} + 0.8\mathbb{1}_{2 \leq t < 3} + 0.7\mathbb{1}_{3 \leq t < 4},$$
$$\bar{v}(t) = 0.21\mathbb{1}_{0 \leq t < 1} + 0.14\mathbb{1}_{1 \leq t < 2} + 0.12\mathbb{1}_{2 \leq t < 3} + 0.08\mathbb{1}_{3 \leq t < 4},$$
$$\gamma(t) = 0.3\mathbb{1}_{0 \leq t < 1} + 0.2\mathbb{1}_{1 \leq t < 2} + 0.15\mathbb{1}_{2 \leq t < 3} + 0.1\mathbb{1}_{3 \leq t < 4},$$
$$\rho_{x,v}(t) = -0.7\mathbb{1}_{0 \leq t < 1} - 0.5\mathbb{1}_{1 \leq t < 2} - 0.2\mathbb{1}_{2 \leq t < 3} - 0.1\mathbb{1}_{3 \leq t < 4}.$$

a. Derive the characteristic function for this model with piece-wise constant parameters.

b. Implement the corresponding characteristic function to recover the probability density function.

c. Take the COS method with as the number of expansion terms $N = 500$, the integration domain $[-16, 16]$ and price a put option for strike prices in the range $K = \{50, 55, \ldots, 295, 300\}$.

d. Determine the put prices by means of the put-call parity.

e. Implement a Monte Carlo simulation (the Euler scheme) for 100.000 paths and 100 time-steps. In order to avoid complex numbers consider the following truncation scheme $v(t_i) = \max(v(t_i), 0)$ at each time-step t_i. The details of this Monte Carlo scheme are presented in the next chapter. Compare your results against ones obtained by the COS method.

Exercise 8.9 Show that for a CIR process $v(t)$ given by:
$$dv(t) = \kappa\left(v(t_0) - v(t)\right)dt + \gamma\sqrt{v(t)}dW(t)$$
the following equality holds true,
$$\mathbb{E}\left[v(t)v(s)\right] = v_0^2 + v_0 e^{-\kappa(t)}\gamma^2 \frac{e^{\kappa s} - e^{-\kappa s}}{2\kappa}, \quad \text{for} \quad s \leq t. \tag{8.61}$$

Exercise 8.10 Under the risk-free measure $\mathbb{Q}$, where all underlying assets are martingales w.r.t. the money-savings account $M(t)$, with $dM(t) = rM(t)dt$, the Heston model, in terms of the independent Brownian motions, is defined as,
$$dS(t)/S(t) = rdt + \sqrt{v(t)}d\widetilde{W}_x(t), \tag{8.62}$$
$$dv(t) = \kappa\left(\bar{v} - v(t)\right)dt + \gamma\sqrt{v(t)}\left[\rho_{x,v}d\widetilde{W}_x(t) + \sqrt{1-\rho_{x,v}^2}d\widetilde{W}_v(t)\right].$$

Derive the model dynamics under the stock-measure $\mathbb{Q}^S$, under which all assets are martingales w.r.t. the stock process $S(t)$ in (8.62).

To accomplish this, the following steps need to be taken,

a. Define a process $Y(t) = M(t)/S(t)$ and show that its dynamics are given by,
$$dY(t)/Y(t) = v(t)dt - \sqrt{v(t)}d\widetilde{W}_x(t).$$

b. Define another Brownian motion, $\widetilde{W}_x^S(t)$, for which process $Y(t)$ becomes a martingale, i.e., show that
$$d\widetilde{W}_x^S(t) = \sqrt{v(t)}dt - d\widetilde{W}_x(t).$$

c. Use the derived measure transformation to show that the Heston model under the $\mathbb{Q}^S$-measure reads,
$$dS(t)/S(t) = (r + v(t))dt - \sqrt{v(t)}dW_x^S(t),$$
$$dv(t) = \kappa^*\left(\bar{v}^* - v(t)\right)dt + \gamma\sqrt{v(t)}dW_v^S(t),$$
with a correlation parameter $\rho_{x,v}^* = -\rho_{x,v}$, and $\kappa^* = \kappa - \rho_{x,v}\gamma$, $\bar{v}^* = \kappa\bar{v}/(\kappa - \rho_{x,v}\gamma)$.

d. Using $d\widehat{W}_x^S(t) = -dW_x^S(t)$, show that the model dynamics then transform to,
$$dS(t)/S(t) = (r + v(t))dt + \sqrt{v(t)}d\widehat{W}_x^S(t),$$
$$dv(t) = \kappa^*\left(\bar{v}^* - v(t)\right)dt + \gamma\sqrt{v(t)}d\widehat{W}_v^S(t),$$
with $d\widehat{W}_x^S(t)d\widehat{W}_v^S(t) = \rho_{x,v}dt$, $\kappa^* = \kappa - \rho_{x,v}\gamma$ and $\bar{v}^* = \kappa\bar{v}/(\kappa - \rho_{x,v}\gamma)$.

Exercise 8.11 Use the COS method to recover the probability density function of the Bates model. Compare this Bates model density with the Heston model density. Employ the parameter values for the Heston model as presented in Example 8.4.1.

Vary the Bates jump parameters and draw conclusions about the effect on the tails of the distribution.

CHAPTER 9

Monte Carlo Simulation

In this chapter:

This chapter discusses the Monte Carlo technique and presents methods to *simulate asset paths*. A basic introduction into Monte Carlo discretization is given in **Section 9.1**. Details of the *stochastic Euler and Milstein SDE discretization schemes* are presented in **Section 9.2**, where particularly a detailed derivation of the Milstein scheme is provided. Application of the Euler scheme to *the CIR process* may however lead to undesired and unrealistic path realizations, because the CIR process is guaranteed to be nonnegative whereas a plain Euler discretization is not. In **Section 9.3**, we present *alternative discretizations* for this issue, and in **Section 9.4** we generalize the concept to the *Heston system of stochastic volatility SDEs*. The Monte Carlo computation of the option Greeks is presented in **Section 9.5**.

Keywords: Monte Carlo simulation, stochastic integration, Euler discretization, Milstein discretization, CIR process, exact simulation, QE scheme, Heston SV model.

9.1 Monte Carlo basics

Monte Carlo integration methods are sampling methods, based on probability theory. They rely on trials and randomness to reveal information.

Monte Carlo methods are applicable to a wide range of problems. In this chapter we focus on financial derivative pricing. Boyle [1977] wrote a very early paper advocating the use of Monte Carlo methods for the valuation of options, see [Boyle et al., 1997].

In the computational finance context Monte Carlo methods can basically be described as follows. First of all, stochastic asset paths are generated, under the risk-neutral (or any forward) measure. For each random asset path, the corresponding value of the payoff function can be calculated, based on the asset path's value at the final time T. An approximation of the value of the option at $t = t_0$ is then obtained as the discounted average of these very many simulated payoff values.

From a mathematical point of view, Monte Carlo methods are based on the *central limit theorem* and the *the law of large numbers*. Since the result of an experiment is a random number, the *structure of the error* which is made with this method, has a probabilistic distribution. This means that the solution is typically an expected value, which comes from the Monte Carlo experiment with a variance.

The Monte Carlo method is especially well-suited for pricing *path-dependent options*. Consider an option payoff function $H(T, S)$, which may be dependent on the path of the underlying asset $S(t)$, with $t_0 \leq t \leq T$. Then today's value of a European-style derivative is given, as presented in the Feynman-Kac Theorem, Theorem 3.2.1, by

$$V(t_0, S_0) = e^{-r(T-t_0)} \mathbb{E}^\mathbb{Q} \left[H(T, S) | \mathcal{F}(t_0) \right]$$
$$= e^{-r(T-t_0)} \int_\mathbb{R} H(T, y) f_S(T, y; t_0, S_0) dy, \quad (9.1)$$

where $S_0 = S(t_0)$, the expectation $\mathbb{E}^\mathbb{Q}[\cdot]$ is taken under risk-neutral measure $\mathbb{Q}$, and $f_S(T, y; t_0, S_0)$ represents the probability density function, connected to the general risk-neutral Itô dynamics,

$$dS(t) = \bar{\mu}^\mathbb{Q}(t, S) dt + \bar{\sigma}(t, S) dW^\mathbb{Q}(t), \quad t > t_0, \quad (9.2)$$

as in Equation (3.19). When the PDF is not available in closed-form, the Monte Carlo method is a convenient and valuable pricing method. By approximation and simulation of the asset dynamics, *implicitly* the corresponding PDF is approximated. With a sufficiently large number of asset price paths, it then becomes possible, by means of counting the asset prices at time T, to estimate the PDF by means of a histogram.

Stochastic simulation techniques are also interesting and useful for high-dimensional integration problems, for which deterministic quadrature rules (cubature) cannot be applied, as their computational effort and storage requirements are hampered by the *curse of dimensionality* (i.e. the exponential growth of the number of integration points in tensor-product deterministic grids with increasing problem dimensionality).

Monte Carlo Simulation

A **Monte Carlo algorithm** to approximate an option value can be summarized as follows:

1. Partition the time interval $[0, T]$, so that the following time points, $0 = t_0 < t_1 < \cdots < t_m = T$, are obtained. The time points are defined by $t_i = \frac{i \cdot T}{m}$, $i = 0, \ldots, m$, where $m+1$ represents the number of time steps; The time step is denoted by $\Delta t = t_{i+1} - t_i$, where this time partitioning does not need to be uniform.

2. Generate asset values, $s_{i,j}$, taking the risk-neutral dynamics of the underlying model into account. Note that asset path value $s_{i,j}$ has two indices, $i = 1, \ldots, m$ (the time steps) and $j = 0, \ldots, N$, with N being the number of generated Monte Carlo asset paths.

3. Compute the N payoff values, H_j, and store these results. In the case of European options, we have $H_j = H(T, s_{m,j})$, whereas in the case of path-dependent options we may have $H_j = H(T, s_{i,j})$, $i = 1, \ldots m$.

4. Compute the average:

$$\mathbb{E}^{\mathbb{Q}}\left[H(T, S) | \mathcal{F}(t_0)\right] \approx \frac{1}{N} \sum_{j=1}^{N} H_j.$$

The right-hand side is known as the *Monte Carlo estimate*.

5. Calculate the option value as

$$V(t_0, S) \approx e^{-r(T-t_0)} \frac{1}{N} \sum_{j=1}^{N} H_j.$$

6. Determine the standard error (standard deviation) related to the obtained prices in Step 5.

Remark 9.1.1 (Standard error in Monte Carlo method) *With a Monte Carlo method, we approximate a solution by a sequence of random realizations, i.e., with N Monte Carlo paths we obtain the following approximation:*

$$\mathbb{E}^{\mathbb{Q}}[H(T, S)] \approx \bar{H}_N(T, S) := \frac{1}{N} \sum_{j=1}^{N} H_j.$$

By the strong law of large numbers, we know that, for $N \to \infty$,

$$\lim_{N \to \infty} \bar{H}_N(T, S) = \mathbb{E}^{\mathbb{Q}}[H(T, S)], \quad \text{with probability } 1.$$

In order to estimate the error which occurs due to a finite number of Monte Carlo paths, we need to compute the variance of the estimator $\bar{V}_N(t_0, S)$, as follows,

$$\mathbb{V}ar^{\mathbb{Q}}\left[\bar{H}_N(T,S)\right] = \mathbb{V}ar^{\mathbb{Q}}\left[\frac{1}{N}\sum_{j=1}^{N}H_j\right]$$

$$= \frac{1}{N^2}\sum_{j=1}^{N}\mathbb{V}ar^{\mathbb{Q}}[H_j]$$

$$\approx \frac{1}{N}\mathbb{V}ar^{\mathbb{Q}}[H(T,S)], \qquad (9.3)$$

given that samples $s_{i,j}$ are drawn independently.

In practice, the variance $\mathbb{V}ar^{\mathbb{Q}}[H(T,S)]$ is unknown. However, it can be approximated by the sample variance, $\bar{v}_N^2$, which is an unbiased estimator, i.e.

$$\bar{v}_N^2 := \frac{1}{N-1}\sum_{j=1}^{N}\left(H_j - \bar{H}_N(T,S)\right)^2.$$

Commonly, the standard error, ϵ_N, is defined as:

$$\epsilon_N := \frac{\bar{v}_N}{\sqrt{N}}. \qquad (9.4)$$

From (9.4) we see that when the number of samples increases by a factor of 4, the error reduces only by a factor of 2. ▲

Accurate simulation methods, that are based on Monte Carlo paths, are presented in the subsections to follow.

Unless stated otherwise, in this chapter we will work under the risk-neutral measure when dealing with asset prices.

9.1.1 Monte Carlo integration

As a part of the basic description of Monte Carlo methods, we here consider two stochastic integration example problems, in which a Wiener process $W(t)$ appears.

For a given deterministic function $g(t)$, we first wish to compute numerically,

$$\boxed{\int_0^T g(t)\mathrm{d}W(t).} \qquad (9.5)$$

For this purpose, we define a partition of the integration interval, as follows, $0 = t_0 < t_1 < \cdots < t_m = T$, $t_i = i\frac{T}{m}$, and $\Delta t = t_{i+1} - t_i$.

The numerical approximation within a stochastic integration technique is based on the definition of the Itô integral, as in Definition 1.3.1, particularly in Equation (1.24), where taking the limit $(m \to \infty)$ is omitted, i.e.

$$X(T) := \int_0^T g(t)\mathrm{d}W(t) \approx \sum_{i=0}^{m-1} g(t_i)\left(W(t_{i+1}) - W(t_i)\right). \tag{9.6}$$

We simulate the Wiener increments by using the following relation:

$$W(t_{i+1}) - W(t_i) \stackrel{\mathrm{d}}{=} Z\sqrt{\Delta t}, \text{ with } Z \sim \mathcal{N}(0,1),$$

and recall that $\mathcal{N}(0,1)$ is the standard normal distribution.
Recall that we can express (9.6) as:

$$\mathrm{d}X(t) = g(t)\mathrm{d}W(t), \quad X(0) = X_0, \tag{9.7}$$

for which a basic discretization scheme reads:

$$x_{i+1} = x_i + g(t_i)\left(W(t_{i+1}) - W(t_i)\right), \tag{9.8}$$

with $X(t_i) = x_i$.

However, as already mentioned, Monte Carlo integration is based on the law of large numbers and the central limit theorem. We therefore need to simulate *very many integration paths* between 0 and T. This is indicated by a second index "j", which runs from 1 to N (the number of Monte Carlo paths).
We draw, for each interval and each path, a standard normally distributed random number Z, each time we need to approximate $W(t_{i+1}) - W(t_i), \forall i, j$. This repetition, sampling N Monte Carlo paths, gives rise to two indices, i, j,

$$x_{i+1,j} = x_{i,j} + g(t_i)\left(W(t_{i+1,j}) - W(t_{i,j})\right), \quad j = 1, \ldots N, \tag{9.9}$$

with $X(t_i) = x_i$.

Example 9.1.1 (Monte Carlo integration of $\int_0^1 t^2 \mathrm{d}W(t)$) We will check the numerical results that are obtained by the Monte Carlo method for a concrete example of Equation (9.5), with $g(t) = t^2$ and $T = 1$. Theoretically we know that $X(T)$ is normally distributed, due to the representation in (9.7), with,

$$\mathbb{E}[X(1)] = \mathbb{E}\left[\int_0^1 t^2 \mathrm{d}W(t)\right] = 0,$$

$$\mathbb{V}\mathrm{ar}[X(1)] = \mathbb{V}\mathrm{ar}\left[\int_0^1 t^2 \mathrm{d}W(t)\right] = \mathbb{E}\left[\left(\int_0^1 t^2 \mathrm{d}W(t)\right)^2\right] = \int_0^1 t^4 \mathrm{d}t = 0.2.$$

These results can be derived based on the discussion around the Equations (1.32), (1.33), in Chapter 1.

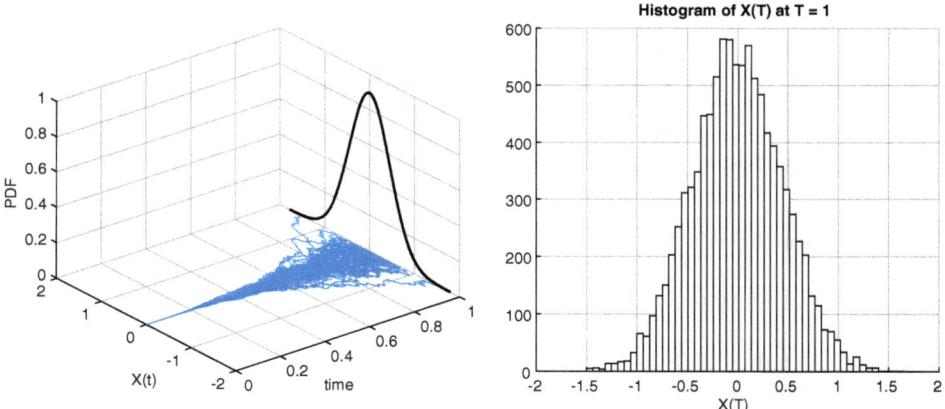

Figure 9.1: *Monte Carlo paths of $X(t)$ and the corresponding histogram at $T=1$ with $g(t)=t^2$ from Example 9.1.1.*

Using the Monte Carlo simulation to solve this stochastic integration problem, with $m=100$ time intervals, and $N=10000$ simulated paths, we obtain the Monte Carlo results as displayed in Figure 9.1. With the parameters chosen, we here find

$$\mathbb{E}\left[\int_0^1 t^2 \mathrm{d}W(t)\right] \approx -0.0037, \quad \mathrm{Var}\left[\int_0^1 t^2 \mathrm{d}W(t)\right] \approx 0.2047,$$

which is a first approximation, which may be crude but already rather accurate, to the analytic solution. The solution will improve with smaller time intervals and when an increasing number of sample paths is selected. ◆

Example 9.1.2 (Monte Carlo integration of $\int_0^2 W(t)\mathrm{d}W(t)$) As another example, we return to Equation (1.41), in Example 1.3.1 of Chapter 1, where the exact solution for a stochastic integral containing Brownian motion $W(t)$ was derived, as

$$X(T) = \int_0^T W(t)\mathrm{d}W(t) = \frac{1}{2}W^2(T) - \frac{1}{2}T.$$

Using as a particular example, $T=2$, we find analytically,

$$\mathbb{E}[X(2)] = \mathbb{E}\left[\int_0^2 W(t)\mathrm{d}W(t)\right] = 0, \ \mathrm{Var}[X(2)] = \mathrm{Var}\left[\int_0^2 W(t)\mathrm{d}W(t)\right] = 2.$$

In this case $X(t)$ is not normally distributed. As the solution of $X(t)$ contains a squared Brownian motion, it will follow a *shifted noncentral chi-squared distribution*. The term $-0.5T$ is the shift that moves the distribution into the negative range.

Monte Carlo Simulation

As before, to approximate the solution of $X(t)$, we take the corresponding SDE and perform an approximation of $\mathrm{d}X(t) = W(t)\mathrm{d}W(t)$, by

$$x_{i+1} = x_i + W(t_i)\left(W(t_{i+1}) - W(t_i)\right), \qquad (9.10)$$

where N such sample paths will be generated. The Monte Carlo approximation of the expectation and the variance, in the form of a histogram, is presented at the right-hand side of Figure 9.2, next to some discrete paths at the left-hand side. ◆

Example 9.1.3 (Expectation and variance of $(W(t_{i+1}) - W(t_i))^2$)
In Remark 2.1.2 in Chapter 2, we gave a heuristic argument for the fact that $\mathrm{d}W^2 = \mathrm{d}t$ can be used. Here, we perform a numerical experiment, in which we measure the expectation and variance of the term $(W(t_{i+1}) - W(t_i))^2$, with respect to a time discretization. The results in Figure 9.3 confirm that the variance converges to 0 much more rapidly than the expectation, and that the expectation converges to $\mathrm{d}t$. ◆

Example 9.1.4 (Smoothness of the payoff and Monte Carlo convergence) In this example we will show that the Monte Carlo convergence strongly depends on the smoothness of the function under consideration. Let's take the following two (nonsmooth and smooth) functions,

$$g_1(x) = \mathbb{1}_{x \geq 0}, \text{ and } g_2(x) = F_{\mathcal{N}(0,1)}(x),$$

and consider the following expectations, $\mathbb{E}[g_i(W(1))|\mathcal{F}(t_0)]$, for $i = 1, 2$, where $W(1)$ is a Brownian motion at time $T = 1$. Both expectations can be solved

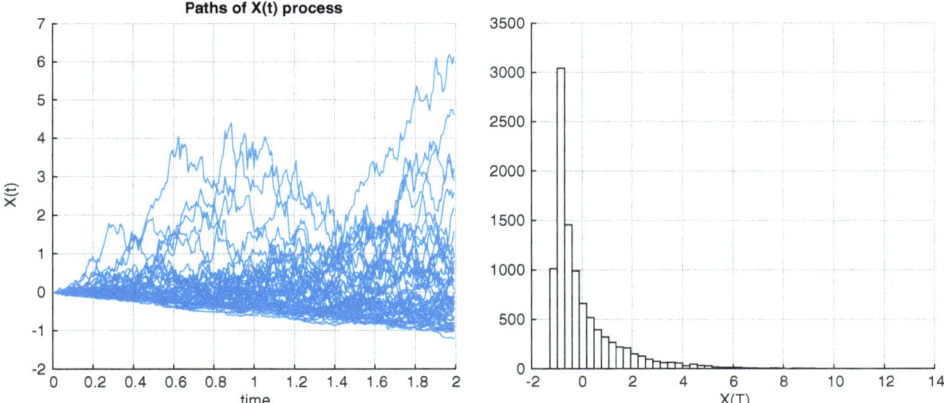

Figure 9.2: *Monte Carlo paths of $X(t)$ from Example 9.1.2, and the corresponding histogram at $T = 2$.*

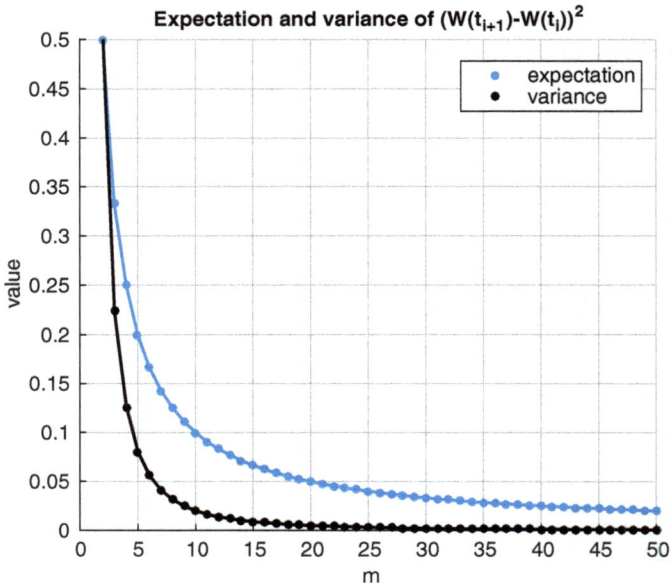

Figure 9.3: *Expectation and variance in the numerical approximation of* $\mathrm{d}W^2$ *versus parameter m.*

analytically and are equal to $\frac{1}{2}$:

$$V_1 := \mathbb{E}[g_1(W(1))|\mathcal{F}(t_0)] = \int_{\mathbb{R}} \mathbb{1}_{x \geq 0} f_{\mathcal{N}(0,1)}(x) = \frac{1}{2},$$

$$V_2 := \mathbb{E}[g_2(W(1))|\mathcal{F}(t_0)] = \int_{\mathbb{R}} F_{\mathcal{N}(0,1)}(x) f_{\mathcal{N}(0,1)}(x) \mathrm{d}x = \int_{\mathbb{R}} F_{\mathcal{N}(0,1)}(x) \mathrm{d}F_{\mathcal{N}(0,T)}(x)$$

$$= \frac{1}{2} F_{\mathcal{N}(0,1)}^2(x) \Big|_{-\infty}^{+\infty} = \frac{1}{2}.$$

We define the following error,

$$\widetilde{c}_N^i = \widetilde{V}_i^N - V_i, \quad \text{with} \quad \widetilde{V}_i^N = \frac{1}{N} \sum_{j=1}^N V_i(w_j),$$

where w_j are samples from $W(1) \sim \mathcal{N}(0,1)$, and the exact solution $V_i = 1/2$. In Figure 9.4 the results are presented, where a larger variance of the error for the nonsmooth function $g_1(x)$ is observed. The results in Figure 9.4 confirm that functions that are of *digital option type*, thus are nonsmooth, may require more paths to achieve the same level of accuracy than for smooth functions. ◆

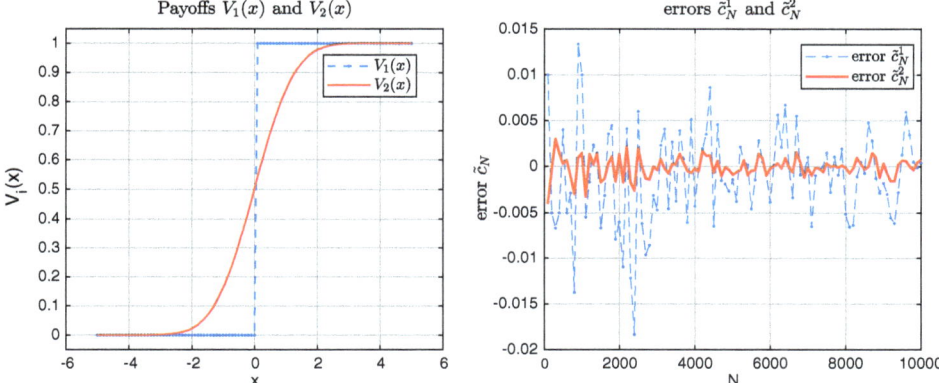

Figure 9.4: *Relation between smoothness of a function and the corresponding Monte Carlo convergence.*

9.1.2 Path simulation of stochastic differential equations

The general SDE involving a state dependent drift term $\bar{\mu}(t, X(t))$ and volatility term $\bar{\sigma}(t, X(t))$ is defined by,

$$dX(t) = \bar{\mu}(t, X(t))dt + \bar{\sigma}(t, X(t))dW(t), \quad t > t_0, \qquad (9.11)$$

and its solution is given by,

$$X(T) = x_0 + \int_{t_0}^{T} \bar{\mu}(t, X(t))dt + \int_{t_0}^{T} \bar{\sigma}(t, X(t))dW(t). \qquad (9.12)$$

Generally, when both sides of (9.12) contain the term $X(t)$, the SDE needs to be solved numerically. Only for a few standard processes, an explicit solution is available.

For this purpose, we define the equidistant grid in time-wise direction between $[0, T]$, as explained earlier, and obtain, for each time point t_i,

$$x_{i+1} = x_i + \int_{t_i}^{t_{i+1}} \bar{\mu}(t, X(t))dt + \int_{t_{i-1}}^{t_i} \bar{\sigma}(t, X(t))dW(t), \qquad (9.13)$$

where $x_i = X(t_i)$. Numerical integration schemes can be employed to approximate the integrals in (9.13). The order of convergence of such an integration scheme is defined as follows:

> **Definition 9.1.1 (Convergence)** Denote by x_m the approximation for $X(T)$, where Δt is the time step size, and m corresponds to the last term in the time discretization, $t_i = i \cdot T/m$, $i = 0, \ldots, m$. Then, the approximation x_m converges in a strong sense to $X(T)$, with order $\alpha > 0$, if
>
> $$\epsilon^s(\Delta t) := \mathbb{E}^{\mathbb{Q}}[|x_m - X(T)|] = O(h^\alpha).$$
>
> For a sufficiently smooth function $g(\cdot)$, the approximation x_m converges in a weak sense to $X(T)$, with respect to $g(\cdot)$, with order $\beta > 0$, if
>
> $$\epsilon^w(\Delta t) := \left|\mathbb{E}^{\mathbb{Q}}[g(x_m)] - \mathbb{E}^{\mathbb{Q}}[g(X(T))]\right| = O(h^\beta).$$
>
> In other words, a numerical integration method converges in a strong sense, if the asset prices converge, and weak convergence implies a convergent approximation of the probability distribution of $X(T)$, for a given time T. The convergence then concerns only the marginal distribution of $X(T)$. ◂

9.2 Stochastic Euler and Milstein schemes

In this section, we discuss in more detail the stochastic Euler and Milstein discretization schemes for SDEs. Next to the derivations of these methods, we also focus on their convergence properties.

9.2.1 Euler scheme

The *stochastic Euler, or Euler-Maruyama, scheme* is a basic numerical integration method for computing stochastic integrals as defined in (9.13). This scheme approximates (9.13), as follows:

> $$x_{i+1} \approx x_i + \int_{t_i}^{t_{i+1}} \bar{\mu}(t_i, x_i) dt + \int_{t_i}^{t_{i+1}} \bar{\sigma}(t_i, x_i) dW(t).$$
>
> So, in the Euler scheme the integrands are approximated by their values *at the left-side boundary* of the integration interval. Or, each integrand is approximated (or *frozen*) at its initial value, giving
>
> $$\begin{aligned} x_{i+1} &\approx x_i + \bar{\mu}(t_i, x_i)(t_{i+1} - t_i) + \bar{\sigma}(t_i, x_i)\left(W(t_{i+1}) - W(t_i)\right) \\ &\stackrel{d}{=} x_i + \bar{\mu}(t_i, x_i)\Delta t + \bar{\sigma}(t_i, x_i) W(\Delta t), \end{aligned} \qquad (9.14)$$
>
> where x_i denotes the approximation of $X(t_i)$ and $x_0 = X(t_0)$.

The quantity $W(\Delta t)$ is normally distributed with mean zero and variance Δt, i.e. $W(\Delta t) \sim \mathcal{N}(0, \Delta t)$. A sample from this distribution is obtained by computing

$Z \cdot \sqrt{\Delta t}$, where Z is drawn from a standard normal distribution, i.e. $Z \sim \mathcal{N}(0,1)$. The discretized SDE in (9.14) can therefore be simulated, by means of

$$x_{i+1} \approx x_i + \bar{\mu}(t_i, x_i)\Delta t + \bar{\sigma}(t_i, x_i)\sqrt{\Delta t}Z. \tag{9.15}$$

Example 9.2.1 (Euler scheme for the GBM process) Within the Monte Carlo simulation, with many paths, the Euler discretization for the ith timestep and j^{th} path, reads:

$$s_{i+1,j} \approx s_{i,j} + rs_{i,j}\Delta t + \sigma s_{i,j}\left(W_{i+1,j} - W_{i,j}\right), \tag{9.16}$$

with $\Delta t = t_{i+1} - t_i$, for any $i = 1, \ldots m$, $s_0 = S(t_0)$ and $j = 1, \ldots N$.

The GBM process with dynamics, $\mathrm{d}S(t) = rS(t)\mathrm{d}t + \sigma S(t)\mathrm{d}W(t)$, has as exact solution in the time interval $[t_i, t_{i+1}]$,

$$S(t_{i+1}) = S(t_i)\exp\left((r - \frac{1}{2}\sigma^2)\Delta t + \sigma\left(W(t_{i+1}) - W(t_i)\right)\right). \tag{9.17}$$

As an example, we set $S(t_0) = 50$, $r = 0.06$, $\sigma = 0.3$, $T = 1$ and determine the strong convergence error at the maturity time T, i.e.

$$\epsilon^s(\Delta t) = \frac{1}{N}\sum_{j=1}^{N}|S_j(T) - s_{m,j}| = \frac{1}{N}\sum_{j=1}^{N}|S(t_0)e^{(r-\frac{1}{2}\sigma^2)T + \sigma W_{m,j}} - s_{m,j}|,$$

for different Δt-values, and also the weak convergence error:

$$\epsilon^w(\Delta t) = \left|\frac{1}{N}\sum_{j=1}^{N}S_j(T) - \frac{1}{N}\sum_{j=1}^{N}s_{m,j}\right|$$

$$= \left|S(t_0)\frac{1}{N}\sum_{j=1}^{N}e^{(r-\frac{1}{2}\sigma^2)T + \sigma W_{m,j}} - \frac{1}{N}\sum_{j=1}^{N}s_{m,j}\right|,$$

where m stands for time $t_m \equiv T$ and the index j indicates the path number at which solutions from (9.17) and (9.16) are evaluated. Note that in this experiment it is crucial to use the *same Brownian motion* for both equations (9.17) and (9.16). If the Brownian motions wouldn't be the same, we wouldn't be able to measure the strong convergence, as the random paths would be different.

For different mesh widths Δt, the results are presented in Figure 9.5. We postulate that

$$\epsilon^s(\Delta t) \leq C \cdot (\Delta t)^{\frac{1}{2}} = \mathcal{O}((\Delta t)^{\frac{1}{2}}).$$

At the right-hand side of Figure 9.5, it becomes clear that the Euler scheme converges in a strong sense with order $\frac{1}{2}$ and in a weak sense with order 1, for the Black-Scholes model. For a proof of this numerical finding, we refer to [Kloeden

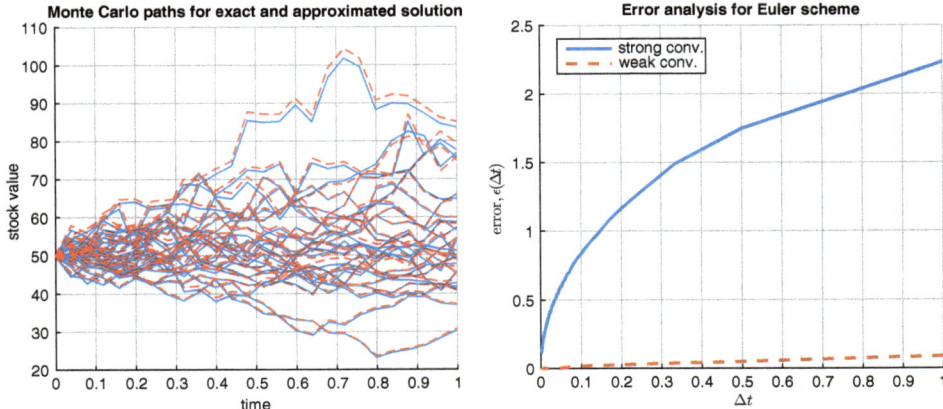

Figure 9.5: *Left: generated paths with exact simulation (9.17) versus Euler approximation (9.16); Right: error against value of time step Δt for the Euler discretization.* ♦

and Platen, 1992, 1995, 1999]. The differences in the paths between the exact scheme (9.17) and the Euler approximation (9.16) are shown at the left-hand side picture. ♦

In the case of deterministic differential equations, one may employ the *Taylor expansion* to define discretizations by which we may obtain a higher order of convergence. For stochastic differential equations a similar approach is available, which is based on the stochastic Taylor expansion, or the so-called *Itô-Taylor expansion*. The stochastic Euler approximation is based on the first two terms of this expansion.

For the Itô process SDE, $dX(t) = \bar{\mu}(t, X(t))dt + \bar{\sigma}(t, X(t))dW(t)$, the discretization under the *Milstein scheme* is obtained by adding a *third term* to the Euler discretization, i.e.,

$$x_{i+1} = x_i + \int_{t_i}^{t_{i+1}} \bar{\mu}(t, X(t))dt + \int_{t_i}^{t_{i+1}} \bar{\sigma}(t, X(t))dW(t)$$

$$\approx x_i + \int_{t_i}^{t_{i+1}} \bar{\mu}(t_i, x_i)dt + \int_{t_i}^{t_{i+1}} \bar{\sigma}(t_i, x_i)dW(t)$$

$$+ \frac{1}{2}\bar{\sigma}(t_i, x_i)(W^2(\Delta t) - \Delta t)\frac{\partial \bar{\sigma}}{\partial x}(t_i, x_i),$$

with $x_0 = X(t_0)$.

In the case of the risk neutral GBM process, with $\bar{\mu}^{\mathbb{Q}}(t, S(t)) = rS(t)$ and $\bar{\sigma}(t, S(t)) = \sigma S(t)$, the discretization reads,

$$s_{i+1} \approx s_i + rs_i \Delta t + \sigma s_i \left(W(t_{i+1}) - W(t_i)\right)$$
$$+ \frac{1}{2}\sigma^2 s_i \left((W(t_{i+1}) - W(t_i))^2 - \Delta t\right)$$
$$\stackrel{d}{=} s_i + rs_i \Delta t + \sigma s_i \sqrt{\Delta t} Z + \frac{1}{2}\sigma^2 s_i \left(\Delta t Z^2 - \Delta t\right).$$

The additional correction term in the Milstein scheme improves the *speed of convergence* compared to the Euler discretization for scalar SDEs. For the Black-Scholes model, as well as for the local volatility model, this scheme exhibits both a strong and weak convergence of order 1. For a proof we refer to [Kloeden and Platen, 1992, 1995, 1999].

Although the Milstein Scheme is definitely manageable in the one-dimensional case, *its extension to multi-dimensional SDE problems is far from trivial.*

9.2.2 Milstein scheme: detailed derivation

We will give a detailed derivation of the Milstein discretization scheme here, and start with the general Itô process, as in (9.11) and its discretization (9.13). For notational convenience, we will sometimes use the following notation in this section,

$$a_1(t) \equiv \bar{\mu}(t) := \bar{\mu}(t, X(t)), \text{ and } a_2(t) \equiv \bar{\sigma}(t) := \bar{\sigma}(t, X(t)).$$

As $\bar{\mu}(t)$ and $\bar{\sigma}(t)$ both depend on $X(t)$ as in (9.12), they are stochastic variables and we may therefore derive their dynamics.

$$da_k(t) = \frac{\partial a_k}{\partial t} dt + \frac{\partial a_k}{\partial X} dX(t) + \frac{1}{2}\frac{\partial^2 a_k}{\partial X^2}(dX(t))^2, \quad k = 1, 2. \quad (9.18)$$

By integrating the processes (9.18), for $k = 1$ and $k = 2$ in the time interval $[t_i, t]$ with $t \leq t_{i+1}$, we find:

$$a_k(t) = a_k(t_i) + \int_{t_i}^{t} \underbrace{\left(\frac{\partial a_k}{\partial z} + \bar{\mu}(z)\frac{\partial a_k}{\partial X} + \frac{1}{2}\bar{\sigma}^2(z)\frac{\partial^2 a_k}{\partial X^2}\right)}_{A_k(z)} dz$$
$$+ \int_{t_i}^{t} \bar{\sigma}(z)\frac{\partial a_k}{\partial X} dW(z), \quad (9.19)$$

where we also defined the quantities $A_k(z), k = 1, 2$. After substitution of (9.19) into Equation (9.13), we have,

$$x_{i+1} = x_i + \int_{t_i}^{t_{i+1}} \left[\bar{\mu}(t_i) + \int_{t_i}^{t} A_1(z)dz + \int_{t_i}^{t} \bar{\sigma}(z)\frac{\partial \bar{\mu}(z)}{\partial X} dW(z)\right] dt$$
$$+ \int_{t_i}^{t_{i+1}} \left[\bar{\sigma}(t_i) + \int_{t_i}^{t} A_2(z)dz + \int_{t_i}^{t} \bar{\sigma}(z)\frac{\partial \bar{\sigma}(z)}{\partial X} dW(z)\right] dW(t).$$

Neglecting all terms of higher order, i.e. $(dt)^2 = 0$ and $dt \cdot dW(t) = 0$, as in Table 2.1 in Chapter 2, the dynamics simplify to

$$x_{i+1} = x_i + \int_{t_i}^{t_{i+1}} \bar{\mu}(t_i)dt + \int_{t_i}^{t_{i+1}} \bar{\sigma}(t_i)dW(t)$$
$$+ \int_{t_i}^{t_{i+1}} \left(\int_{t_i}^{t} \bar{\sigma}(z) \frac{\partial \bar{\sigma}(z)}{\partial X} dW(z) \right) dW(t). \quad (9.20)$$

The first two integrals in the expression above can be simplified, i.e.,

$$x_{i+1} = x_i + \bar{\mu}(t_i)\Delta t + \bar{\sigma}(t_i)\left(W(t_{i+1}) - W(t_i)\right)$$
$$+ \int_{t_i}^{t_{i+1}} \left(\int_{t_i}^{t} \bar{\sigma}(z) \frac{\partial \bar{\sigma}(z)}{\partial X} dW(z) \right) dW(t). \quad (9.21)$$

For the inner integral in the last term, we apply the Euler discretization, as follows

$$\int_{t_i}^{t} \bar{\sigma}(z) \frac{\partial \bar{\sigma}(z)}{\partial X} dW(z) \approx \bar{\sigma}(t_i) \frac{\partial \bar{\sigma}}{\partial X}(t_i) \left(W(t) - W(t_i)\right).$$

As the expression under the outer integral in (9.21) is now constant in the interval $[t_i, t_{i+1}]$, it is placed outside the integral, giving

$$\bar{\sigma}(t_i) \frac{\partial \bar{\sigma}}{\partial X}(t_i) \int_{t_i}^{t_{i+1}} \left(W(t) - W(t_i)\right) dW(t). \quad (9.22)$$

The stochastic integral at the right side can be simplified with the help of the solution in Example 9.1.2, i.e.,

$$\int_{t_i}^{t_{i+1}} \left(W(t) - W(t_i)\right) dW(t) = \frac{1}{2} \left(W(t_{i+1}) - W(t_i)\right)^2 - \frac{1}{2}\Delta t.$$

After substitution, Equation (9.22) reads,

$$\int_{t_i}^{t_{i+1}} \left(\int_{t_i}^{t_{i+1}} \bar{\sigma}(z) \frac{\partial \bar{\sigma}(z)}{\partial X} dW(z) \right) dW(t) \approx \frac{1}{2} \bar{\sigma}(t_i) \frac{\partial \bar{\sigma}}{\partial X}(t_i) \left((W(\Delta t))^2 - \Delta t\right).$$

The following discrete dynamics then result:

$$x_{i+1} \approx x_i + \bar{\mu}(t_i, x_i)\Delta t + \bar{\sigma}(t_i, x_i)W(\Delta t)$$
$$+ \frac{1}{2} \bar{\sigma}(t_i, x_i) \left(W^2(\Delta t) - \Delta t\right) \frac{\partial \bar{\sigma}}{\partial X}(t_i, x_i),$$

where we used that $W(t_{i+1}) - W(t_i) \stackrel{d}{=} \sqrt{\Delta t} Z$, with $Z \sim \mathcal{N}(0,1)$. This is the well-known Milstein discretization!

Example 9.2.2 (Test case for the Milstein scheme) We set again, $S(t_0) = 50$, $r = 0.06$, $\sigma = 0.3$, $T = 1$, and measure the strong and weak

Monte Carlo Simulation 271

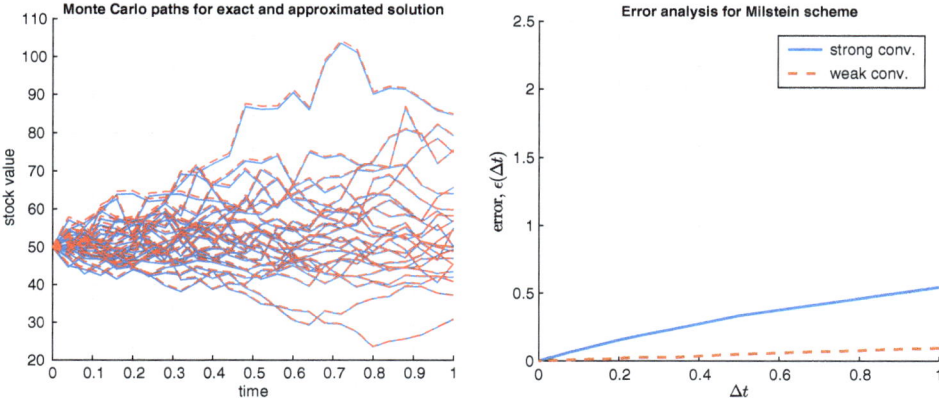

Figure 9.6: *Left: generated paths exact vs. approximation; Right: error against value of step size Δt for the Milstein discretization*

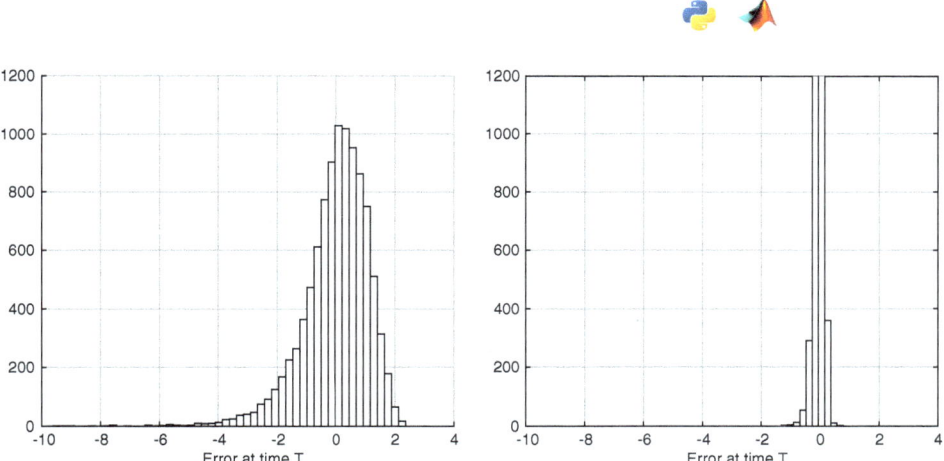

Figure 9.7: *Error comparison between Euler (left) and Milstein (right) discretization schemes, for $T = 1$ with $\Delta t = 0.1$.*

convergence, just as in Example 9.2.1. The results are presented in Figure 9.6. The distribution of the errors, generated by the Euler and Milstein schemes, is compared in the Figure 9.7. Based on the results in Figures 9.6, we confirm that

$$\epsilon^s(\Delta t) \leq C \cdot \Delta t = \mathcal{O}(\Delta t),$$

for the Milstein scheme in this test.

Table 9.1: *Call option prices in dependence of the number of Monte Carlo paths for the Euler and Milstein discretizations.*

Type:	N	100	1000	5000	10000	50000	100000	BS
Europ. call	Euler	0.7709	0.7444	0.7283	0.7498	0.7328	0.7356	0.7359
	Milstein	0.7690	0.7438	0.7283	0.7497	0.7327	0.7356	
Digital call	Euler	2.8253	2.4062	2.4665	2.4411	2.4502	2.4469	2.4483
	Milstein	2.8253	2.4062	2.4674	2.4406	2.4515	2.4462	

If we compare the Euler with the Milstein trajectories, we cannot see any significant differences by mere eye, especially not for small values of the volatility. ♦

Example 9.2.3 (Analytic versus Monte-Carlo method for BS model)
We take $S(t_0) = 5$, $\sigma = 0.3$, $r = 0.06$, $T = 1$, the number of Monte Carlo time steps $m = 1000$ (so that $\Delta t = 1/1000$) and compare the obtained call option values with the analytic solutions from the Black-Scholes formula, for the strike price $K = S(t_0)$, i.e. $V_c(t_0, S(t_0)) = 0.7359$. Table 9.1 shows that both methods require a similar number of paths to achieve satisfactory pricing results. This is an indication that the Euler and Milstein schemes exhibit the *same order of weak convergence*.

We display in the same table the convergence for the digital option, with its exact solution in Equation (3.26). The Black-Scholes digital option price is equal to 2.4483. Also here a very fine convergence is achieved in the number of Monte Carlo paths N. ♦

Monte Carlo simulation with bivariate samples

Multi-dimensional samples can also easily be generated by means of a Monte Carlo simulation. This can most easily be done in the formulation of a system of SDEs with the independent Brownian motions, where random numbers can be drawn independently of each other.

Here, a 2D experiment is presented, where bivariate Monte Carlo samples are shown, with the corresponding (1D) marginal distributions, in Figure 9.8. In the experiment, for illustration purposes, 5000 Monte Carlo samples have been generated with the following set of parameters, $\mathbf{X} = [X, Y]^T$, with:

$$\mathbf{X} \sim \mathcal{N}\left(\begin{bmatrix} 0 \\ 0 \end{bmatrix}, \begin{bmatrix} \sigma_1^2 & \rho\sigma_1\sigma_2 \\ \rho\sigma_1\sigma_2 & \sigma_2^2 \end{bmatrix}\right), \text{ with } \sigma_1 = 1, \sigma_2 = 0.5, \rho = 0.8.$$

Monte Carlo simulation and jumps

Here, we comment on the discretization scheme in the case of a *jump diffusion process*. In Section 5.1 we have studied the jump diffusion asset dynamics,

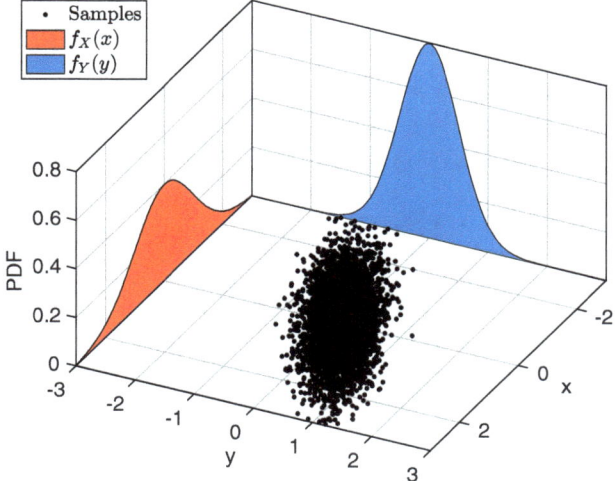

Figure 9.8: *Monte Carlo samples (black dots) and the corresponding marginal distributions.*

written down in Equation (5.10). Under the log-transformation, $X(t) = \log S(t)$, the dynamics were given by, see also Equation (5.11),

$$dX(t) = \left(r - \xi_p \mathbb{E}[e^J - 1] - \frac{1}{2}\sigma^2\right)dt + \sigma dW^{\mathbb{Q}}(t) + J dX_{\mathcal{P}}^{\mathbb{Q}}(t). \quad (9.23)$$

By assuming independence between the jump magnitude J, Poisson process $X_{\mathcal{P}}(t)$ and Brownian motion $W(t)$, the solution of (9.23) is given by

$$X(T) = X_0 + \left(r - \xi_p \mathbb{E}[e^J - 1] - \frac{1}{2}\sigma^2\right)T + \sigma W^{\mathbb{Q}}(T) + \sum_{k=1}^{X_{\mathcal{P}}^{\mathbb{Q}}(T)} J$$

$$= X_0 + \left(r - \xi_p \mathbb{E}[e^J - 1] - \frac{1}{2}\sigma^2\right)T + \sigma W^{\mathbb{Q}}(T) + J X_{\mathcal{P}}^{\mathbb{Q}}(T).$$

Using $S(t) = \exp(X(t))$, we thus find,

$$S(T) = S(t_0)\exp\left(\left(r - \xi_p \mathbb{E}[e^J - 1] - \frac{1}{2}\sigma^2\right)T + \sigma W^{\mathbb{Q}}(T) + J X_{\mathcal{P}}^{\mathbb{Q}}(T)\right)$$

$$= S(t_0)\exp\left(\left(r - \xi_p \mathbb{E}[e^J - 1] - \frac{1}{2}\sigma^2\right)T + \sigma W^{\mathbb{Q}}(T)\right)\prod_{k=1}^{X_{\mathcal{P}}^{\mathbb{Q}}(T)} e^J.$$

Discretizing the process in (9.23) by the Euler discretization, gives us,

$$x_{i+1} \approx x_i + \left(r - \xi_p \mathbb{E}[e^J - 1] - \frac{1}{2}\sigma^2\right)\Delta t + \sigma\left(W^{\mathbb{Q}}(t_{i+1}) - W^{\mathbb{Q}}(t_i)\right)$$
$$+ J\left(X_{\mathcal{P}}^{\mathbb{Q}}(t_{i+1}) - X_{\mathcal{P}}^{\mathbb{Q}}(t_i)\right).$$

The increments $X_{\mathcal{P}}^{\mathbb{Q}}(t_{i+1}) - X_{\mathcal{P}}^{\mathbb{Q}}(t_i)$ represent the number of jumps that will take place in the time interval (t_i, t_{i+1}). As defined in Equation (5.2), the number of jumps in an interval Δt will follow a Poisson distribution with parameter ξ_p. Depending on the distribution of the jump magnitude variable J, the expectation in the drift will be different, but it will be a deterministic quantity.

Recall that $X_{\mathcal{P}}^{\mathbb{Q}}(t)$ is a Poisson process (see Definition 5.1.2), so that the simulation of the process is given by:

$$x_{i+1} \stackrel{d}{=} x_i + \left(r - \xi_p \mathbb{E}[e^J - 1] - \frac{1}{2}\sigma^2\right)\Delta t + \sigma\sqrt{\Delta t}Z + JX_{\mathcal{P}}, \quad (9.24)$$

where $Z \sim \mathcal{N}(0,1)$, and $X_{\mathcal{P}}$ is a Poisson distributed random variable with

$$\mathbb{Q}[X_{\mathcal{P}} = k] = e^{-\xi_p \Delta t}(\xi_p \Delta t)^k / k!$$

Summarizing, a Monte Carlo simulation for a jump diffusion process based on the Euler discretization scheme is certainly feasible.

Remark 9.2.1 (Jumps and the number of Monte Carlo asset paths)
Very often jump processes have relatively small jump intensities, meaning that jumps are rare to occur. This implies that within a Monte Carlo simulation very many Monte Carlo paths need to be generated to get decent statistics, regarding the jump occurrence.

When a Monte Carlo simulation for GBM would require N paths to get accurate and satisfactory results, it is not uncommon to require $10N$ paths, when rarely occurring jumps are added to the stochastic process. ▲

9.3 Simulation of the CIR process

Standard discretization schemes, such as the Euler and Milstein schemes, are well-suited for simulating a wide range of stochastic processes. However, they typically do not perform well for processes that are, by definition, positive and have a probability mass around zero, as we will discuss in the subsections to follow.

9.3.1 Challenges with standard discretization schemes

A typical example of a process with probability mass around zero is the CIR process, which was discussed to model the variance for the Heston stochastic volatility model in Chapter 8, with the following dynamics,

$$\boxed{dv(t) = \kappa(\bar{v} - v(t))dt + \gamma\sqrt{v(t)}dW(t), \quad v(t_0) > 0.} \quad (9.25)$$

Monte Carlo Simulation

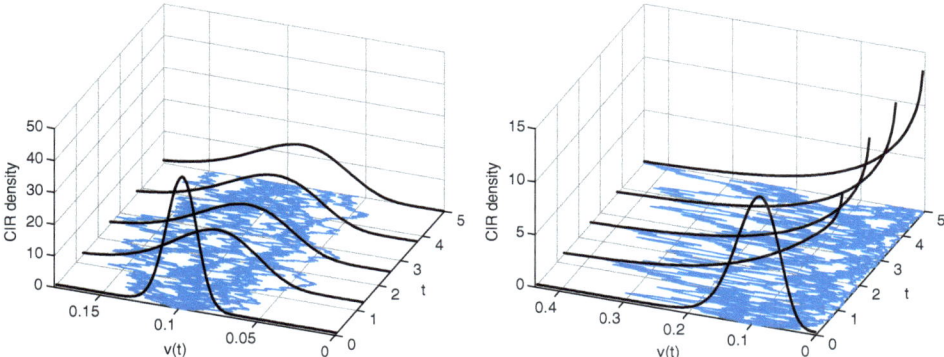

Figure 9.9: *Paths and the corresponding PDF for the CIR process (9.25) in the cases where the Feller condition is satisfied and is not satisfied. Simulations were performed with $\kappa = 0.5$, $v_0 = 0.1$, $\bar{v} = 0.1$. Left: $\gamma = 0.1$; Right: $\gamma = 0.35$.*

It is well-known that if the Feller condition, $2\kappa\bar{v} > \gamma^2$, is satisfied, the process $v(t)$ cannot reach zero, and if this condition does not hold the origin is accessible and strongly reflecting. In both cases, the $v(t)$ process cannot become negative. In Figure 9.9, some paths for these two cases are presented. For the case where the Feller condition is not satisfied, the paths are *accumulating* around the zero region.

The nonnegativity problem becomes apparent when a standard discretization is employed. If we apply, for example, the Euler discretization to the process in (9.25), i.e.,

$$v_{i+1} = v_i + \kappa(\bar{v} - v_i)\Delta t + \gamma\sqrt{v_i}\sqrt{\Delta t}Z.$$

and assume that $v_i > 0$, we may calculate the *probability* that a next realization, v_{i+1}, becomes negative, i.e. $\mathbb{P}[v_{i+1} < 0]$.

$$\mathbb{P}[v_{i+1} < 0 | v_i > 0] = \mathbb{P}[v_i + \kappa(\bar{v} - v_i)\Delta t + \gamma\sqrt{v_i \Delta t}Z < 0 | v_i > 0]$$
$$= \mathbb{P}[\gamma\sqrt{v_i \Delta t}Z < -v_i - \kappa(\bar{v} - v_i)\Delta t | v_i > 0],$$

which equals,

$$\mathbb{P}[v_{i+1} < 0 | v_i > 0] = \mathbb{P}\left[Z < -\frac{v_i + \kappa(\bar{v} - v_i)\Delta t}{\gamma\sqrt{v_i \Delta t}} \middle| v_i > 0\right] > 0.$$

Since Z is a normally distributed random variable, it is unbounded. Therefore the probability of v_i becoming negative, is positive under the Euler discretization, implying $\mathbb{P}[v_{i+1} < 0 | v_i > 0] > 0$.

Especially when the Feller condition is not satisfied most of the probability mass of the variance is concentrated around zero. This case gives rise to a high probability of the variance becoming negative when using the Euler discretization scheme. This is clearly undesirable. A similar exercise can be defined for the Milstein scheme.

> Application of the Euler scheme to the CIR-type process may lead to undesired and unrealistic path realizations. Whereas the mean-reverting CIR process is guaranteed to be nonnegative, the Euler discretization is not.

Example 9.3.1 (Bounded Jacobi process) As another process with similar interesting properties, we discuss *the bounded Jacobi process*, which is often used for modeling variables for which the paths need to be bounded in a certain region. The bounded Jacobi process is for example used to model a *stochastic correlation*. All realizations should then be bounded in the interval $[-1, 1]$, and the dynamics are given by,

$$\mathrm{d}\rho(t) = \kappa_\rho \left(\mu_\rho - \rho(t)\right) \mathrm{d}t + \gamma_\rho \sqrt{1 - \rho^2(t)} \mathrm{d}W(t). \tag{9.26}$$

Note that the Jacobi process has two boundaries, at $\rho(t) = -1$ and at $\rho(t) = 1$. Given the following parameter constraint,

$$\kappa_\rho > \max\left(\frac{\gamma_\rho^2}{1 - \mu_\rho}, \frac{\gamma_\rho^2}{1 + \mu_\rho}\right), \tag{9.27}$$

the boundaries $\rho(t) = -1$ or $\rho(t) = 1$ are not attainable.

By the Euler discretization, the discrete process for (9.26) is given by:

$$\rho_{i+1} = \rho_i + \kappa_\rho \left(\mu_\rho - \rho_i\right) \Delta t + \gamma_\rho \sqrt{1 - \rho_i^2} \sqrt{\Delta t} Z, \tag{9.28}$$

again with $Z \sim \mathcal{N}(0, 1)$. Similar to the example for the CIR process, it is now possible to show that, due to the Euler discretization,

$$\mathbb{P}[|\rho_{i+1}| > 1 \mid |\rho_i| < 1] > 0.$$

We will not discuss the concept of stochastic correlation in more detail in this book. ♦

9.3.2 Taylor-based simulation of the CIR process

As the mean-reverting process is a continuous process, it may only become negative when it first reaches zero. Notice that equation (9.25) is deterministic for $v(t) = 0$. As soon as the variance reaches zero, it will immediately be positive afterwards. In contrast, the Euler discretization is not continuous. In other words, the discrete form can become negative immediately.

In this section, we discuss possible solutions to this problem.

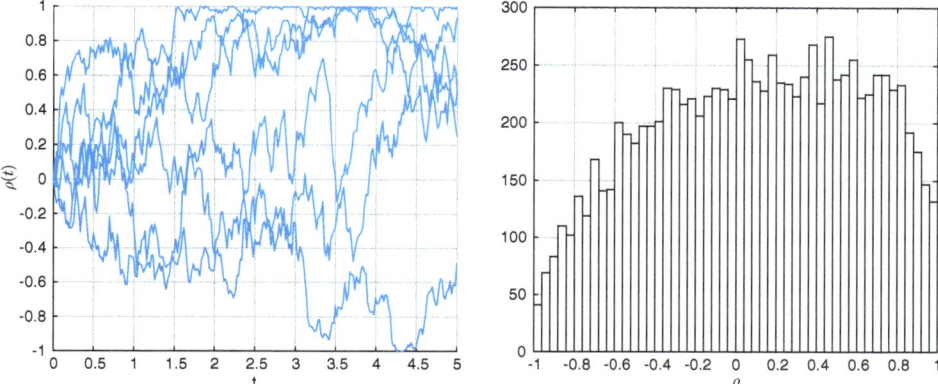

Figure 9.10: *Monte Carlo Euler discretization paths and a histogram at $T = 5$ for the bounded Jacobi process with $\rho(t_0) = 0$, $\kappa = 0.25$, $\mu_\rho = 0.5$ and $\gamma_\rho = 0.6$.*

Truncated Euler scheme

The truncated Euler scheme, as the name says, is based on the previously introduced Euler discretization scheme. In order to prevent the process to cross the axis, we need to deal with possible negative path realizations v_{i+1}. The *truncated Euler scheme* offers a basic approach to handle this problem (discussed in [Lord et al., 2010]).

It can be summarized, as follows,

$$\begin{cases} \hat{v}_{i+1} = v_i + \kappa(\bar{v} - v_i)\Delta t + \gamma\sqrt{v_i \Delta t}Z, \\ v_{i+1} = \max(\hat{v}_{i+1}, 0). \end{cases}$$

Although the scheme certainly provides paths that are nonnegative, the accuracy of this scheme is parameter-dependent, meaning that, when the Feller condition is not satisfied and the density should accumulate around zero, the *adjusted paths*, due to the truncation, may be highly biased. In essence, by the truncation a different process than the original CIR process is represented numerically. When truncation takes place for many Monte Carlo paths, the accuracy may be limited as the discretization bias could be high.

Reflecting Euler scheme

One of the important properties of the CIR process is that the origin is attainable, however, it is *not* an absorbing boundary. This means that the Monte Carlo paths may *reach* the boundary $v = 0$, but they cannot stay at the boundary – they should immediately move away from it. When using the truncated Euler scheme from above, using $v_{i+1} = \max(\hat{v}_{i+1}, 0)$, the paths that attain negative values

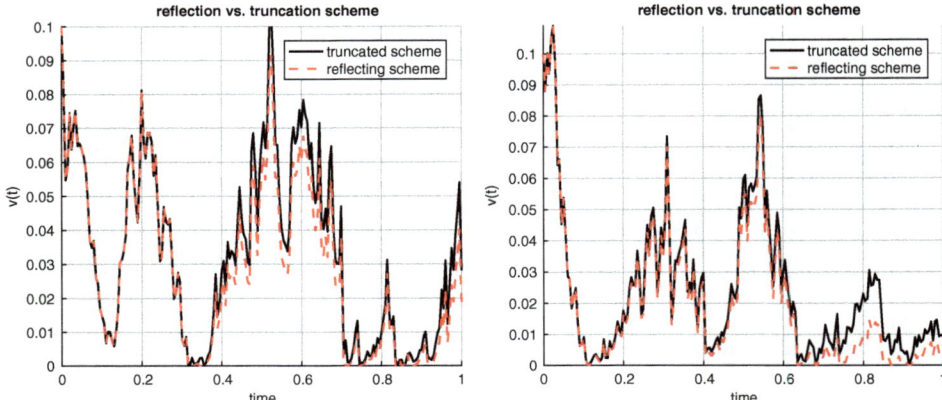

Figure 9.11: *Comparison of the reflecting and truncated Euler schemes for two specific Monte Carlo paths.*

are projected at the origin. Another possible modification for the simulation of the CIR process, which is particularly useful when the Feller condition is not satisfied, is to use the so-called *reflecting principle*, where the variance paths are forced to move upwards again.

The reflecting scheme is given by the following adjustment of the Euler scheme,

$$\begin{cases} \hat{v}_{i+1} = v_i + \kappa(\bar{v} - v_i)\Delta t + \gamma\sqrt{v_i \Delta t}Z, \\ v_{i+1} = |\hat{v}_{i+1}|. \end{cases} \quad (9.29)$$

In Figure 9.11, some paths that were generated by the reflecting and truncated Euler schemes are compared. The paths that are generated by both schemes are essentially the same, until they reach the boundary at $v(t) = 0$. After hitting the origin, the reflecting scheme will have paths that are above those that are generated by the truncation scheme. Notice that by the reflecting scheme we however *do not improve* the quality of the Euler scheme.

9.3.3 Exact simulation of the CIR model

A different approach for simulating paths from the CIR process takes into account the fact that variance process $v(t)$, follows the *noncentral chi-squared distribution*. Details of the noncentral chi-squared distribution have been given already in Definition 8.1.1 in the context of the Heston stochastic volatility model. Conditional on state $v(s)$, $s < t$, the distribution at time t is given by,

$$v(t)|v(s) \sim \bar{c}(t,s) \cdot \chi^2(\delta, \bar{\kappa}(t,s)), \quad (9.30)$$

with

$$\bar{c}(t,s) = \frac{\gamma^2}{4\kappa}\left(1 - e^{-\kappa(t-s)}\right), \quad \delta = \frac{4\kappa\bar{v}}{\gamma^2}, \quad \bar{\kappa}(t,s) = \frac{4\kappa e^{-\kappa(t-s)}}{\gamma^2(1 - e^{-\kappa(t-s)})}v(s).$$

Equation (9.30) may form the basis for an *exact simulation scheme* for the path realizations of the CIR process, as, for $i = 0, \ldots, m-1$,

$$\bar{c}(t_{i+1}, t_i) = \frac{\gamma^2}{4\kappa}\left(1 - e^{-\kappa(t_{i+1}-t_i)}\right),$$

$$\bar{\kappa}(t_{i+1}, t_i) = \frac{4\kappa e^{-\kappa(t_{i+1}-t_i)}}{\gamma^2(1 - e^{-\kappa(t_{i+1}-t_i)})}\boxed{v_i}.$$

$$\boxed{v_{i+1}} = \bar{c}(t_{i+1}, t_i)\chi^2(\delta, \bar{\kappa}(t_{i+1}, t_i)),$$

with a constant parameter $\delta = 4\kappa\bar{v}/\gamma^2$, and some initial value $v(t_0) = v_0$.

By the simulation scheme above, one can simulate CIR paths without paying special attention to the Feller condition. The scheme relies however on an efficient sampling from the noncentral chi-squared distribution. Although this distribution is quite popular, many standard computational packages may not yet include a highly efficient sampling algorithm from it, particularly not when dealing with the case in which the Feller condition is not satisfied or in the case of very large scale simulations.

For this reason, in the next section, the *Quadratic Exponential (QE) scheme* is explained, which enables, by means of sophisticated approximations, an *efficient and robust* sampling scheme from the noncentral chi-squared distribution.

9.3.4 The Quadratic Exponential scheme

In [Andersen, 2008], the so-called Quadratic Exponential (QE) scheme was proposed for simulating the Heston variance process $v(t)$, which is driven by the CIR dynamics (9.25).

In essence, the QE scheme for sampling from $v(t)$ consists of two different sampling algorithms, and the switching between these algorithms is dependent on the parameter values of the CIR process. Two different cases are considered: one for which the density of the process $v(t)$ *is far* from the zero region, and another where the distribution is *close to* the origin.

In the first case, the analysis is based on the observation that the noncentral chi-squared distribution can, for moderate or high levels of the noncentrality parameter $\bar{\kappa}$, be well approximated by a power function which is applied to a Gaussian variable [Patnaik, 1949], i.e.,

$$\boxed{v(t) \approx v_1(t) = a(b + Z_v)^2,} \qquad (9.31)$$

with some constants a and b and $Z_v \sim \mathcal{N}(0,1)$.

To determine the constants a and b, the *moment matching technique* is applied, by which the first two moments from the two distributions are equated. As explained in Definition 8.1.1, the mean $\bar{m}$ and variance $\bar{s}^2$ of the CIR-type process $v(t)$ are known, and given by

$$\begin{cases} \bar{m} := \mathbb{E}\left[v(t)|\mathcal{F}(0)\right] = \bar{c}(t,0)(\delta + \bar{\kappa}(t,0)), \\ \bar{s}^2 := \mathbb{V}\text{ar}\left[v(t)|\mathcal{F}(0)\right] = \bar{c}^2(t,0)(2\delta + 4\bar{\kappa}(t,0)), \end{cases} \quad (9.32)$$

with the parameters $\delta, \bar{c}(t,0), \bar{\kappa}(t,0)$ as in Definition 8.1.1. Equation (9.31) essentially indicates that $v(t)$ is distributed as "a" times the noncentral chi-squared distribution with one degree of freedom and with noncentrality parameter b^2, i.e.,

$$v_1(t) \sim a \cdot \chi^2(1, b^2),$$

implying the following expectation and variance,

$$\mathbb{E}\left[v_1(t)|\mathcal{F}(0)\right] = a(1 + b^2), \quad \mathbb{V}\text{ar}\left[v_1(t)|\mathcal{F}(0)\right] = 2a^2(1 + 2b^2). \quad (9.33)$$

By equating these two sets of equations, $\mathbb{E}[v(t)] = \mathbb{E}[v_1(t)]$ and $\mathbb{V}\text{ar}[v(t)] = \mathbb{V}\text{ar}[v_1(t)]$, we find:

$$\bar{m} = a(1 + b^2), \quad \bar{s}^2 = 2a^2(1 + 2b^2). \quad (9.34)$$

From the first equation, it follows that $a = \bar{m}/(1 + b^2)$, and the second equation gives:

$$b^4 - 2\frac{\bar{m}^2}{\bar{s}^2} + 1 + 2b^2\left(1 - 2\frac{\bar{m}^2}{\bar{s}^2}\right) = 0. \quad (9.35)$$

Setting $z := b^2$, we obtain a quadratic equation in z, as follows,

$$z^2 - 2\frac{\bar{m}^2}{\bar{s}^2} + 1 + 2z\left(1 - 2\frac{\bar{m}^2}{\bar{s}^2}\right) = 0, \quad (9.36)$$

which has solutions if $\bar{s}^2/\bar{m}^2 \leq 2$. Under this condition, the solution for b^2 is given by,

$$b^2 = 2\frac{\bar{m}^2}{\bar{s}^2} - 1 + \sqrt{2\frac{\bar{m}^2}{\bar{s}^2}}\sqrt{2\frac{\bar{m}^2}{\bar{s}^2} - 1} \geq 0, \quad \text{and} \quad a = \frac{\bar{m}}{1 + b^2}. \quad (9.37)$$

The condition $\bar{s}^2/\bar{m}^2 \leq 2$ corresponds to the case for which the density of the corresponding variance process $v(t)$ is far from zero (which can also be seen from the fact that the mean $\bar{m}$ is significantly larger than the variance $\bar{s}^2$). As shown in [Andersen, 2008], the representation in (9.31) appears very accurate for large values of $v(t)$, however, when the probability mass of $v(t)$ accumulates around zero the approximation in (9.31) may become inaccurate.

> When the representation in (9.31) does not provide a sufficiently accurate result, it is suggested to approximate the cumulative distribution function of $v(t)$ by the following *exponential function*,
>
> $$F_{v(t)}(x) := \mathbb{P}\left[v(t) \leq x\right] \approx F_{v_2(t)}(x) = c + (1-c)\left(1 - e^{-dx}\right), \quad \text{for} \quad x \geq 0,$$
>
> and the density by,
>
> $$f_{v(t)}(x) \approx f_{v_2(t)}(x) = c\delta(0) + d(1-c)e^{-dx}, \quad \text{for} \quad x \geq 0,$$
>
> with constant parameters $c \in [0,1]$ and $d \in \mathbb{R}$.

From the approximation above, it is again possible to derive the corresponding mean and the variance, i.e.,

$$\mathbb{E}[v_2(t)|\mathcal{F}(t_0)] = \frac{1-c}{d}, \quad \mathrm{Var}[v_2(t)|\mathcal{F}(t_0)] = \frac{1-c^2}{d^2}, \tag{9.38}$$

To determine the parameters c and d, we apply once more the moment matching technique, which results in the following system of equations:

$$\bar{m} = \frac{1-c}{d}, \quad \bar{s}^2 = \frac{1-c^2}{\left(\frac{1-c}{\bar{m}}\right)^2}.$$

By solving this system for c and d and taking into account the condition $c \in [0,1]$, we find:

$$c = \frac{\frac{\bar{s}^2}{\bar{m}^2} - 1}{\frac{\bar{s}^2}{\bar{m}^2} + 1}, \quad \text{and} \quad d = \frac{1-c}{\bar{m}} = \frac{2}{\bar{m}\left(\frac{\bar{s}^2}{\bar{m}^2} + 1\right)}. \tag{9.39}$$

It is clear that, in order to keep $c \in [0,1]$, we need to assume $\frac{\bar{s}^2}{\bar{m}^2} \geq 1$.

The main advantage of such an exponential representation lies in the fact that $F_{v_2(t)}(\cdot)$ is invertible, so that sampling can take place directly from $v_2(t)$ via

$$v_2(t) = F_{v_2(t)}^{-1}(u), \quad \text{for} \quad u \sim \mathcal{U}([0,1]). \tag{9.40}$$

The inverse of $F_{v_2(t)}(x)$ is given by,

$$F_{v_2(t)}^{-1}(u) := F_{v_2(t)}^{-1}(u; c, d) = \begin{cases} 0, & 0 \leq u \leq c, \\ d^{-1}\log\left(\frac{1-c}{1-u}\right), & c < u \leq 1, \end{cases} \tag{9.41}$$

with c and d as in (9.39).

> The description of the overall QE scheme can be concluded by a rule for switching between the two different sampling algorithms. The first sampling algorithm was well-defined for $\bar{s}^2/\bar{m}^2 \leq 2$, while the second was well-defined for $\bar{s}^2/\bar{m}^2 \geq 1$. As these two intervals for $\bar{s}^2/\bar{m}^2$ overlap, it is suggested to choose some point $a^* \in [1, 2]$. When $\bar{s}^2/\bar{m}^2 \leq a^*$, one should use the first sampling algorithm, using Equation (9.31), and when $\bar{s}^2/\bar{m}^2 > a^*$, the second one should be employed, using Equation (9.41).
>
> The combined algorithms, with (9.31) and (9.41), is called *the Quadratic Exponential scheme*.

Example 9.3.2 In this numerical experiment the performance of the QE scheme is analyzed and compared to exact simulation. We set $a^* = 1.5$ and consider two cases,

$$\bar{s}^2/\bar{m}^2 < a^*,$$

with parameters $\gamma = 0.2$, $\bar{v} = 0.05$, $v(0) = 0.3$ and $\kappa = 0.5$, and the second test case,

$$\bar{s}^2/\bar{m}^2 \geq a^*,$$

with parameters $\gamma = 0.6$, $\bar{v} = 0.05$, $v(0) = 0.3$ and $\kappa = 0.5$. After the samples have been generated with the help of the QE scheme, the corresponding CDFs for the two test cases are computed and presented in Figure 9.12.

A very good agreement between the CDF originating from the QE scheme and the one obtained by directly inverting the noncentral chi-square distribution can be observed. ♦

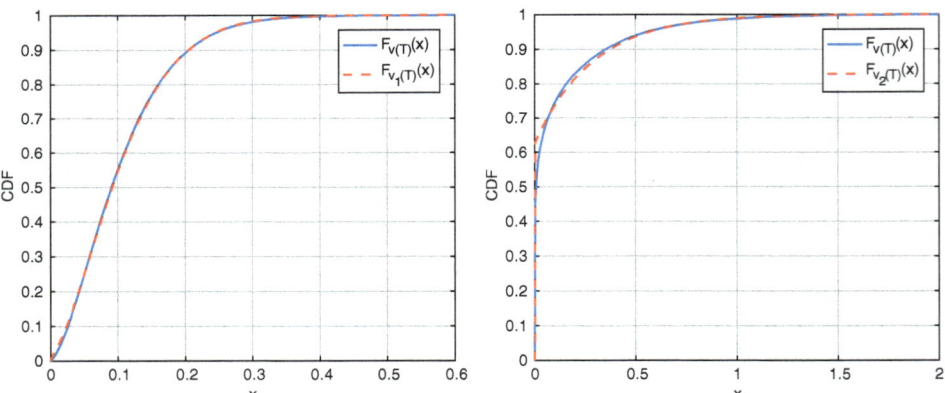

Figure 9.12: *QE scheme performance and comparison to the exact simulation in terms of the corresponding CDFs. Left: First test case. Right: Second test case.*

9.4 Monte Carlo scheme for the Heston model

Here, we generalize the concepts from the previous section towards a system of SDEs. We first discuss the concept of *conditional sampling* for a simplified SDE model, after which we will derive an efficient Monte Carlo scheme for the Heston stochastic volatility model. The conditional sampling is connected to a reformulation of the SDE system in which the *time integrated variance* process will appear.

9.4.1 Example of conditional sampling and integrated variance

The concept of conditional sampling from an integrated process is connected to the attempt to choose possibly *large time steps* Δt in a Monte Carlo simulation. In the Euler and Milstein schemes, convergence was expressed in terms of the time step Δt. Here, we consider a simulation approach for which, in principle, we don't need to take any intermediate time steps in the simulation of the stochastic process.

To clarify the idea, we consider, under the log transformation $X(t) = \log S(t)$, the following basic model:

$$\begin{cases} dX(t) = -\frac{1}{2}v(t)dt + \sqrt{v(t)}dW(t), \\ dv(t) = \sqrt{v(t)}dW(t), \end{cases} \quad (9.42)$$

where, for illustrative purposes, we assume that the processes are driven by the same Brownian motion $W(t)$.

To obtain samples for $X(T)$, at time T, we integrate the SDEs (9.42), and obtain:

$$\begin{cases} X(T) = X(s) - \frac{1}{2}\int_s^T v(t)dt + \int_s^T \sqrt{v(t)}dW(t), \\ v(T) = v(s) + \int_s^T \sqrt{v(t)}dW(t). \end{cases} \quad (9.43)$$

From the second equation in (9.43), it follows that

$$\int_s^T \sqrt{v(t)}dW(t) = v(T) - v(s),$$

which we may substitute in the first SDE in (9.43), to get,

$$X(T) = X(s) - v(s) + v(T) - \frac{1}{2}\int_s^T v(t)dt. \quad (9.44)$$

So, in order to obtain samples for $X(T)$, we need samples for $v(s)$, $v(T)$, but also samples from $\int_s^T v(t)dt$. The samples from $v(s)$ and $v(T)|v(s)$, that are needed

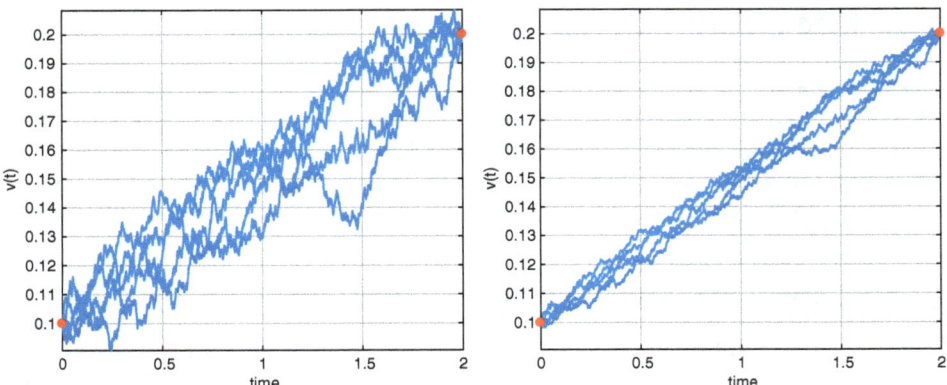

Figure 9.13: *The concept of sampling from $Y(t)$, given realizations for $v(s) = 0.1$ and $v(t)|v(s) = 0.2$, marked by the red dots, is based on the Brownian bridge technique.*

in (9.44), may be obtained by the techniques described in Sections 9.3.3 or 9.3.4. However, also samples from the *integrated variance, given the samples of $v(s)$ and $v(T)|v(s)$*, need to be determined.

For this, we consider the *integrated CIR process*, Y, as follows,

$$\mathrm{d}Y(t) = v(t)\mathrm{d}t, \quad Y(t_0) = 0, \tag{9.45}$$

which is equivalent to,

$$Y(t) = \int_{t_0}^{t} v(z)\mathrm{d}z, \quad t_0 < t. \tag{9.46}$$

For any two realizations, $v(s)$ and $v(t)|v(s)$, we need the distribution of $Y(t)$. Figure 9.13 sketches the main concept of conditional sampling.

The objective here is to derive the *conditional characteristic function*, for $Y(t)$, given the initial $v(s)$, and the end point $v(t)$, i.e., $\phi_{Y(t)|v(s),v(t)}(u;s,t)$. For the derivation of this conditional characteristic function of the integrated CIR process, we make use of the well-known fact that for the squared Bessel process of dimension δ, as defined by the following SDE,

$$\mathrm{d}y(t) = \delta \mathrm{d}t + 2\sqrt{|y(t)|}\mathrm{d}W_y(t), \quad y(t_0) = y_0, \tag{9.47}$$

with $\delta \geq 0$, the following conditional expectation is known analytically (see [Pitman and Yor, 1982]):

$$\mathbb{E}\left[e^{-\frac{c^2}{2}\int_s^t y(u)du}\Big|y(s),y(t)\right] = \frac{c\cdot\tau}{\sinh(c\cdot\tau)}$$

$$\times \exp\left(\frac{y(s)+y(t)}{2\tau}(1-c\cdot\tau\coth(c\cdot\tau))\right)$$

$$\times I_{\frac{\delta}{2}-1}\left(\frac{c\sqrt{y(s)y(t)}}{\sinh(c\cdot\tau)}\right) \Big/ I_{\frac{\delta}{2}-1}\left(\frac{\sqrt{y(s)y(t)}}{\tau}\right), \tag{9.48}$$

for some constant $c \in \mathbb{R}$ and $\tau = t - s$.

The squared Bessel process has been discussed in the previous chapter, in Equation (8.13). The expectation in Equation (9.48) can be used for deriving the *conditional expectation for the CIR process*, when we express the dynamics of the CIR process in terms of the Bessel process (9.47).

9.4.2 The integrated CIR process and conditional sampling

The aim here is to employ the known characteristic function (9.48) in the context of the Heston CIR variance process.

Integrating the Heston variance process (9.25) over a time interval $[s,t]$, results in,

$$v(t) = v(s) + \int_s^t \kappa(\bar{v}-v(z))\mathrm{d}z + \int_s^t \gamma\sqrt{v(z)}\mathrm{d}W(z). \tag{9.49}$$

The Bessel process in (9.47) has a volatility coefficient equal to 2, while in the CIR process (9.25) the volatility coefficient equals γ. We apply a *time change* (see Remark 8.1.1) to the last term in Equation (9.49), which can be expressed as:

$$\int_s^t \gamma\sqrt{v(z)}\mathrm{d}W(z) = 2\int_s^t \frac{\gamma}{2}\sqrt{v(z)}\mathrm{d}W(z)$$

$$= 2\int_s^t \sqrt{v\left(\frac{\gamma^2}{4}\frac{4z}{\gamma^2}\right)}\mathrm{d}W\left(\frac{\gamma^2 z}{4}\right).$$

The process in (9.49) can thus be written as,

$$v(t) = v(s) + \int_s^t \kappa\left[\bar{v}-v\left(\frac{\gamma^2 z}{4}\frac{4}{\gamma^2}\right)\right]\mathrm{d}z + 2\int_s^t \sqrt{v\left(\frac{\gamma^2 z}{4}\frac{4}{\gamma^2}\right)}\mathrm{d}W\left(\frac{\gamma^2 z}{4}\right).$$

With a new variable, $w = \frac{\gamma^2}{4}z$, so that $\mathrm{d}w = \frac{\gamma^2}{4}\mathrm{d}z$, we find,

$$v(t) = v(s) + \frac{4}{\gamma^2}\int_{\frac{\gamma^2}{4}s}^{\frac{\gamma^2}{4}t} \kappa\left[\bar{v}-v\left(\frac{4w}{\gamma^2}\right)\right]\mathrm{d}w + 2\int_{\frac{\gamma^2}{4}s}^{\frac{\gamma^2}{4}t} \sqrt{v\left(\frac{4w}{\gamma^2}\right)}\mathrm{d}W(w).$$

By defining another process, $\nu(w) := v\left(\frac{4w}{\gamma^2}\right)$, with $\nu_0 = v_0$, the corresponding dynamics are given by:

$$\nu\left(\frac{\gamma^2 t}{4}\right) = \nu\left(\frac{\gamma^2 s}{4}\right) + \frac{4}{\gamma^2}\int_{\frac{\gamma^2 s}{4}}^{\frac{\gamma^2 t}{4}} \kappa\left(\bar{v} - \nu(w)\right)\mathrm{d}w + 2\int_{\frac{\gamma^2 s}{4}}^{\frac{\gamma^2 t}{4}} \sqrt{\nu(w)}\mathrm{d}W(w), \quad (9.50)$$

so that,

$$\nu(\bar{t}) = \nu(\bar{s}) + \int_{\bar{s}}^{\bar{t}} (2a\nu(w) + b)\,\mathrm{d}w + 2\int_{\bar{s}}^{\bar{t}} \sqrt{\nu(w)}\mathrm{d}W(w), \quad (9.51)$$

with $a = -2\kappa/\gamma^2$, $b = 4\kappa\bar{v}/\gamma^2$, $\bar{s} = \gamma^2 s/4$, $\bar{t} = \gamma^2 t/4$, and where the corresponding $\nu(t)$-dynamics are given by:

$$\mathrm{d}\nu(t) = (2a\nu(t) + b)\,\mathrm{d}t + 2\sqrt{\nu(t)}\mathrm{d}W(t), \quad \nu_0 = v_0. \quad (9.52)$$

The volatility of the process in (9.52) is of the desired form, like in (9.47). However, an additional drift term appears in the dynamics in (9.52), as compared to the drift in Equation (9.47).

Generally, the drift of a process can be adapted by a change of measure, i.e. by changing measures the drift changes while the volatility of the process will remain the same. This suggests us to define, under a new Brownian motion $\widehat{W}(t)$, a modified process for $\nu(t)$, as follows,

$$\mathrm{d}\nu(t) = b\,\mathrm{d}t + 2\sqrt{\nu(t)}\mathrm{d}\widehat{W}(t), \quad (9.53)$$

for some Brownian motion $\widehat{W}(t)$. We match the right-hand sides of Equations (9.52) and (9.53), to determine this Brownian motion $\widehat{W}(t)$,

$$(2a\nu(t) + b)\,\mathrm{d}t + 2\sqrt{\nu(t)}\mathrm{d}W(t) = b\,\mathrm{d}t + 2\sqrt{\nu(t)}\mathrm{d}\widehat{W}(t),$$

which yields the following *change of measure*,

$$\mathrm{d}\widehat{W}(t) = \mathrm{d}W(t) + a\sqrt{\nu(t)}\mathrm{d}t.$$

The proposed measure change implies the following *Radon-Nikodym derivative*:

$$\left.\frac{\mathrm{d}\mathbb{Q}}{\mathrm{d}\widehat{\mathbb{Q}}}\right|_{\mathcal{F}(t)} = \exp\left(-\frac{a^2}{2}\int_s^t \nu(z)\mathrm{d}z + a\int_s^t \sqrt{\nu(z)}\mathrm{d}\widehat{W}(z)\right).$$

The last integral in this expression can, with the help of Equation (9.53), be expressed, as

$$\int_s^t \sqrt{\nu(z)}\mathrm{d}\widehat{W}(z) = \frac{1}{2}\int_s^t (\mathrm{d}\nu(z) - b\,\mathrm{d}z) = \frac{1}{2}[\nu(t) - \nu(s) - b\cdot(t-s)].$$

so that the Radon-Nikodym derivative can be written as,

$$\left.\frac{\mathrm{d}\mathbb{Q}}{\mathrm{d}\widehat{\mathbb{Q}}}\right|_{\mathcal{F}(t)} = \exp\left(-\frac{a^2}{2}\int_s^t \nu(z)\mathrm{d}z + \frac{a}{2}[\nu(t) - \nu(s) - b\cdot(t-s)]\right). \quad (9.54)$$

By the measure transformation, we can change measures based on the Bayes' formula, i.e.

$$\mathbb{E}^A[X] = \mathbb{E}^B\left[X\frac{\mathrm{d}\mathbb{Q}^A}{\mathrm{d}\mathbb{Q}^B}\bigg|_{\mathcal{F}}\right] \bigg/ \mathbb{E}^B\left[\frac{\mathrm{d}\mathbb{Q}^A}{\mathrm{d}\mathbb{Q}^B}\right], \tag{9.55}$$

for any two equivalent measures $\mathbb{Q}^A$ and $\mathbb{Q}^B$ and random variable $X \in \mathcal{F}$.

Changing variables and conditional sampling

For a given constant $c \in \mathbb{R}$, the conditional expectation for the integrated variance $Y(t)$ in (9.46), in terms of the process $\nu(t)$, is equal to:

$$\mathbb{E}\left[e^{-cY(t)}\big|v(s), v(t)\right] := \mathbb{E}\left[\exp\left(-c\int_s^t v(z)\mathrm{d}z\right)\bigg|v(s), v(t)\right]$$

$$= \mathbb{E}\left[\exp\left(-c\int_s^t \nu\left(\frac{\gamma^2 z}{4}\right)\mathrm{d}z\right)\bigg|\nu\left(\frac{\gamma^2 s}{4}\right), \nu\left(\frac{\gamma^2 t}{4}\right)\right].$$

By a further change of variables, using $s_* = \frac{\gamma^2}{4}s$ and $t_* = \frac{\gamma^2}{4}t$, we find:

$$\mathbb{E}\left[e^{-cY(t)}\big|v(s), v(t)\right] = \mathbb{E}\left[\exp\left(-\frac{4c}{\gamma^2}\int_{s_*}^{t_*}\nu(w)\,\mathrm{d}w\right)\bigg|\nu(s_*), \nu(t_*)\right].$$

The last equality can be solved with the help of the measure change in (9.54), i.e.

$$\mathbb{E}\left[e^{-cY(t)}\big|v(s), v(t)\right] = \frac{\mathbb{E}^{\widehat{\mathbb{Q}}}\left[\exp\left(-\frac{4c}{\gamma^2}\int_{s_*}^{t_*}\nu(z)\,\mathrm{d}z\right)\frac{\mathrm{d}\mathbb{Q}}{\mathrm{d}\widehat{\mathbb{Q}}}\bigg|\nu(s_*), \nu(t_*)\right]}{\widehat{\mathbb{E}}\left[\frac{\mathrm{d}\mathbb{Q}}{\mathrm{d}\widehat{\mathbb{Q}}}\bigg|\nu(s_*), \nu(t_*)\right]} =: \frac{\omega_1}{\omega_2}.$$

(9.56)

For simplicity, we denote $\mathbb{E}^{\widehat{\mathbb{Q}}}[\cdot|\nu(s_*), \nu(t_*)] =: \widehat{\mathbb{E}}[\cdot|s_*, t_*]$, so that the expectation in the denominator reads, with (9.54),

$$\omega_2 = \widehat{\mathbb{E}}\left[\exp\left(-\frac{a^2}{2}\int_{s_*}^{t_*}\nu(z)\mathrm{d}z + \frac{a}{2}(\nu(t_*) - \nu(s_*) - b\tau_*)\right)\bigg|s_*, t_*\right]$$

$$= e^{\frac{a}{2}(\nu(t_*) - \nu(s_*) - b\tau_*)}\widehat{\mathbb{E}}\left[\exp\left(-\frac{a^2}{2}\int_{s_*}^{t_*}\nu(z)\mathrm{d}z\right)\bigg|s_*, t_*\right],$$

with $\tau_* = t_* - s_*$. The expectation in the numerator is equal to:

$$\omega_1 = \widehat{\mathbb{E}}\left[\exp\left(-\left(\frac{4c}{\gamma^2} + \frac{a^2}{2}\right)\int_{s_*}^{t_*}\nu(z)\mathrm{d}z + \frac{a}{2}(\nu(t_*) - \nu(s_*) - b\tau_*)\right)\bigg|s_*, t_*\right]$$

$$= e^{\frac{a}{2}(\nu(t_*) - \nu(s_*) - b\tau_*)}\widehat{\mathbb{E}}\left[\exp\left(-\left(\frac{4c}{\gamma^2} + \frac{a^2}{2}\right)\int_{s_*}^{t_*}\nu(z)\mathrm{d}z\right)\bigg|s_*, t_*\right].$$

As the first exponential terms in ω_1 and ω_2 are identical, the expression (9.56) simplifies, to

$$\mathbb{E}\left[\exp\left(-c\int_s^t v(z)\mathrm{d}z\right)\Big|v(s),v(t)\right] = \frac{\hat{\mathbb{E}}\left[\exp\left(-\left(\frac{4c}{\gamma^2}+\frac{a^2}{2}\right)\int_{s_*}^{t_*}\nu(z)\mathrm{d}z\right)\Big|s_*,t_*\right]}{\hat{\mathbb{E}}\left[\exp\left(-\frac{a^2}{2}\int_{s_*}^{t_*}\nu(z)\mathrm{d}z\right)\Big|s_*,t_*\right]}, \quad (9.57)$$

where $\nu(t)$ is driven by the SDE (9.53).

After substitution of (9.48) to both expectations (in the numerator and the denominator) and further simplifications, we find a somewhat involved expression, which is however a closed-form expression,

$$J(c) := \mathbb{E}\left[\mathrm{e}^{-cY(t)}\Big|v(s),v(t)\right] = \frac{\psi(c)\mathrm{e}^{-0.5(\psi(c)-\kappa)\tau}(1-\mathrm{e}^{-\kappa\tau})}{\kappa(1-\mathrm{e}^{-\psi(c)\tau})}$$

$$\times \exp\left[\frac{v(s)+v(t)}{\gamma^2}\left(\frac{\kappa(1+\mathrm{e}^{-\kappa\tau})}{1-\mathrm{e}^{-\kappa\tau}}-\frac{\psi(c)(1+\mathrm{e}^{-\psi(c)\tau})}{1-\mathrm{e}^{-\psi(c)\tau}}\right)\right]$$

$$\times \frac{I_{\frac{b}{2}-1}\left(\frac{4\psi(c)\sqrt{v(t)v(s)}}{\gamma^2}\frac{\mathrm{e}^{-0.5\psi(c)\tau}}{1-\mathrm{e}^{\psi(c)\tau}}\right)}{I_{\frac{b}{2}-1}\left(\frac{4\kappa\sqrt{v(t)v(s)}}{\gamma^2}\frac{\mathrm{e}^{-0.5\kappa\tau}}{1-\mathrm{e}^{\kappa\tau}}\right)}, \qquad (9.58)$$

with $\tau = t-s$ and $\psi(c) = \sqrt{\kappa^2 + 2\gamma^2 c}$.

By setting $c = -iu$, this gives us the requested conditional characteristic function

$$\boxed{\phi_{Y(t)|v(s),v(t)}(u;s,t) = J(-iu).}$$

With the conditional sampling technique presented, we can generate the required samples for an $X(t)$ process. Although the technique is considered *exact*, it is computationally expensive. Generation of each sample requires the inversion of the CDF, which is given in terms of the characteristic function. An alternative approach for sampling of the stochastic volatility model is presented below, in the context of the Heston model.

9.4.3 Almost exact simulation of the Heston model

We now focus on the Heston stochastic volatility model, which, under the log transformation reads,

$$\begin{cases} \mathrm{d}X(t) = \left(r - \frac{1}{2}v(t)\right)\mathrm{d}t + \sqrt{v(t)}\left[\rho_{x,v}\mathrm{d}\widetilde{W}_v(t) + \sqrt{1-\rho_{x,v}^2}\mathrm{d}\widetilde{W}_x(t)\right], \\ \mathrm{d}v(t) = \kappa\left(\bar{v}-v(t)\right)\mathrm{d}t + \gamma\sqrt{v(t)}\mathrm{d}\widetilde{W}_v(t), \end{cases} \quad (9.59)$$

with the parameters as given for Equation (8.18).

Monte Carlo Simulation

It is crucial here that the variance process is driven by an independent Brownian motion, while the process for $X(t) := \log S(t)$ is now correlated to $v(t)$. The motivation for this is that we are able to use the properties of the marginal distribution of $v(t)$.

After integration of both processes in (9.59) in a certain time interval $[t_i, t_{i+1}]$, the following discretization scheme is obtained:

$$x_{i+1} = x_i + \int_{t_i}^{t_{i+1}} \left(r - \frac{1}{2}v(t)\right) dt + \rho_{x,v} \boxed{\int_{t_i}^{t_{i+1}} \sqrt{v(t)} d\widetilde{W}_v(t)}$$

$$+ \sqrt{1 - \rho_{x,v}^2} \int_{t_i}^{t_{i+1}} \sqrt{v(t)} d\widetilde{W}_x(t),$$

and

$$v_{i+1} = v_i + \kappa \int_{t_i}^{t_{i+1}} (\bar{v} - v(t)) dt + \gamma \boxed{\int_{t_i}^{t_{i+1}} \sqrt{v(t)} d\widetilde{W}_v(t)}.$$

Notice that the two integrals with $\widetilde{W}_v(t)$ in the SDEs above are the same, and in terms of the variance realizations they are given by:

$$\int_{t_i}^{t_{i+1}} \sqrt{v(t)} d\widetilde{W}_v(t) = \frac{1}{\gamma}\left(v_{i+1} - v_i - \kappa \int_{t_i}^{t_{i+1}} (\bar{v} - v(t)) dt\right). \quad (9.60)$$

The variance v_{i+1} can then be simulated, for given value of v_i, by means of the CIR process, or, equivalently, by either the noncentral chi-squared distribution or by the QE scheme (the techniques described in the Sections 9.3.3 or 9.3.4, respectively).

As a final step in the Heston model simulation, the discretization for x_{i+1} is given by:

$$x_{i+1} = x_i + \int_{t_i}^{t_{i+1}} \left(r - \frac{1}{2}v(t)\right) dt + \frac{\rho_{x,v}}{\gamma}\left(v_{i+1} - v_i - \kappa \int_{t_i}^{t_{i+1}} (\bar{v} - v(t)) dt\right)$$

$$+ \sqrt{1 - \rho_{x,v}^2} \int_{t_i}^{t_{i+1}} \sqrt{v(t)} d\widetilde{W}_x(t). \quad (9.61)$$

We approximate all integrals appearing in the expression above by their left integration boundary values of the integrand, as in the Euler discretization scheme:

$$x_{i+1} \approx x_i + \int_{t_i}^{t_{i+1}} \left(r - \frac{1}{2}v_i\right) dt + \frac{\rho_{x,v}}{\gamma}\left(v_{i+1} - v_i - \kappa \int_{t_i}^{t_{i+1}} (\bar{v} - v_i) dt\right)$$

$$+ \sqrt{1 - \rho_{x,v}^2} \int_{t_i}^{t_{i+1}} \sqrt{v_i} d\widetilde{W}_x(t). \quad (9.62)$$

The calculation of the integrals is now trivial and results in:

$$x_{i+1} \approx x_i + \left(r - \frac{1}{2}v_i\right)\Delta t + \frac{\rho_{x,v}}{\gamma}\left(v_{i+1} - v_i - \kappa(\bar{v} - v_i)\Delta t\right)$$

$$+ \sqrt{1 - \rho_{x,v}^2}\sqrt{v_i}\left(\widetilde{W}_x(t_{i+1}) - \widetilde{W}_x(t_i)\right). \quad (9.63)$$

After collection of all terms and using the well-known property of normally distributed random variables, $\widetilde{W}_x(t_{i+1}) - \widetilde{W}_x(t_i) \stackrel{d}{=} \sqrt{\Delta t} Z_x$, with $Z_x \sim \mathcal{N}(0,1)$, we find:

$$\begin{cases} x_{i+1} \approx x_i + k_0 + k_1 v_i + k_2 v_{i+1} + \sqrt{k_3 v_i} Z_x, \\ v_{i+1} = \bar{c}(t_{i+1}, t_i) \chi^2(\delta, \bar{\kappa}(t_{i+1}, t_i)), \end{cases}$$

as in (9.30), with

$$\bar{c}(t_{i+1}, t_i) = \frac{\gamma^2}{4\kappa}(1 - e^{-\kappa(t_{i+1}-t_i)}), \quad \delta = \frac{4\kappa \bar{v}}{\gamma^2},$$

$$\bar{\kappa}(t_{i+1}, t_i) = \frac{4\kappa e^{-\kappa \Delta t}}{\gamma^2(1 - e^{-\kappa \Delta t})} v_i,$$

and $\chi^2(\delta, \bar{\kappa}(\cdot, \cdot))$ the noncentral chi-squared distribution with δ degrees of freedom and noncentrality parameter $\bar{\kappa}(t_{i+1}, t_i)$. The remaining constants are known as

$$k_0 = \left(r - \frac{\rho_{x,v}}{\gamma} \kappa \bar{v}\right) \Delta t, \quad k_1 = \left(\frac{\rho_{x,v} \kappa}{\gamma} - \frac{1}{2}\right) \Delta t - \frac{\rho_{x,v}}{\gamma},$$

$$k_2 = \frac{\rho_{x,v}}{\gamma}, \quad k_3 = (1 - \rho_{x,v}^2) \Delta t.$$

Example 9.4.1 (Heston model, Monte Carlo experiment) The simulation scheme, which is described in Subsection 9.4.3, is called here an *"Almost Exact Simulation" (AES) scheme* for the Heston SV model. It is not the Quadratic Exponential scheme, because we employ the direct sampling from the noncentral chi-squared distribution to simulate the variance process.

In this simulation experiment, European call options are computed by means of the AES scheme, with $S(t_0) = 100$, for three different strike prices: $K = 100$ (ATM), $K = 70$ (ITM) and $K = 140$ (OTM). The results obtained are compared to an Euler discretization with the truncated variance technique as described in Section 9.3. In the experiment different time steps are used, varying from one time step per year to 64 time steps per year. The model parameters are chosen as:

$$\kappa = 0.5, \gamma = 1, \bar{v} = 0.04, v_0 = 0.04, r = 0.1, \rho_{x,v} = -0.9.$$

The numerical results are based on 500.000 Monte Carlo paths, They are presented in Table 9.2. The reference results have been generated by the COS method, from Section 8.4. The experiment shows that with smaller time steps Δt, the numerical results improve when using the AES scheme. This pattern is not observed when the truncated Euler discretization is employed, for which the error doesn't seem to converge. This lack of convergence is due to the bias which is generated by the truncation. ◆

Remark 9.4.1 *It is clear that this chapter serves as an introduction into the Monte Carlo simulation, describing the road towards an accurate simulation for asset price models with stochastic volatility and the CIR process. We have*

Monte Carlo Simulation

Table 9.2: *Errors of Heston model Monte Carlo simulation results. AES stands for "Almost Exact Simulation", Euler stands for "Truncated Euler Scheme" in Section 9.3".*

	$K=100$		$K=70$		$K=140$	
Δt	Euler	AES	Euler	AES	Euler	AES
1	0.94 (0.023)	-1.00 (0.012)	-0.82 (0.028)	-0.53 (0.016)	1.29 (0.008)	0.008 (0.001)
1/2	2.49 (0.022)	-0.45 (0.011)	-0.11 (0.030)	-0.25 (0.016)	1.03 (0.008)	-0.0006 (0.001)
1/4	2.40 (0.016)	-0.18 (0.010)	0.37 (0.027)	-0.11 (0.016)	0.53 (0.005)	0.0005 (0.001)
1/8	2.08 (0.016)	-0.10 (0.010)	0.43 (0.025)	-0.07 (0.016)	0.22 (0.003)	0.0009 (0.001)
1/16	1.77 (0.015)	-0.03 (0.010)	0.40 (0.023)	-0.03 (0.016)	0.08 (0.001)	0.0002 (0.001)
1/32	1.50 (0.014)	-0.03 (0.009)	0.34 (0.022)	-0.01 (0.016)	0.03 (0.001)	-0.002 (0.001)
1/64	1.26 (0.013)	-0.001 (0.009)	0.27 (0.021)	-0.005 (0.016)	0.02 (0.001)	0.001 (0.001)

discussed the Taylor-based Monte Carlo path discretization Euler and Milstein schemes, as well as the exact simulation scheme for CIR type processes. The latter scheme is typically more involved, but they are more accurate. Large time-steps can be used in the simulation, thus requiring fewer time-steps for satisfactory accuracy. The starting point for such schemes has been the exact simulation in [Broadie and Kaya, 2006]. Other examples of exact or almost exact simulation include the Stochastic Collocation Monte Carlo method (SCMC) from [Grzelak et al., 2018] or the SABR simulation methods in [Chen et al., 2012; Leitao et al., 2017b,a]. Closely connected to the conditional sampling schemes in Section 9.4 are Brownian bridge techniques, by which Monte Carlo convergence can be accelerated. ▲

9.4.4 Improvements of Monte Carlo simulation

Convergence acceleration of the plain Monte Carlo method's convergence can be achieved in several ways. *Antithetic sampling* is one of the basic techniques to accelerate the convergence of the Monte Carlo method, as it reduces the variance of the sampling by employing, next to a random $Z \sim \mathcal{N}(0,1)$-sample within a simulation, also the sample $-Z$ to generate another path. It is well-known that if $Z \sim \mathcal{N}(0,1)$, then also $-Z \sim \mathcal{N}(0,1)$. This property can be used to drastically reduce the number of paths needed in a Monte Carlo simulation. Suppose that V_i^N represents the approximation obtained by the MC method, and $\hat{V}_i^N$ is obtained by using $-Z$. By taking the average, $\bar{V}_i^N = 1/2(V_i^N + \hat{V}_i^N)$, another approximation is obtained. Since V_i^N and $\hat{V}_N^i$ are both random variables we aim at $\mathrm{Var}[\bar{V}_i^N] < \mathrm{Var}[V_i^N] + \mathrm{Var}[\hat{V}_i^N]$. We have:

$$\mathrm{Var}[\bar{V}_i^N] = \frac{1}{4}\mathrm{Var}[V_i^N + \hat{V}_i^N] = \frac{1}{4}(\mathrm{Var}[V_i^N] + \mathrm{Var}[\hat{V}_i^N]) + \frac{1}{2}\mathrm{Cov}[V_i^N, \hat{V}_i^N]$$

So, when assuming a negative correlation between V_i^N and $\hat{V}_i^N$, it follows that $\mathrm{Var}[\bar{V}_i^N] \leq \frac{1}{2}(\mathrm{Var}[V_i^N + \hat{V}_i^N])$.

Other key techniques regarding variance reduction in Monte Carlo methods include importance sampling, Quasi-Monte Carlo (QMC) methods, stratified sampling, and control variates. In the latter variance reduction technique, a very similar function, or model, to the quantity of interest is employed, for which essential (analytical) properties are known that give rise to a significant reduction of the Monte Carlo variance.

Multilevel Monte Carlo methods (MLMC), as proposed by Giles in [Giles, 2007, 2008; Giles and Reisinger, 2012], form another interesting recent variance reduction technique, which is based on the additivity of the expectation operator. All these variance reduction techniques have their successful applications for specific tasks in Computational Finance, and combinations of these techniques may sometimes further improve the convergence rate.

Importance sampling

In this section, we present an intuitive way to understand the *Radon-Nikodym derivative*, in the form of an importance sampling variance reduction technique, in the context of Monte Carlo simulation. We would like to calculate the expectation $\mathbb{E}^X[g(X)]$ of some function $g(\cdot)$ of random variable X, with probability density $f_X(x)$, i.e.

$$\mathbb{E}^X[g(X)] = \int_\mathbb{R} g(x)f_X(x)\mathrm{d}x = \int_\mathbb{R} g(x)\mathrm{d}F_X(x). \tag{9.64}$$

To reduce the variance, we consider another random variable Y with probability density $f_Y(x)$, such that $f_X(x) \neq 0$ for all x with $f_Y(x) \neq 0$, and write,

$$\int_\mathbb{R} g(x)\mathrm{d}F_X(x) = \int_\mathbb{R} g(x)f_X(x)\mathrm{d}x = \int_\mathbb{R} g(x)\frac{f_X(x)}{f_Y(x)}f_Y(x)\mathrm{d}x$$

$$= \int_\mathbb{R} g(x)\frac{f_X(x)}{f_Y(x)}\mathrm{d}F_Y(x), \tag{9.65}$$

which implies that

$$\mathbb{E}^X[g(X)] = \mathbb{E}^Y[g(Y)L(Y)], \quad \text{with} \quad L(x) = \frac{f_X(x)}{f_Y(x)}. \tag{9.66}$$

Function $L(x)$ is often referred to as the *score function* in Monte Carlo simulation, in statistics it is the *likelihood ratio*, and in probability theory the *Radon-Nikodym derivative*.

With the help of the Radon-Nikodym derivative, we can derive the following expression,

$$\mathbb{E}^X[g(X)] \approx \frac{1}{N}\sum_{j=1}^N g(x_j) \approx \frac{1}{N}\sum_{j=1}^N g(y_j)L(y_j), \tag{9.67}$$

where y_j are the samples of Y. This representation is called *importance sampling*.

Example 9.4.2 (Importance sampling for Monte Carlo simulation)
We perform a Monte Carlo simulation to approximate $\mathbb{P}[X > 3]$, which is equivalent to calculating $\mathbb{E}[\mathbb{1}_{X>3}]$ with $X \sim \mathcal{N}(0,1)$. It is trivial to find that $\mathbb{P}[X > 3] = 0.001349898031630$.

Now, we consider the following four distributions as the possible candidates for importance sampling variance reduction, $\mathcal{U}([0,1])$, $\mathcal{U}([0,4])$, $\mathcal{N}(0,0.25)$ and $\mathcal{N}(0,9)$.

From Figure 9.14 it is clear that the best performing importance sampling variable is $\mathcal{N}(0,9)$, and the reasons can be explained as follows. The domains of the distributions $\mathcal{U}([0,1])$ and $\mathcal{U}([0,4])$ are insufficiently wide to cover the whole domain of $\mathcal{N}(0,1)$. Figure 9.14 shows however that $\mathcal{U}([0,4])$ performs better, but it generates a bias because there cannot be samples > 4. Furthermore, variable $\mathcal{N}(0,0.25)$ does not fit the tails of $\mathcal{N}(0,1)$ sufficiently well. ♦

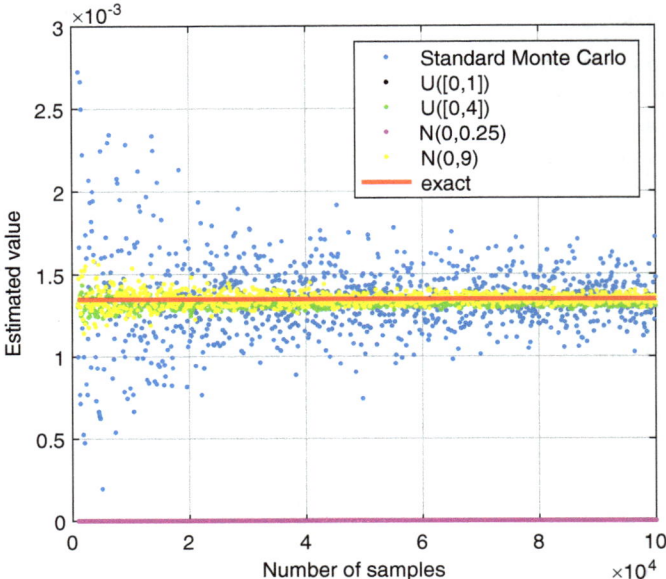

Figure 9.14: *Different samples in dependence of the variable used for importance sampling.*

9.5 Computation of Monte Carlo Greeks

Within the banks, the computation of the option Greeks is a very important task, which is performed for all derivatives in the trading books. Banks report, in their everyday operations, the so-called P&L attribution,[1] which explains the impact of the daily fluctuations on the value the derivative. To illustrate this, we consider a portfolio $\Pi(t,S)$ consisting of an option $V(t,S)$ on an underlying stock $S(t)$, and Δ stocks that are used for hedging, $\Pi(t,S) = V(t,S) - \Delta S(t)$. Discretization gives us,

$$\Pi_{i+1} = \Pi_i + V_{i+1} - V_i - \Delta(S_{i+1} - S_i). \qquad (9.68)$$

At each time point t_{i+1} the value of the option V_{i+1} is substituted as well as stock value S_{i+1}, and the value of the portfolio at time t_{i+1}, Π_{i+1}, is evaluated. When the left- and right-hand sides of the equation are sufficiently close, the P&L is said to be well-explained. The P&L attribution is used to verify whether all risk factors modeled are sufficient to materially explain the changes in the portfolio values. On the other hand, when the two sides of (9.68) differ substantially, this is an indication that the hedge has not been performed correctly, where some

---

[1] also often called "P&L Explained".

risk factors are most likely not taken into account properly in the hedge strategy. The option Greeks form the basis of these attributions.

In practice, asset price models are calibrated to financial market option prices, yielding an optimal set of model parameters, $\boldsymbol{\theta} = (\theta_1, \ldots, \theta_N)$. From a risk management perspective, it is then crucial to understand how financial derivative prices would evolve if the asset model parameters change. When the option price varies significantly with changing parameters, this would imply a frequent hedging re-balancing strategy and therefore an increase in the hedging costs. On the other hand, if the sensitivity of the derivative price with respect to the model parameter is not significant, this means a low re-balancing frequency and therefore a reduction of the costs.

In the context of Monte Carlo pricing techniques, there are basically three methods for approximation of the option Greeks: finite difference techniques (bump-and-revalue), pathwise sensitivity and likelihood ratio methods, that are discussed next.

9.5.1 Finite differences

In this section, we will use the following notation,

$$V(\theta) \equiv V(t_0, S; \theta) = \mathbb{E}^{\mathbb{Q}}\left[\frac{V(T, S; \theta)}{M(T)}\bigg|\mathcal{F}(t_0)\right],$$

with one parameter θ. The option's sensitivity with respect to parameter θ can be approximated by a forward difference approximation,

$$\frac{\partial V}{\partial \theta} \approx \frac{V(\theta + \Delta\theta) - V(\theta)}{\Delta\theta}. \tag{9.69}$$

With $V(\theta)$ continuous and at least twice differentiable, we have for any $\Delta\theta > 0$,

$$V(\theta + \Delta\theta) = V(\theta) + \frac{\partial V}{\partial \theta}\Delta\theta + \frac{1}{2}\frac{\partial^2 V}{\partial \theta^2}\Delta\theta^2 + \cdots. \tag{9.70}$$

So, for forward differencing, we find,

$$\frac{\partial V}{\partial \theta} = \frac{V(\theta + \Delta\theta) - V(\theta)}{\Delta\theta} - \frac{1}{2}\frac{\partial^2 V}{\partial \theta^2}\Delta\theta + \mathcal{O}(\Delta\theta^2). \tag{9.71}$$

In the Monte Carlo framework, the sensitivity to θ is estimated as follows,

$$\frac{\partial V}{\partial \theta} \approx \frac{\bar{V}(\theta + \Delta\theta) - \bar{V}(\theta)}{\Delta\theta}, \tag{9.72}$$

where $\bar{V}(\theta) = \frac{1}{N}\sum_{i=1}^{N} V_i(\theta)$.

Since the expectation is an unbiased estimator of the mean value, i.e., $\mathbb{E}[V(\theta)] \equiv \bar{V}(\theta)$, we have as the forward difference error,

$$\epsilon_1(\Delta\theta) := \mathbb{E}\left[-\frac{1}{2}\frac{\partial^2 V}{\partial \theta^2}\Delta\theta + \mathcal{O}(\Delta\theta^2)\right] = \mathcal{O}(\Delta\theta).$$

An alternative approximation of this derivative is based on *central differences*,

$$\frac{\partial V}{\partial \theta} \approx \frac{V(\theta + \Delta\theta) - V(\theta - \Delta\theta)}{2\Delta\theta}. \tag{9.73}$$

With the help of the Taylor expansion, we find, regarding central differencing,

$$\frac{\partial V}{\partial \theta} = \frac{V(\theta + \Delta\theta) - V(\theta - \Delta\theta)}{2\Delta\theta} + \mathcal{O}(\Delta\theta^2). \tag{9.74}$$

The central differences error is estimated by,

$$\epsilon_2(\Delta\theta) = \mathcal{O}(\Delta\theta^2).$$

Central differences generally yield a smaller bias than the forward differences approximation.

Example 9.5.1 (Black-Scholes delta and vega with finite differences)
In this numerical experiment, we analyze the errors when estimating the Monte Carlo Greeks for the Black-Scholes model with forward and central finite differences. For a call option the Black-Scholes delta, $\frac{\partial V}{\partial S(t_0)}$, and vega, $\frac{\partial V}{\partial \sigma}$ are simulated and compared to the analytic expressions.

We confirm in Figure 9.15 that central differencing yields better convergence properties than forward differencing. Interestingly, for the vega approximation, it is shown that when the *shock size* $\Delta\theta$ is large, forward differencing may

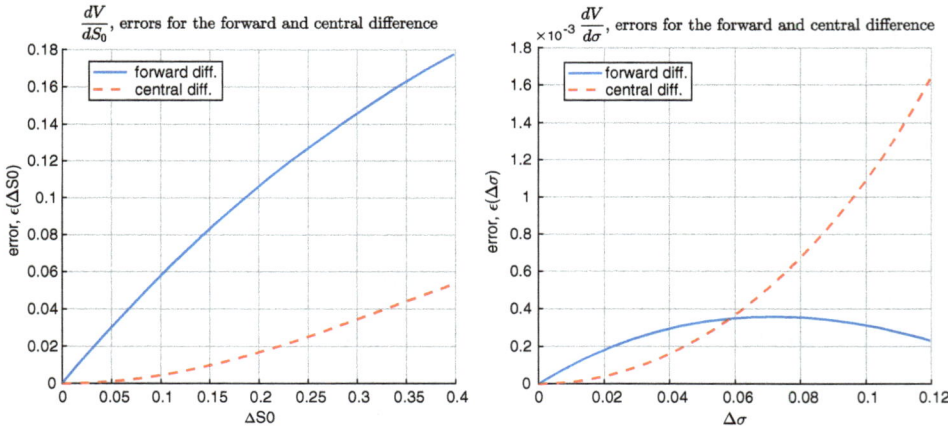

Figure 9.15: *The Black-Scholes delta (left) and vega (right) estimated with forward and with central differences. The governing parameters, $S(t_0) = 1$, $r = 0.06$, $\sigma = 0.3$, $T = 1$, $K = S(t_0)$.*

outperform central differencing. Notice further the differences in the sizes of the errors for the sensitivities. For an approximately 40% shock size for delta the maximum error is 0.18, whereas the same magnitude of shock for vega gives an error of about 0.0016. ♦

9.5.2 Pathwise sensitivities

A second technique within the Monte Carlo context to compute the Greeks is by the *pathwise sensitivity method*. This method was developed particularly for computing sensitivities of discrete models. The aim is to efficiently estimate the sensitivities at time t_0, like with respect to $S(t_0)$ and also to other model parameters. Compared to the finite difference methods, it is not necessary to re-evaluate and simulate the "shocked" Monte Carlo paths, where in the forward difference discretization we need one additional simulation, and in the central difference discretization two additional simulations.

The pathwise sensitivity method is applicable when we deal with continuous functions of the parameters of interest. It is based on interchanging the differentiation and expectation operators, as follows,

$$\frac{\partial V}{\partial \theta} = \frac{\partial}{\partial \theta} \mathbb{E}^{\mathbb{Q}}\left[\frac{V(T,S(T);\theta)}{M(T)}\bigg|\mathcal{F}(t_0)\right] = \mathbb{E}^{\mathbb{Q}}\left[\frac{\partial}{\partial \theta}\frac{V(T,S(T);\theta)}{M(T)}\bigg|\mathcal{F}(t_0)\right]. \quad (9.75)$$

The method cannot easily be applied in the case of nonsmooth payoff functions, as the differentiation of a non-smooth payoff, like a barrier or digital option is not straightforward.

Assuming a constant interest rate, we find,

$$\frac{\partial V}{\partial \theta} = e^{-r(T-t_0)}\mathbb{E}^{\mathbb{Q}}\left[\frac{\partial V(T,S(T);\theta)}{\partial S(T)}\frac{\partial S(T)}{\partial \theta}\bigg|\mathcal{F}(t_0)\right]. \quad (9.76)$$

At time T, $V(T,S(T);\theta)$ is a stochastic quantity, which depends on the distribution of $S(T)$. To compute the derivative with respect to θ, the probability space of $S(T)$ is *fixed*, which means that if the process $S(T)$ is driven by a Brownian motion $W(t)$, the paths of $W(t)$ will not be influenced by a change of parameter θ.

Example 9.5.2 (Black-Scholes delta and vega, pathwise method)
As an example, we apply the pathwise sensitivity methodology from Equation (9.76) to a call option under the Black-Scholes model, i.e.,

$$V(T,S;\theta) = \max(S(T) - K, 0) \text{ with } S(T) = S(t_0)e^{(r-\frac{1}{2}\sigma^2)(T-t_0)+\sigma(W(T)-W(t_0))}.$$

The derivative of the payoff with respect to $S(T)$ is given by,

$$\frac{\partial V}{\partial S(T)} = \mathbb{1}_{S(T)>K}, \quad (9.77)$$

and the necessary derivatives with respect to $S(t_0)$ and σ are as follows,

$$\frac{\partial S(T)}{\partial S(t_0)} = \mathrm{e}^{(r-\frac{1}{2}\sigma^2)(T-t_0)+\sigma(W(T)-W(t_0))},$$

$$\frac{\partial S(T)}{\partial \sigma} = S(T)\left(-\sigma(T-t_0)+W(T)-W(t_0)\right).$$

So, for the estimates of delta and vega we obtain,

$$\frac{\partial V}{\partial S(t_0)} = \mathrm{e}^{-r(T-t_0)}\mathbb{E}^{\mathbb{Q}}\left[\mathbb{1}_{S(T)>K}\mathrm{e}^{(r-\frac{1}{2}\sigma^2)(T-t_0)+\sigma(W(T)-W(t_0))}\Big|\mathcal{F}(t_0)\right]$$

$$= \mathrm{e}^{-r(T-t_0)}\mathbb{E}^{\mathbb{Q}}\left[\frac{S(T)}{S(t_0)}\mathbb{1}_{S(T)>K}\Big|\mathcal{F}(t_0)\right],$$

and

$$\frac{\partial V}{\partial \sigma} = \mathrm{e}^{-r(T-t_0)}\mathbb{E}^{\mathbb{Q}}\left[\mathbb{1}_{S(T)>K}S(T)\left(-\sigma(T-t_0)+W(T)-W(t_0)\right)\Big|\mathcal{F}(t_0)\right]$$

$$= \frac{\mathrm{e}^{-r(T-t_0)}}{\sigma}\mathbb{E}^{\mathbb{Q}}\left[S(T)\left(\log\left(\frac{S(T)}{S(t_0)}\right)-\left(r+\frac{1}{2}\sigma^2\right)(T-t_0)\right)\mathbb{1}_{S(T)>K}\Big|\mathcal{F}(t_0)\right].$$

We refer to Figure 9.16 for the method's convergence in terms of the number of Monte Carlo asset paths. ♦

Example 9.5.3 (Heston model's delta, pathwise sensitivity) We check here the sensitivity of the call option value under a stock price process which is governed by the Heston SV model. The pathwise sensitivity for Heston's delta

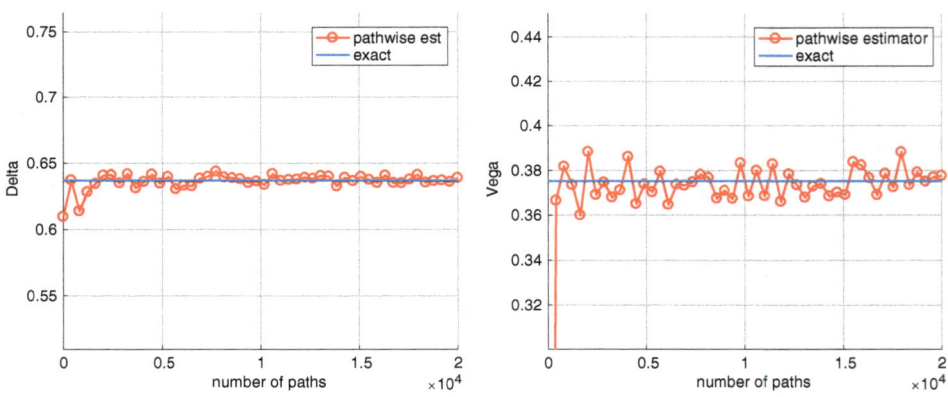

Figure 9.16: *The Black-Scholes delta (left) and vega (right) estimated by the pathwise sensitivity method. The parameters are $S(t_0) = 1$, $r = 0.06$, $\sigma = 0.3$, $T = 1$, $K = S(t_0)$.*

parameter is given by,

$$\frac{\partial V}{\partial S(t_0)} = e^{-r(T-t_0)} \mathbb{E}^{\mathbb{Q}} \left[\frac{\partial V(T, S; S(t_0))}{\partial S(T)} \frac{\partial S(T)}{\partial S(t_0)} \bigg| \mathcal{F}(t_0) \right]$$

The derivative of the payoff function with respect to $S(T)$ reads,

$$\frac{\partial V(T, S; \theta)}{\partial S(T)} = \mathbb{1}_{S(T) > K}, \qquad (9.78)$$

and the solution for the Heston process is then found to be,

$$S(T) = S(t_0) \exp\left[\int_{t_0}^{T} \left(r - \frac{1}{2}v(t)\right) dt + \int_{t_0}^{T} \sqrt{v(t)} dW_x(t) \right],$$

so that the sensitivity to $S(t_0)$ is given by,

$$\frac{\partial S(T)}{\partial S(t_0)} = \frac{S(T)}{S(t_0)}. \qquad (9.79)$$

As in the Black-Scholes model, the option Greek delta is given by the following expression,

$$\frac{\partial V}{\partial S(t_0)} = e^{-r(T-t_0)} \mathbb{E}^{\mathbb{Q}} \left[\frac{S(T)}{S(t_0)} \mathbb{1}_{S(T) > K} \bigg| \mathcal{F}(t_0) \right].$$

The method's convergence for this quantity is presented in Figure 9.17. ◆

Example 9.5.4 (Asian option's delta and vega, pathwise sensitivity)
We now consider an Asian option, see also Exercise 5.9, with the payoff function defined as,

$$V(T, S) = \max\left(A(T) - K, 0\right), \quad A(T) = \frac{1}{m} \sum_{i=1}^{m} S(t_i),$$

and the stock price process is governed by the GBM model,

$$\frac{dS(t)}{S(t)} = r dt + \sigma dW^{\mathbb{Q}}(t),$$

with the solution,

$$S(t_i) = S(t_0) e^{\left(r - \frac{1}{2}\sigma^2\right)(t_i - t_0) + \sigma(W^{\mathbb{Q}}(t_i) - W^{\mathbb{Q}}(t_0))}. \qquad (9.80)$$

Since the payoff function depends on m stock values in time, the pathwise sensitivity to $S(t_0)$ is given by,

$$\frac{\partial V}{\partial S(t_0)} = e^{-r(T-t_0)} \mathbb{E}^{\mathbb{Q}} \left[\sum_{i=1}^{m} \frac{\partial V(T, S(t_i); S(t_0))}{\partial S(t_i)} \frac{\partial S(t_i)}{\partial S(t_0)} \bigg| \mathcal{F}(t_0) \right].$$

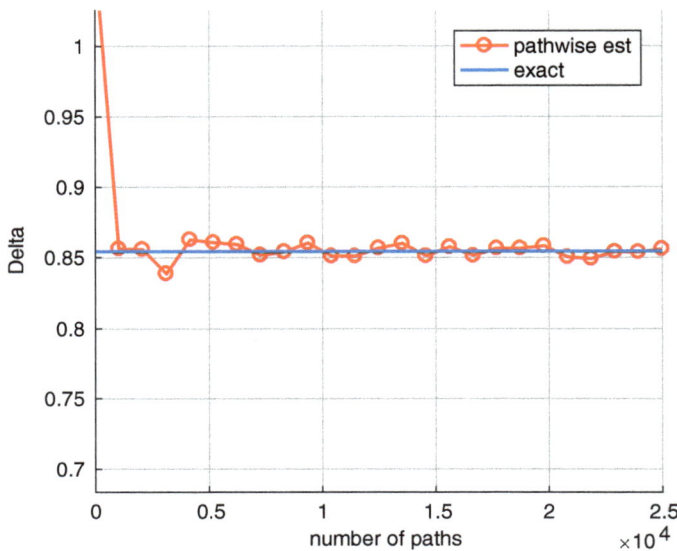

Figure 9.17: *Convergence of the Heston model's delta, estimated by the pathwise sensitivity method, with parameters $\gamma = 0.5$, $\kappa = 0.5$, $\bar{v} = 0.04$, $\rho_{x,v} = -0.9$, $v_0 = 0.04$, $T = 1.0$, $S(t_0) = 100.0$, $r = 0.1$.*

The sensitivity with respect to volatility σ then reads,

$$\frac{\partial V}{\partial \sigma} = e^{-r(T-t_0)} \mathbb{E}^{\mathbb{Q}} \left[\sum_{i=1}^{m} \frac{\partial V(T, S(t_i); \sigma)}{\partial S(t_i)} \frac{\partial S(t_i)}{\partial \sigma} \bigg| \mathcal{F}(t_0) \right].$$

The derivative of the payoff with respect to the stock price at time t_i is given by,

$$\frac{\partial V}{\partial S(t_i)} = \frac{1}{m} \mathbb{1}_{A(T) > K}. \tag{9.81}$$

Moreover, for the sensitivity of the stock with respect to the initial stock and the volatility, we obtain,

$$\frac{\partial S(t_i)}{\partial S(t_0)} = \frac{S(t_i)}{S(t_0)}, \quad \text{and} \quad \frac{\partial S(t_i)}{\partial \sigma} = S(t_i) \left(-\sigma(t_i - t_0) + W^{\mathbb{Q}}(t_i) - W^{\mathbb{Q}}(t_0) \right).$$

Summarizing, for the Asian option we find the following estimate for delta,

$$\frac{\partial V}{\partial S(t_0)} = e^{-r(T-t_0)} \frac{1}{m} \mathbb{E}^{\mathbb{Q}} \left[\mathbb{1}_{A(T) > K} \sum_{i=1}^{m} \frac{S(t_i)}{S(t_0)} \bigg| \mathcal{F}(t_0) \right]$$

$$= e^{-r(T-t_0)} \mathbb{E}^{\mathbb{Q}} \left[\mathbb{1}_{A(T) > K} \frac{A(T)}{S(t_0)} \bigg| \mathcal{F}(t_0) \right],$$

and for vega,

$$\frac{\partial V}{\partial \sigma} = \frac{e^{-r(T-t_0)}}{m} \mathbb{E}^{\mathbb{Q}} \left[1_{A(T)>K} \sum_{i=1}^{m} S(t_i) \left(-\sigma(t_i - t_0) + W^{\mathbb{Q}}(t_i) - W^{\mathbb{Q}}(t_0) \right) \Big| \mathcal{F}(t_0) \right].$$

By Equation (9.80), we find,

$$S(t_i) \left(-\sigma(t_i - t_0) + W^{\mathbb{Q}}(t_i) - W^{\mathbb{Q}}(t_0) \right)$$
$$= \frac{1}{\sigma} S(t_i) \left(\log \left(\frac{S(t_i)}{S(t_0)} \right) - \left(r + \frac{1}{2}\sigma^2 \right)(t_i - t_0) \right).$$

The vega estimate then reads,

$$\frac{\partial V(S(t_0))}{\partial \sigma} = \frac{e^{-r(T-t_0)}}{m} \mathbb{E}^{\mathbb{Q}}$$
$$\times \left[1_{A(T)>K} \sum_{i=1}^{m} \frac{S(t_i)}{\sigma} \left(\log \left(\frac{S(t_i)}{S(t_0)} \right) - \left(r + \frac{1}{2}\sigma^2 \right)(t_i - t_0) \right) \Big| \mathcal{F}(t_0) \right].$$

The method's convergence is displayed in Figure 9.18. ◆

9.5.3 Likelihood ratio method

A third technique to determine the Monte Carlo Greeks is the *likelihood ratio method*. Note that the main principles of the likelihood ratio method have

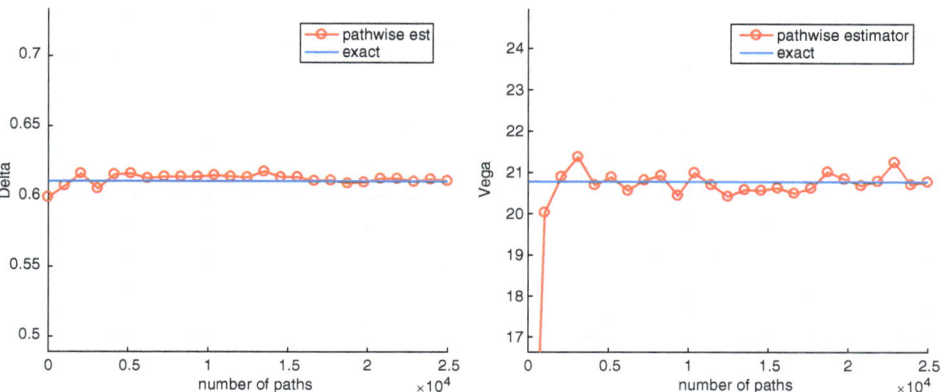

Figure 9.18: *The delta (left) and vega (right) estimated with the pathwise method for an Asian option. The parameters are: $S(t_0) = 100$, $\sigma = 0.15$, $T = 1$, $r = 0.05$, $K = S(t_0)$, $m = 10$ (equally spaced time points).*

already been discussed, in Section 9.4.4, where some techniques to improve the convergence of Monte Carlo simulation were discussed, see Equation (9.66).

The basis of the likelihood ratio method is formed by a differentiation of the probability density function, instead of the payoff function which was the basis for the pathwise sensitivity method. To explain the likelihood ratio method, we start with the option pricing formula in integral formulation (assuming constant interest rates),

$$V(\theta) \equiv V(t_0, S(t_0); \theta) = e^{-r(T-t_0)} \int_{\mathbb{R}} V(T, z) f_{S(T)}(z; \theta) dz, \qquad (9.82)$$

where, as usual, $V(T, z) = H(T, z)$ is the payoff function and $f_{S(T)}(s; \theta)$ the probability density function of $S(T)$.

In order to compute the sensitivity with respect to parameter θ, the integration and differentiation operators are interchanged, i.e.,

$$\frac{\partial V}{\partial \theta} = e^{-r(T-t_0)} \frac{\partial}{\partial \theta} \int_{\mathbb{R}} V(T, z) f_{S(T)}(z; \theta) dz$$

$$= e^{-r(T-t_0)} \int_{\mathbb{R}} V(T, z) \frac{\partial}{\partial \theta} f_{S(T)}(z; \theta) dz. \qquad (9.83)$$

The interchange of the integration and differentiation operators is typically not problematic, since the probability density functions in finance are usually smooth. The above expression is written in terms of the expectation operator, by multiplying and dividing the integrand by $f_{S(T)}(z; \theta)$,

$$\frac{\partial V}{\partial \theta} = e^{-r(T-t_0)} \int_{\mathbb{R}} V(T, z) \frac{\partial f_{S(T)}(z; \theta)}{\partial \theta} \frac{f_{S(T)}(z; \theta)}{f_{S(T)}(z; \theta)} dz$$

$$= e^{-r(T-t_0)} \int_{\mathbb{R}} V(T, z) \frac{\frac{\partial}{\partial \theta} f_{S(T)}(z; \theta)}{f_{S(T)}(z; \theta)} f_{S(T)}(z; \theta) dz$$

$$= e^{-r(T-t_0)} \mathbb{E}^{\mathbb{Q}} \left[V(T, z) \frac{\frac{\partial}{\partial \theta} f_{S(T)}(z; \theta)}{f_{S(T)}(z; \theta)} \bigg| \mathcal{F}(t_0) \right]. \qquad (9.84)$$

The ratio in (9.84) can be expressed as a logarithm, as follows,

$$\frac{\partial V}{\partial \theta} = e^{-r(T-t_0)} \mathbb{E}^{\mathbb{Q}} \left[V(T, z) \frac{\partial}{\partial \theta} \log f_{S(T)}(z; \theta) \bigg| \mathcal{F}(t_0) \right]$$

$$= e^{-r(T-t_0)} \mathbb{E}^{\mathbb{Q}} \left[V(T, z) L(z; \theta) \bigg| \mathcal{F}(t_0) \right].$$

The expression $L(z; \theta)$ under the expectation operator is again the *score function* or *likelihood ratio weight*.

Since here we need to compute the first derivative of the density function, and not of the payoff, we can replace different payoff functions relatively easily as the score function will remain the same. This simplifies the use of the method. On the downside, the computation of the first derivatives with respect to the parameters will be nontrivial when the probability density function of the asset price model is not known in closed-form, like for the Heston model. We conclude this section with some examples, where we show that the likelihood ratio method exhibits a larger bias than the pathwise sensitivity method.

Example 9.5.5 (Black-Scholes delta and vega, likelihood ratio method)
Recall the lognormal probability density function under the Black-Scholes model,

$$f_{S(T)}(x) = \frac{1}{\sigma x \sqrt{2\pi(T-t_0)}} \exp\left[-\frac{\left(\log \frac{x}{S(t_0)} - \left(r - \frac{1}{2}\sigma^2\right)(T-t_0)\right)^2}{2\sigma^2(T-t_0)}\right].$$

The first derivatives with respect to $S(t_0)$ and σ are found to be,

$$\frac{\partial \log f_{S(T)}(x)}{\partial S(t_0)} = \frac{\beta(x)}{S(t_0)\sigma^2(T-t_0)}, \text{ with } \beta(x) = \log \frac{x}{S(t_0)} - \left(r - \frac{1}{2}\sigma^2\right)(T-t_0),$$

and

$$\frac{\partial \log f_{S(T)}(x)}{\partial \sigma} = -\frac{1}{\sigma} + \frac{1}{\sigma^3(T-t_0)}\beta^2(x) - \frac{1}{\sigma}\beta(x). \tag{9.85}$$

Therefore, the estimate for delta is given by,

$$\frac{\partial V}{\partial S(t_0)} = \frac{\mathrm{e}^{-r(T-t_0)}}{S(t_0)\sigma^2(T-t_0)}\mathbb{E}^{\mathbb{Q}}\left[\max(S(T)-K,0)\beta(S(T))\Big|\mathcal{F}(t_0)\right],$$

while the vega estimate reads,

$$\frac{\partial V}{\partial \sigma} = \mathrm{e}^{-r(T-t_0)}\mathbb{E}^{\mathbb{Q}}$$
$$\times \left[\max(S(T)-K,0)\left(-\frac{1}{\sigma} + \frac{1}{\sigma^3(T-t_0)}\beta^2(S(T)) - \frac{1}{\sigma}\beta(S(T))\right)\Big|\mathcal{F}(t_0)\right].$$

In Figure 9.19, we compare the convergence of the pathwise sensitivity and the likelihood ratio method. The results show a superior convergence of the pathwise sensitivity method. The approximated delta and vega, as functions of the option expiry, are presented in Figure 9.20. Clearly, the pathwise sensitivity method appears to provide a better convergence behavior with smaller variance.♦

Remark 9.5.1 (Monte Carlo methods, early-exercise options) *We have not yet discussed the Monte Carlo techniques that are well-suited for options with early-exercise opportunities, like Bermudan options and American options. These techniques combine a forward path simulation stage, with a backward*

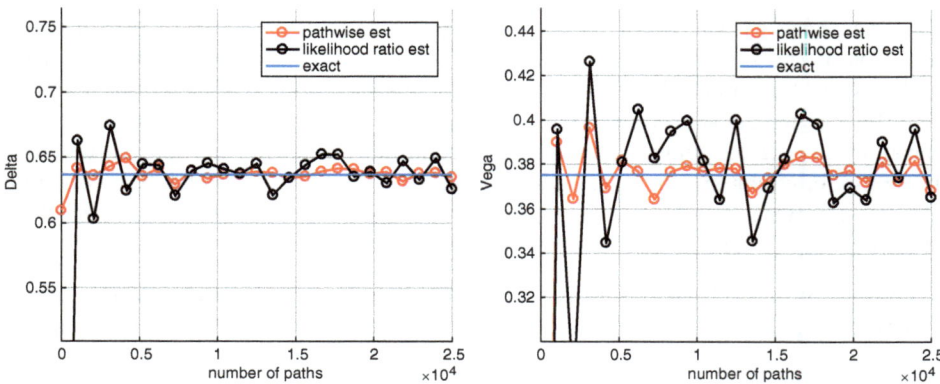

Figure 9.19: *Convergence results for the pathwise and likelihood ratio sensitivity methods for delta (left) and vega (right). The parameters are $S(t_0) = 1$, $r = 0.06$, $\sigma = 0.3$, $T = 1$, $K = S(t_0)$.*

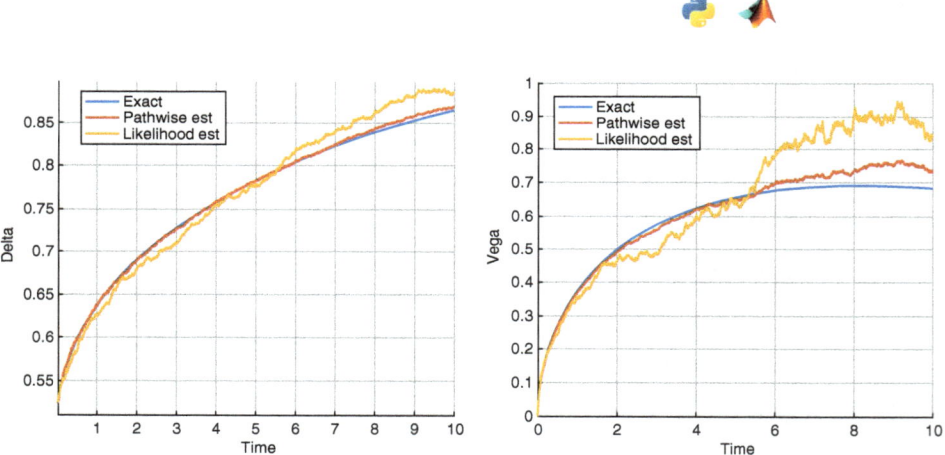

Figure 9.20: *Delta (left) and vega (right) estimated by the pathwise and likelihood ratio sensitivity methods as functions of the option expiry. The parameters in the experiment are $S(t_0) = 1$, $r = 0.06$, $\sigma = 0.3$, $T = 1$, $K = S(t_0)$, number of paths 25000 with 1000 time-steps.*

stage in which conditional expectations need to be approximated. This is done by some form of regression of Monte Carlo sample points onto a functional form which is composed of basis functions. A prominent example in this field is the Longstaff-Schwartz method (LSM) [Longstaff and Schwartz, 2001]. Also the Stochastic Grid Bundling Method (SGBM) [Jain and Oosterlee, 2015], with *localized regression in bundled partitions of asset points, is very suitable for such options and generalizes to an accurate computation of the option's Greeks and also towards Credit Value Adjustment (CVA), see the Chapters 12, 13 and 15.*

Of course, there is much more to say about this fascinating numerical technique. There is ample literature about the Monte Carlo method, with efficient pricing versions for a variety of financial derivatives contracts, and also for risk management purposes. Monographs with detailed Monte Carlo methods content, include [Kloeden and Platen, 1992, 1995, 1999; Glasserman, 2003; Jäckel, 2002], and Monte Carlo methods are also discussed in many textbooks, such as [Seydel, 2017; Higham, 2004]. ▲

9.6 Exercise set

Exercise 9.1 For a given deterministic function $g(t) = \cos(t)$ and for $T = 2$, determine theoretically,
$$\int_0^T g(t)\mathrm{d}W(t).$$
Confirm your answer by means of a Monte Carlo experiment.

Exercise 9.2 For a given deterministic function $g(t) = \exp(t)$ and for $T = 2$, determine theoretically,
$$\int_0^T g(t)\mathrm{d}W(t).$$
Confirm your answer by means of a Monte Carlo experiment.

Exercise 9.3 For a given function $g(t) = W^4(t)$ and for $T = 2$, determine theoretically,
$$\int_0^T g(t)\mathrm{d}W(t).$$
Confirm your answer by a Monte Carlo experiment.

Exercise 9.4 In Exercise 1.7, we found theoretically that
$$\int_0^T W(z)\mathrm{d}z = \int_0^T (T-z)\mathrm{d}W(z).$$
Take $T = 5$ and confirm by a numerical Monte Carlo experiment that the equality holds true.

Exercise 9.5 Show, by means of a Monte Carlo simulation, that
a. $\mathbb{E}[W(t)|\mathcal{F}(t_0)] = W(t_0)$ for $t_0 = 0$,
b. $\mathbb{E}[W(t)|\mathcal{F}(s)] = W(s)$, with $s < t$. Note that this latter part requires Monte Carlo "sub-simulations".

Exercise 9.6 This exercise is related to Exercise 2.9. With $\mathbf{X}(t) = [X(t), Y(t)]^{\mathrm{T}}$, suppose $X(t)$ satisfies the following SDE,
$$\mathrm{d}X(t) = 0.04X(t)\mathrm{d}t + \sigma X(t)\mathrm{d}W^{\mathbb{P}}(t),$$
and $Y(t)$ satisfies,
$$\mathrm{d}Y(t) = \beta Y(t)\mathrm{d}t + 0.15Y(t)\mathrm{d}W^{\mathbb{P}}(t).$$
Take $\beta = 0.1$, $\sigma = 0.38$, and a money-savings account, $\mathrm{d}M(t) = rM(t)\mathrm{d}t$, with risk-free rate $r = 6\%$, maturity $T = 7y$, and for strike values $K = 0$ to $K = 10$ with steps $\Delta K = 0.5$. Use the Euler discretization to find,
$$V(t, \mathbf{X}) = \mathbb{E}^{\mathbb{Q}}\left[\frac{1}{M(T)}\max\left(\frac{1}{2}X(T) - \frac{1}{2}Y(T), K\right)\bigg|\mathcal{F}(t)\right],$$
and plot the results (strike price versus $V(t, \mathbf{X})$).
Hint: A change of measure is needed to price the option.

Exercise 9.7 For a Wiener process $W(t)$ consider

$$X(t) = W(t) - \frac{t}{T}W(T-t), \text{ for } 0 \le t \le T.$$

For $T = 10$, find analytically $\mathbb{V}\mathrm{ar}[X(t)]$ and perform a numerical simulation to confirm your result. Is the accuracy sensitive to t?

Exercise 9.8 In GBM, with $S(t_0) = 1$, $r = 0.045$ consider the following time-dependent volatility function,

$$\sigma(t) = 0.2 + e^{-0.3 \cdot t}.$$

Implement a Monte Carlo Euler simulation for maturity time $T = 2$, with $N = 100.000$ paths and $m = 200$ time-steps. Investigate the accuracy of the so-called effective[2] parameters (i.e., time-averaged parameters) in the context of pricing a European call option for different strike prices, $K = 0, 0.1, \ldots, 4$.

Compute also the implied volatilities for different maturity times, $T = 0.5, 1.0, 1.5, 2.0$. Describe the observed implied volatilities.

Exercise 9.9 The Euler and Milstein schemes are not well-suited for simulating stochastic processes that are positive and have a probability mass at zero. An example is the CIR process with the following dynamics:

$$dv(t) = \kappa(\bar{v} - v(t))dt + \gamma\sqrt{v(t)}dW(t), \quad v(t_0) > 0. \tag{9.86}$$

When the Feller condition, $2\kappa\bar{v} > \gamma^2$, is satisfied, $v(t)$ can never reach zero, however, if this condition does not hold, the origin is accessible. In both cases, $v(t)$ cannot become negative.

a. Choose a specific parameter set and show that an Euler discretization can give rise to negative $v(t)$ values.

b. Perform the *almost exact simulation* (meaning sampling for the noncentral chi-squared distribution, as in Section 9.3.3) and confirm that the asset paths are not negative when using this technique for the same parameter values as above. Perform two tests in which the time step is varied. Give expectations and variances.

Exercise 9.10 Perform a Monte Carlo simulation for the jump diffusion model. Use both the Kou model and the Merton jump diffusion model, and employ the parameter as given in Exercise 5.11. Simulate the distribution at time T by directly sampling from $S(T)$, i.e.,

1. from the characteristic function one can obtain the CDF for $S(T)$ using the COS method (Exercise 6.4);
2. take uniform samples $u_i = U([0, 1])$;
3. use the Newton-Raphson iteration to compute the samples $s_i = F^{-1}_{S(T)}(u_i)$, where $F_{S(T)}(\cdot)$ is computed using point 1.

Questions:

a. How many asset path should be generated in each test case, so that we obtain reliable option values and corresponding variances?

---

[2] Effective parameters are often time-averaged parameters, see, for example, σ_* in Example 2.2.1.

b. Compare the resulting option values with those obtained by means of the COS method in Exercise 6.6.

c. Use also Kou's jump diffusion model, within the Monte Carlo simulation framework, based on the three test parameter sets

$$\text{Set I: } \xi_p = 8, p_1 = 0.4, p_2 = 0.6, \alpha_1 = 10, \alpha_2 = 5;$$
$$\text{Set II: } \xi_p = 8, p_1 = 0.2, p_2 = 0.8, \alpha_1 = 10, \alpha_2 = 5;$$
$$\text{Set III: } \xi_p = 0.1, p_1 = 0.2, p_2 = 0.8, \alpha_1 = 10, \alpha_2 = 50;$$

Compare the results to those obtained with the help of the COS method, and present 10 relevant asset price paths for both the Merton and the Kou jump diffusion models. (Note, however, that the results are somewhat sensitive towards the choice of dt.)

Exercise 9.11 Perform a Monte Carlo simulation for the infinite activity Lévy VG, CGMY and NIG asset price models. The aim is to approximate as good as possible the option values that have been obtained in Exercise 6.8. Simulate the distribution at time T by directly sampling from $S(T)$, i.e.,

1. from the characteristic function one can obtain the CDF for $S(T)$ using the COS method (Exercise 6.4);
2. take uniform samples $u_i = U([0,1])$;
3. use the Newton-Raphson iteration to compute the samples $s_i = F_{S(T)}^{-1}(u_i)$, where $F_{S(T)}(\cdot)$ is computed using point 1.

Exercise 9.12 We analyze the sensitivity of a modified asset-or-nothing digital option, which was also discussed earlier in (7.40) and is given by,

$$V(\rho) \equiv V(t_0, S_1, S_2; \rho) = e^{-r(T-t_0)} \mathbb{E}^{\mathbb{Q}} \left[S_2(T) \mathbb{1}_{S_1(T) > K} \big| \mathcal{F}(t_0) \right],$$

where the processes, with the help of the Cholesky decomposition, are given by,

$$\frac{\mathrm{d}S_1(t)}{S_1(t)} = r\mathrm{d}t + \sigma_1 \mathrm{d}\widetilde{W}_1^{\mathbb{Q}}(t),$$
$$\frac{\mathrm{d}S_2(t)}{S_2(t)} = r\mathrm{d}t + \sigma_2 \left(\rho \mathrm{d}\widetilde{W}_1^{\mathbb{Q}}(t) + \sqrt{1-\rho^2} \mathrm{d}\widetilde{W}_2^{\mathbb{Q}}(t) \right),$$

with the corresponding solutions,

$$S_1(T) = S_1(t_0) e^{\left(r - \frac{1}{2}\sigma_1^2\right)T + \sigma_1 \widetilde{W}_1^{\mathbb{Q}}(T)},$$
$$S_2(T) = S_2(t_0) e^{\left(r - \frac{1}{2}\sigma_2^2\right)T + \sigma_2 \rho \widetilde{W}_1^{\mathbb{Q}}(T) + \sigma_2 \sqrt{1-\rho^2} \widetilde{W}_2^{\mathbb{Q}}(T)}.$$

Use the pathwise sensitivity method to derive the sensitivity of this option to the correlation parameter ρ, and present the convergence of $\frac{\partial V}{\partial \rho}$ with different numbers of asset paths, for the parameter setting: $S_1(t_0) = 1$, $S_2(t_0) = 1$, $r = 0.06$, $\sigma_1 = 0.3$, $\sigma_2 = 0.2$, $T = 1$, $K = S_1(t_0)$, $\rho = 0.7$.

CHAPTER 10

Forward Start Options; Stochastic Local Volatility Model

In this chapter:

Forward start options are option contracts that start at some time point in the future (*not at "time now"* t_0). These options are path-dependent and an accurate representation of the transition probability density function of the underlying asset price process is important to model the features of forward start options that are observed in the financial market. This brings extra complications that we discuss in this chapter. The pricing of forward start options, *under different asset dynamics*, is discussed in **Section 10.1**. We also consider in this chapter an alternative model where the Heston stochastic volatility model is *augmented by a nonparametric local volatility* component, i.e. the *Heston-Stochastic Local Volatility* (Heston-SLV or H-SLV) model, in **Section 10.2**. We will see that the Heston-SLV model typically yields a *stable hedging performance* and an accurate pricing of the so-called forward volatility sensitive products, like the forward start options.

Keywords: forward start option modeling under different dynamics, stochastic-local volatility models, computation of expectation, Monte Carlo bin method.

10.1 Forward start options

In this section, we discuss the modeling and pricing of forward start options, that are also called *performance options*. Forward start options may be considered as European options, however, with a future start time. As such these options are examples of path-dependent, exotic options. Whereas with European plain vanilla options the initial stock value S_0 is known at initial time t_0, in the case of forward start options the initial stock value is *unknown*, as it will become active at some future time, $t = T_1$. So, forward start options do not directly depend on *today's* value of the underlying asset, but on the performance of the asset over some future time period $[T_1, T_2]$.

10.1.1 Introduction into forward start options

Although the forward start options are seemingly of a rather basic form, they may be nontrivial from a modeling perspective.

We start with some general results regarding the pricing of forward start options that do not depend on any particular asset model. In the subsections to follow, the impact of different alternatives for the underlying asset dynamics on the forward start option prices will be discussed.

With two maturity times T_1 and T_2, with $t_0 < T_1 < T_2$, a forward start option *payoff function* is defined, as follows,

$$V^{\text{fwd}}(T_2, S(T_2)) := \max\left(\frac{S(T_2) - S(T_1)}{S(T_1)} - K, 0\right), \tag{10.1}$$

with a strike price K.

The contract value is determined by the *percentage performance* of the underlying asset $S(t)$, which is measured at the future time points T_1 and T_2. Obviously, with $t_0 = T_1$ the payoff will collapse to a standard, "scaled" European option with the maturity at time T_2, and the payoff then equals to

$$V^{\text{fwd}}(T_2, S(T_2)) = \frac{1}{S_0} \max\left(S(T_2) - S_0 \cdot K^*, 0\right), \quad K^* = K + 1.$$

The forward start call option payoff function in (10.1) can also be expressed as:
$$V^{\text{fwd}}(T_2, S(T_2)) = \max\left(\frac{S(T_2)}{S(T_1)} - K^*, 0\right).$$
with $K^* = K + 1$.

The contract will pay out at time T_2, and *today's value* is given by the following pricing equation:

$$V^{\text{fwd}}(t_0, S_0) = M(t_0)\mathbb{E}^{\mathbb{Q}}\left[\frac{1}{M(T_2)} \max\left(\frac{S(T_2)}{S(T_1)} - K^*, 0\right) \Big| \mathcal{F}(t_0)\right], \tag{10.2}$$

with $M(t)$ the money-savings account.

With a constant interest rate r, the pricing equation reads

$$V^{\text{fwd}}(t_0, S_0) = \frac{M(t_0)}{M(T_2)} \mathbb{E}^{\mathbb{Q}} \left[\max\left(\frac{S(T_2)}{S(T_1)} - K^*, 0\right) \Big| \mathcal{F}(t_0) \right]$$
$$= \frac{M(t_0)}{M(T_2)} \mathbb{E}^{\mathbb{Q}} \left[\max\left(e^{x(T_1, T_2)} - K^*, 0\right) \Big| \mathcal{F}(t_0) \right],$$

where

$$x(T_1, T_2) := \log S(T_2) - \log S(T_1).$$

We can derive the characteristic function of $x(T_1, T_2)$:

$$\phi_x(u) \equiv \phi_x(u, t_0, T_2) = \mathbb{E}^{\mathbb{Q}} \left[e^{iu(\log S(T_2) - \log S(T_1))} \Big| \mathcal{F}(t_0) \right].$$

By iterated expectations, we have

$$\phi_x(u) = \mathbb{E}^{\mathbb{Q}} \left[\mathbb{E}^{\mathbb{Q}} \left[e^{iu(\log S(T_2) - \log S(T_1))} \Big| \mathcal{F}(T_1) \right] \Big| \mathcal{F}(t_0) \right].$$

As $\log S(T_1)$ is measurable with respect to the filtration $\mathcal{F}(T_1)$, we can write

$$\phi_x(u) = \mathbb{E}^{\mathbb{Q}} \left[e^{-iu \log S(T_1)} \mathbb{E}^{\mathbb{Q}} \left[e^{iu \log S(T_2)} \Big| \mathcal{F}(T_1) \right] \Big| \mathcal{F}(t_0) \right].$$

In terms of a *discounted characteristic function*, we need to insert an appropriate discounting term, i.e.

$$\phi_x(u) = \mathbb{E}^{\mathbb{Q}} \left[e^{-iu \log S(T_1)} e^{r(T_2 - T_1)} \mathbb{E}^{\mathbb{Q}} \left[e^{-r(T_2 - T_1)} e^{iu \log S(T_2)} \Big| \mathcal{F}(T_1) \right] \Big| \mathcal{F}(t_0) \right].$$

The inner expectation can be recognized as the discounted characteristic function of $X(T_2) = \log S(T_2)$, and therefore it follows that

$$\phi_x(u) = \mathbb{E}^{\mathbb{Q}} \left[e^{-iuX(T_1)} e^{r(T_2 - T_1)} \psi_X(u, T_1, T_2) \Big| \mathcal{F}(t_0) \right]. \quad (10.3)$$

The function $\psi_X(u, T_1, T_2)$ will be detailed below, for two different asset dynamics, i.e. the Black-Scholes and the Heston dynamics.

Until now, any particular dynamics for the underlying stock $S(t)$ have not been specified, except for the assumption of a constant interest rate. In the follow-up sections, we will consider pricing of forward start options under different asset dynamics, and first discuss forward start options under the Black-Scholes model.

10.1.2 Pricing under the Black-Scholes model

Under the Black-Scholes model, the discounted characteristic function, $\psi_X(u, T_1, T_2)$, for the log-stock $X(t) = \log S(t)$, conditioned on the information available until time T_1, was already derived in Equation (3.29), and has the following form here:

$$\psi_X(u, T_1, T_2) = \exp\left[\left(r - \frac{\sigma^2}{2}\right)iu(T_2 - T_1) - \frac{1}{2}\sigma^2 u^2(T_2 - T_1)\right.$$
$$\left. - r(T_2 - T_1) + iuX(T_1)\right]. \tag{10.4}$$

By substitution of the forward characteristic function of the Black-Scholes model (10.4) into (10.3), the following expression for the *forward characteristic function* $\phi_x(u)$ of $x(T_1, T_2)$ is obtained,

$$\phi_x(u) = \mathbb{E}^{\mathbb{Q}}\left[\left.e^{\left(r - \frac{\sigma^2}{2}\right)iu(T_2 - T_1) - \frac{1}{2}\sigma^2 u^2(T_2 - T_1)}\right| \mathcal{F}(t_0)\right]$$
$$= \exp\left(\left(r - \frac{\sigma^2}{2}\right)iu(T_2 - T_1) - \frac{1}{2}\sigma^2 u^2(T_2 - T_1)\right). \tag{10.5}$$

This is, in fact, the characteristic function of a *normal density* with mean $(r - 1/2\sigma^2)(T_2 - T_1)$ and variance $\sigma^2(T_2 - T_1)$. Expression (10.5) does not depend on $S(t)$, but only on the interest rate and the volatility. It may be quite surprising that the stock has disappeared from the above expression, but this is a consequence of the fact that the ratio $S(T_2)/S(T_1)$, under the lognormal asset dynamics, gives as the solution:

$$\frac{S(T_2)}{S(T_1)} = e^{\left(r - \frac{1}{2}\sigma^2\right)(T_2 - T_1) + \sigma(W(T_2) - W(T_1))}. \tag{10.6}$$

This result implies that we can find a closed-form expression for the value of a forward start option under the Black-Scholes model.

Theorem 10.1.1 (Pricing of forward start option under BS model)
The price for a forward start call option at $t_0 = 0$, as defined in (10.1), under the Black-Scholes dynamics for stock $S(t)$, is given by:

$$V^{fwd}(t_0, S_0) = e^{-rT_2}\mathbb{E}^{\mathbb{Q}}\left[\left.\max\left(\frac{S(T_2)}{S(T_1)} - K^*, 0\right)\right| \mathcal{F}(t_0)\right]$$

with $K^* = K + 1$, and it has a closed-form solution, given by:

$$V^{fwd}(t_0, S_0) = e^{-rT_1} F_{\mathcal{N}(0,1)}(d_1) - K^* e^{-rT_2} F_{\mathcal{N}(0,1)}(d_2), \tag{10.7}$$

with

$$d_1 = \frac{\log\left(\frac{1}{K^*}\right) + \left(r + \frac{1}{2}\sigma^2\right)(T_2 - T_1)}{\sigma\sqrt{T_2 - T_1}}, \quad d_2 = \frac{\log\left(\frac{1}{K^*}\right) + \left(r - \frac{1}{2}\sigma^2\right)(T_2 - T_1)}{\sigma\sqrt{T_2 - T_1}}$$

The proof is given below.

Proof Using Equation (10.6), the pricing equation can be rewritten, as follows:

$$V^{\text{fwd}}(t_0, S_0) = e^{-rT_2}\mathbb{E}^{\mathbb{Q}}\left[\max\left(\frac{S(T_2)}{S(T_1)} - K^*, 0\right)\Big|\mathcal{F}(t_0)\right]$$

$$= e^{-rT_2}\int_{-\infty}^{\infty} \max\left(e^{(r-\frac{1}{2}\sigma^2)(T_2-T_1)+\sigma\sqrt{T_2-T_1}x} - K^*, 0\right) f_{\mathcal{N}(0,1)}(x)\mathrm{d}x,$$

with $f_{\mathcal{N}(0,1)}(x)$ the usual PDF of a standard normal variable.

For a positive strike price, it follows that,

$$V^{\text{fwd}}(t_0, S_0) = e^{-rT_2}\int_{a}^{\infty} e^{(r-\frac{1}{2}\sigma^2)(T_2-T_1)+\sigma\sqrt{T_2-T_1}x} f_{\mathcal{N}(0,1)}(x)\mathrm{d}x$$

$$- K^* e^{-rT_2}\left[1 - F_{\mathcal{N}(0,1)}(a)\right],$$

with $a = \dfrac{1}{\sigma\sqrt{T_2-T_1}}\left(\log K^* - (r - \tfrac{1}{2}\sigma^2)(T_2 - T_1)\right).$

The integral in the last expression can be simplified, as follows:

$$\frac{1}{\sqrt{2\pi}}\int_{a}^{\infty} \exp\left(\left(r - \frac{1}{2}\sigma^2\right)(T_2 - T_1) + \sigma\sqrt{T_2-T_1}x\right)e^{-\frac{x^2}{2}}\mathrm{d}x$$

$$= \frac{e^{(r-\frac{1}{2}\sigma^2)(T_2-T_1)+\frac{1}{2}\sigma^2(T_2-T_1)}}{\sqrt{2\pi}}\int_{a}^{\infty} e^{-\frac{1}{2}(x-\sigma\sqrt{T_2-T_1})^2}\mathrm{d}x$$

$$= e^{r(T_2-T_1)}\left[1 - F_{\mathcal{N}(0,1)}\left(a - \sigma\sqrt{T_2-T_1}\right)\right],$$

which results in the following expression for $V^{\text{fwd}}(t_0, S_0)$,

$$V^{\text{fwd}}(t_0, S_0) = e^{-rT_1}\left[1 - F_{\mathcal{N}(0,1)}\left(a - \sigma\sqrt{T_2-T_1}\right)\right] - K^* e^{-rT_2}\left[1 - F_{\mathcal{N}(0,1)}(a)\right].$$

By using the well-known identity for the standard normal CDF, $F_{\mathcal{N}(0,1)}(a) = 1 - F_{\mathcal{N}(0,1)}(-a)$, we find,

$$V^{\text{fwd}}(t_0, S_0) = e^{-rT_1} F_{\mathcal{N}(0,1)}\left(\sigma\sqrt{T_2-T_1} - a\right) - K^* e^{-rT_2} F_{\mathcal{N}(0,1)}(-a).$$

With,

$$d_1 := \sigma\sqrt{T_2 - T_1} - a$$

$$= \frac{1}{\sigma\sqrt{T_2-T_1}}\left[\log\left(\frac{1}{K^*}\right) + \left(r + \frac{1}{2}\sigma^2\right)(T_2 - T_1)\right],$$

and $d_2 = d_1 - \sigma\sqrt{T_2 - T_1}$, we arrive at formula (10.7). ∎

Remark 10.1.1 (Forward implied volatility) *Based on the Black-Scholes-type equation for the valuation of forward start options, we may determine, for given market prices of forward start options, the corresponding forward implied volatility. In the same way as discussed in Section 4.1.1, the Black-Scholes forward implied volatility σ_{imp}^{fwd} is found by solving,*

$$V^{\text{fwd}}(t_0, K^*, T_1, T_2, \sigma_{imp}^{fwd}) = V^{\text{fwd, mkt}}(K, T).$$

▲

10.1.3 Pricing under the Heston model

In this section, we will price the forward start option under the *Heston stochastic volatility dynamics*.

We have already seen that the pricing of forward start options, under the assumption of constant interest rates, boils down to finding an expression for the following characteristic function, see (10.3),

$$\phi_x(u) = \mathbb{E}^{\mathbb{Q}} \left[e^{-iuX(T_1)} e^{r(T_2-T_1)} \psi_X(u, T_1, T_2) \Big| \mathcal{F}(t_0) \right]. \tag{10.8}$$

To detail this characteristic function under the Heston SV model, an expression for $\psi_X(u, T_1, T_2)$ is required. This characteristic function, with vector $\mathbf{u}^T = (u, 0)^T$, is of the following form:

$$\psi_X(u, T_1, T_2) = e^{\bar{A}(u,\tau) + \bar{B}(u,\tau)X(T_1) + \bar{C}(u,\tau)v(T_1)}, \tag{10.9}$$

with $\tau = T_2 - T_1$ and the complex-valued functions $\bar{A}(u,\tau)$, $\bar{B}(u,\tau)$ and $\bar{C}(u,\tau)$, as presented in Lemma 8.3.2. Recall that for the Heston model, we have $\bar{B}(u,\tau) = iu$, which simplifies the expression for $\phi_x(u)$ in (10.8), as follows,

$$\phi_x(u) = e^{\bar{A}(u,\tau) + r(T_2-T_1)} \mathbb{E}^{\mathbb{Q}} \left[e^{\bar{C}(u,\tau)v(T_1)} \Big| \mathcal{F}(t_0) \right]. \tag{10.10}$$

Just like in the Black-Scholes case, the forward characteristic function under the Heston SV dynamics depends neither on the stock $S(t)$ nor on the log-stock $X(t) = \log S(t)$. It is completely determined in terms of the model's volatility.

To complete the derivations for the Heston model, the expectation in (10.10) should be discussed. This expectation is *a representation of the moment-generating function* of $v(t)$. The theorem below provides us with the corresponding solution.

Theorem 10.1.2 (Moment-generating function for the CIR process)
For a stochastic process $v(t)$, with the dynamics, given by,

$$dv(t) = \kappa(\bar{v} - v(t))dt + \gamma\sqrt{v(t)}dW_v(t), \quad v(t_0) = v_0,$$

for $t \geq t_0$, the moment-generating function (or Laplace transformation) has the following form:

$$\mathbb{E}^{\mathbb{Q}} \left[e^{uv(t)} \Big| \mathcal{F}(t_0) \right] = \left(\frac{1}{1 - 2u\bar{c}(t,t_0)} \right)^{\frac{1}{2}\delta} \exp\left(\frac{u\bar{c}(t,t_0)\bar{\kappa}(t,t_0)}{1 - 2u\bar{c}(t,t_0)} \right),$$

where the parameters, $\bar{c}(t,t_0)$, degrees of freedom δ, and noncentrality $\bar{\kappa}(t,t_0)$, are given by

$$\bar{c}(t,t_0) = \frac{\gamma^2}{4\kappa}(1 - e^{-\kappa(t-t_0)}), \quad \delta = \frac{4\kappa\bar{v}}{\gamma^2}, \quad \bar{\kappa}(t,t_0) = \frac{4\kappa v_0 e^{-\kappa(t-t_0)}}{\gamma^2(1 - e^{-\kappa(t-t_0)})}, \tag{10.11}$$

as is proved below.

Proof From Remark 8.1.1, the density for the noncentral chi-squared distributed random variable $v(t)$ is given by:

$$f_{v(t)}(x) = \frac{1}{\bar{c}(t,t_0)} f_{\chi^2(\delta,\bar{\kappa}(t,t_0))}\left(\frac{x}{\bar{c}(t,t_0)}\right),$$

The density of the noncentral chi-squared distribution $\chi^2(\delta, \bar{\kappa}(t,t_0))$ reads:

$$f_{\chi^2(\delta,\bar{\kappa}(t,t_0))}(x) = \sum_{k=0}^{\infty} \frac{1}{k!} e^{-\frac{\bar{\kappa}(t,t_0)}{2}} \left(\frac{\bar{\kappa}(t,t_0)}{2}\right)^k f_{\chi^2(\delta+2k)}(x),$$

with $\chi^2(\delta + 2k)$ the chi-squared distribution with $\delta + 2k$ degrees of freedom. By the definition of the moment-generating function (see also Section 1.1) we have:

$$\mathcal{M}_{v(t)}(u) := \mathbb{E}^{\mathbb{Q}}\left[e^{uv(t)}\big|\mathcal{F}(t_0)\right]$$

$$= \frac{1}{\bar{c}(t,t_0)} \sum_{k=0}^{\infty} \frac{1}{k!} e^{-\frac{\bar{\kappa}(t,t_0)}{2}} \left(\frac{\bar{\kappa}(t,t_0)}{2}\right)^k \int_0^{\infty} e^{uy} f_{\chi^2(\delta+2k)}\left(\frac{y}{\bar{c}(t,t_0)}\right) dy.$$

Changing variables, i.e. $y = \bar{c}(t,t_0)x$, gives us

$$\mathcal{M}_{v(t)}(u) = \sum_{k=0}^{\infty} \frac{1}{k!} e^{-\frac{\bar{\kappa}(t,t_0)}{2}} \left(\frac{\bar{\kappa}(t,t_0)}{2}\right)^k \int_0^{\infty} e^{u\bar{c}(t,t_0)x} f_{\chi^2(\delta+2k)}(x) dx.$$

The integral in the last expression can be recognized as the *moment-generating function* of the chi-squared distribution, with $\delta + 2k$ degrees of freedom, i.e.

$$\mathcal{M}_{\chi^2(\delta+2k)}(u\bar{c}(t,t_0)) = \int_0^{\infty} e^{u\bar{c}(t,t_0)x} f_{\chi^2(\delta+2k)}(x) dx = \left(\frac{1}{1-2u\bar{c}(t,t_0)}\right)^{\frac{1}{2}\delta+k}.$$

This gives us, by adding and subtracting an exponential term,

$$\mathcal{M}_{v(t)}(u) = \sum_{k=0}^{\infty} \frac{1}{k!} e^{-\frac{\bar{\kappa}(t,t_0)}{2}} \left(\frac{\bar{\kappa}(t,t_0)}{2}\right)^k \left(\frac{1}{1-2u\bar{c}(t,t_0)}\right)^{\frac{1}{2}\delta+k}$$

$$= \left(\frac{1}{1-2\bar{c}(t,t_0)u}\right)^{\frac{1}{2}\delta} \exp\left(\frac{\bar{\kappa}(t,t_0)}{2(1-2u\bar{c}(t,t_0))} - \frac{\bar{\kappa}(t,t_0)}{2}\right)$$

$$\times \sum_{k=0}^{\infty} \frac{1}{k!} e^{-\frac{\bar{\kappa}(t,t_0)}{2(1-2u\bar{c}(t,t_0))}} \left(\frac{\bar{\kappa}(t,t_0)}{2(1-2u\bar{c}(t,t_0))}\right)^k.$$

The expression under the summation represents the probability $\mathbb{P}(Y = k)$ for a *Poisson distribution* with parameter $\hat{\alpha}$, i.e. $Y \sim \text{Poisson}(\hat{\alpha})$, with

$$\mathbb{P}[Y = k] = \frac{1}{k!} e^{-\hat{\alpha}} \hat{\alpha}^k.$$

By using $\hat{\alpha} := \bar{\kappa}(t,t_0)/(2(1-2u\bar{c}(t,t_0)))$, the relation is straightforward:

$$\mathcal{M}_{v(t)}(u) = \left(\frac{1}{1-2u\bar{c}(t,t_0)}\right)^{\frac{1}{2}\delta} \exp\left(\frac{\bar{\kappa}(t,t_0)}{2(1-2u\bar{c}(t,t_0))} - \frac{\bar{\kappa}(t,t_0)}{2}\right) \sum_{k=0}^{\infty} \mathbb{P}[Y=k].$$

As the summation of probabilities over all possible outcomes of the random variable is equal to 1, we thus find:

$$\mathcal{M}_{v(t)}(u) = \mathbb{E}^{\mathbb{Q}}\left[e^{uv(t)}\big|\mathcal{F}(t_0)\right] = \left(\frac{1}{1-2u\bar{c}(t,t_0)}\right)^{\frac{1}{2}\delta} \exp\left(\frac{u\bar{c}(t,t_0)\bar{\kappa}(t,t_0)}{1-2u\bar{c}(t,t_0)}\right),$$

which concludes the proof. ∎

The characteristic function $\phi_x(u)$, as it is given in (10.10) with $\mathbf{u}^{\mathrm{T}} = [u,0]^{\mathrm{T}}$, is now completely determined, as

$$\phi_x(u) = e^{\bar{A}(u,\tau)+r(T_2-T_1)} \mathbb{E}^{\mathbb{Q}}\left[e^{\bar{C}(u,\tau)v(T_1)}\big|\mathcal{F}(t_0)\right]$$

$$= \exp\left(\bar{A}(u,\tau) + r\tau + \frac{\bar{C}(u,\tau)\bar{c}(T_1,t_0)\bar{\kappa}(T_1,t_0)}{1-2\bar{C}(u,\tau)\bar{c}(T_1,t_0)}\right)\left(\frac{1}{1-2\bar{C}(u,\tau)\bar{c}(T_1,t_0)}\right)^{\frac{1}{2}\delta},$$
(10.12)

with $\tau = T_2 - T_1$, parameters δ, $\bar{\kappa}(t,t_0)$, $\bar{c}(t,t_0)$ as in (10.11), $\bar{C}(u,\tau)$ is as in the derivation of this function for the Heston characteristic function, see Equation (8.47) in Chapter 8, and $\bar{A}(u,\tau)$ is given in Lemma 8.3.2.

Example 10.1.1 (Forward implied volatilities under the Heston model)
In this experiment we compute the forward implied volatilities for the Heston model. The following set of model parameters is chosen,

$$r=0, \kappa=0.6, \bar{v}=0.1, \gamma=0.2, \rho_{x,v}=-0.5, v(t_0)=0.05.$$

In a first test, we set $T_1 = 1,2,3,4$ and $T_2 = T_1 + 2$, while in the second test we take $T_1 = 1$ and $T_2 = 2,3,4,5$. In Figure 10.1 the results are shown. Clearly, the Heston model enables forward implied volatilities and these volatilities change with the *performance period* T_1 and T_2. In the first test case, when T_1 and T_2 changing similarly, we may observe how the forward implied volatility moves in time. In the second test case, the aggregated volatilities can be observed. ♦

10.1.4 Local versus stochastic volatility model

In the financial industry, the local volatility (LV) model, as discussed in Section 4.3, is often employed for managing smile and skew risk. An undesirable feature of this model is however the flattening of the *forward implied volatility smile*, which may lead to a significant mispricing for so-called forward volatility sensitive contracts, like the forward start options. The LV model also does not accurately model the implied volatility smile movement when the value of the underlying changes. A result of this may be an unstable hedge position.

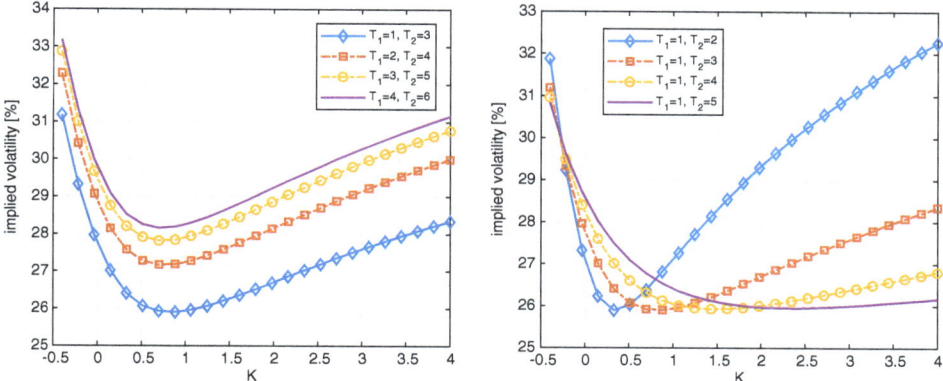

Figure 10.1: *Forward implied volatilities for the Heston model.*

As an example here, we perform a forward start option experiment in which the performance of the local volatility model is compared to the Heston stochastic volatility model. In the experiment, first of all, call option prices are generated by means of the Heston model, and subsequently these generated prices will be used *within the local volatility model*. In the local volatility model, several partial derivatives have to be approximated accurately, as described in Section 4.3.

European call options are used with maturity time $T = 1$, $S_0 = 1$ and strike prices in the range from $0.5 \cdot S_0$ up to $1.7 \cdot S_0$. In the Heston model, the following parameters are used, $\kappa = 1.3$, $\bar{v} = 0.05$, $\gamma = 0.3$, $\rho_{x,v} = -30\%$, $v(0) = 0.1$ and $r = 0$.

In Figure 10.2a the implied Black-Scholes volatilities resulting from the two models are presented. The local volatility model, with the finite difference approximations for the partial derivatives, *accurately reproduces* the implied volatility which was based on the Heston model option values. The local volatility results depend, via the local volatility formula in (4.49), completely on the European option prices at the initial time t_0. This is inherent for the LV model, but it may result in significant pricing differences for financial products with a payoff function, which is based on more than one payment/expiry date.

Because these two asset price models give approximately the same values for the European-style plain vanilla options, the two stock transition densities, from t_0 to T, *should be approximately equal*.

The implied volatility in a local volatility model however has the tendency *to flatten out*. For forward start options, starting at $t = T_1$ with maturity at time $t = T$, one would expect a volatility *with a very similar structure* as the implied volatility observed today. However, with the local volatility model the resulting volatility is *flat* and thus does not resemble the expected structure. We observe

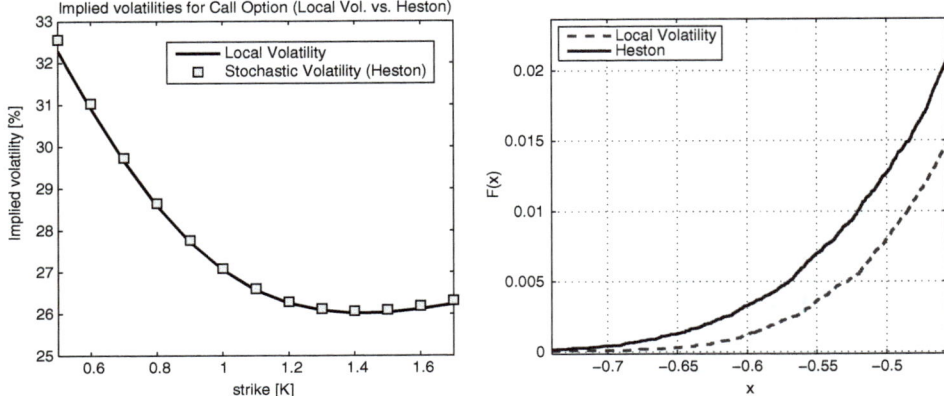

Figure 10.2: *Left: Implied volatilities for the Heston and local volatility models; Right: Zoom of the left tail of the CDFs for $Y(T_1)$, as obtained by the local volatility and the Heston model.*

this for a variable $Y(T_1)$, for $t_0 < T_1 < T$,

$$Y(T_1) := \frac{S(T) - S(T_1)}{S(T_1)},$$

which represents the "forward start" variable. In Figure 10.2b two cumulative distribution functions for $Y(T_1)$, that are generated by the Heston and by the local volatility model, are presented. Although the two models agree in their terminal distribution (the almost perfect match in Figure 10.2a), they are very different when other distributions are involved.

The SV model resembles generally the dynamics of forward start options in the financial market better than a LV model.

Remark 10.1.2 (Building block) *Forward start options are also building blocks for other financial derivatives, such as cliquet options (also called ratchet options). The cliquet options are encountered in the equity and interest rate markets, and they consist of a series of consecutive forward start options.*

Like the forward start option, the cliquet option is also based on the return,

$$R(T_k) = \frac{S(T_k) - S(T_{k-1})}{S(T_{k-1})}.$$

The returns within the cliquet option are often bounded by a cap, "Cap", and a floor "Floor", as follows,

$$R^b(T_k) = \max\left(\min\left(R(T_k), Cap\right), Floor\right).$$

The cliquet option which also comes with a global floor, $Floor_g$, is a product whose payoff is defined by the following expression,

$$V^{cliq}(T) = \max\left(\sum_{k=1}^{m} R_k^b, Floor_g\right), \quad (10.13)$$

with $Floor_g$ a global maximum amount that can be reached in the contract. The lifetime of the cliquet option, from $t = 0$ until $t = T$ is subdivided into m time points $T_k, k = 1, \ldots m$ into reset intervals of size Δt. So, there are m time points at which the bounded return R_k^b is calculated which enters the cliquet option payoff.

A useful feature of the cliquet option contracts is that they enable an investor to periodically lock in profits. If a certain asset performed well a time period $[T_k, T_{k+1}]$, but performed poorly in $[T_{k+1}, T_{k+2}]$, an investor will receive the profit which is accumulated in the first period and does not lose any money in the second period. As the possible profits may be limited as well, i.e. in the contract it can be specified that the profit cannot be larger than, say 20%, cliquet options may be seen as relatively cheap alternatives for plain-vanilla European options. ▲

10.2 Introduction into stochastic-local volatility model

In the previous section, it has been shown that the Heston stochastic volatility model generates an implied volatility forward smile, which has a shape that is very similar to the implied volatility smile which is observed today. This is in line with the observations in the financial market, and it gives rise to accurate pricing methods for forward implied volatility sensitive products. The calibration of the Heston model is however not always an easy task, and moreover not all kinds of implied volatility smiles and skews can be modeled by means of the Heston dynamics. The class of local volatility models, however, is relatively easily calibrated to plain vanilla European options in the market, but it has the drawback of the flattening of the forward implied volatility smile, due to an inaccurate representation of transition probability density functions under this model.

As a combined asset price model, aiming to take advantage of the benefits of the individual models, and remove as much as possible their drawbacks, the class of *stochastic-local volatility models* was developed by Jex et al. [1999] and Lipton [Lipton, 2002; Lipton and McGhee, 2002], amongst others. As pointed out in [Clark, 2011] and [Lipton et al., 2014], for the pricing of forward sensitive options these SLV models are often employed.

The stochastic-local volatility (SLV) model under measure $\mathbb{Q}$ is governed by the following system of SDEs:

$$\begin{cases} dS(t)/S(t) = r dt + \bar{\sigma}(t, S(t))\bar{\xi}(v(t))dW_x(t), \\ dv(t) = a_v(t, v(t))dt + b_v(t, v(t))dW_v(t), \\ dW_x(t)dW_v(t) = \rho_{x,v}dt, \end{cases} \quad (10.14)$$

with correlation parameter $\rho_{x,v}$ between the corresponding Brownian motions, W_x, W_v, and constant interest rate r. The function $\bar{\sigma}(t, S(t))$ is the *local volatility component*, while $\bar{\xi}(v(t))$ controls the *stochastic volatility*, the terms $a_v(t, v(t))$ and $b_v(t, v(t))$ represent the drift and diffusion of the variance process, respectively.

The general SLV model may collapse to either the pure SV model or to the LV model. With local volatility component $\bar{\sigma}(t, S(t)) = 1$, the model boils down to a *pure* stochastic volatility model. On the other hand, if for the stochastic component of the variance, $b_v(t, v(t)) = 0$, the model reduces to the local volatility model.

Two popular stochastic volatility models that fit into this framework are the Heston SV model [Heston, 1993] with the variance process governed by the CIR dynamics [Cox et al., 1985], $\bar{\xi}(v(t)) = \sqrt{v(t)}$, with $a_v(t, v(t)) = \kappa(\bar{v} - v(t))$, $b_v(t, v(t)) = \gamma\sqrt{v(t)}$, and the Schöbel-Zhu model [Schöbel and Zhu, 1999], with $\bar{\xi}(v(t)) = v(t)$ and $a_v(t, v(t)) = \kappa(\bar{v} - v(t))$, $b_v(t, v(t)) = \gamma$. Again, parameter κ controls the speed of mean-reversion, $\bar{v}$ the long-term mean and γ determines the volatility of process $v(t)$.

The SLV model described by the Equations (10.14) is not yet complete as the term $\bar{\sigma}(t, S(t))$ is not yet specified. This function may take different forms. We will discuss a *nonparametric form* for $\bar{\sigma}(t, S(t))$.

In the SLV framework there is thus one free function available, i.e. $\bar{\sigma}(t, S(t))$, and this local volatility component may be defined such that the densities implied from the market and the model are equal. It is well-known that based on European option market data the market implied density $f_{S(T)}(\cdot)$ of the stock $S(T)$ can be determined, see Chapter 4.

Remark 10.2.1 (Parametric form) *The local volatility component $\bar{\sigma}(t, S(t))$ may be prescribed by the constant elasticity of variance (CEV) model, i.e. $\bar{\sigma}(t, S(t)) = \sigma S^{\beta-1}(t)$, which is a well-known parametric form. Choosing a parametric form for $\bar{\sigma}(t, S(t))$ has the undesirable feature that σ and β have to be calibrated. In other words, next to the SV parameters also the LV parameters would be included in the calibration procedure.* ▲

10.2.1 Specifying the local volatility

In the following, an expression for the local volatility component $\bar{\sigma}(t, S(t))$ in the stochastic-local volatility model is derived.

We start with a European call option whose price is given by:

$$V_c(t_0, S_0) = \frac{M(t_0)}{M(t)} \mathbb{E}^{\mathbb{Q}} \left[(S(t) - K)^+ | \mathcal{F}(t_0) \right],$$

where $dM(t) = rM(t)dt$, $M(t_0) = 1$. In the following derivations, the notation $\mathcal{F}(t_0)$ is omitted, for convenience. Itô's lemma is applied to derive the *dynamics of the call option value,* as follows,

$$\begin{aligned}\mathrm{d}V_c(t_0, S_0) &= \left(\mathrm{d}\frac{1}{M(t)}\right) \mathbb{E}\left[(S(t)-K)^+\right] + \frac{1}{M(t)}\mathrm{d}\mathbb{E}\left[(S(t)-K)^+\right]\\ &= -\frac{r}{M(t)}\mathbb{E}\left[(S(t)-K)^+\right]\mathrm{d}t + \frac{1}{M(t)}\mathbb{E}\left[\mathrm{d}(S(t)-K)^+\right],\end{aligned}$$
(10.15)

where the Fubini theorem[1] justifies the equality $\mathrm{d}\left(\mathbb{E}\left[(S(t)-K)^+\right]\right) = \mathbb{E}\left[\mathrm{d}(S(t)-K)^+\right]$. Regarding the right-hand side in (10.15), Itô's lemma cannot be directly applied to determine the dynamics of $\mathrm{d}(S(t)-K)^+$, as the function $g(x) = (x-a)^+$ is not differentiable at the point $x = a$.

A function $g(t,x)$ of t and stochastic variable $X_t = x$ should be in C^2 w.r.t. the space variable x. Various generalizations of the Itô formula have been presented in the literature for functions $g(t,X)$ that are not in C^2 regarding the space variable. A well-known extension is the so-called *Tanaka-Meyer formula*, as derived in [Tanaka, 1963] for $g(t,X) = |X|$, and extended to absolutely continuous $g(t,X)$ functions, with $\frac{\partial g(t,X)}{\partial X}$ of bounded variation, by [Meyer, 1976] and [Wang, 1977]. See also [Protter, 2005; Karatzas and Shreve, 1991].

Theorem 10.2.1 (Tanaka-Meyer formula) *Given a probability space $(\Omega, \mathcal{F}, \mathbb{Q})$, let, for $t_0 \leq t < \infty$, $X(t) = X(t_0) + X_I(t) + X_{II}(t)$ be a semimartingale, where $X_I(t)$ is a continuous local martingale, and $X_{II}(t)$ is a càdlàg adapted process of locally bounded variation, i.e. $X_{II}(t)$ is defined on $\mathbb{R}$ (or a subset) and is right-continuous with left limits almost everywhere,
Then, for the function $g(x) = (x-a)^+$, with $a \in \mathbb{R}$, it follows that,*

$$g(X(t)) = g(X(t_0)) + \int_{t_0}^t \mathbb{1}_{X(z)>a}\mathrm{d}X_I(z) + \int_{t_0}^t \mathbb{1}_{X(z)>a}\mathrm{d}X_{II}(z)$$
$$+ \frac{1}{2}\int_{t_0}^t g''(X(z))(\mathrm{d}X_I(z))^2.$$

A full proof can be found in Tanaka [1963].

Remark 10.2.2 (Local martingale) *A local martingale is a stochastic processes which is locally a martingale. In general, a local martingale is not a martingale, because its expectation may be distorted by large values of a small probability. Itô processes that are described by an SDE without drift term are local martingales, but not always true martingales. The GBM process, with $\mathrm{d}X(t) =$*

---

[1] Fubini's Theorem states that a sufficient condition for the equality, $\mathbb{E}[\int_0^t g(X(z))\mathrm{d}z] = \int_0^t \mathbb{E}[g(X(z))]\mathrm{d}z$ to hold is some regularity of $X(t)$ and g and the finiteness of $\int_0^t \mathbb{E}[|g(X(z))|]\mathrm{d}z$.

$X(t)\mathrm{d}W(t)$, is an example of a local martingale which is also a true martingale, but the CEV model, $\mathrm{d}X(t) = X^\beta(t)\mathrm{d}W(t)$, with an exponent greater than one is a local martingale but it is not a true martingale, because $\mathbb{E}[X(t)|\mathcal{F}_0] < X_0, \forall t > 0$.

$X(t)$ is a strictly local martingale when $\beta > 1$ and $X(t)$ can be shown to be a martingale for $\beta < 1$, as presented in [Lindsay and Brecher, 2012]. ▲

Stopping times & local martingales

The general definition of the local martingale is connected to the concept of the stopping time, as will be defined below.

Definition 10.2.1 (Stopping time) *A nonnegative random variable τ_S is called a stopping time, with respect to the filtration $\mathcal{F}(t)$, if for every $t \geq 0$ the event $\{\tau_S \leq t\}$ is $\mathcal{F}(t)$- measurable, or, in other words, if the following holds true, for the sigma-field,*

$$\sigma(\tau_S \leq t) \in \mathcal{F}(t).$$

If $\tau_S \leq \infty$ a.s., we call τ_S a finite stopping time. ◀

Intuitively, τ_S is a stopping time if for every time t we can determine whether τ_S has occurred before time t on basis of information which we have up to time t.

Example 10.2.1 (Stopping time) Let us consider a Brownian motion $W(t)$ and a stopping time which is defined as $\tau_S = \min\{t : W(t) > 2\}$. The stopping time represents the first time when Brownian motion $W(t)$ reaches the value 2. In Figure 10.3 some paths are presented. The black dots indicate the moments when a path hits the value 2. By the repeating this exercise for many paths, the distribution of the stopping time τ_S may be visualized (see the right-hand side

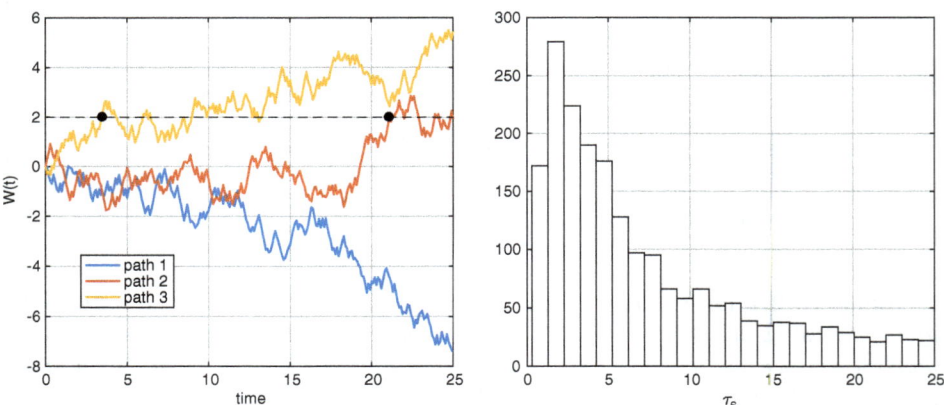

Figure 10.3: *Left: Monte Carlo paths of Brownian motion $W(t)$, and the times at which $W(t) > 2$ for the first time. Right: histogram of the stopping time τ_S obtained with 3000 paths.*

in Figure 10.3). Based on that distribution the most likely time point for which the process will reach a certain level can be determined. ♦

Remark 10.2.3 (Stopped process) *The stopped process $X^{\tau_S}(t)$ is defined by:*

$$X^{\tau_S}(t) := X(t \wedge \tau_S) = \begin{cases} X(t) & t \leq \tau_S, \\ X(\tau_S), & t > \tau_S. \end{cases} \quad (10.16)$$

▲

An important property of the stopped process is the martingale property, which is given in the proposition below.

Proposition 10.2.1 *If a process $X(t)$ is a martingale, then the stopped process $X^{\tau_S}(t) = X(t \wedge \tau_S)$ is also a martingale. In particular, for all $t \geq 0$, we have:*

$$\mathbb{E}\left[X^{\tau_S}(t)|\mathcal{F}(t_0)\right] = \mathbb{E}\left[X(t_0)|\mathcal{F}(t_0)\right] = X(t_0).$$

The proof can be found in [Williams, 1991]- Theorem 10.9. ◀

With the connection of the martingale and the stopping time, we may move forward and define the concept of a local martingale.

Definition 10.2.2 (Local martingale) *A continuous adapted stochastic process $X(t)$ is a local martingale, if a sequence of $\tau_{S,i_{i \in \mathbb{N}}}$ of stopping times exists, such that,*

1. *$\tau_{S,1} \leq \tau_{S,2} \leq \ldots$ and $\tau_{S,i} \to \infty$, a.s.*

2. *the stopped processes $X^{\tau_S,i}$ are martingales for all i.*

The sequence $\tau_{S,i}$ is called a fundamental or localizing sequence. ◀

From the definition above it is clear that any martingale is also a local martingale, but the converse is not always true.

Derivation of stochastic local volatility term

We return to our derivation in Equation (10.15). Applying the Tanaka-Meyer formula, gives us

$$(S(t) - K)^+ = (S(t_0) - K)^+ + \int_{t_0}^{t} \mathbb{1}_{S(z)>K} dS(z) + \frac{1}{2}\int_{t_0}^{t} \delta\left(S(z) - K\right) (dS(z))^2,$$

where δ is the Dirac delta function, see (1.17) and also Example 4.2.2. It is here written, with some abuse of notation (for convenience), in differential form as,

$$d(S(t) - K)^+ = \mathbb{1}_{S(t)>K} dS(t) + \frac{1}{2}\delta\left(S(t) - K\right) (dS(t))^2.$$

Substituting the SLV dynamics of $S(t)$, gives us, to leading order,

$$d\left(S(t) - K\right)^+ = \mathbb{1}_{S(t)>K}\left(rS(t)dt + \bar{\sigma}(t, S(t))\bar{\xi}(v(t))S(t)dW_x(t)\right)$$
$$+ \frac{1}{2}\delta\left(S(t) - K\right)\bar{\sigma}^2(t, S(t))\bar{\xi}^2(v(t))S^2(t)dt.$$

The dynamics of the call option price can now be written as:

$$dV_c(t_0, S_0) = -\frac{r}{M(t)}\mathbb{E}\left[(S(t) - K)^+\right]dt$$

$$+ \frac{1}{M(t)}\mathbb{E}\left[\mathbf{1}_{S(t)>K}\left(rS(t)dt + \bar{\sigma}(t, S(t))\bar{\xi}(v(t))S(t)dW_x(t)\right)\right]$$

$$+ \frac{1}{2M(t)}\mathbb{E}\left[\delta\left(S(t) - K\right)\bar{\sigma}^2(t, S(t))\bar{\xi}^2(v(t))S^2(t)\right]dt.$$

This equation can be simplified by using the equality

$$\mathbb{E}\left[(S(t) - K)^+\right] = \mathbb{E}\left[\mathbf{1}_{S(t)>K}(S(t) - K)\right]$$

$$= \mathbb{E}\left[\mathbf{1}_{S(t)>K}S(t)\right] - K\mathbb{E}\left[\mathbf{1}_{S(t)>K}\right],$$

which leads to the following Result 10.2.1.

Result 10.2.1 *The dynamics of the European call option price $V_c(t_0, S_0)$, with $S(t)$ and $v(t)$ following the dynamics as given in (10.14), are given by*

$$dV_c(t_0, S_0) = \frac{rK}{M(t)}\mathbb{E}\left[\mathbf{1}_{S(t)>K}\right]dt$$

$$+ \frac{1}{2M(t)}\mathbb{E}\left[\delta\left(S(t) - K\right)\bar{\sigma}^2(t, S(t))\bar{\xi}^2(v(t))S^2(t)\right]dt,$$

where the expectations are conditioned on $\mathcal{F}(t_0)$.

In the following, we use another result: Lemma 10.2.1.

Lemma 10.2.1 *The European call option price $V_c(t_0, S_0; K, T) = V_c(t_0, S_0)$, with $S(t)$ and $v(t)$ following the dynamics as in (10.14), satisfies*

$$\frac{\partial V_c(t_0, S_0; K, t)}{\partial K} = -\frac{1}{M(t)}\mathbb{E}\left[\mathbf{1}_{S(t)>K}|\mathcal{F}(t_0)\right], \quad (10.17)$$

and

$$\frac{\partial V_c^2(t_0, S(t_0); K, t)}{\partial K^2} = \frac{f_{S(t)}(K)}{M(t)}, \quad (10.18)$$

where $f_{S(t)}$ is the marginal probability density function of $S(t)$. ◀

The proof of this lemma is very similar to the derivations that are found in Chapter 4.

Returning to the dynamics of the call price in Result 10.2.1, including the results from Lemma 10.2.1, gives,

$$dV_c(t_0, S_0; K, t) = -rK\frac{\partial V_c(t_0, S_0; K, t)}{\partial K}dt \quad (10.19)$$

$$+ \frac{1}{2M(t)}\mathbb{E}\left[\delta\left(S(t) - K\right)\bar{\sigma}^2(t, S(t))\bar{\xi}^2(v(t))S^2(t)\right]dt,$$

which gives us:
$$\Xi(t)dt := \mathbb{E}\left[\delta\left(S(t)-K\right)\bar{\sigma}^2(t,S(t))\bar{\xi}^2(v(t))S^2(t)\right]dt \qquad (10.20)$$
$$= 2M(t)\left(dV_c(t_0,S_0;K,t)+rK\frac{\partial V_c(t_0,S_0;K,t)}{\partial K}dt\right)$$

with,
$$\Xi(t) := \iint_{\mathbb{R}} \delta\left(s-K\right)\bar{\sigma}^2(t,s)\bar{\xi}^2(z)s^2 f_{v(t),S(t)}(z,s)dsdz$$
$$= \int_{\mathbb{R}} \bar{\xi}^2(z)\left(\int_{\mathbb{R}} \delta(s-K)s^2\bar{\sigma}^2(t,s)f_{v(t),S(t)}(z,s)ds\right)dz. \qquad (10.21)$$

Using the equality, $\int_{\mathbb{R}} \delta(s-K)f_{S(t)}(s)ds = f_{S(t)}(K)$, the inner integral simplifies to:
$$\int_{\mathbb{R}} \delta(s-K)s^2\bar{\sigma}^2(t,s)f_{v(t),S(t)}(z,s)ds = K^2\bar{\sigma}^2(t,K)f_{v(t),S(t)}(z,K). \qquad (10.22)$$

Then, the expression for the term $\Xi(t)$ is given by
$$\Xi(t) = K^2\bar{\sigma}^2(t,K)\int_{\mathbb{R}} \bar{\xi}^2(z)f_{v(t),S(t)}(z,K)dz, \qquad (10.23)$$

which is equivalent to:
$$\Xi(t) = K^2\bar{\sigma}^2(t,K)f_{S(t)}(K)\mathbb{E}\left[\bar{\xi}^2(v(t))|S(t)=K\right].$$

The dynamics are given by:
$$dV_c(t_0,S_0;K,t) = -rK\frac{\partial V_c(t_0,S_0;K,t)}{\partial K}dt \qquad (10.24)$$
$$+ \frac{1}{2M(t)}K^2\bar{\sigma}^2(t,K)f_{S(t)}(K)\mathbb{E}\left[\bar{\xi}^2(v(t))|S(t)=K\right]dt.$$

By using the second equation in Lemma 10.2.1, we obtain:
$$dV_c(t_0,S_0;K,t) = \left(-rK\frac{\partial V_c(t_0,S_0;K,t)}{\partial K}\right. \qquad (10.25)$$
$$\left.+ \frac{1}{2}K^2\bar{\sigma}^2(t,K)\mathbb{E}\left[\bar{\xi}^2(v(t))|S(t)=K\right]\frac{\partial^2 V_c(t_0,S_0;K,t)}{\partial K^2}\right)dt,$$

which can be expressed as:
$$\bar{\sigma}^2(t,K)\mathbb{E}\left[\bar{\xi}^2(v(t))|S(t)=K\right] = \frac{\frac{\partial V_c(t_0,S_0;K,t)}{\partial t}+rK\frac{\partial V_c(t_0,S_0;K,t)}{\partial K}}{\frac{1}{2}K^2\frac{\partial^2 V_c(t_0,S_0;K,t)}{\partial K^2}}$$
$$=: \sigma_{\text{LV}}^2(t,K),$$

where $\sigma_{\text{LV}}(t,K)$ denotes the *Dupire local volatility term* [Dupire, 1994].

> The following relation is now obtained, in the context of the SLV model,
>
> $$\bar{\sigma}^2(t,K) = \frac{\sigma_{\text{LV}}^2(t,K)}{\mathbb{E}\left[\bar{\xi}^2(v(t))|S(t)=K\right]}. \qquad (10.26)$$
>
> The SLV local volatility component $\bar{\sigma}^2(t,K)$ thus consists of two components, a deterministic local volatility $\sigma_{\text{LV}}(t,K)$ and a conditional expectation $\mathbb{E}[\bar{\xi}^2(v(t))|S(t)=K]$.

Numerical evaluation of the term $\sigma_{\text{LV}}(t,K)$ is well-established, see Section 4.3, and [Andreasen and Huge, 2011; Coleman et al., 1999; De Marco et al., 2013]. The efficient computation of the conditional expectation in (10.26) needs to be discussed next. The stock process $S(t)$ contains a local volatility component $\sigma_{\text{LV}}(t,S)$, which is not known analytically. The difficulty lies however in the fact that the joint distribution of the variance v and the stock S, i.e. $f_{v(t),S(t)}$, is not known.

Remark 10.2.4 (Literature) *In an SLV framework a conditional expectation thus needs to be determined. The exact form of it depends on the choice of the particular stochastic process. The conditional expectation cannot be extracted directly from the market quotes, which is thus different from the LV model model situation. A common approach consists of solving the Kolmogorov forward partial differential equation (PDE) [Deelstra and Rayée, 2012; Ren et al., 2007; Clark, 2011], approximating the conditional expectation and the stochastic-local volatility component simultaneously. PDE discretization techniques are common practice in the financial industry in a hybrid local volatility context.*

The Markovian projection technique has also been applied in an SLV context, for which we refer to [Piterbarg, 2007; Henry-Labordère, 2009]. Although this method is generally applicable, it involves several conditional expectations that need to be approximated.

In [Lorig et al., 2015] for a general class of stochastic-local volatility models a family of asymptotic expansions for European-style option prices and implied volatilities is derived. Further, in [Pascucci and Mazzon, 2017] the authors derive an asymptotic expansion for forward start options in a multi-factor local-stochastic volatility model, which results in explicit approximation formulas for the forward implied volatility.

In [van der Stoep et al., 2014], in a Monte Carlo setting, a nonparametric method was introduced for the evaluation of the problematic conditional expectation, which relies on splitting the Monte Carlo realizations in bins. A similar technique is presented in [Guyon and Henry-Labordère, 2012; Jourdain and Sbai, 2012], based on kernel estimators in an interacting particle system. ▲

10.2.2 Monte Carlo approximation of SLV expectation

We concentrate on the Monte Carlo evaluation of the SLV model.

First of all, by an Euler discretization, we may simulate the SLV model (10.14), as follows,

$$s_{i+1,j} = s_{i,j} + rs_{i,j}\Delta t + \sqrt{\frac{\sigma_{\text{LV}}^2(t_i, s_{i,j})}{\mathbb{E}\left[\bar{\xi}^2(v(t_i))|S(t_i) = s_{i,j}\right]}} s_{i,j}\bar{\xi}(v_{i,j})\sqrt{\Delta t}Z_x, \quad (10.27)$$
$$v_{i+1,j} = v_{i,j} + a_v(t_i, v_{i,j})\Delta t + b_v(t_i, v_{i,j})\sqrt{\Delta t}Z_v.$$

with $j = 1, \ldots, N$ (the number of Monte Carlo paths) and $i = 0, \ldots, m$ (the number of time steps), $Z_x = Z_1$, $Z_v = \rho_{x,v}Z_1 + \sqrt{1 - \rho_{x,v}^2}Z_2$, Z_1 and Z_2 independent standard normal variables. The time step $\Delta t = i \cdot \frac{T}{m}$.

To determine the values of the asset paths for the *next* time step, t_{i+1}, two components, $\sigma_{\text{LV}}^2(t_i, s_{i,j})$ and $\mathbb{E}\left[\bar{\xi}^2(v(t_i))|S(t_i) = s_{i,j}\right]$ should be accurately approximated. Efficient evaluation of $\sigma_{\text{LV}}^2(t_i, s_{i,j})$ is already explained, see Chapter 4, but the conditional expectation needs to be discussed. A difficulty is that the conditioning for the evaluation should be performed for each individual stock realization, $s_{i,j}$, i.e., in a discretization for (S, v), each realization of $s_{i,j}$ has exactly one corresponding realization of the variance $v_{i,j}$ and this makes the evaluation of the conditional expectation difficult.

In the next subsection, a nonparametric method is presented for the conditional expectation.

Nonparametric method with bins

Suppose that for a given discretization in time, with t_i, $i = 1, \ldots, m$, there are N pairs of Monte Carlo realizations, $(s_{i,1}, v_{i,1})$, $(s_{i,2}, v_{i,2})$, $\ldots$, $(s_{i,N}, v_{i,N})$, for which the conditional expectation in (10.27) should be approximated. For each $s_{i,j}$-value, we have only one $v_{i,j}$-value, and the conditional expectation will then be equal to $\bar{\xi}^2(v_{i,j})$, which is inaccurate. This is however a consequence of the discretization of the continuous system (S, v). An accurate estimate would obviously require an *infinite set of paths*, which is practically unfeasible.

A technique to overcome this problem is based on *grouping pairs of realizations into bundles*, which may then provide a more accurate estimate for the desired expectation. Let's subdivide the range of $S(t_i)$-values into l mutually exclusive bins, $(b_1, b_2], (b_2, b_3], \ldots, (b_l, b_{l+1}]$, with $b_1 \geq 0$ and $b_{l+1} < \infty$.

For any particular stock realization $s_{i,j}$, for which $s_{i,j} \in (b_k, b_{k+1}]$ for some $k \in \{1, 2, \ldots, l\}$, the following approximation is then proposed,

$$\mathbb{E}\left[\bar{\xi}^2(v(t_i))|S(t_i) = s_{i,j}\right] \approx \mathbb{E}\left[\bar{\xi}^2(v(t_i))|S(t_i) \in (b_k, b_{k+1}]\right]. \quad (10.28)$$

By defining the left and right boundaries of $(b_k, b_{k+1}]$ to be $s_{i,j} - \varepsilon$ and $s_{i,j} + \varepsilon$, respectively, we would obtain,

$$\mathbb{E}\left[\bar{\xi}^2(v(t_i))|S(t_i) = s_{i,j}\right] = \lim_{\varepsilon \to 0^+} \mathbb{E}\left[\bar{\xi}^2(v(t_i))|S(t_i) \in (s_{i,j} - \varepsilon, s_{i,j} + \varepsilon]\right]$$

$$= \lim_{\varepsilon \to 0^+} \frac{\mathbb{E}\left[\bar{\xi}^2(v(t_i))\mathbb{1}_{S(t_i) \in (s_{i,j} - \varepsilon, s_{i,j} + \varepsilon]}\right]}{\mathbb{Q}\left[S(t_i) \in (s_{i,j} - \varepsilon, s_{i,j} + \varepsilon]\right]}. \quad (10.29)$$

In the limiting case, where the two boundaries of the bin are equal to $s_{i,j}$, the approximation of the conditional expectation would converge to its exact value. This is an indication for the suitability of the approximation in (10.28). *Proper bin boundaries* b_k for $k = 1, \ldots, l+1$, are obtained by ordering the stock path values, $s_{i,1}, s_{i,2}, \ldots, s_{i,N}$, to get the increasing sequence, $\bar{s}_{i,1} \leq \bar{s}_{i,2} \leq \cdots \leq \bar{s}_{i,N}$, with $\bar{s}_{i,1}$ and $\bar{s}_{i,N}$ the minimum and maximum values at time step i, respectively. The bins are defined such that each bin contains approximately the same number of Monte Carlo paths, The bin boundaries, $b_{i,k}$, $k = 1, \ldots, l+1$, thus depend on the number of paths per bin.

$$b_{i,1} = \bar{s}_{i,1}, \ b_{i,l+1} = \bar{s}_{i,N}, \ b_{i,k} = \bar{s}_{i,(k-1)N/l}, \ k = 2 \ldots l. \quad (10.30)$$

Each pair $(s_{i,j}, v_{i,j})$ is now assigned to a bin according to its $s_{i,j}$-value. With the path numbers corresponding to the kth bin, at time t_i, denoted by $\mathcal{J}_{i,k}$, i.e. $\mathcal{J}_{i,k} := \{j|(s_{i,j}, v_{i,j}) \in (b_{i,k}, b_{i,k+1}]\}$, and the number of paths in the kth bin, $N_k := |\mathcal{J}_{i,k}|$, the conditional expectation is approximated by,

$$\mathbb{E}\left[\bar{\xi}^2(v(t_i))\middle| S(t_i) = s_{i,j}\right] \approx \frac{\mathbb{E}\left[\bar{\xi}^2(v(t_i))\mathbb{1}_{S(t_i) \in (b_{i,k}, b_{i,k+1}]}\right]}{\mathbb{Q}\left[S(t_i) \in (b_{i,k}, b_{i,k+1}]\right]}$$

$$\approx \frac{\frac{1}{N}\sum_{j=1}^{N} \bar{\xi}^2(v_{i,j})\mathbb{1}_{s_{i,j} \in (b_{i,k}, b_{i,k+1}]}}{\mathbb{Q}\left[S(t_i) \in (b_{i,k}, b_{i,k+1}]\right]}$$

$$= \frac{1}{N\alpha(k)} \sum_{j \in \mathcal{J}_{i,k}} \bar{\xi}^2(v_{i,j}), \quad (10.31)$$

where $\alpha(k) := \mathbb{Q}\left[S(t_i) \in (b_{i,k}, b_{i,k+1}]\right]$ represents the probability that the stock value is in the kth bin.

The second approximation step is by approximation of the expectation by an average which is based on a finite number of $(s_{i,j}, v_{i,j})$-pairs. The value of $\alpha(k)$ depends on the way the bins are chosen, i.e. $\alpha(k) = 1/l$.

Remark 10.2.5 *The choice of bins affects the convergence of the nonparametric method. When the bins are defined according to (10.30), this implies that bins close to the mean value of the joint density are much smaller in size than the bins in the tails. This is desirable as the region close to the mean contains many more observations, requiring a high accuracy and thus small bin sizes.* ▲

Algorithm 10.1 Nonparametric method

For each time step $t_i, i = 1 \ldots m$
  Generate N pairs of observations $(s_{i,j}, \bar{\xi}^2(v_{i,j}))$, $j = 1 \ldots N$.
  Sort the elements $\bar{s}_{i,j}$: $\bar{s}_{i,1} \leq \bar{s}_{i,2} \leq \cdots \leq \bar{s}_{i,N}$.
  Determine the boundaries of l bins $(b_{i,k}, b_{i,k+1}]$, $k = 1 \ldots l$, according to (10.30).
  For the kth bin: approximate
  $$\mathbb{E}\left[\bar{\xi}^2(v(t_i)) \middle| S(t_i) \in (b_{i,k}, b_{i,k+1}]\right] \approx \frac{1}{N\alpha(k)} \sum_{j \in \mathcal{J}_{i,k}} \bar{\xi}^2(v_{i,j}),$$
  with $\mathcal{J}_{i,k}$ the path numbers j with $s_{i,j} \in (b_k, b_{k+1})$
  and $\alpha(k)$ the probability of the stock being in the kth bin.

We summarize the nonparametric method in Algorithm 10.1.

Example 10.2.2 A first example is based on the pure Heston SV model, where $\bar{\xi}(x) = \sqrt{x}$ and the conditional expectation is given by $\mathbb{E}\left[v(t_i)|S(t_i) = s_{i,j}\right]$. For the *pure* Heston model the conditional expectation can also be computed highly accurately by the 2D version of COS method, from [Ruijter and Oosterlee, 2012].

In Figure 10.4 the results of the computation of the conditional expectation by the bin method and the reference function, obtained by Fourier expansions are presented. Each figure contains a contour plot of the recovered joint PDF, the corresponding conditional expectation, and its approximation by means of the bins. In the simulations, 10^5 Monte Carlo paths, with 5 and 20 bins, respectively, are used. The approximation obtained by the algorithm introduced converges to the reference expectation. ♦

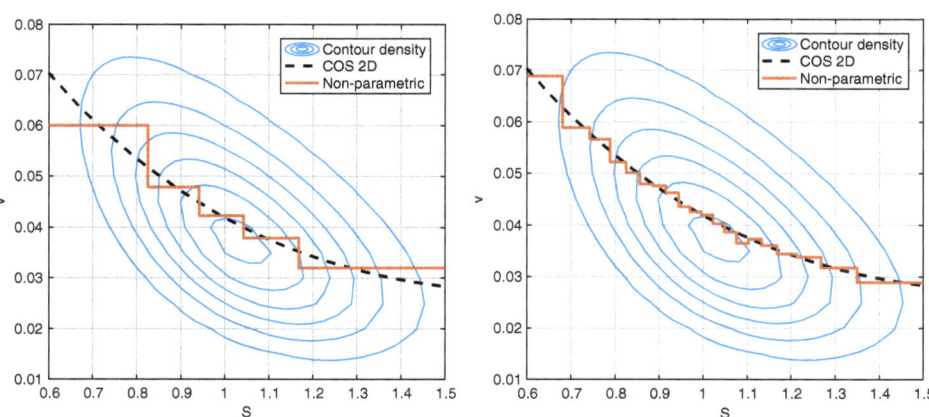

Figure 10.4: *The approximation obtained by the nonparametric bin method converges to the conditional expectation obtained by the COS method when the number of bins increases (5 and 20, respectively).*

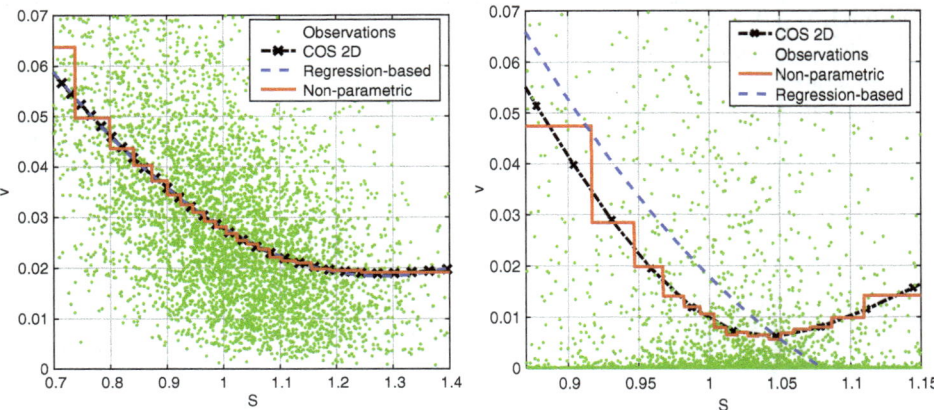

Figure 10.5: *A continuous approximation ('CA') gives a better fit to the theoretical conditional expectation, which is recovered by the COS method (left), and a zoomed plot (right).*

As the conditional expectation is a continuous function, the approximation should satisfy this property too. Furthermore, at the left boundary the fit of the nonparametric approximation to the reference can be improved. To obtain a continuous approximation the mid-points of the approximations of the nonparametric method are connected by interpolation, see Figure 10.5.

10.2.3 Monte Carlo AES scheme for SLV model

The *Almost Exact Simulation scheme* (AES), which is based on the QE scheme [Andersen, 2008], see Section 9.3.4, is generalized to the simulation of the Heston SLV model. The main difference in Monte Carlo simulation between the pure Heston SV model and the Heston SLV model is found in the fact that the variance of the latter is also governed by a state-dependent local volatility component. This requires an additional *approximation* (i.e. freezing of coefficients) to be explained. Numerical experiments show that with this additional approximation an accurate simulation result is still obtained.

Recall the dynamics of the Heston SLV model, which is expressed in terms of independent Brownian motions,

$$\begin{cases} \mathrm{d}S(t)/S(t) = r\mathrm{d}t + \bar{\sigma}(t, S(t))\sqrt{v(t)}\left(\rho_{x,v}\mathrm{d}\widetilde{W}_v(t) + \sqrt{1-\rho_{x,v}^2}\mathrm{d}\widetilde{W}_x(t)\right), \\ \mathrm{d}v(t) = \kappa(\bar{v} - v(t))\mathrm{d}t + \gamma\sqrt{v(t)}\mathrm{d}\widetilde{W}_v(t), \end{cases}$$

with $\rho_{x,v}$ the correlation parameter between the $S(t)$ and $v(t)$ processes. The discretization for $X(t) = \log(S(t))$, with (some abuse of notation)

$\bar{\sigma}(t, X(t)) := \bar{\sigma}(t, e^{X(t)})$, reads,

$$X(t + \Delta t) = X(t) + \int_t^{t+\Delta t} \left(r - \frac{1}{2}\bar{\sigma}^2(z, X(z))v(z)\right) dz$$

$$+ \rho_{x,v} \int_t^{t+\Delta t} \bar{\sigma}(z, X(z))\sqrt{v(z)}d\widetilde{W}_v(z)$$

$$+ \sqrt{1 - \rho_{x,v}^2} \int_t^{t+\Delta t} \bar{\sigma}(z, X(z))\sqrt{v(z)}d\widetilde{W}_x(z). \quad (10.32)$$

Recall from Section 8.1.2 that the variance process follows a scaled noncentral chi-squared distribution, i.e.

$$v(t + \Delta t)|v(t) \sim \bar{c}(t + \Delta t, t)\chi^2(\delta, \bar{\kappa}(t + \Delta t, t)), \quad (10.33)$$

with $\chi^2(\delta, \bar{\kappa}(t + \Delta t, t))$ the noncentral chi-squared distribution with δ degrees of freedom and noncentrality parameter $\bar{\kappa}(t + \Delta t, t)$, and,

$$\bar{c}(t + \Delta t, t) = \frac{\gamma^2}{4\kappa}(1 - e^{-\kappa\Delta t}), \quad \delta = \frac{4\kappa\bar{v}}{\gamma^2}, \quad \bar{\kappa}(t + \Delta t, t) = \frac{4\kappa e^{-\kappa\Delta t}}{\gamma^2(1 - e^{-\kappa\Delta t})}v(t).$$

Integrating the variance process, see also Equation (9.49), gives,

$$\int_t^{t+\Delta t} \sqrt{v(z)}d\widetilde{W}_v(z) = \frac{1}{\gamma}\left(v(t + \Delta t) - v(t) - \kappa\bar{v}\Delta t + \kappa \int_t^{t+\Delta t} v(z)dz\right). \quad (10.34)$$

In the last integral in (10.32) the local and stochastic volatilities are coupled. This complicates the simulation as the integrated variance, as in (10.34), cannot be directly used. Since Monte Carlo simulation with a local volatility component typically involves many time steps, $\bar{\sigma}(s, X(s))$ in (10.32) is *frozen locally*, i.e.,

$$\int_t^{t+\Delta t} \bar{\sigma}(z, X(z))\sqrt{v(z)}d\widetilde{W}_v(z) \approx \bar{\sigma}(t, X(t)) \int_t^{t+\Delta t} \sqrt{v(z)}d\widetilde{W}_v(z). \quad (10.35)$$

Due to the approximation in (10.35), it is possible to use (10.34) in (10.32):

$$X(t + \Delta t) \approx X(t) + \int_t^{t+\Delta t} \left(r - \frac{1}{2}\bar{\sigma}^2(z, X(z))v(z)\right) dz$$

$$+ \frac{\rho_{x,v}\bar{\sigma}(t, X(t))}{\gamma}\left(v(t + \Delta t) - v(t) - \kappa\bar{v}\Delta t + \kappa \int_t^{t+\Delta t} v(z)dz\right)$$

$$+ \sqrt{1 - \rho_{x,v}^2} \int_t^{t+\Delta t} \bar{\sigma}(z, X(z))\sqrt{v(z)}d\widetilde{W}_x(z).$$

With an Euler discretization for all integrals w.r.t. time, the discretized process for $X(t)$ reads,

$$X(t+\Delta t) \approx X(t) + r\Delta t - \frac{1}{2}\bar{\sigma}^2(t, X(t))v(t)\Delta t$$

$$+ \frac{1}{\gamma}\rho_{x,v}\bar{\sigma}(t, X(t))\left(v(t+\Delta t) - v(t) - \kappa\bar{v}\Delta t + \kappa v(t)\Delta t\right)$$

$$+ \sqrt{1-\rho_{x,v}^2}\int_t^{t+\Delta t}\bar{\sigma}(z, X(z))\sqrt{v(z)}\mathrm{d}\widetilde{W}_x(z).$$

Furthermore, by the Itô isometry, we have

$$\int_t^{t+\Delta t}\bar{\sigma}(z, X(z))\sqrt{v(z)}\mathrm{d}\widetilde{W}_x(z) \sim \widetilde{Z}_x\sqrt{\int_t^{t+\Delta t}\bar{\sigma}^2(z, X(z))v(z)\mathrm{d}z}, \qquad (10.36)$$

where $\widetilde{Z}_x \sim N(0,1)$. The integral at the right side of (10.36) is again approximated by the Euler discretization, i.e. $\int_t^{t+\Delta t}\bar{\sigma}^2(z, X(z))v(z)\mathrm{d}z \approx \bar{\sigma}^2(t, X(t))v(t)\Delta t$, so that the discretization scheme reads

$$v_{i+1,j} \sim \bar{c}(t+\Delta t, t)\chi^2(\delta, \bar{\kappa}(t_{i+1}, t_i)),$$

$$x_{i+1,j} = x_{i,j} + r\Delta t - \frac{1}{2}\bar{\sigma}^2(t_i, x_{i,j})v_{i,j}\Delta t + \frac{\rho_{x,v}}{\gamma}\bar{\sigma}(t_i, x_{i,j})\left(v_{i+1,j} - \kappa\bar{v}\Delta t\right.$$

$$\left. + v_{i,j}(\kappa\Delta t - 1)\right) + \sqrt{1-\rho_{x,v}^2}\sqrt{\bar{\sigma}^2(t_i, x_{i,j})v_{i,j}\Delta t}\widetilde{Z}_x,$$

with

$$\bar{\sigma}^2(t_i, x_{i,j}) \stackrel{\text{def}}{=} \bar{\sigma}^2(t_i, e^{x_{i,j}}) = \frac{\sigma_{\mathrm{LV}}^2(t_i, s_{i,j})}{\mathbb{E}\left[v(t_i)|S(t_i) = s_{i,j}\right]}. \qquad (10.37)$$

In (10.37), Dupire's local volatility component is computed,

$$\sigma_{\mathrm{LV}}^2(t_i, s_{i,j}) = \left.\frac{\frac{\partial V_c(t_0, S_0; s, t)}{\partial t} + rs\frac{\partial V_c(t_0, S_0; s, t)}{\partial s}}{\frac{1}{2}s^2\frac{\partial^2 V_c(t_0, S_0; s, t)}{\partial s^2}}\right|_{s=s_{i,j}, t=t_i}$$

by using the usual finite difference approximations, with time step Δt and asset price step Δs,

$$\left.\frac{\partial V_c(t, s_{i,j})}{\partial t}\right|_{t=t_i} \approx \frac{V_c(t_i + \Delta t, s_{i,j}) - V_c(t_i, s_{i,j})}{\Delta t},$$

$$\left.\frac{\partial V_c(t_i, s)}{\partial s}\right|_{s=s_{i,j}} \approx \frac{V_c(t_i, s_{i,j} + \Delta s) - V_c(t_i, s_{i,j})}{\Delta s},$$

and

$$\left.\frac{\partial^2 V_c(t_i, s)}{\partial s^2}\right|_{s=s_{i,j}} \approx \frac{V_c(t_i, s_{i,j} + \Delta s) - 2V_c(t_i, s_{i,j}) + V_c(t_i, s_{i,j} - \Delta s)}{\Delta s^2}. \qquad (10.38)$$

For stability reasons, the derivatives are often expressed in terms of implied volatilities, see Section 4.3.1, and also [Deelstra and Rayée, 2012]. As a continuum of European call prices in time-to-maturity and strike is not available, interpolation is required, as explained in Section 4.3.3, and [Andreasen and Huge, 2011].

Example 10.2.3 (Efficient simulation scheme) The presented AES scheme is compared to the Euler discretization scheme here, for the Heston SLV model. We generate our own "market data", by means of the plain Heston SV model, with the following parameter values (which is Case III in [Andersen, 2008]),

$$\kappa = 1.05, \ \gamma = 0.95, \ \bar{v} = 0.0855, \ v_0 = 0.0945, \ \rho_{x,v} = -0.315, r = 0,$$

and $T = 5$.

Then, we will employ the Heston SLV model to resemble the "market observed implied volatilities". We will use the following SLV parameters for this purpose, $\gamma = 0.7125$, $\kappa = 1.3125$, $\rho_{x,v} = -0.3937$, $\bar{v} = 0.0641$ and $v_0 = 0.1181$. The local volatility component is used to optimally fit the implied volatilities.

For different time steps Δt and strike prices K, the absolute error in the implied volatilities $|\sigma_{\text{imp}}^{\text{mkt}} - \sigma_{\text{imp}}^{\text{SLV}}|$ is computed, where $\sigma_{\text{imp}}^{\text{mkt}}$ and $\sigma_{\text{imp}}^{\text{SLV}}$ denote the implied volatilities that are implied by the market and the HSLV model, respectively. Twenty Monte Carlo simulations, with 5×10^4 paths each are performed. The number of bins is set to 20.

The corresponding results are presented in Table 10.1.

Table 10.1: *Average error $|\sigma_{imp}^{mkt} - \sigma_{imp}^{SLV}|$ from Monte Carlo simulations of the HSLV model with the Euler and efficient ('low-bias') schemes using 20 random seeds, for multiple time step sizes Δt and strike prices K. Numbers in parentheses are standard deviations over the seeds.*

	Error (%): $|\sigma_{imp}^{mkt} - \sigma_{imp}^{SLV}|$					
K	70%		100%		150%	
Δt	Euler	Low-bias	Euler	Low-bias	Euler	Low-bias
1	0.34 (0.12)	3.15 (0.10)	0.80 (0.13)	2.68 (0.11)	1.89 (0.19)	2.27 (0.17)
1/2	0.32 (0.13)	1.24 (0.11)	1.03 (0.15)	0.90 (0.11)	1.59 (0.20)	0.66 (0.15)
1/4	0.40 (0.16)	0.34 (0.12)	0.82 (0.15)	0.13 (0.11)	1.09 (0.20)	0.06 (0.13)
1/8	0.28 (0.13)	0.02 (0.12)	0.50 (0.13)	0.11 (0.14)	0.62 (0.19)	0.13 (0.20)
1/16	0.16 (0.15)	0.02 (0.15)	0.30 (0.15)	0.07 (0.16)	0.35 (0.18)	0.06 (0.21)
1/32	0.05 (0.13)	0.02 (0.14)	0.11 (0.15)	0.03 (0.14)	0.14 (0.21)	0.00 (0.19)

The AES Heston SLV simulation outperforms the Euler scheme, as it results in a higher accuracy and a faster convergence to the reference value. ◆

European option results under the SLV model

The performance of the Heston SLV model is analyzed with respect to the quality of a pre-calibrated Heston model. In the case the Heston model is well-calibrated, a limited contribution of the local volatility component is expected [van der Stoep et al., 2014]. On the other hand, if the Heston model is not sufficiently well-calibrated, the local volatility contribution should be more pronounced. The quality of the fit may then be related to the approximation of the conditional expectation expression.

With the AES scheme from the previous section, European call options under the Heston SLV model are priced within the Monte Carlo framework. The results are compared against the Heston model and the standard local volatility model. As the benchmark, we use a Heston model, with parameters,

$$\gamma = 0.5, \kappa = 0.3, \rho_{x,v} = -0.6, r = 0, v_0 = 0.04, \bar{v} = 0.05.$$

The Feller condition is not satisfied, as $2\kappa\bar{v}/\gamma^2 = 0.12$.

The European-style options that are considered are with maturity times (in years), $T = \{0.5, 2, 5, 8, 10\}$. The Monte Carlo simulation is performed with 5×10^5 paths, 100 time steps per year, and 20 bins to approximate the conditional expectation.

In Figure 10.6 (left side) we display the results obtained with an experiment, in which the Heston model is insufficiently calibrated. In this case the local volatility term can compensate for the large distance between the market and the Heston model implied volatilities.

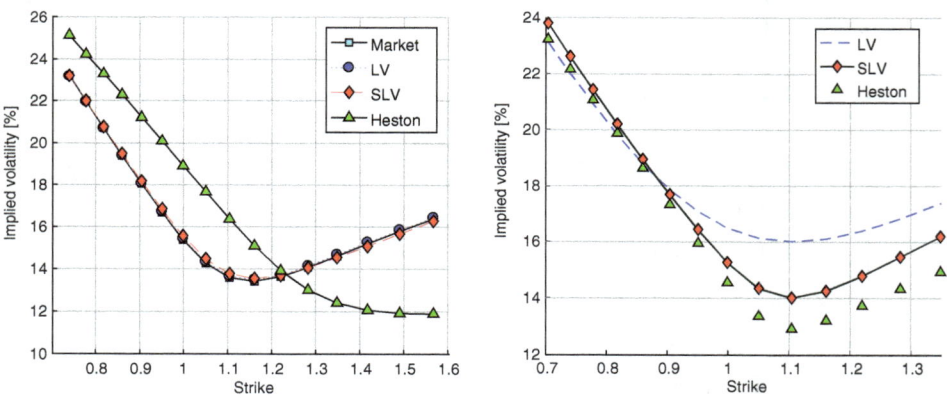

Figure 10.6: *Implied volatility $T = 2$, with an insufficiently calibrated Heston model; Left: European call option, right: forward start option, with $T_1 = 2$, $T_2 = 4$. In detail: The "market implied volatility" is generated by the Heston SV model with $\gamma = 0.5$, $\kappa = 0.3$, $\rho_{x,v} = -0.6$, $\bar{v} = 0.05$ and $v_0 = 0.04$, whereas the Heston SLV model parameters are given by $\gamma = 0.3750$, $\kappa = 0.3750$, $\rho_{x,v} = -0.75$, $\bar{v} = 0.0375$, $v_0 = 0.05$.*

The local volatility term in the Heston SLV model thus acts *a compensator*, which may bridge the gap between the market and calibrated Heston SV prices, in the case of an unsatisfactory calibration of a model.

At the right side of Figure 10.6, forward smiles from a forward start option simulation are presented, for the case in which the Heston model is insufficiently calibrated. We observe that the SLV model provides a forward implied volatility smile which is located *between the volatilities implied by the Heston model and the LV model*. As in [Engelmann et al., 2011], the forward implied volatilities do not become flat as for the local volatility model and preserve a shape very similar to the Heston model.

The results by the SLV model appear to represent an "advanced interpolation", between the Heston and the local volatility models.

In the numerical experiments, 10^5 Monte Carlo paths and 20 bins have been used, so that the calculation of the conditional expectation by the nonparametric method takes less than 0.025 seconds per time step.

In the Monte Carlo pricing under the SLV dynamics, a bias is introduced, which is basically due to three error sources. The discretization of the SLV dynamics by means of the AES discretization, introduces a discretization error. Another error is due to the calculation of the Dupire local volatility term (10.37). In particular, finite differences for the three derivatives in (10.38) introduce an error. Finally, at each time step, $\mathbb{E}\left[v(t)|S(t)=s\right]$ is approximated by means of the nonparametric method. The corresponding error analysis for this technique has been presented in [van der Stoep et al., 2014].

10.3 Exercise set

Exercise 10.1 In Theorem 10.1.1 the price of a forward start call option has been derived. Derive, in the same fashion, the price of a forward start put option.

Exercise 10.2 Price a forward start option by the COS method, choosing suitable parameters yourself.

  a. Under Black-Scholes dynamics for the underlying (so that you can verify the answer with the help of the solution in Theorem 10.1.1).
  b. Under the Heston dynamics, based on the characteristic function given in (10.10).
  c. Perform a Monte Carlo simulation with these dynamics and compare your COS method solutions from above with the Monte Carlo results.

Exercise 10.3 For the process,

$$\frac{\mathrm{d}F(t)}{F(t)} = \bar{\sigma}_F(t)\mathrm{d}W_F(t), \quad F(t_0) = \frac{S_0}{P(t_0, T_2)}, \tag{10.39}$$

the dynamics of the log-forward $X(t) := \log F(t)$ are governed by a one-dimensional SDE.
Show that the corresponding characteristic function, $\phi^{T_2} := \phi_X^{T_2}(u, X, t, T)$, is given by,[2]

$$\frac{\partial \phi^{T_2}}{\partial t} - \frac{1}{2}\bar{\sigma}_F^2(t)\frac{\partial \phi^{T_2}}{\partial X} + \frac{1}{2}\bar{\sigma}_F^2(t)\frac{\partial^2 \phi^{T_2}}{\partial X^2} = 0. \tag{10.40}$$

As the log-forward process is affine (it includes only a deterministic, time-dependent volatility) the model belongs to the affine class of processes. Moreover, its characteristic function is of the following form,

$$\phi_X^{T_2}(u, X, t, T) = \mathrm{e}^{\bar{A}(u,\tau) + \bar{B}(u,\tau)X(t)}, \tag{10.41}$$

with $\tau = T - t$ and $X(t) := \log F(t)$.
Show that, in (10.41), we have,

$$\bar{B}(u, \tau) = iu, \quad \bar{A}(u, \tau) = -\frac{1}{2}u(i+u)\int_0^\tau \bar{\sigma}_F^2(T-\tau)\mathrm{d}\tau.$$

Exercise 10.4 Consider the following 2D model, with two correlated geometric Brownian motions,

$$\mathrm{d}Y_1(t) = \sigma_1 Y_1(t)\mathrm{d}W_1(t), \quad Y_1(0) = y_{10}, \tag{10.42}$$
$$\mathrm{d}Y_2(t) = \sigma_2 Y_2(t)\mathrm{d}W_2(t), \quad Y_2(0) = y_{20}, \tag{10.43}$$

with $\mathrm{d}W_1(t)\mathrm{d}W_2(t) = \rho\mathrm{d}t$. Show that the expectation of $Y_2(t)$, conditional on the event $Y_1(t) = y_1$, is given by,

$$\mathbb{E}[Y_2(t)|Y_1(t) = y_1] = y_{20}\left(\frac{y_1}{y_{10}}\right)^{\rho\frac{\sigma_2}{\sigma_1}} \mathrm{e}^{t\left(\frac{1}{2}\rho\sigma_1\sigma_2 - \frac{1}{2}\sigma_2^2\rho^2\right)}. \tag{10.44}$$

Exercise 10.5 Related to the previous exercise 10.4, we prescribe $y_{10} = 1$, $y_{20} = 0.05$, $\rho = -0.5$ and $t = 5$ and consider two sets of volatility parameters, i.e. Set 1 where $\sigma_1 = \sigma_2 = 0.3$ and a more extreme Set 2 with $\sigma_1 = \sigma_2 = 0.9$. With these two parameter sets,

---

[2] Note that the characteristic function resembles the one for the Black-Scholes model with $r = 0$ and a time-dependent volatility $\bar{\sigma}_F(t)$.

a. Perform a Monte Carlo simulation, with different numbers of paths, using $y_1 = 1.75$.

b. Compare the analytic expression in (10.44) with the numerical approximation using the "*nonparametric method with bins*", as proposed in Section 10.2.2, with $l = 50$ bins. Vary the number of bins and assess the quality of the approximations.

Exercise 10.6 Consider the Schöbel-Zhu model, Equation (8.2), with the following set of parameters:

$$S(t_0) = 5, \sigma(t_0) = 0.1, \kappa = 0.1, \bar{\sigma} = 0.25, \gamma = 0.25, \rho_{x,\sigma} = -0.5.$$

Compute the forward Black-Scholes implied volatilities for $T_1 = 2$, $T_2 = 3$. Analyze the impact on the implied volatilities of variation of the parameters, κ, $\bar{\sigma}$, γ and $\rho_{x,\sigma}$. *Hint:* Compute the corresponding option prices with the Monte Carlo method.

Exercise 10.7 Generalize the Heston model derivations, as presented in Section 10.1.3, towards the Bates model and compute the corresponding forward implied volatilities for the parameters specified in Example 10.1.1 and with $\mu_J = 0.05$, $\sigma_J = 0.2$, $\xi_p = 0.1$. Vary the parameters and discuss the impact.

Exercise 10.8 In Equation (10.26) the local volatility function $\bar{\sigma}^2(t, K)$ for the stochastic local volatility model is found. Derive this function when jumps are also included in the dynamics of the stock process $S(t)$. *Hint:* Consider the Bates model as an example.

Exercise 10.9 Perform a hedge experiment, as in Section 3.3 (for the Black-Scholes model) or in Section 5.2.3 (for the Merton model), but here the market should be simulated with the Heston SV model. Choose the Heston SV parameters as used in Section 8.2.2, that give rise to skew and/or smile implied volatility patterns. Show that delta hedging is not sufficient to hedge all risk, if we deal with a market which is governed by an implied volatility smile or skew.

Exercise 10.10 Perform a hedge experiment, as in Section 3.3 (for the Black-Scholes model) or in Section 5.2.3 (for the Merton model), but here the market should be simulated with the Bates SV model. Choose the Bates SV parameters as used in Example 8.4.2, that give rise to steep skew and/or smile implied volatility patterns. Show that delta hedging is not sufficient to hedge all risk, if we deal with a market which is governed by an implied volatility smile or skew.

CHAPTER 11

Short-Rate Models

In this chapter:

We will move into the world of interest rates, and discuss the modeling of *stochastic interest rates*. After a brief introduction of interest rates, markets and products, in **Section 11.1**, the instantaneous *short-rate* is defined, in **Section 11.2**, as the interest rate one earns on a riskless investment over an infinitesimal period of time. The modeling of the short-rate is the focus in this chapter. The *Heath-Jarrow-Morton (HJM) framework* [Heath et al., 1992] represents a class of models, which present the dynamics of the *instantaneous forward rates* directly.

Among many models, the practically well-used short-rate models are the models developed by Vašiček [Vašiček, 1977], Cox-Ingersoll-Ross [Cox et al., 1985] and, in particular, the Hull-White model [Hull and White, 1990] is discussed in detail in **Section 11.3**. We will present the framework and show how, by changing the instantaneous volatility, different interest rate models can be derived.

Changing measures is very convenient in the interest rates context, as shown in **Section 11.4**.

Keywords: short-rates, Heath-Jarrow-Morton framework, affine models, Hull-White model.

11.1 Introduction to interest rates

We will start here the discussion of the mathematical modeling of interest rates to price interest rate related financial products. Particularly, we will encounter the so-called *short-rate models* in this chapter. Before we dive into the models, first a brief introduction into basic products and the terminology is presented.

11.1.1 Bond securities, notional

Bond securities are financial products that pay a regular interest, which is called a *coupon*, on a predefined amount of money. The main issuers of bonds in global financial markets are central and local governments, whereas companies may issue so-called *corporate bonds*.

The main reason for issuing bonds is to borrow money from investors on the financial market. If one believes that company A is more reliable (the chance of bankruptcy is smaller) than company B, one would request a higher interest rate on money lent to company B. Bonds are paid back in terms of coupons. A bond can be seen as a credit derivative as the issuer may fail to pay the obligations. In such an event, called a *default*, the investor does not, or only partially, get back the investment.

The bond market is one of the largest of the financial markets, and bonds come in many varieties. They do not only differ with respect to their issuers and the types of coupons, but also the maturity at which final payments are made may vary. Traditionally, bonds that were issued by governments were considered riskless. History has shown, however, a dozen of cases[1] where countries were not able to pay their financial obligations. It is maybe more appropriate to state that major governments are less likely to default compared to companies. The risk of a counterparty default is typically associated with the market believing in the likeliness to fail paying financial obligations. Practically, the assessment of the default probability is done by rating agencies that give ratings to the quality of the financial products issued. The lower the rating, the lower the value of the bond typically is, and therefore the higher the interest rate an investor may receive when investing in this derivative (note the relation between the bond's price and its rating).

We can distinguish basically two types of coupons in the bonds issued. There are coupons paid in the future that are predetermined (fixed), so-called *fixed-income securities*, and floating coupons, that are reset on a regular basis, giving rise to the term *float-rate note*. For interest rate derivatives that are quoted as a percentage or in basis points, it is important to know on which *reference* amount this percentage is based. The amount which is used to determine actual coupon payments is called *the notional amount*. So, if an investor receives, say 5%, on a notional of €1000 the actual coupon received is €50.

---

[1] An overview of sovereign defaults in the 19th and 20th century is presented at http://en.wikipedia.org/wiki/Sovereign_default.

Short-Rate Models

> **Definition 11.1.1** *A basic interest rate product is the zero-coupon bond, $P(t,T)$, which pays 1 currency unit at maturity time T, i.e. $P(T,T) = 1$. We are interested in its value at a time $t < T$.*
>
> *The fundamental theorem of asset pricing states that the price at time t of any contingent claim with payoff, $H(T)$, is given by:*
>
> $$V(t) = \mathbb{E}^{\mathbb{Q}}\left[e^{-\int_t^T r(z)dz} H(T)\,\Big|\,\mathcal{F}(t)\right], \qquad (11.1)$$
>
> *where the expectation is taken under the risk-neutral measure $\mathbb{Q}$.*
> *The price of a zero-coupon bond at time t with maturity T is thus given by:*
>
> $$P(t,T) = \mathbb{E}^{\mathbb{Q}}\left[e^{-\int_t^T r(z)dz}\,\Big|\,\mathcal{F}(t)\right], \qquad (11.2)$$
>
> *since $H(T) = V(T) = P(T,T) \equiv 1$.* ◂

A zero-coupon bond is a contract with price $P(t,T)$, at time $t < T$, to deliver at time T, $P(T,T) = €1$, see a sketch in Figure 11.1.

11.1.2 Fixed-rate bond

In a very basic version of a bond, *the fixed-rate bond*, $V^{FRB}(t)$, there is no stochasticity in the future payments, as the interest rate is fixed throughout the contract period.

For a given fixed rate r, a notional amount N and a set of payment dates $T_1, T_2, \ldots, T_m$, a fixed interest rate bond is an investment with several coupon payments, that are defined by:

$$V_i^{\text{fix}}(T_i) \equiv H_i^{\text{fix}}(T_i) = \begin{cases} rN\tau_i, & i \in \{1, 2, \ldots, m-1\} \\ rN\tau_m + N, & i = m, \end{cases} \qquad (11.3)$$

with $\tau_i = T_i - T_{i-1}$. Since this bond is a sum of payments, each cash flow can be priced separately,

$$V_i^{\text{fix}}(t_0) = M(t_0)\mathbb{E}^{\mathbb{Q}}\left[\frac{1}{M(T_i)} V_i^{\text{fix}}(T_i)\Big|\mathcal{F}(t_0)\right] = P(t_0, T_i) V_i^{\text{fix}}(T_i), \qquad (11.4)$$

Figure 11.1: *Cash flow for a zero-coupon bond, $P(t,T)$, with the payment at time T.*

as each payment of the fixed-rate bond is "predetermined". We will have, however, $M(t_0) = 1$. A schematic representation of the payment structure is then found in Figure 11.2.

Overall, the price of the fixed-rate bond at time t_0 yields:

$$V^{FRB}(t_0) = \sum_{i=1}^{m} \mathbb{E}^{\mathbb{Q}} \left[\frac{1}{M(T_i)} V_i^{\text{fix}}(T_i) \Big| \mathcal{F}(t_0) \right] = \sum_{i=1}^{m} P(t_0, T_i) V_i^{\text{fix}}(T_i).$$

Figure 11.3 shows an early (historical) fixed-rate bond from the state of Louisiana.

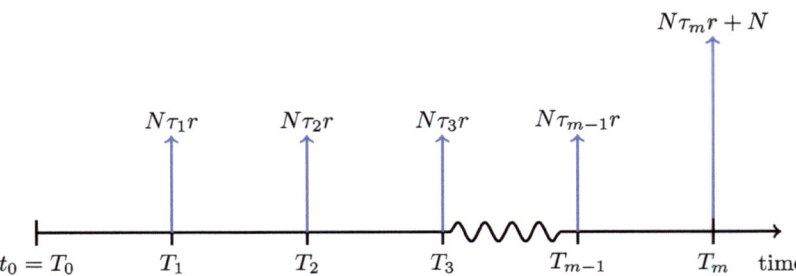

Figure 11.2: *Cash flows for a fixed-rate bond with m payments of $r\%$ and notional N.*

Figure 11.3: *An 1878 \$5 State of Louisiana Bank note. These were named "baby bonds" due to a girl's portrait on the note. Since then the name "baby bond" is linked to fixed income securities in small denominations. After the US civil war, the federal government imposed tax on state bank notes. The state of Louisiana issued small bank notes and sold them as interest-bearing bonds (thus avoiding the tax). The bond has four payments of 7.5% each, paid semi-annually from Aug 1884 until Feb 1886.*

11.2 Interest rates in the Heath-Jarrow-Morton framework

In many cases, interest rates are modeled by a basic deterministic function, for example, of time. The assumption of deterministic interest rates is generalized when rates are modeled by *a stochastic* instantaneous spot-rate process $r(t)$.

> The instantaneous short-rate is defined as the interest rate one earns on a riskless investment over an infinitesimal period of time $\mathrm{d}t$.

Among many models, the most successful short-rate models (due to their simple structure) are the models developed by Vašiček [1977], Cox *et al.* [1985] and Hull and White [1990], the latter two are extensions of the Vašiček model.

11.2.1 The HJM framework

The Heath-Jarrow-Morton (HJM) framework [Heath *et al.*, 1992] represents a class of interest rate (IR) models, that represent the dynamics of the so-called *instantaneous forward rates* directly. We will present the HJM framework and show how, by changing the instantaneous volatility, different interest rate models can be derived. For this we need the definition of the forward rate.

> **Definition 11.2.1 (Forward and instantaneous forward rates)**
> Suppose that at time t we enter into a forward contract to deliver at time T_1, $t < T_1 < T_2$, a bond which expires at maturity time T_2. Let the forward price of the bond at time t be denoted by $P_f(t, T_1, T_2)$. Suppose that, at the same time, a zero-coupon bond, which matures at time T_1, $P(t, T_1)$, and a bond, which matures at time T_2, $P(t, T_2)$, are purchased. Based on the assumption of no-arbitrage and market completeness, the following equality has to be satisfied,
>
> $$\boxed{P_f(t, T_1, T_2) = \frac{P(t, T_2)}{P(t, T_1)}.} \quad (11.5)$$
>
> The implied forward rate at time t for the period $[T_1, T_2]$, i.e. $r_F(t, T_1, T_2)$, is defined as,
>
> $$P_f(t, T_1, T_2) = \mathrm{e}^{-(T_2-T_1)r_F(t,T_1,T_2)}. \quad (11.6)$$
>
> Equating Equations (11.5) and (11.6) gives us
>
> $$\mathrm{e}^{-(T_2-T_1)r_F(t,T_1,T_2)} = \frac{P(t, T_2)}{P(t, T_1)},$$

Forward rate $r_F(t, T_1, T_2)$ is defined by,

$$r_F(t, T_1, T_2) := -\frac{\log P(t, T_2) - \log P(t, T_1)}{T_2 - T_1}.$$

In the limit $T_2 - T_1 \to 0$, we arrive at the definition of the *instantaneous forward rate*

$$f^r(t, T_2) \stackrel{\text{def}}{=} \lim_{T_1 \to T_2} r_F(t, T_1, T_2) = -\frac{\partial}{\partial T_2} \log P(t, T_2).$$

◀

Within the HJM framework the dynamics of instantaneous forward rate $f^r(t, T)$ are of the main interest.

The point of departure is the assumption that, for a fixed maturity $T \geq 0$, the instantaneous forward rate $f^r(t, T)$, under the real-world measure $\mathbb{P}$, is governed by the following dynamics:

$$df^r(t, T) = \alpha^{\mathbb{P}}(t, T)dt + \bar{\eta}(t, T)dW^{\mathbb{P}}(t), \quad f^r(0, T) = f^r_{0,T}, \quad (11.7)$$

for any time $t < T$, with corresponding drift term $\alpha^{\mathbb{P}}(t, T)$ and volatility term $\bar{\eta}(t, T)$. Here,

$$f^r_{0,T} = -\frac{\partial}{\partial T} \log P(0, T).$$

In the HJM framework, the short-rate $r(t)$ is defined as the limit of the instantaneous forward rate $r(t) \equiv f^r(t, t)$.

We also define the money-savings account, by

$$M(t) := \exp\left(\int_0^t r(z)dz\right) = \exp\left(\int_0^t f^r(z, z)dz\right). \quad (11.8)$$

The zero-coupon bond $P(t, T)$, with maturity T, is a tradable asset and should therefore be a martingale. Its value is thus determined by

$$P(t, T) = M(t)\mathbb{E}^{\mathbb{Q}}\left[\frac{1}{M(T)} \cdot 1 \Big| \mathcal{F}(t)\right] = \mathbb{E}^{\mathbb{Q}}\left[\exp\left(-\int_t^T r(z)dz\right) \Big| \mathcal{F}(t)\right]. \quad (11.9)$$

Measure $\mathbb{Q}$ is the risk-neutral measure associated with the money-savings account $M(t)$ as the numéraire. A change of measure, from the real-world measure $\mathbb{P}$ to the risk-neutral measure $\mathbb{Q}$, will affect the drift terms in the diffusion SDE model (11.7), so that the dynamics of $f^r(t, T)$ under measure $\mathbb{Q}$ are given by

$$df^r(t, T) = \alpha^{\mathbb{Q}}(t, T)dt + \bar{\eta}(t, T)dW^{\mathbb{Q}}(t), \quad f^r(0, T) = f^r_{0,T}, \quad (11.10)$$

where $\alpha^{\mathbb{Q}}(t, T)$ is a state-dependent function, associated with measure $\mathbb{Q}$.

Remark 11.2.1

Although the ZCB, $P(0,T)$, can be priced as in Equation (11.9), its value can be directly related to today's yield curve via:

$$f^r(0,T) = -\frac{\partial}{\partial T}\log P(0,T). \tag{11.11}$$

The construction of a yield curve will be presented in Section 12.1.5.

To estimate the instantaneous forward rate $f^r(t,T)$, based on zero-coupon bond prices $P(0,T)$, one needs to approximate the derivative in (11.11) by means of a finite difference discretization, i.e. for $t = t_0 = 0$:

$$f^r(0,T) \approx -\frac{\log(P(0,T+\Delta T)) - \log(P(0,T-\Delta T))}{2\Delta T}.$$

Alternatively, we can write

$$f^r(0,T) = -\frac{\partial}{\partial T}\log P(0,T) = -\frac{1}{P(0,T)}\frac{\partial}{\partial T}P(0,T),$$

where discretization of the latter representation may be numerically stable.

In Figure 11.4 zero-coupon bond curves $P(0,T)$ and instantaneous forward rates $f^r(0,T)$ are presented. Given the discrete nature of market quotes, the obtained $f^r(0,T)$ may become nonsmooth. In practice, a smoothing procedure (like averaging) is often applied. ▲

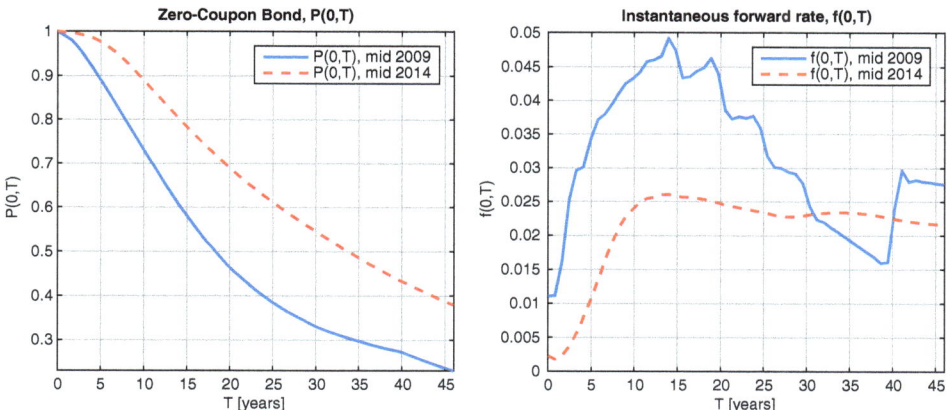

Figure 11.4: Zero-coupon bond $P(0,T)$ and instantaneous forward rate $f^r(0,T)$, obtained from market data from the years 2009 and 2014.

The HJM condition for an arbitrage-free model

An important result within the HJM framework is the no-arbitrage relation between drift $\alpha^{\mathbb{Q}}(t,T)$ and the volatility structure $\bar{\eta}(t,T)$. In this section we show how this relation is constructed.

The ZCB is directly related to the instantaneous forward rates $f^r(t,T)$, via the relation:

$$P(t,T) = \exp\left(-\int_t^T f^r(t,z)\mathrm{d}z\right), \qquad (11.12)$$

which can be derived from (11.11).

We derive the dynamics under measure $\mathbb{Q}$ of the discounted ZCB, as follows,

$$\mathrm{d}\Pi_p(t) := \mathrm{d}\left(\frac{P(t,T)}{M(t)}\right) = \mathrm{d}\left[\exp\left(-\int_t^T f^r(t,z)\mathrm{d}z - \int_0^t r(z)\mathrm{d}z\right)\right]. \quad (11.13)$$

This tradable financial derivative, $P(t,T)$, which is discounted by the money-savings account, is thus a martingale, following the discussion in [Harrison and Kreps, 1979], and its dynamics should be free of any drift terms. This implies that the dynamics of $f^r(t,T)$ should be of a specific form. In the lemma below, the HJM arbitrage-free conditions are presented.[2]

Lemma 11.2.1 (HJM arbitrage-free condition) *For the instantaneous forward rates, modeled by the following SDE:*

$$\mathrm{d}f^r(t,T) = \alpha^{\mathbb{Q}}(t,T)\mathrm{d}t + \bar{\eta}(t,T)\mathrm{d}W^{\mathbb{Q}}(t),$$

the no-arbitrage drift condition is given by

$$\alpha^{\mathbb{Q}}(t,T) = \bar{\eta}(t,T)\int_t^T \bar{\eta}(t,z)\mathrm{d}z. \qquad (11.14)$$

◀

The proof is based on the derivation of the dynamics of $\Pi_p(t)$ and setting the drift term equal to zero.

The dynamics of $f^r(t,T)$ are fully determined by the volatility structure, $\bar{\eta}(t,T)$. This is a crucial observation, as it indicates that by the volatility $\bar{\eta}(t,T)$, the complete, risk-free model can be derived.

---

[2] Note, however, that the discounted short-rate process, or the discounted volatility, are *not martingales*, as these stochastic quantities are not tradable (i.e. we cannot directly buy a short-rate or a volatility).

11.2.2 Short-rate dynamics under the HJM framework

Based on the relation $f^r(t,t) = r(t)$, and by integrating the system in (11.10), we obtain the *short-rate dynamics* in the HJM framework, that are of the form:

$$f^r(t,T) = f^r(0,T) + \int_0^t \alpha^{\mathbb{Q}}(z,T)\mathrm{d}z + \int_0^t \bar{\eta}(z,T)\mathrm{d}W^{\mathbb{Q}}(z). \quad (11.15)$$

Using $T = t$, this implies that,

$$\boxed{r(t) \equiv f^r(t,t) = f^r(0,t) + \int_0^t \alpha^{\mathbb{Q}}(z,t)\mathrm{d}z + \int_0^t \bar{\eta}(z,t)\mathrm{d}W^{\mathbb{Q}}(z).} \quad (11.16)$$

Proposition 11.2.1 (Short-rate dynamics under the HJM framework)
Suppose that $f^r(0,t)$, $\alpha^{\mathbb{Q}}(t,T)$ and $\bar{\eta}(t,T)$ are differentiable in their second argument, with

$$\int_0^T \left|\frac{\partial}{\partial t} f^r(0,t)\right| < \infty,$$

then the short-rate process under the HJM framework is given by,

$$\mathrm{d}r(t) = \bar{\zeta}(t)\mathrm{d}t + \bar{\eta}(t,t)\mathrm{d}W^{\mathbb{Q}}(t), \quad (11.17)$$

or, equivalently, in integral form:

$$r(t) = r(0) + \int_0^t \bar{\zeta}(z)\mathrm{d}z + \int_0^t \bar{\eta}(z,z)\mathrm{d}W^{\mathbb{Q}}(z), \quad (11.18)$$

with

$$\bar{\zeta}(t) = \alpha^{\mathbb{Q}}(t,t) + \frac{\partial}{\partial t}f^r(0,t) + \int_0^t \frac{\partial}{\partial t}\alpha^{\mathbb{Q}}(z,t)\mathrm{d}z + \int_0^t \frac{\partial}{\partial t}\bar{\eta}(z,t)\mathrm{d}W^{\mathbb{Q}}(z),$$

see the proof below. ◂

Proof Recall that,

$$r(t) = f^r(0,t) + \int_0^t \alpha^{\mathbb{Q}}(z,t)\mathrm{d}z + \int_0^t \bar{\eta}(z,t)\mathrm{d}W^{\mathbb{Q}}(z). \quad (11.19)$$

The last integral can be expressed in a different manner, i.e.

$$\int_0^t \bar{\eta}(z,t)\mathrm{d}W^{\mathbb{Q}}(z) = \int_0^t \bar{\eta}(z,z)\mathrm{d}W^{\mathbb{Q}}(z) + \int_0^t \left(\bar{\eta}(z,t) - \bar{\eta}(z,z)\right)\mathrm{d}W^{\mathbb{Q}}(z)$$

$$= \int_0^t \bar{\eta}(z,z)\mathrm{d}W^{\mathbb{Q}}(z) + \int_0^t \left(\int_z^t \frac{\partial}{\partial u}\bar{\eta}(z,u)\mathrm{d}u\right)\mathrm{d}W^{\mathbb{Q}}(z).$$

Applying Fubini's theorem to the last integral, gives,

$$\int_0^t \left(\int_z^t \frac{\partial}{\partial u}\bar{\eta}(z,u)\mathrm{d}u\right)\mathrm{d}W^{\mathbb{Q}}(z) = \int_0^t \left(\int_0^u \frac{\partial}{\partial u}\bar{\eta}(z,u)\mathrm{d}W^{\mathbb{Q}}(z)\right)\mathrm{d}u, \quad (11.20)$$

and

$$\int_0^t \alpha^{\mathbb{Q}}(z,t)\mathrm{d}z = \int_0^t \alpha^{\mathbb{Q}}(z,z)\mathrm{d}z + \int_0^t \left(\int_0^u \frac{\partial}{\partial u}\alpha^{\mathbb{Q}}(z,u)\mathrm{d}z\right)\mathrm{d}u. \quad (11.21)$$

Furthermore,

$$f^r(0,t) = f^r(0,0) + f^r(0,t) - f^r(0,0)$$
$$= r(0) + \int_0^t \frac{\partial}{\partial z}f^r(0,z)\mathrm{d}z,$$

since $f^r(0,0) = r(0)$. By collecting all terms, Equation (11.19) becomes,

$$r(t) = f^r(0,t) + \int_0^t \alpha^{\mathbb{Q}}(z,t)\mathrm{d}z + \int_0^t \bar{\eta}(z,t)\mathrm{d}W^{\mathbb{Q}}(z)$$
$$= r(0) + \int_0^t \frac{\partial}{\partial z}f^r(0,z)\mathrm{d}z$$
$$+ \left(\int_0^t \alpha^{\mathbb{Q}}(z,z)\mathrm{d}z + \int_0^t \left(\int_0^u \frac{\partial}{\partial u}\alpha^{\mathbb{Q}}(z,u)\mathrm{d}z\right)\mathrm{d}u\right)$$
$$+ \int_0^t \left(\int_0^u \frac{\partial}{\partial u}\bar{\eta}(z,u)\mathrm{d}W^{\mathbb{Q}}(z)\right)\mathrm{d}u + \int_0^t \bar{\eta}(z,z)\mathrm{d}W^{\mathbb{Q}}(z).$$

So, we get,

$$r(t) = r(0) + \int_0^t \bar{\zeta}(z)\mathrm{d}z + \int_0^t \bar{\eta}(z,z)\mathrm{d}W^{\mathbb{Q}}(z),$$

with

$$\bar{\zeta}(t) = \frac{\partial}{\partial t}f^r(0,t) + \alpha^{\mathbb{Q}}(t,t) + \int_0^t \frac{\partial}{\partial t}\alpha^{\mathbb{Q}}(z,t)\mathrm{d}z + \int_0^t \frac{\partial}{\partial t}\bar{\eta}(z,t)\mathrm{d}W^{\mathbb{Q}}(z). \quad \blacksquare$$

A question which remains is regarding the dynamics of the zero-coupon bond. We have already seen that $P(t,T)$ can be defined in terms of the forward rates, and that the ZCB can be obtained, for a given time t, by integrating the yield curve, which is known at time t. The lemma below details the *dynamics of* $P(t,T)$, for a given HJM instantaneous forward rate $f^r(t,T)$:

Lemma 11.2.2 (ZCB dynamics under risk-neutral measure) *The risk-free dynamics of the ZCB are given by:*

$$\mathrm{d}P(t,T) = r(t)P(t,T)\mathrm{d}t - P(t,T)\left(\int_t^T \bar{\eta}(z,T)\mathrm{d}z\right)\mathrm{d}W^{\mathbb{Q}}(t). \quad (11.22)$$

◀

Short-Rate Models

Proof In (11.12) the ZCB is given by $P(t,T) = \exp(Z(t,T))$, with $Z(t,T) := -\int_t^T f^r(t,s)\mathrm{d}s$. By Itô's lemma the dynamics of $P(t,T)$ are then given by:

$$\frac{\mathrm{d}P(t,T)}{P(t,T)} = \mathrm{d}Z(t,T) + \frac{1}{2}(\mathrm{d}Z(t,T))^2. \quad (11.23)$$

The missing element in the dynamics for the ZCB in (11.23) is the Itô differential of process $Z(t,T)$. We start its derivation by standard calculus, i.e. from

$$\mathrm{d}Z(t,T) = f^r(t,t)\mathrm{d}t - \int_t^T \mathrm{d}f^r(t,z)\mathrm{d}z. \quad (11.24)$$

By the definition of the short-rate, $r(t) := f^r(t,t)$ as in (11.16), and the definition of the instantaneous forward rate $f^r(t,T)$, under the $\mathbb{Q}$ measure (11.15), we find:

$$\mathrm{d}Z(t,T) = \left(r(t) - \int_t^T \alpha^{\mathbb{Q}}(t,z)\mathrm{d}z\right)\mathrm{d}t - \left(\int_t^T \bar{\eta}(t,z)\mathrm{d}z\right)\mathrm{d}W^{\mathbb{Q}}(t),$$

so that the dynamics for the ZCB under the general HJM model can be written as:

$$\frac{\mathrm{d}P(t,T)}{P(t,T)} = D(t,T)\mathrm{d}t - \left(\int_t^T \bar{\eta}(t,z)\mathrm{d}z\right)\mathrm{d}W^{\mathbb{Q}}(t),$$

where

$$D(t,T) := r(t) - \int_t^T \alpha^{\mathbb{Q}}(t,z)\mathrm{d}z + \frac{1}{2}\left(\int_t^T \bar{\eta}(t,z)\mathrm{d}z\right)^2. \quad (11.25)$$

The remaining part of the proof is to show that $D(t,T) = r(t)$. From the HJM drift condition under the risk-neutral measure, we find, by differentiation of the last integral with respect to T,

$$\frac{\mathrm{d}}{\mathrm{d}T}\left[\frac{1}{2}\left(\int_t^T \bar{\eta}(t,z)\mathrm{d}z\right)^2\right] = \bar{\eta}(t,T)\int_t^T \bar{\eta}(t,z)\mathrm{d}z =: \alpha^{\mathbb{Q}}(t,T), \quad (11.26)$$

where the last equality is due to (11.14) in Lemma 11.2.1.

Integrating both sides of (11.26) gives us

$$\int_t^T \alpha^{\mathbb{Q}}(t,z)\mathrm{d}z = \frac{1}{2}\left(\int_t^T \bar{\eta}(t,z)\mathrm{d}z\right)^2,$$

which implies that $D(t,T)$ in (11.25) equals $r(t)$ and this completes the proof. ∎

11.2.3 The Hull-White dynamics in the HJM framework

We consider a specific short-rate model, which is generated by a HJM volatility given by:

$$\boxed{\bar{\eta}(t,T) = \eta \cdot \mathrm{e}^{-\lambda(T-t)}.}$$

In the past, the requirement "$\lambda > 0$" was added to the equation above, however, it should be mentioned that this parameter is typically obtained in the industry by calibration. Then, it is not uncommon to also encounter negative λ-values.

By Lemma 11.2.1, we find:

$$\alpha^{\mathbb{Q}}(z,t) = \eta e^{-\lambda(t-z)} \int_z^t \eta e^{-\lambda(\bar{z}-z)} d\bar{z} = -\frac{\eta^2}{\lambda} e^{-\lambda(t-z)}\left(e^{-\lambda(t-z)} - 1\right),$$

which implies that $\alpha^{\mathbb{Q}}(t,t) = 0$. The remaining terms in (11.16) are as follows:

$$\int_0^t \frac{\partial}{\partial t} \alpha^{\mathbb{Q}}(z,t) dz = \frac{\eta^2}{\lambda} e^{-\lambda t}(e^{\lambda t} - 1),$$

and

$$\frac{\partial}{\partial t} \bar{\eta}(z,t) = -\lambda \eta e^{-\lambda(t-z)} = -\lambda \bar{\eta}(z,t), \qquad (11.27)$$

with $\bar{\eta}(t,t) = \eta$. The dynamics for $r(t)$ are therefore given by:

$$dr(t) = \left[\frac{\partial}{\partial t} f^r(0,t) + \int_0^t \frac{\partial}{\partial t} \alpha^{\mathbb{Q}}(z,t) dz - \lambda \int_0^t \bar{\eta}(z,t) dW^{\mathbb{Q}}(z)\right] dt + \eta dW^{\mathbb{Q}}(t), \qquad (11.28)$$

where the Brownian motion $W^{\mathbb{Q}}(t)$ appears twice. To determine the solution for the integral $\int_0^t \bar{\eta}(z,t) dW^{\mathbb{Q}}(z)$, we can use (11.16), which yields:

$$r(t) = f^r(0,t) + \int_0^t \alpha^{\mathbb{Q}}(z,t) dz + \int_0^t \bar{\eta}(z,t) dW^{\mathbb{Q}}(z),$$

so that

$$\int_0^t \bar{\eta}(z,t) dW^{\mathbb{Q}}(z) = r(t) - f^r(0,t) - \int_0^t \alpha^{\mathbb{Q}}(z,t) dz.$$

Since

$$\int_0^t \alpha^{\mathbb{Q}}(z,t) dz = \frac{\eta^2}{2\lambda^2} e^{-2\lambda t}\left(e^{\lambda t} - 1\right)^2,$$

we obtain the following dynamics for process $r(t)$:

$$dr(t) = \left(\frac{\partial}{\partial t} f^r(0,t) - \lambda r(t) + \lambda f^r(0,t) + \frac{\eta^2}{2\lambda}\left(1 - e^{-2\lambda t}\right)\right) dt + \eta dW^{\mathbb{Q}}(t)$$

$$= \lambda \left(\frac{1}{\lambda} \frac{\partial}{\partial t} f^r(0,t) + f^r(0,t) + \frac{\eta^2}{2\lambda^2}\left(1 - e^{-2\lambda t}\right) - r(t)\right) dt + \eta dW^{\mathbb{Q}}(t).$$

Short-Rate Models

> By defining,
>
> $$\theta(t) := \frac{1}{\lambda}\frac{\partial}{\partial t}f^r(0,t) + f^r(0,t) + \frac{\eta^2}{2\lambda^2}\left(1 - e^{-2\lambda t}\right), \qquad (11.29)$$
>
> the dynamics of process $r(t)$ are given by
>
> $$dr(t) = \lambda(\theta(t) - r(t))dt + \eta dW^{\mathbb{Q}}(t), \qquad (11.30)$$
>
> which is known by the name *Hull-White short-rate process*.
> The Hull-White process is *free of arbitrage* if and only if function $\theta(t)$ is given by (11.29), with the instantaneous forward rate
>
> $$f^r(0,t) = -\frac{\partial}{\partial t}\log P_{mkt}(0,t),$$
>
> where $P_{mkt}(0,t)$ is a ZCB from the financial market.
> This is an immediate consequence of a model being in the HJM framework. It implies that the expectation $\mathbb{E}^{\mathbb{Q}}\left[e^{-\int_0^t r(s)ds}|\mathcal{F}(0)\right]$, by construction, should be equal to $P_{mkt}(0,t)$, which is observed at $t_0 = 0$ and has maturity time t. It is used in the computation of $\theta(t)$ in (11.29).

The dynamics of the ZCB under the Hull-White model

In Lemma 11.2.2, the general dynamics for the zero-coupon bond, driven by the short-rate process $r(t)$, in the HJM framework were presented:

$$\frac{dP(t,T)}{P(t,T)} = r(t)dt - \left(\int_t^T \bar{\eta}(z,T)dz\right)dW^{\mathbb{Q}}(t).$$

For the Hull-White model, the HJM volatility is given by,

$$\bar{\eta}(t,T) = \eta e^{-\lambda(T-t)}.$$

> This implies the following dynamics for the ZCB under the Hull-White model,
>
> $$\frac{dP(t,T)}{P(t,T)} = r(t)dt - \left(\int_t^T \eta e^{-\lambda(T-z)}dz\right)dW^{\mathbb{Q}}(t)$$
> $$= r(t)dt + \eta \bar{B}_r(t,T)dW^{\mathbb{Q}}(t), \qquad (11.31)$$
>
> with $\bar{B}_r(t,T) = \frac{1}{\lambda}\left(e^{-\lambda(T-t)} - 1\right).$

We will describe the main properties of the Hull-White model in more detail below.

11.3 The Hull-White model

The Hull-White model [Hull and White, 1990] is a single-factor, no-arbitrage yield curve model in which the short-term interest rate is driven by an extended Ornstein-Uhlenbeck (OU) mean reverting process, i.e.,

$$\boxed{dr(t) = \lambda \left(\theta(t) - r(t)\right) dt + \eta dW_r(t), \quad r(0) = r_0,} \qquad (11.32)$$

where $\theta(t)$ is a time-dependent drift term, which is used to fit the mathematical bond prices to the yield curve observed in the market and $W_r(t) \equiv W_r^{\mathbb{Q}}(t)$ is the Brownian motion under measure $\mathbb{Q}$. This Hull-White model SDE, including the definition of parameter $\theta(t)$, we have encountered in the previous section, in Equation (11.30).

Parameter η determines the overall level of the volatility and λ is the reversion rate parameter. A large value of λ causes short-term rate movements to dampen out rapidly, so that the long-term volatility is reduced.

11.3.1 The solution of the Hull-White SDE

We will present the solution, as well as the characteristic function, for this interest rate process.

To obtain the solution to SDE (11.32), we apply Itô's lemma to a process $y(t) := e^{\lambda t} r(t)$, i.e.

$$dy(t) = \lambda y(t) dt + e^{\lambda t} dr(t). \qquad (11.33)$$

After substitution of (11.32), we find:

$$dy(t) = \lambda y(t) dt + e^{\lambda t} \left[\lambda \left(\theta(t) - r(t)\right) dt + \eta dW_r(t)\right],$$

and arrive at the following system of equations:

$$\begin{cases} y(t) = e^{\lambda t} r(t), \\ dy(t) = \lambda \theta(t) e^{\lambda t} dt + \eta e^{\lambda t} dW_r(t). \end{cases} \qquad (11.34)$$

The right-hand side of the $y(t)$-dynamics does not depend on state vector $y(t)$, and we can therefore determine the solution of $y(t)$ by integrating both sides of the SDE,

$$\int_0^t dy(z) = \lambda \int_0^t \theta(z) e^{\lambda z} dz + \eta \int_0^t e^{\lambda z} dW_r(z),$$

which gives us,

$$y(t) = y(0) + \lambda \int_0^t \theta(z) e^{\lambda z} dz + \eta \int_0^t e^{\lambda z} dW_r(z).$$

By the definition of $y(t)$, with $y(0) = r_0$, the solution for process $r(t)$ reads:

$$r(t) = e^{-\lambda t} r_0 + \lambda \int_0^t \theta(z) e^{-\lambda(t-z)} dz + \eta \int_0^t e^{-\lambda(t-z)} dW_r(z).$$

Short-Rate Models 353

> Interest rate $r(t)$ is thus *normally distributed* with
> $$\mathbb{E}\left[r(t)|\mathcal{F}(t_0)\right] = r_0 \mathrm{e}^{-\lambda t} + \lambda \int_0^t \theta(z)\mathrm{e}^{-\lambda(t-z)}\mathrm{d}z,$$
> using $t_0 = 0$, and
> $$\mathbb{V}\mathrm{ar}\left[r(t)|\mathcal{F}(t_0)\right] = \frac{\eta^2}{2\lambda}\left(1 - \mathrm{e}^{-2\lambda t}\right).$$
> Moreover, for $\theta(t)$ constant, i.e., $\theta(t) \equiv \theta$ (in which case we deal with the *Vašiček model* [Vašiček, 1977]), we have
> $$\lim_{t \to \infty} \mathbb{E}\left[r(t)|\mathcal{F}(t_0)\right] = \theta.$$
> This means that the first moment of the process converges to the mean reverting level θ, for large values of t.

11.3.2 The HW model characteristic function

To derive the Hull-White process' characteristic function, we use a decomposition, based on the proposition to follow, see also Arnold [1973], Øksendal [2000], Pelsser [2000].

> **Proposition 11.3.1 (Hull-White decomposition)**
> *The Hull-White short-rate process (11.32) can be decomposed as*
> $$r(t) = \widetilde{r}(t) + \psi(t),$$
> where
> $$\psi(t) = r_0 \mathrm{e}^{-\lambda t} + \lambda \int_0^t \theta(z)\mathrm{e}^{-\lambda(t-z)}\mathrm{d}z,$$
> and
> $$\mathrm{d}\widetilde{r}(t) = -\lambda \widetilde{r}(t)\mathrm{d}t + \eta \mathrm{d}W_r(t), \text{ with } \widetilde{r}_0 = 0. \quad (11.35)$$
> ◂

Proof The proof is by Itô's lemma.

An advantage of this transformation is that the stochastic process $\widetilde{r}(t)$ in (11.35) is a *basic OU mean reverting process*, determined only by λ and η, and independent of function $\psi(t)$. It is therefore relatively easy to analyze this model.

We determine here the *discounted characteristic function* of the spot interest rate $r(t)$ in the Hull-White model,

$$\phi_{\mathrm{HW}}(u;t,T) := \mathbb{E}^{\mathbb{Q}}\left[\mathrm{e}^{-\int_t^T r(z)\mathrm{d}z + iur(T)}|\mathcal{F}(t)\right],$$

which, by the decomposition of $r(t)$ in Proposition 11.3.1, can be expressed as:

$$\phi_{\text{HW}}(u;t,T) = \mathbb{E}^{\mathbb{Q}}\left[e^{iu\psi(T)-\int_t^T \psi(z)dz} \cdot e^{iu\widetilde{r}(T)-\int_t^T \widetilde{r}(z)dz}\Big|\mathcal{F}(t)\right]$$
$$= e^{iu\psi(T)-\int_t^T \psi(z)dz} \cdot \phi_{\widetilde{\text{HW}}}(u;t,T),$$

and we can use the property that process $\widetilde{r}(t)$ is *affine*. Hence, the discounted characteristic function, which is written as $\phi_{\widetilde{\text{HW}}}(u;t,T)$, with $u \in \mathbb{C}$, for the affine short-rate model is of the following form:

$$\phi_{\widetilde{\text{HW}}}(u;t,T) = \exp\left(\bar{A}(u,\tau) + \bar{B}(u,\tau)\widetilde{r}(t)\right). \tag{11.36}$$

The initial condition at $\tau = 0$, accompanying (11.36), is

$$\phi_{\widetilde{\text{HW}}}(u;T,T) = \exp\left(iu\widetilde{r}(T)\right),$$

so that $\bar{A}(u,0) = 0$ and $\bar{B}(u,0) = iu$. The solutions for $\bar{A}(u,\tau)$ and $\bar{B}(u,\tau)$ are provided by the following lemma:

Lemma 11.3.1 (Coefficients for HW discounted ChF) *The functions $\bar{A}(u,\tau)$ and $\bar{B}(u,\tau)$ in (11.36) are given by:*

$$\bar{A}(u,\tau) = \frac{\eta^2}{2\lambda^3}\left(\lambda\tau - 2\left(1 - e^{-\lambda\tau}\right) + \frac{1}{2}\left(1 - e^{-2\lambda\tau}\right)\right) - iu\frac{\eta^2}{2\lambda^2}\left(1 - e^{-\lambda\tau}\right)^2$$
$$- \frac{1}{2}u^2\frac{\eta^2}{2\lambda}\left(1 - e^{-2\lambda\tau}\right),$$
$$\bar{B}(u,\tau) = iue^{-\lambda\tau} - \frac{1}{\lambda}\left(1 - e^{-\lambda\tau}\right).$$

Note that $\bar{B}(0,\tau) \equiv \bar{B}_r(t,T)$, *as in (11.31)* ◀

The proof is by means of the techniques that have been presented in Chapter 7, in Section 7.3.

By inserting $u = 0$, the risk-free pricing formula for a *zero-coupon bond* $P(t,T)$ is given by:

$$P(t,T) = \phi_{\text{HW}}(0;t,T) = \mathbb{E}^{\mathbb{Q}}\left[e^{-\int_t^T r(z)dz} \cdot 1\Big|\mathcal{F}(t)\right]$$
$$= \exp\left(-\int_t^T \psi(z)dz + \bar{A}(0,\tau) + \bar{B}(0,\tau)\widetilde{r}(t)\right).$$
$$=: \exp\left(-\int_t^T \psi(z)dz + \bar{A}_r(t,T) + \bar{B}_r(t,T)\widetilde{r}(t)\right).$$

Short-Rate Models 355

> The zero-coupon bond can thus be written as the product of a deterministic factor and the bond price in a Vašiček model with zero mean, under measure $\mathbb{Q}$. Recall that process $\widetilde{r}(t)$ at time $t = 0$ is equal to 0, so that
>
> $$P(0,T) = \exp\left(-\int_0^T \psi(z)\mathrm{d}z + \bar{A}_r(t,T)\right).$$
>
> This gives us
>
> $$\psi(T) = -\frac{\mathrm{d}}{\mathrm{d}T}\log P(0,T) + \frac{\mathrm{d}}{\mathrm{d}T}\bar{A}_r(t,T) = f^r(0,T) + \frac{\eta^2}{2\lambda^2}\left(1 - \mathrm{e}^{-\lambda T}\right)^2,$$
>
> where $f^r(t,T)$ is again the instantaneous forward rate.

This shows that function $\psi(t)$ can be obtained from the initial forward curve $f^r(0,T)$. The other time-invariant parameters, λ and η, have to be estimated using market prices of certain interest rate derivatives. From the Hull-White decomposition in Proposition 11.3.1, we have

$$\theta(t) = \frac{1}{\lambda}\frac{\partial}{\partial t}\psi(t) + \psi(t),$$

which gives,

$$\theta(t) = f^r(0,t) + \frac{1}{\lambda}\frac{\partial}{\partial t}f^r(0,t) + \frac{\eta^2}{2\lambda^2}\left(1 - \mathrm{e}^{-2\lambda t}\right). \tag{11.37}$$

Characteristic function $\phi_{\mathrm{HW}}(u;t,T)$ for the Hull-White model can be obtained by integration of $\psi(z)$ over the interval $[t,T]$.

> **Example 11.3.1 (The Vašiček model)** With $\theta(t) \equiv \theta$ in the Hull-White decomposition from Proposition 11.3.1, we can derive the characteristic function for the Vašiček model. Straightforward integration gives us the following representation for function $\psi(t)$ in the Vašiček setting:
>
> $$\psi(t) = \theta + (r_0 - \theta)\mathrm{e}^{-\lambda t}, \tag{11.38}$$
>
> which results in the following characteristic function,
>
> $$\phi_{\mathrm{Vas}}(u;t,T) = \mathbb{E}^{\mathbb{Q}}\left[\exp\left(-\int_t^T r(z)\mathrm{d}z + iur(T)\right)\Big|\mathcal{F}(t)\right]$$
>
> $$= \exp\left(-\int_t^T \psi(z)\mathrm{d}z + iu\psi(T)\right) \cdot \phi_{\widetilde{\mathrm{HW}}}(u;t,T),$$
>
> with $\phi_{\widetilde{\mathrm{HW}}}(u;t,T)$ given by (11.36). By integrating $\psi(t)$ in (11.38), we find:
>
> $$\exp\left(-\int_t^T \psi(z)\mathrm{d}z + iu\psi(T)\right) = \exp\left[\frac{1}{\lambda}(r_0 - \theta)\left(\mathrm{e}^{-\lambda T} - \mathrm{e}^{-\lambda t}\right) - (T-t)\theta\right]$$
>
> $$\times \exp\left[iu\left(\theta + (r_0 - \theta)\mathrm{e}^{-\lambda T}\right)\right]. \quad \blacklozenge$$

Remark 11.3.1 (Function $\theta(t)$) *In the Hull-White model (11.32), the function $\theta(t)$ is given by,*

$$\theta(t) := \frac{1}{\lambda}\frac{\partial}{\partial t}f^r(0,t) + f^r(0,t) + \frac{\eta^2}{2\lambda^2}\left(1 - e^{-2\lambda t}\right). \qquad (11.39)$$

This function is determined by a differentiation of the instantaneous forward rate $f^r(0,t)$. This may be unstable because $f^r(0,t)$ is also obtained by differentiation of the ZCB, in Equation (11.11).

A stable representation can be found based on the function $\psi(t)$ in the Hull-White decomposition (Proposition 11.3.1), $r(t) = \tilde{r}(t) + \psi(t)$,

$$\psi(t) := r_0 e^{-\lambda t} + \lambda \int_0^t \theta(z)e^{-\lambda(t-z)}dz \qquad (11.40)$$

By substituting (11.39) into (11.40), $\psi(t)$ simplifies,

$$\psi(t) = e^{-\lambda t}\left[r_0 + e^{\lambda t}f^r(0,t) - f^r(0,0) + \frac{\eta^2}{\lambda^2}(\cosh(\lambda t) - 1)\right], \qquad (11.41)$$

where $\cosh(x) = \frac{e^x + e^{-x}}{2}$.

The expression for $\psi(t)$ in (11.41) depends only on the instantaneous forward rate $f^r(0,t)$ at $t_0 = 0$ and final time t. It is not based on integration or differentiation of the function $\theta(t)$, which makes this representation computationally attractive. This representation is thus also preferred from a stability perspective. ▲

Alternative derivation of the ZCB

In the derivation of the price of the ZCB, one does not necessarily have to split off the time-dependent term $\theta(t)$, as in the Hull-White decomposition in Proposition 11.3.1. It is possible to directly derive the ZCB value. Based on the Hull-White dynamics, the present value of $P(t,T)$, which pays 1 currency unit at maturity time T, is given by:

$$P(t,T) = M(t)\mathbb{E}^{\mathbb{Q}}\left[\frac{1}{M(T)}\bigg|\mathcal{F}(t)\right] = \mathbb{E}^{\mathbb{Q}}\left[e^{-\int_t^T r(z)dz}\bigg|\mathcal{F}(t)\right],$$

with $dM(t) = r(t)M(t)dt$.

In the Hull-White model the volatility is constant and the drift is a linear function of state variable $r(t)$, so that the model belongs to the class of affine processes. Its *discounted* characteristic function can therefore directly be derived.

11.3.3 The CIR model under the HJM framework

An alternative to the Hull-White model is found in the CIR short-rate model, which gives rise to nonnegative short rates. The dynamics of a Cox-Ingersoll-Ross (CIR) process for short-rate $r(t)$, are given by the following SDE:

$$\boxed{dr(t) = \lambda(\theta - r(t))dt + \gamma\sqrt{r(t)}dW_r(t),} \qquad (11.42)$$

Short-Rate Models 357

with λ the speed of mean reversion, θ the long-term mean and where γ controls the volatility. The square-root process, in (11.42), precludes negative values for $r(t)$, and if $r(t)$ reaches zero it may subsequently become positive. The square-root process has been discussed in detail in the context of the variance process of the Heston model in Section 8.1.2.

In this case, the *Feller condition* is defined by $2\lambda\theta \geq \gamma^2$, which guarantees that $r(t)$ stays positive; otherwise, it may reach zero.

Short-rate $r(t)$ under the CIR dynamics is distributed as

$$r(t)|r(s) \sim \bar{c}(t,s)\chi^2\left(\delta, \bar{\lambda}(t,s)\right), \quad 0 < s < t, \tag{11.43}$$

with

$$\bar{c}(t,s) = \frac{1}{4\lambda}\gamma^2(1 - e^{-\lambda(t-s)}), \quad \delta = \frac{4\lambda\theta}{\gamma^2}, \quad \bar{\lambda}(t,s) = \frac{4\lambda r(s)e^{-\lambda(t-s)}}{\gamma^2(1 - e^{-\lambda(t-s)})}. \tag{11.44}$$

The cumulative distribution function (CDF) is found to be as in Equation (8.6).

The corresponding density function is given, as in Equation (8.9), with the modified Bessel function of the first kind as in (8.10).

The CIR process, with constant parameters λ, θ and γ, is of the affine form, as its drift is a linear function of $r(t)$ and the instantaneous covariance equals $\gamma^2 r(t)$, which is also linear in $r(t)$. Based on the results in Section 7.3, the corresponding *discounted characteristic function* reads:

$$\phi_{\text{CIR}}(u;t,T) = e^{\bar{A}(u,\tau) + \bar{B}(u,\tau)r(t)}, \tag{11.45}$$

By $u = 0$ in (11.45), we find the price of the ZCB, i.e.

$$\phi_{\text{CIR}}(0;t,T) = \mathbb{E}^{\mathbb{Q}}\left[\exp\left(-\int_t^T r(z)\mathrm{d}z\right)\bigg|\mathcal{F}(t)\right]$$

$$= e^{\bar{A}(0,\tau) + \bar{B}(0,\tau)r(t)} =: P(t,T). \tag{11.46}$$

Solving the corresponding Riccati ODEs, gives

$$\bar{B}(0,\tau) = \frac{2(1 - e^{\tau a})}{(a+\lambda)(e^{\tau a} - 1) + 2a}, \quad \bar{A}(0,\tau) = \frac{2\lambda\theta}{\gamma^2}\log\left(\frac{2ae^{\frac{1}{2}\tau(a+\lambda)}}{2a + (a+\lambda)(e^{a\tau} - 1)}\right),$$

with $a = \sqrt{(\lambda^2 + 2\gamma^2)}$, and remaining parameters constant.

Using the notation, with $u = 0$, $\bar{A}(0,\tau) \equiv \bar{A}_{\text{CIR}}(t,T)$, $\bar{B}(0,\tau) \equiv \bar{B}_{\text{CIR}}(t,T)$, by Itô's lemma, the SDE of the zero-coupon bond price $P(t,T)$ in (11.46) is given by:

$$\frac{\mathrm{d}P(t,T)}{P(t,T)} = \left[\frac{\mathrm{d}\bar{A}_{\text{CIR}}(t,T)}{\mathrm{d}t} + r(t)\frac{\mathrm{d}\bar{B}_{\text{CIR}}(t,T)}{\mathrm{d}t} + \lambda\bar{B}_{\text{CIR}}(t,T)(\theta - r(t))\right.$$

$$\left. + \frac{1}{2}\gamma^2\bar{B}_{\text{CIR}}^2(t,T)r(t)\right]\mathrm{d}t + \gamma\bar{B}_{\text{CIR}}(t,T)\sqrt{r(t)}\mathrm{d}W_r(t). \tag{11.47}$$

At the same time, by Lemma 11.2.2, we know that in the general HJM framework the dynamics of the zero-coupon bond, under the risk-neutral measure, are given by:

$$\frac{\mathrm{d}P(t,T)}{P(t,T)} = r(t)\mathrm{d}t - \left(\int_t^T \bar{\eta}(t,z)\mathrm{d}z\right)\mathrm{d}W_r(t). \tag{11.48}$$

By equating the diffusion terms of the SDEs in (11.47) and (11.48), the following relation should be satisfied

$$\int_t^T \bar{\eta}(t,z)\mathrm{d}z = -\gamma \bar{B}_{\mathrm{CIR}}(t,T)\sqrt{f^r(t,t)}, \tag{11.49}$$

with $f^r(t,t) \equiv r(t)$ and $\bar{B}(0,\tau) \equiv \bar{B}_{\mathrm{CIR}}(t,T)$ (indicating that the function $\bar{B}_{\mathrm{CIR}}$ explicitly depends on final time T). After differentiation of (11.49) with respect to parameter T, we arrive at the following HJM volatility generator:

$$\bar{\eta}(t,T) = -\gamma\sqrt{f^r(t,t)}\frac{\mathrm{d}\bar{B}_{\mathrm{CIR}}(t,T)}{\mathrm{d}T}$$
$$= \frac{4\gamma a^2 \mathrm{e}^{a(T-t)}}{\left(\lambda(\mathrm{e}^{a(T-t)}-1)+a(\mathrm{e}^{a(T-t)}+1)\right)^2}\sqrt{f^r(t,t)}.$$

Example 11.3.2 In the Figures 11.5 and 11.6, the impact of parameter variation on typical Monte Carlo sample paths is presented. Figure 11.5 shows the impact for the Hull-White model, whereas Figure 11.6 displays the variation for the CIR short rate model. Before the year 2008 it seemed obvious to consider

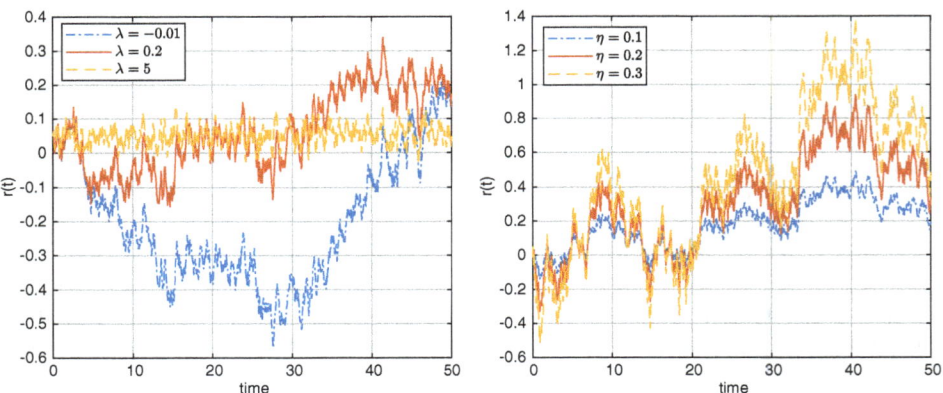

Figure 11.5: *Within the HW model context, the impact of variation of mean-reversion λ, and of the volatility parameter η on the Monte Carlo paths.*

only positive interest rates, however, after the serious financial crisis of 2007, negative rates have been encountered in the financial markets. Negative rates are discussed in some detail in Section 14.4.1 in a forthcoming chapter.

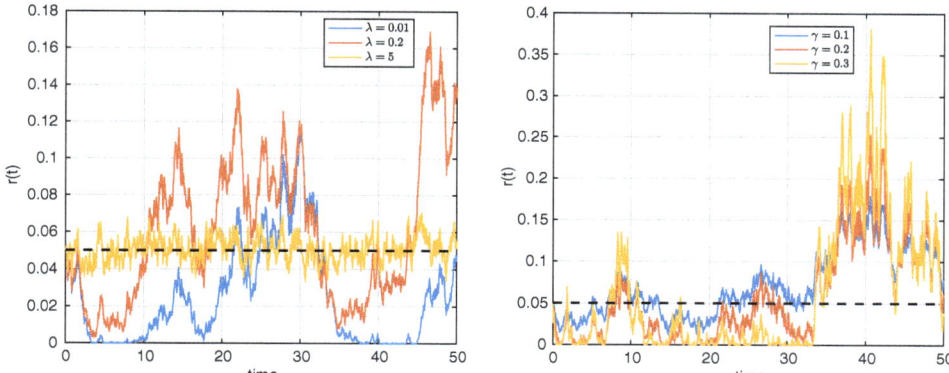

Figure 11.6: *For the CIR model, the impact of variation of mean-reversion λ, and of the volatility parameter γ on the Monte Carlo paths.*

11.4 The HJM model under the *T*-forward measure

So far, we have considered the HJM model under the risk-neutral measure, generated by money-savings account $M(t)$ as the numéraire. In this section, an extension of the HJM framework is explained, so that forward rates under the T_i−forward measure can be used. The derivation will be particularly useful in the context of the Libor market model (LMM), where the forward rates are modeled by a multi-dimensional system of SDEs. The details on the modeling of the LMM will be given in Section 14.1.

A change of measure, from the risk-neutral measure $\mathbb{Q}$ which is implied by the money-savings account $M(t)$, to a measure implied by the zero-coupon bond $P(t, T_i)$, with $t_0 < t < T_i$, requires the following Radon-Nikodym derivative:

$$\boxed{\lambda_{\mathbb{Q}}^i(t) = \frac{\mathrm{d}\mathbb{Q}^{T_i}}{\mathrm{d}\mathbb{Q}}\bigg|_{\mathcal{F}(t)} = \frac{P(t, T_i)}{P(t_0, T_i)} \frac{M(t_0)}{M(t)}.} \qquad (11.50)$$

By Itô's lemma, the dynamics of $\lambda_{\mathbb{Q}}^i(t)$ are given by

$$\mathrm{d}\lambda_{\mathbb{Q}}^i(t) = \frac{M(t_0)}{P(t_0, T_i)} \mathrm{d}\left(\frac{P(t, T_i)}{M(t)}\right)$$

$$= \frac{M(t_0)}{P(t_0, T_i)} \left[\frac{1}{M(t)} \mathrm{d}P(t, T_i) - \frac{P(t, T_i)}{M^2(t)} \mathrm{d}M(t)\right]. \qquad (11.51)$$

Remember, however, that $M(t_0) = 1$. With the help of Lemma 11.2.2, we substitute the HJM arbitrage-free dynamics for the ZCB, i.e.,

$$\frac{dP(t,T_i)}{P(t,T_i)} = r(t)dt - \left(\int_t^{T_i} \bar{\eta}(z,T_i)dz\right)dW^{\mathbb{Q}}(t), \tag{11.52}$$

with volatility function $\bar{\eta}(z,T_i)$, as defined in Lemma 11.2.1.

After substitution of (11.52) into (11.51), the SDE for $\lambda_{\mathbb{Q}}^i(t)$ is given by:

$$\frac{d\lambda_{\mathbb{Q}}^i(t)}{\lambda_{\mathbb{Q}}^i(t)} = -\left(\int_t^{T_i} \bar{\eta}(z,T_i)dz\right)dW^{\mathbb{Q}}(t). \tag{11.53}$$

By (Girsanov's) Theorem 7.2.2, the measure transformation, from measure $\mathbb{Q}$ to the T_i-forward measure $\mathbb{Q}^{T_i}$, is given by:

$$dW^i(t) = \left(\int_t^{T_i} \bar{\eta}(z,T_i)dz\right)dt + dW^{\mathbb{Q}}(t). \tag{11.54}$$

This transformation can be used to determine the dynamics of the instantaneous forward rate $f^r(t,T)$, with $t < T < T_i$, under measure $\mathbb{Q}^{T_i}$. Recall the main result from Lemma 11.2.1, i.e. that the arbitrage-free dynamics of the forward rate $f^r(t,T)$ are given by

$$df^r(t,T) = \bar{\eta}(t,T)\left(\int_t^T \bar{\eta}(t,z)dz\right)dt + \bar{\eta}(t,T)dW^{\mathbb{Q}}(t).$$

By the equality in (11.54), the dynamics for $f^r(t,T)$ are given by:

$$df^r(t,T) = \bar{\eta}(t,T)\left(\int_t^T \bar{\eta}(t,z)dz\right)dt + \bar{\eta}(t,T)\left[dW^i(t) - \left(\int_t^{T_i} \bar{\eta}(z,T_i)dz\right)dt\right],$$

which simplifies to:

$$df^r(t,T) = -\bar{\eta}(t,T)\left(\int_T^{T_i} \bar{\eta}(t,z)dz\right)dt + \bar{\eta}(t,T)dW^i(t), \tag{11.55}$$

for $T < T_i$.

11.4.1 The Hull-White dynamics under the T-forward measure

We discuss the dynamics of the Hull-White short-rate model under the T−forward measure. The advantages of the short-rate model under the changed measure will become clear in particular in the context of equity hybrid models, in Chapter 13.

Recall the Hull-White model under risk-neutral measure $\mathbb{Q}$, given by SDE (11.32). To change between measures, from $\mathbb{Q}$ governed by the money-savings account $M(t)$, to the T-forward measure $\mathbb{Q}^T$, implied by the zero-coupon bond $P(t,T)$, we make use again of the Radon-Nikodym derivative,

$$\lambda_{\mathbb{Q}}^T(t) = \frac{\mathrm{d}\mathbb{Q}^T}{\mathrm{d}\mathbb{Q}}\bigg|_{\mathcal{F}(t)} = \frac{P(t,T)}{P(t_0,T)} \frac{M(t_0)}{M(t)}, \quad \text{for } t > t_0. \tag{11.56}$$

Based on the dynamics for the ZCB from (11.31) and the definition of the money-savings account $M(t)$, the dynamics of $\lambda_{\mathbb{Q}}^T(t)$ read,

$$\mathrm{d}\lambda_{\mathbb{Q}}^T(t) = \frac{M(t_0)}{P(t_0,T)} \left(\frac{1}{M(t)} \mathrm{d}P(t,T) - \frac{P(t,T)}{M^2(t)} \mathrm{d}M(t) \right)$$

$$= \frac{M(t_0)}{P(t_0,T)} \frac{P(t,T)}{M(t)} \eta \bar{B}_r(t,T) \mathrm{d}W_r^{\mathbb{Q}}(t).$$

with $\bar{B}_r(t,T) = \frac{1}{\lambda}\left(\mathrm{e}^{-\lambda(T-t)} - 1\right)$. Using the definition of $\lambda_{\mathbb{Q}}^T(t)$ from (11.56), we thus find,

$$\frac{\mathrm{d}\lambda_{\mathbb{Q}}^T(t)}{\lambda_{\mathbb{Q}}^T(t)} = \eta \bar{B}_r(t,T) \mathrm{d}W_r^{\mathbb{Q}}(t). \tag{11.57}$$

This representation gives us the Girsanov kernel, which enables us to change between measures, from the risk-neutral measure $\mathbb{Q}$ to the T-forward measure $\mathbb{Q}^T$, i.e.

$$\mathrm{d}W_r^T(t) = -\eta \bar{B}_r(t,T) \mathrm{d}t + \mathrm{d}W_r^{\mathbb{Q}}(t). \tag{11.58}$$

The measure transformation defines the following short-rate dynamics under the $\mathbb{Q}^T$ measure,

$$\mathrm{d}r(t) = \lambda(\theta(t) - r(t))\mathrm{d}t + \eta \mathrm{d}W_r^{\mathbb{Q}}(t)$$

$$= \lambda \left(\theta(t) + \frac{\eta^2}{\lambda} \bar{B}_r(t,T) - r(t) \right) \mathrm{d}t + \eta \mathrm{d}W_r^T(t). \tag{11.59}$$

The process in (11.59) can further be rewritten into

$$\mathrm{d}r(t) = \lambda \left(\hat{\theta}(t,T) - r(t) \right) \mathrm{d}t + \eta \mathrm{d}W_r^T(t), \tag{11.60}$$

with

$$\hat{\theta}(t,T) = \theta(t) + \frac{\eta^2}{\lambda} \bar{B}_r(t,T), \quad \text{where} \quad \bar{B}_r(t,T) = \frac{1}{\lambda}\left(\mathrm{e}^{-\lambda(T-t)} - 1\right).$$

11.4.2 Options on zero-coupon bonds under Hull-White model

In this final section we discuss a European option with expiry at time T, on a zero-coupon bond $P(T, T_S)$ with *maturity time* T_S, $T < T_S$. Although this product is very basic and does not require extensive analysis, it is important to discuss the product, as it forms an important building block for pricing *swaption products*, in the follow-up chapter. A European-style option is defined by the following equation:

$$V^{\text{ZCB}}(t_0, T) = \mathbb{E}^{\mathbb{Q}} \left[\frac{M(t_0)}{M(T)} \max \left(\bar{\alpha}(P(T, T_S) - K), 0 \right) \Big| \mathcal{F}(t_0) \right],$$

with $\bar{\alpha} = 1$ for a call and $\bar{\alpha} = -1$ for a put option, strike price K and $dM(t) = r(t)M(t)dt$. By a measure change, from the risk-free to the T-forward measure, the pricing equation is given by,

$$V^{\text{ZCB}}(t_0, T) = P(t_0, T) \mathbb{E}^T \left[\max \left(\bar{\alpha}(P(T, T_S) - K), 0 \right) \Big| \mathcal{F}(t_0) \right].$$

> When dealing with an *affine* short-rate model $r(t)$, the zero-coupon bond $P(T, T_S)$ is an exponential function, and the pricing equation can be expressed as:
>
> $$V^{\text{ZCB}}(t_0, T) = P(t_0, T) \mathbb{E}^T \left[\max \left(\bar{\alpha} \left(e^{\bar{A}_r(\tau) + \bar{B}_r(\tau) r(T)} - K \right), 0 \right) \Big| \mathcal{F}(t_0) \right]$$
>
> $$= P(t_0, T) e^{\bar{A}_r(\tau)} \mathbb{E}^T \left[\max \left(\bar{\alpha} \left(e^{\bar{B}_r(\tau) r(T)} - \hat{K} \right), 0 \right) \Big| \mathcal{F}(t_0) \right], \tag{11.61}$$
>
> with $\tau = T_S - T$, $\hat{K} = K e^{-\bar{A}_r(\tau)}$, and $r(T)$ is the short-rate process at time T under T-forward measure $\mathbb{Q}^T$. $\bar{A}_r(\tau)$ and $\bar{B}_r(\tau)$ are given by
>
> $$\bar{A}_r(\tau) := \lambda \int_0^\tau \theta(T_S - z) \bar{B}_r(z) dz + \frac{\eta^2}{4\lambda^3} \left(e^{-2\lambda\tau}(4e^{\lambda\tau} - 1) - 3 \right) + \frac{\eta^2}{2\lambda^2} \tau,$$
>
> $$\bar{B}_r(\tau) := \frac{1}{\lambda} \left(e^{-\lambda\tau} - 1 \right). \tag{11.62}$$

The Hull-White dynamics of the short-rate $r(t)$ under the T-forward measure are given by (11.60). The Hull-White process is normally distributed, $r(T) \sim \mathcal{N}\left(\mu_r(T), v_r^2(T)\right)$, with mean $\mu_r(T)$ and variance $v_r^2(T)$, given by:

$$\mu_r(T) = r_0 e^{-\lambda(T-t_0)} + \lambda \int_{t_0}^T \hat{\theta}(z, T) e^{-\lambda(T-z)} dz,$$

$$v_r^2(T) = \frac{\eta^2}{2\lambda} \left(1 - e^{-2\lambda(T-t_0)} \right), \tag{11.63}$$

with $\hat{\theta}(z, T)$ as in (11.60). This implies that $\bar{B}_r(\tau) r(T)$ is also normally distributed, with

$$z(T) := \bar{B}_r(\tau) r(T) \sim \mathcal{N}\left(\bar{B}_r(\tau) \mu_r(T), \bar{B}_r^2(\tau) v_r^2(T) \right),$$

yielding the following expression for a European call option on the ZCB in Equation (11.61):

$$V_c^{\text{ZCB}}(t_0, T) = P(t_0, T) e^{\bar{A}_r(\tau)} \mathbb{E}^T \left[\max\left(e^{z(T)} - \hat{K}, 0\right) \Big| \mathcal{F}(t_0) \right] \quad (11.64)$$

$$= P(t_0, T) e^{\bar{A}_r(\tau)} \int_{-\infty}^{a} \left(e^{\bar{B}_r(\tau)\mu_r(T) + \bar{B}_r(\tau)v_r(T)x} - \hat{K} \right) f_{\mathcal{N}(0,1)}(x) \mathrm{d}x,$$

with $f_{\mathcal{N}(0,1)}(x)$ the standard normal PDF, and[3]

$$a = \frac{\log \hat{K} - \bar{B}_r(\tau)\mu_r(T)}{\bar{B}_r(\tau)v_r(T)}. \quad (11.65)$$

The integral can be decomposed into two integrals, i.e.

$$\int_{-\infty}^{a} \left(e^{\bar{B}_r(\tau)\mu_r(T) + \bar{B}_r(\tau)v_r(T)x} - \hat{K} \right) f_{\mathcal{N}(0,1)}(x) \mathrm{d}x = \psi_1 + \psi_2,$$

where

$$\psi_1 = e^{\bar{B}_r(\tau)\mu_r(T)} \int_{-\infty}^{a} \frac{1}{\sqrt{2\pi}} e^{\bar{B}_r(\tau)v_r(T)x} e^{-\frac{x^2}{2}} \mathrm{d}x$$

which, after simplification, is found to be,

$$\psi_1 = \exp\left(\frac{1}{2}\bar{B}_r^2(\tau)v_r^2(T) + \bar{B}_r(\tau)\mu_r(T)\right) F_{\mathcal{N}(0,1)}(a - \bar{B}_r(\tau)v_r(T)),$$

and

$$\psi_2 := -\hat{K} \int_{-\infty}^{a} f_{\mathcal{N}(0,1)}(x) \mathrm{d}x = -\hat{K} F_{\mathcal{N}(0,1)}(a),$$

with $F_{\mathcal{N}(0,1)}(x)$ the standard normal CDF, and a in (11.65).

By setting $d_1 = a - \bar{B}_r(\tau)v_r(T)$ and $d_2 = d_1 + \bar{B}_r(\tau)v_r(T)$, we find

$$\psi_1 = \exp\left(\frac{1}{2}\bar{B}_r^2(\tau)v_r^2(T) + \bar{B}_r(\tau)\mu_r(T)\right) F_{\mathcal{N}(0,1)}(d_1),$$

$$\psi_2 = -\hat{K} F_{\mathcal{N}(0,1)}(d_2).$$

The pricing equation is given by,

$$\frac{V_c^{\text{ZCB}}(t_0, T)}{P(t_0, T)} = \exp\left(\bar{A}_r(\tau)\right) \left[\exp\left(\frac{1}{2}\bar{B}_r^2(\tau)v_r^2(T) + \bar{B}_r(\tau)\mu_r(T)\right) F_{\mathcal{N}(0,1)}(d_1) \right.$$

$$\left. - \hat{K} F_{\mathcal{N}(0,1)}(d_2) \right], \quad (11.66)$$

with $\tau = T_S - T$, $d_1 = a - \bar{B}_r(\tau)v_r(T)$, $d_2 = d_1 + \bar{B}_r(\tau)v_r(T)$, with a in (11.65), $\bar{A}_r(\tau)$ and $\bar{B}_r(\tau)$ in (11.62), $\mu_r(T)$ and $v_r(T)$ in (11.63) and $\hat{K} = K e^{-\bar{A}_r(\tau)}$.

---

[3] Note that $\bar{B}_r(\tau) \leq 0$.

Pricing of European put options can be done analogously, or with the help of the put-call parity which also holds for the interest rate products.

This concludes the sections with derivations for the short-rate processes and ZCBs, under different measures. Next to the basic ZCB interest rate products, other interest rate derivatives exist. They will be presented in the chapter to follow.

11.5 Exercise set

Exercise 11.1 Show that for any stochastic process $r(t)$ with the dynamics given by:
$$\mathrm{d}r(t) = \lambda(\hat{\theta}(t, T_2) - r(t))\mathrm{d}t + \eta \mathrm{d}W_r^{T_2}(t), \quad r(t_0) = r_0,$$
and
$$\hat{\theta}(t, T_2) = \theta(t) + \frac{\eta^2}{\lambda}\bar{B}_r(T_2 - t),$$
for any time t, the mean and variance are given by,
$$\mathbb{E}^{T_2}[r(t)|\mathcal{F}(t_0)] = r_0 e^{-\lambda t} + \lambda \int_0^t \hat{\theta}(z, T_2) e^{-\lambda(t-z)} \mathrm{d}z,$$
$$\mathrm{Var}^{T_2}[r(t)|\mathcal{F}(t_0)] = \frac{\eta^2}{2\lambda}\left(1 - e^{-2\lambda t}\right).$$

Show that the moment generating function (the Laplace transform) is of the following, closed form:
$$\mathbb{E}^{T_2}\left[e^{ur(t)}\Big|\mathcal{F}(t_0)\right] = \exp\left[u\left(r_0 e^{-\lambda t} + \lambda \int_0^t \hat{\theta}(z, T_2) e^{-\lambda(t-z)} \mathrm{d}z\right) + \frac{1}{2}u^2 \frac{\eta^2}{2\lambda}\left(1 - e^{-2\lambda t}\right)\right].$$

Exercise 11.2 Give a proof for the HJM arbitrage-free condition, which is stated in Lemma 11.2.1.

Exercise 11.3 Give a proof for the Hull-White decomposition, which is stated in Proposition 11.3.1.

Exercise 11.4 Derive the formulas for the coefficients of the discounted characteristic function for the Hull-White model in Lemma 11.3.1.

Exercise 11.5 Show that the integration of the function $\psi(t)$ in (11.40) results in
$$\psi(t) = r_0 e^{-\lambda t} + \lambda \int_0^t \left(\frac{1}{\lambda}\frac{\partial}{\partial u} f^r(0, u) + f^r(0, u) + \frac{\eta^2}{2\lambda^2}\left(1 - e^{-2\lambda u}\right)\right) e^{-\lambda(t-u)} \mathrm{d}u$$
$$= e^{-\lambda t}\left[r_0 + e^{\lambda t}f^r(0, t) - f^r(0, 0) + \frac{\eta^2}{\lambda^2}(\cosh(\lambda t) - 1)\right],$$
where $\cosh(x) = \frac{e^x + e^{-x}}{2}$. Determine the intermediate steps.

Exercise 11.6 Derive the characteristic function for the HW model, directly, i.e. without the Hull-White decomposition.

Exercise 11.7 Show that the European put option on the ZCB can be expressed as,
$$V^{\mathrm{ZCB}}(t_0, T) = P(t_0, T)e^{\bar{A}_r(\tau)}\left[\hat{K}F_{\mathcal{N}(0,1)}(d_2) - e^{\frac{1}{2}\bar{B}_r^2(\tau)v_r^2(T) + \bar{B}_r(\tau)\mu_r(T)}F_{\mathcal{N}(0,1)}(d_1)\right]$$
with $\tau = T - t_0$, $d_1 = \bar{B}_r(\tau)v_r(T) - a$, $d_2 = -a$ with
$$a = \frac{\log \hat{K} - \bar{B}_r(\tau)\mu_r(T)}{\bar{B}_r(\tau)v_r(T)},$$
$\bar{A}_r(\tau)$ and $\bar{B}_r(\tau)$ are given in (11.62), $\mu_r(T)$ and $v_r(T)$ are in defined in (11.63) and (11.63) and $\hat{K} = Ke^{\bar{A}_r(\tau)}$.
Hint: The derivations can be done analogously to the derivations in Section 11.4.2.

Exercise 11.8 The so-called *Ho-Lee short-rate model* is defined by means of the HJM volatility, as follows,
$$\bar{\eta}(t, T) = \eta.$$
Show that:
$$dr(t) = \theta(t)dt + \eta dW^{\mathbb{Q}}(t), \tag{11.67}$$
where $\theta(t) = \frac{\partial}{\partial t} f^r(0, t) + \eta^2 t$.

Exercise 11.9 Consider the Vašiček short-rate model,
$$dr(t) = \lambda \left(\theta - r(t)\right) dt + \eta dW(t),$$
with parameters $\lambda = 0.05$, $\theta = 0.02$ and $\eta = 0.1$ and initial rate $r(t_0) = 0$. At time t_0, we wish to hedge a position in a 10y zero-coupon bond, $P(t_0, 10y)$, using two other bonds, $P(t_0, 1y)$ and $P(t_0, 20y)$.

  a. Determine two weights, ω_1 and ω_2, such that $\omega_1 + \omega_2 = 1$ and $\omega_1 P(t_0, 1y) + \omega_2 P(t_0, 20y) = P(t_0, 10y)$.
  b. Perform a minimum variance hedge and determine the weights ω_1 and ω_2, such that
  $$\operatorname{Var}\left[\int_0^{10y} \omega_1 P(t, 1y) + \omega_2 P(t, 20y) dt\right] = \operatorname{Var}\left[\int_0^{10y} P(t, 10y) dt\right],$$
  while $\omega_1 + \omega_2 = 1$. What can be said about this type of hedge compared to the point addressed?
  c. Change the measure, to the $T = 10y$-forward measure, and, for the given weights determined earlier, check whether the variance of the estimator increases.

Exercise 11.10 Assume that the short-rate follows the CIR model with constant parameters under the risk-free measure $\mathbb{Q}$,
$$dr(t) = \lambda\left(\theta - r(t)\right) dt + \gamma \sqrt{r(t)} dW^{\mathbb{Q}}(t).$$
Write a code for the simulation of this model to price a European call option with expiration time $T = 3$ on the zero-coupon bond $P(1, 3)$, with strike price $K = 0.75$. In the simulation use the parameter settings, $r_0 = 0.03$, $\lambda = 0.5$, $\theta = 0.1$ and $\gamma = 0.13$.

Exercise 11.11 Consider the Hull-White model,
$$dr(t) = \lambda(\theta(t) - r(t))dt + \eta dW_r^{\mathbb{Q}}(t),$$
with constant, positive, parameters λ and η. A zero-coupon bond is given by
$$P_{mkt}(0, t) = e^{1-e^{at}}, \quad a = 0.4. \tag{11.68}$$

  a. Use Equation (11.68) to determine $\theta(t)$.
  b. Use $r_0 = f^r(0, \epsilon)$, $\epsilon \to 0$, $\lambda = 0.4$ and $\eta = 0.075$ and price the model bond,
  $$P_{model}(0, t) = \mathbb{E}^{\mathbb{Q}}\left[e^{-\int_0^t r(s) ds}\right].$$

Compare $P_{mkt}(0, t)$ with $P_{model}(0, t)$ for $t \in [0, 10]$ and discuss the differences.

CHAPTER 12

Interest Rate Derivatives and Valuation Adjustments

In this chapter:

We discuss financial products that are available in the money market, and also we explain a specific form of risk management related to counterparties of such contracts. We start the presentation with some *well-known interest rate derivatives* in two sections, **Section 12.1** and **Section 12.2**. *Forward rate agreements, caps, floors, swaps, and swaptions* are discussed in quite some detail. These are all heavily traded interest rate products. Many of these products are based on so-called Libor rates, and can be re-written in terms of zero-coupon bond prices. When dealing with *not too complicated* products the bond prices may be modeled by means of the short-rate models from the previous chapter. They typically give accurate prices, especially when at-the-money options are considered.

Next to this, we will focus on a specific risk management issue within financial institutions. *Adjustments* are added to the fair values of derivatives, in order to deal with the possible default of a counterparty. The probability of a counterparty default is taken into account, by means of an extra charge to the derivative's price. In certain trades, collateral is exchanged, and then the risk of a loss due to a defaulting counterparty is reduced. Adjustments to the option values are thus common for over-the-counter trades in which collateral is not exchanged. *Credit Valuation Adjustment* (CVA) is explained in detail in **Section 12.3**. This section explains the modeling of the *exposure to the counterparty*, which requires very similar modeling and pricing techniques as financial option valuation. The techniques and models explained so far are thus also useful in the context of CVA.

Keywords: interest rate derivatives, Hull-White model, valuation adjustment, expected exposure, potential future exposure, counterparty default.

12.1 Basic interest rate derivatives and the Libor rate

Many interest rate products are traded heavily every day in the financial money markets. It is often the zero-coupon bond (ZCB), $P(t,T)$, which forms the basis of such interest rate derivative contracts. When dealing with *not too complicated* interest rate products, the corresponding bond prices may be modeled by means of the arbitrage-free short-rate models from the previous chapter. Here we will discuss a few of the most common interest rate products, but first we define the Libor rate.

12.1.1 Libor rate

The London-InterBank Offered Rate (Libor) is an interest rate, which is determined, or fixed (as the process of determining the rate is called *fixing*), each working day at 11AM London time. The legal entity handling the Libor rate assessment is the British Banker's Association (BBA), which is the trade association for UK banking. This association unites approximately 250 members world-wide from 180 different counties. In the procedure to determine the Libor rate, a number of international banks (the pull) is asked to submit their Libor quotes to the BBA. These rates then represent the rate at which the banks are able to borrow money from other banks. In each assessment stage the top and the bottom quantiles from all submissions are eliminated and the average of the remaining quotes is agreed to represent the *Libor fixing*. If a bank is able to borrow at a very low interest rate, this is also an indication for the condition of the financial system, as banks are then willing to lend out money at a low cost. In the case of a crisis, banks, like other institutions, are more reluctant regarding the lending money at low rates. For fair business, it is crucial that Libor quotes and the submissions are fair and represent the actual funding rate.

We have already mentioned the two types of coupons in bond products, those that are fixed and those that are reset on a regular basis (the floating bonds). Commonly, the floating rate notes are tied to a reference benchmark, such as to treasury bills, Libor rates or the so-called consumer price indices, CPIs, that are related to changes in the inflation. If a coupon is connected to a Libor rate, which is quoted in terms of basis points (100 bp= 1% interest), the bond may be quoted at "*Libor* + x *bp*", which implies that the bond costs the Libor rate plus $x/100$. In this case, x represents a *spread*, which is determined at the time of issuance of the bond and it will remain fixed until the bond matures.

Let us assume there are two counterparties, A and B, where counterparty A will pay to counterparty B 1€ at time T_1 and at time T_2 counterparty A will receive back 1€ and will also receive the interest rate K over the accrual time $T_2 - T_1$. The cash flows are presented in Figure 12.1.

Interest Rate Derivatives and Valuation Adjustments

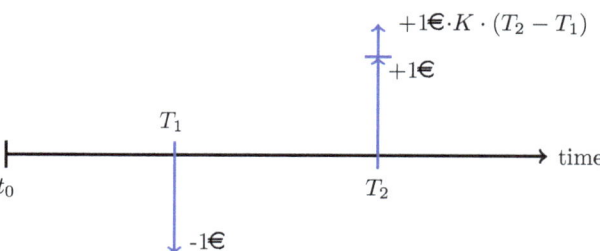

Figure 12.1: *Cash flows between two counterparties.*

The fair value of this contract is calculated as follows,

$$V(t_0) = \mathbb{E}^{\mathbb{Q}}\left[\frac{-1}{M(T_1)} + \frac{1 + K \cdot (T_2 - T_1)}{M(T_2)}\bigg|\mathcal{F}(t_0)\right]$$
$$= -P(t_0, T_1) + (1 + (T_2 - T_1) \cdot K)P(t_0, T_2).$$

The interest rate K for which the value of this contract at time t_0 equals 0, meaning no payments at inception, is given by

$$K = \frac{1}{(T_2 - T_1)}\left(\frac{P(t_0, T_1)}{P(t_0, T_2)} - 1\right).$$

Generally, the fair rate K for interbank lending with trade date t, starting date T_{i-1} and maturity date T_i with tenor $\tau_i = T_i - T_{i-1}$, and denoted by $K \equiv \ell_i(t) := \ell(t; T_{i-1}, T_i)$, equals,

$$\ell(t; T_{i-1}, T_i) = \frac{1}{\tau_i}\left(\frac{P(t, T_{i-1})}{P(t, T_i)} - 1\right).$$

We deal with a trading time horizon $[0, T^*]$ and a set of times $\{T_i, i = 1, \ldots, m\}, m \in \mathbb{N}$, such that $0 \leq T_0 < T_i < \cdots < T_m \leq T^*$, and define a so-called *tenor*, $\tau_i := T_i - T_{i-1}$.

Definition 12.1.1 (Simple compounded forward Libor rate) *For a given tenor τ_i, and the risk-free ZCB maturing at time T_i with a nominal of 1 (unit of currency), $P(t, T_i)$, the Libor forward rate $\ell(t; T_{i-1}, T_i)$ for a period $[T_{i-1}, T_i]$ is defined as,*

$$\boxed{\ell_i(t) \equiv \ell(t; T_{i-1}, T_i) = \frac{1}{\tau_i}\frac{P(t, T_{i-1}) - P(t, T_i)}{P(t, T_i)}.} \quad (12.1)$$

◀

For notational convenience, we use $\ell_i(t) \equiv \ell(t; T_{i-1}, T_i)$, so that $\ell_i(T_{i-1}) = \ell(T_{i-1}; T_{i-1}, T_i)$.

12.1.2 Forward rate agreement

We start the products overview of the different interest rate products with the forward rate agreement. A forward rate agreement (FRA), or simply "forward contract", is a financial product by which a contract party can *"lock in"* an interest rate for a given period of time. In the interest rate market, it is common to fix an interest rate at a future time T_{i-1} which is then accrued over a future time period $[T_{i-1}, T_i]$. For this period $[T_{i-1}, T_i]$, the two parties agree to *exchange a fixed rate K* for a payment of the (floating) Libor rate, which is observed at time T_{i-1}. Typically, these payments are exchanged at time T_i (which is often two business days after time T_{i-1}).

> The payoff of the FRA contract at time T_{i-1} is given by:
>
> $$V^{\text{FRA}}(T_{i-1}) = H^{\text{FRA}}(T_{i-1}) = \frac{\tau_i \left(\ell(T_{i-1}; T_{i-1}, T_i) - K\right)}{1 + \tau_i \ell(T_{i-1}; T_{i-1}, T_i)}, \quad (12.2)$$
>
> with the tenor $\tau_i = T_i - T_{i-1}$ and we assume a unit notional amount, $N = 1$.

Using the definition of the Libor rate $\ell(T_{i-1}; T_{i-1}, T_i)$, we can directly connect the denominator in (12.2) with the ZCB $P(T_{i-1}, T_i)$, as follows,

$$P(T_{i-1}, T_i) = \frac{1}{1 + \tau_i \ell(T_{i-1}; T_{i-1}, T_i)},$$

so that the FRA's payoff function can also be expressed as,

$$V^{\text{FRA}}(T_{i-1}) = \tau_i P(T_{i-1}, T_i) \left(\ell(T_{i-1}; T_{i-1}, T_i) - K\right). \quad (12.3)$$

The current price of the FRA contract is determined by,

$$V^{\text{FRA}}(t_0) = M(t_0) \cdot \mathbb{E}^{\mathbb{Q}} \left[\frac{1}{M(T_{i-1})} \tau_i P(T_{i-1}, T_i) \left(\ell(T_{i-1}; T_{i-1}, T_i) - K\right) \Big| \mathcal{F}(t_0) \right]$$

$$= \mathbb{E}^{\mathbb{Q}} \left[\frac{1 - P(T_{i-1}, T_i)}{M(T_{i-1})} - \tau_i K \frac{P(T_{i-1}, T_i)}{M(T_{i-1})} \Big| \mathcal{F}(t_0) \right], \quad (12.4)$$

since $M(t_0) = 1$.

> Since the bonds $P(T_{i-1}, T_i)$ are traded assets, the discounted bonds should be *martingales*, which gives us,
>
> $$V^{\text{FRA}}(t_0) = P(t_0, T_{i-1}) - P(t_0, T_i) - \tau_i K P(t_0, T_i)$$
> $$= \tau_i P(t_0, T_i) \left(\ell(t_0; T_{i-1}, T_i) - K\right). \quad (12.5)$$
>
> Note that, by definition,
>
> $$P(t_0, T_{i-1}) := \mathbb{E}^{\mathbb{Q}} \left[\frac{1}{M(T_{i-1})} \Big| \mathcal{F}(t_0) \right].$$

Most commonly, the FRAs are traded at zero value, which implies that for the fixed rate we should have, $K = \ell(t_0; T_{i-1}, T_i)$.

12.1.3 Floating rate note

Another heavily traded product is the floating rate note (FRN). Given the floating Libor rate, $\ell_i(T_i) := \ell(T_{i-1}; T_{i-1}, T_i)$, and a notional amount N, the FRN is an instrument with coupon payments, which is defined as,

$$V_i^{\text{FRN}}(T_i) = \begin{cases} N\tau_i\ell(T_{i-1}; T_{i-1}, T_i), & i \in \{1, 2, \ldots, m-1\}, \\ N\tau_m\ell(T_{m-1}; T_{m-1}, T_m) + N, & i = m. \end{cases} \quad (12.6)$$

The bond thus consists of a sum of payments, see also Figure 12.2, and each cash flow can be priced separately, as

$$\begin{aligned} V_i^{\text{FRN}}(t_0) &= \mathbb{E}^{\mathbb{Q}}\left[\frac{1}{M(T_i)}V_i^{\text{FRN}}(T_i)\Big|\mathcal{F}(t_0)\right] \\ &= P(t_0, T_i)\mathbb{E}^{T_i}\left[V_i^{\text{FRN}}(T_i)\big|\mathcal{F}(t_0)\right], \end{aligned} \quad (12.7)$$

by a measure change, from the risk-neutral measure $\mathbb{Q}$ to the T_i forward measure. Now, $\ell(T_{i-1}; T_{i-1}, T_i)$ is a traded quantity, so it should again be a martingale, and therefore we have:

$$\mathbb{E}^{T_i}\left[\ell(T_{i-1}; T_{i-1}, T_i)\big|\mathcal{F}(t_0)\right] = \ell(t_0; T_{i-1}, T_i),$$

resulting in,

$$\mathbb{E}^{T_i}\left[V_i^{\text{FRN}}(T_i)\big|\mathcal{F}(t_0)\right] = \begin{cases} N\tau_i\ell(t_0; T_{i-1}, T_i), & i \in \{1, 2, \ldots, m-1\}, \\ N\tau_m\ell(t_0; T_{m-1}, T_m) + N, & i = m. \end{cases} \quad (12.8)$$

12.1.4 Swaps

Swaps are financial products that enable their holders to *swap two sets of interest rate payments*. For example, a company may have a financial contract in which,

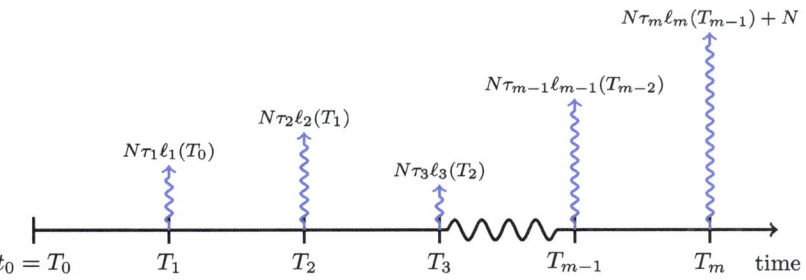

Figure 12.2: *Cash flows for a floating rate note with $\ell_i(T_{i-1}) := \ell(T_{i-1}; T_{i-1}, T_i)$ and each coupon defined as $N\tau_i\ell(T_{i-1}; T_{i-1}, T_i)$, $\tau_i = T_i - T_{i-1}$.*

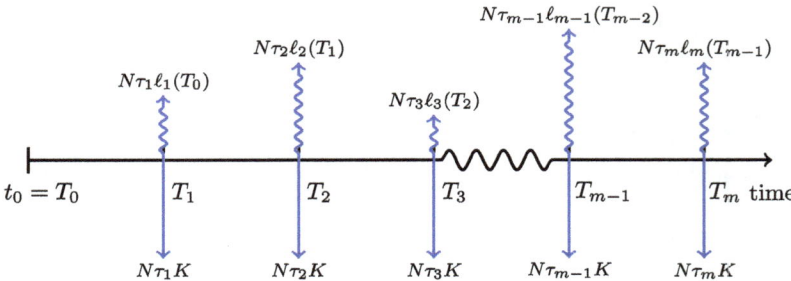

Figure 12.3: *Cash flows for a swap with* $\ell_i(T_{i-1}) := \ell(T_{i-1}; T_{i-1}, T_i)$, $\tau_i = T_i - T_{i-1}$.

in each time period, a floating interest rate is received which changes each period, and depends on the rates at that time. It may however be desirable to secure the cash flows by receiving *a fixed interest rate* in these time periods.

A swap is an agreement between two parties to exchange financial instruments (usually cash flows) in the future. The swap can be purchased to swap the floating interest rate with a fixed rate, or vice versa.

Swaps are OTC contracts, that were traded for the first time in the nineteen-eighties. The most commonly known type of swap is the *plain vanilla interest rate swap*, where one party agrees to pay the fixed cash flows that are equal to the interest at a predetermined, fixed rate on a notional amount and where the other party pays a floating interest on the same notional amount. The payments are made at predetermined future dates for a predetermined number of years.

The plain vanilla interest rate swap consists of two legs, a fixed leg with a series of fixed rate payments, at a fixed rate K, at the future times $T_{i+1}, \ldots, T_m$ and the float leg consisting of a series of *floating* Libor rates, see Figure 12.3.

The payment of the fixed leg at each time T_i is equal to $\tau_i N K$, with N the notional amount and K is the fixed interest rate. The float leg equals $N\tau_i \ell(T_{i-1}; T_{i-1}, T_i)$, with $\ell(T_{i-1}; T_{i-1}, T_i)$ the Libor rate over the period $[T_{i-1}, T_i]$, which resets at time T_{i-1}.

In financial jargon, the two terms related to interest rate swaps that are often used, are *interest rate swap payer* and *interest rate swap receiver*. Those terms distinguish between the cases when one receives a float leg and pays the fixed leg (swap payer), and the case where one pays the float leg and receives the fixed leg (swap receiver).

The payoff of the interest rate (*payer* or *receiver*) swap is given by:

$$V^{\text{PS,RS}}(T_i, \ldots, T_m) = \bar{\alpha} H^S(T_i, \ldots, T_m) = \bar{\alpha} \sum_{k=i+1}^{m} \tau_k N \big(\ell_k(T_{k-1}) - K\big), \quad (12.9)$$

with $\bar{\alpha} = 1$ for a payer and $\bar{\alpha} = -1$ for a receiver swap, and where we again use the short-hand notation $\ell_k(T_{k-1}) := \ell(T_{k-1}; T_{k-1}, T_k)$, which indicates that the Libor rate is determined (expires) at time T_{k-1}.

Because this IR swap starts in the future, at time T_i, it is also sometimes called a forward swap.

In order to determine today's t_0-value of the swap, we evaluate the corresponding expectation of the discounted future cash flows, i.e., each payment which takes place at the time points, $T_{i+1}, \ldots, T_m$ needs to be discounted to today,

$$V^{\text{PS,RS}}(t_0) = \bar{\alpha} \cdot N \cdot M(t_0) \mathbb{E}^{\mathbb{Q}} \left[\sum_{k=i+1}^{m} \frac{1}{M(T_k)} \tau_k \big(\ell_k(T_{k-1}) - K\big) \Big| \mathcal{F}(t_0) \right]$$

$$= \bar{\alpha} \cdot N \cdot M(t_0) \sum_{k=i+1}^{m} \tau_k \mathbb{E}^{\mathbb{Q}} \left[\frac{1}{M(T_k)} \big(\ell_k(T_{k-1}) - K\big) \Big| \mathcal{F}(t_0) \right].$$

From here on, we will focus on the payer swap, $\bar{\alpha} = 1$, and use the notation $V^{\text{PS}} = V^{\text{S}}$.

The measure change, from the risk-neutral measure $\mathbb{Q}$ to the T_k-forward measure $\mathbb{Q}^{T_k}$, as in Equation (11.61), gives,

$$V^{\text{S}}(t_0) = N \sum_{k=i+1}^{m} \tau_k P(t_0, T_k) \Big(\mathbb{E}^{T_k} \big[\ell_k(T_{k-1}) | \mathcal{F}(t_0) \big] - K \Big)$$

$$= N \sum_{k=i+1}^{m} \tau_k P(t_0, T_k) \Big(\ell_k(t_0) - K \Big), \qquad (12.10)$$

where the last step is because the Libor rate $\ell_k(t)$, under its own natural measure $\mathbb{Q}^{T_k}$, is a martingale, i.e. $\mathbb{E}^{T_k}[\ell_k(t)|\mathcal{F}(t_0)] = \ell_k(t_0)$.

Equation (12.10) is rewritten, as follows:

$$V^{\text{S}}(t_0) = N \sum_{k=i+1}^{m} \tau_k P(t_0, T_k) \ell_k(t_0) - NK \sum_{k=i+1}^{m} \tau_k P(t_0, T_k). \qquad (12.11)$$

The first summation in (12.11) can be further simplified, by the definition of the Libor rate (12.1), i.e.,

$$\sum_{k=i+1}^{m} \tau_k P(t_0, T_k) \ell_k(t_0) = \sum_{k=i+1}^{m} \tau_k P(t_0, T_k) \left(\frac{1}{\tau_k} \frac{P(t_0, T_{k-1}) - P(t_0, T_k)}{P(t_0, T_k)} \right)$$

$$= \sum_{k=i+1}^{m} \Big(P(t_0, T_{k-1}) - P(t_0, T_k) \Big)$$

$$= P(t_0, T_i) - P(t_0, T_m),$$

where the last step is due to a telescopic summation.

The price of the (payer) swap is now given, as follows:

$$V^{\text{S}}(t_0) = N \left(P(t_0, T_i) - P(t_0, T_m) \right) - NK \sum_{k=i+1}^{m} \tau_k P(t_0, T_k). \qquad (12.12)$$

Definition 12.1.2 (Annuity factor) *The annuity factor, is defined as,*

$$A_{i,m}(t) := \sum_{k=i+1}^{m} \tau_k P(t, T_k). \qquad (12.13)$$

In fact, the annuity $A_{i,m}(t)$ is nothing but a linear combination of zero-coupon bonds. ◀

As each of these zero-coupon bonds is a tradable asset, the linear combination, and therefore the annuity, is tradable as well. Therefore, we may consider the *annuity function as a numéraire* when pricing the corresponding derivatives.

Typically, interest rate swaps are considered *perfect interest rate products*, where two trading parties can hedge their particular exposures. Standard practice in determining the strike price K for the interest rate swaps is that the value of the swap at the initial time t_0 should equal zero. Entering such a financial deal is then *for free*. Moreover, the strike value for which the swap equals zero is called the *swap rate*, and it is indicated by $S_{i,m}(t_0)$.

By setting (12.12) equal to zero, we find,

$$S_{i,m}(t_0) = \frac{P(t_0, T_i) - P(t_0, T_m)}{\sum_{k=i+1}^{m} \tau_k P(t_0, T_k)} = \frac{P(t_0, T_i) - P(t_0, T_m)}{A_{i,m}(t_0)}, \qquad (12.14)$$

which, alternatively, may be written as

$$S_{i,m}(t_0) = \frac{1}{A_{i,m}(t_0)} \sum_{k=i+1}^{m} \tau_k P(t_0, T_k) \ell_k(t_0) = \sum_{k=i+1}^{m} \omega_{i,k}(t_0) \ell_k(t_0),$$

with $\omega_{i,k}(t_0) = \tau_k \frac{P(t_0, T_k)}{A_{i,m}(t_0)}$ and $\ell_k(t_0) := \ell(t_0; T_{k-1}, T_k)$.

> By using Equation (12.14), the value of the (payer) swap (12.12) is expressed, as,
>
> $$V^S(t_0) = N \cdot A_{i,m}(t_0) \big(S_{i,m}(t_0) - K \big). \qquad (12.15)$$
>
> This compact representation is convenient, since all the components have a specific meaning: the annuity $A_{i,m}(t_0)$ represents the present value of a basis point of the swap; $S_{i,m}(t)$ is the swap rate and K is the strike price. Obviously, with $K = S_{i,m}(t_0)$, today's value of this swap is equal to zero.

Remark 12.1.1 *Regarding the basic interest rate swap, the pricing can be performed without any assumptions about the underlying model. The pricing can simply be done by using interest rate products that are available in the market.*

Notice that the same is true for the FRA's and the FRN's. ▲

12.1.5 How to construct a yield curve

Instead of directly working with the bond prices, it is common in the interest rates world to use the term *yield* at which bonds trade. So, the *bond yield* represents the interest rates at which bonds are traded. Yields are different for different maturity times and this gives rise to the existence of a *yield curve*, which indicates today's yield (interest rate) for any given maturity. An investor who wishes to invest €1 in bonds today, can, with the help of the yield curve, assess the future expected market value. Usually yields increase at longer maturities, so that investors are rewarded for the additional risk related to holding a bond for a longer time.

A yield curve is thus an important element to determine the present value of future cash flows. The yield curve forms, for example, the basis for the forward rates that are needed for pricing interest rate derivatives. The idea of a yield curve is to map the market quotes of liquid interest rate-based products to a unified curve, which then represents the expectation of the future rates.

The concept by itself is like an implied volatility problem, however, here discount factors are implied from multiple interest rates derivatives instead of volatilities from option prices. A discrete set of products π is mapped to a discrete set of input nodes on a yield curve Ω_{yc}. Typically, the set π will contain cash products, and, depending on the market, different interest rate products, and Ω_{yc} will consist of discount factors.

So, a yield curve is represented by a set of nodes,

$$\Omega_{yc} = \{(t_1, p(t_1)), (t_2, p(t_2)), \ldots, (t_n, p(t_n))\}, \quad (12.16)$$

where the discount factor $p(t_i)$ is defined as:

$$p_i \equiv p(t_i) := P(t_0, t_i) = \mathbb{E}^{\mathbb{Q}}\left[1 \cdot e^{-\int_{t_0}^{t_i} r(z)dz}\bigg|\mathcal{F}(t_0)\right].$$

Since discount factors, as opposed to the short-rate $r(t)$, are deterministic, they can be expressed by the simply compounded rate r_i, i.e. $p(t_i) = e^{-r_i(t_i - t_0)}$. Often, Ω_{yc} in (12.16) is called the set of *spine points* of the yield curve. The spine points are thus directly implied from the financial products that are used for its calibration. Based on a finite number of spine points p_i, a continuous function $P(t_0, t)$ is defined, where at each time $t = t_i$, we have $P(t_0, t_i) = p_i$. Points between spine points can be calculated by some interpolation scheme.

An accurate yield curve is crucial, not only for pricing interest rate derivatives but also for hedging purposes, e.g. once the price of an interest rate product is determined, the product's sensitivity with respect to the products that are used to construct the yield curve will be used for hedging and risk management.

In [Hagan and West, 2008] the criteria for the curve construction and its interpolation were given, such as

- the yield curve should be able to price back the products that were used to construct it,

- the forward rates implied from the curve should be continuous,

- the interpolation used should be "as local as possible", meaning that a small change in a node of the curve should not affect nodes that are "far away" from it,
- the hedge should be local too, i.e., if one of the products in the curve is hedged then the hedge shouldn't be based on many other products.

With these criteria, the suitability of a yield curve configuration, in terms of the products used and the interpolation, for the purpose of interest rate derivative pricing, can be assessed.

Each calibration product, which is of plain vanilla type, is a function of the spine points. The i-th product is defined as $V_i(t_0) := V_i(t_0, \Omega_{yc})$, and the product is assumed to be quoted on the market, with market quote V_i^{mkt}.

During the calibration process, we search for a set of spine discount factors $\mathbf{p} := [p_1, p_2, \ldots, p_n]^T$, for which

$$d_i := V_i(t_0, \mathbf{p}) - V_i^{mkt} = 0, \ i = 1, \ldots, n, \quad (12.17)$$

where d_i stands for the defect, i.e., the present value of a derivative minus the market quote.

The problem of finding implied volatilities is similar, but here the situation is a bit more involved as the yield curve will be based on multiple products, so that the calibration problem becomes multidimensional. We take $\mathbf{d} = [d_1, d_2, \ldots, d_n]^T$ and determine the optimal spine points, by solving $\mathbf{d} = 0$, by means of a Newton iteration, i.e,

$$\mathbf{d}(\mathbf{p} + \boldsymbol{\Delta p}) = \mathbf{d}(\mathbf{p}) + \frac{\partial \mathbf{d}(\mathbf{p})}{\partial \mathbf{p}} \boldsymbol{\Delta p} + O\left(\boldsymbol{\Delta p}^2\right)$$

$$=: \mathbf{d}(\mathbf{p}) + \mathbf{J}(\mathbf{p}) \boldsymbol{\Delta p} + O\left(\boldsymbol{\Delta p}^2\right), \quad (12.18)$$

with Jacobian $\mathbf{J}(\mathbf{p})$. This, in matrix notation, is equivalent to:

$$\begin{bmatrix} d_1(\mathbf{p}+\boldsymbol{\Delta p}) \\ d_2(\mathbf{p}+\boldsymbol{\Delta p}) \\ \vdots \\ d_n(\mathbf{p}+\boldsymbol{\Delta p}) \end{bmatrix} = \begin{bmatrix} d_1(\mathbf{p}) \\ d_2(\mathbf{p}) \\ \vdots \\ d_n(\mathbf{p}) \end{bmatrix} + \begin{bmatrix} \frac{\partial d_1}{\partial p_1} & \frac{\partial d_1}{\partial p_2} & \cdots & \frac{\partial d_1}{\partial p_n} \\ \frac{\partial d_2}{\partial p_1} & \frac{\partial d_2}{\partial p_2} & \cdots & \frac{\partial d_2}{\partial p_n} \\ \vdots & & \ddots & \vdots \\ \frac{\partial d_n}{\partial p_1} & \frac{\partial d_n}{\partial p_2} & \cdots & \frac{\partial d_n}{\partial p_n} \end{bmatrix} \begin{bmatrix} \Delta p_1 \\ \Delta p_2 \\ \vdots \\ \Delta p_n \end{bmatrix} + O\left(\boldsymbol{\Delta p}^2\right).$$

Ignoring higher-order terms, aiming for a zero defect, i.e., $\mathbf{d}(\mathbf{p} + \boldsymbol{\Delta p}) = 0$, and solving this linear equation system for $\boldsymbol{\Delta p}$, gives us,

$$\boldsymbol{\Delta p} = -\mathbf{J}^{-1}(\mathbf{p}) \mathbf{d}(\mathbf{p}). \quad (12.19)$$

With a superscript k, indicating the iteration number, we approximate the solution by the following Newton iteration:

$$\mathbf{p}^{(k+1)} = \mathbf{p}^{(k)} - \mathbf{J}^{-1}\left(\mathbf{p}^{(k)}\right) \mathbf{d}\left(\mathbf{p}^k\right), \quad (12.20)$$

for $k = 1, \ldots$. A simple example will be presented next.

Interest Rate Derivatives and Valuation Adjustments 377

Example 12.1.1 (Yield curve calculation without interpolation)
With some of the basic interest rate products being defined in the earlier sections of this chapter, we give an example of the yield curve construction. The basic yield curve here will be based on two products, a Forward Rate Agreement (FRA) and a swap agreement. It will give us a rather trivial, but insightful, yield curve. In practice, the number of products for the curve construction may vary from just a few to ten or more, depending on the market. The present value of a FRA contract is given by:

$$V^{FRA}(t_0) = P(t_0, T_1) \left(\frac{\tau(\ell(t_0; T_2, T_3) - K_1)}{1 + \tau \ell(t_0; T_2, T_3)} \right), \quad (12.21)$$

with $\tau = T_3 - T_2$ and T_1, T_2, T_3 being the settlement date, the fixing date and the maturity date, respectively. For simplicity, we set $t_0 = T_1 = T_2$, and $T_3 = 1$, giving

$$V^{FRA}(t_0) = \frac{\ell(t_0; t_0, 1) - K_1}{1 + \ell(t_0; t_0, 1)}. \quad (12.22)$$

So, d_1 in Equation (12.17) is given by:

$$d_1 = \frac{\ell(t_0; t_0, 1) - K_1}{1 + \ell(t_0; t_0, 1)} - V_1^{mkt} = 0. \quad (12.23)$$

By the definition of the Libor rate, i.e.

$$\ell(t_0; T_1, T_2) = \frac{1}{T_2 - T_1} \left(\frac{P(t_0, T_1) - P(t_0, T_2)}{P(t_0, T_2)} \right),$$

we can express d_1 in terms of the spine discount factors, as follows

$$d_1 = \frac{\left(\frac{1}{p_1} - 1\right) - K_1}{1 + \left(\frac{1}{p_1} - 1\right)} - V_1^{mkt} = 0, \quad (12.24)$$

which simplifies to,

$$d_1 = 1 - (1 + K_1)p_1 - V_1^{mkt} = 0. \quad (12.25)$$

The swap, with a set of payment dates, $T_i, \ldots, T_m$, has the following present value:

$$V^S(t_0) = P(t_0, T_i) - P(t_0, T_m) - K_2 \sum_{k=i+1}^{m} \tau_k P(t_0, T_k). \quad (12.26)$$

If we take $T_i = 1$ and $T_m = 2$, this gives us the following d_2-value, in terms of the spine discount factors, p_1 and p_2,

$$d_2 = p_1 - p_2 - K_2 p_2 - V_2^{mkt} = 0. \quad (12.27)$$

With these results, we calibrate the yield curve spine discount factors, $\mathbf{p} = [p_1, p_2]^T$, as follows,

$$\begin{bmatrix} d_1 \\ d_2 \end{bmatrix} = \begin{bmatrix} 1 - (1+K_1)p_1 - V_1^{mkt} \\ p_1 - p_2(1+K_2) - V_2^{mkt} \end{bmatrix}, \quad (12.28)$$

and

$$\mathbf{J} := \begin{bmatrix} \frac{\partial d_1}{\partial p_1} & \frac{\partial d_1}{\partial p_2} \\ \frac{\partial d_2}{\partial p_1} & \frac{\partial d_2}{\partial p_2} \end{bmatrix} = \begin{bmatrix} -(1+K_1) & 0 \\ 1 & -(1+K_2) \end{bmatrix}, \qquad (12.29)$$

with the inverse of the Jacobian:

$$\mathbf{J}^{-1} = \frac{1}{(1+K_1)(1+K_2)} \begin{bmatrix} -(1+K_2) & 0 \\ -1 & -(1+K_1) \end{bmatrix}. \qquad (12.30)$$

Based on the quotes from the market, V_1^{mkt}, V_2^{mkt}, and the corresponding strike prices K_1 and K_2, by Equation (12.20), the optimal spine discount factors p_1 and p_2 can be obtained. Often in the financial market, the strike prices K_1 and K_2 are set such that $V_1^{mkt} = 0$ and $V_2^{mkt} = 0$ at inception t_0, i.e., the most common contracts are those for which the value at the start of the contract is equal to zero.◆

12.2 More interest rate derivatives

Here, we will discuss a few interest rate derivatives with a somewhat more involved contract definition.

12.2.1 Caps and floors

Somewhat more involved interest rate products are caps and floors. They are based on simpler building blocks.

An interest rate cap has been defined to provide an insurance to the holder of a loan, which is based on a floating rate, against a floating rate which possibly might increase above a pre-defined level, i.e. *the cap-rate K*. The interest rate floor provides an insurance against a floating rate which might decrease below a pre-defined level, which is the floor-rate K in this case. Caps and floors can be decomposed into sums of basic interest rate contracts. In the case of the cap, these are called *caplets*, and they are defined as follows,

---

Definition 12.2.1 (Caplet/Floorlet) *Given two future time points*, $T_{i-1} < T_i$, *with* $\tau_i = T_i - T_{i-1}$, *the* T_{i-1}-*caplet/floorlet with rate* K_i *and nominal amount* N_i *is a contract which pays, at time* T_i, *the amount,*

$$V_i^{CPL}(T_i) = H_i^{CPL}(T_i) = \tau_i N_i \max\left(\ell_i(T_{i-1}) - K_i, 0\right),$$

$$V_i^{FL}(T_i) = H_i^{FL}(T_i) = \tau_i N_i \max\left(K_i - \ell_i(T_{i-1}), 0\right),$$

with $\ell_i(T_{i-1}) := \ell(T_{i-1}; T_{i-1}, T_i)$. ◀

At time T_{i-1}, we can already *observe* the determined Libor rate $\ell_i(T_{i-1})$ in the financial money market, however, the payment will take place only at time T_i.

A cap consists of m caplets with the same strike price K and the same notional amount N. The value of the cap at time t is the sum of the values of the individual caplets at time t.

Caplets and floorlets are thus basically *European options on the interest rate*, which is accrued from time T_{i-1} to T_i. It is an important notion that the contract pays at time T_i, whereas the rate is reset already at time T_{i-1}. So, at time T_{i-1} the rate, which will be used for the payment at time T_i, is known. We should model the Libor rate $\ell_i(t)$ as a stochastic quantity, which, in financial terminology, *resets at time T_{i-1}*.

The caplet price under the T_i-forward measure is given by:

$$V_i^{\text{CPL}}(t) = \mathbb{E}^{\mathbb{Q}}\left[\frac{N_i \tau_i}{M(T_i)} \max\left(\ell_i(T_{i-1}) - K_i, 0\right) \Big| \mathcal{F}(t)\right]$$
$$= N_i \tau_i P(t, T_i) \mathbb{E}^{T_i}\left[\max\left(\ell_i(T_{i-1}) - K_i, 0\right) \Big| \mathcal{F}(t)\right]. \quad (12.31)$$

The floorlet price under the T_i-forward measure is given by:

$$V_i^{\text{FL}}(t) = \mathbb{E}^{\mathbb{Q}}\left[\frac{N_i \tau_i}{M(T_i)} \max\left(K_i - \ell_i(T_{i-1}), 0\right) \Big| \mathcal{F}(t)\right]$$
$$= N_i \tau_i P(t, T_i) \mathbb{E}^{T_i}\left[\max\left(K_i - \ell_i(T_{i-1}), 0\right) \Big| \mathcal{F}(t)\right]. \quad (12.32)$$

Note that depending on the interest rate model of choice, the expectations in (12.31) will differ, see, for example, in Chapter 14.

Pricing of caplets/floorlets under the Hull-White model

The Libor rate can be written in terms of bond prices, and these bond prices can also be modeled by means of the short-rate models that were discussed in the previous chapter.

Here, we will determine the pricing formula of a caplet under the Hull-White short-rate model. The derivation for floorlets goes similarly.

Details Hull-White model

We briefly repeat the Hull-White dynamics, and show that the characteristic function can also be derived without the Hull-White decomposition.

The single-factor Hull-White model [Hull and White, 1990] is given by the following dynamics, see also (11.32),

$$dr(t) = \lambda\left(\theta(t) - r(t)\right) dt + \eta dW_r^{\mathbb{Q}}(t), \quad r(t_0) = r_0, \quad (12.33)$$

with $\theta(t) \in \mathbb{R}$, $t \in \mathbb{R}^+$, the drift term, which is used to fit the bond prices to the yield curve, η determines the level of the volatility and λ is the reversion rate parameter.

The HW process is an affine process, see also Section 11.3.1, and its *discounted characteristic function* is given by:

$$\phi_{r_{\text{HW}}}(u; t, T) = \mathbb{E}^{\mathbb{Q}}\left[e^{-\int_t^T r(z)dz + iur(T)} \Big| \mathcal{F}(t)\right] = e^{\bar{A}(u,\tau) + \bar{B}(u,\tau)r(t)}, \quad (12.34)$$

for $\tau = T - t$ and initial condition,

$$\phi_{r_{\text{HW}}}(u; T, T) = \mathbb{E}^{\mathbb{Q}}\left[e^{-\int_T^T r(z)dz + iur(T)} \Big| \mathcal{F}(T)\right] = e^{iur(T)},$$

implying $\bar{A}(u, 0) = 0$ and $\bar{B}(u, 0) = iu$.

We solve the following set of ODEs for $\bar{A}(u, \tau)$ and $\bar{B}(u, \tau)$,

$$\frac{d\bar{A}}{d\tau} = \lambda\theta(T - \tau)\bar{B} + \frac{1}{2}\eta^2\bar{B}^2,$$

$$\frac{d\bar{B}}{d\tau} = -1 - \lambda\bar{B}.$$

For $\bar{B}(u, \tau)$, which does not depend on $\bar{A}(u, \tau)$, we find,

$$\frac{d}{d\tau}\left(e^{\lambda\tau}\bar{B}\right) = -e^{\lambda\tau},$$

with corresponding solution:

$$\bar{B}(u, \tau) = \bar{B}(u, 0)e^{-\lambda\tau} - \frac{1}{\lambda}\left(1 - e^{-\lambda\tau}\right).$$

Using $\bar{B}(u, 0) = iu$, we get,

$$\bar{B}(u, \tau) = iue^{-\lambda\tau} - \frac{1}{\lambda}\left(1 - e^{-\lambda\tau}\right), \quad (12.35)$$

and integrating the equation for $\bar{A}(u, \tau)$ gives us,

$$\bar{A}(u, \tau) = \bar{A}(u, 0) + \lambda\int_0^\tau \theta(T - z)\bar{B}(u, z)dz + \frac{1}{2}\eta^2\int_0^\tau \bar{B}^2(u, z)dz.$$

Including $\bar{A}(u, 0) = 0$ and the solution for $\bar{B}(u, \tau)$ in (12.35), gives us

$$\bar{A}(u, \tau) = \lambda\int_0^\tau \theta(T - z)\bar{B}(u, z)dz + \frac{1}{2}\eta^2\int_0^\tau \bar{B}^2(u, z)dz$$

$$= \lambda\int_0^\tau \theta(T - z)\bar{B}(u, z)dz + \frac{\eta^2}{4\lambda^3}(iu\lambda + 1)$$

$$\times \left(e^{-2\lambda T}\left(4e^{\lambda T} - 1 - \lambda iu\right) + (iu\lambda - 3)\right) + \frac{\eta^2\tau}{2\lambda^2}.$$

With $u = 0$ and $\bar{A}_r(\tau) \equiv \bar{A}(0, \tau)$, $\bar{B}_r(\tau) \equiv \bar{B}(0, \tau)$ in the characteristic function (12.34) for the Hull-White model, we obtain the price of the ZCB, as

$$P(t, T) := \phi_{r_{\text{HW}}}(0; t, T) = \mathbb{E}^{\mathbb{Q}}\left[e^{-\int_t^T r(z)dz} \Big| \mathcal{F}(t)\right] = e^{\bar{A}_r(\tau) + \bar{B}_r(\tau)r(t)}, \quad (12.36)$$

with,

$$\bar{B}_r(\tau) = \frac{1}{\lambda}\left(e^{-\lambda\tau} - 1\right),$$

$$\bar{A}_r(\tau) = \lambda\int_0^\tau \theta(T - z)\bar{B}_r(z)dz + \frac{\eta^2}{4\lambda^3}\left[e^{-2\lambda\tau}\left(4e^{\lambda\tau} - 1\right) - 3\right] + \frac{\eta^2\tau}{2\lambda^2}.$$

Pricing the caplet

The price of a caplet, with a strike price K, is given by:

$$V^{\mathrm{CPL}}(t_0) = N\tau_i \mathbb{E}^{\mathbb{Q}} \left[\frac{1}{M(T_i)} \max\left(\ell_i(T_{i-1}) - K, 0\right) \Big| \mathcal{F}(t_0) \right]$$
$$= N\tau_i P(t_0, T_i) \mathbb{E}^{T_i} \left[\max\left(\ell_i(T_{i-1}) - K, 0\right) \Big| \mathcal{F}(t_0) \right]. \quad (12.37)$$

By the definition of the Libor rate in (12.1), the (scaled) caplet valuation formula[a] can be written as,

$$\frac{V^{\mathrm{CPL}}(t_0)}{P(t_0, T_i)} = N\tau_i \mathbb{E}^{T_i} \left[\max\left(\frac{1}{\tau_i}\left(\frac{1}{P(T_{i-1}, T_i)} - 1 \right) - K, 0 \right) \Big| \mathcal{F}(t_0) \right]$$
$$= N \cdot \mathbb{E}^{T_i} \left[\max\left(e^{-\bar{A}_r(\tau_i) - \bar{B}_r(\tau_i) r(T_{i-1})} - 1 - \tau_i K, 0 \right) \Big| \mathcal{F}(t_0) \right]$$
$$= N \cdot e^{-\bar{A}_r(\tau_i)} \mathbb{E}^{T_i} \left[\max\left(e^{-\bar{B}_r(\tau_i) r(T_{i-1})} - \hat{K}, 0 \right) \Big| \mathcal{F}(t_0) \right],$$
$$(12.38)$$

with $\hat{K} = (1 + \tau_i K) e^{\bar{A}_r(\tau_i)}$, and using the results from (12.35). The value of a caplet under the Hull-White model can thus be priced in a similar fashion as the option on a zero-coupon bond $P(T_{i-1}, T_i)$, which was discussed in Section 11.4.2.

---
[a] We scale only for notational convenience.

Tower property and caplet pricing

An elegant way to derive the caplet prices is by means of the *tower property of expectations*. The price of a caplet with a strike price K can be written as,

$$V^{\mathrm{CPL}}(t_0) = N\tau_i \mathbb{E}^{\mathbb{Q}} \left[\mathbb{E}^{\mathbb{Q}} \left[\frac{1}{M(T_i)} \max\left(\ell_i(T_{i-1}) - K, 0\right) \Big| \mathcal{F}(T_{i-1}) \right] \Big| \mathcal{F}(t_0) \right]$$
$$= N\tau_i \mathbb{E}^{\mathbb{Q}} \left[\frac{1}{M(T_{i-1})} \mathbb{E}^{\mathbb{Q}} \left[\frac{M(T_{i-1})}{M(T_i)} \max\left(\ell_i(T_{i-1}) - K, 0\right) \Big| \mathcal{F}(T_{i-1}) \right] \Big| \mathcal{F}(t_0) \right]. \quad (12.39)$$

After a change of measure, the inner expectation can be expressed as,

$$\mathbb{E}^{\mathbb{Q}} \left[\frac{M(T_{i-1})}{M(T_i)} \max\left(\ell_i(T_{i-1}) - K, 0\right) \Big| \mathcal{F}(T_{i-1}) \right]$$
$$= P(T_{i-1}, T_i) \max\left(\ell_i(T_{i-1}) - K, 0\right).$$

The caplet value can therefore be found as,

$$V^{\mathrm{CPL}}(t_0) = N\tau_i \mathbb{E}^{\mathbb{Q}}\left[\frac{1}{M(T_{i-1})}P(T_{i-1},T_i)\max\left(\ell_i(T_{i-1})-K,0\right)\Big|\mathcal{F}(t_0)\right]. \tag{12.40}$$

For the (scaled) caplet value, this results in

$$\begin{aligned}
\frac{V^{\mathrm{CPL}}(t_0)}{P(t_0,T_{i-1})} &= N\tau_i \cdot \mathbb{E}^{T_{i-1}}\left[P(T_{i-1},T_i)\right.\\
&\quad \times \left.\max\left(\frac{1}{\tau_i}\left(\frac{1}{P(T_{i-1},T_i)}-1\right)-K,0\right)\Big|\mathcal{F}(t_0)\right]\\
&= \hat{N}\cdot\mathbb{E}^{T_{i-1}}\left[\max\left(\frac{1}{\hat{K}}-P(T_{i-1},T_i)\right)\Big|\mathcal{F}(t_0)\right], \tag{12.41}
\end{aligned}$$

with $\hat{N} = N(1+\tau_i K)$ and $\hat{K} = 1+\tau_i K$.

The last equation is, in fact, the value of a put option on a ZCB with expiry date T_{i-1}, maturity date T_i and strike price $\hat{K}^{-1}$.

Example 12.2.1 (Caplet implied volatility) In Figure 12.4 the impact of the different Hull-White model parameters, λ, η, on the caplet implied volatility is presented. The mean reversion parameter λ appears to have a much smaller effect on the implied volatilities than the volatility parameter η. In practice, therefore, λ is often kept fixed whereas η is determined in a calibration process.

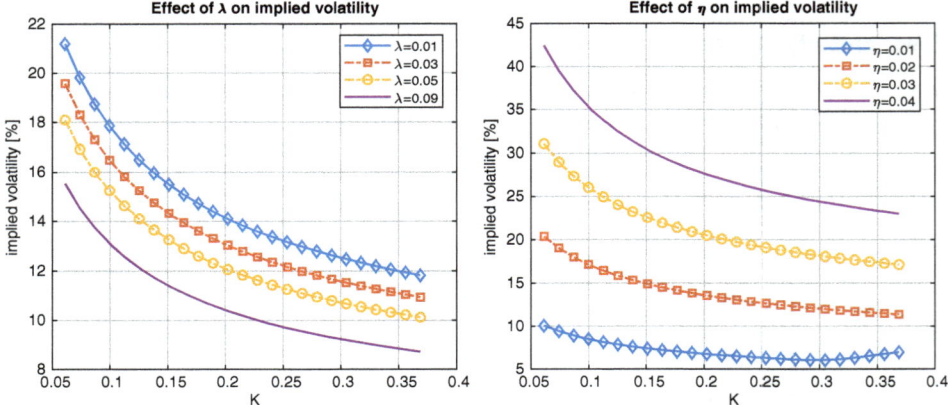

Figure 12.4: *Effect of λ and η in the Hull-White model on the caplet implied volatilities.*

12.2.2 European swaptions

We continue the discussion of some interest rate products with the swaption. Swaptions are the options on the interest rate swaps. A holder of a European swaption has the right, but not an obligation, to enter into a swap contract at a future date for a predetermined strike price K. Similarly as for the swaps, the swaptions have payer and receiver versions, that either pay the fixed rate and receive the float leg (the payer swaption) or vice versa (the receiver swaption). The strike price of the swaption determines the fixed rate of the underlying swap. The fixed rate which values the forward starting swap to "par" is called the *forward swap rate* and swaptions struck at this rate are called at-the-money (ATM). Note that European-style swaptions are OTC contracts, and therefore their terms may vary to fit specific investor's needs. The most commonly traded swaptions are contracts based on par interest rate swap.

In the standard setting, the swaption's maturity time coincides with the first reset date of the underlying interest swap, i.e. $T_0 = T_i$. With time t_0 today's date, T_i, the first reset date, is some future date at which a swaption holder has the option to enter into a swap deal.

The value of the swaption product at time T_i is then given by:

$$V^{\text{Swpt}}(T_i) = H^{\text{Swpt}}(T_i) = \max\left(V^S(T_i), 0\right)$$

$$= N \cdot \max\left(\sum_{k=i+1}^{m} \tau_k P(T_i, T_k)\left(\ell(T_i; T_{k-1}, T_k) - K\right), 0\right), \quad (12.42)$$

and today's discounted value is equal to:

$$V^{\text{Swpt}}(t_0) = N \cdot \mathbb{E}^{\mathbb{Q}}\left[\frac{M(t_0)}{M(T_i)}\right.$$

$$\left.\times \max\left(\sum_{k=i+1}^{m} \tau_k P(T_i, T_k)\left(\ell(T_i; T_{k-1}, T_k) - K\right), 0\right) \bigg| \mathcal{F}(t_0)\right].$$

$$(12.43)$$

Using the representation from (12.15), the value of the swaption also equals,

$$V^{\text{Swpt}}(t_0) = N \cdot \mathbb{E}^{\mathbb{Q}}\left[\frac{M(t_0)}{M(T_i)} \max\left(A_{i,m}(T_i)(S_{i,m}(T_i) - K), 0\right) \bigg| \mathcal{F}(t_0)\right]$$

$$= N \cdot \mathbb{E}^{\mathbb{Q}}\left[\frac{A_{i,m}(T_i) M(t_0)}{M(T_i)} \max\left(S_{i,m}(T_i) - K, 0\right) \bigg| \mathcal{F}(t_0)\right],$$

$$(12.44)$$

with $M(t_0) = 1$, $t_0 < T_i < T_{i+1}$, with T_{i+1} the first payment date.

As discussed, the annuity $A_{i,m}(T_i)$ is a combination of tradable ZCBs and can be considered as a numéraire. The corresponding Radon-Nikodym derivative for changing measures, from the risk-neutral measure $\mathbb{Q}$ to the new annuity measure (also known as the swap measure) $\mathbb{Q}^{i,m}$, associated with annuity $A_{i,m}(t)$, is given by,

$$\lambda_{\mathbb{Q}}^{i,m}(T_i) = \frac{d\mathbb{Q}^{i,m}}{d\mathbb{Q}}\bigg|_{\mathcal{F}(T_i)} = \frac{A_{i,m}(T_i)}{A_{i,m}(t_0)}\frac{M(t_0)}{M(T_i)}.$$

The value of a swaption is thus given by:

$$V^{\text{Swpt}}(t_0) = N \cdot \mathbb{E}^{i,m}\left[\frac{A_{i,m}(T_i)M(t_0)}{M(T_i)}\frac{A_{i,m}(t_0)}{A_{i,m}(T_i)}\frac{M(T_i)}{M(t_0)}\right.$$

$$\left.\times \max\left(S_{i,m}(T_i) - K, 0\right)\bigg|\mathcal{F}(t_0)\right]$$

$$= N \cdot A_{i,m}(t_0)\mathbb{E}^{i,m}\left[\max\left(S_{i,m}(T_i) - K, 0\right)\bigg|\mathcal{F}(t_0)\right].$$

To avoid arbitrage, the swap rate $S_{i,m}(T_i)$, which is defined as,

$$S_{i,m}(t) = \frac{P(t,T_i) - P(t,T_m)}{A_{i,m}(t)},$$

has to be a martingale under the swap measure associated with the annuity $A_{i,m}(t)$. This implies that the dynamics of the swap rate $S_{i,m}(t)$ under the swap measure $\mathbb{Q}^{i,m}$ have to be free of drift terms.

Swaptions under the Hull-White model

In this subsection, we will price the European swaptions under the Hull-White short-rate model as an illustrative example. In this case, swaption pricing can be performed semi-analytically. The expression (12.43), after the change of measure, from measure $\mathbb{Q}$ to the T_i-forward measure $\mathbb{Q}^{T_i}$, gives us,

$$V^{\text{Swpt}}(t_0) = N \cdot P(t_0, T_i)$$

$$\times \mathbb{E}^{T_i}\left[\max\left(\sum_{k=i+1}^{m}\tau_k P(T_i, T_k)\Big(\ell(T_i; T_{k-1}, T_k) - K\Big), 0\right)\bigg|\mathcal{F}(t_0)\right].$$

Using Equation (12.12), the following relation holds:

$$\sum_{k=i+1}^{m}\tau_k P(T_i, T_k)\Big(\ell_k(T_i) - K\Big) = 1 - P(T_i, T_m) - K\sum_{k=i+1}^{m}\tau_k P(T_i, T_k)$$

$$= 1 - \sum_{k=i+1}^{m} c_k P(T_i, T_k), \quad (12.45)$$

with $c_k = K\tau_k$ for $k = i+1, \ldots, m-1$ and $c_m = 1 + K\tau_m$.

This gives us the following expression for the swaption price:

$$V^{\text{Swpt}}(t_0) = N \cdot P(t_0, T_i)\mathbb{E}^{T_i}\left[\max\left(1 - \sum_{k=i+1}^{m} c_k P(T_i, T_k), 0\right)\Big|\mathcal{F}(t_0)\right].$$

As the ZCB $P(T_i, T_k)$ in the case of an affine short-rate process $r(t)$ can be expressed as

$$P(T_i, T_k) = \exp\left(\bar{A}_r(\bar{\tau}_k) + \bar{B}_r(\bar{\tau}_k)r(T_i)\right),$$

with $\bar{\tau}_k := T_k - T_i$, the (scaled) pricing equation reads:

$$\frac{V^{\text{Swpt}}(t_0)}{P(t_0, T_i)} = N \cdot \mathbb{E}^{T_i}\left[\max\left(1 - \sum_{k=i+1}^{m} c_k e^{\bar{A}_r(\bar{\tau}_k) + \bar{B}_r(\bar{\tau}_k)r(T_i)}, 0\right)\Big|\mathcal{F}(t_0)\right].$$
(12.46)

For the *"maximum of the sum"* in the above expression, a *"sum of the maximum"* expression has been derived in [Jamshidian, 1989], which is presented in the following result.

Result 12.2.1 (Maximum of a sum) *We wish to evaluate,*

$$A = \max\left(K - \sum_k \psi_k(r), 0\right),$$
(12.47)

where $\psi_k(r)$, $\mathbb{R} \to \mathbb{R}^+$, is a monotonically increasing or decreasing sequence of functions.

An approach for the transformation of the calculation of the maximum of a sum to an expression for the sum of certain maxima was presented in [Jamshidian, 1989].

Since each $\psi_k(r)$ is monotonically increasing, $\sum_k \psi_k(r)$ is monotonically increasing as well. This implies that a value $r = r^*$ exists, such that

$$K - \sum_k \psi_k(r^*) = 0,$$
(12.48)

where r^* may be determined by some root-finding algorithm, like the Newton-Raphson iteration, which was discussed in Subsection 4.1.1. Using Equation (12.48), Equation (12.47) can be written as,

$$\bar{M} = \max\left(\sum_{k=1}^m \psi_k(r^*) - \sum_k \psi_k(r), 0\right) = \max\left(\sum_k (\psi_k(r^*) - \psi_k(r)), 0\right).$$

The expression for the quantity $\bar{M}$ is then given by,

$$\bar{M} = \max\left(\sum_k (\psi_k(r^*) - \psi_k(r)), 0\right) = \sum_k (\psi_k(r^*) - \psi_k(r))\mathbb{1}_{r > r^*},$$

which can be expressed as,
$$\bar{M} = \sum_k \max\left(\psi_k(r^*) - \psi_k(r), 0\right),$$

concluding the result.

With the help of Result 12.2.1, Equation (12.46) can be re-written as:
$$\frac{V^{\text{Swpt}}(t_0)}{P(t_0, T_i)} = N \cdot \sum_{k=i+1}^{m} c_k \mathbb{E}^{T_i}\left[\max\left(\hat{K}_k - e^{\bar{A}_r(\bar{\tau}_k) + \bar{B}_r(\bar{\tau}_k)r(T_i)}, 0\right)\right],$$

with $\hat{K}_k := \exp\left(\bar{A}_r(\bar{\tau}_k) + \bar{B}_r(\bar{\tau}_k)r^*\right)$, where parameter r^* is chosen such that,
$$\sum_{k=i+1}^{m} c_k \exp\left(\bar{A}_r(\bar{\tau}_k) + \bar{B}_r(\bar{\tau}_k)r^*\right) = 1.$$

The pricing formula for the swaption is not yet complete, as the sum of the expectations above still needs to be determined. Notice that each element of this sum represents a European put option on a zero-coupon bond, i.e.,
$$V_p^{\text{ZCB}}(t_0, T_i) = P(t_0, T_i)\mathbb{E}^{T_i}\left[\max\left(\hat{K}_k - e^{\bar{A}_r(\bar{\tau}_k) + \bar{B}_r(\bar{\tau}_k)r(T_i)}, 0\right)\right],$$

with $\bar{\tau}_k = T_k - T_i$. Pricing of European-type options on the ZCB under the Hull-White model gives us a closed-form solution.

The *pricing formula for a European swaption* reads:
$$V^{\text{Swpt}}(t_0) = N \cdot \sum_{k=i+1}^{m} c_k V_p^{\text{ZCB}}(t_0, T_k), \qquad (12.49)$$

with the strike price $\hat{K}_k := \exp\left(\bar{A}_r(\bar{\tau}_k) + \bar{B}_r(\bar{\tau}_k)r^*\right)$, and where r^* is determined by solving of the following equation:
$$1 - \sum_{k=i+1}^{m} c_k \exp\left(\bar{A}_r(\bar{\tau}_k) + \bar{B}_r(\bar{\tau}_k)r^*\right) = 0, \qquad (12.50)$$

where $c_k = \hat{K}_k \bar{\tau}_k$ for $k = i+1, \ldots, m-1$ and $c_m = 1 + \hat{K}_k \bar{\tau}_m$.

12.3 Credit Valuation Adjustment and Risk Management

So far in this book, we have priced different financial derivatives, assuming (implicitly) that the counterparty of the derivative would meet the payment obligations.

Let us focus on a basic interest rate swap, $V^S(t_0)$, as in Section 12.1.4, between two counterparties A and B. The contract is basically an exchange of fixed against floating rate payments, on a notional N. Imagine a situation in which interest rates moved substantially, so that party A will have significant financial obligations towards the counterparty B. Depending on the financial situation, it may happen that party A encounters difficulties to meet the payment obligations. Moreover, counterparty B may endure a significant financial loss due to default of party A. If the counterparty defaults, the loss will be the replacement costs of the contract (i.e. the current market value).

Counterparty credit risk

In financial jargon the situation described above is defined as *Counterparty Credit Risk*.

> **Definition 12.3.1 (Counterparty credit risk (CCR))**
> *Counterparty credit risk (CCR) is related to the situation where a counterparty will default prior to the expiration of the contract and is unable make all payments required by the financial contract.* ◄

In the year 2007, a financial crisis occurred, which originated in the United States' credit and housing market, and spread around the world, from the financial markets into the real economy. Financial institutions with a high reputation went bankrupt or were bailed out, including the investment bank Lehman Brothers (founded in 1850).

In the worst times of that crisis, the bankruptcy of large financial institutions triggered a widespread propagation of so-called *default risk* through the financial network. This initiated a thorough review of the standards and methodologies for the valuation of financial derivatives. Policies, rules and regulations in the financial world changed drastically in the wake of that crisis. An important area of financial risk which required special attention referred to the *counterparty credit risk (CCR)*. This is the risk that a party of a financial contract is not able to fulfill the payment duties that are agreed upon in the contract, which is also known as *a default*.

Since then, the *probability of default* of the counterparty of a financial contract has been incorporated in the prices of financial derivatives, and thus plays a prominent role in the pricing context. Counterparties are charged an additional premium, which is added to the fair price of the derivative, due to the probability of default. This way the risk that the counterparty would miss payment obligations is compensated for the other party in the contract. The total amount of trades of complex, and thus risky, financial derivatives has significantly reduced in the wake of the financial crisis. The lack of confidence in the financial system may have resulted in a drastic reduction of complexity, simply because risk of basic financial products is easier to estimate, and also just to keep money in the pocket.

As a consequence of CCR and its effect on the value of a financial derivative, a derivative contract with a defaultable counterparty may be considered to be worth less than a contract with a free-of-risk counterparty; the lower the creditworthiness of a counterparty, the lower the market value of the derivative contract may be.

From the perspective of derivatives pricing, it may be clear that pricing a derivative under the risk-neutral measure is not sufficient regarding the risk related to the possible default of a counterparty. Typically, the probability of default of a counterparty can be assessed either in an *implied* fashion, meaning that the relevant information is extracted from the market quotes of credit derivatives, like Credit Default Swaps (CDS), that give an indication of a companies' or countries creditworthiness, or it can be inferred from scores and insights on the credit quality of companies and countries, from, for example, rating agencies.

The pricing of a derivative under CCR is related to the *exposure in the future to a specific counterparty*, as this future exposure gives us an indication about the possible sizes of losses in the event of a counterparty default.

Mathematically, the (positive) exposure, $E(t)$, is defined as,

$$E(t) := \max(V(t), 0), \quad V(t) := \mathbb{E}^{\mathbb{Q}}\left[\frac{M(t)}{M(T)} H(T, \cdot) \Big| \mathcal{F}(t)\right], \quad (12.51)$$

where $V(t)$ represents the value of a derivative contract at time t, and where T is the maturity time of the contract.

Example 12.3.1 (Simulation of exposure) As an example, we consider an interest rate swap, based on a notional N, with the price being defined by Equation (12.10), i.e.

$$V^S(t) = N \cdot \sum_{k=i+1}^{m} \tau_k P(t, T_k) \left(\ell(t; T_{k-1}, T_k) - K\right), \quad (12.52)$$

with $\tau_k = T_k - T_{k-1}$ and $T_{i+1}, \ldots T_m$ are the payment dates.

As already discussed, the value of a swap today (at time t_0) is completely determined by the zero-coupon bonds $P(t_0, T_k)$, for $k = i, i+1, \ldots, m$. These bonds are available in the money market, so seemingly we do not need a stochastic interest rate model for the valuation of the swap. On the other hand, if we wish to calculate the value *at any future time*, $t > t_0$, in order to assess the exposure, we need to choose a stochastic model which models the uncertainty related to the future swap values.

A popular model used for the simulation of exposure is the Hull-White short-rate model, as in (12.33), with the term-structure parameter $\theta(t)$, and ZCBs $P(t, T)$ as in Equation (12.36).

To estimate the exposure $E(t)$ in (12.51) for any time t until the last payment date of the swap T_m, we simulate the Hull-White process (12.33) by means of the Monte Carlo method, and obtain realizations for $r(t)$. With these simulated paths, any ZCB $P(t, T_k)$, as defined in (12.36) can be valued. In Figure 12.5 typical profiles for the swap value and the exposure profile are presented. $V^S(t_0)$ equals 0 at the inception time t_0. Moreover, as t approaches T, being the final payment of the swap, the value of the swap converges to 0 (as after the last payment there is no exposure). ◆

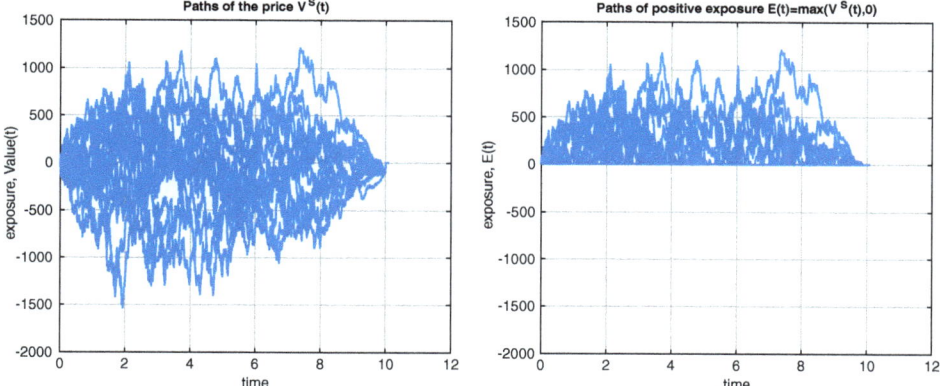

Figure 12.5: *Monte Carlo simulation of a swap $V^S(t)$ and the exposure paths $E(t)$ over time.*

> Given the exposure profiles $E(t)$, we wish to calculate the *expected (positive) exposure*, which is defined as follows,
>
> $$\text{EE}(t_0, t) = \mathbb{E}^{\mathbb{Q}}\left[\frac{M(t_0)}{M(t)} E(t) \Big| \mathcal{F}(t_0)\right], \quad (12.53)$$

where $E(T)$ is the positive exposure in (12.51) and $M(t)$ is the money-savings account; $dM(t) = r(t)M(t)dt$, with $r(t)$ the Hull-White short-rate process. The concept of expected exposure is particularly important for the computation of the so-called *credit value adjustment (CVA)*, which will also be discussed.

Potential Future Exposure (PFE)

Another measure which contributes to the valuation of CCR, especially in the context of risk management, is called the *Potential Future Exposure* (PFE). PFE is the maximum credit exposure calculated at some confidence level. In risk management and for setting trading limits to traders, PFE is often considered as the *worst-case scenario exposure*.

> PFE, at time t, i.e. $\text{PFE}(t_0, t)$, is defined as a quantile of the exposure $E(t)$,
>
> $$\text{PFE}(t_0, t) = \inf\left\{x \in \mathbb{R} : p \leq F_{E(t)}(x)\right\}, \quad (12.54)$$
>
> where p is the significance level and $F_{E(t)}(x)$ is the CDF of the exposure at time t.

Before the appearance in the Basel II accords [Bank for International Settlements, 2004], the concepts EE and PFE had already emerged and were commonly used

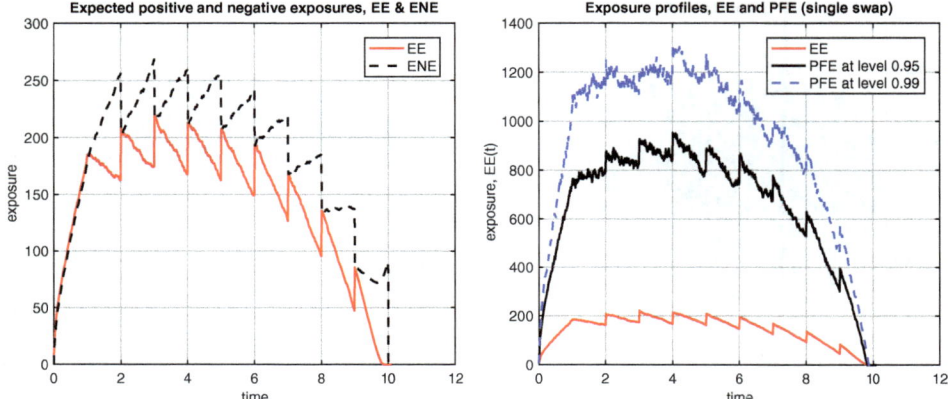

Figure 12.6: *Left: the (positive) Expected Exposure, EE, and the Expected Negative Exposure, ENE, which is defined as* $ENE(t) = \max(-V(t), 0)$; *Right: The EE and two PFE quantities, with levels, 0.95 and 0.99, respectively.*

as representative metrics for credit exposure [Gregory]. EE thus represents the *average* expected loss in the future, while PFE may manifest the *worst* exposure given a certain confidence level. These two quantities indicate the loss from both a pricing and risk management perspective [Gregory], respectively. There has been a debate on the computation of PFE, whether to compute it under the real-world or the risk-neutral measure. It is argued that PFE should be computed based on simulations under the real-world measure, reflecting the future developments in the market realistically, from a risk management perspective [Kenyon *et al.*, 2015].

In Figure 12.6, the time-dependent values of $EE(t_0, t)$ and $PFE(t_0, t)$ for the example from Figure 12.5 are plotted. The rugged pattern of these quantities in the figure is due to the fact that each time point a payment takes place, so that the values, and also the exposures, suddenly drop in value.

Types of exposure

In Example 12.3.1, we basically considered one financial derivative, i.e. the interest rate swap. In practice, however, *portfolios with derivatives* are much bigger, ranging from a few up to millions of traded products. The corresponding exposure can then be placed into three main categories:

1. *Contract-level exposure*, which is the exposure for one individual contract, as in the earlier example. For a derivative with value $V_1(t)$, the contract-level exposure is given by:
$$E(t) = \max(V_1(t), 0).$$

2. *Counterparty-level exposure*, where the exposure is based on all derivatives that are traded with a specific counterparty. In the case of two derivatives

with the counterparty, with values $V_1(t)$ and $V_2(t)$, respectively, the exposure at the counterparty-level, $E_c(t)$, is given by:

$$E_c(t) = \max(V_1(t), 0) + \max(V_2(t), 0) =: E_1(t) + E_2(t).$$

3. *Netting exposure*, which includes offsetting the values of multiple trades between two, or more, counterparties. As above, for two contracts with values $V_1(t)$ and $V_2(t)$, the netted exposure, $E_n(t)$, is defined by:

$$E_n(t) = \max(V_1(t) + V_2(t), 0).$$

Since $\max(x + y, 0) \leq \max(x, 0) + \max(y, 0)$, the netting exposure is typically lower than the counterparty-level exposure. From the exposure perspective, the concept of netting is beneficial to reduce the risk to a counterparty. However, not all trades can be used in netting- because netting is only applicable to so-called *homogeneous trades* that can be legally netted (such as those trades that are specified in the ISDA master agreement, where ISDA stands for International Swaps and Derivatives Association, see https://www.isda.org/).

Example 12.3.2 (Closed-form exposure profile of stock contract) Consider a contract which will pay at time T, the value of a stock $S(T)$. The contract's value today is simply:

$$V(t_0) = M(t_0)\mathbb{E}^{\mathbb{Q}}\left[\frac{S(T)}{M(T)}\Big|\mathcal{F}(t_0)\right] = S(t_0), \quad (12.55)$$

since the discounted stock process is a martingale under risk-neutral measure $\mathbb{Q}$. By definition, the expected exposure at time t is given by:

$$\mathrm{EE}(t_0, t) = \mathbb{E}^{\mathbb{Q}}\left[\frac{M(t_0)}{M(t)}\max\left(V(t), 0\right)\Big|\mathcal{F}(t_0)\right]. \quad (12.56)$$

Inserting (12.55) into (12.56) gives us:

$$\mathrm{EE}(t_0, t) = \mathbb{E}^{\mathbb{Q}}\left[\frac{M(t_0)}{M(t)}\max\left(M(t)\mathbb{E}^{\mathbb{Q}}\left[\frac{S(T)}{M(T)}\Big|\mathcal{F}(t)\right], 0\right)\Big|\mathcal{F}(t_0)\right]$$

$$= M(t_0)\mathbb{E}^{\mathbb{Q}}\left[\frac{1}{M(t)}\max\left(S(t), 0\right)\Big|\mathcal{F}(t_0)\right] = S(t_0). \quad (12.57)$$

So, the exposure of a payoff, which is the stock value $S(T)$ at some future time T, is equivalent to a call option on the stock with zero strike price, $K = 0$, which is, in fact, equal to $S(t_0)$.

♦

Example 12.3.3 (Exposure profile of an interest rate swap) The exposure profile of an interest rate swap can be found in closed-form. Following the derivations in Subsection 12.1.4, today's value on an interest rate swap is given by Equation (12.52). The expected exposure at time t is now given by,

$$\text{EE}(t_0, t) = \mathbb{E}^{\mathbb{Q}} \left[\frac{M(t_0)}{M(t)} \max\left(V^S(t), 0\right) \Big| \mathcal{F}(t_0) \right]$$

$$= \mathbb{E}^{\mathbb{Q}} \left[\frac{M(t_0)}{M(t)} \max\left(N \sum_{k=i+1}^{m} \tau_k P(t, T_k) \left(\ell(t; T_{k-1}, T_k) - K\right), 0 \right) \Big| \mathcal{F}(t_0) \right],$$

for $t \geq T_i$.

> Notice that the expected exposure $\text{EE}(t_0, t)$ is equivalent to the value of a swaption $V^{\text{Swpt}}(t)$ with strike price K, as defined in Subsection 12.2.2:
>
> $$\text{EE}(t_0, t) = \mathbb{E}^{\mathbb{Q}} \left[\frac{1}{M(t)} V^{\text{Swpt}}(t, K) \Big| \mathcal{F}(t_0) \right],$$
>
> since $M(t_0) = 1$.

♦

12.3.1 Unilateral Credit Value Adjustment

We show how to derive *an adjustment to the usual risk-neutral pricing*, which is known as Credit Value Adjustment (CVA). Let us consider transactions from the point of view of a *safe investor*, i.e., a default-free company which may face counterparty risk. By $\bar{V}^D(t, T)$ we denote a time t discounted payoff (with final time T), which is subject to counterparty default risk and by $\bar{V}(t, T)$[1] the same quantity where counterparty risk is not involved (i.e. a so-called risk-free contract).

By $\bar{V}^D(t, T)$ we denote the sum of discounted future cash flows, $C(T_i)$, that occur at times T_i, between times t and T, and that are discounted to time t, i.e.,

$$\bar{V}^D(t, T) = \mathbb{E}^{\mathbb{Q}} \left[\sum_{i=1}^{N} \frac{M(t)}{M(T_i)} C(T_i) \Big| \mathcal{F}(t) \right] = \sum_{i=1}^{N} \mathbb{E}^{\mathbb{Q}} \left[\frac{M(t)}{M(T_i)} C(T_i) \Big| \mathcal{F}(t) \right]. \quad (12.58)$$

Note that with the definition above, for any time $s < T$, the following relation holds,

$$\bar{V}^D(t, T) = \bar{V}^D(t, s) + \mathbb{E}^{\mathbb{Q}} \left[\frac{M(t)}{M(s)} \bar{V}^D(s, T) \Big| \mathcal{F}(t) \right]$$

$$= \bar{V}^D(t, s) + \mathbb{E}^{\mathbb{Q}} \left[\frac{M(t)}{M(s)} \sum_{i=j+1}^{N} \mathbb{E}^{\mathbb{Q}} \left[\frac{M(s)}{M(T_i)} C(T_i) \Big| \mathcal{F}(s) \right] \Big| \mathcal{F}(t) \right] \quad (12.59)$$

---

[1] We use a different notation for these option values, with the bar, and with two time arguments, as this appears necessary with the default time t_D.

where j indicates the index for which $T_j = s$. From the tower property, we obtain,

$$\bar{V}^D(t,T) = \bar{V}^D(t,s) + \mathbb{E}^{\mathbb{Q}}\left[\frac{M(t)}{M(s)}\sum_{i=j+1}^{N}\frac{M(s)}{M(T_i)}C(T_i)\bigg|\mathcal{F}(t)\right]$$

$$= \mathbb{E}^{\mathbb{Q}}\left[\sum_{i=1}^{j}\frac{M(t)}{M(T_i)}C(T_i)\bigg|\mathcal{F}(t)\right] + \mathbb{E}^{\mathbb{Q}}\left[\sum_{i=j+1}^{N}\frac{M(t)}{M(T_i)}C(T_i)\bigg|\mathcal{F}(t)\right]$$

$$= \sum_{i=1}^{N}\mathbb{E}^{\mathbb{Q}}\left[\frac{M(t)}{M(T_i)}C(T_i)\bigg|\mathcal{F}(t)\right], \tag{12.60}$$

which confirms the definition in (12.58).

There are different scenarios regarding the default time t_D of the counterparty,

1. If the default of a counterparty happens *after the final payment* of the derivative, i.e. $t_D > T$, the contract value at time t is simply $\mathbb{1}_{t_D > T}\bar{V}(t,T)$,

2. If the default occurs prior to the maturity time, i.e. $t_D < T$, then

    (a) We will receive/pay all payments until the default time, $\mathbb{1}_{t_D \leq T}\bar{V}(t,t_D)$,

    (b) Depending on the counterparty, we may recover some of the future payments, assuming the recovery fraction to be R_c, the value equals $\mathbb{1}_{t_D \leq T}R_c\max(\bar{V}(t_D,T),0)$,

    (c) On the other hand, if we owe money to the defaulting counterparty, we cannot keep the money but have to pay it back completely, i.e., $\mathbb{1}_{t_D \leq T}\min(\bar{V}(t_D,T),0)$.

Based on all these situations and corresponding derivatives values, the price of a *risky* derivative $\bar{V}^D(t,T)$ is given by:

$$\bar{V}^D(t,T) = \mathbb{E}^{\mathbb{Q}}\bigg[\mathbb{1}_{t_D>T}\bar{V}(t,T) + \mathbb{1}_{t_D\leq T}\bar{V}(t,t_D)$$
$$+ \frac{M(t)}{M(t_D)}\mathbb{1}_{t_D\leq T}R_c\max(\bar{V}(t_D,T),0)$$
$$+ \frac{M(t)}{M(t_D)}\mathbb{1}_{t_D\leq T}\min(\bar{V}(t_D,T),0)\bigg|\mathcal{F}(t)\bigg].$$

Since $x = \max(x,0) + \min(x,0)$, a simplified version is given by,

$$\bar{V}^D(t,T) = \mathbb{E}^{\mathbb{Q}}\bigg[\mathbb{1}_{t_D>T}\bar{V}(t,T) + \mathbb{1}_{t_D\leq T}\bar{V}(t,t_D) + \frac{M(t)}{M(t_D)}\mathbb{1}_{t_D\leq T}\bar{V}(t_D,T)$$
$$+ \frac{M(t)}{M(t_D)}\mathbb{1}_{t_D\leq T}(R_c-1)\max(\bar{V}(t_D,T),0)\bigg|\mathcal{F}(t)\bigg].$$

The first three terms in the expression at the right-hand side yield,

$$\mathbb{E}^{\mathbb{Q}}\left[\mathbb{1}_{t_D>T}\bar{V}(t,T) + \mathbb{1}_{t_D\leq T}\bar{V}(t,t_D) + \frac{M(t)}{M(t_D)}\mathbb{1}_{t_D\leq T}\bar{V}(t_D,T)\Big|\mathcal{F}(t)\right]$$

$$= \mathbb{E}^{\mathbb{Q}}\left[\mathbb{1}_{t_D>T}\bar{V}(t,T) + \mathbb{1}_{t_D\leq T}\bar{V}(t,T)\Big|\mathcal{F}(t)\right]$$

$$= \bar{V}(t,T)\mathbb{E}^{\mathbb{Q}}\left[\mathbb{1}_{t_D>T} + \mathbb{1}_{t_D\leq T}\Big|\mathcal{F}(t)\right]$$

$$= \bar{V}(t,T).$$

Here, we used again the fact that $\bar{V}(t,T)$ represents the sum of discounted payments that take place between t and T. Then, we can use

$$\mathbb{E}[\bar{V}(t,s) + \frac{M(t)}{M(s)}\bar{V}(s,T)] = \mathbb{E}[\bar{V}(t,T)],$$

where we add two sums.

By assuming a constant recovery rate R_c, the value of the risky asset $\bar{V}^D(t,T)$ is given by,

$$\bar{V}^D(t,T) = \mathbb{E}^{\mathbb{Q}}\left[\bar{V}(t,T) + \frac{M(t)}{M(t_D)}\mathbb{1}_{t_D\leq T}(R_c - 1)\max(\bar{V}(t_D,T),0)\Big|\mathcal{F}(t)\right]$$

$$= \bar{V}(t,T) + \mathbb{E}^{\mathbb{Q}}\left[\frac{M(t)}{M(t_D)}\mathbb{1}_{t_D\leq T}(R_c - 1)\max(\bar{V}(t_D,T),0)\Big|\mathcal{F}(t)\right]$$

$$= \bar{V}(t,T) - (1 - R_c)\mathbb{E}^{\mathbb{Q}}\left[\frac{M(t)}{M(t_D)}\mathbb{1}_{t_D\leq T}\max(\bar{V}(t_D,T),0)\Big|\mathcal{F}(t)\right]$$

$$=: \bar{V}(t,T) - \text{CVA}(t,T),$$

So,

$$\text{CVA}(t,T) = (1 - R_c)\mathbb{E}^{\mathbb{Q}}\left[\frac{M(t)}{M(t_D)}\mathbb{1}_{t_D\leq T}\max(\bar{V}(t_D,T),0)\Big|\mathcal{F}(t)\right].$$

Generally, we find, for the corresponding option values:

risky derivative = risk-free derivative - CVA.

CVA can thus be interpreted as the price of counterparty risk, i.e. the expected loss due to a future counterparty default. With R_c the recovery rate, $(1 - R_c)$ is called the *loss-given-default* (LGD).

Interest Rate Derivatives and Valuation Adjustments

12.3.2 Approximations in the calculation of CVA

For the CVA charge, the joint distribution of the default time t_D and the exposure, at time t, is needed. By the tower property of expectations, we find:

$$\text{CVA}(t,T) = (1 - R_c)\, \mathbb{E}^{\mathbb{Q}} \left[\frac{M(t)}{M(t_D)} \mathbb{1}_{t_D \leq T} \max(\bar{V}(t_D, T), 0) \Big| \mathcal{F}(t) \right]$$

$$= (1 - R_c)\, \mathbb{E}^{\mathbb{Q}} \left[\mathbb{E}^{\mathbb{Q}} \left[\frac{M(t)}{M(t_D)} \mathbb{1}_{t_D \leq T} \max(\bar{V}(t_D, T), 0) \Big| \mathcal{F}(t_D) \right] \Big| \mathcal{F}(t) \right]$$

$$= (1 - R_c)\, \mathbb{E}^{\mathbb{Q}} \left[\mathbb{1}_{t_D \leq T} \mathbb{E}^{\mathbb{Q}} \left[\frac{M(t)}{M(t_D)} \max(\bar{V}(t_D, T), 0) \Big| \mathcal{F}(t_D) \right] \Big| \mathcal{F}(t) \right].$$

Under the assumption of *independence* between the default time t_D and the exposure, the following approximation can be derived,

$$\text{CVA}(t,T) = (1 - R_c) \int_t^T \mathbb{E}^{\mathbb{Q}} \left[\frac{M(t)}{M(t_D)} \max(\bar{V}(t_D, T), 0) \Big| t_D = z \right] f_{t_D}(z)\mathrm{d}z$$

$$= (1 - R_c) \int_t^T \mathbb{E}^{\mathbb{Q}} \left[\frac{M(t)}{M(t_D)} \max(\bar{V}(t_D, T), 0) \Big| t_D = z \right] \mathrm{d}F_{t_D}(z)$$

$$\approx (1 - R_c) \sum_{k=1}^m \text{EE}(t, T_k) \bar{q}(T_{k-1}, T_k). \tag{12.61}$$

With the *expected positive exposure*

$$\text{EE}(t, T_k) := \mathbb{E}^{\mathbb{Q}} \left[\frac{M(t)}{M(T_k)} \max(\bar{V}(T_k, T), 0) \Big| \mathcal{F}(t) \right],$$

and the *probability of default*, in (T_{k-1}, T_k), is given by,

$$\bar{q}(T_{k-1}, T_k) := F_{t_D}(T_k) - F_{t_D}(T_{k-1}) = \mathbb{E}\left[\mathbb{1}_{T_{k-1} < t_D \leq T_k}\right]. \tag{12.62}$$

As a result, the *CVA charge* can be approximated by,

$$\text{CVA} \approx \underbrace{\text{LGD}}_{\text{loss given default}} \times \underbrace{\text{PD}}_{\text{probability of default}} \times \underbrace{\text{EE}}_{\text{expected positive exposure}}.$$

In the above derivations the *independence between the counterparties' risk factors*, that are modeled by $\mathbb{1}_{t_D \leq T}$ and the risk associated with the underlying exposure, which is governed by $\bar{V}(t_D, T)$, is assumed. This assumption of independence may lead an *underestimation* of risk under certain circumstances, which may result in a potential significant loss. The phenomenon which is then not modeled is commonly referred to as *Wrong Way Risk* (WWR), as compared to the so-called Right Way Risk. WWR is defined as the risk which occurs when the "exposure to a counterparty is adversely correlated with the credit quality of the counterparty". An example of WWR is when fluctuations in the interest rate would cause changes in the value of certain derivative transactions, and also impact, at the same time, the creditworthiness of the counterparty. The counterparty is affected twice in such a situation by the specific phenomenon.

Suppose a bank enters a swap contract with an oil company where the bank pays a fixed amount and receives a floating amount, and the floating amount would be linked to the oil price. When the oil price moves upwards in this example, the bank would expect to receive money from the oil producer. The producer's credit quality, on the other hand, may also deteriorate as the costs may increase due to the price movement. Under normal circumstances, the oil producer would look for protection against a rise in the oil prices by entering a long oil derivative contract, instead of paying an amount which is linked to the oil price.

12.3.3 Bilateral Credit Value Adjustment (BCVA)

In the unilateral CVA case, as discussed in the previous section, we have seen a generalization of the concept of risk-free derivatives pricing, to a situation where a counterparty may default. However, there may be a *symmetry problem* in the logic followed there. Since in any financial transaction there are (at least) two counterparties, investor I and counterparty C, the CVA charge may differ, and may depend on the perspective. An investor will calculate the price of the derivative as,

$$\bar{V}_I(t_0, T) = \bar{V}(t_0, T) - \text{CVA}_I(t_0, T),$$

with $\text{CVA}_I(t_0, T)$ the charge computed by assuming that the counterparty may default. On the other hand, counterparty C would calculate the price of the derivative as,

$$\bar{V}_C(t_0, T) = \bar{V}(t_0, T) - \text{CVA}_C(t_0, T).$$

In general, $\text{CVA}_I(t_0, T) \neq \text{CVA}_C(t_0, T)$, because the two counterparties may be subject to a different credit risk value, so that the adjusted value which is calculated by investor I is not the opposite of the adjusted value calculated by counterparty C. In other words, the two parties may *not agree* on the adjusted price of the risky derivative.

The above reasoning motivates to use a *generalized version of the unilateral CVA case*, which is called *bilateral CVA* (BCVA). In BCVA, the fact that both counterparties may default is included in the modeling. When deriving the price of a "risky asset", we do not only include the CVA component, but also a component which accounts for the risk associated with the *own* default risk (which is known under the name *"debt value adjustment"*, $\text{DVA}(t_0, T)$). Depending on the perspective, the $\text{CVA}(t_0, T)$ charge for the investor is equivalent to the $\text{DVA}(t_0, T)$ for the counterparty, i.e., $\text{CVA}_C(t_0, T) = \text{DVA}_I(t_0, T)$ and $\text{CVA}_I(t_0, T) = \text{DVA}_C(t_0, T)$.

For the two counterparties, we would have the following pricing equations:

$$\bar{V}_I(t_0, T) = \bar{V}(t_0, T) - \text{CVA}_I(t_0, T),$$
$$\bar{V}_C(t_0, T) = \bar{V}(t_0, T) - \text{CVA}_C(t_0, T).$$

Interest Rate Derivatives and Valuation Adjustments

Since the investor and the counterparty should agree on the price, the condition $\bar{V}_I(t_0, T) = \bar{V}_C(t_0, T)$ should be imposed, which implies the following adjustments,

$$\bar{V}_I(t_0, T) = \bar{V}(t_0, T) - \text{CVA}_I(t_0, T) + \text{CVA}_C(t_0, T)$$
$$=: \bar{V}(t_0, T) - \text{CVA}_I(t_0, T) + \text{DVA}_I(t_0, T),$$

and, equivalently, for the counterparty,

$$\bar{V}_C(t_0, T) = \bar{V}(t_0, T) - \text{CVA}_C(t_0, T) + \text{CVA}_I(t_0, T)$$
$$=: \bar{V}(t_0, T) - \text{CVA}_C(t_0, T) + \text{DVA}_C(t_0, T).$$

This means that $\text{DVA}(t_0, T)$ is given by:

$$\text{DVA}(t, T) = (1 - R_c) \, \mathbb{E}^{\mathbb{Q}} \left[\frac{M(t)}{M(\hat{t}_D)} \mathbb{1}_{\hat{t}_D \leq T} \max(-\bar{V}(\hat{t}_D, T), 0) \middle| \mathcal{F}(t) \right],$$

where $\hat{t}_D$ indicates the investor's own default time.

Here, the *Negative Exposure* is the negative part of the default-free contract value.

Modeling CVA and DVA is commonly referred to as *bilateral CVA* (BCVA), which is defined, as follows,

$$\text{BCVA}(t, T) = \text{CVA}(t, T) - \text{DVA}(t, T).$$

Note that $\text{DVA}(t, T)$ is a positive quantity, and it can therefore be seen as a *profit upon self-default*, which is a problematic aspect.

12.3.4 Exposure reduction by netting

In any modern financial institution, it is crucial nowadays to properly manage the credit exposure. Proper management of the credit exposure in the case of a counterparty default may not only reduce potential losses but it may even save a financial institution from a bankruptcy. There are a number of possible steps an institution may undertake to reduce the counterparty risk. One measure which is powerful is *netting*. In the case of a default of one of the counterparties, a close-out netting agreement facilitates the legal framework to aggregate all the financial transactions with the counterparty which has defaulted.

Example 12.3.4 (Impact of netting) To illustrate the netting principle, consider two counterparties with only two transactions. Deal 1 has a value of €100 and Deal 2 has a loss of -€50. As it was discussed earlier, in the case of a counterparty default, the other party is obliged to pay any remaining financial obligations, and in the case of a positive trade value the party would only obtain a certain percentage of the full amount recovered. Let us assume for the recovery

Table 12.1: *Total gain and losses for a basic example, under a scenario without netting and a scenario with netting*

	scenario without netting	scenario with netting
deal 1	40% × €100 = €40	€100 = €100
deal 2	-€50=-€50	-€50=-€50
total	-€10	40% × €50 = €20

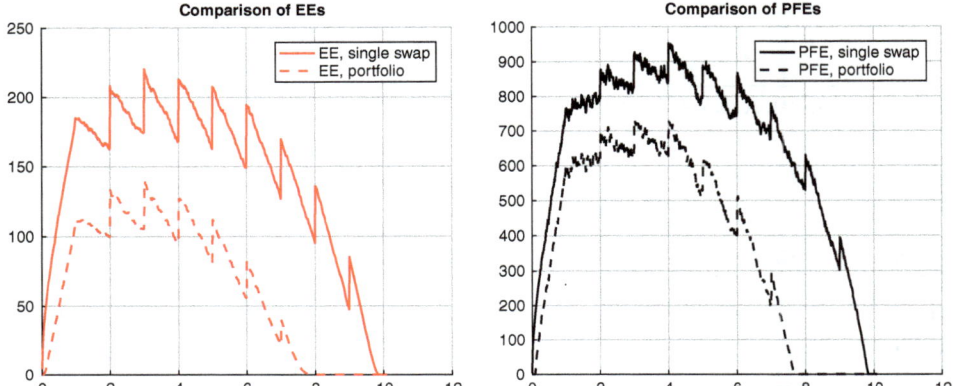

Figure 12.7: *The effect of netting on the reduction of exposure. Left: two exposure profiles (for a single swap and for a portfolio). Right: The corresponding PFEs (for a single swap and for a portfolio).*

rate $R_c = 40\%$. In Table 12.1, two different scenarios, one based on netting and the other without netting, are presented, where it can be observed that the netting may have a significant effect on the value of the portfolio, in the case of a default of the counterparty.

The main difference between the two scenarios lies in the appearance of the recovery rate. In the case without any netting, the recovery rate is applied on the basis of the individual deals. When the netting is taken into account, the recovery applies to the overall portfolio. ♦

Example 12.3.5 (Exposure reduction for interest rate swap)
This example is an extension of Example 12.3.1, where the exposure of an interest rate swap $V^S(t)$ was computed. To illustrate the effect of netting, the portfolio consisting of one swap is extended with a short position of an interest rate swap with only one coupon whose payment takes place only once, at time T_m. Its value is given by,

$$V_2^S(t_0) = N\tau_m P(t_0, T_m)\ell(t_0; T_{m-1}, T_m). \tag{12.63}$$

The complete portfolio is now given by $V(t_0) = V^S(t_0) - V_2^S(t_0)$, with $V^S(t_0)$ as in Equation (12.52) and $V_2^S(t_0)$ is defined in (12.63). In Figure 12.7, the (positive) effect of the additional swap in the portfolio on the exposure profiles is presented.

We see in the left-hand figure that the expected loss in the case of counterparty default is approximately 200, but, as indicated by the PFE, this loss can potentially be twice this size. The exposure is reduced in the case of two swaps (right-hand figure), but the pattern is essentially the same. ♦

12.4 Exercise set

Exercise 12.1 Derive the value of a caplet under the Hull-White model, departing from Equation (12.38).

Exercise 12.2 Consider the problem of pricing a European call option on the zero-coupon bond $P(T_1, T_2)$, with $T_0 = 0$, $T_1 = 2$ and $T_2 = 6$ years, where the initial zero-coupon bond curve is given by $P(0,t) = e^{-0.03t^2 - 0.1t}$.

For the Hull-White model,

$$dr(t) = \lambda(\theta(t) - r(t))dt + \eta dW_r^{\mathbb{Q}}(t), \quad r(0) \approx f^r(0,0),$$

with $\lambda = 0.2$, $\eta = 0.1$ and where $\theta(t)$ is expressed in terms of the instantaneous forward rate $f^r(0,t)$,

$$\theta(t) = f^r(0,t) + \frac{1}{\lambda}\frac{\partial}{\partial t}f^r(0,t) + \frac{\eta^2}{2\lambda^2}\left(1 - e^{-2\lambda t}\right),$$

perform the following tasks,

a. As the instantaneous forward rates $f^r(0,t)$ are expressed in terms of the zero-coupon bond curve via the relation, $f^r(0,t) = -\frac{\partial}{\partial t}\log P(0,t)$, determine analytically the long-term short-rate mean $\theta(t)$.

b. For 100.000 paths and 300 time-steps and the following Euler discretization,

$$r(t + \Delta t) = r(t) + \lambda(\theta(t) - r(t))\Delta t + \eta\sqrt{\Delta t}Z, \quad Z \sim \mathcal{N}(0,1),$$

price an option on the zero-coupon bond, given by,

$$V(0) = \mathbb{E}^{\mathbb{Q}}\left[\frac{M(0)}{M(T_1)}\max(P(T_1, T_2) - K, 0)\right]$$

$$= \mathbb{E}^{\mathbb{Q}}\left[\frac{M(0)}{M(T_1)}\max\left(e^{\bar{A}_r(4) + \bar{B}_r(4)r(2)} - K, 0\right)\right],$$

for the strike prices, $K = \{0, 0.1, 0.2, \ldots, 1.3\}$, with $\tau = T_2 - T_1$ and $\bar{A}_r(\tau)$, $\bar{B}_r(\tau)$.

c. Compare the Monte Carlo pricing results to the analytic expression. What can be said about the Monte Carlo accuracy in dependence of the number of paths and time-steps?

Exercise 12.3 Consider the problem of pricing a European swaption, with $T_0 = 0$, swap maturity $T_m \equiv T_3 = 3$ and the tenor of the swap $T_n \equiv T_6 = 6$ years (with annual frequency). The initial zero-coupon bond curve is given by $P(0,t) = e^{-0.02t^2 - 0.06t}$.

For the Hull-White model, with $\lambda = 1.5$, $\eta = 0.07$ and where $\theta(t)$ expressed in terms of instantaneous forward rate $f^r(0,t)$. Complete the following tasks.

a. For 100.000 paths and 300 time-steps perform the following Euler discretization, i.e.:

$$r(t + \Delta t) = r(t) + \lambda(\theta(t) - r(t))\Delta t + \eta\sqrt{\Delta t}Z, \quad Z \sim \mathcal{N}(0,1),$$

and price the following swaption:

$$V^{\text{Swpt}}(0) = \mathbb{E}^{\mathbb{Q}}\left[\frac{1}{M(T_3)}\max\left(\sum_{k=4}^{6}\tau_k P(T_3, T_k)\Big(\ell(T_3, T_{k-1}, T_k) - K\Big), 0\right)\Big|\mathcal{F}(0)\right],$$

with $\tau_k = 1$ and for strikes $K = \{0.01, 0.02, \ldots, 0.25\}$.

Hint: Express the Libor rate $\ell(T_3, T_{k-1}, T_k)$ in terms of ZCBs and for each bond $P(T_3, T_k)$ use the equality $P(T_3, T_k) = e^{\bar{A}(T_3, T_k) + \bar{B}(T_3, T_k)r(T_3)}$.

b. The Hull-White model under the T_3 measure is governed by the following SDE:

$$dr(t) = \lambda \left(\hat{\theta}(t, T_3) - r(t) \right) dt + \eta W_r^{T_3}(t), \quad r(0) \approx f^r(0, 0), \quad (12.64)$$

with $\hat{\theta}(t, T_3) = \theta(t) + \frac{\eta^2}{\lambda^2}(e^{-\lambda(3-t)} - 1)$. Implement a Monte Carlo simulation and perform the following swaption pricing exercise:

$$V^{\text{Swpt}}(0) = N \cdot P(0, T_3) \mathbb{E}^{T_3} \left[\max \left(1 - \sum_{k=4}^{6} c_k P(T_3, T_k), 0 \right) \Big| \mathcal{F}(0) \right],$$

with the appropriate coefficients c_k.

Hint: As the expectation is taken under the T_3-forward measure, to evaluate

$$P(T_3, T_k) = e^{\bar{A}_r(T_3, T_k) + \bar{B}_r(T_3, T_k) r(T_3)},$$

$r(T_3)$ needs to be simulated under the T_3-forward measure using the dynamics in (12.64).

c. Implement the Newton-Rapshon algorithm to determine r^* in (12.50) and perform analytically the swaption pricing, see Section 12.2.2. Compare your results with the Monte Carlo simulation.

Exercise 12.4 A so-called *bullet-type mortgage* is one of the simplest mortgages a client of a bank may receive. In this type of contract, the borrower receives N_0 at the time of settling, t_0, and the notional is redeemed only at the end of the last period, in one single period (from which the name "bullet" results). At the end of each period, only the interest part is paid back to the loaner, so that the notional remains constant until T_m, i.e., $N(T_i) = N_0 \mathbb{1}_{T_i < T_m}$. So, for the fixed interest rate K, the individual payments are equal to $C(T_i) = K N_0 \tau_i$ and the total amount of the payments $I = \sum_{i=1}^{m} C(T_i)$. Assume a *constant mortgage prepayment rate* P (prepayments are extra payments by the mortgagor during the life of the mortgage. These are deviations from the scheduled mortgage payments) and perform the following tasks:

a. Show that under constant prepayment rate the total amount of payments is equal to,

$$I = K N_0 \frac{1 - (1 - P)^m}{P}.$$

b. Take a bullet mortgage with $T_m = 10$ and $K = 3\%$. Generate the following two figures: 1) for $P = 0\%$ and $P = 10\%$ plot the prepayment amounts, the notional payments and the interest rate payments; 2) plot the outstanding notional in time under three different prepayment levels, $P = 0\%$, $P = 4\%$ and $P = 12\%$.

Exercise 12.5 An annuity-type mortgage is a somewhat more advanced mortgage contract, because, contrary to the bullet mortgage, it also involves *repayments*, $Q(T_i)$, of the outstanding notional, i.e.,

$$N(T_{i+1}) = N(T_i) - \Delta T_i Q(T_i) = N(T_i) - \Delta T_i \left(C(T_i) - I(T_i) \right),$$

where $C(T_i)$ indicates an individual payment (which consists of two payments, $Q(T_i)$ the amount of money to repay the initial amount N_0 and $I(T_i)$ the interest payment paid at time T_i), and $\Delta T_i = T_{i+1} - T_i$. Assume a constant installment, $C(T_i) = C$, and show that for the discounted annuity-based mortgage the following holds true,

$$V^{An}(t_0; K) = \sum_{i=1}^{m} \frac{C}{(1+K)^{T_i}} = \frac{C}{K} \left(1 - \frac{1}{(1+K)^{T_m}} \right).$$

On the other hand, the annuity mortgage needs to be equal to the initial mortgage value. In order to find C in the above equation, we thus have one more equation, i.e.,

$$V^{An}(t_0; K) = N_0,$$

where N_0 is the initial mortgage value.

a. To determine C, the discounted cashflows should equal the initial mortgage value. Show that this implies,

$$Q(T_i) = C(T_i) - I(T_i) = \frac{KN_0}{1 - (1+K)^{-T_m}} - KN(T_{i-1}).$$

b. As for the bullet mortgage exercise, we will introduce a prepayment (an additional reduction of the notional, next to the contractual repayments),

$$N(T_{i+1}) = N(T_i) - Q(T_i) - PN(T_i),$$

and show that individual payments should be equal to,

$$C(T_i) = \frac{KN(T_i)}{1 - (1+K)^{-(T_m - T_i)}}.$$

c. Plot the following two graphs, based on an annuity mortgage with $T_m = 10$ and $K = 3\%$ under the following different scenarios. 1) the installment composition in the cases $P = 0\%$ and $P = 12\%$. 2) The outstanding notional against time, under different levels of prepayment, $P = 0\%$, $P = 4\%$ and $P = 12\%$.

Exercise 12.6 Based on the notation in Equation (12.9), we have three interest rate payer swap contracts, $V_1(T_1) := V^{PS}(T_0, T_1)$, $V_2(T_2) = V^{PS}(T_0, T_1, T_2)$, $V_3(T_3) = V^{PS}(T_0, T_1, T_2, T_3)$. Assume an equally spaced tenor structure, $T_{i+1} - T_i = 1, \forall i$, so that $[T_1, T_2, T_3] = [1, 2, 3]$. The par values for the swap contracts are given in the table below:

Table 12.2: *Swaps and the corresponding fixed rates K_i. In other words, for these fixed rates the values of the corresponding swaps equal 0.*

Swap	Par fixed rate K
$V_1(t_0)$	0.01
$V_2(t_0)$	0.0214
$V_3(t_0)$	0.036

The objective of this exercise is to derive a yield curve which will be calibrated to the above instruments, and thus to find the spine discount factors, $p_i := P(t_0, T_i)$, such that all instruments in Table 12.2 will be priced back to their par values. The spine discount factors will be set at the maturity times of the swaps, i.e. $[1, 2, 3]$.

Perform the following tasks:

- Implement a function which prices an interest rate swap for a given maturity, for a strip of discount factors and a fixed rate K_i.

- Implement a function which solves the Jacobian matrix of the following form,

$$\mathbf{J} = \begin{bmatrix} \frac{\partial V_1(t_0)}{\partial p_1} & \frac{\partial V_1(t_0)}{\partial p_2} & \frac{\partial V_1(t_0)}{\partial p_3} \\ \frac{\partial V_2(t_0)}{\partial p_1} & \frac{\partial V_2(t_0)}{\partial p_2} & \frac{\partial V_2(t_0)}{\partial p_3} \\ \frac{\partial V_3(t_0)}{\partial p_1} & \frac{\partial V_3(t_0)}{\partial p_2} & \frac{\partial V_3(t_0)}{\partial p_3} \end{bmatrix} \quad (12.65)$$

Hint: To compute the partial derivatives use finite differences with $\Delta p_i = 10^{-5}$.

- Implement the multi-dimensional Newton-Raphson algorithm discussed in Section 12.1.5 to determine the optimal spine points.
- Solve this problem without using the Newton-Raphson algorithm as well.

Exercise 12.7 This exercise is based on Exercise 12.6 with the following modification: Assume that $T_{i+1} - T_i = 0.5$ (so that we have twice the frequency of payments). This modification requires the specification of an interpolation between the spine points. Consider linear interpolation based on the *"effective rate"*, $r_i := -\frac{\log P_i(t_0, T_i)}{T_i}$.

Exercise 12.8 In Example 12.3.2 it was shown that the EE profile is equivalent to a European option price, with strike price $K = 0$. Consider the GBM process for stock $S(t)$ (choose the model parameters yourself) and confirm these findings numerically.

Exercise 12.9 Modify the computer codes to generate the results in Figure 12.7 by using CIR dynamics instead of the Hull-White short rate dynamics. Note that the ZCBs under the CIR dynamics have been presented in Section 11.3.3.

What is the impact of varying the mean reversion and the volatility parameters on the computed profiles?

For which parameter settings would you expect very similar results for the CIR and HW dynamics?

Exercise 12.10 Use the code from Figure 12.7, and add two additional swaps to the portfolio, where one of them has a maturity time of 4 years and the other one of 12 years. Plot the corresponding EE and PFE profiles. Modify the notional amounts of these swaps and report the impact of this change to the results.

CHAPTER 13

Hybrid Asset Models, Credit Valuation Adjustment

In this chapter:

Financial contracts that involve *multiple asset classes* require well-developed pricing models.

In **Section 13.1**, we will discuss affine hybrid equity interest rate models, like the *Black-Scholes Hull-White* and *Schöbel-Zhu Hull-White equity models*, for which the *characteristic function* can be determined relatively easily. Hybrid models can be used for hybrid payoffs which have a *limited sensitivity to the interest rate smile*. However, these models are also important in the context of risk management, particularly for *Credit Valuation Adjustment (CVA)*, where the implied volatility smile and the stochastic interest rate may have a prominent effect, for example on the Potential Future Exposure (PFE) quantity. We pay attention to a *hybrid extension of the Heston stochastic volatility model*, which is called the *Heston Hull-White model*. This hybrid model, in **Section 13.2**, combines two correlated asset classes, i.e. equity and interest rates. We present an *approximation* of the full-scale model so that the model fits in the *class of affine diffusion processes*, for which a closed-form solution of the characteristic function can be derived.

By defining the affine hybrid Heston model under the *forward measure*, we can price financial derivative products in a similar way as under the plain Heston model. The chapter is concluded with a CVA risk management application under the Heston Hull-White hybrid model in **Section 13.3**.

Keywords: hybrid models, correlation between asset classes, Black-Scholes Hull-White model, Schöbel-Zhu Hull-White model, Heston-type hybrid models, Heston Hull-White model, multi-factor Gaussian interest rate processes, characteristic function, risk management CVA.

13.1 Introduction to affine hybrid asset models

> In the case of financial turbulence, stocks may go down and investors may escape from the market to reduce their losses. Central banks may then decrease the interest rates in order to increase cash flows: this may again lead to an increase in stock values, since it becomes less attractive for investors to keep their money in bank accounts. Movements in the interest rate market may thus have an influence on the behavior of stock prices, especially in the long run. This is taken into account in so-called *hybrid models*.

We will present a number of hybrid models which can be used for pricing the corresponding hybrid derivatives, as well as for risk management.

Hybrid models can be expressed by a system of SDEs, for example for stock, volatility and interest rate, with a full correlation matrix.

By correlating these SDEs from the different asset classes, one can define the hybrid models. Even if each of the individual SDEs yields a closed-form solution, a nonzero correlation structure between the processes may cause difficulties for efficient valuation and calibration.

Typically, however, a closed-form solution for hybrid models is not known, and a numerical approximation by means of a Monte Carlo (MC) simulation or a discretization of the corresponding PDEs has to be employed. The speed of pricing European derivative products is crucial, especially for the calibration of the SDEs. Several theoretically attractive SDE models, that cannot fulfill the speed requirements, are not used in practice, because of the fact that the calibration stage is too time-consuming.

> Although hybrid models can relatively easily be defined, these models are only used, when they provide a satisfactory fit to market implied volatility structure and when it is possible to set a nonzero correlation structure among the processes from the different asset classes. Furthermore, highly efficient valuation and calibration are mandatory.
>
> For this reason, a focus in this chapter is to derive the characteristic function of the hybrid SDE system. With *a characteristic function* available, highly efficient pricing of European options may take place, for example, by the COS method.

13.1.1 Black-Scholes Hull-White (BSHW) model

As a starting point, we extend the standard Black-Scholes model [Black and Scholes, 1973] by the Hull-White [Hull and White, 1990] short-rate model. This model is called the Black-Scholes Hull-White hybrid (BSHW) model, which is often considered as a benchmark for modeling Foreign-Exchange (FX) [German and Kohlhagen, 1983], inflation-indexed derivatives (based on the Consumer-Price-Index, CPI) [Jarrow and Yildirim, 2003] or long-maturity options [Brigo and Mercurio, 2007].

> Under the risk-neutral measure $\mathbb{Q}$, the dynamics of the model with $\mathbf{X}(t) = [S(t), r(t)]^T$ are given by the following system of SDEs:
>
> $$\begin{aligned} \mathrm{d}S(t)/S(t) &= r(t)\mathrm{d}t + \sigma \mathrm{d}W_x(t), \quad S(t_0) = S_0 > 0, \\ \mathrm{d}r(t) &= \lambda(\theta(t) - r(t))\mathrm{d}t + \eta \mathrm{d}W_r(t), \quad r(t_0) = r_0, \end{aligned} \quad (13.1)$$
>
> where $W_x(t)$ and $W_r(t)$ are two *correlated* Brownian motions with $\mathrm{d}W_x(t)\mathrm{d}W_r(t) = \rho_{x,r}\mathrm{d}t$, and $|\rho_{x,r}| < 1$ is the instantaneous correlation parameter between the asset price and the short-rate process. Parameters σ and η determine the volatility of equity and interest rate, respectively; $\theta(t)$ is a deterministic function (as defined in Equation (11.37)) and λ determines the speed of mean reversion.

After a transformation into log-coordinates, $X(t) = \log S(t)$, it is easy to see that the model satisfies the affinity conditions from (7.52), (7.53) and (7.54), so that the corresponding characteristic function, $\phi_{\mathrm{BSHW}}(u;t,T)$, can easily be derived.

> For the state vector $\mathbf{X}(t) = [X(t), r(t)]^T$, the discounted characteristic function, with $\mathbf{u} = [u, 0]^T$ and $\tau := T - t$, reads:
>
> $$\phi_{\mathrm{BSHW}}(u;t,T) = \exp\left(\bar{A}(u,\tau) + \bar{B}(u,\tau)X(t) + \bar{C}(u,\tau)r(t)\right), \quad (13.2)$$
>
> with final condition, $\phi_{\mathrm{BSHW}}(u;T,T) = \exp(iuX(T))$. The functions $\bar{A}(u,\tau)$, $\bar{B}(u,\tau)$ and $\bar{C}(u,\tau)$ are found to be:
>
> $$\bar{B}(u,\tau) = iu,$$
>
> $$\bar{C}(u,\tau) = \frac{1}{\lambda}(iu - 1)(1 - e^{-\lambda\tau}),$$
>
> $$\bar{A}(u,\tau) = \frac{1}{2}\sigma^2 iu(iu-1)\tau + \frac{\rho_{x,r}\sigma\eta}{\lambda}iu(iu-1)\left(\tau + \frac{1}{\lambda}\left(e^{-\lambda\tau} - 1\right)\right)$$
>
> $$+ \frac{\eta^2}{4\lambda^3}(i+u)^2\left(3 + e^{-2\lambda\tau} - 4e^{-\lambda\tau} - 2\lambda\tau\right)$$
>
> $$+ \lambda\int_0^\tau \theta(T-z)\bar{C}(u,z)\mathrm{d}z.$$

The expression for $\bar{A}(u,\tau)$ contains an integral over the deterministic function $\theta(t)$, which may be calibrated to the current market interest rate yield. This integral can be determined analytically.

The characteristic function for the BSHW model can thus be determined and Fourier inversion techniques, like the COS method, can be used for a number of payoff functions.

Moreover, the valuation of plain vanilla equity options can be performed analytically under the T-forward measure [Brigo and Mercurio, 2007; Overhaus et al., 2007], just like in the standard Black-Scholes model, as we will show in the section to follow.

13.1.2 BSHW model and change of measure

The BSHW model represents an example for which it is beneficial to change measures. The process, under the risk-neutral measure, has the following representation in the terms of *independent* Brownian motions:

$$\begin{bmatrix} dr(t) \\ dS(t)/S(t) \end{bmatrix} = \begin{bmatrix} \lambda(\theta(t) - r(t)) \\ r(t) \end{bmatrix} dt + \begin{bmatrix} \eta & 0 \\ \sigma\rho_{x,r} & \sigma\sqrt{1-\rho_{x,r}^2} \end{bmatrix} \begin{bmatrix} d\widetilde{W}_r(t) \\ d\widetilde{W}_x(t) \end{bmatrix}. \tag{13.3}$$

To determine the distribution of $S(t)$, we perform a *measure transformation*, i.e. we switch between tradable numéraires. Instead of using the money-savings account $M(t)$, the *optimal* candidate for the numéraire is here the zero-coupon bond $P(t,T)$, for which its value at maturity T is equal to one unit of currency. At time T, we therefore do not have any randomness with respect to this numéraire, which is beneficial because of the interest in the stock distribution at time T. ZCB $P(t,T)$ is governed by the dynamics in (11.31) under the Hull-White model. In the dynamics of the zero-coupon bond, we also consider the independent Brownian motion $\widetilde{W}_r(t)$, instead of $W_r(t)$.

We employ the measure transformation, and define the Radon-Nikodym derivative,

$$\lambda_{\mathbb{Q}}^T(t) = \frac{d\mathbb{Q}^T}{d\mathbb{Q}}\bigg|_{\mathcal{F}(t)} = \frac{P(t,T)}{P(0,T)}\frac{M(0)}{M(t)}. \tag{13.4}$$

By Itô's lemma, the dynamics of $\lambda_{\mathbb{Q}}^T(t)$ are given by

$$d\lambda_{\mathbb{Q}}^T(t) = \frac{1}{M(t)}dP(t,T) - \frac{P(t,T)}{M^2(t)}dM(t).$$

After substitutions and simplifications, it becomes:

$$\frac{d\lambda_{\mathbb{Q}}^T(t)}{\lambda_{\mathbb{Q}}^T(t)} = \eta \bar{B}_r(t,T) d\widetilde{W}_r(t),$$

see also (11.31).

> This representation gives us the Girsanov kernel, which describes the transition from risk-neutral measure $\mathbb{Q}$, to the T-forward measure $\mathbb{Q}^T$, i.e.,
>
> $$\begin{cases} d\widetilde{W}_r(t) = \eta \bar{B}_r(t,T) dt + d\widetilde{W}_r^T(t), \\ d\widetilde{W}_x(t) = d\widetilde{W}_x^T(t). \end{cases} \tag{13.5}$$

The measure transformation brings us the following dynamics for the ZCB $P(t,T)$, under the T-forward measure,

$$\frac{dP(t,T)}{P(t,T)} = r(t)dt + \eta \bar{B}_r(t,T)\left(\eta \bar{B}_r(t,T)dt + d\widetilde{W}_r^T(t)\right)$$

$$= \left(r(t) + \eta^2 \bar{B}_r^2(t,T)\right)dt + \eta \bar{B}_r(t,T) d\widetilde{W}_r^T(t), \tag{13.6}$$

where the short-rate process $r(t)$ is, under the same measure, given by Equation (11.60).

We return to the dynamics of the stock process $S(t)$, in terms of the independent Brownian motions (see Equation (13.3)),

$$\frac{\mathrm{d}S(t)}{S(t)} = r(t)\mathrm{d}t + \sigma \left(\rho_{x,r} \mathrm{d}\widetilde{W}_r(t) + \sqrt{1 - \rho_{x,r}^2} \mathrm{d}\widetilde{W}_x(t) \right). \tag{13.7}$$

After application of the measure transformation in (13.5), the dynamics of $S(t)$, under the T-forward measure, are given by

$$\frac{\mathrm{d}S(t)}{S(t)} = \left(r(t) + \rho_{x,r} \eta \sigma \bar{B}_r(t,T) \right) \mathrm{d}t + \sigma \left(\rho_{x,r} \mathrm{d}\widetilde{W}_r^T(t) + \sqrt{1 - \rho_{x,r}^2} \mathrm{d}\widetilde{W}_x^T(t) \right).$$

The BSHW model under the T-forward measure is thus governed by the following system of SDEs:

$$\begin{cases} \dfrac{\mathrm{d}S(t)}{S(t)} = \left(r(t) + \rho_{x,r} \eta \sigma \bar{B}_r(t,T) \right) \mathrm{d}t + \sigma \mathrm{d}W_x^T(t), \\ \mathrm{d}r(t) = \lambda \left(\theta(t) + \dfrac{\eta^2}{\lambda} \bar{B}_r(t,T) - r(t) \right) \mathrm{d}t + \eta \mathrm{d}W_r^T(t), \end{cases} \tag{13.8}$$

with $\bar{B}_r(t,T) = \frac{1}{\lambda} \left(\mathrm{e}^{-\lambda(T-t)} - 1 \right)$ and $\mathrm{d}W_x^T(t) \mathrm{d}W_r^T(t) = \rho_{x,r}\mathrm{d}t$.

Option pricing under the *T*-forward measure

The change of measure, from the risk-neutral to the T-forward measure, does not really simplify the dynamics of the underlying BSHW model – it even seems to make it somewhat more involved. The advantages of the measure transformation come from the payoff, when pricing options.

The pricing problem of a European-style payoff function, $H(T,S)$, can be expressed as

$$V(t_0, S) = M(t_0)\mathbb{E}^{\mathbb{Q}} \left[\frac{1}{M(T)} H(T,S) \Big| \mathcal{F}(t_0) \right] = M(t_0) \int_\Omega \frac{1}{M(T)} H(T,S) \mathrm{d}\mathbb{Q},$$

where $M(t_0) = 1$. From (13.4), $\mathrm{d}\mathbb{Q} = P(t_0,T)M(T)\mathrm{d}\mathbb{Q}^T$, and since $M(t_0) = P(T,T) = 1$, pricing under the T-forward measure results in

$$V(t_0, S) = \int_\Omega \frac{1}{M(T)} H(T,S) P(t_0,T) M(T) \mathrm{d}\mathbb{Q}^T$$
$$= P(t_0,T)\mathbb{E}^T \left[H(T,S) \big| \mathcal{F}(t_0) \right].$$

As the zero-coupon bond $P(t_0, T)$ is $\mathcal{F}(t_0)$-measurable (as its value is known at time t_0) and $M(t_0) = 1$, we obtain

$$V(t_0, S) = P(t_0, T)\mathbb{E}^T\left[H(T, S)|\mathcal{F}(t_0)\right]. \quad (13.9)$$

Under the $\mathbb{Q}$-measure, the stock, discounted by the money-savings account, is a martingale, however, this is not the case under the T-forward measure. Under T-forward measure $\mathbb{Q}^T$, the numéraire is the zero-coupon bond $P(t, T)$ and the process $\frac{S(t)}{P(t,T)}$ thus has to be a martingale. With the *forward stock price*, defined as

$$S_F(t, T) := \frac{S(t)}{P(t, T)},$$

and $S(t)$ given in (13.8), $P(t, T)$ from (13.6), it follows, by Itô's lemma, that

$$dS_F(t, T) = \frac{1}{P(t, T)}dS(t) - \frac{S(t)}{P(t, T)}dP(t, T) + \frac{S(t)}{P^3(t, T)}(dP(t, T))^2$$
$$- \frac{1}{P^2(t, T)}dP(t, T)dS(t),$$

which can be simplified to

$$\frac{dS_F(t, T)}{S_F(t, T)} = \sigma dW_x^T(t) - \eta \bar{B}_r(t, T)dW_r^T(t). \quad (13.10)$$

Process $S_F(t, T)$ does not contain any dt-terms, so $S_F(t, T)$ is a martingale under the T-forward measure, $\mathbb{Q}^T$.

We further simplify the SDE in (13.10), recalling that for two correlated Brownian motions, $W_1(t)$ and $W_2(t)$, with correlation $\rho_{1,2}$ and positive constants a, b, the following equality holds in distributional sense,

$$aW_1(t) + bW_2(t) \stackrel{d}{=} \sqrt{a^2 + b^2 + 2ab\rho_{1,2}}W_3(t),$$

where $W_3(t)$ is also a Brownian motion. With this insight, the SDE in (13.10) can be rewritten as:

$$\frac{dS_F(t, T)}{S_F(t, T)} = \bar{\sigma}_F(t)dW_F(t), \quad S_F(t_0, T) = \frac{S_0}{P(t_0, T)}, \quad (13.11)$$

with $\bar{\sigma}_F(t) = \sqrt{\sigma^2 + \eta^2 \bar{B}_r^2(t, T) - 2\rho_{x,r}\sigma\eta\bar{B}_r(t, T)}$.

Equation (13.11) resembles the SDE for GBM in the Black-Scholes model, except that the volatility parameter $\bar{\sigma}_F(t)$ is time-dependent. For a European-style option with only a payment at time T, the distribution of the underlying asset at time T should be known. In the case of the time-dependent volatility it is possible to determine a constant volatility, σ_c, so that the average volatility, accumulated until maturity time T match with σ_c. We illustrated this in Subsection 2.2.3, Equation (2.25).

The COS method and stochastic discounting

The COS method from Chapter 6 can be generalized towards asset dynamics with a stochastic interest rate process, based on the following risk-neutral pricing formula:

$$V(t_0, S) = \mathbb{E}^{\mathbb{Q}}\left[e^{-\int_{t_0}^{T} r(z)\mathrm{d}z} V(T, S) \big| \mathcal{F}(t_0)\right]$$

$$= \int_{\mathbb{R}} V(T, e^y) f_{\mathbf{X}}(T, y; t_0, x) \mathrm{d}y, \quad (13.12)$$

where $\mathbf{X}(t) = [X(t), r(t), \ldots]^{\mathrm{T}}$, $X(t_0) \equiv \log S(t_0)$, and $f_{\mathbf{X}}(T, y; t_0, x) := \int_{\mathbb{R}} e^z f_{X,z}(T, y, z; t_0, x) \mathrm{d}z$, with $z(t) = -\int_{t_0}^{T} r(z) \mathrm{d}z$.

Assuming a rapid decay of the density function, the following approximation will be made,

$$V(t_0, S) \approx \int_a^b V(T, e^y) f_{\mathbf{X}}(T, y; t_0, x) \mathrm{d}y, \quad (13.13)$$

where a and b are appropriate integration boundaries (as discussed in Section 6.2.4). The discounted characteristic function is now given by:

$$\phi_{\mathbf{X}}(\mathbf{u}; t_0, T) = \mathbb{E}^{\mathbb{Q}}\left[e^{-\int_{t_0}^{T} r(z)\mathrm{d}z + i\mathbf{u}^{\mathrm{T}} \mathbf{X}(T)} \big| \mathcal{F}(t_0)\right],$$

which, for $\tau = T - t_0$, $\mathbf{u} = [u, 0, \ldots, 0]^{\mathrm{T}}$, reads

$$\phi_{\mathbf{X}}(u; t_0, T) = \iint_{\mathbb{R}} e^{z + iuy} f_{X,z}(T, y, z; t_0, x) \mathrm{d}z \mathrm{d}y$$

$$= \int_{\mathbb{R}} e^{iuy} f_{\mathbf{X}}(T, y; t_0, x) \mathrm{d}y. \quad (13.14)$$

The integration in (13.14) represents the *Fourier transform* of $f_{\mathbf{X}}(T, y; t_0, x)$. It can be approximated on a bounded domain $[a, b]$, by

$$\phi_{\mathbf{X}}(u; t, T) \approx \int_a^b e^{iuy} f_{\mathbf{X}}(T, y; t_0, x) \mathrm{d}y =: \hat{\phi}_{\mathbf{X}}(u; t, T). \quad (13.15)$$

The multi-variate density function $f_{\mathbf{X}}(T, y; t_0, x)$ can be connected to its characteristic function, via the following result:

Result 13.1.1 *For a given bounded domain $[a,b]$, and with N the number of terms in the Fourier cosine series expansion, the probability density function $f_{\mathbf{X}}(T, y; t_0, x)$ given by (13.12) can be approximated by,*

$$f_{\mathbf{X}}(T, y; t_0, x) \approx \frac{2}{b-a} \sum_{k=0}^{N-1}{}' \operatorname{Re}\left\{\hat{\phi}_{\mathbf{X}}\left(\frac{k\pi}{b-a}; t, T\right) \exp\left(-i\frac{ka\pi}{b-a}\right)\right\}$$

$$\times \cos\left(k\pi \frac{y-a}{b-a}\right),$$

with $\operatorname{Re}\{\cdot\}$ the real part of the argument in brackets; $\hat{\phi}_{\mathbf{X}}(u; t, T)$ is the corresponding characteristic function and the remaining settings are as in the COS method, from Chapter 6.

Implied volatility and models with stochastic interest rates

When generalizing the Black-Scholes model by means of stochastic interest rates, a natural question is how to calculate the implied volatilities within the Black-Scholes formula? In the standard procedure to determine the implied volatility, a constant interest rate is inserted in the formulas. Which r-value should be chosen when the interest rates are modeled by a stochastic process? The answers are related to the previous derivations, where we changed between the risk-neutral and T-forward measure. It was shown that the stock dynamics in the Black-Scholes Hull-White model can be written as,

$$\frac{\mathrm{d}S(t)}{S(t)} = r(t)\mathrm{d}t + \sigma \mathrm{d}W_x(t),$$

with $r(t)$ governed by the Hull-White model. A measure transformation led to the dynamics that were free of drift terms for the forward stock $S_F(t,T) = S(t)/P(t,T)$, see (13.11),

$$\frac{\mathrm{d}S_F(t,T)}{S_F(t,T)} = \bar{\sigma}_F(t)\mathrm{d}W_F(t).$$

This is an important observation as this way an alternative formula for pricing European options can be derived. The pricing formula is then independent of the interest rates (it does however still depend on the volatility), so we can work with any model in which the interest rate is not constant.

The equality at final time, $S(T) = S_F(T,T)$, implies that for the valuation of a contract with a fixed maturity time, the drift-free stochastic process $S_F(t,T)$ may be employed. Because process $S_F(t,T)$ does not contain $r(t)$, it is significantly easier to value the derivatives.

With the pricing formula in (13.9) and payoff $H(T,S) = \max(S(T) - K, 0)$, we find

$$V(t_0, S) = P(t_0, T)\mathbb{E}^T\left[\max(S_F(T,T) - K, 0)|\mathcal{F}(t_0)\right] \quad (13.16)$$
$$= S_{F,0}P(t_0, T)F_\mathcal{N}(d_1) - KP(t_0, T)F_\mathcal{N}(d_2). \quad (13.17)$$

with

$$d_1 = \frac{\log\left(\frac{S_{F,0}}{K}\right) + \frac{1}{2}\sigma_c^2(T - t_0)}{\sigma_c\sqrt{T - t_0}}, \quad d_2 = \frac{\log\left(\frac{S_{F,0}}{K}\right) - \frac{1}{2}\sigma_c^2(T - t_0)}{\sigma_c\sqrt{T - t_0}}.$$

Furthermore, $S_{F,0} := \frac{S_0}{P(t_0,T)}$, $F_\mathcal{N}(\cdot)$ is the standard normal CDF, σ_c is as in (2.25), so

$$\sigma_c^2 = \frac{1}{T - t_0}\int_{t_0}^T \bar{\sigma}_F^2(z)\mathrm{d}z.$$

As mentioned, this derived formula does not directly contain interest rate r, so it can also be applied for the computation of the implied volatilities for models in which the interest rate is stochastic.[1]

By prescribing a certain time-dependent volatility function in the Black-Scholes model, one has control over the ATM implied volatilities. These time-dependent equity volatilities are known by the name *volatility term structure*.

Example 13.1.1 (BSHW model and implied volatility) In Figure 13.1 possible shapes of the ATM implied volatilities are presented. Clearly, a time-dependent volatility function is insufficient to generate implied volatility smiles, but it is sufficient to describe the implied volatility term structure which can be observed in the interest rate market.

The numerical experiment was performed with the following set of BSHW parameters $\sigma = 0.2$, $\lambda = 0.1$, $\eta = 0.01$ and $\rho_{x,r} = 0.3$. Each parameter is varied individually, keeping the remaining parameters fixed. All parameters have a significant effect on the model's implied ATM volatilities.

As with the plain Black-Scholes model, we encounter flat implied volatilities for the BSHW model. By adding the stochastic interest rate, the implied volatility skew is not impacted, however, the interest rate parameters influence the implied volatility term-structure. ♦

13.1.3 Schöbel-Zhu Hull-White (SZHW) model

In this section we present a first *stochastic volatility (SV) equity hybrid model*, which contains a *stochastic interest rate* process and a *full matrix of correlations* between the underlying Brownian motions. Also here particularly the Hull-White stochastic interest rate process [Hull and White, 1990] is added to the SV model.

---

[1] Of course, the model is still interest rate dependent via the zero-coupon bond $P(t_0, T)$, which has to be calculated and used in the pricing formula.

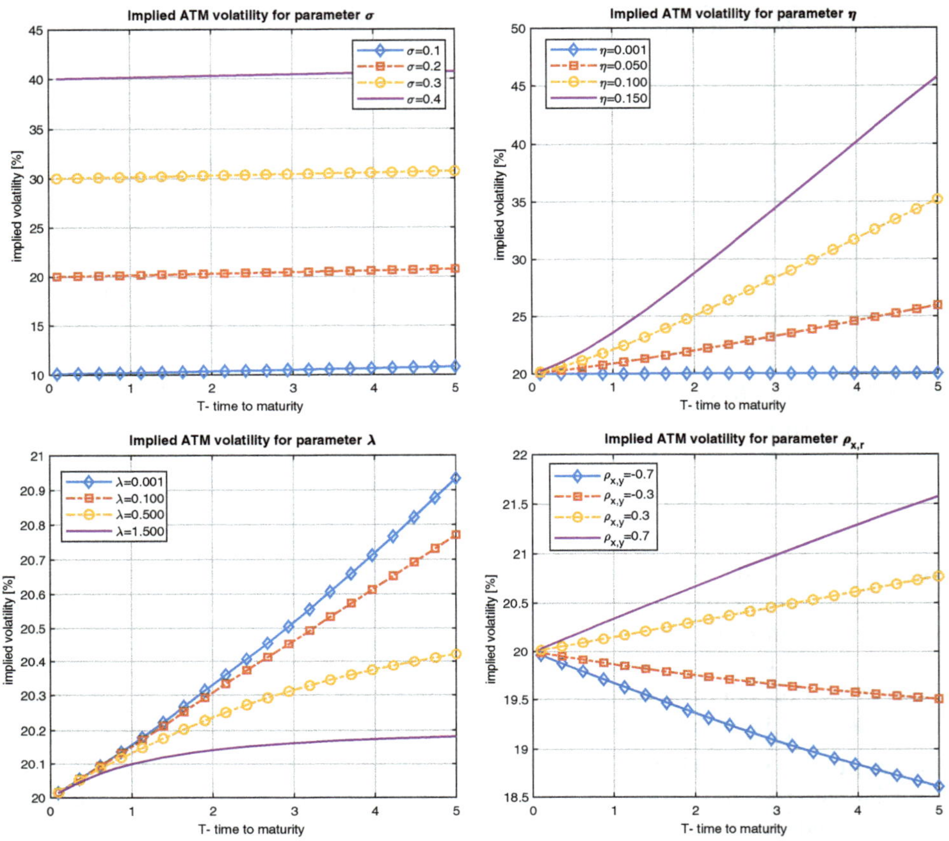

Figure 13.1: *Implied volatility term structure in the Black-Scholes Hull-White model (equivalent to a Black-Scholes model with time-dependent volatility).*

For state vector $\mathbf{X}(t) = [S(t), r(t), \sigma(t)]^{\mathrm{T}}$, we fix a probability space $(\Omega, \mathcal{F}, \mathbb{Q})$ and a filtration $\mathcal{F} = \{\mathcal{F}(t) : t \geq 0\}$, which satisfies the usual conditions.

Under the risk-neutral measure $\mathbb{Q}$, consider a 3D system of stochastic differential equations, of the following form:

$$\begin{aligned}
\mathrm{d}S(t)/S(t) &= r(t)\mathrm{d}t + \sigma^p(t)\mathrm{d}W_x(t), \\
\mathrm{d}r(t) &= \lambda(\theta(t) - r(t))\mathrm{d}t + \eta\mathrm{d}W_r(t), \\
\mathrm{d}\sigma(t) &= \kappa\left(\bar{\sigma} - \sigma(t)\right)\mathrm{d}t + \gamma\sigma^{1-p}(t)\mathrm{d}W_\sigma(t),
\end{aligned} \qquad (13.18)$$

where p is an exponent, κ and λ control the speed of mean reversion, η represents the interest rate volatility, and $\gamma\sigma^{1-p}(t)$ determines the volatility of the $\sigma(t)$ process. Depending on parameter p, $\sigma(t)$ denotes either the volatility ($p = 1$) or the variance ($p = \frac{1}{2}$). Parameters $\bar{\sigma}$ and $\theta(t)$ are the long-run mean of the volatility, and the interest rate processes, respectively. $W_k(t)$ with $k = \{x, r, \sigma\}$ are correlated Wiener processes, also governed by the instantaneous correlation matrix:

$$\mathbf{C} := \begin{bmatrix} 1 & \rho_{x,\sigma} & \rho_{x,r} \\ \rho_{\sigma,x} & 1 & \rho_{\sigma,r} \\ \rho_{r,x} & \rho_{r,\sigma} & 1 \end{bmatrix}. \tag{13.19}$$

The system in (13.18) for $p = \frac{1}{2}$ is the *Heston Hull-White hybrid model*, to be discussed in the section to follow.

If we set $p = \frac{1}{2}$ and set $r(t)$ constant, we obtain the plain *Heston model* [Heston, 1993],

$$\begin{cases} \mathrm{d}S(t)/S(t) = r\mathrm{d}t + \sqrt{\sigma(t)}\mathrm{d}W_x(t), \\ \mathrm{d}\sigma(t) = \kappa^H \left(\bar{\sigma}^H - \sigma(t)\right)\mathrm{d}t + \gamma^H \sqrt{\sigma(t)}\mathrm{d}W_\sigma(t), \end{cases}$$

where $\sigma(t) \equiv v(t)$, i.e. the variance process is of CIR-type [Cox et al., 1985].

For $p = 1$ and $r(t)$ constant, the model is, in fact, the generalized Stein-Stein [Stein and Stein, 1991] model, which is identical to the *Schöbel-Zhu [Schöbel and Zhu, 1999] model*:

$$\begin{cases} \mathrm{d}S(t)/S(t) = r\mathrm{d}t + \sqrt{v(t)}\mathrm{d}W_x(t), \\ \mathrm{d}v(t) = 2\kappa \left(\frac{\bar{\sigma}}{2\kappa}\sigma(t) + \frac{\gamma^2}{2\kappa} - v(t)\right)\mathrm{d}t + 2\gamma\sqrt{v(t)}\mathrm{d}W_\sigma(t), \end{cases} \tag{13.20}$$

in which the squared volatility, $v(t) = \sigma^2(t)$, represents the variance of the instantaneous stock return.

The plain Schöbel-Zhu model is a particular case of the original Heston model, as stated in [Heston, 1993] and [Schöbel and Zhu, 1999]: for $\bar{\sigma} = 0$ in Equation (13.20), the Schöbel-Zhu model equals the Heston model in which $\kappa^H = 2\kappa$, $\bar{\sigma}^H = \gamma^2/2\kappa$, and $\gamma^H = 2\gamma$. This relation gives a direct connection between their discounted characteristic functions (see [Lord and Kahl, 2006]). Finally, if we set $r(t)$ constant and $p = 0$ in the system of Equations (13.18), plus zero correlations, the model collapses to the standard Black-Scholes model [Black and Scholes, 1973].

Affinity of SZHW model

In this section, the parameters in the Equations (13.18) will be set, so that the Schöbel-Zhu Hull-White (SZHW) model originates.

The Schöbel-Zhu Hull-White hybrid model can be expressed by the following 3D system of SDEs

$$\begin{cases} \mathrm{d}S(t)/S(t) = r(t)\mathrm{d}t + \sigma(t)\mathrm{d}W_x(t), \\ \mathrm{d}r(t) = \lambda \left(\theta(t) - r(t)\right)\mathrm{d}t + \eta\mathrm{d}W_r(t), \\ \mathrm{d}\sigma(t) = \kappa(\bar{\sigma} - \sigma(t))\mathrm{d}t + \gamma\mathrm{d}W_\sigma(t), \end{cases} \tag{13.21}$$

with the parameters as in Equations (13.18), with $p=1$, and the correlations $\mathrm{d}W_x(t)\mathrm{d}W_\sigma(t) = \rho_{x,\sigma}\mathrm{d}t$, $\mathrm{d}W_x(t)\mathrm{d}W_r(t) = \rho_{x,r}\mathrm{d}t$, $\mathrm{d}W_r(t)\mathrm{d}W_\sigma(t) = \rho_{r,\sigma}\mathrm{d}t$. System (13.21) is not affine in its present form. The technique to make the system affine, is by adding an additional equation to the originally 3D system.

By extending the state vector (as in [Cheng and Scaillet, 2007] or [Pelsser]) with another, latent stochastic variable, defined by $v(t) := \sigma^2(t)$, and using $X(t) = \log S(t)$, we obtain the following 4D system of SDEs,

$$\begin{cases} \mathrm{d}X(t) = \left(\widetilde{r}(t) + \psi(t) - \frac{1}{2}v(t)\right)\mathrm{d}t + \sqrt{v(t)}\mathrm{d}W_x(t), \\ \mathrm{d}\widetilde{r}(t) = -\lambda\widetilde{r}(t)\mathrm{d}t + \eta\mathrm{d}W_r(t), \\ \mathrm{d}v(t) = \left(-2v(t)\kappa + 2\kappa\bar{\sigma}\sigma(t) + \gamma^2\right)\mathrm{d}t + 2\gamma\sqrt{v(t)}\mathrm{d}W_\sigma(t), \\ \mathrm{d}\sigma(t) = \kappa(\bar{\sigma} - \sigma(t))\mathrm{d}t + \gamma\mathrm{d}W_\sigma(t), \end{cases} \quad (13.22)$$

where we used $r(t) = \widetilde{r}(t) + \psi(t)$, as in Subsection 11.3, and where $\theta(t)$ is included in $\psi(t)$.
Model (13.22) is *affine in the extended state vector* $\mathbf{X}(t) = [X(t), \widetilde{r}(t), v(t), \sigma(t)]^{\mathrm{T}}$.

By the extension of the vector space, we have obtained an affine model which enables us to apply the results to derive the corresponding *characteristic function*. We will however not do that here. The detailed derivation can, for example, be found in [Grzelak et al., 2012].

13.1.4 Hybrid derivative product

Hybrid products are financial contracts that combine different market sectors, assets and instruments. Hybrid products arise from the need of an investor to benefit from profits in different market sectors. Contracts can be based on the best performing sector, for example, with a guarantee that an investment cannot decrease significantly in value. Hybrid contracts are however not so often encountered anymore in the financial industry, since the 2007/2008 financial crisis.

A diversification product (performance basket)

Hybrid products in strategic trading are so-called diversification products. These products are based on sets of assets with different expected returns and risk levels. Proper construction of such products may give reduced risk compared to any single asset, and an expected return that is greater than that of the least risky asset [Hunter and Picot, 2006]. A simple example is a portfolio with two assets: a stock with a high risk and high return and a bond with a low risk and low return. If one introduces an equity component in a pure bond portfolio the expected return will increase. However, because of a nonperfect correlation between these two assets also a risk reduction is expected. If the percentage of the equity in the portfolio is increased, it eventually starts to dominate and the risk may increase with a higher impact for a low or negative correlation [Hunter and Picot, 2006].

An example is a financial product, defined in the following way:

$$V^d(t_0, S, r) = \mathbb{E}^{\mathbb{Q}}\left[\frac{M(t_0)}{M(T)} \max\left(0, \omega_d \cdot \frac{S(T)}{S_0} + (1 - \omega_d) \cdot \frac{P(T, T_1)}{P(t_0, T_1)}\right)\bigg|\mathcal{F}(t_0)\right],$$

where $S(T)$ is the underlying asset at time T, $M(t_0) = 1$, $P(t, T)$ is the zero-coupon bond, ω_d represents a percentage ratio.

Figure 13.2 shows the pricing results for the SZHW model discussed.

The product pricing is performed for different correlations $\rho_{x,r}$ and the remaining parameters have been calibrated to market data, see the data in Table 13.1.

For $\omega_d \in [0\%, 100\%]$ the max disappears from the payoff and only a sum of discounted expectations remains. Figure 13.2 shows that a positive correlation between the products in the basket significantly increases the contract value, while negative correlation has a reversed effect. The absolute difference between the models increases with the percentage ω_d.

13.2 Hybrid Heston model

The Heston equity model with a deterministic interest rate has established itself as one of the benchmark models for equity derivatives. The assumption of deterministic interest rates in the plain Heston model is harmless when equity products with a short time to maturity need to be priced. For long-term equity, foreign-exchange, or equity-interest rate products, however, a deterministic interest rate may be inaccurate.

Table 13.1: *Calibration results for the SZHW and the Heston model, as defined in (13.21). The experiment was done with $\kappa = 0.5$, and different correlations $\rho_{x,r}$ for the SZHW model. In the simulation for the Heston model a constant interest rate $r = 0.033$ was used.*

			model	λ	η		
			Hull-White	1.12	0.02		

model	$\rho_{x,r}$	$\bar{\sigma}$	γ	$\rho_{x,\sigma}$	$\rho_{r,\sigma}$	σ_0/v_0
	-70%	0.137	0.236	-0.381%	-0.339%	0.084
SZHW	0%	0.167	0.2	-85.0%	-0.8%	0.035
	70%	0.102	0.211	-85.0%	-34.0%	0.01
Heston	–	0.0770	0.3500	-66.22%	–	0.0107

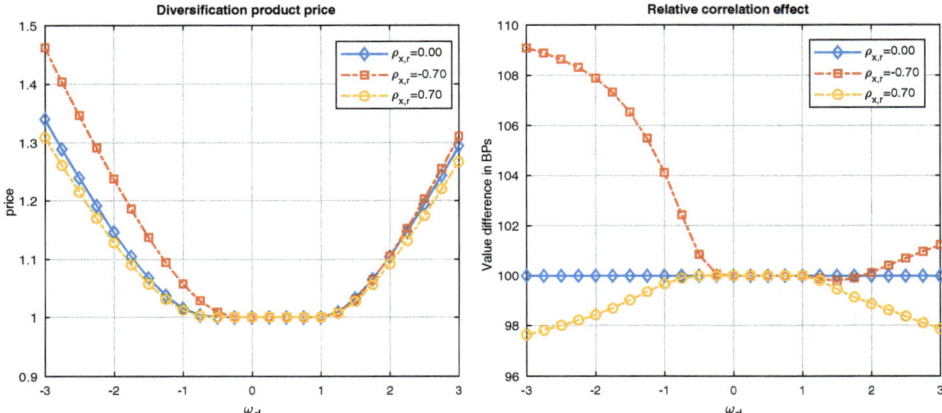

Figure 13.2: *Left: Pricing of a diversification hybrid product, with $T = 9$ and $T_1 = 10$, under different correlations $\rho_{x,r}$. The remaining parameters are as in Table 13.1. Right: Price differences with respect to the model with $\rho_{x,r} = 0\%$ expressed in basis points (BPs) for different correlations $\rho_{x,r}$.*

In Section 13.2.1 we therefore discuss a Heston hybrid model with a stochastic interest rate process. Section 13.2.2 presents a deterministic approximation of the Heston Hull-White hybrid model, together with the corresponding characteristic function.

13.2.1 Details of Heston Hull-White hybrid model

In Section 13.1.3, we briefly discussed the SZHW model [Grzelak et al., 2012; van Haastrecht et al., 2009]. A full matrix of correlations was imposed directly on the driving Brownian motions. The model was in the class of AD processes, but since the SZHW model is based on a Vašiček-type process [Vašiček, 1977] for the stochastic volatility, the volatilities may become negative.

In this section, we will discuss the *Heston Hull-White (HHW) hybrid model* and a model *approximation*, so that we can obtain the characteristic function.

We extend the Heston model state vector by a stochastic interest rate process, i.e., $\mathbf{X}(t) = [S(t), v(t), r(t)]^{\mathrm{T}}$. In particular, we add the Hull-White (HW) interest rate [Hull and White, 1990]. The HHW model is presented in the following way, under the $\mathbb{Q}$-measure:

$$\begin{cases} \mathrm{d}S(t)/S(t) = r(t)\mathrm{d}t + \sqrt{v(t)}\mathrm{d}W_x(t), & S(0) > 0, \\ \mathrm{d}v(t) = \kappa(\bar{v} - v(t))\mathrm{d}t + \gamma\sqrt{v(t)}\mathrm{d}W_v(t), & v(0) > 0, \\ \mathrm{d}r(t) = \lambda(\theta(t) - r(t))\mathrm{d}t + \eta\mathrm{d}W_r(t), & r(0) \in \mathbb{R}, \end{cases} \qquad (13.23)$$

For the HHW model, the correlations are given by $\mathrm{d}W_x(t)\mathrm{d}W_v(t) = \rho_{x,v}\mathrm{d}t$, $\mathrm{d}W_x(t)\mathrm{d}W_r(t) = \rho_{x,r}\mathrm{d}t$, and $\mathrm{d}W_v(t)\mathrm{d}W_r(t) = \rho_{v,r}\mathrm{d}t$, and κ, γ and $\bar{v}$ are as in (8.18); $\lambda > 0$ determines the speed of mean reversion for the interest rate process; $\theta(t)$, as described in Section 11.2, is the interest rate term-structure and η controls the volatility of the interest rate.

System (13.23) is not in the affine form, not even for $X(t) = \log S(t)$. The *symmetric* instantaneous covariance matrix is given by

$$\bar{\boldsymbol{\sigma}}(\mathbf{X}(t))\bar{\boldsymbol{\sigma}}(\mathbf{X}(t))^{\mathrm{T}} = \begin{bmatrix} v(t) & \rho_{x,v}\gamma v(t) & \rho_{x,r}\eta\sqrt{v(t)} \\ * & \gamma^2 v(t) & \rho_{r,v}\gamma\eta\sqrt{v(t)} \\ * & * & \eta^2 \end{bmatrix}_{(3\times 3)}. \qquad (13.24)$$

Setting the correlation $\rho_{r,v}$ to zero would still not make the system affine. Matrix (13.24) is of the linear form with respect to state vector $[X(t) = \log S(t), v(t), r(t)]^{\mathrm{T}}$ if $\rho_{r,v}$ and $\rho_{x,r}$ are set to zero, as covered in [Muskulus et al., 2007].

Since for pricing and for risk management CVA computations a nonzero correlation between stock and interest rate may be important (see, for example, [Hunter and Picot, 2006]), an alternative approximation to the Heston hybrid models needs to be formulated so that correlations can be imposed.

As mentioned, Zhu in [Zhu, 2000] presented a hybrid model which modeled a skew pattern for equity and included a stochastic (but uncorrelated) interest rate process. Generalizations were then presented by Giese [2006] and Andreasen [2006], where the Heston stochastic volatility model [Heston, 1993] was used, combined with an indirectly correlated interest rate process. Correlation was modeled by including additional terms in the SDEs.

Remark 13.2.1 (Gaussian multi-factor short-rate model) *We may also consider the Gaussian multi-factor short-rate model (Gn++) [Brigo and Mercurio, 2007], which is known as the multi-factor Hull-White model [Hul]. For a given state vector, $\mathbf{R}(t) = [r(t), \varsigma_1(t), \ldots, \varsigma_{n-1}(t)]^{\mathrm{T}}$, this model is defined by the following system of SDEs:*

$$\begin{cases} \mathrm{d}r(t) = \left(\theta(t) + \sum_{k=1}^{n-1}\varsigma_k(t) - \beta r(t)\right)\mathrm{d}t + \eta\mathrm{d}W_r(t), & r(0) > 0, \\ \mathrm{d}\varsigma_k(t) = -\lambda_k\varsigma_k(t)\mathrm{d}t + \varsigma_k\mathrm{d}W_{\varsigma_k}(t), & \varsigma_k(0) = 0, \end{cases} \qquad (13.25)$$

where

$$\mathrm{d}W_r(t)\mathrm{d}W_{\varsigma_k}(t) = \rho_{r,\varsigma_k}\mathrm{d}t, \ k=1,\ldots,n-1, \quad \mathrm{d}W_{\varsigma_i}(t)\mathrm{d}W_{\varsigma_j}(t) = \rho_{\varsigma_i,\varsigma_j}\mathrm{d}t, \ i \neq j,$$

with $\beta > 0$, $\lambda_k > 0$ the mean reversion parameters; $\eta > 0$ and parameters ς_k determine the volatility magnitude of the interest rate. In the system above,

$\theta(t) > 0$, $t > 0$, is the long-term interest rate (which is usually calibrated to the current yield curve).

The $Gn++$ model provides a satisfactory fit to ATM humped structures of the volatility of the instantaneous forward rate. Moreover, the easy construction of the model (based on a multivariate normal distribution) provides a closed-form solution for caps and swaptions, enabling fast calibration. We will stay here with the one-factor HW process for the stochastic short-rate. ▲

In order to calibrate the HHW model, we need to determine a suitable characteristic function. To derive a closed-form characteristic function with a nonzero correlation between the equity process $S(t)$ and the interest rate $r(t)$, we will assume that the Brownian motions for the interest rate $r(t)$ and the variance $v(t)$ are not correlated.

Regarding the calibration of such hybrid models, typically the interest rate model is pre-calibrated to available interest rate derivatives. Subsequently, the parameters for the interest rate process are considered as an input for the hybrid model. In the second part of the calibration, the stock price parameters, together with the correlations between the asset classes, are estimated. In an idealized world, one would determine the correlation parameters from preferably correlation-type financial products or hybrid products. However, this is very difficult (or impossible) because these products are not traded sufficiently often (they have limited market liquidity).

13.2.2 Approximation for Heston hybrid models

Since the full-scale HHW model is not affine, it is not possible to directly derive a characteristic function. We therefore *linearize* the Heston hybrid model, to provide an approximation for the expressions in (13.24). The corresponding characteristic function is subsequently derived.

An approximation for the term $\eta \rho_{x,r} \sqrt{v(t)}$ in matrix (13.24) is found by *replacing it by its expectation*, i.e.,

$$\boxed{\eta \rho_{x,r} \sqrt{v(t)} \approx \eta \rho_{x,r} \mathbb{E}\left[\sqrt{v(t)}\right].} \qquad (13.26)$$

By taking the expectation of the stochastic variable, the model becomes affine, and a corresponding characteristic function for the approximate model is available.

In Lemma 13.2.1 the closed-form expressions for the expectation and variance of $\sqrt{v(t)}$ (a CIR-type process) are presented.

Lemma 13.2.1 (Expectation and variance for CIR-type process)
For a given time $t > 0$, the expectation and variance of $\sqrt{v(t)}$, where $v(t)$ is a CIR-type process, as in (8.18), are given by

$$\mathbb{E}\left[\sqrt{v(t)}|\mathcal{F}(0)\right] = \sqrt{2\bar{c}(t,0)}e^{-\bar{\kappa}(t,0)/2}\sum_{k=0}^{\infty}\frac{1}{k!}\left(\frac{\bar{\kappa}(t,0)}{2}\right)^k \frac{\Gamma\left(\frac{1+\delta}{2}+k\right)}{\Gamma\left(\frac{\delta}{2}+k\right)}, \tag{13.27}$$

and

$$\mathbb{V}ar\left[\sqrt{v(t)}|\mathcal{F}(0)\right] = \bar{c}(t,0)(\delta + \bar{\kappa}(t,0))$$

$$- 2\bar{c}(t,0)e^{-\bar{\kappa}(t,0)}\left(\sum_{k=0}^{\infty}\frac{1}{k!}\left(\frac{\bar{\kappa}(t,0)}{2}\right)^k \frac{\Gamma\left(\frac{1+\delta}{2}+k\right)}{\Gamma\left(\frac{\delta}{2}+k\right)}\right)^2, \tag{13.28}$$

where

$$\bar{c}(t,0) = \frac{1}{4\kappa}\gamma^2(1-e^{-\kappa t}), \quad \delta = \frac{4\kappa\bar{v}}{\gamma^2}, \quad \bar{\kappa}(t,0) = \frac{4\kappa v(0)e^{-\kappa t}}{\gamma^2(1-e^{-\kappa t})}, \tag{13.29}$$

with $\Gamma(k)$ being the Gamma function, as in (5.50). ◂

Proof It was shown in [Cox et al., 1985; Broadie and Kaya, 2006], that, $v(t)|v(0)$, $t > 0$ is distributed as $\bar{c}(t,0)$ times a noncentral chi-squared random variable, $\chi^2(\delta, \bar{\kappa}(t,0))$, with δ the "degrees of freedom" parameter and noncentrality parameter $\bar{\kappa}(t,0)$, i.e.,

$$v(t)|v(0) \sim \bar{c}(t,0)\chi^2(\delta, \bar{\kappa}(t,0)), \quad t > 0. \tag{13.30}$$

The corresponding CDF and PDF are known and presented in (8.6) and (8.9), respectively. The density function for $v(t)$ can be expressed as:

$$f_{v(t)}(x) := \frac{d}{dx}F_{v(t)}(x) = \frac{1}{\bar{c}(t,0)}f_{\chi^2(\delta,\bar{\kappa}(t,0))}\left(x/\bar{c}(t,0)\right).$$

By [Dufresne, 2001], it follows that:

$$\mathbb{E}\left[\sqrt{v(t)}|\mathcal{F}(0)\right] := \int_0^{\infty} \frac{\sqrt{x}}{\bar{c}(t,0)} f_{\chi^2(\delta,\bar{\kappa}(t,0))}\left(\frac{x}{\bar{c}(t,0)}\right) dx$$

$$= \sqrt{2\bar{c}(t,0)}\frac{\Gamma\left(\frac{1+\delta}{2}\right)}{\Gamma\left(\frac{\delta}{2}\right)}{}_1F_1\left(-\frac{1}{2},\frac{\delta}{2},-\frac{\bar{\kappa}(t,0)}{2}\right), \tag{13.31}$$

where ${}_1F_1(a;b;z)$ is a so-called *confluent hyper-geometric function*, which is also known as Kummer's function [Kummer, 1936] of the first kind, given by:

$${}_1F_1(a;b;z) = \sum_{k=0}^{\infty}\frac{(a)_k}{(b)_k}\frac{z^k}{k!}, \tag{13.32}$$

where $(a)_k$ and $(b)_k$ are *Pochhammer symbols* of the form:

$$(a)_k = \frac{\Gamma(a+k)}{\Gamma(a)} = a(a+1)\cdots(a+k-1). \tag{13.33}$$

Using the principle of Kummer (see [Koepf, 1998] pp.42), we find:

$$_1F_1\left(-\frac{1}{2}, \frac{\delta}{2}, -\frac{\bar{\kappa}(t,0)}{2}\right) = e^{-\bar{\kappa}(t,0)/2} \,_1F_1\left(\frac{1+\delta}{2}, \frac{\delta}{2}, \frac{\bar{\kappa}(t,0)}{2}\right). \tag{13.34}$$

Therefore, by (13.32) and (13.34), Equation (13.31) can be written as:

$$\mathbb{E}\left[\sqrt{v(t)}\Big| \mathcal{F}(0)\right] = \sqrt{2\bar{c}(t,0)} e^{-\bar{\kappa}(t,0)/2} \frac{\Gamma\left(\frac{1+\delta}{2}\right)}{\Gamma\left(\frac{\delta}{2}\right)} \,_1F_1\left(\frac{1+\delta}{2}, \frac{\delta}{2}, \frac{\bar{\kappa}(t,0)}{2}\right)$$

$$= \sqrt{2\bar{c}(t,0)} e^{-\bar{\kappa}(t,0)/2} \sum_{k=0}^{\infty} \frac{1}{k!} (\bar{\kappa}(t,0)/2)^k \frac{\Gamma\left(\frac{1+\delta}{2}+k\right)}{\Gamma\left(\frac{\delta}{2}+k\right)},$$

which concludes the proof for the expectation.

Furthermore, the mean and variance of process $v(t)$ of the noncentral chi-squared distribution are well-known, see Equation (8.12). This combined with the results for $\mathbb{E}\left[\sqrt{v(t)}\right]$ concludes the proof. ∎

This analytic expression for the expectation $\sqrt{v(t)}$ in (13.27) is involved and requires rather expensive numerical operations.

We therefore provide details of a simplified approximation.

An approximations for expectation $\mathbb{E}\left[\sqrt{v(t)}\right]$

In order to find a first-order approximation to (13.27), the so-called *delta method* (see for example [Amstrup et al., 2006; Oehlert, 1992]) may be applied, which states that a function $g(X)$ can be approximated by a first-order Taylor expansion at $\mathbb{E}[X]$, for a given random variable X with expectation $\mathbb{E}[X]$ and variance $\mathbb{Var}[X]$. Assuming function g to be sufficiently smooth and the first two moments of X to exist, a first-order Taylor expansion gives,

$$g(X) \approx g(\mathbb{E}[X]) + (X - \mathbb{E}[X])\frac{\partial g}{\partial X}(\mathbb{E}[X]). \tag{13.35}$$

Since the variance of $g(X)$ can be approximated by the variance of the right-hand side of (13.35) we have

$$\mathbb{Var}[g(X)] \approx \mathbb{Var}\left[g(\mathbb{E}[X]) + (X - \mathbb{E}[X])\frac{\partial g}{\partial X}(\mathbb{E}[X])\right]$$

$$= \left(\frac{\partial g}{\partial X}(\mathbb{E}[X])\right)^2 \mathbb{Var}[X]. \tag{13.36}$$

Using this result for function $g(v(t)) = \sqrt{v(t)}$, we find

$$\text{Var}\left[\sqrt{v(t)}\right] \approx \left(\frac{1}{2}\frac{1}{\sqrt{\mathbb{E}[v(t)]}}\right)^2 \text{Var}[v(t)] = \frac{1}{4}\frac{\text{Var}[v(t)]}{\mathbb{E}[v(t)]}. \tag{13.37}$$

However, from the definition of the variance, we also have

$$\text{Var}\left[\sqrt{v(t)}\right] = \mathbb{E}[v(t)] - \left(\mathbb{E}\left[\sqrt{v(t)}\right]\right)^2, \tag{13.38}$$

and combining Equations (13.37) and (13.38) gives the approximation:

$$\mathbb{E}\left[\sqrt{v(t)}\right] \approx \sqrt{\mathbb{E}[v(t)] - \frac{1}{4}\frac{\text{Var}[v(t)]}{\mathbb{E}[v(t)]}}. \tag{13.39}$$

As $v(t)$ is a square-root process,

$$v(t) = v(0)e^{-\kappa t} + \bar{v}(1 - e^{-\kappa t}) + \gamma \int_0^t e^{\kappa(z-t)}\sqrt{v(z)}dW_v(z). \tag{13.40}$$

The expectation $\mathbb{E}[v(t)|\mathcal{F}(0)] = \bar{c}(t,0)(\delta + \bar{\kappa}(t,0))$ and the variance, $\text{Var}[v(t)|\mathcal{F}(0)] = \bar{c}^2(t,0)(2\delta + 4\bar{\kappa}(t,0))$, with $\bar{c}(t,0)$, δ, and $\bar{\kappa}(t,0)$ given by (13.29). This gives us the following approximation:

Result 13.2.1 *The expectation $\mathbb{E}\left[\sqrt{v(t)}\right]$, with stochastic process $v(t)$ given by Equation (13.23), can be approximated by*

$$\mathbb{E}\left[\sqrt{v(t)}|\mathcal{F}(0)\right] \approx \sqrt{\bar{c}(t,0)(\bar{\kappa}(t,0) - 1) + \bar{c}(t,0)\delta + \frac{\bar{c}(t,0)\delta}{2(\delta + \bar{\kappa}(t,0))}} =: \mathcal{E}(t), \tag{13.41}$$

with $\bar{c}(t,0)$, δ, and $\bar{\kappa}(t,0)$ given in Lemma 13.2.1, and where κ, $\bar{v}$, γ and $v(0)$ are the parameters given in Equation (13.23).

Since Result 13.2.1 provides an explicit approximation in terms of a deterministic function for $\mathbb{E}\left[\sqrt{v(t)}\right]$, we are, in principle, able to derive the corresponding characteristic function.

Limits of the approximation for $\mathbb{E}[\sqrt{v(t)}]$

The parameters for which the expression under the square root in approximation (13.41) is nonnegative follow from:

$$\bar{c}(t,0)(\bar{\kappa}(t,0) - 1) + \bar{c}(t,0)\delta + \frac{\bar{c}(t,0)\delta}{2(\delta + \bar{\kappa}(t,0))} \geq 0. \tag{13.42}$$

Division by $\bar{c}(t,0) > 0$, gives $2\left(\bar{\kappa}(t,0) + \delta\right)^2 - 2(\bar{\kappa}(t,0) + \delta) + \delta \geq 0$. By setting $y = \bar{\kappa}(t,0) + \delta$, we find $2y^2 - 2y + \delta \geq 0$. The parabola is nonnegative for the discriminant $4 - 4 \cdot 2 \cdot \delta \leq 0$, so that the expression in (13.41) is nonnegative for $\delta = 4\kappa\bar{v}/\gamma^2 \geq \frac{1}{2}$.

> If the *Feller condition* is satisfied, the expression under the square-root is well-defined. If $8\kappa\bar{v}/\gamma^2 \geq 1$ but the Feller condition is not satisfied, the approximation is also valid. If the expression under the square-root in (13.41) becomes negative, it is beneficial for accuracy reasons to employ the approximations in Lemma 13.2.1 instead.

The approximation for $\mathbb{E}\left[\sqrt{v(t)}\right]$ in (13.41) is still nontrivial, and may cause difficulties when deriving the corresponding characteristic function. In order to find the coefficients of the characteristic function, a routine for numerically solving the corresponding ODEs has to be incorporated. Numerical integration, however, slows down the option valuation and would make the SDE model less attractive. Aiming to derive a closed-form expression for the characteristic function, expectation $\mathbb{E}\left[\sqrt{v(t)}\right]$ in (13.41) can be further approximated by a function of the following form:

$$\mathbb{E}\left[\sqrt{v(t)}\right] \approx a + be^{-ct} =: \widetilde{\mathcal{E}}(t), \qquad (13.43)$$

with constant values a, b, and c that are obtained by matching the functions $\mathcal{E}(t)$ in (13.41), and $\widetilde{\mathcal{E}}(t)$ in (13.43), for $t \to +\infty$, $t \to 0$, and $t = 1$,

$$\lim_{t \to +\infty} \mathcal{E}(t) = \sqrt{\bar{v} - \frac{\gamma^2}{8\kappa}} = a = \lim_{t \to +\infty} \widetilde{\mathcal{E}}(t),$$

$$\lim_{t \to 0} \mathcal{E}(t) = \sqrt{v(0)} = a + b = \lim_{t \to 0} \widetilde{\mathcal{E}}(t), \qquad (13.44)$$

$$\lim_{t \to 1} \mathcal{E}(t) = \mathcal{E}(1) = a + be^{-c} = \lim_{t \to 1} \widetilde{\mathcal{E}}(t).$$

The values a, b and c can be found as,

$$a = \sqrt{\bar{v} - \frac{\gamma^2}{8\kappa}}, \quad b = \sqrt{v(0)} - a, \quad c = -\log\left(b^{-1}(\mathcal{E}(1) - a)\right), \qquad (13.45)$$

where $\mathcal{E}(t)$ is given by (13.41). Further details on the validity of this approximation may be found in [Grzelak and Oosterlee, 2011].

Example 13.2.1 To assess numerically the quality of approximation (13.45) to $\mathbb{E}\left[\sqrt{v(t)}\right]$ in (13.27), a numerical experiment is performed (see the results in Figure 13.3).

For randomly chosen sets of parameters the approximation (13.45) resembles $\mathbb{E}\left[\sqrt{v(t)}\right]$ in (13.27) very well. ♦

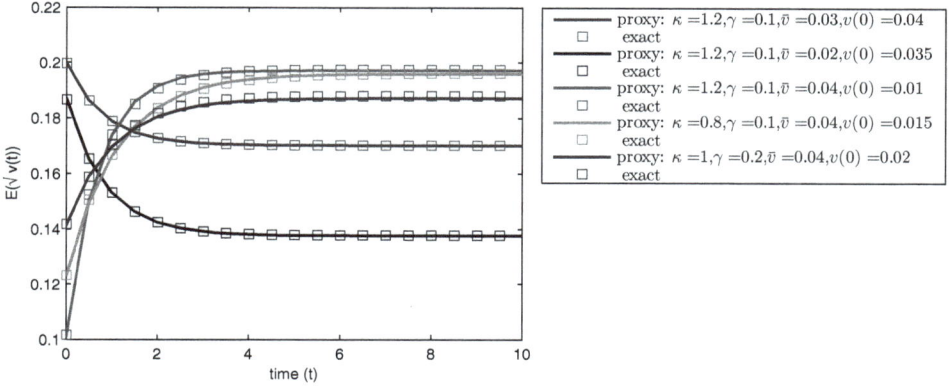

Figure 13.3: *Approximation* $\mathbb{E}\left[\sqrt{v(t)}\right] \approx a + be^{-ct}$ *(continuous line) versus the exact solution given in (13.27) (squares) for 5 random sets of* κ, γ, $\bar{v}$ *and* $v(0)$.

Characteristic function for the H1-HW model

The resulting approximate model, based on (13.26), (13.43), is called here *the H1-HW model*. Here, we derive the characteristic function for the H1-HW approximation of the HHW model. The nonaffine term in matrix (13.24) equals $\eta \rho_{x,r} \sqrt{v(t)}$ and it will be approximated by $\eta \rho_{x,r} \mathbb{E}\left[\sqrt{v(t)}\right]$.

For simplicity, we assume that the term-structure of the interest rate, $\theta(t)$, is constant, $\theta(t) = \theta$, see Equation (11.37), which is the Vašiček model in Section 11.3.1. The discounted characteristic function for the H1-HW model is of the form:

$$\phi_{\text{H1-HW}}(u; t, T) = \exp\left(\bar{A}(u, \tau) + \bar{B}(u, \tau) X(t) + \bar{C}(u, \tau) r(t) + \bar{D}(u, \tau) v(t)\right),$$
(13.46)

with initial conditions $\bar{A}(u, 0) = 0$, $\bar{B}(u, 0) = iu$, $\bar{C}(u, 0) = 0$, $\bar{D}(u, 0) = 0$. The characteristic function for the H1-HW model can be derived in closed-form, with the help of the following lemmas.

Lemma 13.2.2 (ODEs related to the H1-HW model) *The functions $\bar{B}(u, \tau)$, $\bar{C}(u, \tau)$, $\bar{D}(u, \tau)$, and $\bar{A}(u, \tau)$ for $u \in \mathbb{C}$ and $\tau \geq 0$ in (13.46) for the H1-HW model satisfy the following system of ODEs:*

$$\frac{d\bar{B}}{d\tau} = 0, \quad \bar{B}(u, 0) = iu,$$

$$\frac{d\bar{C}}{d\tau} = -1 - \lambda \bar{C} + \bar{B}, \quad \bar{C}(u, 0) = 0,$$

$$\frac{d\bar{D}}{d\tau} = \bar{B}(\bar{B}-1)/2 + (\gamma\rho_{x,v}\bar{B} - \kappa)\bar{D} + \gamma^2\bar{D}^2/2, \ \bar{D}(u,0) = 0,$$

$$\frac{d\bar{A}}{d\tau} = \lambda\theta\bar{C} + \kappa\bar{v}\bar{D} + \eta^2\bar{C}^2/2 + \eta\rho_{x,r}\mathbb{E}\left[\sqrt{v(t)}\right]\bar{B}\bar{C},$$

$$\bar{A}(u,0) = 0,$$

with $\tau = T - t$, and where κ, λ, and θ and η, $\rho_{x,r}$, and $\rho_{x,v}$ correspond to the parameters in the HHW model (13.23). ◂

The different terms can be found by straightforward application of the theory for AD processes. The following lemma gives the closed-form solution for the functions $\bar{B}(u,\tau)$, $\bar{C}(u,\tau)$, $\bar{D}(u,\tau)$ and $\bar{A}(u,\tau)$ in (13.46).

Lemma 13.2.3 (Characteristic function for the H1-HW model) *The solution of the ODE system in Lemma 13.2.2 is given by:*

$$\bar{B}(u,\tau) = iu,$$

$$\bar{C}(u,\tau) = (iu - 1)\lambda^{-1}(1 - e^{-\lambda\tau}),$$

$$\bar{D}(u,\tau) = \frac{1 - e^{-D_1\tau}}{\gamma^2(1 - ge^{-D_1\tau})}(\kappa - \gamma\rho_{x,v}iu - D_1),$$

$$\bar{A}(u,\tau) = \lambda\theta I_1(\tau) + \kappa\bar{v}I_2(\tau) + \frac{1}{2}\eta^2 I_3(\tau) + \eta\rho_{x,r}I_4(\tau),$$

with $D_1 = \sqrt{(\gamma\rho_{x,v}iu - \kappa)^2 - \gamma^2 iu(iu-1)}$, *and where* $g = \dfrac{\kappa - \gamma\rho_{x,v}iu - D_1}{\kappa - \gamma\rho_{x,v}iu + D_1}$.

The integrals $I_1(\tau)$, $I_2(\tau)$, and $I_3(\tau)$ admit an analytic solution, and $I_4(\tau)$ admits a semi-analytic solution:

$$I_1(\tau) = \frac{1}{\lambda}(iu - 1)\left(\tau + \frac{1}{\lambda}(e^{-\lambda\tau} - 1)\right),$$

$$I_2(\tau) = \frac{\tau}{\gamma^2}(\kappa - \gamma\rho_{x,v}iu - D_1) - \frac{2}{\gamma^2}\log\left(\frac{1 - ge^{-D_1\tau}}{1 - g}\right),$$

$$I_3(\tau) = \frac{1}{2\lambda^3}(i+u)^2\left(3 + e^{-2\lambda\tau} - 4e^{-\lambda\tau} - 2\lambda\tau\right),$$

$$I_4(\tau) = iu\int_0^\tau \mathbb{E}\left[\sqrt{v(T-z)}\right]\bar{C}(u,z)dz$$

$$= -\frac{1}{\lambda}(iu + u^2)\int_0^\tau \mathbb{E}\left[\sqrt{v(T-z)}\right](1 - e^{-\lambda z})dz. \quad \blacktriangleleft$$

Proof Due to the initial condition $\bar{B}(u,0) = iu$, obviously $\bar{B}(u,\tau) = iu$. For the second ODE, multiplying both sides by $e^{\lambda\tau}$, gives,

$$\frac{d}{d\tau}\left(e^{\lambda\tau}\bar{C}\right) = (iu - 1)e^{\lambda\tau}, \tag{13.47}$$

by integrating both sides and using the condition $\bar{C}(u,0) = 0$,
$$\bar{C}(u,\tau) = (iu-1)\lambda^{-1}\left(1-e^{-\lambda\tau}\right).$$
By using $\bar{a} := -\frac{1}{2}(u^2+iu)$, $\bar{b} := \gamma\rho_{x,v}iu - \kappa$, and $\bar{c} := \frac{1}{2}\gamma^2$, the ODEs for $\bar{D}(u,\tau)$ and $I_2(\tau)$ are given by the following Riccati equation:

$$\frac{d\bar{D}}{d\tau} = \bar{a} + \bar{b}\bar{D} + \bar{c}\bar{D}^2, \quad \bar{D}(u,0) = 0, \tag{13.48}$$

$$I_2(\tau) = \kappa\bar{v}\int_0^\tau \bar{D}(u,z)dz. \tag{13.49}$$

Equations (13.48) and (13.49) are of the same form as those in the original Heston dynamics [Heston, 1993]. Their solutions are thus given by:

$$\bar{D}(u,\tau) = \frac{-\bar{b}-D_1}{2c(1-Ge^{-D_1\tau})}(1-e^{-D_1\tau}),$$

$$I_2(\tau) = \frac{1}{2\bar{c}}\left((-\bar{b}-D_1)\tau - 2\log\left(\frac{1-Ge^{-D_1\tau}}{1-G}\right)\right),$$

with $D_1 = \sqrt{\bar{b}^2 - 4\bar{a}\bar{c}}$, $G = \frac{-\bar{b}-D_1}{-\bar{b}+D_1}$.

The evaluation of the integrals $I_1(\tau)$, $I_3(\tau)$ and $I_4(\tau)$ is straightforward. The proof is finished by the corresponding substitutions. ∎

Note that by taking $\mathbb{E}\left[\sqrt{v(T-s)}\right] \approx a + be^{-c(T-s)}$, with a, b, and c as given in (13.43), a closed-form expression is obtained:

$$I_4(\tau) = -\frac{1}{\lambda}(iu+u^2)\left[\frac{b}{c}\left(e^{-ct} - e^{-cT}\right) + a\tau + \frac{a}{\lambda}\left(e^{-\lambda\tau}-1\right) \right.$$
$$\left. + \frac{b}{c-\lambda}e^{-cT}\left(1-e^{-\tau(\lambda-c)}\right)\right]. \tag{13.50}$$

The above derivation for the H1-HW model is based on a zero correlation between the variance and interest rate processes. A generalization of the H1-HW model to a *full matrix of nonzero correlations* between the processes can be done is a similar way, by similar approximations of the nonaffine covariance matrix terms by means of their respective expectations. However, it is important to quantify approximation errors that will be made, and their impact on the option prices.

Remark 13.2.2 (Error analysis for the approximate Heston model)
A better theoretical understanding of the difference between the full-scale and the approximate H1-HW model may be based on the corresponding option pricing PDEs, where classical PDE error analysis to examine the quality of the approximation has been performed in [Guo et al., 2013]. In the case of these Heston Hull-White models, the option pricing PDE will be three-dimensional. By changing measures, from the spot measure to the T-forward measure, the pricing PDE reduces to a two-dimensional PDE to facilitate the analysis. Elements from probability theory as well as PDE theory and numerical analysis were employed.▲

13.2.3 Monte Carlo simulation of hybrid Heston SDEs

We consider the Monte Carlo simulation of the following general hybrid system of SDEs:

$$\mathrm{d}S(t)/S(t) = r(t)\mathrm{d}t + b_x(t, v(t))\mathrm{d}W_x(t),$$
$$\mathrm{d}v(t) = a_v(t, v(t))\mathrm{d}t + b_v(t, v(t))\mathrm{d}W_v(t), \quad (13.51)$$
$$\mathrm{d}r(t) = a_r(t, r(t))\mathrm{d}t + b_r(t, r(t))\mathrm{d}W_r(t),$$

with $S(t_0) > 0$, $v(t_0) > 0$, $r(t_0) \in \mathbb{R}$, $\mathrm{d}W_x(t)\mathrm{d}W_v(t) = \rho_{x,v}\mathrm{d}t$, $\mathrm{d}W_x(t)\mathrm{d}W_r(t) = \rho_{x,r}\mathrm{d}t$, $\mathrm{d}W_r(t)\mathrm{d}W_v(t) = 0$, and functions $b_x(\cdot)$, $a_v(\cdot)$, $b_v(\cdot)$, $a_r(\cdot)$ and $b_r(\cdot)$ satisfying the usual growth conditions (see, for example, [Shreve, 2004]).

Remark 13.2.3 (Correlation) *With a zero correlation parameter, sampling from a joint distributions can be performed by sampling the marginal distributions independently, i.e., we don't need to sample from the joint distribution.*

Within the SZHW model, the variance and the interest rate are governed by mean-reverting, normally distributed stochastic processes and sampling from their joint distribution is therefore rather easy [van Haastrecht et al., 2014].

However, sampling from the joint distribution is not at all standard, when we deal with a lognormal variance distribution which is correlated with a normally distributed interest rate process. ▲

Remark 13.2.4 (Characteristic function) *The model setting in (13.51) is rather general and many choices for the volatility and drift coefficients can be made. On the other hand, if we wish to make, next to Monte Carlo simulation, use of a characteristic function and thus Fourier techniques for pricing, the number of choices for the model parameters is limited. Staying with the class of affine models would require that squares of the volatility coefficients and all possible combinations, $b_i^2(t, v(t))$, $b_i(t, v(t))b_j(t, v(t))$, $i \neq j$, $i, j \in \{x, v, r\}$, are linear in the state variables. This restriction is severe, as only a few models would be available, and this also motivates the H1-HW model, defined earlier in this chapter.* ▲

We first define a time grid, $t_i = i\frac{T}{m}$, with $i = 0, \ldots, m$, and $\Delta t_i = t_{i+1} - t_i$. A very basic strategy to simulate the system in (13.51) is by means of an Euler discretization. This gives rise, using $X(t) = \log(S(t))$, to the following, discrete system:

$$x_{i+1} = x_i + \left[r_i - \frac{1}{2}b_x^2(t_i, v_i)\right]\Delta t_i + b_x(t_i, v_i)\sqrt{\Delta t_i}Z_x,$$
$$v_{i+1} = v_i + a_v(t_i, v_i)\Delta t_i + b_v(t_i, v_i)\sqrt{\Delta t_i}Z_v, \quad (13.52)$$
$$r_{i+1} = r_i + a_r(t_i, r_i)\Delta t_i + b_r(t_i, r_i)\sqrt{\Delta t_i}Z_r,$$

with $Z_i \sim \mathcal{N}(0, 1)$, that are correlated as the Brownian motions in (13.51).

Although the discrete system in (13.52) is commonly used for pricing by Monte Carlo simulation, issues may be encountered when one of the processes is of CIR-type, like in the Heston model. The Euler disretization scheme may lead to negative values, making the Monte Carlo simulation results unrealistic.

A more general framework for the Monte Carlo simulation of the system (13.51) relies on the assumption that the marginal distributions of the variance process $v(t)$ and the interest rate $r(t)$ are known. Sampling from these distributions can then be performed efficiently using available software packages.

A formulation in terms of the independent Brownian motions, based on the Cholesky decomposition of correlation matrix $\mathbf{C}$, as $\mathbf{C} = \mathbf{L}\mathbf{L}^T$, and

$$\mathbf{C} = \begin{bmatrix} 1 & 0 & \rho_{x,r} \\ 0 & 1 & \rho_{x,v} \\ \rho_{x,r} & \rho_{x,v} & 1 \end{bmatrix}, \quad \mathbf{L} = \begin{bmatrix} 1 & 0 & 0 \\ 0 & 1 & 0 \\ \rho_{x,r} & \rho_{x,v} & \sqrt{1 - \rho_{x,r}^2 - \rho_{x,v}^2} \end{bmatrix}, \quad (13.53)$$

reformulates, for $X(t) = \log S(t)$, the system in (13.51), as:

$$dr(t) = a_r(t, r(t))dt + b_r(t, r(t))d\widetilde{W}_r(t),$$
$$dv(t) = a_v(t, v(t))dt + b_v(t, v(t))d\widetilde{W}_v(t),$$

and the log-stock dynamics,

$$dX(t) = \left(r(t) - \frac{1}{2}b_x^2(t, v(t))\right) dt + \rho_{x,r} b_x(t, v(t))d\widetilde{W}_r(t)$$
$$+ \rho_{x,v} b_x(t, v(t))d\widetilde{W}_v(t) + \sqrt{1 - \rho_{x,r}^2 - \rho_{x,v}^2}\, b_x(t, v(t))d\widetilde{W}_x(t).$$

> One may wonder whether the expression under the square root in Equation (13.53) may become negative. In such a case, however, the correlation matrix is not positive definite, and is therefore not a valid correlation matrix.

Integration over a time interval $[t_i, t_{i+1}]$, gives,

$$x_{i+1} = x_i + \int_{t_i}^{t_{i+1}} \left(r(z) - \frac{1}{2}b_x^2(z, v(z))\right) dz$$
$$+ \rho_{x,r} \int_{t_i}^{t_{i+1}} b_x(z, v(z))d\widetilde{W}_r(z) + \rho_{x,v} \int_{t_i}^{t_{i+1}} b_x(z, v(z))d\widetilde{W}_v(z)$$
$$+ \sqrt{1 - \rho_{x,r}^2 - \rho_{x,v}^2} \int_{t_i}^{t_{i+1}} b_x(z, v(z))d\widetilde{W}_x(z). \quad (13.54)$$

The resulting representation for x_{i+1} forms the basis for a general Monte Carlo simulation procedure.

The Heston Hull-White model

By specific choices for the functions in (13.51), well-known SDE systems can be recognized. With the following set:

$$b_x(t, v(t)) = \sqrt{v(t)},$$
$$a_v(t, v(t)) = \kappa(\bar{v} - v(t)), \quad b_v(t, v(t)) = \gamma\sqrt{v(t)}, \quad (13.55)$$
$$a_r(t, r(t)) = \lambda(\theta(t) - r(t)), \quad b_r(t, r(t)) = \eta,$$

System (13.51) is the Heston Hull-White model from (13.23). The Monte Carlo technique for the standard Heston model presented in (9.60) can directly be employed for the discretization of x_{i+1} in (13.54), i.e.,

$$x_{i+1} = x_i + \int_{t_i}^{t_{i+1}} \left(r(z) - \frac{1}{2}v(z)\right) dz + \rho_{x,r} \int_{t_i}^{t_{i+1}} \sqrt{v(z)} d\widetilde{W}_r(z)$$
$$+ \frac{\rho_{x,v}}{\gamma}\left(v_{i+1} - v_i - \kappa\bar{v}\Delta t_i + \kappa \int_{t_i}^{t_{i+1}} v(z) dz\right)$$
$$+ \sqrt{1 - \rho_{x,r}^2 - \rho_{x,v}^2} \int_{t_i}^{t_{i+1}} \sqrt{v(z)} d\widetilde{W}_x(z).$$

Collecting the different terms of the discretization for x_i gives:

$$x_{i+1} = x_i + k_0 + k_1 \int_{t_i}^{t_{i+1}} v(z) dz + \int_{t_i}^{t_{i+1}} r(z) dz + k_2(v_{i+1} - v_i)$$
$$+ \rho_{xr} \int_{t_i}^{t_{i+1}} \sqrt{v(z)} d\widetilde{W}_r(z) + k_3 \int_{t_i}^{t_{i+1}} \sqrt{v(z)} d\widetilde{W}_x(z), \quad (13.56)$$

with

$$k_0 = -\frac{\rho_{x,v}}{\gamma}\kappa\bar{v}\Delta t_i, \quad k_1 = \kappa k_2 - \frac{1}{2}, \quad k_2 = \frac{\rho_{x,v}}{\gamma}, \quad k_3 = \sqrt{1 - \rho_{x,r}^2 - \rho_{x,v}^2}.$$

The integrals in the discretization above are difficult to determine analytically. We therefore apply the Euler discretization scheme to (13.56), resulting in the following approximation:

$$x_{i+1} \approx x_i + k_0 + k_1 v_i \Delta t_i + r_i \Delta t_i + k_2(v_{i+1} - v_i)$$
$$+ \rho_{x,r}\sqrt{v_i}\left(\widetilde{W}_r(t_i) - \widetilde{W}_r(t_i)\right) + k_3\sqrt{v_i}\left(\widetilde{W}_x(t_{i+1}) - \widetilde{W}_x(t_i)\right). \quad (13.57)$$

with $\widetilde{W}_r(t_{i+1}) - \widetilde{W}_r(t_i) \stackrel{d}{=} \sqrt{\Delta t_i}\widetilde{Z}_r$ and $\widetilde{W}_x(t_{i+1}) - \widetilde{W}_x(t_i) \stackrel{d}{=} \sqrt{\Delta t_i}\widetilde{Z}_x$, with $\widetilde{Z}_r$ and $\widetilde{Z}_x$ i.i.d. $\mathcal{N}(0,1)$ variables,

$$x_{i+1} \approx x_i + k_0 + k_1 v_i \Delta t_i + r_i \Delta t_i + k_2(v_{i+1} - v_i)$$
$$+ \rho_{xr}\sqrt{v_i \Delta t_i}\widetilde{Z}_r + k_3\sqrt{v_i \Delta t_i}\widetilde{Z}_x. \quad (13.58)$$

For the dynamics of the interest rate process,
$$dr(t) = \lambda(\theta(t) - r(t))dt + \eta d\widetilde{W}_r(t),$$
the Euler discretization gives rise to the following approximation:
$$r_{i+1} \approx r_i + \lambda\theta(t_i)\Delta t_i - \lambda r_i \Delta t_i + \eta\sqrt{\Delta t_i}\widetilde{Z}_r. \tag{13.59}$$

Summarizing, the Monte Carlo scheme for the HHW model can be discretized as follows:
$$v_{i+1} = \bar{c}(t_{i+1}, t_i)\chi^2(\delta, \bar{\kappa}(t_{i+1}, t_i)),$$
$$r_{i+1} \approx r_i + \lambda\theta(t_i)\Delta t_i - \lambda r_i \Delta t_i + \eta\sqrt{\Delta t_i}\widetilde{Z}_r,$$
$$x_{i+1} \approx x_i + k_0 + (k_1\Delta t_i + k_2)v_i + r_i\Delta t_i + k_2 v_{i+1}$$
$$+ \sqrt{v_i \Delta t_i}(\rho_{x,r}\widetilde{Z}_r + k_3 \widetilde{Z}_x),$$

with,
$$k_0 = -\frac{\rho_{x,v}}{\gamma}\kappa\bar{v}\Delta t_i, \quad k_1 = \kappa k_2 - \frac{1}{2}, \quad k_2 = \frac{\rho_{x,v}}{\gamma}, \quad k_3 = \sqrt{1 - \rho_{x,r}^2 - \rho_{x,v}^2},$$

and
$$\bar{c}(t_{i+1}, t_i) = \frac{\gamma^2}{4\kappa}(1 - e^{-\kappa(t_{i+1} - t_i)}), \quad \delta = \frac{4\kappa\bar{v}}{\gamma^2},$$
$$\bar{\kappa}(t_{i+1}, t_i) = \frac{4\kappa e^{-\kappa(t_{i+1} - t_i)}}{\gamma^2(1 - e^{-\kappa(t_{i+1} - t_i)})}v_i,$$

with $\widetilde{Z}_r, \widetilde{Z}_x$ independent $\mathcal{N}(0,1)$ random variables; $\chi^2(\delta, \bar{\kappa}(t_{i+1}, t_i))$ a random variable with noncentral chi-squared distribution with δ degrees of freedom and noncentrality parameter $\bar{\kappa}(t_{i+1}, t_i)$.

The simulation scheme depends on the availability of the noncentral chi-squared distribution, which may be available in statistical packages.[2] Alternatively, the QE scheme from Section 9.3.4 may be employed.

13.2.4 Numerical experiment, HHW versus SZHW model

We compare the H1-HW model, with (13.26), (13.43), to the SZHW model (presented in Section 13.1.3). For both models the interest rate process $r(t)$ is driven by a correlated normally distributed short-rate model, so that we only need to focus on the differences between the volatility processes.

The volatility in the Schöbel-Zhu model is governed by the normally distributed Ornstein-Uhlenbeck volatility process $\sigma(t)$, whereas in the Heston

---

[2]In MATLAB it can be used via ncx2rnd().

model the volatility is governed by $\sqrt{v(t)}$, with $v(t)$ distributed as $\bar{c}(t,0)$ times a noncentral chi-squared random variable $\chi^2(\delta, \bar{\kappa}(t,0))$.

We first determine under which conditions $\sqrt{v(t)}$ is approximately a normal distribution, as $\sigma(t)$ in the Schöbel-Zhu model is normally distributed.

For any time, $t > 0$, the square-root of the variance process $v(t)$ can be approximated by:

$$\sqrt{v(t)} \approx \mathcal{N}\left(\mathcal{E}(t), \bar{c}(t,0) - \frac{\bar{c}(t,0)\delta}{2(\delta + \bar{\kappa}(t,0))}\right), \qquad (13.60)$$

with $\bar{c}(t,0)$, δ and $\bar{\kappa}(t,0)$ from (13.29) and $\mathcal{E}(t)$ from (13.41).

The normal approximation (13.60) is a satisfactory approximation for either a large number of degrees of freedom δ, or a large noncentrality parameter $\bar{\kappa}(t,0)$. Particularly, $\delta \gg 0$ implies that $4\kappa\bar{v} \gg \gamma^2$, which is closely related to the Feller condition, $2\kappa\bar{v} > \gamma^2$. The Heston model thus has a similar volatility structure as the Schöbel-Zhu model, when the Feller condition is satisfied.

Figure 13.4 (left side picture) confirms this observation. The volatilities for the Heston and Schöbel-Zhu models differ significantly, when the Feller condition does not hold, as the volatility in the Heston model gives rise to much heavier tails than those in the Schöbel-Zhu model. This may have an effect when calibrating the models to the market data with significant implied volatility smile or skew (right side figure).

We also check the model's performance during calibration to real financial market data. The SZHW and the H1-HW model (i.e. the affine Heston with the Hull-White short-rate process) are calibrated to the implied volatilities from

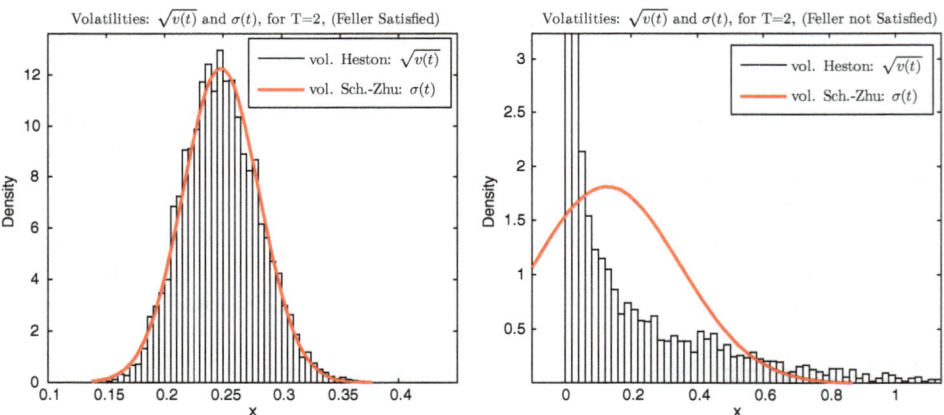

Figure 13.4: *Histogram for $\sqrt{v(t)}$ (the Heston model) and density for $\sigma(t)$ (the Schöbel-Zhu model), with $T = 2$. Left: Feller condition satisfied, i.e. $\kappa = 1.2$, $v(0) = \bar{v} = 0.063$, $\gamma = 0.1$; Right: Feller condition not satisfied, $\kappa = 0.25$, $v(0) = \bar{v} = 0.063$, $\gamma = 0.63$, as in [Antonov, 2007].*

Table 13.2: *Calibration results for the Schöbel-Zhu hybrid model (SZHW) and the H1-HW hybrid.*

T	Strike	Market	SZHW	H1-HW	err.(SZHW)	err.(H1-HW)
	40%	57.61	54.02	57.05	3.59 %	-0.56 %
T=6m	100%	22.95	25.21	21.57	-2.26 %	-1.38 %
	120%	15.9	18.80	16.38	-2.90 %	0.48 %
	40%	48.53	47.01	48.21	1.52 %	0.32 %
T=1y	100%	24.49	24.97	24.28	-0.48 %	0.21 %
	120%	19.23	19.09	19.14	0.14 %	0.09 %
	40%	36.76	36.15	36.75	0.61 %	0.01 %
T=10y	100%	29.18	29.47	29.18	-0.29 %	0.00 %
	120%	27.66	27.93	27.62	-0.27 %	0.04 %

the S&P500 (27/09/2010) with spot price at 1145.88. For both models, the correlation between the stock and interest rates, $\rho_{x,r}$, is set to +30%.

The calibration results, in Table 13.2, confirm that the H1-HW model is more *flexible* than the Schöbel-Zhu Hull-White model. The difference is pronounced for large strikes at which the error for the affine Heston hybrid model is up to 20 times lower than for the SZHW hybrid model.

13.3 CVA exposure profiles and hybrid models

As mentioned in Section 12.3, credit *exposure* is defined as the potential future losses without any recovery, see (12.51),

$$E(t) := \max(V(t), 0), \quad V(t) := \mathbb{E}^{\mathbb{Q}}\left[\frac{M(t)}{M(T)} H(T, \cdot) \Big| \mathcal{F}(t)\right],$$

Exposure evolves over time as the market moves with volatility, and it often cannot be expressed in closed-form.

In order to quantify the exposure, the metrics *expected exposure*, EE (12.53), and *potential future exposure*, PFE (12.54), of the future losses in practice, the exposure profile needs to be computed for a large number of scenarios on a set of time steps. This is one of the involved parts in computing the Credit Value Adjustment, CVA. Numerical methods are therefore used to estimate, by means of a MC simulation, the values, the exposure, and their distribution during the life of the financial contracts. These numerical methods contain essentially three stages, i.e. a forward stage for generating future asset path scenarios, a backward stage to value the financial derivatives (the MtM values of the contract) and a backward stage to calculate the exposure along the generated future asset paths.

A general Monte Carlo framework is formulated by Pykhtin and Zhu [2007] for the computation of exposure profiles for over-the-counter (OTC) derivative products.

The forward Monte Carlo method typically generates the discrete asset paths from initial time up to maturity time. Along these paths, the option values are subsequently determined, at each *CVA monitoring date*. Calculation of the exposure profiles asks for efficient numerical methods, as the computational demand grows rapidly with respect to the number of MC paths. The *finite difference method* [de Graaf et al., 2014], approximating numerical solutions to the governing partial differential equation, may be suitable as it typically results in approximate option prices on a grid of underlying values. This feature may be exploited in the EE context, as all grid points can then be used to generate option probability density functions.

Computational complexity is an issue for the computation of the CVA of a whole portfolio, as there are then multiple financial derivatives in the exposed portfolio. Additional market factors in the asset dynamics, such as a *stochastic asset volatility and a stochastic interest rate*, also increase the computational effort. Monte Carlo variants that are based on regression to approximate expectation values at future time points, see, for example, [Longstaff and Schwartz, 2001; Jain and Oosterlee, 2015; Feng and Oosterlee, 2017; Feng et al., 2016; Feng, 2017] are most often used in practice and appear very useful because of the potential high-dimensionality of this valuation problem in risk management.

13.3.1 CVA and exposure

As explained in Section 12.3, CVA represents the price of counterparty credit risk. It is based on an expected value (the expected exposure, EE) which is computed under the risk-neutral measure. We will focus on the computation of CVA, and we will also compute EE and PFE under the risk-neutral measure. A computation of PFE under the real-world measure can be found, for example, in [Feng et al., 2016].

The three key elements in the calculation of CVA are the loss given default, the discounted exposure and the survival/default probability of the counterparty (remember the definition of survival probability in Definition 1.1.1). By assuming independence among them, the calculation formula of CVA is given by Equation (12.61), i.e.,

$$\mathrm{CVA}(t,T) \approx \underbrace{(1-R)}_{\mathrm{LGD}} \sum_{i=1}^{M} \mathrm{EE}(t,T_i) \cdot \underbrace{(F_{t_D}(T_i) - F_{t_D}(T_{i-1}))}_{\mathrm{PD}},$$

where the LGD is assumed to be a fixed ratio based on market information, and the probability of default (PD) can often be obtained via the survival probability curve on the CDS market [Brigo and Mercurio, 2007].

13.3.2 European and Bermudan options example

We will study the CVA, EE and PFE for two different options. *Bermudan options* have not yet been discussed in detail in this book. These options can be exercised early, i.e. prior to the maturity time T, at certain, pre-specified dates in the future. We denote this set of early-exercise dates by $\mathcal{T}_E$.

The received payoff from *immediate exercise* of the option at a time point t_i is given by

$$H(t_i, S(t_i)) := \max(\bar{\alpha}(S(t_i) - K), 0), \quad \text{with} \quad \begin{cases} \bar{\alpha} = 1, & \text{for a call;} \\ \bar{\alpha} = -1, & \text{for a put,} \end{cases} \quad (13.61)$$

where K is the strike price and $S(t_i)$ is the underlying asset value at time t_i.

The option holder determines the exercise strategy, aiming for "an optimal profit". We assume here that the option holder is not influenced by the credit quality of the option writer when making the decision to exercise early. This can be generalized, see, for example, [Feng, 2017].

> The *optimal stopping time*, see the definition in Definition 10.2.1, which is denoted by t_s, is the optimal time for the holder to exercise the option. Exercising the option early should maximize the expected payoff at time $t = 0$, i.e.
> $$V^{\text{Berm}}(t_0) = \max_{t_s \in \mathcal{T}_E} \mathbb{E}\left[\frac{M(t_0)}{M(t_s)} \cdot H(t_s, S(t_s)) \Big| \mathcal{F}(t_0)\right]. \quad (13.62)$$

The technique for pricing Bermudan options by a Monte Carlo method is to determine the optimal exercise strategy for each generated asset path. At each exercise date t_i, the Monte paths will reach some asset value $S(t_i)$. The option holder compares the possible payoff from the immediate exercise at t_i, with *the expected future payoff*, which is based on the *continuation of the option*. This way, the optimal exercise strategy, as for each time point t_i the specific asset value $S^*(t_i)$ is determined where for higher asset values a different decision ("exercise", or "not exercise") is made than for the lower asset values.

The *continuation value* of the option, at time t_i, can be expressed as the *conditional expectation of the discounted option value* at a later time t_{i+1}. The continuation value of the option can thus be written as

$$c(t_i) := \mathbb{E}\left[\frac{M(t_i)}{M(t_{i+1})} V(t_{i+1}) \Big| \mathcal{F}(t_i)\right], \quad (13.63)$$

where $V(t_{i+1})$ is the option value at time t_{i+1}.

> The pricing of Bermudan options, by means of so-called backward dynamic programming, gives rise to a *backward induction algorithm*, as in [Longstaff and Schwartz, 2001], and can be expressed by:
> $$V^{\text{Berm}}(t_i) = \begin{cases} H(t_i, S(t_m)), & \text{for } t_m = T, \\ \max\{c(t_i), H(t_i, S(t_i))\}, & \text{for } t_i \in \mathcal{T}_E, \\ c(t_i), & \text{for } t_i \in \mathcal{T} \setminus \mathcal{T}_E. \end{cases} \quad (13.64)$$

Similar to pricing Bermudan options, the exposure profile at future time points of *European options* can be determined based on simulation. The European option value at the expiry time T equals the payoff function, $V(t_m) = H(t_m, S(t_m))$. At the time points $t_i < T$, the value of the European option is equal to the discounted conditional expected payoff, i.e.,

$$V(t_i) := \mathbb{E}\left[\frac{M(t_i)}{M(t_m)} \cdot H(t_m, S(t_m)) \big| \mathcal{F}(t_i)\right], \quad (13.65)$$

where $H(t_m, S(t_m))$ is the payoff at time $t_m = T$.

> By the tower property of expectations, the future option value at $t_i < t_m$, can also be calculated in a backward iteration as,
>
> $$V(t_i) = \mathbb{E}\left[\frac{M(t_i)}{M(t_{i+1})} \cdot \mathbb{E}\left[\frac{M(t_{i+1})}{M(t_m)} \cdot H(t_m, S(t_m)) \big| \mathcal{F}(t_{i+1})\right] \big| \mathcal{F}(t_i)\right]$$
>
> $$= \mathbb{E}\left[\frac{M(t_i)}{M(t_{i+1})} \cdot V(t_{i+1}) \big| \mathcal{F}(t_i)\right] = c(t_i). \quad (13.66)$$

Impact of stochastic volatility and interest rate

Much research, in terms of efficient Monte Carlo methods [Longstaff and Schwartz, 2001; Jain and Oosterlee, 2015] and in terms of efficient PDE methods [Zvan et al., 1998; d'Halluin et al., 2004; Forsyth and Vetzal, 2002; In 't Hout, 2017] has been dedicated to valuing Bermudan options in an accurate and efficient way.

The exposure at path j and time point t_i is computed by

$$E_i^j = \begin{cases} 0, & \text{if the option is exercised,} \\ c(t_i), & \text{if the option is continued.} \end{cases}$$

The values of the EE and PFE at time t_i can be approximated by,

$$\text{EE}(t_0, t_i) \approx \frac{1}{N}\sum_{j=1}^{N} E_i^j,$$

$$\text{PFE}(t_0, t_i) \approx \inf\left\{x \in \mathbb{R} : 0.974 \leq F_{E(t_i)}(x)\right\},$$

with $\text{PFE}(t_0, t_i)$ as in (12.54). For paths at which the option has been exercised at some time point t_i, the exposure on these paths for $t > t_i$ equals zero, as the option has terminated.

We analyze the impact of the stochastic volatility and the stochastic interest rate on the exposure profiles and on the CVA, see also [Feng and Oosterlee, 2017; Feng, 2017]. We consider the effect of including the stochastic volatility and stochastic interest rate on the metrics of the future losses (i.e. CVA, EE, PFE). A stochastic volatility process may help to explain the implied volatility surface in the financial derivatives market (such as the volatility smile), and randomness in the interest rate process may give a significant contribution to the prices, especially of long-term financial derivatives [Lauterbach and Schultz,

1990]. The models chosen here are the Black-Scholes (BS), the Heston model, but also the Heston Hull-White (HHW) model [Grzelak and Oosterlee, 2011], and the Black-Scholes Hull-White (BSHW) model. The latter two processes have been discussed in this chapter.

Let the collection of equally-spaced discrete monitoring dates be,

$$\mathcal{T} = \{0 = t_0 < t_1 < \cdots < t_m = T, \Delta t = t_{i+1} - t_i\}.$$

The Quadratic Exponential (QE) scheme is employed within the MC method for an accurate simulation of the Heston-type models.

CVA is computed here via formula (12.61), with LGD= 1. The options will be valued at the monitoring dates to determine the exposure profiles. Simulation is done with $N = 10^6$ MC paths and $\Delta t = 0.1$. We employ a small time step when simulating the asset dynamics to enhance the accuracy of the CVA calculation, and assume that the Bermudan option can only be exercised at certain dates, i.e. $\mathcal{T}_E \subset \mathcal{T}$.

The survival probability is assumed to the independent of the exposure with a constant intensity, expressed by

$$F_{t_D}(t) = 1 - \exp(-0.03t),$$

for the period $t \in [0, T]$.

The following parameters for the models are used, in [Feng and Oosterlee, 2017; Feng, 2017],

$$S_0 = 100, v_0 = 0.05, r_0 = 0.02, \kappa = 0.3, \gamma = 0.6, \bar{v} = 0.05, \lambda = 0.01, \eta = 0.01,$$
$$\theta = 0.02, \rho_{x,v} = -0.3, \rho_{x,r} = 0.6. \tag{13.67}$$

We set the parameters of the models such that the values of European put options with expiry date T have the same prices under all models. For example, under the Black-Scholes model the *implied interest rate*, i.e. $r_T = -\log(P(0,T))/T$ is used, and the *implied volatility* is also computed. Under the Heston model, the parameters of the Heston process are the above mentioned parameters, and the corresponding interest rate is computed. Under the BSHW model, the parameters of the Hull-White process are the same as above, and a suitable volatility is determined.

As in [Feng and Oosterlee, 2017] we use a *percentage CVA*, defined as $\left(100 \cdot \frac{\text{CVA}(t_0)}{V(0)}\right)\%$. Table 13.3 presents the percentage CVA for European put options with two maturity times, $T = 1$, $T = 5$, for the strike prices $K = \{80, 100, 120\}$. It can be observed that the CVA percentage does not change much with the strike price. However, European options with maturity $T = 5$ exhibit a higher CVA percentage than those with maturity $T = 1$. With the chosen parameters, there is only a small impact of the stochastic volatility and the stochastic interest rate on the CVA percentage.

Figure 13.5 presents the EE and PFE values in time for a Bermudan put option, which is at-the-money. In Figure 13.5(b) the PFE values for the HHW model are relatively close to those of the Heston model, and the PFE values

Table 13.3: *CVA(%) of European options with $T = 1$, $T = 5$ and strike values $K = \{80, 100, 120\}$.*

	K/S_0	European option, CVA (%)			
		BS	Heston	BSHW	HHW
$T=1$	80%	2.951 (0.010)	2.959 (0.003)	2.953 (0.005)	2.949 (0.005)
	100%	2.956 (0.011)	2.958 (0.003)	2.952 (0.002)	2.952 (0.002)
	120%	2.955 (0.002)	2.959 (0.001)	2.953 (0.001)	2.952 (0.001)
$T=5$	80%	13.93 (0.036)	13.94 (0.021)	13.88 (0.016)	13.93 (0.027)
	100%	13.95 (0.039)	13.96 (0.010)	13.90 (0.003)	13.94 (0.018)
	120%	13.92 (0.010)	13.95 (0.007)	13.90 (0.005)	13.94 (0.010)

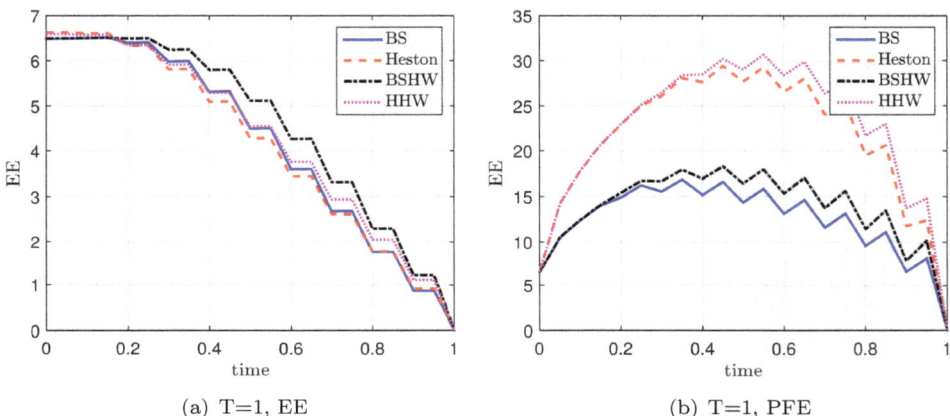

(a) T=1, EE   (b) T=1, PFE

Figure 13.5: *Impact of stochastic volatility and interest rate on EE and PFE with different asset dynamics, at the money $K = 100$.*

for the BSHW model are very similar to those of the BS model. Under the model assumptions and parameters, the stochastic volatility has a significant contribution to the PFE values, compared to the stochastic interest rate. Adding stochastic volatility thus has more impact on the right-side tails of the exposure profiles than on the EE values.

Figures 13.5(a) and (b) show that the stochastic interest rate increases the future EE values of Bermudan options, while the stochastic volatility has the opposite effect.

The stochastic interest rate plays a significant role in the case of a longer maturity derivative, and results in increasing PFE profiles; stochastic asset volatility appears to have an effect on PFE values at the early stages of a contract.

13.4 Exercise set

Exercise 13.1 In this exercise, we will apply the COS method to value a plain vanilla call option under the Black-Scholes Hull-White model (BSHW). The state vector $[S, r]^T$ is assumed to satisfy, under the risk neutral measure,

$$dS(t) = r(t)S(t)dt + \sigma S(t)dW_x(t),$$
$$dr(t) = \lambda(\theta(t) - r(t))dt + \eta dW_r(t),$$

with parameters σ, λ, η, further, $dW_x(t), dW_r(t) = \rho dt$ and a deterministic function $\theta(t)$ given.

a. Apply the transformations $\tilde{r}(t) = r(t) - \psi(t)$, where

$$\psi(t) = e^{-\lambda t}r_0 + \lambda \int_0^t e^{-\lambda(t-z)}\theta(z)dz. \tag{13.68}$$

Use the following variables, $X(t) = \log(S(t)), \tilde{x}(t) = X(t) - \Psi(t)$, where $\Psi(t) = \int_0^t \psi(z)dz$, to show that

$$d\tilde{x}(t) = \left(\tilde{r}(t) - \frac{1}{2}\sigma^2\right)dt + \sigma dW_x(t),$$
$$d\tilde{r}(t) = -\lambda \tilde{r}(t)dt + \eta dW_r(t).$$

b. Deduce that the discounted characteristic function for $\mathbf{X}(T) = [X(T), r(T)]^T$, with starting point $\mathbf{X}(t) = [X(t), r(t)]^T$ at t, has the form:

$$\phi(\mathbf{u}, \mathbf{X}, t, T) = e^{-\int_t^T \psi(z)dz + i\mathbf{u}^T[\Psi(T), \psi(T)]^T} e^{\bar{A}(\mathbf{u}, \tau) + \bar{B}_x(\mathbf{u}, \tau)\tilde{x}(t) + \bar{B}_r(\mathbf{u}, \tau)\tilde{r}(t)},$$

where $\tau = T - t$, $\mathbf{u} = [u, 0]^T$, and show that $\bar{A}$ and $\bar{B} = [\bar{B}_x, \bar{B}_r]^T$ satisfy a system of ODEs of the form:

$$\frac{dA}{d\tau} = -r_0 + \bar{B}^T a_0 + \frac{1}{2}\bar{B}^T c_0 \bar{B},$$
$$\frac{dB}{d\tau} = -r_1 + a_1^T \bar{B} + \frac{1}{2}\bar{B}^T c_1 \bar{B},$$

with boundary conditions $A(u, 0) = 0, B_x(u, 0) = iu, B_r(u, 0) = 0$. Define all terms.

c. Solve this system of ODEs, and derive the characteristic function for constant $\theta(t) = \theta$.

d. Implement the COS method, for the BSHW model for a European call option with payoff $\max(S(T) - K, 0)$. Plot the values of the options as a function of the strike $K \in [S_0/2, 2S_0]$, for the following parameters: $\lambda = 1, T = 0.5, \theta = 0.1, \rho = -0.6, \eta = 0.1, \sigma = 0.3, r_0 = 0.2, S_0 = 1$.

e. Determine the analytic solution of a European call option under the BSHW model, and compare the solution with the numerical solution obtained.

Exercise 13.2 For a state vector $X(t) = [S(t), r(t)]^T$ consider the BSHW model, with correlation $dW_x(t)dW_r(t) = \rho dt$. Choose $\rho = 0.5, \theta = 0.1, \eta = 0.3, \sigma = 0.3, S_0 = 1, r_0 = \theta, T = 1$, and for lambda choose either $\lambda = 1.0$ or $\lambda = 0.2$.

a. Show that the solution of $S(t), t > 0$ can be expressed as:

$$S(t) = S_0 \exp\left(\int_0^t \left(r(z) - \frac{1}{2}\sigma^2\right) dz + \sigma W_x(t)\right)$$

b. Construct a self-financing portfolio and derive the pricing PDE.
   Hint: The construction of the PDE is similar to the one for the Heston model.

c. Perform a Monte-Carlo simulation to price a put option with $K = 1.1$, with the Milstein and Euler discretization methods, for the two λ-values, and discuss the results obtained. How can one improve the Monte Carlo results?

d. For this system of SDEs, determine the characteristic function, and use the COS method, to recover the density function for the BSHW system given here.

e. Employ the COS method to price the put option with $K = 1.1$ and compare your results with those obtained in Exercise 13.2b.

Exercise 13.3 The Black-Scholes Hull-White model is, under the risk-free measure $\mathbb{Q}$, defined as,

$$\frac{dS(t)}{S(t)} = r(t)dt + \sigma dW_x^{\mathbb{Q}}(t),$$

$$dr(t) = \lambda(\theta(t) - r(t))dt + \eta dW_r^{\mathbb{Q}}(t),$$

with all parameters positive and the correlation $dW_x^{\mathbb{Q}}(t)dW_r^{\mathbb{Q}}(t) = \rho_{x,r}dt$. Show that, for the T_2-forward price $S_F(t, T_2) := S(t)/P(t, T_2)$, this gives rise to the following dynamics,

$$\frac{dS_F(t, T_2)}{S_F(t, T_2)} = \bar{\sigma}_F(t)dW_F(t), \quad S_F(t_0, T_2) = \frac{S_0}{P(t_0, T_2)}, \tag{13.69}$$

with $\bar{\sigma}_F(t) = \sqrt{\sigma^2 + \eta^2 \bar{B}_r^2(T_2 - t) - 2\rho_{x,r}\sigma\eta\bar{B}_r(T_2 - t)}$, and $\bar{B}_r(\tau) = \frac{1}{\lambda}\left(e^{-\lambda\tau} - 1\right)$.

Exercise 13.4 Consider the Black-Scholes Hull-White hybrid model, with parameters λ, σ, η and correlation coefficient $\rho_{x,r}$.

Show that for $X(t) = \log S(t)$ the following holds true,

$$\mathbb{E}^{\mathbb{Q}}\left[\frac{r(t)}{M(t)}\mathbf{1}_{S(t)>K}\right] = P(0,t)\mathbb{E}^t\left[r(t)| X > \log(K)\right](1 - F_X(\log(K)))$$

$$= P(0,t)\left(\mu_r^t(t) + \sigma_r^t(t)\mathbb{E}^t\left[Z| X > \log(K)\right]\right)\left(1 - F_{S(t)}^t(K)\right).$$

where $\mathbb{E}^t$ indicates the expectation under the t-forward measure (with the numeraire a ZCB with maturity time t). Further, $r(t) \sim \mu_r^t(t) + \sigma_r^t(t)Z$, $Z \sim \mathcal{N}(0, 1)$, where

$$\mu_r^t(t) = r(0)e^{-\lambda t} + \int_0^t \tilde{\theta}(u)e^{-\lambda(t-u)}du, \quad \tilde{\theta}(u) := \lambda\theta(u) + \frac{\eta^2}{\lambda}\left(e^{-\lambda(t-u)} - 1\right),$$

and

$$\sigma_r^t(t) = \left(\frac{\eta^2}{2\lambda}\left(1 - e^{-2\lambda t}\right)\right)^{1/2}.$$

Exercise 13.5 Consider the BSHW model with $S(t_0) = 1$, $\sigma = 0.08$ and $\lambda = 0.5$, $\eta = 0.03$ and $P(0, t) = \exp(-0.01t)$. The correlation $\rho_{x,r} = -0.5$. Consider a portfolio consisting of:

- 1 stock;
- 1 payer swap with expiry date $T_1 = 1y$, maturity date $T_2 = 10y$, with a frequency of $1y$ (the swap rate K is chosen such that the swap value today equals 0) and unit notional $N = 1$;
- 1 call option on the stock with strike price $K = 1.2$ and maturity date $T = 12y$.

a. Set the probability of default equal to 0.2 for all swap payment dates T_i and the recovery rate $R_c = 0.5$. Compute the Expected Exposure profile, and the PFE with quantile 0.05,

b. By means of a Monte Carlo simulation, calculate the value of the portfolio and the CVA. Discuss the results.

c. Which financial derivative should be added to the portfolio (think of netting), so that the CVA charge would be reduced?

Exercise 13.6 Implement a Monte Carlo Euler method, the Monte Carlo AES scheme for the HHW model, as well as the COS method for the approximate H1-HW model.

a. Compare the results of these codes, for the same set of parameters in terms of convergence and required computational time to reach a specific accuracy of the option prices.

b. For which parameter setting would you expect differences between the two Monte Carlo based codes? Confirm your intuition by means of the corresponding results.

c. For which parameters would you expect differences between the HHW results and the H1-HW results? Show the corresponding results.

Exercise 13.7 Similarly as the approximation of the nonaffine terms in the instantaneous covariance matrix of the Heston hybrid model from Section 13.2.2, include the additional correlation, $\rho_{r,v}$, between the interest rate, $r(t)$, and the stochastic variance, $v(t)$.

a. Show that, for $\mathbf{X}(t) = [x(t), v(t), r(t)]^{\mathrm{T}}$ the model has the following symmetric instantaneous covariance matrix:

$$\mathbf{\Sigma} := \sigma(\mathbf{X}(t))\sigma(\mathbf{X}(t))^{\mathrm{T}} = \begin{bmatrix} v(t) & \rho_{x,v}\gamma v(t) & \rho_{x,r}\eta\sqrt{v(t)} \\ * & \gamma^2 v(t) & \rho_{r,v}\gamma\eta\sqrt{v(t)} \\ * & * & \eta^2 \end{bmatrix}_{(3\times 3)}. \quad (13.70)$$

b. Determine the associated Kolmogorov backward equation for the characteristic function, with the terminal condition.

c. The affinity issue arises in two terms of matrix (13.70), namely, in elements $(1,3)$ and $(2,3)$: $\mathbf{\Sigma}_{(1,3)} = \rho_{x,r}\eta\sqrt{v(t)}$, $\mathbf{\Sigma}_{(2,3)} = \rho_{r,v}\gamma\eta\sqrt{v(t)}$.

Use the deterministic approximations $\mathbf{\Sigma}_{(1,3)} \approx \rho_{x,r}\eta\mathbb{E}\left[\sqrt{v(t)}\right]$ and $\mathbf{\Sigma}_{(2,3)} \approx \rho_{r,v}\gamma\eta\mathbb{E}\left[\sqrt{v(t)}\right]$.

The representations of the Heston Hull-White model and the model in (13.23) with $\rho_{r,v} \neq 0$ for $p = 0$ are closely related. Specify this relation in terms of the coefficients of the corresponding ChF.

Exercise 13.8 The dynamics for the stock, $S(t)$, in the so-called Heston-CIR model read:

$$\begin{cases} dS(t)/S(t) = & r(t)dt + \sqrt{v(t)}dW_x^{\mathbb{Q}}(t), \; S(0) > 0 \\ dv(t) = & \kappa(\bar{v} - v(t))dt + \gamma\sqrt{v(t)}dW_v^{\mathbb{Q}}(t), \; v(0) > 0, \\ dr(t) = & \lambda(\theta(t) - r(t))dt + \eta\sqrt{r(t)}dW_r^{\mathbb{Q}}(t), \; r(0) > 0, \end{cases} \quad (13.71)$$

with $dW_x^{\mathbb{Q}}(t)dW_v^{\mathbb{Q}}(t) = \rho_{x,v}dt$, $dW_x^{\mathbb{Q}}(t)dW_r^{\mathbb{Q}}(t) = \rho_{x,r}dt$ and $dW_v^{\mathbb{Q}}(t)dW_r^{\mathbb{Q}}(t) = 0$.

Assume that the nonaffine term in the pricing PDE, $\Sigma_{(1,3)}$, can be approximated, as

$$\Sigma_{(1,3)} \approx \eta\rho_{x,r}\mathbb{E}\left[\sqrt{r(t)}\sqrt{v(t)}\right] \stackrel{!}{=} \eta\rho_{x,r}\mathbb{E}\left[\sqrt{r(t)}\right] \cdot \mathbb{E}\left[\sqrt{v(t)}\right]. \quad (13.72)$$

Since the processes involved are of the same type, the expectations in (13.72) can be determined as presented in Section 13.2.2.

a. Determine the characteristic function for the log-stock, $X(t) = \log S(t)$, and the corresponding Riccati ODEs

b. Find the solution for these Riccati ODEs.

Exercise 13.9 Derive the PDE under the HHW model under the T-forward measure. In order to reduce the complexity, it is convenient to switch between two pricing measures, i.e. we focus on the *forward stock price*, $S_F(t,T)$, defined by:

$$S_F(t,T) = \frac{S(t)}{P(t,T)}, \quad (13.73)$$

with $P(t,T)$ the zero-coupon bond expiring at time T, which pays 1 unit of currency at time T. By switching from the risk-neutral measure, $\mathbb{Q}$, to the T−forward measure, $\mathbb{Q}^T$, discounting will be *decoupled* from the expectation operator, i.e.,

$$V(t,S,v) = P(t,T)\mathbb{E}^T\left[\max(S_F(T,T) - K, 0)|\mathcal{F}(t)\right], \quad (13.74)$$

where independent variable r does not appear as an argument anymore.

Determine the dynamics of the forward $S_F(t,T)$ in (13.73) by Itô's formula.

Implement the Euler SDE scheme for the HHW model, and compare pricing under the $\mathbb{Q}$ and T-forward measure.

Exercise 13.10 In this exercise we focus on the SZHW model.

a. Develop a Monte Carlo Euler simulation for the SZHW model.

b. Check the convergence of the Monte Carlo simulation by decreasing the time step Δt and by increasing the number of Monte Carlo paths.

c. Derive the characteristic function of the SZHW model.

d. The code for the COS method based on the characteristic function for the SZHW model is added. Compare the computational speed and the method's convergence with the Monte Carlo scheme.

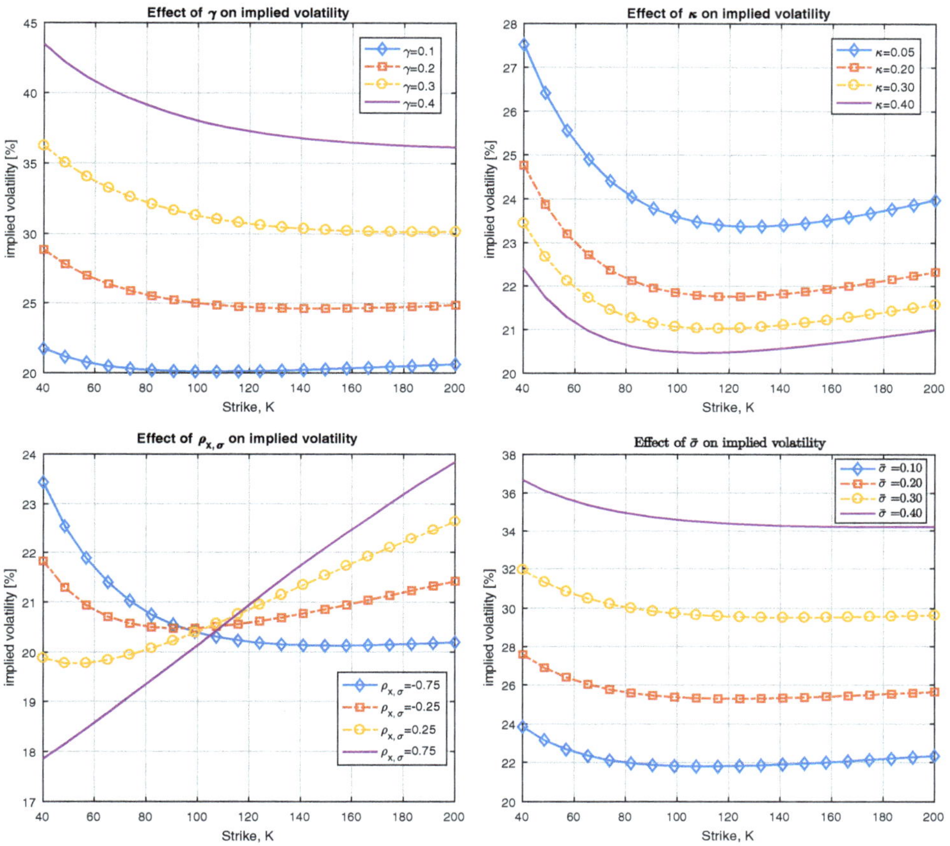

Hybrid Asset Models, Credit Valuation Adjustment

Figure 13.6: *Impact of changing model parameters on the SZHW implied volatility; the baseline model parameters are $\sigma_0 = 0.1$, $\gamma = 0.11$, $\rho_{r,\sigma} = 0.32$, $\rho_{x,\sigma} = -0.42$, $\rho_{x,r} = 0.3$, $\kappa = 0.4$, $\bar{\sigma} = 0.05$, $T = 5$, $\lambda = 0.425$, $\eta = 0.1$, and $P(0,T) = e^{-0.025T}$.*

e. Confirm the impact of parameters changes on the implied volatility curve as presented in Figure 13.6.

f. Investigate the effect of the correlations between the different SDEs in the SZHW model on the implied volatilities, for large maturity times T.

CHAPTER 14

Advanced Interest Rate Models and Generalizations

In this chapter:

We continue with the modeling of *stochastic interest rates*. When dealing with *not too complicated* interest rate derivatives, the short-rate models are often accurate and perform well, especially for at-the-money options. However, the assumption of instantaneous rates is debatable, see, for example, [Musiela and Rutkowski, 1997]. A more general class of mathematical models, that describe the dynamics of the *forward interest rates* has been introduced in the literature as well. These models are called *the market models*. The Libor market model will be discussed in quite some detail in **Section 14.1**. The Libor market model dynamics can also be placed in the Heath-Jarrow-Morton (HJM) framework.

The well-known and well-established lognormal Libor market model, which will be presented under different pricing measures, is explained in **Section 14.2**. This lognormal market model has its limitations, as it is not able to model the interest rate smile/skew shaped implied volatility curves, that are observed also in the money markets. We will also explain the *convexity correction* concept, which should be incorporated in a market model in the case of payments delays. We discuss the CEV and displaced diffusion Libor market models in **Section 14.3**, where we present the stochastic volatility extension of the Libor market model.

This chapter is concluded in **Section 14.4**, with an explanation of modern risk management, in the form of modeling of *negative interest rates*, that are observed in the current money market and of the multi-curve setting.

Keywords: Libor rates, Libor market model, implied volatility smile, displaced diffusion, negative interest rates.

14.1 Libor market model

Since its introduction by Brace *et al.* [1997] and Jamshidian [1997], the Libor market model (LMM) framework enjoys popularity among practitioners from the financial industry, mainly due to the fact that the model primitives can be directly related to observed products and quantities in the money market, e.g. the forward Libor rates and caplet implied Black volatilities. In this framework, closed-form solutions for caps and European swaptions (although not in the same formulation) can be obtained, when the LMM is based on the assumption that the discrete forward Libor rate follows a lognormal distribution, under its own numéraire. The *forward measure* pricing methodology, by Jamshidian [1987; 1989], and Geman *et al.* [1995], is convenient for the valuation of various interest rate contracts. The LMM modeling approach is also contained in the HJM framework. The fundamental techniques to price interest rate derivatives stem from the original work of Heath, Jarrow and Morton (HJM) in the late 1980s, which is considered to be a modern modeling framework for interest rates.

Complex fixed-income products, such as various kinds of swaptions, usually involve cash flows at different points in time. As the valuation of these products can not be decomposed into a sequence of independent payments, it is a challenging task to develop a model, which is mathematically consistent and agrees with the well–established market formulas like with variants of the Black-Scholes model for the pricing of interest rate derivatives. For this, more general LMM dynamics will be presented, that are different from the lognormal LMM dynamics.

14.1.1 General Libor market model specifications

The interest rate model of interest should provide arbitrage-free dynamics for all Libor rates and has to facilitate the pricing of caplets and floorlets in a similar fashion as the market conventional Black-Scholes formula, under the assumption of lognormality, while, on the other hand, we should be able to price complex interest rate derivatives, like swaptions.

The model which satisfies these requirements is the well-known Libor market model (LMM), also known as the BGM model [Brace *et al.*, 1997; Jamshidian, 1997; Miltersen *et al.*, 1997]. Let us recall the definition of the Libor rate (12.1):

For $t \leq T_{i-1}$, the definition of the *Libor rate* has been given by:

$$\ell(t; T_{i-1}, T_i) = \frac{1}{\tau_i} \left(\frac{P(t, T_{i-1}) - P(t, T_i)}{P(t, T_i)} \right)$$

$$= \frac{1}{\tau_i} \left(\frac{P(t, T_{i-1})}{P(t, T_i)} - 1 \right), \qquad (14.1)$$

with $P(t, T_i)$ the ZCB which pays out at time T_i. We will use the short-hand notation $\ell_i(t) := \ell(t; T_{i-1}, T_i)$, for convenience.

Advanced Interest Rate Models and Generalizations 447

The definition of the Libor rate can be connected to the HJM framework. In Equation (11.5) the forward rate has been defined as,

$$P_f(t, T_1, T_2) = \frac{P(t, T_2)}{P(t, T_1)}.$$

Furthermore, in Definition 11.2.1 a continuous compounding has been applied. When assuming a simple compounding, we obtain,

$$P_f(t, T_1, T_2) = \frac{1}{1 + (T_2 - T_1)R(t, T_1, T_2)},$$

using $\tau_2 = T_2 - T_1$ and $\ell(t; T_1, T_2) := R(t, T_1, T_2)$, it follows that,

$$\frac{1}{1 + \tau_2 \ell(t; T_1, T_2)} = \frac{P(t, T_2)}{P(t, T_1)}.$$

Therefore, we find (14.1), with $i = 2$.

The Libor rates, $\ell(t_0; T_{i-1}, T_i)$, are forward rates that are thus defined by three moments in time, the present time t_0 at which the rate is modeled by an SDE, its "expiry date", T_{i-1} (which is also called the "fixing date" or the "reset date") when this Libor rate is fixed and thus known, and its maturity date T_i at which the Libor rate terminates, with $t_0 \leq T_{i-1} \leq T_i$. Libor rate $\ell(t; T_{i-1}, T_i)$ is thus determined at its reset date T_{i-1}, after which the Libor rate is known and does not depend on any volatility anymore. The Libor rate $\ell(t_0; T_{i-1}, T_i)$ may be interpreted as a rate which will be accrued over the period $[T_{i-1}, T_i]$, but is observed "today" at t_0.

Another element of a forward rate is the *accrual period*, denoted by $\tau_i = T_i - T_{i-1}$, which is the time between expiry and maturity date, see Figure 14.1. In practice, the accrual period is not necessarily exactly the period between T_{i-1} and T_i, as typically an additional *"reset delay"* is involved. This reset delay is often equal to a few business days, and it depends on the currency for which the forward rate is calculated (in Euros, it is 2 business days).

Another date which is important in the contract evaluations is the *pay delay*, τ^*, $T_p := T_i + \tau^*$, as in Figure 14.2. The pay date of a contract typically is a few days after the maturity of the contract. However, τ^* is often neglected in the evaluations,

$$\mathbb{E}^{\mathbb{Q}}\left[\frac{\ell(T_{i-1}; T_{i-1}, T_i)}{M(T_i + \tau^*)}\right] \approx \mathbb{E}^{\mathbb{Q}}\left[\frac{\ell(T_{i-1}; T_{i-1}, T_i)}{M(T_i)}\right], \qquad (14.2)$$

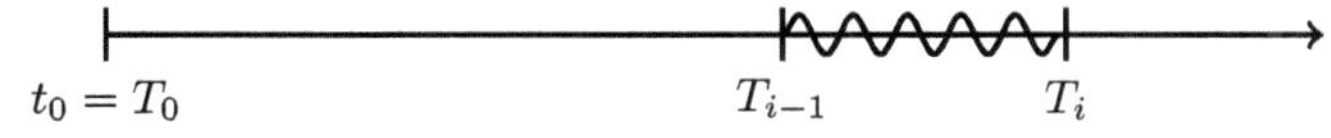

├─────────────────────────┼──────∿∿∿∿──┤──────→
$t_0 = T_0$ T_{i-1} T_i

Figure 14.1: *A forward rate, $\ell(t_0; T_{i-1}, T_i)$, visualized.*

├─────────────────────────┼──────∿∿∿∿──┤──┤──→
$t_0 = T_0$ T_{i-1} T_i $T_{i+\tau^*}$

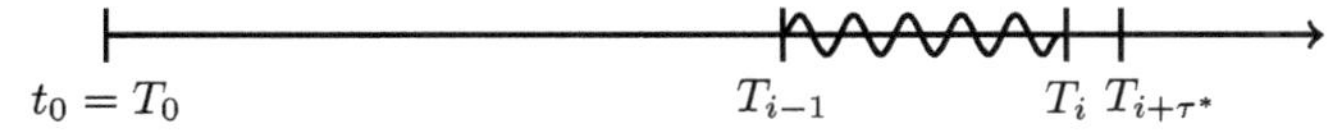

Figure 14.2: *Forward rate $\ell(t_0; T_{i-1}, T_i)$ visualized with pay delay.*

as τ^* is typically only a few business days. In the case that the pay delay τ^* tends to be a long time period, a *convexity correction* needs to be taken into account (to be discussed in Section 14.2.4).

For Libor rate $\ell_i(t)$, we define the dynamics, as follows:

$$\mathrm{d}\ell_i(t) = \bar{\mu}_i^{\mathbb{P}}(t)\mathrm{d}t + \bar{\sigma}_i(t)\mathrm{d}W_i^{\mathbb{P}}(t), \quad \text{for} \quad i = 1, \ldots, m, \quad (14.3)$$

with a certain, possibly stochastic, volatility function $\bar{\sigma}_i(t)$ and with $W_i^{\mathbb{P}}(t)$ Brownian motion under measure $\mathbb{P}$, which can be correlated according to:

$$\mathrm{d}W_i^{\mathbb{P}}(t)\mathrm{d}W_j^{\mathbb{P}}(t) = \rho_{i,j}\mathrm{d}t.$$

Correlation $\rho_{i,j}$ can also be time-dependent.

We denote by $\mathbb{Q}^i$ the T_i-forward measure, associated with the ZCB $P(t, T_i)$ as the numéraire, and $\mathbb{E}^{T_i}$ is the corresponding expectation operator under this measure.

By the results of Harrison and Kreps [1979], it is known that for a given arbitrage-free market, for any strictly positive numéraire financial product whose price is given by $g_1(t)$, there exists a measure for which $\frac{g_2(t)}{g_1(t)}$ is a martingale for all product prices $g_2(t)$. This implies that the following *martingale equality* holds:

$$\mathbb{E}^{T_i}\left[\frac{P(T_{i-1}, T_{i-1})}{P(T_{i-1}, T_i)}\bigg|\mathcal{F}(t)\right] = \frac{P(t, T_{i-1})}{P(t, T_i)}. \quad (14.4)$$

By the equality in (14.1), the left- and right-hand sides in Equation (14.4) can be rewritten so that

$$\mathbb{E}^{T_i}\left[1 + \tau_i\ell(T_{i-1}; T_{i-1}, T_i)\bigg|\mathcal{F}(t)\right] = 1 + \tau_i\ell(t; T_{i-1}, T_i),$$

or,

$$\mathbb{E}^{T_i}\left[\ell(T_{i-1}; T_{i-1}, T_i)|\mathcal{F}(t)\right] = \ell(t; T_{i-1}, T_i).$$

This result can be explained by a simple analogy. Since $\frac{1}{P(T_{i-1}, T_i)}$ is a martingale under the $\mathbb{Q}^i$-measure, the same holds true for Libor rate $\ell(t; T_{i-1}, T_i)$. To see that $\ell(t; T_{i-1}, T_i)$ is indeed a martingale, we may look at the right-hand side of (14.1), which represents the price of a traded asset (the spread between two zero-coupon bonds with nominal $\frac{1}{\tau_i}$). The left-hand side of the same equation therefore also has to be tradable. If we consider the measure $\mathbb{Q}^i$, the T_i-forward measure associated with numéraire $P(t, T_i)$, the forward rate, $\ell(t; T_{i-1}, T_i)$, should be a martingale, so it should be free of drift terms. This implies that it should be possible to transform SDE (14.3) to,

$$\mathrm{d}\ell_i(t) = \bar{\sigma}_i(t)\mathrm{d}W_i^i(t), \quad \text{for } t < T_{i-1}, \quad (14.5)$$

with $\bar{\mu}_i^i(t) = 0$, and $\bar{\sigma}_i(t)$ an instantaneous volatility of the forward rate $\ell(t; T_{i-1}, T_i)$. $W_i^i(t)$ is the Brownian motion of $\ell_i(t)$ under the T_i-forward measure. So, the subscript in W_i^i refers to the specific Libor rate and the superscript to the measure.

We explicitly mention, however, that *only* the i-th Libor rate, $\ell_i(t)$, is a martingale under the $\mathbb{Q}^i$ forward measure. If we, for example, would represent the dynamics of Libor rate $\ell_i(t)$ under the T_j-forward measure (with $i \neq j$), the dynamics of $\ell_i(t)$ would be given by

$$d\ell_i(t) = \bar{\mu}_i^j(t)dt + \bar{\sigma}_i(t)dW_i^j(t),$$

with a certain nonzero drift term $\bar{\mu}_i^j(t)$. The explicit form of the drift $\mu_i^j(t)$, which depends on the measure $\mathbb{Q}^j$, will be presented in the subsections to follow.

14.1.2 Libor market model under the HJM framework

In this section, we show that the Libor market model also falls into the class of HJM models. As explained in Definition 11.2.1, any ZCB $P(t,T)$ can be directly related to the instantaneous forward rate $f^r(t,T)$, via the following relation,

$$P(t,T) = \exp\left(-\int_t^T f^r(t,z)dz\right).$$

By this, we can also link $\ell_i(t)$ to the instantaneous forward rate $f^r(t,T)$, i.e.

$$\tau_i \ell_i(t) + 1 = \frac{P(t, T_{i-1})}{P(t, T_i)} = \exp\left(\int_{T_{i-1}}^{T_i} f^r(t,z)dz\right), \tag{14.6}$$

and thus the dynamics for (14.6), are given by

$$d\left(\tau_i \ell_i(t)\right) = d\left[\exp\left(\int_{T_{i-1}}^{T_i} f^r(t,z)dz\right)\right]. \tag{14.7}$$

By the notation $\xi(t) := \int_{T_{i-1}}^{T_i} f^r(t,z)dz$, and the application of Itô's lemma, we obtain:

$$de^{\xi(t)} = e^{\xi(t)}d\xi(t) + \frac{1}{2}e^{\xi(t)}\left(d\xi(t)\right)^2. \tag{14.8}$$

The integration boundaries in (14.7) do not depend on t, and therefore the $\xi(t)$-dynamics are given by:

$$d\xi(t) = d\left[\int_{T_{i-1}}^{T_i} f^r(t,z)dz\right] = \int_{T_{i-1}}^{T_i} df^r(t,z)dz. \tag{14.9}$$

After collecting all terms in (14.8), (14.9), and returning to (14.7), we find,

$$d\left(\tau_i \ell_i(t)\right) = \exp\left(\int_{T_{i-1}}^{T_i} f^r(t,z)dz\right)\left[\int_{T_{i-1}}^{T_i} df^r(t,z)dz + \frac{1}{2}\left(\int_{T_{i-1}}^{T_i} df^r(t,z)dz\right)^2\right],$$

which by (14.6) can be simplified as:

$$\frac{\mathrm{d}(\tau_i \ell_i(t))}{\tau_i \ell_i(t) + 1} = \int_{T_{i-1}}^{T_i} \mathrm{d}f^r(t,z)\mathrm{d}z + \frac{1}{2}\left(\int_{T_{i-1}}^{T_i} \mathrm{d}f^r(t,z)\mathrm{d}z\right)^2. \tag{14.10}$$

From the definition of the instantaneous forward rate, $f^r(t,T)$, under the T_i-forward measure, we have:

$$\mathrm{d}f^r(t,T) = \alpha^{T_i}(t,T)\mathrm{d}t + \bar{\eta}(t,T)\mathrm{d}W^i(t), \tag{14.11}$$

with

$$\alpha^{T_i}(t,T) = -\bar{\eta}(t,T)\left(\int_T^{T_i} \bar{\eta}(t,z)\mathrm{d}z\right), \tag{14.12}$$

where $\bar{\eta}(t,T)$ is the general notation for the volatility of the instantaneous forward rate $f^r(t,T)$ in the HJM framework. The integral at the right-hand side of (14.10) can now be written as

$$\int_{T_{i-1}}^{T_i} \mathrm{d}f^r(t,z)\mathrm{d}z = \int_{T_{i-1}}^{T_i} \alpha^{T_i}(t,z)\mathrm{d}t\mathrm{d}z + \int_{T_{i-1}}^{T_i} \bar{\eta}(t,z)\mathrm{d}W^i(t)\mathrm{d}z$$

$$= \left(\int_{T_{i-1}}^{T_i} \alpha^{T_i}(t,z)\mathrm{d}z\right)\mathrm{d}t + \left(\int_{T_{i-1}}^{T_i} \bar{\eta}(t,z)\mathrm{d}z\right)\mathrm{d}W^i(t). \tag{14.13}$$

By substitution of (14.13) into (14.10) we obtain the following dynamics for the Libor rate $\ell_i(t)$:

$$\frac{\mathrm{d}\ell_i(t)}{\ell_i(t) + \frac{1}{\tau_i}} = \left[\int_{T_{i-1}}^{T_i} \alpha^{T_i}(t,z)\mathrm{d}z + \frac{1}{2}\left(\int_{T_{i-1}}^{T_i} \bar{\eta}(t,z)\mathrm{d}z\right)^2\right]\mathrm{d}t$$

$$+ \left(\int_{T_{i-1}}^{T_i} \bar{\eta}(t,z)\mathrm{d}z\right)\mathrm{d}W^i(t).$$

Using (14.12), we arrive at:

$$\frac{\mathrm{d}\ell_i(t)}{\ell_i(t) + \frac{1}{\tau_i}} = \left[\frac{1}{2}E^2(t; T_{i-1}, T_i) - \int_{T_{i-1}}^{T_i} \bar{\eta}(t,z)E(t;z,T_i)\mathrm{d}z\right]\mathrm{d}t$$

$$+ E(t; T_{i-1}, T_i)\mathrm{d}W^i(t), \tag{14.14}$$

with $E(t; a, b) = \int_a^b \bar{\eta}(t,z)\mathrm{d}z$.

The choice of instantaneous volatility function $\bar{\eta}(t,z)$ in (14.14) can still be defined, and different choices of $\bar{\eta}(t,z)$ give rise to different dynamics of the Libor market model.

14.2 Lognormal Libor market model

In (14.5) the general form of Libor rate $\ell_i(t)$ under the T_i-forward measure was introduced. At the same time, in (14.14), for the same Libor rate, the dynamics in the HJM framework were derived.

In the lognormal Libor market model, the choice of volatility, $\bar{\sigma}_i(t)$ in (14.1), is given by

$$\bar{\sigma}_i(t) = \sigma_i(t)\ell_i(t),$$

with a time-dependent volatility parameter $\sigma_i(t)$. The dynamics of the *lognormal Libor rate* $\ell_i(t)$ then read:

$$\frac{\mathrm{d}\ell_i(t)}{\ell_i(t)} = \sigma_i(t)\mathrm{d}W_i^i(t), \quad \text{for} \quad t < T_{i-1}.$$

The dynamics of the lognormal LMM can also be obtained by a particular choice of the HJM volatility $\bar{\eta}(t,T)$, in (14.14).

By choosing $\bar{\eta}(t,z) = \eta(t)$, which does *not* depend on z, and is given by:

$$\bar{\eta}(t,z) = \eta(t) = \frac{\ell_i(t)\sigma_i(t)}{1+\tau_i\ell_i(t)}, \tag{14.15}$$

and which implies $E(t;a,b) = \eta(t)(b-a)$, the drift term in (14.14) equals:

$$\frac{1}{2}E^2(t;T_{i-1},T_i) - \int_{T_{i-1}}^{T_i} \eta(t)E(t;z,T_i)\mathrm{d}z$$

$$= \frac{1}{2}\eta^2(t)(T_i - T_{i-1})^2 - \eta(t)\int_{T_{i-1}}^{T_i} E(t;z,T_i)\mathrm{d}z$$

$$= \frac{1}{2}\eta^2(t)(T_i - T_{i-1})^2 - \eta^2(t)\int_{T_{i-1}}^{T_i} (T_i - z)\mathrm{d}z$$

$$= 0.$$

With zero-drift, the lognormal Libor rate dynamics under the HJM framework are given by:

$$\mathrm{d}\ell_i(t) = \left(\ell_i(t) + \frac{1}{\tau_i}\right)E(t;T_{i-1},T_i)\mathrm{d}W^i(t),$$

which, with $\bar{\eta}(t,z)$ in (14.15), thus equals:

$$\mathrm{d}\ell_i(t) = \sigma_i(t)\ell_i(t)\mathrm{d}W_i^i(t), \quad \text{for } t \leq T_{i-1}. \tag{14.16}$$

In this setting, the Libor rate $\ell_i(t)$ is written as:

$$d\ell_i(t) = \sigma_i(t)\ell_i(t)\mathbf{L}_i(t)d\widetilde{\mathbf{W}}(t),$$

with $\mathbf{L}_i(t)$ the i-th row of matrix $\mathbf{L}(t)$, with the HJM generator given by:

$$\bar{\eta}(t,z) = \frac{\ell_i(t)\sigma_i(t)}{1+\tau_i\ell_i(t)}\mathbf{L}_i(t). \quad (14.17)$$

14.2.1 Change of measure in the LMM

In (14.5) the dynamics of the Libor rate under the T_i-forward measure have been described by the SDE, $d\ell_i(t) = \bar{\sigma}_i(t)dW_i^i(t)$, for $t < T_{i-1}$, where $\bar{\sigma}_i(t)$ is a certain state-dependent process. In this section, we will change measures and determine the dynamics of the Libor rate $\ell_i(t)$ *under the T_{i-1}-forward measure*. This result will be generalized to the dynamics *under any measure*, T_j, with $j \neq i$.

As explained in detail in Chapter 7, to perform a measure transformation the dynamics of the Radon-Nikodym derivative $\lambda_i^{i-1}(t)$, for a measure change from the T_i-forward to the T_{i-1}-forward measure, are needed, i.e.

$$\lambda_i^{i-1}(t) = \frac{d\mathbb{Q}^{i-1}}{d\mathbb{Q}^i}\bigg|_{\mathcal{F}(t)} := \frac{P(t,T_{i-1})}{P(t_0,T_{i-1})}\frac{P(t_0,T_i)}{P(t,T_i)}. \quad (14.18)$$

By the definition of the Libor rate in (14.1), the Radon-Nikodym derivative in (14.18) can be expressed in terms of $\ell_i(t)$, as

$$\lambda_i^{i-1}(t) = \frac{P(t_0,T_i)}{P(t_0,T_{i-1})}(\tau_i\ell_i(t)+1), \quad (14.19)$$

with the corresponding dynamics given by:

$$d\lambda_i^{i-1}(t) = \frac{P(t_0,T_i)}{P(t_0,T_{i-1})}\tau_i d\ell_i(t). \quad (14.20)$$

With $d\ell_i(t)$ in (14.5) and by using (14.18), the dynamics of $\lambda_i^{i-1}(t)$ are given by:

$$d\lambda_i^{i-1}(t) = \lambda_i^{i-1}(t)\frac{\tau_i\bar{\sigma}_i(t)}{\tau_i\ell_i(t)+1}dW_i^i(t). \quad (14.21)$$

Based on the *Girsanov Theorem* 7.2.2, the dynamics of $\lambda_i^{i-1}(t)$ define the following measure transformation, i.e.

$$\boxed{dW_i^{i-1}(t) = -\frac{\tau_i\bar{\sigma}_i(t)}{\tau_i\ell_i(t)+1}dt + dW_i^i(t).} \quad (14.22)$$

> So, the dynamics of Libor rate $\ell_i(t)$, under the T_{i-1}-forward measure, read:
> $$d\ell_i(t) = \bar{\sigma}_i(t)dW_i^i(t)$$
> $$= \bar{\sigma}_i(t)\frac{\tau_i\bar{\sigma}_i(t)}{\tau_i\ell_i(t)+1}dt + \bar{\sigma}_i(t)dW_i^{i-1}(t),$$
> with some volatility function $\bar{\sigma}_i(t)$.

This formula can be used for other "one time-step measure changes" as well.

> In particular, when moving from the T_i- to the T_{i+1}-measure, the following measure transformation is required,
> $$dW_i^i(t) = -\frac{\tau_{i+1}\bar{\sigma}_{i+1}(t)}{\tau_{i+1}\ell_{i+1}(t)+1}dt + dW_i^{i+1}(t), \qquad (14.23)$$
> using (14.22) with i and $i+1$.
> The dynamics of the Libor rate $\ell_i(t)$ under the T_{i+1}-forward measure can be determined accordingly.

Function $\bar{\sigma}_i(t)$ may contain a stochastic volatility term, correlations, or the Libor rate itself (as in the lognormal case). Two standard formulations to express the dynamics of the Libor market model are particularly practical. Their differences are found in the measure under which the models are presented. Commonly, the LMM is derived under either the spot or the terminal measure. These variants will be discussed in the subsections to follow.

14.2.2 The LMM under the terminal measure

In the previous section, we saw how to change the Libor rate dynamics from the T_i- to the T_{i-1}-forward measure, and also from the T_i- to the T_{i+1}-forward measure. Now, we generalize the concept and show how the dynamics of the Libor rate $\ell_i(t)$ are defined under the T_m-forward measure, where $i < m$. The Libor market model under the terminal measure should be defined in such a way that all Libor rates are expressed under the T_m-forward measure. The term *"terminal measure"* is an indication for the measure associated with a *final* zero-coupon bond, which is based on the last Libor rate in the contract. For a given tenor structure, $0 \leq T_0 < T_i < \cdots < T_m < T^*$, the corresponding Libor rates are $\ell_i(t)$, $i \in \{1,\ldots,m-1\}$, as well as $\ell_m(t)$, which indicates the last Libor rate.

Based on the transition between the forward measures of two consecutive points in time, T_i and T_{i+1}, in (14.23) we perform a one time-step recursion, and find for the relation between T_i and T_{i+2}:

$$dW_i^i(t) = -\frac{\tau_{i+1}\bar{\sigma}_{i+1}(t)}{\tau_{i+1}\ell_{i+1}(t)+1}dt + \left(-\frac{\tau_{i+2}\bar{\sigma}_{i+2}(t)}{\tau_{i+2}\ell_{i+2}(t)+1}dt + dW_i^{i+2}(t)\right)$$

$$= -\sum_{k=i+1}^{i+2}\frac{\tau_k\bar{\sigma}_k(t)}{\tau_k\ell_k(t)+1}dt + dW_i^{i+2}(t).$$

This result can be generalized to the terminal measure T_m, as follows,

$$\mathrm{d}W_i^i(t) = -\sum_{k=i+1}^{m} \frac{\tau_k \bar{\sigma}_k(t)}{\tau_k \ell_k(t) + 1} \mathrm{d}t + \mathrm{d}W_i^m(t),$$

implying the following dynamics for Libor rate $\ell_i(t)$ under the T_m-forward measure, for any $i < m$,

$$\mathrm{d}\ell_i(t) = \bar{\sigma}_i(t)\mathrm{d}W_i^i(t) = \bar{\sigma}_i(t)\left(-\sum_{k=i+1}^{m} \frac{\tau_k \bar{\sigma}_k(t)}{\tau_k \ell_k(t) + 1} \mathrm{d}t + \mathrm{d}W_i^m(t)\right).$$

14.2.3 The LMM under the spot measure

As indicated in [Brace et al., 1997; Musiela and Rutkowski, 1997], the main problem with market models is that they *do not provide* a continuous time dynamics for any bond in the tenor structure. The well-known, continuously re-balanced, money-savings account, in terms of the instantaneous short rate, is given by $\mathrm{d}M(t) = r(t)M(t)\mathrm{d}t$, with $M(t_0) = 1$. The use of a continuous account as a numéraire in the Libor market model, however, does not fit well to the discrete set-up of the Libor market model, from Section 14.1.1.

Let us consider a discrete tenor structure $\mathcal{T}$ and the Libor rates $\ell_i(t)$. A numéraire for the Libor model should preferably be based on a preassigned maturity and on the tenor structure. We may use a *discretely re-balanced money-savings account*, where re-balancing takes place at predefined maturity dates, based on the following strategy,

- At time t_0, we start with 1 currency unit, and we buy $\frac{1}{P(0,T_1)}$ T_1-bonds.

- At time T_1, we receive the amount $\frac{1}{P(0,T_1)}$, as the owned bonds all pay out 1, and we buy the amount of $\frac{1}{P(0,T_1)} \cdot \frac{1}{P(T_1,T_2)}$ T_2-bonds.

- At time T_2, we thus receive the amount $\frac{1}{P(0,T_1)P(T_1,T_2)}$, and buy ..., etc.

This strategy shows that, between the time points T_0 and T_{i+1}, the spot-Libor rate portfolio contains the following amount of T_{i+1}-bonds:

$$\prod_{k=1}^{i+1} \frac{1}{P(T_{k-1}, T_k)}.$$

With $\bar{m}(t) := \min(i : t \leq T_i)$, as the next reset moment, the value of the portfolio at time t is found from the following definition:

Advanced Interest Rate Models and Generalizations 455

Definition 14.2.1 (Spot-Libor measure) *The spot-Libor measure, $\mathbb{Q}^{\bar{m}(t)}$, is associated to a money-savings account, which is defined as*

$$M(t) := \frac{P\left(t, T_{\bar{m}(t)}\right)}{\prod_{k=1}^{\bar{m}(t)} P(T_{k-1}, T_k)}, \qquad (14.24)$$

with $\bar{m}(t) = \min\left(i : t \leq T_i\right)$. ◀

This definition shows that, in essence, the money-savings account under the Libor market model is governed by *the next zero-coupon bond*, plus the amount of money accumulated so far.

It is now possible to relate this money-savings account to the Libor rates. The following relation holds for any Libor rate $\ell(t; T_i, T_j)$, with $i < j$, using (14.1), with the well-known tenor structure and $\tau_i = T_i - T_{i-1}$,

$$1 + \ell(t; T_i, T_j)(T_j - T_i) = \frac{P(t, T_i)}{P(t, T_j)} \equiv \frac{P(t, T_i) P(t, T_{i+1}) \cdots P(t, T_{j-1})}{P(t, T_{i+1}) \cdots P(t, T_{j-1}) P(t, T_j)}$$

$$= \prod_{k=i+1}^{j} \frac{P(t, T_{k-1})}{P(t, T_k)} = \prod_{k=i+1}^{j} \left(1 + \ell(t; T_{k-1}, T_k)\tau_k\right).$$

Herewith, the evolution of all zero coupon bond prices, with maturity T_i, and all forward Libor rates for the periods $[T_i, T_j]$, with $j > i$, can be determined.

On the other hand, for $\bar{m}(t) = \min(i : t \leq T_i)$ and by the definition of the Libor rate, the price of a bond $P(t, T_j)$, is given by

$$P(t, T_j) = \frac{P\left(t, T_{\bar{m}(t)}\right)}{1 + \ell(t; T_{\bar{m}(t)}, T_j)\left(T_j - T_{\bar{m}(t)}\right)} = \frac{P\left(t, T_{\bar{m}(t)}\right)}{\prod_{k=\bar{m}(t)+1}^{j} \left(1 + \ell(t; T_{k-1}, T_k)\tau_k\right)}.$$
(14.25)

By analyzing the ratio $P(t, T_j)/M(t)$ and using (14.25) and (14.24), we find,

$$\frac{P(t, T_j)}{M(t)} = \frac{P(t, T_{\bar{m}(t)})}{\prod_{k=\bar{m}(t)+1}^{j} (1 + \tau_k \ell(t; T_{k-1}, T_k))} \cdot \frac{\prod_{k=1}^{\bar{m}(t)} P(T_{k-1}, T_k)}{P(t, T_{\bar{m}(t)})}$$

$$= \prod_{k=1}^{\bar{m}(t)} \left(1 + \tau_k \underbrace{\ell(T_{k-1}; T_{k-1}, T_k)}_{\text{fwd Libor rates}}\right)^{-1} \cdot \prod_{k=\bar{m}(t)+1}^{j} \left(1 + \tau_k \underbrace{\ell(t; T_{k-1}, T_k)}_{\text{init. yield curve}}\right)^{-1}.$$

Thus, the *relative bond prices* with respect to the money-savings account are uniquely determined by the initial yield curve and by a collection of forward

Libor rates. This suggests us that the money-savings account may be another viable choice for the numéraire.

The dynamics of Libor rate $\ell_i(t)$, under the Libor spot measure, still needs to be determined. From Definition 14.2.1, it is known that, for a given time t, the money-savings account only depends on the volatility of the *next* zero-coupon bond, $P(t, T_{\bar{m}(t)})$, while the zero-coupon bonds $P(T_{j-1}, T_j)$, for $j \in \{1, \ldots, \bar{m}(t)\}$ are supposed to be known, i.e. deterministic, and do not affect the volatility of $M(t)$.

This defines the following Radon-Nikodym derivative for changing the measures, from the $T_{\bar{n}(t)}$-forward, with $\bar{n}(t) = \bar{m}(t)+1$, to the spot-Libor measure $\mathbb{Q}^M$, generated by the money-savings account $M(t)$ (from Definition 14.2.1):

$$\lambda_{\bar{n}(t)}^M(t) = \left.\frac{d\mathbb{Q}^M}{d\mathbb{Q}^{\bar{n}(t)}}\right|_{\mathcal{F}(t)} = \frac{M(t)}{M(t_0)} \frac{P(t_0, T_{\bar{n}(t)})}{P(t, T_{\bar{n}(t)})}.$$

With the definition of the money-savings account under the Libor market model, this gives

$$\lambda_{\bar{n}(t)}^M(t) = \left.\frac{d\mathbb{Q}^M}{d\mathbb{Q}^{\bar{n}(t)}}\right|_{\mathcal{F}(t)} = \frac{P(t, T_{\bar{m}(t)})}{P(t, T_{\bar{n}(t)})} \frac{P(t_0, T_{\bar{n}(t)})}{M(t_0)} \prod_{k=1}^{\bar{m}(t)} \frac{1}{P(T_{k-1}, T_k)}$$

$$= \frac{P(t, T_{\bar{m}(t)})}{P(t, T_{\bar{n}(t)})} \cdot \bar{P} \qquad (14.26)$$

The term named $\bar{P}$ in (14.26) consists of zero-coupon bonds *from the past*. This suggests that the spot-Libor measure is a measure associated with the zero-coupon bond $P(t, T_{\bar{m}(t)})$, which is scaled by some constant factor. We may indicate the spot-Libor measure either by $\mathbb{Q}^M$ or by $\mathbb{Q}^{\bar{m}(t)}$ with as the corresponding numéraire the zero-coupon bond $P(t, T_{\bar{m}(t)})$. Using the definition of Libor rate $\ell_i(t)$, we get:

$$\lambda_{\bar{n}(t)}^M(t) = (1 + \tau_{\bar{n}(t)} \ell_{\bar{n}(t)}(t)) \cdot \bar{P},$$

with $\bar{P}$ from (14.26) and the dynamics given by:

$$d\lambda_{\bar{n}(t)}^M(t) = \tau_{\bar{n}(t)} \cdot \bar{P} \cdot d\ell_{\bar{n}(t)}(t) = \tau_{\bar{n}(t)} \cdot \bar{P} \cdot \bar{\sigma}_{\bar{n}(t)}(t) dW_{\bar{n}(t)}^{\bar{n}(t)}(t).$$

These can be expressed as

$$d\lambda_{\bar{n}(t)}^M(t) = \lambda_{\bar{n}(t)}^M(t) \frac{\tau_{\bar{n}(t)} \bar{\sigma}_{\bar{n}(t)}(t)}{\tau_{\bar{n}(t)} \ell_{\bar{n}(t)}(t) + 1} dW_{\bar{n}(t)}^{\bar{n}(t)}(t).$$

From Girsanov's theorem, we may conclude that the measure transformation is defined by:

$$dW_{\bar{n}(t)}^M(t) = -\frac{\tau_{\bar{n}(t)} \bar{\sigma}_{\bar{n}(t)}(t)}{\tau_{\bar{n}(t)} \ell_{\bar{n}(t)}(t) + 1} dt + dW_{\bar{n}(t)}^{\bar{n}(t)}(t).$$

The change of measure, from the $T_{\bar{n}(t)}$-forward to the $\mathbb{Q}^M$ measure, gives rise to a similar transformation as when changing forward measures from T_{i-1} to T_i

(as in (14.22)). This is a consequence of the fact that the spot-Libor measure is, in essence, a forward measure associated with a zero-coupon bond, $P(t, T_{\bar{m}(t)})$, with $\bar{m}(t)$ the nearest reset date.

> We end up with the following dynamics for the Libor rate $\ell_i(t)$, $i > \bar{m}(t)$, under the spot-Libor measure $\mathbb{Q}^M$:
>
> $$d\ell_i(t) = \bar{\sigma}_i(t) \sum_{k=\bar{m}(t)+1}^{i} \frac{\tau_k \bar{\sigma}_k(t)}{\tau_k \ell_k(t)+1} dt + \bar{\sigma}_i(t) dW_i^M(t). \qquad (14.27)$$

14.2.4 Convexity correction

The term *convexity* is often used in finance. Here, we will explain a specific scenario for which a *convexity correction* must be incorporated in a Libor market model. However, the concept of convexity adjustment is required for all asset classes when there are payment delays or when the moments of payment do not correspond to the payment date of the numeraire. Generally, if we have a maturity date T but payment takes place at time $T + \tau^*$, convexity has to be taken into account. The higher the uncertainty in the market (high volatility) the more pronounced the effect of the convexity will become.

Let us consider a basic interest rate payoff function, which pays a percentage of a notional N, and the percentage paid will be determined by the Libor rate $\ell(T_{i-1}; T_{i-1}, T_i)$, at time T_i. The price of such a contract is given by:

$$V(t_0) = N \cdot M(t_0) \mathbb{E}^{\mathbb{Q}} \left[\frac{\ell(T_{i-1}; T_{i-1}, T_i)}{M(T_i)} \Big| \mathcal{F}(t_0) \right]$$
$$= N \cdot P(t_0, T_i) \mathbb{E}^{T_i} \left[\ell(T_{i-1}; T_{i-1}, T_i) | \mathcal{F}(t_0) \right], \qquad (14.28)$$

and since $\ell(T_{i-1}; T_{i-1}, T_i)$ is a martingale under the T_i-forward measure, we have,

$$V(t_0) = N \cdot P(t_0, T_i) \ell(t_0; T_{i-1}, T_i). \qquad (14.29)$$

Suppose now that we consider the same contract, however, the payment will take place at some earlier time $T_{i-1} < T_i$. The current value of the contract is then given by:

$$V(t_0) = N \cdot M(t_0) \mathbb{E}^{\mathbb{Q}} \left[\frac{\ell(T_{i-1}; T_{i-1}, T_i)}{M(T_{i-1})} \Big| \mathcal{F}(t_0) \right]. \qquad (14.30)$$

When changing measures, to the T_{i-1}-forward measure, we work with the following Radon-Nikodym derivative:

$$\frac{d\mathbb{Q}^{T_{i-1}}}{d\mathbb{Q}} \Big|_{\mathcal{F}(T_{i-1})} = \frac{P(T_{i-1}, T_{i-1})}{P(t_0, T_{i-1})} \frac{M(t_0)}{M(T_{i-1})},$$

so that,

$$V(t_0) = N \cdot M(t_0) \mathbb{E}^{T_{i-1}} \left[\frac{P(t_0, T_{i-1})}{P(T_{i-1}, T_{i-1})} \frac{M(T_{i-1})}{M(t_0)} \frac{\ell(T_{i-1}; T_{i-1}, T_i)}{M(T_{i-1})} \Big| \mathcal{F}(t_0) \right]$$
$$= N \cdot P(t_0, T_{i-1}) \mathbb{E}^{T_{i-1}} \left[\ell(T_{i-1}; T_{i-1}, T_i) | \mathcal{F}(t_0) \right]. \quad (14.31)$$

Although the Libor rate $\ell(T_{i-1}; T_{i-1}, T_i)$ is a martingale under the T_i-forward measure, *it is however not a martingale* under the T_{i-1} forward measure, i.e.,

$$\mathbb{E}^{T_{i-1}} \left[\ell(T_{i-1}; T_{i-1}, T_i) | \mathcal{F}(t_0) \right] \neq \mathbb{E}^{T_i} \left[\ell(T_{i-1}; T_{i-1}, T_i) | \mathcal{F}(t_0) \right] = \ell(t_0; T_{i-1}, T_i).$$

The difference between these two expectations is commonly referred to as a *convexity*. By the change of measure technique, we can simplify the expressions above, to some extent. By changing to the T_i-forward measure, we find:

$$\frac{d\mathbb{Q}^i}{d\mathbb{Q}^{i-1}} \Big|_{\mathcal{F}(T_{i-1})} = \frac{P(T_{i-1}, T_i)}{P(t_0, T_i)} \frac{P(t_0, T_{i-1})}{P(T_{i-1}, T_{i-1})},$$

so that the value of the derivative is equal to:

$$V(t_0) = N \cdot P(t_0, T_{i-1}) \mathbb{E}^{T_{i-1}} \left[\ell(T_{i-1}; T_{i-1}, T_i) | \mathcal{F}(t_0) \right]$$
$$= N \cdot P(t_0, T_{i-1}) \mathbb{E}^{T_i} \left[\ell(T_{i-1}; T_{i-1}, T_i) \frac{P(t_0, T_i)}{P(T_{i-1}, T_i)} \frac{P(T_{i-1}, T_{i-1})}{P(t_0, T_{i-1})} \Big| \mathcal{F}(t_0) \right]$$
$$= N \cdot \mathbb{E}^{T_i} \left[\ell(T_{i-1}; T_{i-1}, T_i) \frac{P(t_0, T_i)}{P(T_{i-1}, T_i)} \Big| \mathcal{F}(t_0) \right],$$

which can be written as,

$$V(t_0) = N \cdot \mathbb{E}^{T_i} \left[\ell(T_{i-1}; T_{i-1}, T_i) | \mathcal{F}(t_0) \right]$$
$$+ N \cdot \mathbb{E}^{T_i} \left[\ell(T_{i-1}; T_{i-1}, T_i) \left(\frac{P(t_0, T_i)}{P(T_{i-1}, T_i)} - 1 \right) \Big| \mathcal{F}(t_0) \right].$$

The last equality holds by simply adding and subtracting $\ell(T_{i-1}; T_{i-1}, T_i)$.

After further simplifications, we find,

$$V(t_0) = N \cdot (\ell(t_0; T_{i-1}, T_i) + \text{cc}(T_{i-1}, T_i)),$$

with the *convexity correction*, $\text{cc}(T_{i-1}, T_i)$, given by:

$$\text{cc}(T_{i-1}, T_i) = \mathbb{E}^{T_i} \left[\ell(T_{i-1}; T_{i-1}, T_i) \left(\frac{P(t_0, T_i)}{P(T_{i-1}, T_i)} - 1 \right) \Big| \mathcal{F}(t_0) \right]. \quad (14.32)$$

and the convexity correction reads:

$$\text{cc}(T_{i-1}, T_i) = P(t_0, T_i) \mathbb{E}^{T_i} \left[\frac{\ell(T_{i-1}; T_{i-1}, T_i)}{P(T_{i-1}, T_i)} \Big| \mathcal{F}(t_0) \right] - \ell(t_0; T_{i-1}, T_i). \quad (14.33)$$

From the definition of the Libor rate $\ell(T_{i-1}; T_{i-1}, T_i)$, we know,

$$P(T_{i-1}, T_i) = \frac{1}{1 + \tau_i \ell(T_{i-1}; T_{i-1}, T_i)} =: \frac{1}{1 + \tau_i \ell_i(T_{i-1})}. \tag{14.34}$$

By using (14.34), the expectation in (14.33) can be written as:

$$\mathbb{E}^{T_i}\left[\frac{\ell_i(T_{i-1})}{P(T_{i-1}, T_i)}\Big|\mathcal{F}(t_0)\right] = \ell_i(t_0) + \tau_i \mathbb{E}^{T_i}\left[\ell_i^2(T_{i-1})\big|\mathcal{F}(t_0)\right]. \tag{14.35}$$

It is important to note that although $\ell_i(T_{i-1})$ is a martingale under the T_i-forward measure, the quantity $\ell_i^2(T_{i-1})$ is *not* a martingale under the same measure! To see this, let us consider the dynamics

$$d\ell_i(t) = \sigma \ell_i(t) dW_i^i(t),$$

and apply Itô's Lemma to $\ell_i^2(t)$. This will give us,

$$d\ell_i^2(t) = \frac{1}{2}\sigma^2 \ell_i^2(t) dt + 2\sigma \ell_i^2(t) dW_i^i(t).$$

This SDE has a drift term, so it is not a martingale.

We have different choices to determine the expectation at the right-hand side in the above expression. Since Libor rate $\ell_i(t) := \ell(t; T_{i-1}, T_i)$ is a martingale under the T_i-forward measure, the dynamics should have no drift term. A basic choice for the Libor rate is then to define it as a lognormal process, as follows

$$d\ell_i(t) = \sigma \ell_i(t) dW_i^i(t), \tag{14.36}$$

with the solution given by, using $t_0 = 0$,

$$\ell_i(T_{i-1}) = \ell_i(t_0) e^{-\frac{1}{2}\sigma^2 T_{i-1} + \sigma W_i^i(T_{i-1})}. \tag{14.37}$$

Taking the expectation of the squared Libor rate gives us,

$$\mathbb{E}^{T_i}\left[\ell^2(T_{i-1})\big|\mathcal{F}(t_0)\right] = \ell^2(t_0) e^{-\sigma^2 T_{i-1}} \mathbb{E}^{T_i}\left[e^{2\sigma W_i^i(T_{i-1})}\big|\mathcal{F}(t_0)\right] = \ell^2(t_0) e^{\sigma^2 T_{i-1}}.$$

Equation (14.33) then reads,

$$\mathrm{cc}(T_{i-1}, T_i) = P(t_0, T_i)\left(\ell_i(t_0) + \tau_i \ell_i^2(t_0) e^{\sigma^2 T_{i-1}}\right) - \ell_i(t_0), \tag{14.38}$$

In practice, however, this specification is not optimal, as it is not clear how to specify the parameter σ in the dynamics in (14.36). One may consider to choose σ to be the caplet ATM volatility, however this is not a unique choice.

A more reliable approach for the calculation of the second moment of the Libor rate $\ell_i(T_{i-1})$ is to use all strike prices available in the market. This can be

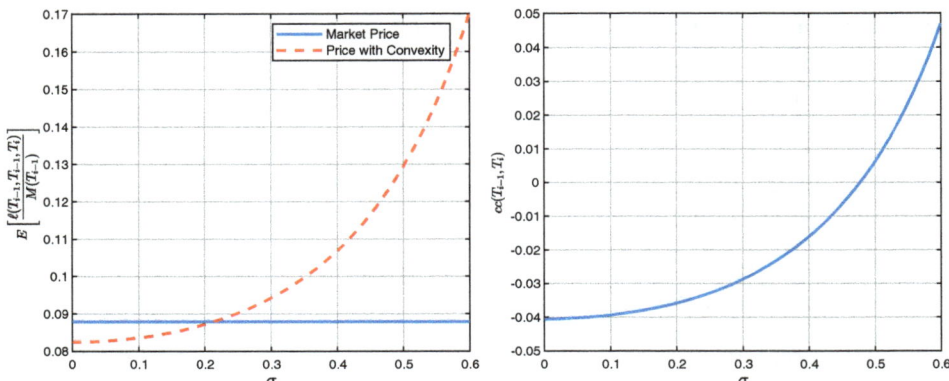

Figure 14.3: The effect of the convexity adjustment $cc(T_{i-1}, T_i)$, as a function of the volatility σ. Left: the effect of the convexity on the derivatives price; Right: the effect of the volatility on the convexity.

done by employing *the Breeden-Litzenberger method* from Section 4.2, i.e.,

$$\mathbb{E}^{T_i}\left[\ell_i^2(T_{i-1})\big|\mathcal{F}(t_0)\right] = \ell_i^2(t_0) + 2\int_0^{\ell_i(t_0)} V_p(t_0, \ell_i(t_0); y, T_{i-1})dy$$

$$+ 2\int_{\ell_i(t_0)}^{\infty} V_c(t_0, \ell_i(t_0); y, T_{i-1})dy, \qquad (14.39)$$

where $V_p(t_0, \ell_i(t_0); y, T)$ and $V_c(t_0, \ell_i(t_0); y, T)$ are values (without discounting) of put and call options (i.e., floorlets $V_i^{\mathrm{FL}}(t_0)$ and caplets $V_i^{\mathrm{CPL}}(t_0)$) on the rate $\ell_i(T_{i-1})$ with strike price y. In Figure 14.3 it is shown how the convexity is affected by the volatility.

14.3 Parametric local volatility models

From the point of view of a complex derivatives trader, see [Rebonato, 2005], an important requirement of an interest rate model is its ability to replicate changes in the future implied volatility surfaces. In this section, we present the *stochastic volatility extension* of the LMM.

14.3.1 Background, motivation

For many years, the lognormal LMM [Brace *et al.*, 1997; Jamshidian, 1997; Miltersen *et al.*, 1997] established itself as a benchmark for interest rate derivatives. The lognormal market models have their limitations, however, as they are not able to model interest rate smile/skew shaped implied volatility curves, that are also observed in the money markets. If, after a calibration of a lognormal market model, the fit is accurate for at-the-money (ATM) financial products, one cannot always accurately price interest rate options with a strike that is different for the ATM level. Without enhancements this model is not able to present strike-dependent volatilities of fixed-income derivatives, such as caps and swaptions. The volatility *skew and smile* have become more apparent in the money market quotes, and model extensions have been introduced to model those features by means of stochastic volatility models.

Before we start with stochastic volatility-based models, let us discuss how an implied volatility can be incorporated in a model. It is well-known, for example, from [Brigo and Mercurio, 2007] that the addition of an uncorrelated stochastic volatility process to the forward rates generates a smile which has its minimum for ATM options. However, an interest skew is also present for many fixed-income products, and to model this feature a stochastic volatility model with an uncorrelated Brownian motion is not sufficient. In the literature there are three popular ways of generating a skew in stochastic volatility Libor market (SV-LMM) models:

- By a nonlinear local volatility function (with the CEV model as a prominent example),

- By the so-called displaced diffusion (DD) dynamics,

- By imposing nonzero correlation between the forward rates and a stochastic volatility process.

A number of stochastic volatility extensions of the LMM have been presented, see e.g., Brigo and Mercurio [2007]. An important step forward in the modeling were the local volatility-type [Andersen and Andreasen, 2000], and the stochastic volatility extensions [Andersen and Andreasen, 2000, 2002; Rebonato, 2002], by which a model can be fitted reasonably well to market data, while the model's stability can be maintained.

We will discuss a few models in more detail below, starting with the CEV LMM.

14.3.2 Constant Elasticity of Variance model (CEV)

An interesting candidate, which is based on a *parametric local volatility model*, is the Constant Elasticity of Variance (CEV) model, which is defined as:

$$\mathrm{d}\ell_i(t) = \bar{\sigma}_i \ell_i^\beta(t) \mathrm{d}W_i^i(t), \quad (14.40)$$

with an exponent β. With $0 < \beta < 1$, the CEV model exhibits the following properties:

- All the solutions are nonexplosive (i.e. bounded),

- For $\beta > \frac{1}{2}$, a solution exists and it is unique,

- For $\beta = 1$, this is the usual lognormal distribution. The value $\ell_i(t) = 0$ is an unattainable barrier, i.e., the forward process always remains strictly positive,

- For $0 < \beta < 1$, the value $\ell_i(t) = 0$ is an attainable barrier,

- For $0 < \beta < \frac{1}{2}$, the solution is not unique, unless one prescribes a separate boundary condition for $\ell_i(t) = 0$.

Process $\ell_i(t)$ is a GBM Libor rate for $\beta = 1$ (which has been discussed already); for $\beta = 0$ it is normally distributed and for $\beta = 0.5$ it follows the Cox-Ingersoll-Ross (CIR) process.

The price of a caplet/floorlet, under these dynamics, is given by the following pricing equation:

$$V_i^{\text{CPL/FL}}(t_0) = \mathbb{E}^{\mathbb{Q}}\left[\frac{M(t_0)}{M(T_i)} \max\left(\bar{\alpha}\left(\ell(T_{i-1}; T_{i-1}, T_i) - K\right), 0\right) \Big| \mathcal{F}(t_0)\right]$$

$$= P(t_0, T_i)\mathbb{E}^{T_i}\left[\max\left(\bar{\alpha}\left(\ell(T_{i-1}; T_{i-1}, T_i) - K\right), 0\right) \Big| \mathcal{F}(t_0)\right],$$

with $M(t_0) = 1$, $\bar{\alpha} = 1$ for caplets and $\bar{\alpha} = -1$ for floorlets. The explicit solution for the caplet and floorlet prices is given in Theorem 14.3.1.

Theorem 14.3.1 (Caplet and floorlet prices under the CEV model) *A closed-form expression exists for caplet/floorlet solutions under the CEV process. With elasticity parameter $\beta \in \mathbb{R}^+$, the CEV process caplets/floorlets at time t_0 are given by:*

- For $\beta \in (0, 1)$:

$$V_i^{\text{CPL}}(t_0) = P(t_0, T_i)\left(\ell_i(t_0)\left(1 - F_{\chi^2(b+2,c)}(a)\right) - KF_{\chi^2(b,a)}(c)\right),$$
$$V_i^{\text{FL}}(t_0) = P(t_0, T_i)\left(K\left(1 - F_{\chi^2(b,a)}(c)\right) - \ell_i(t_0)F_{\chi^2(b+2,c)}(a)\right).$$

- For $\beta > 1$:

$$V_i^{\text{CPL}}(t_0) = P(t_0, T_i)\left(\ell_i(t_0)\left(1 - F_{\chi^2(-b,a)}(c)\right) - KF_{\chi^2(2-b,c)}(a)\right),$$
$$V_i^{\text{FL}}(t_0) = P(t_0, T_i)\left(K\left(1 - F_{\chi^2(2-b,c)}(a)\right) - \ell_i(t_0)F_{\chi^2(-b,a)}(c)\right),$$

where $F_{\chi^2(a,b)}(c) = \mathbb{P}\left[\chi^2(a,b) \leq c\right]$. Here, $F_{\chi^2(a,b)}(c)$ is the noncentral chi-squared cumulative distribution function with degrees of freedom parameter a, noncentrality parameter b, which is evaluated at value c.

The parameters are given by:

$$a = \frac{K^{2(1-\beta)}}{(1-\beta)^2\bar{\sigma}^2(T_i - t_0)}, \quad b = \frac{1}{1-\beta}, \quad c = \frac{(\ell_i(t_0))^{2(1-\beta)}}{(1-\beta)^2\bar{\sigma}^2(T_i - t_0)}.$$

A proof can be found in Schröder [1989].

We focus on parameter β and its effect on the model's implied volatility. Two distinct patterns can be modeled. For $\beta < 1$ the volatility increases when the stock price decreases, and for $\beta > 1$ the volatility increases when the stock price increases. For $\beta < 1$, one may expect the model to generate an *implied volatility skew*.

Example 14.3.1 (CEV model and implied volatility pattern) To calculate the implied volatilities under the LMM CEV model, we choose specific β and $\bar{\sigma}(t) = \sigma$ (constant) parameter values and compute the corresponding call and put option prices for a set of strike prices $\{K_1, \ldots, K_m\}$. Each obtained option price is substituted in the implied volatility iteration, as explained in Section 4.1.1.

In Figure 14.4 the effect of varying the parameters β and σ in the CEV model on the implied volatility is presented. In this numerical experiment, $\ell_i(t_0) = 1$, $r = 0$ and $T = 1$. When exponent β is varied, $\sigma = 0.1$ is fixed, and when σ is varied, $\beta = 0.4$ is fixed. The experiments indicate that different values of parameter β give rise to different implied volatility skews. By varying parameter σ, the level of the ATM volatility changes. These experiments also indicate that the model may not be able to generate an implied volatility smile, which is however often observed in the financial market data.

It was shown in [Schröder, 1989] that the CEV process, equals a space transformed *squared Bessel* process (see also Chapter 8). Prescribing an *absorbing* boundary for the squared Bessel process is the specification which is in agreement with the arbitrage-free constraints, and thus results in a martingale process. Accurate handling of the absorbing boundary behavior is however nontrivial, as the transition density of the absorbed process does not integrate to unity and the moments are not known in closed-form (see also the derivations in Result 8.1.2).

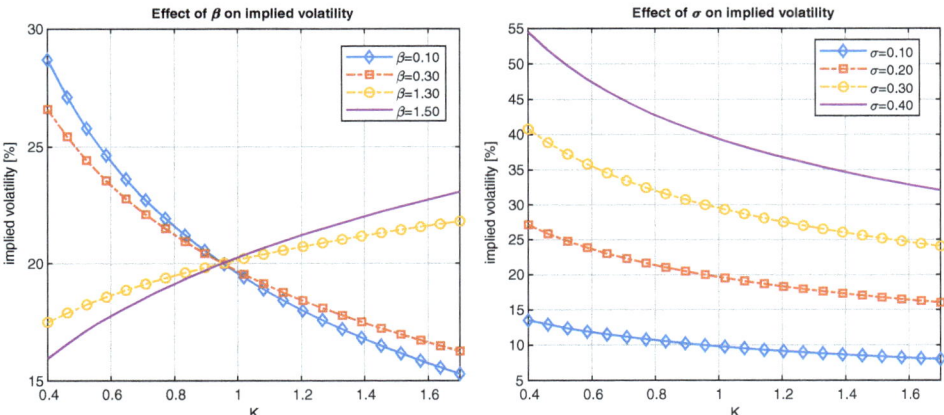

Figure 14.4: *Effect of variation of parameters β and σ on the implied volatilities in the CEV model.*

Intermezzo: The distribution of the CEV process

Let $(\Omega, \mathcal{F}, \mathcal{F}(t), \mathbb{P})$ be a filtered probability space generated by a one-dimensional Brownian motion $W(t) : 0 \leq t < T$. For all $0 \leq t < T$, the CEV process is described by SDE (14.40). Volatility $\bar{\sigma}(t)$ is set to be a constant, i.e. $\bar{\sigma}(t) \equiv \sigma$. Following [Schröder, 1989], consider an invertible transformation, $X(t) = \ell_i^{1-\beta}(t)/(1-\beta)$, for $\beta \neq 1$. Application of Itô's lemma gives us the following SDE for $X(t)$, which can be recognized as the time-changed Bessel process, see also Remark 8.1.1,

$$dX(t) = (1-\beta)\frac{\ell_i^{-\beta}(t)}{1-\beta}\sigma\ell_i^\beta(t)dW(t) - \frac{1}{2}\beta(1-\beta)\frac{\ell_i^{-1-\beta}(t)}{1-\beta}\sigma^2\ell_i^{2\beta}(t)dt$$

$$= \sigma dW(t) - \frac{\beta\sigma^2}{(2-2\beta)X(t)}dt. \tag{14.41}$$

A second transformation, $\bar{X}(t) = X^2(t)$, results in the time-changed squared Bessel process of dimension $\delta := (1-2\beta)/(1-\beta)$, that thus satisfies the following SDE:

$$d\bar{X}(t) = 2\sqrt{|\bar{X}(t)|}\sigma dW(t) + \delta\sigma^2 dt. \tag{14.42}$$

Let $\bar{\nu}(t)$ be a time-changed function (see Remark 8.1.1 for more details about time changes), so that $\bar{\nu}(t) = \sigma^2 t$. Then, $\bar{X}(t) = Y(\bar{\nu}(t))$, where $Y(t)$ is a δ-dimensional squared Bessel process, i.e., the strong solution of the SDE:

$$\boxed{dY(t) = 2\sqrt{|Y(t)|}dW(t) + \delta dt,} \tag{14.43}$$

with degree of freedom δ. The squared Bessel process is a Markov process and its transition densities are known explicitly. They have already been detailed in the context of the CIR process, in Section 8.1.2.

By solving a series of inequalities (see the results in Table 14.1), we find essentially three different parameter ranges which determine the behavior of the CEV process at the boundary and the form of the transition densities:

1. For $\beta > 1$, SDE (14.40) has a unique solution and the boundary zero is not attainable. The density function integrates to one over $\ell_i(t) \in (0, \infty)$ for all $t \geq 0$ and the process $\ell_i(t)$ is a strict *local martingale*.

2. For $\beta < \frac{1}{2}$, SDE (14.40) does not have a unique solution, unless a separate boundary condition is specified for the boundary behavior at $\ell_i(t) = 0$.

Table 14.1: *The mapping of three parameter ranges.*

CEV exponent	Squared Bessel δ
$0 < \beta < \frac{1}{2}$	$0 < \delta < 2$
$\frac{1}{2} \leq \beta < 1$	$-\infty < \delta \leq 0$
$\beta > 1$	$2 < \delta < \infty$

- The density integrates to unity, if the boundary is *reflecting* and process $\ell_i(t)$ is a strict *sub-martingale* (briefly discussed in Section 2.3.2, Chapter 2).

- The density will not integrate to unity, if the boundary at $\ell_i(t) = 0$ is *absorbing*[1] and process $\ell_i(t)$ is a *true martingale*.

3. For $\frac{1}{2} \leq \beta < 1$, a unique strong solution to SDE (14.40) exists, and boundary value zero is absorbing. The density function does not integrate to unity for $t > 0$ and process $\ell_i(t)$ is a *true martingale*.

For most financial applications, parameter β ranges between 0 and 1, which is like in the Cases 2 and 3 in the list above. We therefore focus on these two cases.

Based on the transition density of the squared Bessel diffusions in Y-space, as already presented in Result 8.1.2, it is easy to obtain the transition density for the CEV process (14.40). First of all, note that,

$$\ell_i(T) = \left((1-\beta)\sqrt{|Y(\bar{\nu}(T))|}\right)^{\frac{1}{1-\beta}}.$$

The following map,

$$h(z) := \left((1-\beta)\sqrt{z}\right)^{\frac{1}{1-\beta}}, \quad z \geq 0,$$

is defined, with its inverse

$$h^{-1}(y) = \frac{y^{2(1-\beta)}}{(1-\beta)^2}, \quad y \geq 0.$$

So, $\ell_i(T) = h(Y(\bar{\nu}(T)))$ and initial value $Y_0 = h^{-1}(\ell_i(t_0)) = \ell_i(t_0)^{2(1-\beta)}/(1-\beta)^2$. $Y(\bar{\nu}(T))$ has density function $f_B(\bar{\nu}(T), y)$ and it follows that the transition density for $\ell_i(T)$ is given by

$$f_{CEV}(T, \ell_i(T); t_0, \ell_i(t_0)) = f_B(\bar{\nu}(T); h^{-1}(y))\frac{\mathrm{d}h^{-1}(y)}{\mathrm{d}y},$$

where we use $f_{CEV}(T, \ell_i(T); t_0, \ell_i(t_0))$ to denote the *conditional transition density for the LMM CEV process*. By combining the two cases considered, the related transition densities for the CEV process $\ell_i(t)$, in Equation (14.40), are of the following form:

---

[1] There is a degenerate part with an atom in the boundary and an absolutely continuous part over $(0, \infty)$.

1. For $0 < \beta < \frac{1}{2}$ with absorption at zero, and for $\frac{1}{2} \leq \beta < 1$:

$$f_{CEV}(T, \ell_i(T); t_0, \ell_i(t_0)) = \frac{1}{\bar{\nu}(T)} \left(\frac{\ell_i(T)}{\ell_i(t_0)}\right)^{-\frac{1}{2}}$$

$$\times \exp\left(-\frac{\ell_i(T)^{2(1-\beta)} + \ell_i(t_0)^{2(1-\beta)}}{2(1-\beta)^2 \bar{\nu}(T)}\right)$$

$$\times I_{|\frac{\bar{\beta}-2}{2}|}\left[\frac{(\ell_i(t_0)\ell_i(T))^{1-\beta}}{\bar{\nu}(T)(1-\beta)^2}\right] \frac{\ell_i^{1-2\beta}(T)}{1-\beta}, \tag{14.44}$$

where $\bar{\nu}(T) = \sigma^2 T$ and $\bar{\beta} = \frac{1-2\beta}{1-\beta}$.

2. For $0 < \beta < \frac{1}{2}$ with a reflecting boundary at $\ell_i(t) = 0$:

$$f_{CEV}(T, \ell_i(T); t_0, \ell_i(t_0)) = \frac{1}{\bar{\nu}(T)} \left(\frac{\ell_i(T)}{\ell_i(t_0)}\right)^{-\frac{1}{2}}$$

$$\times \exp\left(-\frac{\ell_i^{2(1-\beta)}(T) + \ell_i(t_0)^{2(1-\beta)}}{2(1-\beta)^2 \bar{\nu}(T)}\right)$$

$$\times I_{\frac{\bar{\beta}-2}{2}}\left[\frac{(\ell_i(t_0)\ell_i(T))^{1-\beta}}{\bar{\nu}(T)(1-\beta)^2}\right] \frac{\ell_i^{1-2\beta}(T)}{1-\beta}. \tag{14.45}$$

By integrating these identities, we find the following CDFs.

Result 14.3.1 *The CDFs of the LMM CEV price process, as in Equation (14.40), are given by the following formulas:*

1. For $0 < \beta < \frac{1}{2}$ with absorption at zero, and for $\frac{1}{2} \leq \beta < 1$:

$$\mathbb{P}[\ell_i(T) \leq x | \ell_i(t_0)] = 1 - F_{\chi^2}(a; b, c(x)). \tag{14.46}$$

2. For $0 < \beta < \frac{1}{2}$ with a reflecting boundary at $\ell_i(t_0) = 0$:

$$\mathbb{P}[\ell_i(T) \leq x | \ell_i(t_0)] = F_{\chi^2}(c(x); 2-b, a), \tag{14.47}$$

with the following parameters:

$$a = \frac{\ell_i(t_0)^{2(1-\beta)}}{(1-\beta)^2 \bar{\nu}(T)}, \quad b = \frac{1}{1-\beta}, \quad c(x) = \frac{x^{2(1-\beta)}}{(1-\beta)^2 \bar{\nu}(T)}, \quad \bar{\nu}(T) = \sigma^2 T,$$

and $F_{\chi^2}(x; \delta, \lambda)$ is the noncentral chi-squared cumulative distribution function with noncentrality parameter λ and degree of freedom δ.

The proofs of these results can be found in Schroder [Schröder, 1989] using classical results for Bessel processes. An alternative proof based on the Green's function theory can be found in [Lesniewski, 2009].

As already stated in Result 8.1.1, the probability density function will not integrate to one when the boundary is absorbing. The shortage in the total probability mass is the probability which is absorbed at $\ell_i(t) = 0$. Following the result from Result 14.3.1, a formula for the absorption probability can be obtained:

Corollary 14.3.2 *For $0 < \beta < 1$, the probability of $\ell_i(T)$, which is governed by SDE (14.40) and conditional on $\ell_i(t_0)$, reads*

$$\mathbb{P}[\ell_i(T) = 0 | \ell_i(t_0)] = 1 - \gamma\left(\frac{1}{2(1-\beta)}, \frac{\ell_i(t_0)^{2(1-\beta)}}{2(1-\beta)^2 \bar{\nu}(T)}\right) \Big/ \Gamma\left(\frac{1}{2(1-\beta)}\right), \tag{14.48}$$

where $\gamma(a, z)$ is the lower incomplete Gamma function and $\Gamma(z)$ is the Gamma function.

Andersen and Andreasen [2010] and Rebonato [2009] argue that if the Libor rate follows a LMM CEV process under a certain measure, there is only one acceptable boundary condition at zero to ensure the arbitrage-free conditions, which is the *absorption condition*. With a reflecting boundary at zero, an investor would wait until the value zero is reached (which would happen with a strictly positive probability). When the price is zero, the investor would take a long position and would sell it immediately when the boundary value 0 has reflected the price process, thus realizing a risk-free profit.

14.3.3 Displaced diffusion model

Although the CEV model appears interesting from a theoretical point of view, it gives rise to difficulties during the calibration phase. Rapid and accurate pricing of plain vanilla options is nontrivial within the LMM CEV model context.

One way to deal with these numerical issues is to *replace* the model by the so-called displaced diffusion (DD) model. DD processes were introduced by Rubinstein [1983]. We follow the definition of the DD process in [Rebonato, 2002].

Consider the following deterministic function, $g(\ell_i(t)) = \ell_i^\beta(t)$. A first-order Taylor expansion of $g(\ell_i(t))$ around $\ell_i(t_0)$, by the *delta method*, see also Equation (13.35), is given by,

$$g(\ell_i(t)) = g(\ell_i(t_0)) + \frac{\mathrm{d}g(\ell_i(t))}{\mathrm{d}\ell_i(t)}\bigg|_{\ell_i(t)=\ell_i(t_0)} (\ell_i(t) - \ell_i(t_0))$$
$$+ \mathcal{O}((\ell_i(t) - \ell_i(t_0))^2),$$

which is equivalent to,

$$\ell_i^\beta(t) \approx \ell_i^\beta(t_0) + \beta \ell_i^{\beta-1}(t_0)(\ell_i(t) - \ell_i(t_0))$$
$$= \vartheta\left((1-\beta)\ell_i(t_0) + \beta \ell_i(t)\right),$$

with $\vartheta := \ell_i^{\beta-1}(t_0)$. After substitution in the LMM CEV model (14.40), we find:
$$\mathrm{d}\ell_i(t) = \bar{\sigma}_i \vartheta \left((1-\beta)\ell_i(t_0) + \beta \ell_i(t)\right) \mathrm{d}W_i^i(t). \tag{14.49}$$

The *displaced diffusion model* is then described by the following SDE:
$$\mathrm{d}\ell_i(t) = \sigma_i \left(\beta \ell_i(t) + (1-\beta)\ell_i(t_0)\right) \mathrm{d}W_i^i(t), \quad \ell_i(t_0) > 0, \tag{14.50}$$
with volatility, $\sigma_i = \vartheta \bar{\sigma}_i$ and the so-called *displacement parameter* β.

Definition 14.3.1 (Alternative formulation) *Another formulation of the DD process, different from the one introduced above, was presented by Marris [1999]. Equation (14.50) can be written as,*
$$\mathrm{d}\ell_i(t) = \left(\ell_i(t) + \frac{1}{\beta}(1-\beta)\ell_i(t_0)\right) \beta \sigma_i \mathrm{d}W_i^i(t),$$

so that we get:

$$\frac{\mathrm{d}(\ell_i(t) + a)}{\ell_i(t) + a} = \tilde{\sigma}_i \mathrm{d}W_i^i(t), \tag{14.51}$$

with $a := \frac{1}{\beta}(1-\beta)\ell_i(t_0)$ and $\tilde{\sigma}_i := \beta \sigma_i$.

◀

The stochastic process in (14.49) can thus be recognized as a *shifted lognormal process*, with shift parameter equal to a. A shifted lognormal process may thus be considered as a first-order approximation to the CEV dynamics.

European option prices on the Libor rate can be efficiently calculated using the well-known *Black-76 valuation formula*, with an adjusted value for the strike.

Theorem 14.3.2 (European options under displaced diffusion model) In the DD model, with a constant skew parameter β and volatility σ_i, we can price European call options, as follows,
$$V_c(t_0) = P(t_0, T_i) \left[\frac{\ell_i(t_0)}{\beta} \cdot F_{\mathcal{N}(0,1)}(d_1) - \left(K + \frac{1-\beta}{\beta}\ell_i(t_0)\right) \cdot F_{\mathcal{N}(0,1)}(d_2)\right], \tag{14.52}$$

with
$$d_1 = \frac{\log\left(\frac{\ell_i(t_0)}{\beta K + (1-\beta)\ell_i(t_0)}\right) + \frac{1}{2}\sigma_i^2 \beta^2 (T-t_0)}{\sigma_i \beta \sqrt{T-t_0}}, \quad d_2 = d_1 - \sigma_i \beta \sqrt{T_i - t_0}.$$

Proof The process in (14.51) can be reformulated as,

$$\frac{\mathrm{d}\left(\ell_i(t) + \frac{1-\beta}{\beta}\ell_i(t_0)\right)}{\ell_i(t) + \frac{1-\beta}{\beta}\ell_i(t_0)} = \sigma_i \beta \mathrm{d}W_i^i(t). \tag{14.53}$$

We define a process $\hat{\ell}_i(t)$, as follows,

$$\hat{\ell}_i(t) = \ell_i(t) + \frac{(1-\beta)}{\beta}\ell_i(t_0). \tag{14.54}$$

Since the shift in process $\ell_i(t)$ is constant, it follows that $\mathrm{d}\hat{\ell}_i(t) = \mathrm{d}\ell_i(t)$, with initial value $\hat{\ell}_i(t_0) = \frac{1}{\beta}\ell_i(t_0)$. By substitution, we obtain,

$$\mathrm{d}\hat{\ell}_i(t) = \mathrm{d}\ell_i(t)$$
$$= \sigma_i\left[\beta\ell_i(t) + (1-\beta)\ell_i(t_0)\right]\mathrm{d}W_i^i(t) = \sigma_i\beta\hat{\ell}_i(t)\mathrm{d}W_i^i(t).$$

Based on this result, it can be concluded that $\hat{\ell}_i(t)$ is governed by the standard Black model with volatility $\hat{\sigma}_i = \beta\sigma_i$, so that the standard pricing formula can be employed.

To price a European call option on the Libor rate $\ell_i(T_i) \equiv \ell(T_{i-1}, T_{i-1}, T_i)$ with payment at time T_i, the following expectation needs to be determined:

$$V_c(t_0) = \mathbb{E}^{\mathbb{Q}}\left[\frac{M(t_0)}{M(T_i)}\max(\ell_i(T_i) - K, 0)\Big|\mathcal{F}(t_0)\right]$$

$$= P(t_0, T_i)\mathbb{E}^{T_i}\left[\max\left(\hat{\ell}_i(T_i) - \frac{(1-\beta)}{\beta}\ell_i(t_0) - K, 0\right)\Big|\mathcal{F}(t_0)\right]$$

$$= P(t_0, T_i)\mathbb{E}^{T_i}\left[\max\left(\hat{\ell}_i(T_i) - K^*, 0\right)\Big|\mathcal{F}(t_0)\right],$$

with $K^* = \frac{(1-\beta)}{\beta}\ell_i(t_0) + K$.

This implies that a European call option under the DD model can be priced by the standard Black-Scholes model with interest rate $r = 0$, which is often referred to as the *Black-76 model*, and where the results are multiplied by $P(t_0, T_i)$ with volatility $\hat{\sigma}_i := \beta\sigma_i$, strike price $K^* := \frac{(1-\beta)}{\beta}\ell_i(t_0) + K$, and initial value $\hat{\ell}_i(t_0) = \ell_i(t_0)/\beta$.

By the Black-Scholes model, the expectation reads,

$$\mathbb{E}^{T_i}\left[\max\left(\hat{\ell}_i(T_i) - K^*, 0\right)\Big|\mathcal{F}(t_0)\right] = \hat{\ell}_i(t_0)F_{\mathcal{N}(0,1)}(d_1) - K^*F_{\mathcal{N}(0,1)}(d_2), \tag{14.55}$$

with

$$d_1 = \frac{\log\frac{\hat{\ell}_i(t_0)}{K^*} + \frac{1}{2}\hat{\sigma}_i(T_i - t_0)}{\hat{\sigma}_i\sqrt{T - t_0}}, \quad d_2 = d_1 - \hat{\sigma}_i\sqrt{T_i - t_0}.$$

By multiplying the results with $P(t_0, T_i)$ and substitutions, the proof is finished. ∎

> The concept of displacing (i.e. shifting) a distribution is common to handling interest rates when a negative interest rate should be incorporated within a model. We will discuss this in Section 14.4.1.

Example 14.3.2 (DD smile and skew) By the displaced diffusion model, an implied volatility skew can also be modeled. In Figure 14.5, some examples of implied volatilities for different β and σ values are presented. Figure 14.5 shows very similar effects regarding β and σ as in the CEV model, in Figure 14.4. ♦

DD processes are thus of interest, since they can be used as simplified approximate models of the CEV model. DD models also have some drawbacks, however, like the requirement that $\ell_i(t) + a$ in Definition 14.3.1 has to be positive, which implies that $\ell_i(t)$ may only vary in $(-a, +\infty)$.

One of the advanced models is the displaced diffusion stochastic volatility (DD-SV) model, as developed in [Andersen and Andreasen, 2002], and the paper [Piterbarg, 2005b] connected time-dependent model volatilities and skews for the Libor rates and swap rates to the market implied quantities.

14.3.4 Stochastic volatility LMM

The 2007/2008 banking crisis also brought essential changes to the interest rate swaption volatilities, for which a highly flexible modeling framework seems to be

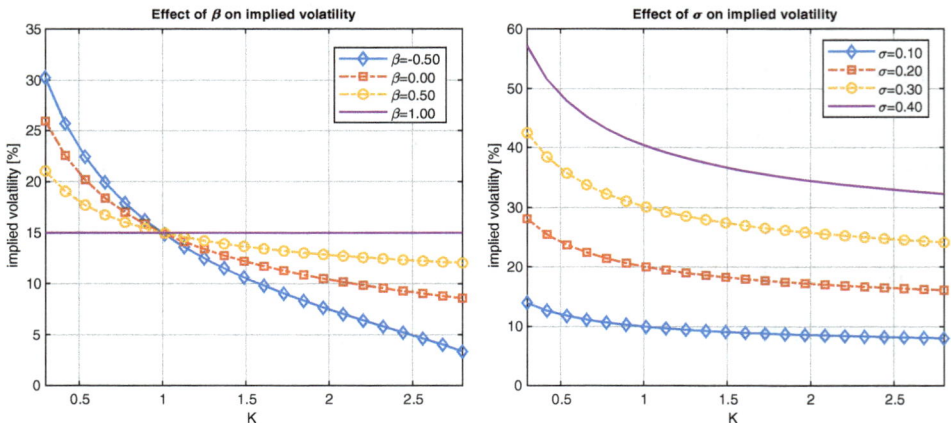

Figure 14.5: *The effect of β and σ on the implied volatilities under the DD model with $\ell_i(t_0) = 1$, $T = 2$. In the left-hand figure, $\sigma = 0.15$; in the right-hand figure, $\beta = 0.5$.*

favorable. Flexible models are usually of multi-factor form. Notable examples are multi-factor short-rate models, multi-factor Quasi-Gaussian models or Libor market model extensions. The CEV process, as introduced by Cox [1975], is an important building block of the CEV Libor Market Model (CEV LMM) by Andersen and Andreasen [2010].

As the general dynamics for the Libor rate, Heston-type dynamics are considered for each Libor rate with a nonzero correlation between the forward rates and the stochastic volatility. *This model, however, is not well-suited for accurately and efficiently modeling Libor rate dynamics, as can be seen below.*

For this, we first consider, a Libor rate $\ell_i(t)$, which under its own measure, is governed by the following dynamics,

$$\begin{cases} d\ell_i(t) = \sigma_i \ell_i(t) \sqrt{\nu(t)} dW_i^i(t), \\ d\nu(t) = \lambda(\nu_0 - \nu(t))dt + \eta\sqrt{\nu(t)} dW_\nu^i(t), \end{cases} \quad (14.56)$$

with the correlation structure given by:

$$dW_i^i(t) dW_l^i(t) = \rho_{i,l} dt, \quad dW_i^i(t) dW_\nu^i(t) = \rho_{i,\nu} dt.$$

Volatility process $\nu(t)$ is modeled to be mean-reverting, with the long term mean ν_0, and the volatility of the variance process is denoted by η. For any two Libor rates, $\ell_i(t)$ and $\ell_l(t)$, the correlation is given by $\rho_{i,l}$ and each Libor rate $\ell_i(t)$ is correlated to the volatility process $\nu(t)$ with $\rho_{i,\nu}$.

We will discuss a measure change for $\ell_i(t)$. First of all, the system in (14.56) reads, in terms of independent Brownian motions,

$$\begin{cases} d\ell_i(t) = \sigma_i \ell_i(t) \sqrt{\nu(t)} d\widetilde{W}_i^i(t), \\ d\nu(t) = \lambda(\nu_0 - \nu(t))dt + \eta\sqrt{\nu(t)} \left(\rho_{i,\nu} d\widetilde{W}_i^i(t) + \sqrt{1-\rho_{i,\nu}^2} d\widetilde{W}_\nu^i(t) \right). \end{cases}$$

Based on the results from Equation (14.22), the measure change from the T_i- to the T_{i-1}-forward measure gives rise to the following adjustment of the Brownian motion,

$$d\widetilde{W}_i^i(t) = \frac{\tau_i \bar{\sigma}_i(t,\ell)}{\tau_i \ell_i(t)+1} dt + d\widetilde{W}_i^{i-1}(t),$$

with $\bar{\sigma}_i(t,\ell) = \sigma_i \ell_i(t)\sqrt{\nu(t)}$. The Brownian motion $\widetilde{W}_\nu^i(t)$ of the variance process $\nu(t)$ is assumed to be independent of the Libor rates, which implies that this Brownian motion is also invariant under measure changes, i.e. $d\widetilde{W}_\nu^i(t) = d\widetilde{W}_\nu^{i-1}(t)$.

The dynamics of the Libor rate $\ell_i(t)$, under the $\mathbb{Q}^{i-1}$-forward measure, now read

$$d\ell_i(t) = \sigma_i \ell_i(t) \sqrt{\nu(t)} \left(\frac{\tau_i \bar{\sigma}_i(t,\ell)}{\tau_i \ell_i(t)+1} dt + d\widetilde{W}_i^{i-1}(t) \right),$$

and the dynamics for the variance process are given by,

$$d\nu(t) = \lambda(\nu_0 - \nu(t))dt$$
$$+ \eta\sqrt{\nu(t)}\left[\rho_{i,\nu}\left(\frac{\tau_i\bar{\sigma}_i(t,\ell)}{\tau_i\ell_i(t)+1}dt + d\widetilde{W}_i^{i-1}(t)\right) + \sqrt{1-\rho_{i,\nu}^2}d\widetilde{W}_\nu^{i-1}(t)\right].$$

After rewriting, we find:

$$d\nu(t) = \lambda\left(\nu_0 - \nu(t) + \rho_{i,\nu}\frac{\eta}{\lambda}\frac{\tau_i\sigma_i\ell_i(t)}{\tau_i\ell_i(t)+1}\nu(t)\right)dt$$
$$+ \eta\sqrt{\nu(t)}\left[\rho_{i,\nu}d\widetilde{W}_i^{i-1}(t) + \sqrt{1-\rho_{i,\nu}^2}d\widetilde{W}_\nu^{i-1}(t)\right].$$

The derivation above shows that a measure change which is applied to the Heston-type Libor model will affect the drift term of the Libor rate ℓ_i, as well as the variance process $\nu(t)$. The Libor rates and stochastic volatility process are correlated, which has a significant impact on the dynamics of the processes under a measure change. The authors in [Wu and Zhang, 2008] showed that the stochastic volatility process $\nu(t)$ evolves only as a square-root process under a reference measure and the mean-reversion property disappears when changing numéraires. Moreover, when the original definition of the volatility process changes, we cannot easily use the Fourier-based pricing methods for the efficient pricing of caplets and swaptions. With *parameter freezing techniques* an approximate model may be derived, which is to a square-root type model (see [Wu and Zhang, 2008]).

Here, the correlation parameter between the Libor rate and the volatility is set *equal to zero*. Because the variance process is now not correlated to the Libor rates, a measure change will not affect the form of the dynamics.

This leads to the following *displaced diffusion* version,

$$\begin{cases} d\ell_i(t) = \sigma_i\left(\beta\ell_i(t) + (1-\beta)\ell_i(t_0)\right)\sqrt{\nu(t)}dW_i^i(t), \\ d\nu(t) = \lambda(\nu_0 - \nu(t))dt + \eta\sqrt{\nu(t)}dW_\nu(t), \end{cases} \quad (14.57)$$

with the correlations,

$$dW_i^i(t)dW_l^i(t) = \rho_{il}dt, \quad dW_i^i dW_\nu(t) = 0.$$

Typically, a model with zero correlation is able to generate different volatility smile patterns, but it cannot represent skew-shaped implied volatilities. The present model is however also able to generate an implied volatility skew shape, due to the displacement parameter β, see Figure 14.5, and the stochastic volatility process. The model is called *the Displaced Diffusion Stochastic Volatility Libor Market Model*, or SV-LMM for short. A connection with the standard Heston model is presented below.

Remark 14.3.1 (Generalizations in the industry) *Another displaced diffusion model under the T_i-forward measure, with time-dependent parameters, which is also extended by a process for stochastic volatility, is given by,*

$$\begin{cases} d\ell_i(t) = \sigma_i(t)\left(\beta_i(t)\ell_i(t) + (1-\beta_i(t))\ell_i(t_0)\right)\sqrt{\nu(t)}dW_i^i(t), \\ d\nu(t) = \lambda(\nu(t_0) - \nu(t))dt + \eta\sqrt{\nu(t)}dW_v^i(t), \quad \nu(t_0) = 1, \end{cases}$$

with $\beta_i(t)$ the time-dependent displacement coefficients, $\sigma_i(t)$ the time-dependent volatility functions, $\nu(t)$ the variance processes with parameters λ, η and independent Brownian motion $W_\nu^i(t)$. In this setting, there is one variance process, $\nu(t)$, for all Libor rates.

The following instantaneous HJM volatility function results for this model,

$$\bar{\eta}(t,z) = \frac{\sigma_i(t)\Big(\beta_i(t)\ell_i(t) + (1-\beta_i(t))\ell_i(t_0)\Big)}{1+\tau_i\ell_i(t)}\sqrt{\nu(t)}. \quad (14.58)$$

The concept in [Piterbarg, 2005b] of effective skew and effective volatility enables the calibration of the volatility smiles for a whole set of swaptions. By changing the measure, from the risk-neutral to the forward measure associated with the zero-coupon bond as the numéraire, the dimensionality of the approximating characteristic function can be reduced. This, combined with freezing the Libor rates (keeping them to specific values) and linearizations of the nonaffine terms arising in the corresponding instantaneous covariance matrix, are some key issues to efficient model evaluation by approximate models. See, for details, [Andersen and Piterbarg, 2010]. ▲

Relation with the Heston model

Here, the relation between the SV-LMM in (14.57) and a *pure* standard Heston model is presented.

The Libor rate dynamics in (14.57) can be written as,

$$\mathrm{d}\ell_i(t) = \beta\sigma_i\left(\ell_i(t) + (1-\beta)\frac{\ell_i(t_0)}{\beta}\right)\sqrt{\nu(t)}\mathrm{d}W_i^i(t),$$

which is equivalent to,

$$\frac{\mathrm{d}\ell_i(t)}{\ell_i(t) + (1-\beta)\frac{\ell_i(t_0)}{\beta}} = \beta\sigma_i\sqrt{\nu(t)}\mathrm{d}W_i^i(t). \quad (14.59)$$

Since $\mathrm{d}(\ell_i(t) + a) = \mathrm{d}\ell_i(t)$, for constant values of a, the process in (14.59) equals,

$$\frac{\mathrm{d}\left(\ell_i(t) + (1-\beta)\frac{\ell_i(t_0)}{\beta}\right)}{\ell_i(t) + (1-\beta)\frac{\ell_i(t_0)}{\beta}} = \beta\sigma_i\sqrt{\nu(t)}\mathrm{d}W_i^i(t).$$

By another process, $\jmath_i(t)$, which is defined as,

$$\jmath_i(t) := \ell_i(t) + (1-\beta)\frac{\ell_i(t_0)}{\beta}, \quad (14.60)$$

with $\jmath_i(t_0) = \ell_i(t_0)/\beta$, it is easy to see that, with β and $\ell_i(t_0)$ constant,

$$\mathrm{d}\jmath_i(t) = \mathrm{d}\ell_i(t).$$

The dynamics for $J_i(t)$ are then given by:

$$\frac{dJ_i(t)}{J_i(t)} = \beta\sigma_i\sqrt{\nu(t)}dW_i^i(t) = \sqrt{\beta^2\sigma_i^2\nu(t)}dW_i^i(t).$$

By defining $\hat{\nu}(t) := \beta^2\sigma_i^2\nu(t)$, the process for $J_i(t)$ can be written as

$$\frac{dJ_i(t)}{J_i(t)} = \sqrt{\hat{\nu}(t)}dW_i^i(t).$$

The final step is the derivation of the dynamics for $\hat{\nu}(t)$. By Itô's lemma, we find:

$$d\hat{\nu}(t) = \beta^2\sigma_i^2 d\nu(t)$$
$$= \beta^2\sigma_i^2\lambda(\nu_0 - \nu(t))dt + \beta^2\sigma_i^2\eta\sqrt{\nu(t)}dW_\nu(t)$$
$$= \lambda(\beta^2\sigma_i^2\nu_0 - \beta^2\sigma_i^2\nu(t))dt + \beta\sigma_i\eta\sqrt{\beta^2\sigma_i^2\nu(t)}dW_\nu(t),$$

which gives us,

$$d\hat{\nu}(t) = \lambda(\hat{\nu}_0 - \hat{\nu}(t))dt + \hat{\eta}\sqrt{\hat{\nu}(t)}dW_\nu(t),$$

with $\hat{\nu}_0 = \beta^2\sigma_i^2\nu_0$ and $\hat{\eta} = \beta\sigma_i\eta$.

The resulting dynamics for $J_i(t)$ then read:

$$\begin{cases} \dfrac{dJ_i(t)}{J_i(t)} = \sqrt{\hat{\nu}(t)}dW_i^i(t), \\ d\hat{\nu}(t) = \lambda(\hat{\nu}_0 - \hat{\nu}(t))dt + \hat{\eta}\sqrt{\hat{\nu}(t)}dW_\nu(t), \end{cases} \quad (14.61)$$

which resemble the dynamics of a standard Heston model, with appropriately shifted parameters. This implies that the model can be calibrated like the standard Heston model.

Pricing of caplets under the SV-LMM model

For a given Libor rate, with the money-savings account $M(t)$ as the númeraire, the price of a caplet or floorlet option, for $t \leq T_{i-1} \leq T_i$, is given by:

$$\frac{V_i^{\text{CPL/FL}}(t_0)}{M(t_0)} = N_i\tau_i\mathbb{E}^{\mathbb{Q}}\left[\frac{1}{M(T_i)}\max(\bar{\alpha}(\ell_i(T_{i-1}) - K), 0)\Big|\mathcal{F}(t_0)\right], \quad (14.62)$$

with $\bar{\alpha} = 1$ for a caplet and $\bar{\alpha} = -1$ for a floorlet and $M(t_0) = 1$.

By changing measures, from the risk-neutral $\mathbb{Q}$-measure associated with money-savings account $M(t)$ to the T_i-forward measure $\mathbb{Q}^{T_i}$, where the numéraire is the ZCB $P(t, T_i)$, we find,

$$\frac{d\mathbb{Q}^{T_i}}{d\mathbb{Q}}\bigg|_{\mathcal{F}(t)} = \frac{P(t, T_i)}{P(t_0, T_i)}\frac{M(t_0)}{M(t)}, \quad (14.63)$$

and the price in (14.62) can be connected to the derivation leading to Equation (12.31). With basic manipulations, this price is equal to:

$$V_i^{\mathrm{CPL/FL}}(t_0) = N_i \tau_i P(t_0, T_i) \mathbb{E}^{T_i} \left[\max(\bar{\alpha}(\ell_i(T_{i-1}) - K), 0) \big| \mathcal{F}(t_0) \right].$$
(14.64)

We recall that Libor rate, $\ell_i(t)$, is a martingale under its natural measure, i.e. under the T_i-forward measure.

Caplets and floorlets are priced under this SV-LMM model, by means of the connection of the displaced diffusion model with the Heston model dynamics. For this, certain parameter modifications need to be taken into account.

> The SV-LMM model is related to the Heston model via Equation (14.60),
> $$\jmath_i(t) = \ell_i(t) + (1-\beta)\frac{\ell_i(t_0)}{\beta},$$

The pricing equation is then given by,

$$V(t_0) = N_i \tau_i P(t_0, T_i) \mathbb{E}^{T_i} \left[\max \left(\bar{\alpha} \left(\jmath_i(T_{i-1}) - (1-\beta)\frac{\ell_i(t_0)}{\beta} - K \right), 0 \right) \Big| \mathcal{F}(t_0) \right],$$
$$= N_i \tau_i P(t_0, T_i) \mathbb{E}^{T_i} \left[\max(\bar{\alpha}(\jmath_i(t) - K^*), 0) \big| \mathcal{F}(t_0) \right],$$

with $K^* = (1-\beta)\frac{\ell_i(t_0)}{\beta} + K$, and process $\jmath_i(t)$ given by (14.61) with initial $\jmath_i(t_0) = \ell_i(t_0)/\beta$.

14.4 Risk management: The impact of a financial crisis

Due to the financial crisis, in the second decade of the 21st Century, the money market was in a difficult state, especially in Europe. Financial institutions were reluctant to invest and to lend out money because of a lack of confidence in the credibility of countries and companies. In order to re-gain confidence in the financial system, in particular the European central banks decided to intervene and to stimulate the monetary supply and demand. An exceptional measure undertaken was to *lower the interest rates*. It was expected that such measures would encourage investors to borrow money at a low rate and invest it into the economy, which, as a result, would again be stimulated.

In the context of financial risk management, important changes in the modeling of interest rate derivatives have taken place in the wake of the financial crisis. In this section, we will highlight two of these changes, i.e. the fact that *negative interest rates* are quoted in the financial markets and the fact that the frequency of payments has an impact on the price of an interest rate derivative. We start with the negative interest rates.

14.4.1 Valuation in a negative interest rates environment

Over multiple years in the above mentioned decade, the interest rates have been lowered, so that in 2014, for the first time in the Eurozone history, the *interest rates became negative in value* (in detail, minus 10 basis points). Keeping money in a bank account would give rise to a loss and not to a gain due to the interest rates. In Figure 14.6 a comparison of the interest rates (the yield curve) in the years 2008 and 2017 is presented. The yield curve in the graphs is based on the discount factors $P(0,t)$, $t > 0$, i.e.,

$$P(0,t) = e^{-r(t)t}, \quad r(t) = -\frac{1}{t} \log P(0,t). \tag{14.65}$$

$r(t)$ is then often called the "zero rate". As can be observed in Figure 14.6 (right side), up to 7 years the expected yield will remain negative. This means that investments with maturity times shorter than 7 years effectively will give rise to costs rather than income.

Different suitable model adaptations to deal with the possible negativity of the rates have been proposed in the industry. Moreover, interest rate models that would give rise to negative interest rates, under certain scenarios, were considered not sufficiently realistic, for many years. An example is the Hull-White (HW) model, which was discussed in detail in Section 11.3. Due to the availability of closed-form solutions for several interest rate products under this model, the HW model has however become a standard model for interest rates.

Some interest rate paths under the Hull-White model, which is calibrated to market data from pre-crisis and post-crisis times, are presented in Figure 14.7. It

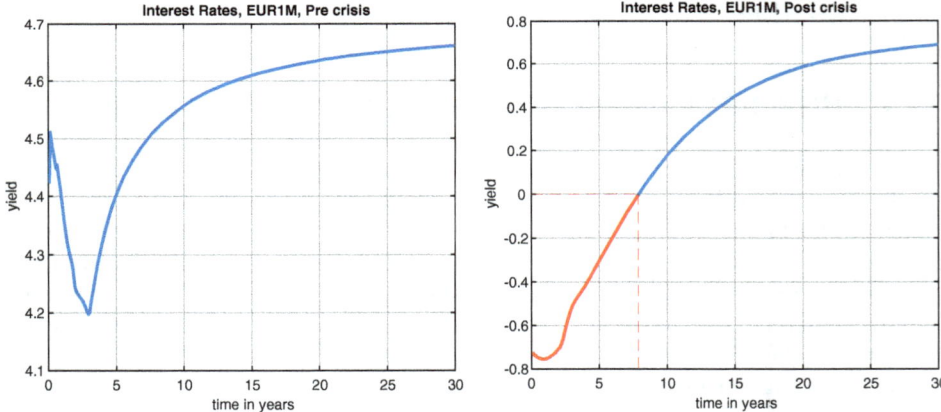

Figure 14.6: *The yield obtained from the 1M EUR curve, left: yield curve from the pre crisis and (right) from the post crisis times.*

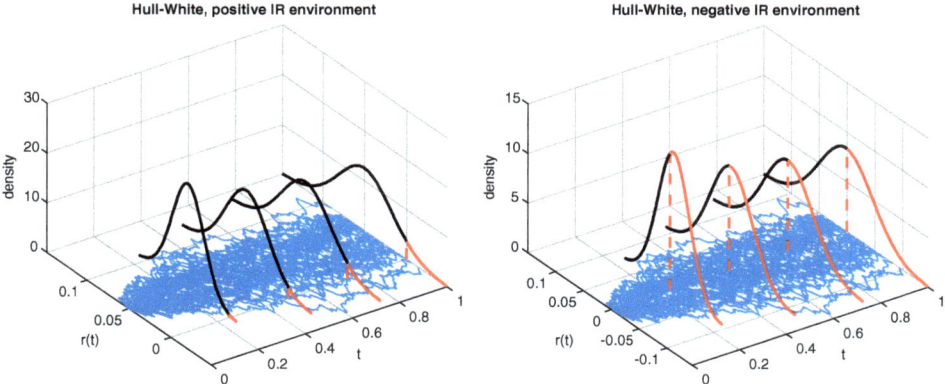

Figure 14.7: *Monte Carlo paths under the Hull-White model in positive and negative interest rate environments.*

is clear that in the post-crisis market data, a significant number of Monte Carlo paths take negative values.

Negative rates have become true, observable phenomena in the interest rate markets, whereas historically this was not the case. The negativity should thus be regarded as a *desirable property* of an interest rate model in post-crisis times. However, the accurate pricing of interest rate options requires a proper analysis since the fact that also forward rates can become negative needs to be modeled appropriately too.

An obvious example for which the pricing under negative rates needs to be modified is the *valuation of caplets*, as presented in Section 12.1. Based on the assumption of lognormal Libor rates, $\ell_i(t) := \ell(t; T_{i-1}, T_i)$, with the corresponding dynamics,

$$\mathrm{d}\ell_i(t) = \sigma_i \ell_i(t) \mathrm{d}W_i^i(t), \tag{14.66}$$

the pricing equation of a caplet is given by:

$$V_i^{\mathrm{CPL}}(t_0) = N_i \tau_i \mathbb{E}^{\mathbb{Q}} \left[\frac{M(t_0)}{M(T_i)} \max\left(\ell_i(T_{i-1}) - K, 0\right) \Big| \mathcal{F}(t_0) \right]$$

$$= N_i \tau_i P(t_0, T_i) \mathbb{E}^{T_i} \left[\max\left(\ell_i(T_{i-1}) - K, 0\right) \Big| \mathcal{F}(t_0) \right], \tag{14.67}$$

see (12.31), and the solution is given by

$$V_i^{\mathrm{CPL}}(t_0) = N_i \tau_i P(t_0, T_i) \left[\ell_i(t_0) N(d_1) - K_i N(d_2) \right], \tag{14.68}$$

with,

$$d_1 = \frac{\log\left(\frac{\ell_i(t_0)}{K}\right) + \frac{1}{2}\sigma_i^2(T_i - t_0)}{\sigma_i\sqrt{T_i - t_0}}, \quad d_2 = d_1 - \sigma_i\sqrt{T_i - t_0},$$

with $\ell_i(t_0) = \ell(t_0; T_{i-1}, T_i)$.

In a negative interest rate environment, the ZCBs $P(t_0, T_i)$ may get values that are higher than one unit of currency. This would imply that the Libor rate $\ell(t_0; T_{i-1}, T_i)$ would be negative. When the Libor rate is negative, the above pricing equation is not valid anymore, as the logarithm of a negative value is not well-defined.

Caplets can be priced under a negative interest rate environment by an *adaptation* of the underlying dynamics of the Libor rate in (14.66). Instead of GBM dynamics, the arithmetic Brownian motion (ABM) dynamics, that also give rise to *negative realizations* could, for example, be chosen. Such a solution, although straight-forward, has a significant disadvantage, however, as the normal distribution has much flatter distribution tails, compared to a lognormal process. Instead of completely changing the underlying dynamics, the industrial standard for dealing with the negativity has become to "shift" the original process.

This shifted process is defined, as follows,

$$\hat{\ell}_i(t) = \ell_i(t) + \theta_i, \qquad (14.69)$$

where the process $\hat{\ell}_i(t)$ is governed by a lognormal process, with the following dynamics,

$$\mathrm{d}\hat{\ell}_i(t) = \hat{\sigma}_i\hat{\ell}_i(t)\mathrm{d}W_i^i(t). \qquad (14.70)$$

The concept of shifting a distribution is very similar to the idea of "displacement", as discussed in Section 14.3.3.

The Libor rate $\ell_i(t)$, as observed in the market, is simply given by $\ell_i(t) := \hat{\ell}_i(t) - \theta_i$.

Shifting of the processes is equivalent to "moving" the probability density along the x-axis. As in Figure 14.8, different shift parameters θ_i will have a similar effect on $\hat{\ell}_i(t)$. The pricing of *caplets* is now given by,

$$V_i^{\mathrm{CPL}}(t_0) = N_i\tau_i P(t_0, T_i)\mathbb{E}^{T_i}\left[\max\left(\ell_i(T_{i-1}) - K, 0\right)\Big|\mathcal{F}(t_0)\right]$$

$$= N_i\tau_i P(t_0, T_i)\mathbb{E}^{T_i}\left[\max\left(\hat{\ell}_i(T_{i-1}) - \theta_i - K, 0\right)\Big|\mathcal{F}(t_0)\right]$$

$$= N_i\tau_i P(t_0, T_i)\mathbb{E}^{T_i}\left[\max\left(\hat{\ell}_i(T_{i-1}) - \hat{K}, 0\right)\Big|\mathcal{F}(t_0)\right], \qquad (14.71)$$

with $\hat{K} = K + \theta_i$.

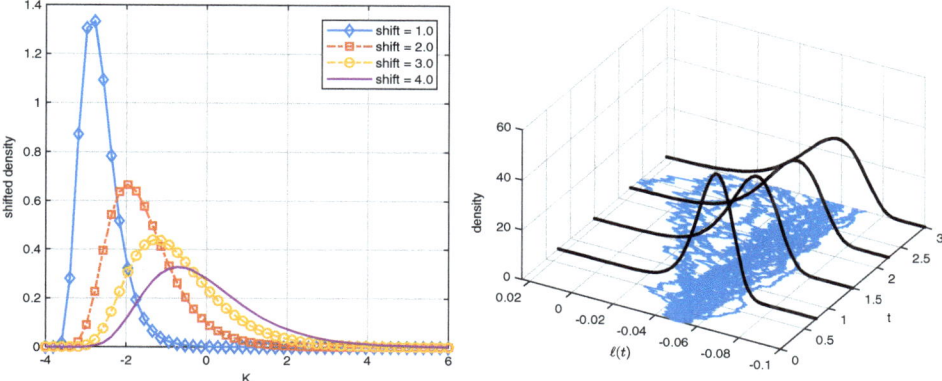

Figure 14.8: *Shifted lognormal distributions used for pricing in a negative interest rate environment.*

> Option pricing under shifted distributions is very convenient. The corresponding *solution* is in accordance with the unshifted variant,
>
> $$V_i^{\text{CPL}}(t_0) = N_i \tau_i P(t_0, T_i) \left[\hat{\ell}_i(t_0) N(d_1) - \hat{K}_i N(d_2) \right], \quad (14.72)$$
>
> with,
>
> $$d_1 = \frac{\log\left(\frac{\hat{\ell}_i(t_0)}{\hat{K}}\right) + \frac{1}{2}\sigma_i^2 (T_i - t_0)}{\sigma_i \sqrt{T_i - t_0}}, \quad d_2 = d_1 - \sigma_i \sqrt{T_i - t_0},$$
>
> with $\hat{K} = K + \theta_i$ and $\hat{\ell}_i(t_0) = \ell_i(t_0) + \theta_i$.

14.4.2 Multiple curves and the Libor rate

The market uses multiple curves when pricing financial products, which we will explain below.

We use an equally spaced tenor structure, $0 \leq T_0 < T_1 < \cdots < T_{m-1} < T_m$, where the spacing between dates T_{i-1} and T_i is typically related to a product's payment structure, like 3 or 6 months periods, depending on the currency.

Consider an investment over a period $[T_{i-1}, T_{i+1}]$ having two different tenor structures, i.e. with $\tau_i = T_i - T_{i-1}$ and $\hat{\tau}_i = \tau_i + \tau_{i-1} \equiv T_i - T_{i-2}$. Basically, we then have two possible strategies. Under the first tenor structure, we can invest for $[T_{i-1}, T_i]$ and re-invest for $[T_i, T_{i+1}]$, while under the second structure we invest over the entire period $[T_{i-1}, T_{i+1}]$, see Figure 14.9. Based on arbitrage arguments, the strategy involving re-investment should give exactly the same result as one

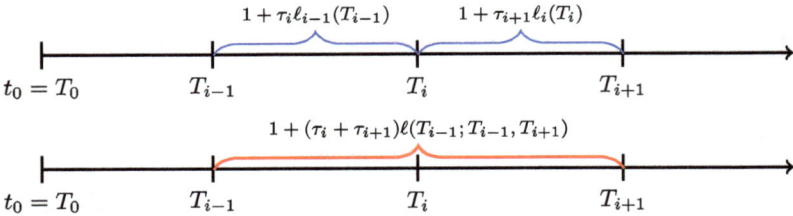

Figure 14.9: *Investment strategy from T_{i-1} until T_{i+1}, using the Libor rates under different tenor structures. Top: tenor structure $T_i - T_{i-1}$; bottom: tenor structure $T_{i+1} - T_{i-1}$.*

investment over the whole period, so that the following equality should hold,

$$\mathbb{E}^{\mathbb{Q}}\left[\frac{1}{M(T_{i+1})}\left(1+\tau_i \ell_i(T_{i-1})\right)\left(1+\tau_{i+1}\ell_{i+1}(T_i)\right)\right]$$
$$= \mathbb{E}^{\mathbb{Q}}\left[\frac{1}{M(T_{i+1})}(1+\hat{\tau}_{i+1}\ell(T_{i-1};T_{i-1},T_{i+1}))\right], \qquad (14.73)$$

with $\hat{\tau}_{i+1} = \tau_i + \tau_{i+1} = T_{i+1} - T_{i-1}$. Writing out the right-hand side of the equality, gives us

$$\mathbb{E}^{\mathbb{Q}}\left[\frac{1}{M(T_{i+1})}(1+\hat{\tau}_{i+1}\ell(T_{i-1};T_{i-1},T_{i+1}))\right]$$
$$= P(t_0,T_{i+1})\mathbb{E}^{\mathbb{Q}}\left[(1+\hat{\tau}_{i+1}\ell(T_{i-1};T_{i-1},T_{i+1}))\right].$$

Using the definition of the Libor rate, we find,

$$\mathbb{E}^{\mathbb{Q}}\left[\frac{1}{M(T_{i+1})}(1+\hat{\tau}_{i+1}\ell(T_{i-1};T_{i-1},T_{i+1}))\right]$$
$$= P(t_0,T_{i+1}) + \hat{\tau}_{i+1}P(t_0,T_{i+1})\ell(t_0;T_{i-1},T_{i+1})$$
$$= P(t_0,T_{i-1}).$$

Also the left-hand side of (14.73) equals $P(t_0, T_{i-1})$.

Equation (14.73) implies that if we trade in a swap with two float legs with different payment frequency, the value of this derivative should equal zero, i.e.

$$V^S(t_0) = N \cdot M(t_0)\mathbb{E}^{\mathbb{Q}}\left[\sum_{k_1=i+1}^{m_1}\frac{1}{M(T_{k_1})}\tau_{k_1}\ell_{k_1}(T_{k_1-1})\right.$$
$$\left. - \sum_{k_2=i+1}^{m_2}\frac{1}{M(T_{k_2})}\tau_{k_2}\ell_{k_2}(T_{k_2-1})\right]$$
$$= 0, \qquad (14.74)$$

with index k_1 corresponding to payments at $\{T_1, T_2, \ldots, T_{m-1}, T_m\}$ and index k_2 to less frequent payments, at $\{T_2, T_4, \ldots, T_m\}$, as in Figure 14.10. This financial

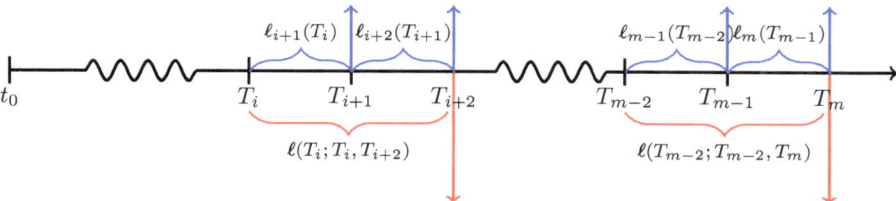

Figure 14.10: *Possible cash flows for a floating interest rate swap.*

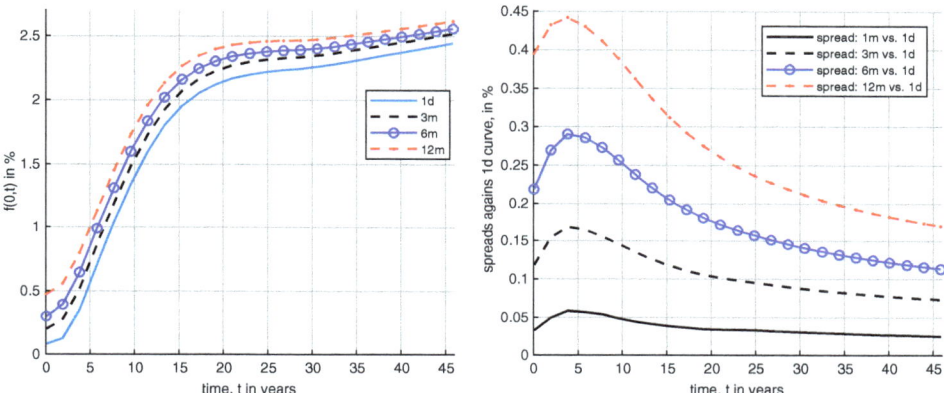

Figure 14.11: *Left: forward rates corresponding to different tenor curves, 1d, 3m, 6m and 12m; Right: spreads between different curves against the 1d curve.*

derivative is commonly known as a *basis swap*, which is a floating-floating interest rate swap. In the case of a *Euribor basis swap* there can be 3 month and 12 month Euribor cash flow exchanges.

Multi-curves and default probabilities

From a *risk management perspective*, however, it is safer to receive frequent payments, as a counterparty *may default* in between payments, and, in the case of frequent payments, there is less money lost if a default occurs.

Until the financial crisis from 2007/2008, the basis spreads based on different tenors, as in Equation (14.74), were negligible.[2] However, interest rate instruments with different tenors are nowadays characterized by different liquidity and credit risk premia, which is reflected in non-zero basis spread values, see Figure 14.11 for an illustration of the current spreads.

Before the mortgage crisis, discounting was based on a single curve which was used for all tenors and also for discounting. The existence of nonzero basis spreads in the market after the mortgage crisis essentially implies that,

---

[2]The reported differences for 3 month versus 6 month EUR basis swaps were up to 0.01% (between 2004–2007).

when modeling interest rate forwards, we need to distinguish forward rates with different tenor structure (with different frequencies).

It is therefore market practice to construct for each tenor a different forward curve, see the left-hand side of Figure 14.11. Each curve is based on a specific selection of interest rate derivatives, that are homogeneous in their tenor (typically 1 month, 3 months, 6 months, 12 months). On the other hand, financial derivatives involving different tenors should be discounted with a unique discount curve. An optimal choice for discounting is a curve which carries the smallest possible credit risk, which suggests that the discounting curve should correspond to the curve with the shortest tenor available on the market (which is typically 1 day). In the Euro zone this is the so-called EONIA (EURO OverNight Index Average) and in the US it is the Fed fund (US Federal Reserve overnight rate).

Notice that, under the multiple curves framework, well-known (single curve) no-arbitrage relations are not valid anymore.

Let us denote discounting by subscript "dc" and forecasting by "fc", then we have,

$$\mathbb{E}^{\mathbb{Q}}\left[\frac{\ell(T_{i-1}; T_{i-1}, T_i)}{M(T_i)}\right] = P_{\text{fc}}(t_0, T_i)\mathbb{E}^{T_i}\left[\ell(T_{i-1}; T_{i-1}, T_i)\right]$$

$$\neq P_{\text{dc}}(t_0, T_i)\mathbb{E}^{T_i}\left[\ell(T_{i-1}; T_{i-1}, T_i)\right]. \quad (14.75)$$

In other words, the measure-change machinery for changing measures from the spot to the forward measure does not comply with the separation of discounting and forecasting. Clearly, extended versions of the Libor rate and measure change are needed to address curve separation.

We introduce the *risk of default* into the lending transactions. Let t_D^B be a random variable indicating the first time probability of default of counterparty B. In Figure 14.12 a generalization of Figure 12.1 is depicted, where the payments take place *only if counterparty B didn't default* prior to the time of the transaction. So, at time t_0, two counterparties agree to make a transaction, if counterparty B didn't go in default prior to time T_1, counterparty A will lend at time T_1 an amount of 1€ to B and, in the case of no default at T_2, counterparty

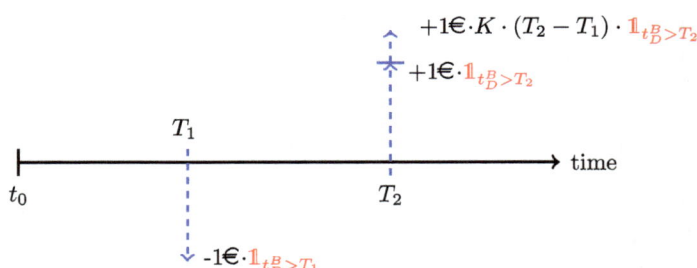

Figure 14.12: *Cash flows between two counterparties including the probability of default.*

B will return 1€ at time T_2 with the additional interest which is proportional to the accruing time $T_2 - T_1$, i.e. 1€ $\cdot K \cdot (T_2 - T_1)$.

Since the payments at times T_1 and T_2 are uncertain, we need to incorporate this information while calculating the fair value of the trade. Assuming independence between the time to default, t_D^B, and the interest rates, we find,

$$V(t_0) = \mathbb{E}^{\mathbb{Q}} \left[\frac{-1}{M(T_1)} \mathbb{1}_{t_D^B > T_1} + \frac{1}{M(T_2)} (1 + K \cdot (T_2 - T_1)) \mathbb{1}_{t_D^B > T_2} \Big| \mathcal{F}(t_0) \right]$$

$$= P(t_0, T_2) (1 + K \cdot (T_2 - T_1)) \mathbb{E}^{\mathbb{Q}} \left[\mathbb{1}_{t_D^B > T_2} \Big| \mathcal{F}(t_0) \right]$$

$$- P(t_0, T_1) \mathbb{E}^{\mathbb{Q}} \left[\mathbb{1}_{t_D^B > T_1} \Big| \mathcal{F}(t_0) \right].$$

The expectations above are now associated with a survival probability, i.e. for $i = 1, 2$,

$$\mathbb{E}^{\mathbb{Q}} \left[\mathbb{1}_{t_D^B > T_i} \Big| \mathcal{F}(t_0) \right] = \mathbb{Q} \left[t_D^B > T_i \right] = 1 - F_{t_D^B}(T_i) =: e^{-\int_{t_0}^{T_i} h(s) ds}, \quad (14.76)$$

where $h(s)$ represents the deterministic *hazard rate*. Typically, the hazard rate for a counterparty can be determined based on credit derivatives, like *Credit Default Swaps, (CDS)*.

The option value $V(t_0)$ can be written as,

$$V(t_0) = P(t_0, T_2)(1 + K \cdot (T_2 - T_1)) D(t_0, T_2) - P(t_0, T_1) D(t_0, T_1), \quad (14.77)$$

with $D(t_0, T_i) = e^{-\int_{t_0}^{T_i} h(s) ds}$. The fair value K, for which the contract in Equation (14.77) equals 0 at the inception time t_0, i.e. $V(t_0) = 0$, is given by,

$$K = \frac{1}{(T_2 - T_1)} \left(\frac{P(t_0, T_1)}{P(t_0, T_2)} \frac{D(t_0, T_1)}{D(t_0, T_2)} - 1 \right). \quad (14.78)$$

As in the single curve setting, the *fair* strike value at which two counterparties agree to exchange funds is called the Libor rate and it is denoted by $\hat{\ell}_i(t) := \hat{\ell}(t; T_{i-1}, T_i)$. In this setting, however, the rate $\hat{\ell}_i(t)$ depends on the creditworthiness of the counterparty and on the tenor $\tau_i = T_i - T_{i-1}$,

$$\hat{\ell}(t; T_{i-1}, T_i) = \frac{1}{\tau_i} \left(\frac{P(t_0, T_{i-1})}{P(t_0, T_i)} \frac{D(t_0, T_{i-1})}{D(t_0, T_i)} - 1 \right), \quad (14.79)$$

Practitioners often associate the hazard rate $h(s)$ in the definition of $D(t_0, T_i)$ as a *spread between the risk-free and the unsecured rate* for different tenors. This implies that $D(t_0, T_i)$ can be interpreted as a discount factor. Equation (14.79) shows that, to determine a fair price of the *unsecured* Libor rate $\hat{\ell}_i(t)$, a *secured*, risk-free, curve is needed, from which the ZCB $P(t_0, T_i)$ should be computed. Given that for each tenor structure, $\tau_i = T_i - T_{i-1}$, a basis spread $D_{\tau_i}(t_0, T_i)$

exists (where the index τ_i highlights that $D_{\tau_i}(t_0, T_i)$ corresponds to a particular tenor), we define a new, unsecured ZCB $P_{\tau_i}(t_0, T_i)$, as follows,

$$P_{\tau_i}(t_0, T_i) := P(t_0, T_i) \cdot D_{\tau_i}(t_0, T_i). \qquad (14.80)$$

Equation (14.80) also indicates how to construct a curve associated with a particular tenor τ_i, i.e. start with the estimation of the risk-free curve (the curve of the shortest tenor, often 1 day), once this *base curve* is established a function $D_{\tau_i}(t_0, T_i)$, corresponding to a particular tenor structure, is determined based on suitable market instruments.

With Equation (14.80) in hand, we rewrite the risky Libor rate, as

$$\hat{\ell}(t; T_{i-1}, T_i) = \frac{1}{\tau_i}\left(\frac{P_{\tau_i}(t, T_{i-1})}{P_{\tau_i}(t, T_i)} - 1\right). \qquad (14.81)$$

With the description above, pricing of an interest rate product given a risk-free curve $P(t_0, T_i)$ and a risky curve $P_{\tau_i}(t_0, T_i)$ can be defined.

14.4.3 Valuation in a multiple curves setting

When pricing interest rate derivatives, the risk of default should be taken into account properly. When a risky Libor rate $\hat{\ell}(t; T_{i-1}, T_i)$ is determined, it is important to discount the future cash flows that depend on this Libor rate. Usually, discounting is based on the so-called *risk-free rate*, however, as discussed, a rate which relates to a risky bond is not representative for the risk-free rate. After the crisis, the consensus has been made that the best approximation for the risk-free rate is the overnight index swap rate (OIS), see Hull and White [2012]. The contractual agreements of the *overnight index swap* are that a counterparty pays a fixed rate and receives the daily-compounded overnight rate.

By combining the concepts of the risky Libor rate and OIS discounting, for the pricing of a basic interest rate derivative, like a (payer/receiver) interest rate swap, see Equation (12.10), under the T_k-forward measure $\mathbb{Q}^{T_k}$, as in Equation (11.61), we find,

$$V^{\text{PS,RS}}(t_0) = \bar{\alpha} \cdot N \sum_{k=i+1}^{m} \tau_k P(t_0, T_k)\left(\mathbb{E}^{T_k}\left[\hat{\ell}_k(T_{k-1})|\mathcal{F}(t_0)\right] - K\right),$$

where $P(t_0, T_k)$ is now the risk-free bond corresponding to the overnight rate.

As illustrated in Bianchetti [2010], the current market consensus is to approximate the above expectation by the forward rate $\hat{\ell}_k(t_0)$, i.e.,

$$V^{\text{PS,RS}}(t_0) \approx \bar{\alpha} \cdot N \sum_{k=i+1}^{m} \tau_k P(t_0, T_k)\left(\hat{\ell}_k(t_0) - K\right), \qquad (14.82)$$

where in the last step of the derivation the following approximation, $\mathbb{E}^{T_k}[\hat{\ell}_k(t)|\mathcal{F}(t_0)] \approx \hat{\ell}_k(t_0)$, was used. Clearly, this is merely an approximation, which is not based on a proper measure transformation. To accomplish the task of consistent derivatives pricing, one can either follow the analogy in the foreign-exchange world described in Bianchetti [2010], where the connection between the risky and risk-free rates has been made via a so-called "quanto correction", or follow the methodology in Mercurio [2010], where a process for $\bar{\psi}_k(t) = \mathbb{E}^{T_k}\left[\hat{\ell}_k(T_{k-1})|\mathcal{F}(t)\right]$ was employed.

14.5 Exercise set

Exercise 14.1 Show that Equation (14.73) is valid by writing out the left-hand side.

Exercise 14.2 Consider two SDEs, one with a time-dependent volatility function, and another with a constant volatility:

$$dS(t) = \sigma(t)(\beta S(t) + (1-\beta)S_0)\sqrt{V(t)}dW_x(t),$$
$$d\hat{S}(t) = \sigma(\beta \hat{S}(t) + (1-\beta)\hat{S}_0)\sqrt{V(t)}dW_x(t),$$

with $t_0 = 0$ and $\hat{S}(0) = \hat{S}_0 = S_0$, and both processes sharing the same variance process,

$$dv(t) = \kappa(v_0 - v(t))dt + \gamma\sqrt{v(t)}dW_v(t),$$

where $dW_x(t)dW_v(t) = 0$.

a. Show that the following equality holds true,

$$\mathbb{E}\left[g\left(\int_0^T \sigma^2(t)v(t)dt\right)\right] = \mathbb{E}\left[g\left(\sigma^2 \int_0^T v(t)dt\right)\right],$$

with,

$$g(x) = \frac{S_0}{\beta}\left(2\phi\left(\frac{1}{2}\beta\sqrt{x}\right) - 1\right),$$

where $\phi(\cdot)$ is a normal cumulative distribution function.

b. Confirm this equality numerically (choosing parameters yourself).

Exercise 14.3 Consider the following dynamics of the stock price process, under the $\mathbb{Q}$-measure,

$$dS(t) = \sigma(\beta S(t) + (1-\beta)S_0)dW^{\mathbb{Q}}(t),$$

with $r = 0$, $S_0 > 0$ and $\beta \in [0,1]$.

a. Find the solution for $S(t)$ (analytically and numerically).

b. Take $\beta = 0.5$, $\sigma = 0.15$, $T = 2$ and $S_0 = 1$ and calculate European option prices for the strike prices K ranging from 0.5 to 2, with steps of 0.1.

c. Compute for these European option prices the corresponding Black-Scholes implied volatilities (using $T = 3$). Draw the results and discuss the shape of the implied volatility w.r.t. variations of the parameters β and σ (try also negative β-values).

Exercise 14.4 "OIS" is an abbreviation which stands for "overnight index swaps". In the standard setting, this overnight index is based on a specified published index of the daily overnight rates. The time of the coupon payments of the OIS may range from 1 week up to 2 years. We denote the daily rate by $\ell(t, T_{i-1}, T_i)$, with $\tau_i = T_i - T_{i-1} = 1d$ (letter "d" indicates "day") show that the daily geometric averaging (often called *geometric compounding*) of the rate over some period $[T_0, T_m]$ equals the forward rate over the same time horizon, i.e., show that,

$$\prod_{i=1}^m (1 + \tau_i \ell(t, T_{i-1}, T_i)) - 1 = (T_m - T_0)\ell(t, T_0, T_m).$$

Table 14.2: *Caplet option prices.*

K	−2%	−1.5%	−0.5%	0%	1%	3%
caplet price	0.0522	0.0468	0.0383	0.0344	0.0302	0.0247
implied volatility [%]						

Exercise 14.5 With $\ell_i(t) := \ell(t, T_{i-1}, T_i)$ and $\tau_i = T_i - T_{i-1}$ show that the following equalities hold,

$$P(t_0, T_{i-1}) = \mathbb{E}^\mathbb{Q}\left[\frac{1}{M(T_{i+1})}\left(1 + \tau_i\ell_i(T_{i-1})\right)\left(1 + \tau_{i+1}\ell_{i+1}(T_i)\right)\big|\mathcal{F}(t_0)\right],$$

$$P(t_0, T_{i-1}) = \mathbb{E}^\mathbb{Q}\left[\frac{1}{M(T_i)}\left(1 + \tau_i\ell_i(T_{i-1})\right)\big|\mathcal{F}(t_0)\right].$$

Exercise 14.6 Consider the problem of pricing a caplet, see Definition 12.2.1, in a negative interest rate environment. For a caplet at time t_0, $V_0^{\text{CPL}}(t_0)$, with notional $N = 1$, accrual $\tau_i = 1$, maturity $T = 5$, forward rate $\ell_i(t_0) = 4\%$, the ZCB $P(t_0, T_i) = 0.85$ and $\theta_i = 4\%$, we observe the following prices (see Table 14.2). Compute the "shifted implied volatilities" and fill in the missing entries in the table.

Exercise 14.7 Consider the so-called *"double-Heston model"* for the forward rate $\ell_i(t) := \ell(t, T_{i-1}, T_i)$, which is described by the following dynamics,

$$d\ell_i(t)/\ell_i(t) = \sqrt{v_1(t)}dW^i_{i,1}(t) + \sqrt{v_2(t)}dW^i_{i,2}(t),$$

$$dv_1(t) = (\bar{v}_1 - v_1(t))\,dt + \sqrt{v_1(t)}dW_{v,1}(t),$$

$$dv_2(t) = (\bar{v}_2 - v_2(t))\,dt + \sqrt{v_2(t)}dW_{v,2}(t),$$

with all the correlation parameters equal to 0.

Find the parameters of the process $\xi(t)$, such that the model can be reformulated into the following model,

$$d\ell_i(t)/\ell_i(t) = \sqrt{\xi(t)}dW^i_*(t),$$

$$d\xi(t) = (\bar{\xi} - \xi(t))\,dt + \sqrt{\xi(t)}dW_\#(t).$$

Exercise 14.8 Consider the Heston model for the Libor rate $\ell_i(t) := \ell(t, T_{i-1}, T_i)$, which is given by,

$$d\ell_i(t)/\ell_i(t) = \sqrt{v(t)}dW_i^{T_i}(t),$$

$$dv(t) = \kappa\left(\bar{v} - v(t)\right)dt + \gamma\sqrt{v(t)}dW_v(t),$$

with some basic model parameters.

Consider the process $\sigma(t) = \sqrt{v(t)}$.

a. Apply Itô's lemma and show that the process $\sigma(t)$ resembles an Ornstein-Uhlenbeck process, with the long-term mean reversion level which is state-dependent and nonstationary. Which problem is encountered when applying Itô's lemma?

b. Discuss the Feller condition for the new process $\sigma(t)$.

Exercise 14.9 Consider the dates $T_{i-1} = 5$, $T_i = 6$, the forward rate $\ell_i(t_0) = 0.13$ and assume the Black-Scholes model for the dynamics of $\ell_i(t)$,

$$d\ell_i(t) = \sigma \ell_i(t) dW_i^i(t).$$

Compute $\mathbb{E}^{T_i}\left[\ell_i^2(T_{i-1})\right]$ using the Breeden-Litzenberger method given in Equation (14.39). Vary the volatility parameter σ and compare the results obtained to those obtained with a Monte Carlo method.

CHAPTER 15

Cross-Currency Models

In this chapter:

In this chapter, the *foreign-exchange (FX) asset class* is the focus. An introduction into the FX world is provided in **Section 15.1**.
Several of the modeling approaches and computational techniques that were used in the previous chapters are generalized to the FX models. The Black-Scholes model for the FX rate is detailed in this section.
Next, systems of SDEs are discussed. Whereas the Black-Scholes Hull-White FX model is included as an exercise, the *Heston-type FX model*, in which the interest rates are short-rate processes that are correlated with the FX process, is derived in **Section 15.2**. In this model, a *full matrix of correlations* is included.
Fast model evaluation is highly desirable for FX options in practice, especially during the calibration of the model. This is the main motivation for the generalization of the linearization techniques, that were presented in Chapter 13, to the world of foreign exchange rates. *Credit valuation adjustment* in the context of an FX swap is also discussed.
The interest rate model in this section *cannot be used to generate implied volatility smiles or skews*, as commonly observed in the interest rate market.
In **Section 15.3** an FX-interest rate hybrid model with the FX modeled by a *Heston model with the interest rates driven by the Libor market model* is presented.
By *changing the measures*, from the risk-neutral to the forward measure, associated with the zero-coupon bond as the numéraire, the dimension of the approximating characteristic function can be significantly reduced.

Keywords: foreign-exchange (FX), correlation between asset classes, characteristic function, Heston-type FX models, Gaussian interest rate processes, Libor market interest rates, forward measure, FX CVA.

15.1 Introduction into the FX world and trading

We start this introduction with a brief historical discussion.

15.1.1 FX markets

The foreign exchange market has grown tremendously over several decades. It is well-known that before WWI the role of the exchange rates were taken by the commodity *gold*. Each country under the gold standard would express its currency in terms of the gold price, e.g., two countries A and B would quote their prices for gold in their local currencies as x and y, respectively, then the exchange rate would effectively be either x/y or y/x, depending on the country. The gold standard as the reference for the value of currencies came to an end in 1933, when president Franklin D. Roosevelt's administration restricted substantial private gold ownership essentially to jewelers. When the gold standard was abandoned, many countries pegged their exchange rates to the U.S. dollar, and the U.S. dollar was in turn directly connected to the price of gold. Since 1971, the U.S. dollar is no longer fixed to the gold price and, as such, currencies are no longer linked to a commodity which can be bought, sold or stored. As a consequence, the value of a currency is mainly driven by the supply and demand, and by the products that are traded in the local markets.

Initially, governments attempted to prescribe the exchange rates themselves to improve a country's trading position, e.g. in a scenario where a country would set the exchange rate lower, relative to the other rates, it would improve the country's export by making the export relatively more affordable, while import from other countries would be relatively more expensive. On multiple occasions, however, such attempts have led to trade wars. On the other hand, for countries that depend on foreign debt, a lower exchange rate would imply higher installments to be paid, especially if loans are taken in a foreign currency. Currently, the major currency exchange rates are determined by supply and demand (i.e., *exchange rates are "floating"*). It is still common practice for countries and their central banks to "fine tune" the exchange rates by keeping gold reserves or foreign currencies, that are then known as foreign exchange reserves. They buy and sell these currencies to stabilize their own currency, when necessary.

Let us consider the euro €, to be the domestic currency and the dollar \$, to be the foreign currency. We set the FX rate to $y_{\$}^{€}(t_0) =: y(t_0)$, expressed in units of the domestic currency, per unit of a foreign currency. If the exchange rate is $y_{\$}^{€}(t_0) = 0.85$ and we wish to exchange \$100 to euros, the calculation is given by:

$$\$100 \cdot y_{\$}^{€}(t_0) = \$100 \cdot 0.85 \frac{€}{\$} = €85.$$

In this example, the amount \$100 is called the *notional in foreign currency*, which will be denoted by N_f. So, a notional in the foreign currency $\$N_f$ is equal to $\$N_f \cdot y(t)$ in the domestic currency (€).

More generally, we will use in this chapter the notation

$$y(t) := y_{\mathrm{f}}^{\mathrm{d}}(t).$$

We exchange a foreign amount of money to our base, domestic, currency.

15.1.2 Forward FX contract

One of the most liquidly traded FX products is the so-called "outright FX forward". This contract is an obligation for a physical exchange of funds at a future date at an agreed rate and there is no payment upfront. Such a forward contract is typically well-suited for hedging the foreign exchange risk at a single payment date (the so-called *"bullet payment"*), as opposed to a stream of FX payments. A stream of FX payments is usually covered by a contract which is called *an FX swap*. In practice, however, forwards are sometimes favored as they are better affordable, albeit less effective, hedging instruments than the swaps.

A *forward contract* is equivalent to borrowing and lending the same amount in two different currencies and converting the proceeds in the domestic currency. Since one borrows and lends the same amount, the initial value has to be zero, as it is usual for a forward contract when the forward price is equal to the forward exchange rate under no arbitrage.

Let us consider a (static) replication strategy for an FX forward contract. We consider two currencies, € and \$, with the corresponding interest rates in their markets, indicated here by $r_€$ and $r_\$$, respectively. In Table 15.1, an example is given of how an FX forward contract can be replicated. At time t_0 borrowing takes place in USD. The interest rate to be paid is $r_\$$, and subsequently this money is exchanged at the spot FX rate to EUR. After the money is exchanged, it is loaned in the EUR market, where it will grow with interest rate $r_€$. At the contract's maturity time T, the above transactions will be reversed. The forward rate $y_F(t_0, T)$ is defined such that the contract value equals zero,

$$\$1 \cdot \mathrm{e}^{r_\$ T} \boxed{y_F(t_0, T)} = \$1 \cdot y(t_0) \cdot \mathrm{e}^{r_€ T}, \qquad (15.1)$$

and therefore,

$$y_F(t_0, T) = y(t_0) \frac{\mathrm{e}^{r_€ T}}{\mathrm{e}^{r_\$ T}} =: y(t_0) \frac{P_\$(t_0, T)}{P_€(t_0, T)}, \qquad (15.2)$$

Table 15.1: *Replication strategy for an FX forward.*

time t_0	time T
borrow in \$: \$1	return: $\$1 \cdot \mathrm{e}^{r_\$ T}$
lend in € : $\$1 \cdot y(t_0)$	obtain: $\$1 \cdot y(t_0) \cdot \mathrm{e}^{r_€ T}$

where the foreign and domestic zero coupon bonds are defined as $P_\$(t,T) = e^{-r_\$ T}$ and $P_{\euro}(t,T) = e^{-r_{\euro} T}$, respectively.

Another way to define the forward FX rate is to connect it to a *future FX rate which is seen from today's perspective*. Suppose we may obtain at a future time T some amount $N_{\text{f}}(T)$ in a foreign currency. There are two possibilities to calculate today's value in the domestic currency d. On the one hand, the notional amount $N_{\text{f}}(T)$ is discounted to today with the foreign currency, giving $N_{\text{f}}(T)/M_{\text{f}}(T)$; subsequently this amount needs to be exchanged at today's (spot) exchange rate to the domestic currency, yielding,

$$V^{\text{FX}}(t_0) = \mathbb{E}^{\mathbb{Q}^{\text{f}}}\left[y(t_0)\frac{N_{\text{f}}(T)}{M_{\text{f}}(T)}\bigg|\mathcal{F}(t_0)\right] = y(t_0)\frac{N_{\text{f}}(T)}{M_{\text{f}}(T)}, \quad (15.3)$$

where $\mathbb{Q}^{\text{f}}$ denotes the foreign currency risk-neutral measure.

On the other hand, one may exchange $N_{\text{f}}(T)$ to the domestic currency at time T based on the exchange rate $y(T)$, and then the amount $N_{\text{f}}(T)y(T)$ will be expressed in the domestic currency. This amount needs to be discounted to time t_0 with a domestic interest rate, i.e.,

$$V^{\text{FX}}(t_0) = \mathbb{E}^{\mathbb{Q}}\left[\frac{y(T)N_{\text{f}}(T)}{M_{\text{d}}(T)}\bigg|\mathcal{F}(t_0)\right] = \frac{N_{\text{f}}(T)}{M_{\text{d}}(T)}\mathbb{E}^{\mathbb{Q}}\left[y(T)\bigg|\mathcal{F}(t_0)\right], \quad (15.4)$$

where $\mathbb{Q} = \mathbb{Q}^{\text{d}}$ denotes the commonly used domestic currency risk-neutral measure (as typically domestic currency is exchanged to foreign currency). The numeraire can be taken out of the expectation as we deal with a deterministic interest rate in this section.

These two ways to calculate today's value of the contract should result in the same value (see Figure 15.1), as otherwise an arbitrage opportunity would occur. Equating (15.3) and (15.4), i.e.

$$y(t_0)\frac{N_{\text{f}}(T)}{M_{\text{f}}(T)} = \frac{N_{\text{f}}(T)}{M_{\text{d}}(T)}\mathbb{E}^{\mathbb{Q}}\left[y(T)|\mathcal{F}(t_0)\right], \quad (15.5)$$

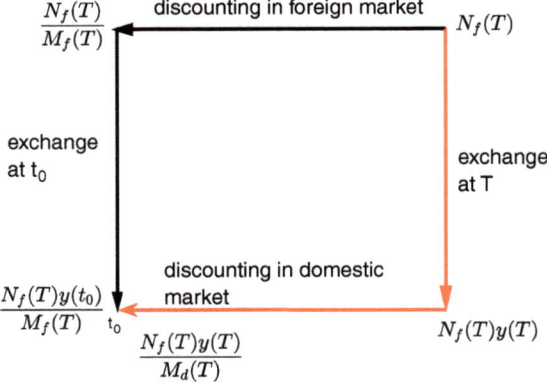

Figure 15.1: *Schematic representation of discounting foreign amount to domestic currency.*

gives,

$$y_F(t_0, T) := \mathbb{E}^{\mathbb{Q}}[y(T)|\mathcal{F}(t_0)] = y(t_0)\frac{N_f(T)}{M_f(T)}\frac{M_d(T)}{N_f(T)} = y(t_0)\frac{M_d(T)}{M_f(T)}$$

$$= y(t_0)\frac{P_f(t_0, T)}{P_d(t_0, T)}. \tag{15.6}$$

In other words, the forward exchange rate $y_F(t_0, T)$ is defined as the expectation of the future spot exchange rate $y(T)$.

15.1.3 Pricing of FX options, the Black-Scholes case

Let us consider the task of pricing a European-type option on the FX rate $y(T)$, with payoff $\max(y(T) - K, 0)$. For simplicity, we assume the foreign and domestic interest rates, r_f and r_d, respectively, to be constant. The stochastic process for the FX rate $y(t)$ is here defined as,

$$dy(t) = \mu y(t)dt + \sigma y(t)dW^{\mathbb{P}}(t). \tag{15.7}$$

These dynamics are under the real-world measure $\mathbb{P}$. To determine the dynamics *under the domestic risk-neutral measure*, the arbitrage-free principle is leading. A money-savings account in the foreign market, $M_f(t)$, is expressed in the domestic currency by means of the exchange rate. Therefore, in the domestic market the foreign savings account is worth $y(t)M_f(t)$.

Any discounted asset in the domestic market needs to be a martingale, so that, $\chi(t) := y(t)\frac{M_f(t)}{M_d(t)}$, needs to a martingale under the domestic risk-neutral measure $\mathbb{Q}^d$. By application of Itô's lemma, we find:

$$d\chi(t) = (r_f - r_d)\frac{M_f(t)}{M_d(t)}y(t)dt + \frac{M_f(t)}{M_d(t)}dy(t)$$

$$= (r_f - r_d)\frac{M_f(t)}{M_d(t)}y(t)dt + \mu y(t)\frac{M_f(t)}{M_d(t)}dt + \sigma\frac{M_f(t)}{M_d(t)}y(t)dW^{\mathbb{P}}(t),$$

and thus

$$d\chi(t)/\chi(t) = (r_f - r_d)dt + \mu dt + \sigma dW^{\mathbb{P}}(t). \tag{15.8}$$

Process $\chi(t)$ will be a martingale if the governing dynamics are free of drift terms. This implies the following change of measure,

$$dW_d^{\mathbb{Q}}(t) = \frac{r_f - r_d + \mu}{\sigma}dt + dW^{\mathbb{P}}(t). \tag{15.9}$$

Returning to (15.7), the following dynamics of the FX process $y(t)$, under domestic risk-neutral measure $\mathbb{Q}^d$, are obtained:

$$dy(t) = \mu y(t)dt + \sigma y(t)dW^{\mathbb{P}}(t)$$

$$= \mu y(t)dt + \sigma y(t)dW_d^{\mathbb{Q}}(t) - \sigma y(t)\frac{r_f - r_d + \mu}{\sigma}dt.$$

After simplifications, the following dynamics for process $y(t)$ are obtained,
$$dy(t) = (r_d - r_f) y(t) dt + \sigma y(t) dW_d^{\mathbb{Q}}(t).$$

Note that $y(t)$ is a *lognormally distributed random variable*, with the density thus given by:

$$f_y(T, y; t_0, y(t_0)) = \frac{1}{\sigma y \sqrt{2\pi(T-t_0)}} \exp\left(-\frac{\left(\log \frac{y}{y(t_0)} - \left(r_d - r_f - \frac{1}{2}\sigma^2\right)(T-t_0)\right)^2}{2\sigma^2(T-t_0)}\right).$$
(15.10)

With the FX process, which is defined under the domestic risk-neutral measure, the value of an FX call option, with deterministic interest rates, is thus given by,

$$V_c^{\text{FX}}(t_0) = \mathbb{E}^{\mathbb{Q}}\left[\frac{M_d(t_0)}{M_d(T)} \max(y(T) - K, 0) \Big| \mathcal{F}(t_0)\right]$$
$$= e^{-r_d(T-t_0)} \mathbb{E}^{\mathbb{Q}}\left[\max(y(T) - K, 0) \Big| \mathcal{F}(t_0)\right]$$
$$= e^{-r_d(T-t_0)} \mathbb{E}^{\mathbb{Q}}\left[y(T) \mathbf{1}_{y(T)>K} \Big| \mathcal{F}(t_0)\right] - e^{-r_d(T-t_0)} K \mathbb{Q}\left[y(T) > K\right],$$

so that the solution reads,

$$V_c^{\text{FX}}(t_0) = e^{-r_f(T-t_0)} y(t_0) F_{\mathcal{N}(0,1)}(d_1) - K e^{-r_d(T-t_0)} F_{\mathcal{N}(0,1)}(d_2),$$

with

$$d_1 = \frac{\log \frac{y(t_0)}{K} + \left(r_d - r_f + \frac{1}{2}\sigma^2\right)(T-t_0)}{\sigma\sqrt{T-t_0}}, \quad d_2 = d_1 - \sigma\sqrt{T-t_0}.$$

Remark 15.1.1 (Advanced FX models) *The literature on modeling FX rates is rich and many stochastic models are available. An industrial standard is a model from [Sippel and Ohkoshi, 2002], where the log-normally distributed FX dynamics were proposed and Gaussian, one-factor, interest rates are employed. This model gives rise to analytic expressions for the prices of basic FX option products for at-the-money options.*

A hybrid FX model which is based on the Heston dynamics for the FX process, in combination with correlated stochastic interest rates, leads to an interesting model. In Section 13.2, this extension of the Heston model with stochastic interest rates was already discussed, where the interest rate processes were short-rate processes, like the Hull-White model.

A Gaussian interest rate model was also used in [Piterbarg, 2006], in which a local volatility model was employed for generating the implied volatility skew which is present in the FX market. Stochastic volatility FX models have been investigated, for example, in [van Haastrecht and Pelsser, 2010], where the Schöbel-Zhu model was applied for pricing FX rates in combination with short-rate processes.

Cross-Currency Models

In [Andreasen, 2006] and [Giese, 2006] an indirectly imposed correlation structure between Gaussian short-rates and the FX process was proposed. The model is intuitively appealing, but it may give rise to large model parameters [Antonov et al., 2008]. An alternative model was given in [Antonov et al., 2008; Antonov and Misirpashaev, 2006], in which calibration formulas were developed by means of Markov projection techniques. ▲

15.2 Multi-currency FX model with short-rate interest rates

Here we discuss the FX Heston-type models, in which the interest rates are stochastic processes, that are correlated with the governing FX process. We first discuss the Heston FX model with Gaussian i.e. Hull-White [Hull and White, 1990], short-rate processes. In this model, a full matrix of correlations is prescribed. Short-rate interest rate models can typically provide a satisfactory fit to at-the-money interest rate products.

In practice, the FX calibration is performed with an a-priori calibrated interest rate model. A highly efficient and fast model evaluation is required. We focus on the efficient evaluation for the vanilla FX options under this hybrid process, and assume that the parameters for the short-rate model have been determined. For plain vanilla European options on a whole *strip of strike prices*, the approximate hybrid model in this section can be evaluated in just milliseconds.

The model describes the spot FX, $y(t)$, which is expressed in units of the domestic currency, per unit of a foreign currency.

The analysis starts with the specification of the underlying short-rate processes, $r_d(t)$ and $r_f(t)$, under their spot measures, i.e. the $\mathbb{Q}$–domestic and $\mathbb{Q}^f$–foreign measures. They are governed by the Hull-White [Hull and White, 1990] one-factor model,

$$\begin{cases} dr_d(t) = \lambda_d(\theta_d(t) - r_d(t))dt + \eta_d dW_d^{\mathbb{Q}}(t), \\ dr_f(t) = \lambda_f(\theta_f(t) - r_f(t))dt + \eta_f dW_f^{\mathbb{Q}^f}(t), \end{cases} \quad (15.11)$$

where $W_d^{\mathbb{Q}}(t)$ and $W_f^{\mathbb{Q}^f}(t)$ are Brownian motions under $\mathbb{Q}$ and $\mathbb{Q}^f$, respectively. Parameters λ_d, λ_f determine the speed of mean reversion to the time-dependent term structure functions $\theta_d(t)$, $\theta_f(t)$, and the parameters η_d, η_f are the respective volatility coefficients.

These processes, under the appropriate measures, are linear in their state variables, so that, for a given maturity T, with $0 \leq t \leq T$, the ZCBs are of the following form:

$$\begin{cases} P_d(t,T) = \exp\left(\bar{A}_d(t,T) + \bar{B}_d(t,T)r_d(t)\right), \\ P_f(t,T) = \exp\left(\bar{A}_f(t,T) + \bar{B}_f(t,T)r_f(t)\right), \end{cases} \quad (15.12)$$

with $\bar{A}_d(t,T)$, $\bar{A}_f(t,T)$ and $\bar{B}_d(t,T)$, $\bar{B}_f(t,T)$ analytically known functions.

The money market accounts in this model are respectively given by,

$$dM_d(t) = r_d(t)M_d(t)dt, \quad dM_f(t) = r_f(t)M_f(t)dt. \tag{15.13}$$

Using the Heath-Jarrow-Morton arbitrage-free argument [Heath et al., 1992], as in Chapter 11, the dynamics for the ZCBs, under their own measures generated by the respective money-savings accounts, are known and given by the following result:

> **Result 15.2.1 (ZCB dynamics under the risk-neutral measure)** *The risk-free dynamics of the ZCBs $P_d(t,T)$ and $P_f(t,T)$, with maturity time T are given by:*
>
> $$\begin{cases} \dfrac{dP_d(t,T)}{P_d(t,T)} = r_d(t)dt - \left(\int_t^T \bar{\sigma}_d(t,z)dz\right) dW_d^{\mathbb{Q}}(t), \\ \dfrac{dP_f(t,T)}{P_f(t,T)} = r_f(t)dt - \left(\int_t^T \bar{\sigma}_f(t,z)dz\right) dW_f^{\mathbb{Q}^f}(t), \end{cases}$$
>
> *where $\bar{\sigma}_d(t,T), \bar{\sigma}_f(t,T)$ are the volatility functions of the instantaneous forward rates $f_d^r(t,T), f_f^r(t,T)$, respectively, that are given by:*
>
> $$\begin{cases} df_d^r(t,T) = \bar{\sigma}_d(t,T) \int_t^T \bar{\sigma}_d(t,z)dzdt + \bar{\sigma}_d(t,T)dW_d^{\mathbb{Q}}(t), \\ df_f^r(t,T) = \bar{\sigma}_f(t,T) \int_t^T \bar{\sigma}_f(t,z)dzdt + \bar{\sigma}_f(t,T)dW_f^{\mathbb{Q}^f}(t). \end{cases}$$

The spot rates at time t are defined by $r_d(t) \equiv f_d^r(t,t)$, $r_f(t) \equiv f_f^r(t,t)$.

By means of the volatility structures, $\bar{\sigma}_d(t,T), \bar{\sigma}_f(t,T)$, one may define a number of different short-rate processes. In the present setting, the volatility functions are chosen to be $\bar{\sigma}_d(t,T) = \eta_d \exp(-\lambda_d(T-t))$ and $\bar{\sigma}_f(t,T) = \eta_f \exp(-\lambda_f(T-t))$. The Hull-White short-rate processes $r_d(t)$ and $r_f(t)$, as in (15.11), are then obtained, as well as the term structure functions, $\theta_d(t)$, and $\theta_f(t)$, that are expressed in terms of instantaneous forward rates, are also known. The choice of the specific volatility determines the dynamics of the ZCBs, as

$$\begin{cases} \dfrac{dP_d(t,T)}{P_d(t,T)} = r_d(t)dt + \eta_d \bar{B}_d(t,T)dW_d^{\mathbb{Q}}(t), \\ \dfrac{dP_f(t,T)}{P_f(t,T)} = r_f(t)dt + \eta_f \bar{B}_f(t,T)dW_f^{\mathbb{Q}^f}(t), \end{cases} \tag{15.14}$$

with $\bar{B}_d(t,T)$ and $\bar{B}_f(t,T)$ as in (15.12), given by:

$$\bar{B}_d(t,T) = \frac{1}{\lambda_d}\left(e^{-\lambda_d(T-t)} - 1\right), \quad \bar{B}_f(t,T) = \frac{1}{\lambda_f}\left(e^{-\lambda_f(T-t)} - 1\right). \tag{15.15}$$

15.2.1 The model with correlated, Gaussian interest rates

The FX-HHW model, with all the processes defined *under the domestic risk-neutral measure* $\mathbb{Q}$, is of the following form:

$$\begin{cases} \mathrm{d}y(t)/y(t) = (r_\mathrm{d}(t) - r_\mathrm{f}(t))\,\mathrm{d}t + \sqrt{v(t)}\mathrm{d}W_y^\mathbb{Q}(t), \\ \mathrm{d}v(t) = \kappa(\bar v - v(t))\mathrm{d}t + \gamma\sqrt{v(t)}\mathrm{d}W_v^\mathbb{Q}(t), \\ \mathrm{d}r_\mathrm{d}(t) = \lambda_\mathrm{d}(\theta_\mathrm{d}(t) - r_\mathrm{d}(t))\mathrm{d}t + \eta_\mathrm{d}\mathrm{d}W_\mathrm{d}^\mathbb{Q}(t), \\ \mathrm{d}r_\mathrm{f}(t) = \left(\lambda_\mathrm{f}(\theta_\mathrm{f}(t) - r_\mathrm{f}(t)) - \eta_\mathrm{f}\rho_{y,\mathrm{f}}\sqrt{v(t)}\right)\mathrm{d}t + \eta_\mathrm{f}\mathrm{d}W_\mathrm{f}^\mathbb{Q}(t), \end{cases} \quad (15.16)$$

with $y(t_0) > 0$, $v(t_0) > 0$, $r_\mathrm{d}(t_0) > 0$ and $r_\mathrm{f}(t_0)$. The parameters κ, λ_d, and λ_f determine the speed of mean reversion of the latter three processes, their long-term mean is given by $\bar v$, $\theta_\mathrm{d}(t)$, $\theta_\mathrm{f}(t)$, respectively. The volatility coefficients for the processes $r_\mathrm{d}(t)$ and $r_\mathrm{f}(t)$ are given by η_d and η_f, and the volatility-of-variance parameter for process $v(t)$ is γ.

In the model, a full matrix of correlations between the Brownian motions $\mathbf{W}(t) = \left[W_y^\mathbb{Q}(t), W_v^\mathbb{Q}(t), W_\mathrm{d}^\mathbb{Q}(t), W_\mathrm{f}^\mathbb{Q}(t)\right]^\mathrm{T}$ is considered,

$$\mathrm{d}\mathbf{W}(t)(\mathrm{d}\mathbf{W}(t))^\mathrm{T} = \begin{pmatrix} 1 & \rho_{y,v} & \rho_{y,\mathrm{d}} & \rho_{y,\mathrm{f}} \\ \rho_{y,v} & 1 & \rho_{v,\mathrm{d}} & \rho_{v,\mathrm{f}} \\ \rho_{y,\mathrm{d}} & \rho_{v,\mathrm{d}} & 1 & \rho_{\mathrm{d},\mathrm{f}} \\ \rho_{y,\mathrm{f}} & \rho_{v,\mathrm{f}} & \rho_{\mathrm{d},\mathrm{f}} & 1 \end{pmatrix} \mathrm{d}t. \quad (15.17)$$

Under the domestic-spot measure, the drift in the short-rate process $r_\mathrm{f}(t)$ gives rise to an additional term, i.e., $-\eta_\mathrm{f}\rho_{y,\mathrm{f}}\sqrt{v(t)}$, which is often called the *quanto correction term*. The r_f-process is thus not a HW process anymore under this measure. When the volatility of the FX model is assumed to be deterministic, then r_f will follow an HW process. Another simplification is based on the assumption $\rho_{y,\mathrm{f}} = 0$, but this is not realistic in practice.

The additional term in the r_f-process ensures the existence of martingales under the domestic spot measure, for the following prices (for more discussion, see [Shreve, 2004]):

$$\chi_1(t) := y(t)\frac{M_\mathrm{f}(t)}{M_\mathrm{d}(t)} \quad \text{and} \quad \chi_2(t) := y(t)\frac{P_\mathrm{f}(t,T)}{M_\mathrm{d}(t)},$$

where $P_\mathrm{f}(t,T)$ is the foreign zero-coupon bond (15.14), and the money-savings accounts $M_\mathrm{d}(t)$ and $M_\mathrm{f}(t)$ are as in (15.13).

To confirm that the processes $\chi_1(t)$ and $\chi_2(t)$ are martingales, one may apply the Itô product rule, which gives:

$$\begin{aligned} \mathrm{d}\chi_1(t)/\chi_1(t) &= \sqrt{v(t)}\mathrm{d}W_y^\mathbb{Q}(t), \\ \mathrm{d}\chi_2(t)/\chi_2(t) &= \sqrt{v(t)}\mathrm{d}W_y^\mathbb{Q}(t) + \eta_\mathrm{f}B_\mathrm{f}(t,T)\mathrm{d}W_\mathrm{f}^\mathbb{Q}(t). \end{aligned}$$

The change of dynamics of the underlying processes, from the foreign-spot to the domestic-spot measure, also influences the dynamics for the associated bonds, that, under the domestic risk-neutral measure $\mathbb{Q}$, with the money-savings account as the numéraire, have the following representations,

$$\begin{cases} \dfrac{dP_d(t,T)}{P_d(t,T)} = r_d(t)dt + \eta_d \bar{B}_d(t,T) dW_d^{\mathbb{Q}}(t), \\ \dfrac{dP_f(t,T)}{P_f(t,T)} = \left(r_f(t) - \rho_{y,f}\eta_f \bar{B}_f(t,T)\sqrt{v(t)}\right)dt + \eta_f \bar{B}_f(t,T) dW_f^{\mathbb{Q}}(t), \end{cases} \quad (15.18)$$

with $\bar{B}_d(t,T)$ and $\bar{B}_f(t,T)$ as in (15.15).

15.2.2 Pricing of FX options

To perform a calibration of the model, we need to price basic options on the FX rate, that are denoted by $V_c^{\mathrm{FX}}(t)$, highly efficiently for a given state vector $\mathbf{X}(t) = [y(t), v(t), r_d(t), r_f(t)]^{\mathrm{T}}$,

$$\boxed{V_c^{\mathrm{FX}}(t) = \mathbb{E}^{\mathbb{Q}}\left[\frac{M_d(t)}{M_d(T)}\max(y(T) - K, 0)\Big|\mathcal{F}(t)\right],}$$

with

$$M_d(t) = \exp\left(\int_0^t r_d(z) dz\right).$$

For this, we consider the *forward price*, $\frac{V_c^{\mathrm{FX}}(t)}{M_d(t)}$, so that,

$$\mathbb{E}^{\mathbb{Q}}\left[\frac{1}{M_d(T)}\max(y(T) - K, 0)\Big|\mathcal{F}(t)\right] = \frac{V_c^{\mathrm{FX}}(t)}{M_d(t)}.$$

By Itô's lemma, we find:

$$d\left(\frac{V_c^{\mathrm{FX}}(t)}{M_d(t)}\right) = \frac{1}{M_d(t)} dV^{\mathrm{FX}}(t) - r_d(t)\frac{V^{\mathrm{FX}}(t)}{M_d(t)} dt. \quad (15.19)$$

The forward price is a martingale, thus its dynamics should be free of drift terms. Including this in (15.19) gives us the following *Fokker-Planck forward equation* for $V_c = V_c^{\mathrm{FX}}(t)$:

$$r_d V_c = \frac{1}{2}\eta_f^2 \frac{\partial^2 V_c}{\partial r_f^2} + \rho_{d,f}\eta_d\eta_f \frac{\partial^2 V_c}{\partial r_d \partial r_f} + \frac{1}{2}\eta_d^2 \frac{\partial^2 V_c}{\partial r_d^2} + \rho_{v,f}\gamma\eta_f\sqrt{v}\frac{\partial^2 V_c}{\partial v \partial r_f}$$

$$+ \rho_{v,d}\gamma\eta_d\sqrt{v}\frac{\partial^2 V_c}{\partial v \partial r_d} + \frac{1}{2}\gamma^2 v \frac{\partial^2 V_c}{\partial v^2} + \rho_{y,f}\eta_f y\sqrt{v}\frac{\partial^2 V_c}{\partial y \partial r_f} + \rho_{y,d}\eta_d y\sqrt{v}\frac{\partial^2 V_c}{\partial y \partial r_d}$$

$$+ \rho_{y,v}\gamma y v \frac{\partial^2 V_c}{\partial y \partial v} + \frac{1}{2}y^2 v \frac{\partial^2 V_c}{\partial y^2} + \left(\lambda_f(\theta_f(t) - r_f) - \rho_{y,f}\eta_f\sqrt{v}\right)\frac{\partial V_c}{\partial r_f}$$

$$+ \lambda_d(\theta_d(t) - r_d)\frac{\partial V_c}{\partial r_d} + \kappa(\bar{v} - v)\frac{\partial V_c}{\partial v} + (r_d - r_f)y\frac{\partial V_c}{\partial y} + \frac{\partial V_c}{\partial t}.$$

This 4D PDE contains nonaffine terms, like square-roots. Finding a numerical solution for this PDE is rather expensive and this PDE formulation is therefore not easily applicable in the context of the model calibration. In the next subsection, we discuss again an *approximation* of the model, which is useful within the calibration.

The FX model under the forward domestic measure

To reduce the complexity of the pricing problem, *a transformation* is performed, from the domestic spot measure, which is generated by the money-savings account in the domestic market $M_d(t)$, to the domestic *forward FX measure*, where the numéraire is the domestic zero-coupon bond $P_d(t,T)$. The forward rate is given by [Musiela and Rutkowski, 1997; Piterbarg, 2006],

$$\boxed{y_F(t,T) = y(t)\frac{P_f(t,T)}{P_d(t,T)},} \qquad (15.20)$$

where $y_F(t,T)$ represents the *forward exchange rate under the domestic T-forward measure*, and $y(t)$ stands for foreign exchange rate under the domestic spot measure.

By switching from the domestic risk-neutral measure $\mathbb{Q}$, to the domestic T-forward measure $\mathbb{Q}^T$, the process of discounting will be decoupled from taking the expectation, i.e.

$$V_c^{\text{FX}}(t) = P_d(t,T)\mathbb{E}^T\left[\max\left(y_F(T,T) - K, 0\right)|\mathcal{F}(t)\right].$$

Itô's formula is applied to determine the dynamics for $y_F(t,T)$ in (15.20), i.e.

$$\mathrm{d}y_F(t,T) = \frac{P_f(t,T)}{P_d(t,T)}\mathrm{d}y(t) + \frac{y(t)}{P_d(t,T)}\mathrm{d}P_f(t,T) - y(t)\frac{P_f(t,T)}{P_d^2(t,T)}\mathrm{d}P_d(t,T)$$

$$+ y(t)\frac{P_f(t,T)}{P_d^3(t,T)}(\mathrm{d}P_d(t,T))^2 + \frac{1}{P_d(t,T)}(\mathrm{d}y(t)\mathrm{d}P_f(t,T))$$

$$- \frac{P_f(t,T)}{P_f^2(t,T)}(\mathrm{d}P_d(t,T)\mathrm{d}y(t)) - \frac{y(t)}{P_d^2(t,T)}\mathrm{d}P_d(t,T)\mathrm{d}P_f(t,T). \quad (15.21)$$

After substitution of the SDEs (15.16), (15.18) into (15.21), we arrive at the following FX forward dynamics:

$$\frac{\mathrm{d}y_F(t,T)}{y_F(t,T)} = \eta_d\bar{B}_d(t,T)\left(\eta_d\bar{B}_d(t,T) - \rho_{y,d}\sqrt{v(t)} - \rho_{d,f}\eta_f\bar{B}_f(t,T)\right)\mathrm{d}t$$

$$+ \sqrt{v(t)}\mathrm{d}W_y^{\mathbb{Q}}(t) - \eta_d\bar{B}_d(t,T)\mathrm{d}W_d^{\mathbb{Q}}(t) + \eta_f\bar{B}_f(t,T)\mathrm{d}W_f^{\mathbb{Q}}(t).$$

$y_F(t,T)$ is a *martingale* under the T-forward domestic measure, meaning,

$$P_d(t,T)\mathbb{E}^T[y_F(T,T)|\mathcal{F}(t)] = P_d(t,T)y_F(t,T) =: P_f(t,T)y(t),$$

the appropriate Brownian motions under the T-forward domestic measure, $\mathrm{d}W_y^T(t)$, $\mathrm{d}W_v^T(t)$, $\mathrm{d}W_d^T(t)$ and $\mathrm{d}W_f^T(t)$, still need to be determined.

A change of measure, from the domestic-spot to the domestic T-forward measure, requires a change of numéraire from money-savings account $M_d(t)$ to zero-coupon bond $P_d(t,T)$. A full matrix of correlations implies that all processes will change their dynamics when changing the measures from spot to forward. Lemma 15.2.1 provides the model dynamics under the domestic T-forward measure $\mathbb{Q}^T$.

Lemma 15.2.1 (FX-HHW model dynamics under the $\mathbb{Q}^T$ measure)
Under the T-forward domestic measure, the model in (15.16), (15.20) is governed by the following dynamics:

$$\frac{dy_F(t,T)}{y_F(t,T)} = \sqrt{v(t)}dW_y^T(t) - \eta_d \bar{B}_d(t,T)dW_d^T(t) + \eta_f \bar{B}_f(t,T)dW_f^T(t), \quad (15.22)$$

where

$$dv(t) = \left(\kappa(\bar{v} - v(t)) + \gamma \rho_{v,d} \eta_d \bar{B}_d(t,T)\sqrt{v(t)}\right)dt + \gamma\sqrt{v(t)}dW_v^T(t),$$

$$dr_d(t) = \left(\lambda_d(\theta_d(t) - r_d(t)) + \eta_d^2 \bar{B}_d(t,T)\right)dt + \eta_d dW_d^T(t),$$

$$dr_f(t) = \left(\lambda_f(\theta_f(t) - r_f(t)) - \eta_f \rho_{y,f}\sqrt{v(t)} + \eta_d \eta_f \rho_{d,f}\bar{B}_d(t,T)\right)dt +$$
$$\eta_f dW_f^T(t),$$

with a full matrix of correlations as in (15.17), and $\bar{B}_d(t,T)$, $\bar{B}_f(t,T)$ in (15.15). ◀

Proof Because the domestic short-rate process $r_d(t)$ is governed by one source of uncertainty, it is convenient to change the order of the state variables to $\mathbf{X}^*(t) = [r_d(t), r_f(t), v(t), y_F(t,T)]^T$ and express the model in terms of the independent Brownian motions $d\widetilde{\mathbf{W}}^{\mathbb{Q}}(t) = [\widetilde{W}_d(t), \widetilde{W}_f(t), \widetilde{W}_v(t), \widetilde{W}_y(t)]^T$, as follows,

$$\begin{bmatrix} dr_d \\ dr_f \\ dv \\ \frac{dy_F}{y_F} \end{bmatrix} = \bar{\mu}(t, \mathbf{X}^*(t))dt + \begin{bmatrix} \eta_d & 0 & 0 & 0 \\ 0 & \eta_f & 0 & 0 \\ 0 & 0 & \gamma\sqrt{v} & 0 \\ -\eta_d \bar{B}_d & \eta_f \bar{B}_f & 0 & \sqrt{v} \end{bmatrix} \mathbf{L} \begin{bmatrix} d\widetilde{W}_d^{\mathbb{Q}} \\ d\widetilde{W}_f^{\mathbb{Q}} \\ d\widetilde{W}_v^{\mathbb{Q}} \\ d\widetilde{W}_y^{\mathbb{Q}} \end{bmatrix},$$

which, equivalently, can be written as:

$$d\mathbf{X}^*(t) = \bar{\mu}(t, \mathbf{X}^*(t))dt + \bar{\mathbf{\Sigma}}\mathbf{L}d\widetilde{\mathbf{W}}^{\mathbb{Q}}(t), \quad (15.23)$$

where $\bar{\mu}(t, \mathbf{X}^*(t))$ represents the drift of the dynamics of $\mathbf{X}^*(t)$ and $\mathbf{L}$ is the Cholesky lower-triangular matrix of the following form:

$$\mathbf{L} = \begin{bmatrix} 1 & 0 & 0 & 0 \\ L_{2,1} & L_{2,2} & 0 & 0 \\ L_{3,1} & L_{3,2} & L_{3,3} & 0 \\ L_{4,1} & L_{4,2} & L_{4,3} & L_{4,4} \end{bmatrix} \stackrel{def}{=} \begin{bmatrix} 1 & 0 & 0 & 0 \\ \rho_{f,d} & L_{2,2} & 0 & 0 \\ \rho_{v,d} & L_{3,2} & L_{3,3} & 0 \\ \rho_{y,d} & L_{4,2} & L_{4,3} & L_{4,4} \end{bmatrix}. \quad (15.24)$$

The representation above appears favorable, since the short-rate process $r_\mathrm{d}(t)$ can be treated *independently* of the other processes.

A matrix representation in terms of independent Brownian motions results in the following dynamics for the domestic short-rate $r_\mathrm{d}(t)$, under the measure $\mathbb{Q}$:

$$\mathrm{d}r_\mathrm{d}(t) = \lambda_\mathrm{d}(\theta_\mathrm{d}(t) - r_\mathrm{d}(t))\mathrm{d}t + \varsigma_1(t)\mathrm{d}\widetilde{\mathbf{W}}^\mathbb{Q}(t),$$

and for the domestic ZCB:

$$\frac{\mathrm{d}P_\mathrm{d}(t,T)}{P_\mathrm{d}(t,T)} = r_\mathrm{d}(t)\mathrm{d}t + \bar{B}_\mathrm{d}(t,T)\varsigma_1(t)\mathrm{d}\widetilde{\mathbf{W}}^\mathbb{Q}(t),$$

with $\varsigma_k(t)$ the kth row vector resulting from multiplying the matrices $\bar{\Sigma}$ and $\mathbf{L}$. For the Hull-White short-rate processes, we have $\varsigma_1(t) := [\eta_\mathrm{d}, 0, 0, 0]$.

The Radon-Nikodym derivative,

$$\lambda_\mathbb{Q}^T(t) = \frac{\mathrm{d}\mathbb{Q}^T}{\mathrm{d}\mathbb{Q}}\bigg|_{\mathcal{F}(t)} = \frac{P_\mathrm{d}(t,T)}{P_\mathrm{d}(0,T)M_\mathrm{d}(t)}. \qquad (15.25)$$

is derived with the help of the Itô derivative of $\lambda_\mathbb{Q}^T(t)$ in (15.25),

$$\frac{\mathrm{d}\lambda_\mathbb{Q}^T(t)}{\lambda_\mathbb{Q}^T(t)} = \bar{B}_\mathrm{d}(t,T)\varsigma_1(t)\mathrm{d}\widetilde{\mathbf{W}}^\mathbb{Q}(t).$$

This implies that the Girsanov kernel for the transition from $\mathbb{Q}$ to $\mathbb{Q}^T$ is given by $\bar{B}_\mathrm{d}(t,T)\varsigma_1(t)$, i.e., the T-bond volatility is given by $\eta_\mathrm{d}\bar{B}_\mathrm{d}(t,T)$, i.e.

$$\lambda_\mathbb{Q}^T(t) = \exp\left(-\frac{1}{2}\int_0^t \bar{B}_\mathrm{d}^2(z,T)\varsigma_1^2(z)\mathrm{d}z + \int_0^t \bar{B}_\mathrm{d}(z,T)\varsigma_1(z)\mathrm{d}\widetilde{\mathbf{W}}^\mathbb{Q}(z)\right).$$

So, we find,

$$\mathrm{d}\widetilde{\mathbf{W}}^T(t) = -\bar{B}_\mathrm{d}(t,T)\varsigma_1^\mathrm{T}(t)\mathrm{d}t + \mathrm{d}\widetilde{\mathbf{W}}^\mathbb{Q}(t).$$

Since the vector $\varsigma_1^T(t)$ is, in fact, of scalar form, the Brownian motions under the T-forward measure are given by:

$$\mathrm{d}\widetilde{\mathbf{W}}^\mathbb{Q}(t) = \left[\mathrm{d}\widetilde{W}_\mathrm{d}^T(t) + \eta_\mathrm{d}\bar{B}_\mathrm{d}(t,T)\mathrm{d}t, \mathrm{d}\widetilde{W}_\mathrm{f}^T(t), \mathrm{d}\widetilde{W}_v^T(t), \mathrm{d}\widetilde{W}_y^T(t)\right]^\mathrm{T}.$$

From the vector representation (15.23), it follows that,

$$\mathbf{L}\mathrm{d}\widetilde{\mathbf{W}}^\mathbb{Q} = \begin{bmatrix} \eta_\mathrm{d}\bar{B}_\mathrm{d}\mathrm{d}t + \mathrm{d}\widetilde{W}_\mathrm{d}^T \\ \rho_{\mathrm{d},\mathrm{f}}\eta_\mathrm{d}\bar{B}_\mathrm{d}\mathrm{d}t + \rho_{\mathrm{d},\mathrm{f}}\mathrm{d}\widetilde{W}_\mathrm{d}^T + \mathbf{L}_{2,2}\mathrm{d}\widetilde{W}_\mathrm{f}^T \\ \rho_{v,\mathrm{d}}\eta_\mathrm{d}\bar{B}_\mathrm{d}\mathrm{d}t + \rho_{v,\mathrm{d}}\mathrm{d}\widetilde{W}_\mathrm{d}^T + \mathbf{L}_{3,2}\mathrm{d}\widetilde{W}_\mathrm{f}^T + \mathbf{L}_{3,3}\mathrm{d}\widetilde{W}_y^T \\ \rho_{y,\mathrm{d}}\eta_\mathrm{d}\bar{B}_\mathrm{d}\mathrm{d}t + \rho_{y,\mathrm{d}}\mathrm{d}\widetilde{W}_\mathrm{d}^T + \mathbf{L}_{4,2}\mathrm{d}\widetilde{W}_\mathrm{f}^T + \mathbf{L}_{4,3}\mathrm{d}\widetilde{W}_y^T + \mathbf{L}_{4,4}\mathrm{d}\widetilde{W}_v^T \end{bmatrix}.$$

Returning to the *dependent* Brownian motions, under the T-forward measure, gives us:

$$\begin{cases} \dfrac{\mathrm{d}y_F(t,T)}{y_F(t,T)} = \sqrt{v(t)}\mathrm{d}W_y^T(t) - \eta_\mathrm{d}\bar{B}_\mathrm{d}(t,T)\mathrm{d}W_\mathrm{d}^T(t) + \eta_\mathrm{f}\bar{B}_\mathrm{f}(t,T)\mathrm{d}W_\mathrm{f}^T(t), \\ \mathrm{d}v(t) = \left(\kappa(\bar{v} - v(t)) + \gamma\rho_{v,\mathrm{d}}\eta_\mathrm{d}\bar{B}_\mathrm{d}(t,T)\sqrt{v(t)}\right)\mathrm{d}t + \gamma\sqrt{v(t)}\mathrm{d}W_v^T(t), \\ \mathrm{d}r_\mathrm{d}(t) = \left(\lambda_\mathrm{d}(\theta_\mathrm{d}(t) - r_\mathrm{d}(t)) + \eta_\mathrm{d}^2\bar{B}_\mathrm{d}(t,T)\right)\mathrm{d}t + \eta_\mathrm{d}\mathrm{d}W_\mathrm{d}^T(t), \\ \mathrm{d}r_\mathrm{f}(t) = \left(\lambda_\mathrm{f}(\theta_\mathrm{f}(t) - r_\mathrm{f}(t)) - \eta_\mathrm{f}\rho_{y,\mathrm{f}}\sqrt{v(t)} + \eta_\mathrm{d}\eta_\mathrm{f}\rho_{\mathrm{d},\mathrm{f}}\bar{B}_\mathrm{d}(t,T)\right)\mathrm{d}t + \eta_\mathrm{f}\mathrm{d}W_\mathrm{f}^T(t), \end{cases}$$

with a full matrix of correlations, given in (15.17). ∎

The system in Lemma 15.2.1 indicates that, after a measure change, from the domestic-spot $\mathbb{Q}$-measure to the domestic T-forward $\mathbb{Q}^T$ measure, the forward exchange rate $y_F(t,T)$ does not depend explicitly on the short-rate processes $r_\mathrm{d}(t)$ and $r_\mathrm{f}(t)$. It does not contain a drift term and it only depends on $W_\mathrm{d}^T(t)$, $W_\mathrm{f}^T(t)$, see (15.22).

Remark 15.2.1 *Since the sum of three correlated, normally distributed random variables, $Q = X + Y + Z$, remains normally distributed, with the mean equal to the sum of the individual means and its variance equal to*

$$v_Q^2 = v_X^2 + v_Y^2 + v_Z^2 + 2\rho_{X,Y}v_Xv_Y + 2\rho_{X,Z}v_Xv_Z + 2\rho_{Y,Z}v_Yv_Z,$$

the forward (15.22) can be represented, by

$$\mathrm{d}y_F/y_F = \left(v + \eta_d^2\bar{B}_d^2 + \eta_f^2\bar{B}_f^2 - 2\rho_{y,d}\eta_d\bar{B}_d\sqrt{v}\right.$$
$$\left. + 2\rho_{y,f}\eta_f\bar{B}_f\sqrt{v} - 2\rho_{d,f}\eta_d\eta_f\bar{B}_d\bar{B}_f\right)^{\frac{1}{2}}\mathrm{d}W_F^T. \qquad (15.26)$$

Although the representation in (15.26) reduces the number of Brownian motions in the dynamics for the y_F, one still needs to find the appropriate cross terms, like $\mathrm{d}W_F^T(t)\mathrm{d}W_v^T(t)$, to determine the covariance terms. It is therefore preferable to stay with the standard formulation. ▲

Approximations and the forward characteristic function

As the dynamics of the forward foreign exchange $y_F(t,T)$, under the domestic forward measure, contain only the Brownian motions $W_\mathrm{d}^T(t)$ and $W_\mathrm{f}^T(t)$ of the short-rate processes, a significant reduction of the pricing problem has now been achieved.

To derive the *forward characteristic function*, as usual, the log-transform of the forward rate $y_F(t,T)$ is considered, $X(t) := \log y_F(t,T)$, which is governed by the following dynamics:

> $$dX(t) = \left(\varsigma\left(t, \sqrt{v(t)}\right) - \frac{1}{2}v(t)\right)dt + \sqrt{v(t)}dW_y^T(t) - \eta_d \bar{B}_d dW_d^T(t) + \eta_f \bar{B}_f dW_f^T(t),$$
> (15.27)
>
> with the variance process $v(t)$ given by:
>
> $$dv(t) = \left(\kappa(\bar{v} - v(t)) + \gamma \rho_{v,d} \eta_d \bar{B}_d \sqrt{v(t)}\right) dt + \gamma \sqrt{v(t)} dW_v^T(t).$$
>
> Here, $\bar{B}_d := \bar{B}_d(t,T)$, $\bar{B}_f := \bar{B}_f(t,T)$, and
>
> $$\varsigma\left(t, \sqrt{v(t)}\right) = \left(\rho_{y,d}\eta_d \bar{B}_d - \rho_{y,f}\eta_f \bar{B}_f\right)\sqrt{v(t)} + \rho_{d,f}\eta_d\eta_f \bar{B}_d \bar{B}_f - \frac{1}{2}\left(\eta_d^2 \bar{B}_d^2 + \eta_f^2 \bar{B}_f^2\right).$$

By the martingale approach and the Feynman-Kac theorem, the following PDE for the forward characteristic function results,

$$\phi_{\mathbf{X}} := \phi_{\mathbf{X}}(u; t, T) = \mathbb{E}^T\left(e^{iuX(T)} \big| \mathcal{F}(t)\right),$$

$$-\frac{\partial \phi_{\mathbf{X}}}{\partial t} = \left(\kappa(\bar{v} - v) + \rho_{v,d}\gamma\eta_d\sqrt{v}\bar{B}_d\right)\frac{\partial \phi_{\mathbf{X}}}{\partial v} + \left(\frac{1}{2}v - \varsigma(t,\sqrt{v})\right)\left(\frac{\partial^2 \phi_{\mathbf{X}}}{\partial X^2} - \frac{\partial \phi_{\mathbf{X}}}{\partial X}\right)$$
$$+ \left(\rho_{y,v}\gamma v - \rho_{v,d}\gamma\eta_d\sqrt{v}\bar{B}_d + \rho_{v,f}\gamma\eta_f\sqrt{v}\bar{B}_f\right)\frac{\partial^2 \phi_{\mathbf{X}}}{\partial X \partial v} + \frac{1}{2}\gamma^2 v \frac{\partial^2 \phi_{\mathbf{X}}}{\partial v^2}.$$

This PDE contains nonaffine $\sqrt{v}$-terms, so that it is nontrivial to find its solution analytically of via Fourier methods. In Section 13.2.1 a method for linearization of these nonaffine square-roots of the square-root process was presented. The method is to *project* the nonaffine square-root terms onto their first moments.

> The approximation of the nonaffine terms in the corresponding PDE is then done by the approximation,
>
> $$\sqrt{v(t)} \approx \mathbb{E}\left[\sqrt{v(t)}\right] =: G(t), \qquad (15.28)$$
>
> with the expectation of the square-root of $v(t)$ as determined[a] in Section 13.2.1.
>
> ---
> [a] In this approximation we also assume that $\mathbb{E}^{\mathbb{Q}}\left[\sqrt{v(t)}\right] \approx \mathbb{E}^T\left[\sqrt{v(t)}\right]$.

Projection of the nonaffine terms onto their first moments allows us to derive the corresponding *forward characteristic function*, $\phi_{\mathbf{X}}$, which is then of the following form:

$$\phi_{\mathbf{X}}(u; t, T) = \exp\left(\bar{A}(\mathbf{u}, \tau) + \bar{B}(\mathbf{u}, \tau)X(t) + \bar{C}(\mathbf{u}, \tau)v(t)\right),$$

where $\tau = T - t$, with

$$\frac{d\bar{B}}{d\tau} = 0,$$

$$\frac{d\bar{C}}{d\tau} = -\kappa\bar{C} + (\bar{B}^2 - \bar{B})/2 + \rho_{y,v}\gamma\bar{B}\bar{C} + \gamma^2\bar{C}^2/2,$$

$$\frac{d\bar{A}}{d\tau} = \kappa\bar{v}\bar{C} + \rho_{v,\mathrm{d}}\gamma\eta_\mathrm{d} G(T-\tau)\bar{B}_\mathrm{d}(\tau)\bar{C} - \varsigma(\tau, G(T-\tau))\left(\bar{B}^2 - \bar{B}\right)$$
$$+ \left(-\rho_{v,\mathrm{d}}\eta_\mathrm{d}\gamma G(T-\tau)\bar{B}_\mathrm{d}(\tau) + \rho_{v,\mathrm{f}}\gamma\eta_\mathrm{f} G(T-\tau)\bar{B}_\mathrm{f}(\tau)\right)\bar{B}\bar{C},$$

with $G(t) = \mathbb{E}\left[\sqrt{v(t)}\right]$ and $\bar{B}_j(\tau) = \lambda_j^{-1}\left(\mathrm{e}^{-\lambda_j \tau} - 1\right)$ for $j \in \{\mathrm{d}, \mathrm{f}\}$. The initial conditions are: $\bar{B}(\mathbf{u}, 0) = iu$, $\bar{C}(\mathbf{u}, 0) = 0$ and $\bar{A}(\mathbf{u}, 0) = 0$.

With $\bar{B}(\mathbf{u}, \tau) = iu$, the complex-valued function $\bar{C}(\mathbf{u}, \tau)$ is of the Heston-type, [Heston, 1993], and its solution reads:

$$\bar{C}(\mathbf{u}, \tau) = \frac{1 - \mathrm{e}^{-d\tau}}{\gamma^2(1 - g\mathrm{e}^{-d\tau})}\left(\kappa - \rho_{y,v}\gamma iu - d\right), \tag{15.29}$$

with $d = \sqrt{(\rho_{y,v}\gamma iu - \kappa)^2 - \gamma^2 iu(iu - 1)}$, $g = \dfrac{\kappa - \gamma\rho_{y,v}iu - d}{\kappa - \gamma\rho_{y,v}iu + d}$.

The parameters κ, γ, $\rho_{y,v}$ are given in (15.16).

Function $\bar{A}(\mathbf{u}, \tau)$ is given by:

$$\bar{A}(\mathbf{u}, \tau) = \int_0^\tau \left(\kappa\bar{v} + \rho_{v,\mathrm{d}}\gamma\eta_\mathrm{d} G(T-z)\bar{B}_\mathrm{d}(z) - \rho_{v,\mathrm{d}}\eta_\mathrm{d}\gamma G(T-z)\bar{B}_\mathrm{d}(z)iu \right.$$
$$\left. + \rho_{v,\mathrm{f}}\gamma\eta_\mathrm{f} G(T-z)\bar{B}_\mathrm{f}(z)iu\right)\bar{C}(z)\mathrm{d}z + (u^2 + iu)\int_0^\tau \varsigma(z, G(T-z))\mathrm{d}z,$$

with $\bar{C}(\mathbf{u}, s)$ in (15.29). It is most convenient to solve $\bar{A}(\mathbf{u}, \tau)$ numerically with, for example, Simpson's quadrature rule. With the correlations $\rho_{v,\mathrm{d}}$, $\rho_{v,\mathrm{f}}$ set equal to zero, a closed-form expression for $\bar{A}(\mathbf{u}, \tau)$ would be available (see Section 13.2.1, for details).

We denote the *approximation*, by means of the linearization of the full-scale FX-HHW model, by *FX-HHW1*. It is clear that efficient pricing with Fourier-based methods can be done with FX-HHW1, and not with FX-HHW.

By the projection of $\sqrt{v(t)}$ onto its first moment in (15.28), the corresponding PDE is affine in its coefficients, and reads:

$$-\frac{\partial\phi_\mathbf{x}}{\partial t} = (\kappa(\bar{v} - v) + \Psi_1)\frac{\partial\phi_\mathbf{x}}{\partial v} + \left(\frac{1}{2}v - \varsigma(t, G(t))\right)\left(\frac{\partial^2\phi_\mathbf{x}}{\partial X^2} - \frac{\partial\phi_\mathbf{x}}{\partial X}\right)$$
$$+ (\rho_{y,v}\gamma v - \Psi_2)\frac{\partial^2\phi_\mathbf{x}}{\partial X\partial v} + \frac{1}{2}\gamma^2 v\frac{\partial^2\phi_\mathbf{x}}{\partial v^2}, \tag{15.30}$$

with $\phi_\mathbf{x} \equiv \phi_\mathbf{x}(u; T, T) = \mathbb{E}^T\left[\mathrm{e}^{iuX(T)}\big|\mathcal{F}(T)\right] = \mathrm{e}^{iuX(T)}$, and

$$\varsigma(t, G(t)) = \Psi_3 + \rho_{\mathrm{d},\mathrm{f}}\eta_\mathrm{d}\eta_\mathrm{f}\bar{B}_\mathrm{d}(t,T)\bar{B}_\mathrm{f}(t,T) - \frac{1}{2}\left(\eta_\mathrm{d}^2\bar{B}_\mathrm{d}^2(t,T) + \eta_\mathrm{f}^2\bar{B}_\mathrm{f}^2(t,T)\right).$$

The three terms, Ψ_1, Ψ_2, and Ψ_3, in PDE (15.30) contain the function $G(t)$:

$$\Psi_1 := \rho_{v,d}\gamma\eta_d \bar{B}_d(t,T)G(t),$$

$$\Psi_2 := \left(\rho_{v,d}\gamma\eta_d \bar{B}_d(t,T) - \rho_{v,f}\gamma\eta_f \bar{B}_f(t,T)\right)G(t),$$

$$\Psi_3 := \left(\rho_{y,d}\eta_d \bar{B}_d(t,T) - \rho_{y,f}\eta_f \bar{B}_f(t,T)\right)G(t).$$

When solving the pricing PDE for $t \to T$, the terms $\bar{B}_d(t,T)$ and $\bar{B}_f(t,T)$ tend to zero, and all terms that contain the approximation will vanish. The case $t \to 0$ is furthermore trivial, since $\sqrt{v(t)} \xrightarrow{t \to 0} \mathbb{E}\left[\sqrt{v(t_0)}\right]$.

Under the T-forward domestic FX measure, the projection of the nonaffine terms onto their first moments is expected to provide high accuracy. In Section 15.2.3, a numerical experiment is performed to confirm this.

15.2.3 Numerical experiment for the FX-HHW model

In this section, the numerical errors resulting from the various approximations in the FX-HHW1 model are analyzed. In a set-up as in [Piterbarg, 2006], the interest rate curves are modeled by ZCBs, using $t_0 = 0$, $P_d(0,T) = \exp(-0.02T)$ and $P_f(0,T) = \exp(-0.05T)$. Furthermore,

$$\eta_d = 0.7\%, \quad \eta_f = 1.2\%, \quad \lambda_d = 1\%, \quad \lambda_f = 5\%.$$

Model parameters that do not satisfy the Feller condition are prescribed,

$$\kappa = 0.5, \quad \gamma = 0.3, \quad \bar{v} = 0.1, \quad v(0) = 0.1.$$

The correlation structure, as defined in (15.17), is given by:

$$\begin{pmatrix} 1 & \rho_{y,v} & \rho_{y,d} & \rho_{y,f} \\ \rho_{y,v} & 1 & \rho_{v,d} & \rho_{v,f} \\ \rho_{y,d} & \rho_{v,d} & 1 & \rho_{d,f} \\ \rho_{y,f} & \rho_{v,f} & \rho_{d,f} & 1 \end{pmatrix} = \begin{pmatrix} 100\% & -40\% & -15\% & -15\% \\ -40\% & 100\% & 30\% & 30\% \\ -15\% & 30\% & 100\% & 25\% \\ -15\% & 30\% & 25\% & 100\% \end{pmatrix}. \quad (15.31)$$

A number of FX options with many expiry times and strike prices, using two different pricing methods for the FX-HHW model, are priced.

The first method is the plain Monte Carlo method, with 50.000 paths and $20 \cdot T_i$ steps, for the full-scale FX-HHW model, without any approximations.

In the second pricing method, the characteristic function, which is based on the approximations in the FX-HHW1 model in Section 15.2.2 is used. Efficient pricing of plain vanilla products is then done by means of the COS method, based on the Fourier cosine series expansion of the probability density function, which is recovered by the characteristic function with 500 Fourier cosine terms.

The experiments are set up with expiry times given by $T_1, \ldots, T_{10}$, and the strike prices are computed by the formula:

$$K_n(T_i) = y_F^{T_i}(0,T) \exp\left(0.1 c_n \sqrt{T_i}\right), \quad \text{with} \quad (15.32)$$

$$c_n = \{-1.5, -1.0, -0.5, 0, 0.5, 1.0, 1.5\},$$

Table 15.2: *Expiries and strike prices of FX options used in the FX-HHW model. Strikes $K_n(T_i)$ were calculated as given in (15.32) with $y(0) = 1.35$.*

T_i	$K_1(T_i)$	$K_2(T_i)$	$K_3(T_i)$	$K_4(T_i)$	$K_5(T_i)$	$K_6(T_i)$	$K_7(T_i)$
6m	1.1961	1.2391	1.2837	1.3299	1.3778	1.4273	1.4787
1y	1.1276	1.1854	1.2462	1.3101	1.3773	1.4479	1.5221
5y	0.8309	0.9291	1.0390	1.1620	1.2994	1.4531	1.6250
10y	0.6224	0.7290	0.8538	1.0001	1.1714	1.3721	1.6071
20y	0.3788	0.4737	0.5924	0.7409	0.9265	1.1587	1.4491
30y	0.2414	0.3174	0.4174	0.5489	0.7218	0.9492	1.2482

and $y_{F^i}^{T_i}(0,T)$ as in (15.20) with $y(0) = 1.35$, the initial spot FX rate (like in dollar \$ per euro €) is set to 1.35. Formula (15.32) for the strike prices is convenient, since for $n = 4$, strike prices $K_4(T_i)$, with $i = 1, \ldots, 10$, are equal to the forward FX rates for time T_i. The strike prices and maturity times are presented in Table 15.2.

The option prices resulting from both models can be expressed in terms of the implied Black volatilities. The approximation model FX-HHW1 appears to be highly accurate for the parameters considered. A maximum error of about 0.1% volatility for at-the-money options with a maturity of 30 years is observed (not reported) and an error smaller than 0.07% for the other options. The FX option prices and the standard deviations are tabulated in Table 15.3. Strike price $K_4(T_i)$ is the at-the-money strike price. In the next subsection, the calibration results to the FX market data are presented.

Calibration to market data

We discuss the calibration of the FX-HHW model to the FX market data. In the simulation the reference market implied volatilities they are taken from [Piterbarg, 2006] and are presented in Table 15.4. In the calibration, the approximate model FX-HHW1 was employed. The correlation structure is as in (15.31). In Figure 15.2 some of the calibration results are presented.

These experiments show that the model can be well calibrated to the market data. For long maturity times and for deep-in-the money options some discrepancy between the market and model prices is observed. This is however typical when dealing with the Heston model (not related to the approximation), since the skew/smile pattern in the FX rate does not flatten for long maturity times. This is sometimes improved by adding jumps to the model (Bates' model, discussed in Section 8.4.3).

Cross-Currency Models

Table 15.3: *Average FX call option prices obtained by the FX-HHW model with 20 Monte Carlo simulations, 50.000 paths and $20 \times T_i$ steps; MC stands for Monte Carlo and COS for the Fourier cosine expansion technique for the FX-HHW1 model with 500 expansion terms. The strike prices $K_n(T_i)$ are tabulated in Table 15.2.*

T_i	method	$K_1(T_i)$	$K_2(T_i)$	$K_3(T_i)$	$K_4(T_i)$	$K_5(T_i)$	$K_6(T_i)$	$K_7(T_i)$
6m	MC	0.1907	0.1636	0.1382	0.1148	0.0935	0.0748	0.0585
	std dev	0.0004	0.0004	0.0005	0.0004	0.0004	0.0004	0.0004
	COS	0.1908	0.1637	0.1382	0.1147	0.0934	0.0746	0.0583
1y	MC	0.2566	0.2209	0.1870	0.1553	0.1264	0.1008	0.0785
	std dev	0.0007	0.0007	0.0007	0.0007	0.0007	0.0007	0.0007
	COS	0.2567	0.2210	0.1870	0.1554	0.1265	0.1008	0.0786
5y	MC	0.4216	0.3709	0.3205	0.2713	0.2246	0.1816	0.1432
	std dev	0.0021	0.0021	0.0021	0.0020	0.0020	0.0019	0.0018
	COS	0.4212	0.3706	0.3203	0.2713	0.2249	0.1822	0.1441
10y	MC	0.4310	0.3871	0.3420	0.2967	0.2521	0.2096	0.1702
	std dev	0.0033	0.0033	0.0033	0.0033	0.0033	0.0031	0.0030
	COS	0.4311	0.3873	0.3423	0.2971	0.2528	0.2106	0.1714
20y	MC	0.3362	0.3109	0.2838	0.2553	0.2260	0.1966	0.1677
	std dev	0.0037	0.0037	0.0037	0.0037	0.0037	0.0036	0.0036
	COS	0.3358	0.3104	0.2833	0.2548	0.2254	0.1960	0.1672
30y	MC	0.2322	0.2191	0.2046	0.1888	0.1720	0.1545	0.1367
	std dev	0.0050	0.0050	0.0050	0.0050	0.0049	0.0048	0.0048
	COS	0.2319	0.2188	0.2042	0.1883	0.1714	0.1539	0.1359

Table 15.4: *Market implied Black volatilities for FX options as given in [Piterbarg, 2006]. The strike prices $K_n(T_i)$ were tabulated in Table 15.2.*

T_i	$K_1(T_i)$	$K_2(T_i)$	$K_3(T_i)$	$K_4(T_i)$	$K_5(T_i)$	$K_6(T_i)$	$K_7(T_i)$
6m	11.41 %	10.49 %	9.66 %	9.02 %	8.72 %	8.66 %	8.68 %
1y	12.23 %	10.98 %	9.82 %	8.95 %	8.59 %	8.59 %	8.65 %
5y	13.44 %	11.84 %	10.38 %	9.27 %	8.76 %	8.71 %	8.83 %
10y	16.43 %	14.79 %	13.34 %	12.18 %	11.43 %	11.07 %	10.99 %
20y	22.96 %	21.19 %	19.68 %	18.44 %	17.50 %	16.84 %	16.46 %
30y	25.09 %	23.48 %	22.17 %	21.13 %	20.35 %	19.81 %	19.48 %

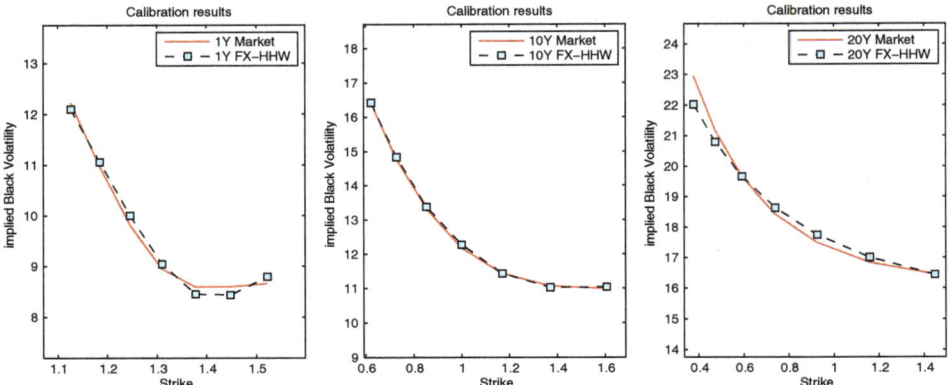

Figure 15.2: *Comparison of implied volatilities from the market and the FX-HHW1 model for FX European call options with maturity times 1, 10 and 20 years. The strike prices are provided in Table 15.2; $y(0) = 1.35$.*

15.2.4 CVA for FX swaps

An FX swap is a financial derivative in which two parties exchange a fixed for a floating FX rate. One party receives a floating FX rate, $y(T)$, and the other party pays the fixed rate K. The amount to be paid is based on some foreign notional amount N_f.

At the maturity date T, the value of the FX swap contract, in the domestic currency, is given by,

$$V^\mathrm{d}(T) = N_\mathrm{f}\left(y(T) - K\right), \tag{15.33}$$

(remember that $y(t) := y_\mathrm{f}^\mathrm{d}(t)$) and the FX swap value at time t is equal to:

$$V^\mathrm{d}(t) = M_\mathrm{d}(t)\mathbb{E}^{\mathbb{Q},\mathrm{d}}\left[\frac{N_\mathrm{f}}{M_\mathrm{d}(T)}\left(y(T) - K\right)\bigg|\mathcal{F}(t)\right]. \tag{15.34}$$

From the definition of the FX forward rate, we have:

$$y_F(t,T) = y(t)\frac{P_\mathrm{f}(t,T)}{P_\mathrm{d}(t,T)},$$

and by the change of measure, from the risk-neutral measure $\mathbb{Q}^\mathrm{d}$ to the forward measure $\mathbb{Q}^{T,\mathrm{d}}$, it follows that,

$$V^\mathrm{d}(t) = M_\mathrm{d}(t)\mathbb{E}^{T,\mathrm{d}}\left[\frac{N_\mathrm{f}}{M_\mathrm{d}(T)}\frac{M_\mathrm{d}(T)}{M_\mathrm{d}(t)}\frac{P_\mathrm{d}(t,T)}{P_\mathrm{d}(T,T)}\left(y(T) - K\right)\bigg|\mathcal{F}(t)\right]$$
$$= P_\mathrm{d}(t,T)N_\mathrm{f}\mathbb{E}^{T,\mathrm{d}}\left[\left(y(T) - K\right)\bigg|\mathcal{F}(t)\right].$$

Using Equation (15.6), the following holds true,

$$\mathbb{E}^{T,\mathrm{d}}\left[y(T)\big|\mathcal{F}(t)\right] = y(t)\frac{P_\mathrm{f}(t,T)}{P_\mathrm{d}(t,T)} =: y_F(t,T),$$

and therefore, we have,

$$V^\mathrm{d}(t) = P_\mathrm{d}(t,T)N_\mathrm{f}\mathbb{E}^{T,\mathrm{d}}\left[(y(T)-K)\big|\mathcal{F}(t)\right]$$
$$= P_\mathrm{d}(t,T)N_\mathrm{f}\left(y_F(t,T)-K\right).$$

Note that Lemma 15.2.1 indicates that, after a measure change, from the domestic-spot measure to the domestic T-forward measure, the forward exchange rate $y_F(t,T)$ does not contain a drift term, and is therefore a martingale.

The expected exposure that we encounter in CVA, see Equation (12.53), in domestic currency, is given by:

$$\mathrm{EE}^\mathrm{d}(t_0,t) = \mathbb{E}^{\mathbb{Q},\mathrm{d}}\left[\frac{M_\mathrm{d}(t_0)}{M_\mathrm{d}(t)}\max(V^\mathrm{d}(t),0)\big|\mathcal{F}(t_0)\right]$$
$$= N_\mathrm{f}\mathbb{E}^{\mathbb{Q},\mathrm{d}}\left[\frac{M_\mathrm{d}(t_0)}{M_\mathrm{d}(t)}P_\mathrm{d}(t,T)\max\left(y_F(t,T)-K,0\right)\big|\mathcal{F}(t_0)\right].$$

Inserting the definition of the ZCB, $P_\mathrm{d}(t,T) = \mathbb{E}^{\mathbb{Q}}\left[M(t)/M(T)\big|\mathcal{F}(t)\right]$, gives,

$$\mathrm{EE}^\mathrm{d}(t_0,t) = N_\mathrm{f}\mathbb{E}^{\mathbb{Q},\mathrm{d}}\left[\frac{M_\mathrm{d}(t_0)}{M_\mathrm{d}(t)}\mathbb{E}^{\mathbb{Q}}\left[\frac{M_\mathrm{d}(t)}{M_\mathrm{d}(T)}\big|\mathcal{F}(t)\right]\max\left(y_F(t,T)-K,0\right)\big|\mathcal{F}(t_0)\right].$$

By means of the tower property of expectations, $\mathbb{E}\left[X\cdot\mathbb{E}\left[Y|\mathcal{F}(t)\right]|\mathcal{F}(t_0)\right] = \mathbb{E}\left[X\cdot Y|\mathcal{F}(t_0)\right]$, we obtain,

$$\mathrm{EE}^\mathrm{d}(t_0,t) = N_\mathrm{f}\mathbb{E}^{\mathbb{Q},\mathrm{d}}\left[\frac{M_\mathrm{d}(t_0)}{M_\mathrm{d}(T)}\max\left(y_F(t,T)-K,0\right)\big|\mathcal{F}(t_0)\right]$$
$$= N_\mathrm{f}P_\mathrm{d}(t_0,T)\mathbb{E}^{T,\mathrm{d}}\left[\max\left(y_F(t,T)-K,0\right)\big|\mathcal{F}(t_0)\right]. \quad (15.35)$$

because $y_F(t,T)$ is a martingale under the domestic T-forward measure.

Now, assuming that the forward rate $y_F(t,T)$, under the domestic T-forward measure, follows a lognormal process,

$$\mathrm{d}y_F(t,T) = \bar{\sigma}y_F(t,T)\mathrm{d}W^{T,\mathrm{d}}(t),$$

the price of a European option equals the expected exposure, and is given by:

$$\mathrm{EE}^\mathrm{d}(t_0,t) := V(t_0) = P_\mathrm{d}(t_0,T)N_\mathrm{f}\left(y_F(t_0,T)\Phi(d_1) - K\Phi(d_2)\right),$$

$$d_1 = \frac{\log\left(\frac{y_F(t_0,T)}{K}\right) + \frac{1}{2}\sigma^2(t-t_0)}{\sigma\sqrt{t-t_0}}, \quad d_2 = d_1 - \sigma\sqrt{t-t_0},$$

which is similar to the pricing of caplets in Section 12.2.1.

Expected exposure for an FX swap can thus be interpreted as a European option on an FX forward rate.

15.3 Multi-currency FX model with interest rate smile

In this section, an extension of the multi-currency model, in which an interest rate smile is incorporated, is presented. This hybrid model, named here *FX-HLMM*, models two types of market implied volatility smiles, i.e. the smile for the FX rate and the smile patterns in the domestic and foreign fixed income markets. This is especially interesting for FX products that are exposed to interest rate smiles. A description of such FX products can be found in the handbook by Hunter [Hunter and Picot, 2006].

Extensions of the FX model from the previous section, on the interest rate side, were presented by Schlögl in [Schlögl, 2002b] or Mikkelsen in [Mikkelsen, 2001], where the short-rate model was replaced by a Libor market model framework. A model for the FX rate with stochastic volatility and Libor interest rates was also proposed in [Takahashi and Takehara, 2008], assuming independence between log-normal Libor rates and the FX variable. In [Kawai and Jäckel, 2007], a displaced diffusion model for the FX rate was combined with the Libor rates.

As in Section 15.2, the stochastic volatility FX is of Heston-type, which, under the domestic risk-neutral measure $\mathbb{Q}$, follows the following dynamics:

$$\begin{aligned} \mathrm{d}y(t)/y(t) &= (\ldots)\mathrm{d}t + \sqrt{v(t)}\mathrm{d}W_y^{\mathbb{Q}}(t), & y(t_0) &> 0, \\ \mathrm{d}v(t) &= \kappa(\bar{v} - v(t))\mathrm{d}t + \gamma\sqrt{v(t)}\mathrm{d}W_v^{\mathbb{Q}}(t), & v(t_0) &> 0, \end{aligned} \quad (15.36)$$

with the parameters as in (15.16). Since the model will be placed under the forward measure, the drift in the first SDE does not need to be specified (the dynamics of domestic-forward FX, $y(t)P_{\mathrm{f}}(t,T)/P_{\mathrm{d}}(t,T)$, do not contain a drift term).

Suppose that the domestic and foreign currencies are independently calibrated to interest rate products that are available in their own markets. For simplicity, assume that the tenor structure for both currencies is the same, i.e., $\mathcal{T}_{\mathrm{d}} \equiv \mathcal{T}_{\mathrm{f}} = \{T_0, T_1, \ldots, T_m \equiv T\}$ and $\tau_i = T_i - T_{i-1}$ for $i = 1\ldots m$. For $t < T_{i-1}$ the Libor rates $\ell_{\mathrm{d},i}(t) := \ell_{\mathrm{d}}(t, T_{i-1}, T_i)$ (domestic) and $\ell_{\mathrm{f},i}(t) := \ell_{\mathrm{f}}(t, T_{i-1}, T_i)$ (foreign), are defined as

$$\boxed{\ell_{\mathrm{d},i}(t) := \frac{1}{\tau_i}\left(\frac{P_{\mathrm{d}}(t, T_{i-1})}{P_{\mathrm{d}}(t, T_i)} - 1\right), \quad \ell_{\mathrm{f},i}(t) := \frac{1}{\tau_i}\left(\frac{P_{\mathrm{f}}(t, T_{i-1})}{P_{\mathrm{f}}(t, T_i)} - 1\right).} \quad (15.37)$$

For each currency, the DD-SV Libor market model is employed for the interest rates, under the T-forward measure generated by the numéraires $P_{\mathrm{d}}(t,T)$ and $P_{\mathrm{f}}(t,T)$, i.e.,

$$\boxed{\begin{cases} \mathrm{d}\ell_{\mathrm{d},i}(t) = \nu_{\mathrm{d},i}\xi_{\mathrm{d},i}(t)\sqrt{\nu_{\mathrm{d}}(t)}\left(\mu_{\mathrm{d}}(t)\sqrt{\nu_{\mathrm{d}}(t)}\mathrm{d}t + \mathrm{d}W_i^{\mathrm{d},T}(t)\right), \\ \mathrm{d}\nu_{\mathrm{d}}(t) = \lambda_{\mathrm{d}}(\nu_{\mathrm{d}}(t_0) - \nu_{\mathrm{d}}(t))\mathrm{d}t + \eta_{\mathrm{d}}\sqrt{\nu_{\mathrm{d}}(t)}\mathrm{d}W_v^{\mathrm{d},T}(t), \end{cases}} \quad (15.38)$$

and

$$\begin{cases} d\ell_{f,i}(t) = \nu_{f,i}\xi_{f,i}(t)\sqrt{\nu_f(t)}\left(\mu_f(t)\sqrt{\nu_f(t)}dt + d\widehat{W}_i^{f,T}(t)\right), \\ d\nu_f(t) = \lambda_f(\nu_f(t_0) - \nu_f(t))dt + \eta_f\sqrt{\nu_f(t)}d\widehat{W}_v^{f,T}(t), \end{cases} \quad (15.39)$$

with

$$\mu_d(t) = -\sum_{k=i+1}^{m} \frac{\tau_k \xi_{d,k}(t)\nu_{d,k}}{1+\tau_k \ell_{d,k}(t)}\rho_{i,k}^d, \quad \mu_f(t) = -\sum_{k=i+1}^{m} \frac{\tau_k \xi_{f,k}(t)\nu_{f,k}}{1+\tau_k \ell_{f,k}(t)}\rho_{i,k}^f,$$

and where

$$\begin{cases} \xi_{d,i} = \vartheta_{d,i}^* \ell_{d,i}(t) + (1-\vartheta_{d,i}^*)\ell_{d,i}(t_0), \\ \xi_{f,i} = \vartheta_{f,i}^* \ell_{f,i}(t) + (1-\vartheta_{f,i}^*)\ell_{f,i}(t_0). \end{cases}$$

Brownian motion $W_i^{d,T}(t)$ corresponds to the i-th domestic Libor rate, $\ell_{d,i}(t)$, under the T-forward domestic measure, and Brownian motion $\widehat{W}_i^{f,T}(t)$ relates to the i-th foreign market Libor rate $\ell_{f,i}(t)$, under the terminal foreign measure T.

In the model, the parameters $\nu_{d,i}(t)$ and $\nu_{f,i}(t)$ determine the level of the interest rate implied volatility smile, the parameters $\vartheta_{d,i}^*$ and $\vartheta_{f,i}^*$ control the slope of the implied volatility smile, and λ_d, λ_f determine the speed of mean reversion for the variance and influence the speed at which the interest rate volatility smile flattens as the swaption expiry increases [Piterbarg, 2005b]. Parameters η_d, η_f determine the curvature of the interest rate smile.

The following correlation structure is imposed, between

FX and its variance process, $v(t)$:	$dW_y^T(t)dW_v^T(t) = \rho_{y,v}dt,$
FX and domestic Libor rates, $\ell_{d,i}(t)$:	$dW_y^T(t)dW_i^{d,T}(t) = \rho_{y,i}^d dt,$
FX and foreign Libor rates, $\ell_{f,i}(t)$:	$dW_y^T(t)d\widehat{W}_i^{f,T}(t) = \rho_{y,i}^f dt,$
Libor rates in the domestic market:	$dW_i^{d,T}(t)dW_j^{d,T}(t) = \rho_{i,j}^d dt,$
Libor rates in the foreign market:	$d\widehat{W}_i^{f,T}(t)d\widehat{W}_j^{f,T}(t) = \rho_{i,j}^f dt,$
Libor rates in dom. & foreign markets:	$dW_i^{d,T}(t)d\widehat{W}_j^{f,T}(t) = \rho_{i,j}^{d,f} dt.$

$$(15.40)$$

A zero correlation is prescribed between the remaining processes, i.e., between

- The Libor rates and their variance process,

$$dW_i^{d,T}(t)dW_\nu^{d,T}(t) = 0, \quad d\widehat{W}_i^{f,T}(t)d\widehat{W}_\nu^{f,T}(t) = 0,$$

- The Libor rates and the FX variance process,

$$dW_i^{d,T}(t)dW_v^T(t) = 0, \quad d\widehat{W}_i^{f,T}(t)dW_v^T(t) = 0,$$

- All variance processes,

$$dW_v^T(t)dW_\nu^{d,T}(t) = 0, \quad dW_v^T(t)d\widehat{W}_\nu^{f,T}(t) = 0, \quad dW_\nu^{d,T}(t)d\widehat{W}_\nu^{f,T}(t) = 0,$$

The FX and the Libor rate variance processes,

$$dW_y^T(t)dW_\nu^{d,T}(t) = 0, \quad dW_y^T(t)d\widehat{W}_\nu^{f,T}(t) = 0.$$

The correlation structure is graphically displayed in Figure 15.3.

We assume that the DD-SV model in (15.38) and (15.39) is already in the *effective* parameter framework, as developed in [Piterbarg, 2005b]. This means that approximate time-homogeneous parameters are used instead of the time-dependent parameters, i.e., $\vartheta_i(t) \equiv \vartheta_i^*$ and $\nu_i(t) \equiv \nu_i^*$.

With the correlation structure as described, the dynamics are derived for the forward FX, which is given by:

$$\boxed{y_F(t,T) = y(t)\frac{P_f(t,T)}{P_d(t,T)},} \quad (15.41)$$

see also (15.20), with $y(t)$ the spot exchange rate and $P_d(t,T)$ and $P_f(t,T)$ zero-coupon bonds.

When deriving the dynamics for (15.41), expressions for the ZCBs, $P_d(t,T)$ and $P_f(t,T)$, are needed. With Equation (15.37), the following expression for the bonds can be obtained:

$$\frac{1}{P_j(t,T)} = \frac{1}{P_j(t,T_{\bar{m}(t)})}\prod_{k=\bar{m}(t)+1}^{m}(1+\tau_k\ell_{j,k}(t)), \quad \text{for } j=\{d,f\}, \quad (15.42)$$

with $T = T_m$ and $\bar{m}(t) = \min(k : t \leq T_k)$ (empty products in (15.42) are defined to be equal to 1). Bond $P_j(t,T_m)$ in (15.42) is fully determined by the Libor rates $\ell_{j,k}(t)$, $k = 1, \ldots, m$ and by the bond $P_j(t,T_{\bar{m}(t)})$. Whereas the Libor rates

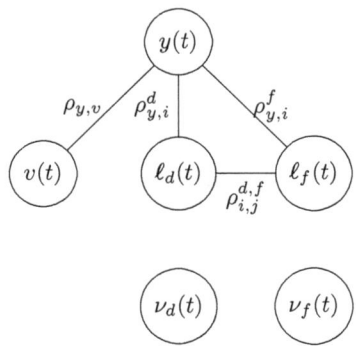

Figure 15.3: *The correlation structure for the FX-HLMM model. Arrows indicate nonzero correlations.*

$\ell_{j,k}(t)$ are defined by the systems in (15.38) and (15.39), the bond $P_j\left(t, T_{\bar{m}(t)}\right)$ is not yet defined in the current framework.

To define the continuous time dynamics for a zero-coupon bond, a linear interpolation scheme, as proposed in [Schlögl, 2002a] is employed, which reads:

$$\frac{1}{P_j(t, T_{\bar{m}(t)})} = 1 + (T_{\bar{m}(t)} - t)\ell_{j,\bar{m}(t)}(T_{\bar{m}(t)-1}), \text{ for } T_{\bar{m}(t)-1} < t < T_{\bar{m}(t)}. \quad (15.43)$$

This basic interpolation technique may perform well within the calibration. By combining (15.43) with (15.42), the domestic and foreign bonds are found to be,

$$\frac{1}{P_{\mathrm{d}}(t,T)} = \left(1 + (T_{\bar{m}(t)} - t)\ell_{\mathrm{d},\bar{m}(t)}(T_{\bar{m}(t)-1})\right) \prod_{k=\bar{m}(t)+1}^{m} (1 + \tau_k \ell_{\mathrm{d},k}(t)),$$

$$\frac{1}{P_{\mathrm{f}}(t,T)} = \left(1 + (T_{\bar{m}(t)} - t)\ell_{\mathrm{f},\bar{m}(t)}(T_{\bar{m}(t)-1})\right) \prod_{k=\bar{m}(t)+1}^{m} (1 + \tau_k \ell_{\mathrm{f},k}(t)).$$

When deriving the dynamics for $y_F(t,T)$ in (15.41), the dt-terms should be set equal to zero, as $y_F(t,T)$ is a martingale under numéraire $P_{\mathrm{d}}(t,T)$.

For each zero-coupon bond, $P_{\mathrm{d}}(t,T)$ or $P_{\mathrm{f}}(t,T)$, the dynamics are determined under the appropriate T-forward measures. For $P_{\mathrm{d}}(t,T)$, it is the domestic T-forward measure, and for $P_{\mathrm{f}}(t,T)$, it is the foreign T-forward measure. The dynamics for the zero-coupon bonds, that are driven by the Libor rate dynamics in (15.38) and (15.39), are now given by:

$$\frac{dP_{\mathrm{d}}(t,T)}{P_{\mathrm{d}}(t,T)} = (\ldots)dt - \sqrt{\nu_{\mathrm{d}}(t)} \sum_{k=\bar{m}(t)+1}^{m} \frac{\tau_k \nu_{\mathrm{d},k} \xi_{\mathrm{d},k}(t)}{1 + \tau_k \ell_{\mathrm{d},k}(t)} dW_k^{\mathrm{d},T}(t),$$

$$\frac{dP_{\mathrm{f}}(t,T)}{P_{\mathrm{f}}(t,T)} = (\ldots)dt - \sqrt{\nu_{\mathrm{f}}(t)} \sum_{k=\bar{m}(t)+1}^{m} \frac{\tau_k \nu_{\mathrm{f},k} \xi_{\mathrm{f},k}(t)}{1 + \tau_k \ell_{\mathrm{f},k}(t)} d\widehat{W}_k^{\mathrm{f},T}(t),$$

and the coefficients were defined in (15.38) and (15.39).

By changing the numéraire, from $P_{\mathrm{f}}(t,T)$ to $P_{\mathrm{d}}(t,T)$, for the foreign bond, only the drift terms will change. Since $y_F(t,T)$ in (15.41) is a martingale under the $P_{\mathrm{d}}(t,T)$ measure, it is *not* necessary to determine the appropriate drift correction.

On the basis of Equation (15.21) for the general dynamics of (15.41) and neglecting all the dt-terms, we get

$$\frac{dy_F(t,T)}{y_F(t,T)} = \sqrt{v(t)}dW_y^T(t) + \sqrt{\nu_{\mathrm{d}}(t)} \sum_{k=\bar{m}(t)+1}^{m} \frac{\tau_k \nu_{\mathrm{d},k} \xi_{\mathrm{d},k}(t)}{1 + \tau_k \ell_{\mathrm{d},k}(t)} dW_k^{\mathrm{d},T}(t)$$

$$- \sqrt{\nu_{\mathrm{f}}(t)} \sum_{k=\bar{m}(t)+1}^{m} \frac{\tau_k \nu_{\mathrm{f},k} \xi_{\mathrm{f},k}(t)}{1 + \tau_k \ell_{\mathrm{f},k}(t)} dW_k^{\mathrm{f},T}(t). \quad (15.44)$$

Note that the *hat* in $\widehat{W}$, has disappeared from the Brownian motion $W_k^{f,T}(t)$ in (15.44), which is an indication for the change of measure, from the foreign to the domestic measure, for the foreign Libor rates.

Since the stochastic volatility process $v(t)$ for the FX is independent of the domestic and foreign Libor rates $\ell_{d,k}(t)$ and $\ell_{f,k}(t)$, the dynamics under the $P_d(t,T)$-measure do not change and are given by:

$$dv(t) = \kappa(\bar{v} - v(t))dt + \gamma\sqrt{v(t)}dW_v^T(t). \qquad (15.45)$$

The model in (15.44), with the stochastic variance in (15.45) and the correlations between the main underlying processes, is however again not affine.

15.3.1 Linearization and forward characteristic function

The model in (15.44) is not of the affine form, as it contains terms like $\xi_{j,k}(t)/(1 + \tau_{j,k}\ell_{j,k}(t))$ with $\xi_{j,k} = \vartheta_{j,k}^*\ell_{j,k}(t) + (1 - \vartheta_{j,k}^*)\ell_{j,k}(t_0)$, for $j \in \{d, f\}$. To derive a characteristic function for an approximate model, we will *freeze* the Libor rates, which is standard practice (see for example [Glasserman and Zhao, 1999; Hull and White, 2000; Jäckel and Rebonato, 2000]), i.e.,

$$\begin{aligned}\ell_{d,k}(t) &\approx \ell_{d,k}(t_0) &\Rightarrow&& \xi_{d,k} &\equiv \ell_{d,k}(t_0),\\ \ell_{f,k}(t) &\approx \ell_{f,k}(t_0) &\Rightarrow&& \xi_{f,k} &\equiv \ell_{f,k}(t_0).\end{aligned} \qquad (15.46)$$

This approximation gives us the following $y_F(t,T)$-dynamics:

$$\boxed{\begin{aligned}\frac{dy_F(t,T)}{y_F(t,T)} &\approx \sqrt{v(t)}dW_y^T(t) + \sqrt{\nu_d(t)}\sum_{k\in\mathcal{A}}\psi_{d,k}dW_k^{d,T}(t) \\ &\quad - \sqrt{\nu_f(t)}\sum_{k\in\mathcal{A}}\psi_{f,k}dW_k^{f,T}(t), \\ dv(t) &= \kappa(\bar{v} - v(t))dt + \gamma\sqrt{v(t)}dW_v^T(t), \\ d\nu_j(t) &= \lambda_j(\nu_j(t_0) - \nu_j(t))dt + \eta_j\sqrt{\nu_j(t)}dW_v^{j,T}(t),\end{aligned}}$$

with $j \in \{d, f\}$, $\mathcal{A} = \{\bar{m}(t) + 1, \ldots m\}$, the correlations are given in (15.40) and

$$\psi_{d,k} := \frac{\tau_k \nu_{d,k}\ell_{d,k}(t_0)}{1 + \tau_k\ell_{d,k}(t_0)}, \quad \psi_{f,k} := \frac{\tau_k \nu_{f,k}\ell_{f,k}(t_0)}{1 + \tau_k\ell_{f,k}(t_0)}. \qquad (15.47)$$

The dynamics for the logarithmic transformation, $X(t) = \log y_F(t,T)$, are derived for which we need to calculate the square of the diffusion coefficients.[1]

With the notation,

$$a := \sqrt{v(t)}dW_y^T(t),\ b := \sqrt{\nu_d(t)}\sum_{k\in\mathcal{A}}\psi_{d,k}dW_k^{d,T}(t),\ c := \sqrt{\nu_f(t)}\sum_{k\in\mathcal{A}}\psi_{f,k}dW_k^{f,T}(t),$$
$$(15.48)$$

---

[1]As in the standard Black-Scholes analysis for $dS(t) = \sigma_1 S(t)dW(t)$, the log-transform gives $d\log S(t) = -\frac{1}{2}\sigma_1^2 dt + \sigma_1 dW(t)$.

we use, for the squared diffusion coefficient $(a+b-c)^2 = a^2+b^2+c^2+2ab-2ac-2bc$. So, the dynamics for $X(t) = \log y_F(t,T)$ can be expressed as:

$$dX(t) \approx -\frac{1}{2}(a+b-c)^2 + \sqrt{v(t)}dW_y^T(t) + \sqrt{\nu_d(t)}\sum_{\mathcal{A}}\psi_{d,k}dW_k^{d,T}(t)$$
$$- \sqrt{\nu_f(t)}\sum_{\mathcal{A}}\psi_{f,k}dW_k^{f,T}(t), \qquad (15.49)$$

with the coefficients a, b and c given in (15.48). Since

$$\left(\sum_{j=1}^m x_j\right)^2 = \sum_{j=1}^m x_j^2 + \sum_{\substack{i,j=1,\ldots,m \\ i\neq j}} x_i x_j, \text{ for } m > 0,$$

it follows that,

$$a^2 = v(t)dt,$$
$$b^2 = \nu_d(t)\left(\sum_{k\in\mathcal{A}}\psi_{d,k}^2 + \sum_{\substack{j,k\in\mathcal{A} \\ j\neq k}}\psi_{d,j}\psi_{d,k}\rho_{j,k}^d\right)dt =: \nu_d(t)A_d(t)dt, \quad (15.50)$$
$$c^2 = \nu_f(t)\left(\sum_{k\in\mathcal{A}}\psi_{f,k}^2 + \sum_{\substack{j,k\in\mathcal{A} \\ j\neq k}}\psi_{f,j}\psi_{f,k}\rho_{j,k}^f\right)dt =: \nu_f(t)A_f(t)dt, \quad (15.51)$$
$$ab = \sqrt{v(t)}\sqrt{\nu_d(t)}\sum_{k\in\mathcal{A}}\psi_{d,k}\rho_{k,x}^d dt,$$
$$ac = \sqrt{v(t)}\sqrt{\nu_f(t)}\sum_{k\in\mathcal{A}}\psi_{f,k}\rho_{k,x}^f dt,$$
$$bc = \sqrt{\nu_d(t)}\sqrt{\nu_f(t)}\sum_{k\in\mathcal{A}}\psi_{d,k}\sum_{j\in\mathcal{A}}\psi_{f,j}\rho_{j,k}^{d,f}dt,$$

with $\rho_{k,x}^d$, $\rho_{k,x}^f$ the correlations between the FX and k^{th} domestic and foreign Libor rates, respectively. The correlation between the jth domestic and k^{th} foreign Libor rate is $\rho_{j,k}^{d,f}$.

By setting $\varsigma\left(t, \sqrt{v(t)}, \sqrt{\nu_d(t)}, \sqrt{\nu_f(t)}\right) := (2ab - 2ac - 2bc)/dt$, the dynamics for $X(t)$ in (15.49) can be expressed as,

$$dX(t) \approx -\frac{1}{2}\left(v(t) + A_d(t)\nu_d(t) + A_f(t)\nu_f(t) + \varsigma\left(t, \sqrt{v(t)}, \sqrt{\nu_d(t)}, \sqrt{\nu_f(t)}\right)\right)dt$$
$$+\sqrt{v(t)}dW_y^T(t) + \sqrt{\nu_d(t)}\sum_{\mathcal{A}}\psi_{d,k}dW_k^{d,T}(t) - \sqrt{\nu_f(t)}\sum_{\mathcal{A}}\psi_{f,k}dW_k^{f,T}(t).$$

The coefficients $\psi_{d,k}$, $\psi_{f,k}$, A_d and A_f in (15.47), (15.50), and (15.51) are deterministic and piecewise constant.

In order to make the model affine, the nonaffine terms in the drift in $\varsigma\left(t, \sqrt{v(t)}, \sqrt{\nu_d(t)}, \sqrt{\nu_f(t)}\right)$ are *linearized* by a projection onto the first moments, i.e.,

$$\varsigma\left(t, \sqrt{v(t)}, \sqrt{\nu_d(t)}, \sqrt{\nu_f(t)}\right)$$
$$\approx \varsigma\left(t, \mathbb{E}\left(\sqrt{v(t)}\right), \mathbb{E}\left(\sqrt{\nu_d(t)}\right), \mathbb{E}\left(\sqrt{\nu_f(t)}\right)\right) =: G(t). \quad (15.52)$$

The variance processes $v(t)$, $\nu_d(t)$ and $\nu_f(t)$ are independent CIR-type processes, so the expectation of their product equals the product of the expectations. Function $G(t)$ can then be determined, with the help of the formulas in Lemma 13.2.1 in Section 13.2.1.

The approximation in (15.52) linearizes all nonaffine terms in the corresponding PDE. As before, the forward characteristic function, $\phi_\mathbf{X} := \phi_\mathbf{X}(u; t, T)$, is defined as the solution of the following backward PDE:

$$0 = \frac{\partial \phi_\mathbf{X}}{\partial t} + \frac{1}{2}\left(v + A_d(t)\nu_d + A_f(t)\nu_f + G(t)\right)\left(\frac{\partial^2 \phi_\mathbf{X}}{\partial X^2} - \frac{\partial \phi_\mathbf{X}}{\partial X}\right)$$
$$+ \lambda_d(\nu_d(t_0) - \nu_d)\frac{\partial \phi_\mathbf{X}}{\partial \nu_d} + \lambda_f(\nu_f(t_0) - \nu_f)\frac{\partial \phi_\mathbf{X}}{\partial \nu_f} + \kappa(\bar{v} - v)\frac{\partial \phi_\mathbf{X}}{\partial v}$$
$$+ \frac{1}{2}\eta_d^2\nu_d\frac{\partial^2 \phi_\mathbf{X}}{\partial \nu_d^2} + \frac{1}{2}\eta_f^2\nu_f\frac{\partial^2 \phi_\mathbf{X}}{\partial \nu_f^2} + \frac{1}{2}\gamma^2 v\frac{\partial^2 \phi_\mathbf{X}}{\partial v^2} + \rho_{y,v}\gamma v\frac{\partial^2 \phi_\mathbf{X}}{\partial X \partial v}, \quad (15.53)$$

with the final condition $\phi_\mathbf{X} \equiv \phi_\mathbf{X}(u; T, T) = e^{iuX(T)}$. Since all coefficients in this PDE are linear, the solution is of the following form:

$$\phi_\mathbf{X}(u; t, T) = \exp\left(\bar{A}(\mathbf{u}, \tau) + \bar{B}(\mathbf{u}, \tau)X(t) + \bar{C}(\mathbf{u}, \tau)v(t)\right.$$
$$\left. + \bar{D}_d(\tau)\nu_d(t) + \bar{D}_f(\tau)\nu_f(t)\right), \quad (15.54)$$

with $\tau := T - t$. Substitution of (15.54) in (15.53) gives us the following system of ODEs,

$$\frac{d\bar{A}}{d\tau} = G(t)\frac{\bar{B}^2 - \bar{B}}{2} + \lambda_d\nu_d(t_0)D_d(\tau) + \lambda_f\nu_f(t_0)D_f(\tau) + \kappa\bar{v}\bar{C},$$

$$\frac{d\bar{B}}{d\tau} = 0,$$

$$\frac{d\bar{C}}{d\tau} = \frac{\bar{B}^2 - \bar{B}}{2} + \left(\rho_{y,v}\gamma\bar{B} - \kappa\right)\bar{C} + \gamma^2\bar{C}^2/2,$$

$$\frac{d\bar{D}_d(\tau)}{d\tau} = A_d(t)\frac{\bar{B}^2 - \bar{B}}{2} - \lambda_d\bar{D}_d(\tau) + \eta_d^2\bar{D}_d^2(\tau)/2,$$

$$\frac{d\bar{D}_f(\tau)}{d\tau} = A_f(t)\frac{\bar{B}^2 - \bar{B}}{2} - \lambda_f\bar{D}_f(\tau) + \eta_f^2\bar{D}_f^2(\tau)/2,$$

with the initial conditions $\bar{A}(\mathbf{u}, 0) = 0$, $\bar{B}(\mathbf{u}, 0) = iu$, $\bar{C}(\mathbf{u}, 0) = 0$, $\bar{D}_d(0) = 0$, $\bar{D}_f(0) = 0$ with $A_d(t)$ and $A_f(t)$ from (15.50), (15.51), respectively, and $G(t)$ as in (15.52).

With $\bar{B}(\mathbf{u},\tau) = iu$, the solution for $\bar{C}(\mathbf{u},\tau)$ is analogous to the solution of the ODE for the FX-HHW1 model in Equation (15.29). As the remaining ODEs contain the piecewise constant functions $A_d(t)$, $A_f(t)$ the solution must be determined iteratively, like for the pure Heston model with piecewise constant parameters in [Andersen and Andreasen, 2000], see also Subsection 8.4.2. For a given grid in time direction, $0 = \tau_0 < \tau_1 < \cdots < \tau_m = \tau$, the functions $\bar{D}_d(\tau)$, $\bar{D}_f(\tau)$ and $\bar{A}(\mathbf{u},\tau)$ can be expressed as:

$$\bar{D}_d(\tau_k) = \bar{D}_d(u,\tau_{k-1}) + \chi_d(u,\tau_k),$$
$$\bar{D}_f(\tau_k) = \bar{D}_f(u,\tau_{k-1}) + \chi_f(u,\tau_k),$$

for $k = 1, \ldots, m$, and

$$\bar{A}(\mathbf{u},\tau_k) = \bar{A}(\mathbf{u},\tau_{k-1}) + \chi_A(u,\tau_k) - \frac{1}{2}(u^2+u)\int_{\tau_{k-1}}^{\tau_k} G(z)\mathrm{d}z,$$

with $G(z)$ in (15.52) and (analytically known) functions $\chi_j(u,\tau_k)$, for $j \in \{\mathrm{d},\mathrm{f}\}$ and $\chi_A(u,\tau_k)$:

$$\chi_j(u,\tau_k) := \left(\lambda_j - \delta_{j,k} - \eta_j^2 D_j(u,\tau_{k-1})\right)\left(1 - \mathrm{e}^{-\delta_{j,k}\Delta\tau_k}\right)/\left(\eta_j^2(1-\ell_{k,j}\mathrm{e}^{-\delta_{j,k}\Delta\tau_k})\right),$$

and

$$\chi_A(u,\tau_k) = \frac{\kappa\bar{v}}{\gamma^2}\left((\kappa - \rho_{y,v}\gamma iu - d_k)\Delta\tau_k - 2\log\left((1 - g_k\mathrm{e}^{-d_k\Delta\tau_k})/(1-g_k)\right)\right)$$
$$+\nu_d(t_0)\frac{\lambda_d}{\eta_d^2}\left((\lambda_d - \delta_{d,k})\Delta\tau_k - 2\log\left((1-\ell_{d,k}\mathrm{e}^{-\delta_{d,k}\Delta\tau_k})/(1-\ell_{d,k})\right)\right)$$
$$+\nu_f(t_0)\frac{\lambda_f}{\eta_f^2}\left((\lambda_f - \delta_{f,k})\Delta\tau_k - 2\log\left((1-\ell_{f,k}\mathrm{e}^{-\delta_{f,k}\Delta\tau_k})/(1-\ell_{f,k})\right)\right),$$

where

$$d_k = \sqrt{(\rho_{y,v}\gamma iu - \kappa)^2 + \gamma^2(iu+u^2)}, \quad g_k = \frac{(\kappa - \rho_{y,v}\gamma iu) - d_k - \gamma^2 C(u,\tau_{k-1})}{(\kappa - \rho_{y,v}\gamma iu) + d_k - \gamma^2 C(u,\tau_{k-1})},$$
$$\delta_{j,k} = \sqrt{\lambda_j^2 + \eta_j^2 A_j(t)(u^2+iu)}, \quad \ell_{j,k} = \frac{\lambda_j - \delta_{j,k} - \eta_j^2 D_j(u,\tau_{k-1})}{\lambda_j + \delta_{j,k} - \eta_j^2 D_j(u,\tau_{k-1})},$$

with $\Delta\tau_k = \tau_k - \tau_{k-1}$, $k = 1,\ldots,m$, $A_d(t)$ and $A_f(t)$ are from (15.50) and (15.51).

The resulting approximation of the full-scale FX-HLMM model is called *FX-LMM1* here.

15.3.2 Numerical experiments with the FX-HLMM model

We consider the FX-HLMM model and check the errors generated by the various approximations that led to the model FX-HLMM1, by some numerical experiments.

Basically, two linearization steps have been performed to define FX-HLMM1: the Libor rates are frozen at their initial values and the nonaffine covariance terms are projected onto a deterministic function.

The following interest rate curves are chosen, with $t_0 = 0$, $P_d(0,T) = \exp(-0.02T)$, $P_f(0,T) = \exp(-0.05T)$, and, as before, for the FX stochastic volatility model,

$$\kappa = 0.5, \quad \gamma = 0.3, \quad \bar{v} = 0.1, \quad v(0) = 0.1.$$

In the simulation, the following parameters for the domestic and foreign markets are chosen:

$$\vartheta^*_{d,k} = 95\%, \quad \nu_{d,k} = 15\%, \quad \lambda_d = 100\%, \quad \eta_d = 10\%,$$
$$\vartheta^*_{f,k} = 50\%, \quad \nu_{f,k} = 25\%, \quad \lambda_f = 70\%, \quad \eta_f = 20\%.$$

In the correlation matrix a number of correlations need to be specified. For the correlations between the Libor rates in each market, large positive values, as frequently observed in fixed income markets (see for example [Brigo and Mercurio, 2007]) are given, $\rho^d_{i,j} = 90\%$, $\rho^f_{i,j} = 70\%$, for $i,j = 1,\ldots,m$ $(i \neq j)$. In order to generate an implied volatility skew for the FX rate, we prescribe a *negative* correlation coefficient between $y_F(t,T)$ and its stochastic volatility process $v(t)$, i.e., $\rho_{y,v} = -40\%$. The correlation coefficients between the FX and the domestic Libor rates are set to $\rho^d_{y,k} = -15\%$, for $k = 1,\ldots,m$, and the correlation between the FX and the foreign Libor rates is $\rho^f_{y,k} = -15\%$. The correlation between the domestic and foreign Libor rates is $\rho^{d,f}_{i,k} = 25\%$ for $i,k = 1,\ldots,m$ $(i \neq k)$. The following block correlation matrix results:

$$\mathbf{C} = \begin{bmatrix} \mathbf{C}_d & \mathbf{C}_{d,f} & \mathbf{C}_{y,d} \\ \mathbf{C}^T_{d,f} & \mathbf{C}_f & \mathbf{C}_{y,f} \\ \mathbf{C}^T_{y,d} & \mathbf{C}^T_{y,f} & 1 \end{bmatrix},$$

with the domestic Libor rate correlations, given by

$$\mathbf{C}_d = \begin{bmatrix} 1 & \rho^d_{1,2} & \cdots & \rho^d_{1,m} \\ \rho^d_{1,2} & 1 & \cdots & \rho^d_{2,m} \\ \vdots & \vdots & \ddots & \vdots \\ \rho^d_{1,m} & \rho^d_{2,m} & \cdots & 1 \end{bmatrix} = \begin{bmatrix} 1 & 90\% & \cdots & 90\% \\ 90\% & 1 & \cdots & 90\% \\ \vdots & \vdots & \ddots & \vdots \\ 90\% & 90\% & \cdots & 1 \end{bmatrix}_{m \times m},$$

the foreign Libor rates correlations, given by:

$$\mathbf{C}_f = \begin{bmatrix} 1 & \rho^f_{1,2} & \cdots & \rho^f_{1,m} \\ \rho^f_{1,2} & 1 & \cdots & \rho^f_{2,m} \\ \vdots & \vdots & \ddots & \vdots \\ \rho^f_{1,m} & \rho^f_{2,m} & \cdots & 1 \end{bmatrix} = \begin{bmatrix} 1 & 70\% & \cdots & 70\% \\ 70\% & 1 & \cdots & 70\% \\ \vdots & \vdots & \ddots & \vdots \\ 70\% & 70\% & \cdots & 1 \end{bmatrix}_{m \times m},$$

the correlation between the Libor rates from the domestic and foreign markets given by:

$$\mathbf{C}_{df} = \begin{bmatrix} 1 & \rho^{d,f}_{1,2} & \cdots & \rho^{d,f}_{1,m} \\ \rho^{d,f}_{1,2} & 1 & \cdots & \rho^{d,f}_{2,m} \\ \vdots & \vdots & \ddots & \vdots \\ \rho^{d,f}_{1,m} & \rho^{d,f}_{2,m} & \cdots & 1 \end{bmatrix} = \begin{bmatrix} 1 & 25\% & \cdots & 25\% \\ 25\% & 1 & \cdots & 25\% \\ \vdots & \vdots & \ddots & \vdots \\ 25\% & 25\% & \cdots & 1 \end{bmatrix}_{m \times m},$$

and the vectors $\mathbf{C}_{y,\mathrm{d}}$ and $\mathbf{C}_{y,\mathrm{f}}$, given by:

$$\mathbf{C}_{\mathbf{y},\mathrm{d}} = \begin{bmatrix} \rho_{y,1}^{\mathrm{d}} \\ \rho_{y,2}^{\mathrm{d}} \\ \vdots \\ \rho_{y,m}^{\mathrm{d}} \end{bmatrix} = \begin{bmatrix} -15\% \\ -15\% \\ \vdots \\ -15\% \end{bmatrix}_{m\times 1}, \mathbf{C}_{\mathbf{y},\mathrm{f}} = \begin{bmatrix} \rho_{y,1}^{\mathrm{f}} \\ \rho_{y,2}^{\mathrm{f}} \\ \vdots \\ \rho_{y,m}^{\mathrm{f}} \end{bmatrix} = \begin{bmatrix} -15\% \\ -15\% \\ \vdots \\ -15\% \end{bmatrix}_{m\times 1}.$$

Since in both markets the Libor rates are assumed to be independent of their variance processes, these correlations are neglected.

We compute the prices of plain vanilla European options on the FX rate in (15.41). The simulation is performed in the same spirit as in Section 15.2.3, where the FX-HHW model was considered. While the prices for the FX-HLMM were obtained by a Monte Carlo simulation (20.000 paths and 20 intermediate points between the dates T_{i-1} and T_i for $i = 1, \ldots, m$), the prices for the FX-HLMM1 model were obtained by the COS method, with 500 cosine series terms.

The strike prices, $K_1(T_i), \ldots, K_7(T_i)$, are tabulated in Table 15.2. The prices and associated standard deviations are presented in Table 15.5.

The FX-HLMM1 model performs very well, as the maximum difference in terms of the implied volatilities is between $0.2\% - 0.5\%$, for this representative parameter setting.

Table 15.5: *Average FX call option prices obtained by the FX-HLMM model with 20 Monte Carlo simulations, 50.000 paths and $20 \times T_i$ steps; MC stands for Monte Carlo and COS for the Fourier cosine expansion technique for the FX-HLMM1 model with 500 expansion terms. Values of the strike prices $K_n(T_i)$ are tabulated in Table 15.2.*

T_i	method	$K_1(T_i)$	$K_2(T_i)$	$K_3(T_i)$	$K_4(T_i)$	$K_5(T_i)$	$K_6(T_i)$	$K_7(T_i)$
2y	MC	0.3336	0.2889	0.2456	0.2046	0.1667	0.1327	0.1030
	std dev	0.0008	0.0009	0.0010	0.0010	0.0011	0.0011	0.0012
	COS	0.3326	0.2880	0.2450	0.2043	0.1667	0.1330	0.1037
5y	MC	0.4243	0.3738	0.3234	0.2743	0.2274	0.1843	0.1457
	std dev	0.0012	0.0013	0.0014	0.0015	0.0016	0.0016	0.0016
	COS	0.4222	0.3717	0.3215	0.2727	0.2265	0.1838	0.1457
10y	MC	0.4363	0.3928	0.3482	0.3031	0.2587	0.2162	0.1764
	std dev	0.0012	0.0016	0.0019	0.0023	0.0026	0.0027	0.0028
	COS	0.4338	0.3905	0.3461	0.3014	0.2576	0.2157	0.1767
20y	MC	0.3417	0.3171	0.2907	0.2629	0.2342	0.2052	0.1768
	std dev	0.0010	0.0013	0.0015	0.0018	0.0021	0.0025	0.0030
	COS	0.3416	0.3176	0.2918	0.2647	0.2367	0.2085	0.1806
30y	MC	0.2396	0.2281	0.2152	0.2011	0.1858	0.1699	0.1534
	std dev	0.0012	0.0015	0.0018	0.0021	0.0024	0.0029	0.0035
	COS	0.2393	0.2279	0.2152	0.2014	0.1866	0.1710	0.1548

Sensitivity to the interest rate skew

Approximation FX-HLMM1 was based on *freezing the Libor rates*. By freezing the Libor rates, i.e., $\ell_{d,i}(t) \equiv \ell_{d,i}(t_0)$ and $\ell_{f,i}(t) \equiv \ell_{f,i}(t_0)$, we have used

$$\begin{cases} \xi_{d,i}(t) = \vartheta^*_{d,i}\ell_{d,i}(t) + (1-\vartheta^*_{d,i})\ell_{d,i}(t_0) = \ell_{d,i}(t_0), \\ \xi_{f,i}(t) = \vartheta^*_{f,i}\ell_{f,i}(t) + (1-\vartheta^*_{f,i})\ell_{f,i}(t_0) = \ell_{f,i}(t_0). \end{cases} \quad (15.55)$$

In the DD-SV model for the Libor rates $\ell_{d,i}(t)$ and $\ell_{f,i}(t)$, for any i, the parameters $\vartheta^*_{d,i}$, $\vartheta^*_{f,i}$ control the slope of the interest rate volatility smiles. Freezing the Libor rates to $\ell_{d,i}(t_0)$ and $\ell_{f,i}(t_0)$ is equivalent to setting $\vartheta^*_{d,i} = 0$ and $\vartheta^*_{f,i} = 0$ in (15.55) in the approximation FX-HLMM1.

By a Monte Carlo simulation, we obtain the FX implied volatilities from the full-scale FX-HLMM model for different values of ϑ^*, and, by comparing them to those from FX-HLMM1 with $\vartheta^* = 0$, the influence of the parameters $\vartheta^*_{d,i}$ and $\vartheta^*_{f,i}$ on the FX is checked. In Table 15.6 the implied volatilities for the FX European call options for the FX-HLMM and FX-HLMM1 models are presented. The experiments are performed for different combinations of the interest rate skew parameters ϑ^*_d and ϑ^*_f.

The experiment indicates that in this case there is only a small impact of the different $\vartheta^*_{d,i}-$ and $\vartheta^*_{f,i}-$values on the FX implied volatilities, implying that the approximate model FX-HLMM1 with $\vartheta^*_{d,i} = \vartheta^*_{f,i} = 0$ is useful for the interest rate modeling, for the parameters studied. With $\vartheta^*_{d,i} \neq 0$ and $\vartheta^*_{f,i} \neq 0$ the implied volatilities that are obtained by the FX-HLMM model appear to be somewhat higher than those obtained by FX-HLMM1, a difference of approximately $0.1\% - 0.15\%$, which is considered highly satisfactory.

Table 15.6: *Implied volatilities of the FX options from the FX-HLMM and FX-HLMM1 models, $T = 10$ and parameters as in Section 15.3.2. The numbers in parentheses are standard deviations (the experiment was performed 20 times with $20 \cdot T$ time steps).*

strike (15.32)	FX-HLMM (Monte Carlo simulation)					FX-HLMM1 (Fourier)
	$\vartheta^*_d = 0$	$\vartheta^*_d = 0.5$		$\vartheta^*_d = 0.5$		$\vartheta^*_d = 0$
	$\vartheta^*_f = 0$	$\vartheta^*_f = 0.5$	$\vartheta^*_f = 1$	$\vartheta^*_f = 0$	$\vartheta^*_f = 1$	$\vartheta^*_f = 0$
0.6224	31.98 % (0.20)	31.91 % (0.17)	31.98 % (0.17)	31.99 % (0.15)	31.96 % (0.18)	31.56 %
0.7290	31.49 % (0.21)	31.43 % (0.16)	31.48 % (0.19)	31.51 % (0.15)	31.46 % (0.18)	31.12 %
0.8538	31.02 % (0.21)	30.96 % (0.17)	31.01 % (0.20)	31.04 % (0.15)	30.97 % (0.18)	30.69 %
1.0001	30.58 % (0.21)	30.53 % (0.17)	30.56 % (0.22)	30.61 % (0.15)	30.52 % (0.17)	30.30 %
1.1714	30.16 % (0.20)	30.11 % (0.17)	30.15 % (0.24)	30.20 % (0.15)	30.08 % (0.16)	29.93 %
1.3721	29.77 % (0.22)	29.73 % (0.16)	29.77 % (0.26)	29.82 % (0.16)	29.68 % (0.17)	29.60 %
1.6071	29.41 % (0.24)	29.38 % (0.17)	29.43 % (0.28)	29.48 % (0.17)	29.31 % (0.18)	29.30 %

This final chapter contained many of the components that we wished to present in this book. An involved SDE system, for which a nontrivial change of measure should take place, is linearized so that we can derive the characteristic function for the approximate model. The COS method can then be used to price basic contracts in a highly efficient manner. Comparison takes place to the solutions found by a Monte Carlo method for the full-scale model.

15.4 Exercise set

Exercise 15.1 Give a proof for the zero-coupon bond dynamics, in Result 15.2.1.

Exercise 15.2 Let r_d and r_f denote the deterministic, domestic and foreign interest rates, respectively, and $M_\mathrm{d}(t)$ and $M_\mathrm{f}(t)$ the corresponding moneyness accounts,

$$\mathrm{d}M_\mathrm{d}(t) = r_\mathrm{d} M_\mathrm{d}(t)\mathrm{d}t, \quad \mathrm{d}M_\mathrm{f}(t) = r_\mathrm{f} M_\mathrm{f}(t)\mathrm{d}t.$$

Let $y(t)$ be the spot FX, which is expressed in units of the domestic currency per unit of the foreign currency. Furthermore,

$$P_\mathrm{d}(t,T) := \mathrm{e}^{-r_\mathrm{d}(T-t)}, \quad P_\mathrm{f}(t,T) := \mathrm{e}^{-r_\mathrm{f}(T-t)},$$

are the domestic and foreign zero-coupon bonds and T is maturity time.

The so-called *time-dependent FX-SABR model* is defined by the following dynamics under the domestic risk-neutral $\mathbb{Q}$-measure,

$$\mathrm{d}y(t) = (r_\mathrm{d} - r_\mathrm{f})\,y(t)\mathrm{d}t + \omega(t)\sigma(t)\left(\frac{P_\mathrm{d}(t,T)}{P_\mathrm{f}(t,T)}\right)^{1-\beta} y^\beta(t)\mathrm{d}W_y^\mathbb{Q}(t),\quad y(0)=y_0,$$

$$\mathrm{d}\sigma(t) = \gamma(t)\sigma(t)\mathrm{d}W_\sigma^\mathbb{Q}(t),\quad \sigma(0)=1,$$

with $\mathrm{d}W_y^\mathbb{Q}(t)\mathrm{d}W_\sigma^\mathbb{Q}(t) = \rho_{y,\sigma}\mathrm{d}t$ and where $\rho_{y,\sigma}$, $\gamma(t)$ and β denote the correlation, volatility-of-volatility parameter and the skew parameter, respectively. The skew parameter is set to $\beta = 0.5$. The volatility dynamics are scaled, which gives rise to a term structure parameter $\omega(t)$.

Show that the FX forward, $y_F(t,T_i) := y(t)\frac{P_\mathrm{f}(t,T_i)}{P_\mathrm{d}(t,T_i)}$, is defined by the following dynamics under the T_i-forward measure,

$$\mathrm{d}y_F(t,T_i) = \left(\frac{P_f(t,T_i)}{P_d(t,T_i)}\right)^\beta \left(\frac{P_f(t,T_i)}{P_d(t,T_i)}\right)^{1-\beta} \omega(t)\sigma(t)\left(\frac{P_d(t,T_i)}{P_f(t,T_i)}\right) y^\beta(t)\mathrm{d}W_y^{T_i}(t),$$

which, by definition of the ZCB under for deterministic interest rates, is defined by $P(T_i,T_N) = \frac{P(t,T_N)}{P(t,T_i)}$, gives:

$$\mathrm{d}y_F(t,T_i) = \omega(t)\sigma(t)\left(\frac{P_d(T_i,T_N)}{P_f(T_i,T_N)}\right)^{1-\beta}(y_F(t,T_i))^\beta\,\mathrm{d}W_y^{T_i}(t).$$

Show that for the scaled process, such that at initial $y(0,T_i) = 1$, we have:

$$\mathrm{d}y_F(t,T_i) = \omega_1(t)\sigma(t)\,(y_F(t,T_i))^\beta\,\mathrm{d}W_y^{T_i}(t),\quad y_F(0,T_i) =: \overline{y}_0^{T_i} = 1,$$
$$\mathrm{d}\sigma(t) = \gamma(t)\sigma(t)\mathrm{d}W_\sigma^{T_i},\quad \sigma(0)=1,$$

with $\mathrm{d}W_y^{T_i}(t)\mathrm{d}W_\sigma^{T_i}(t) = \rho_{y,\sigma}\mathrm{d}t$ and

$$\omega_1(t) := \omega(t)\left(\frac{P_\mathrm{d}(T_i,T_N)}{y_F(0,T_i)P_\mathrm{f}(T_i,T_N)}\right)^{1-\beta}.$$

Exercise 15.3 Under the domestic T_i-forward measure, we deal with the following FX log-normal dynamics, $y_F(t,T_i)$,

$$\mathrm{d}y_F(t,T_i) = \sigma(t)y_F(t,T_i)\mathrm{d}W_y^F(t,T_i),\quad y_F(0,T_i)=1.$$

Show that for the expected ATM payoff at time T_i given by
$$\mathbb{E}\left[(y_F(T_i, T_i) - 1)^+\right] = \mathbb{E}\left[g(x)\right],$$
the function $g(x)$ is given by,
$$g(x) := 2F_{\mathcal{N}(0,1)}\left(\frac{1}{2}\sqrt{x}\right) - 1, \quad x := \int_0^{T_i} \sigma^2(t) \mathrm{d}t,$$
where, as usual, $F_{\mathcal{N}(0,1)}(x)$ denotes the standard normal cumulative distribution function.

Implement a Monte Carlo method and compare the numerical results with the analytic result obtained.

Exercise 15.4 Show that the approximation of the function $g(\cdot)$ in Exercise 15.3 by the corresponding Taylor series yields
$$\mathbb{E}\left[(y_F(T_i, T_i) - 1)^+\right] = \frac{1}{\sqrt{2\pi}}\mathbb{E}\left[\sqrt{x}\right] + \epsilon_{\mathrm{T}}^{(\ell)},$$
with
$$\epsilon_{\mathrm{T}}^{(\ell)} := \frac{2}{\sqrt{\pi}}\left(-\frac{1}{3}\mathbb{E}\left[z^3\right] + \frac{1}{10}\mathbb{E}\left[z^5\right] - \ldots\right), \quad z := \frac{1}{2}\sqrt{x/2}, \quad x := \int_0^{T_i}\sigma^2(t)\mathrm{d}t.$$

Perform a Monte Carlo simulation and compare the obtained numerical result with the results from Exercise 15.3.

Exercise 15.5 Based on the Black-Scholes Hull-White model, the system of SDEs for the FX, under the domestic risk-neutral measure, reads, similar to System (15.16),
$$\mathrm{d}y(t)/y(t) = (r_\mathrm{d}(t) - r_\mathrm{f}(t))\mathrm{d}t + \sigma_y \mathrm{d}W_y^\mathbb{Q}(t),$$
$$\mathrm{d}r_\mathrm{d}(t) = \lambda_\mathrm{d}\left(\theta_\mathrm{d}(t) - r_\mathrm{d}(t)\right)\mathrm{d}t + \eta_\mathrm{d}\mathrm{d}W_\mathrm{d}^\mathbb{Q}(t),$$
$$\mathrm{d}r_\mathrm{f}(t) = \lambda_\mathrm{f}\left(\theta_\mathrm{f}(t) - r_\mathrm{f}(t) - \eta_\mathrm{f}\rho_{y,f}\sigma_y\right)\mathrm{d}t + \eta_\mathrm{f}\mathrm{d}W_\mathrm{f}^\mathbb{Q}(t),$$
with correlation parameters $\mathrm{d}W_y(t)\mathrm{d}W_\mathrm{d}(t) = \rho_{y,d}\mathrm{d}t$, $\mathrm{d}W_y(t)\mathrm{d}W_\mathrm{f}(t) = \rho_{y,f}\mathrm{d}t$ and $\mathrm{d}W_\mathrm{d}(t)\mathrm{d}W_\mathrm{f}(t) = \rho_{d,f}\mathrm{d}t$. Show that the FX-forward rate, under the domestic T-forward measure, which is defined as,
$$y_F(t,T) = y(t)\frac{P_\mathrm{f}(t,T)}{P_\mathrm{d}(t,T)},$$
with $y_F(t,T)$ the forward exchange rate under the T-forward measure, $y(t)$ the foreign exchange rate under the domestic spot measure, and $P_\mathrm{d}(t,T)$, $P_\mathrm{f}(t,T)$ represent the domestic and foreign zero-coupon bonds, respectively, gives rise to the following dynamics,
$$\mathrm{d}y_F(t,T)/y_F(t,T) = \eta_\mathrm{d}\bar{B}_\mathrm{d}(t,T)\left(\eta_\mathrm{d}\bar{B}_\mathrm{d}(t,T) - \rho_{y,d}\sigma_y - \rho_{d,f}\eta_\mathrm{f}\bar{B}_\mathrm{f}(t,T)\right)\mathrm{d}t$$
$$+ \sigma_y\mathrm{d}W_y^\mathbb{Q}(t) - \eta_\mathrm{d}\bar{B}_\mathrm{d}(t,T)\mathrm{d}W_\mathrm{d}^\mathbb{Q}(t) + \eta_\mathrm{f}\bar{B}_\mathrm{f}(t,T)\mathrm{d}W_\mathrm{f}^\mathbb{Q}(t).$$
(15.56)

Exercise 15.6 A change of measure, from the domestic-spot to the domestic T-forward measure, requires a change of numéraire, from the money-savings account $M_\mathrm{d}(t)$ to the zero-coupon bond $P_\mathrm{d}(t,T)$. Show that under the T-forward domestic measure the process in (15.56) reads,
$$\mathrm{d}y_F(t,T)/y_F(t,T) = \sigma_y\mathrm{d}W_y^T(t) - \eta_\mathrm{d}\bar{B}_\mathrm{d}(t,T)\mathrm{d}W_\mathrm{d}^T(t) + \eta_\mathrm{f}\bar{B}_\mathrm{f}(t,T)\mathrm{d}W_\mathrm{f}^T(t). \quad (15.57)$$
with $\bar{B}_\mathrm{d}(t,T)$, $\bar{B}_\mathrm{f}(t,T)$ and remaining parameters as defined in Section 15.2.2.

Exercise 15.7 Determine the value for $\hat{\sigma}_y(t,T)$ for which the forward FX process, $y_F(t,T)$ in (15.57), will be equal in distribution to $\hat{y}_F(t,T)$, which is defined as,

$$\mathrm{d}\hat{y}_F(t,T)/\hat{y}_F(t,T) = \hat{\sigma}_y(t,T)\mathrm{d}W_*^T(t), \tag{15.58}$$

where $W_*^T(t)$ is independent of the other stochastic processes.

Exercise 15.8 Determine the (constant) parameter σ_* for which the process, which is governed by,
$$\mathrm{d}\xi_F(t) = \sigma_* \xi_F(t)\mathrm{d}W_*^T(t),$$
with $\xi_F(0) = \hat{y}_F(0,T)$ will be equal in distribution to $\hat{y}_F(t,T)$ in (15.58) for a time t^*.

Exercise 15.9 Confirm your answers to the Exercises 15.7 and 15.8 by means of a suitable Monte Carlo implementation, where the original and the alternative processes are simulated and the results are compared.

Exercise 15.10 Under the assumption of constant interest rates r_d and r_f, consider the following local volatility model for the FX,

$$\mathrm{d}y(t) = (r_d - r_f)y(t)\mathrm{d}t + \sigma_{LV}(t,y(t))y(t)\mathrm{d}W_d^\mathbb{Q}(t) \tag{15.59}$$

$$=: a(y(t),t)\mathrm{d}t + b(y(t),t)\mathrm{d}W_d^\mathbb{Q}(t),$$

with

$$\sigma_{LV}^2(t,x) = \frac{\frac{\partial V_c^{FX}(t_0,y_0;x,T)}{\partial T} + (r_d - r_f)x\frac{\partial V_c^{FX}(t_0,y_0;x,T)}{\partial x} + r_f V_c^{FX}(t_0,y_0;x,T)}{\frac{1}{2}x^2 \frac{\partial^2 V_c^{FX}(t_0,y_0;x,T)}{\partial x^2}} \tag{15.60}$$

where $V_c^{FX}(t_0,y_0;K,T)$ is a call option price on the FX rate, with strike K and maturity T. Show that the Equations (15.60) and (15.59) satisfy the following Fokker-Planck equation,

$$\frac{\partial f_{y(T)}(x)}{\partial T} + \frac{\partial(a(x,T)f_{y(T)}(x))}{\partial x} - \frac{1}{2}\frac{\partial^2(b^2(x,T)f_{y(T)}(x))}{\partial x^2} = 0.$$

Exercise 15.11 Consider a constant volatility $\sigma = 0.15$. With this volatility parameter the value of an FX call option $V_c^{FX}(t_0)$ for different strike prices K can be computed. Use this to simulate, by a Monte Carlo method, the process in (15.59). Compare the option prices from the Black-Scholes model with $\sigma = 0.15$ to those obtained by simulation of Equation (15.59).

References

M. Abramowitz and I.A. Stegun. *Modified Bessel functions* **I** *and* **K**. Handbook of mathematical functions with formulas, graphs, and mathematical Tables, 9th edition, 1972.

A. Almendral and C.W. Oosterlee. On American options under the variance gamma process. *Applied Mathematical Finance*, 14:131–152, 2007.

S. Amstrup, L. MacDonald, and B. Manly. *Handbook of capture-recapture analysis*. Princeton University Press, 2006.

L.B.G. Andersen. Simple and efficient simulation of the Heston stochastic volatility model. *Journal of Computational Finance*, 11:1–48, 2008.

L.B.G. Andersen and J. Andreasen. Volatility skews and extensions of the Libor Market Model. *Applied Mathematical Finance*, 1(7):1–32, 2000.

L.B.G. Andersen and J. Andreasen. Volatile volatilities. *Risk Magazine*, 15(12): 163–168, 2002.

L.B.G. Andersen and J. Andreasen. Volatility skews and extensions of the Libor market model. *Applied Mathematical Finance*, 7:1:1–32, 2010. doi: https://doi.org/10.1080/135048600450275.

L.B.G. Andersen and V.V. Piterbarg. Moment explosions in stochastic volatility models. *Finance and Stochastics*, 11(1):29–50, Jan 2007. ISSN 1432-1122. doi: 10.1007/s00780-006-0011-7.

L.B.G. Andersen and V.V. Piterbarg. Interest rate modeling. *Atlantic Financial Press*, I-III:416, 2010.

J. Andreasen. Closed form pricing of FX options under stochastic rates and volatility. Presentation at Global Derivatives Conference, Paris, 9-11 May, 2006.

J. Andreasen and B. Huge. Volatility interpolation. *Risk Magazine*, 3:86–89, 2011.

A. Antonov. Effective approximation of FX/EQ options for the hybrid models: Heston and correlated Gaussian interest rates. Presentation at MathFinance Conference, Derivatives and Risk Management in Theory and Practice, Frankfurt, 26-27 May, 2007.

A. Antonov and T. Misirpashaev. Efficient calibration to FX options by Markovian projection in cross-currency Libor Market Models. SSRN working paper, 2006.

A. Antonov, M. Arneguy, and N. Audet. Markovian projection to a displaced volatility Heston model. SSRN working paper, 2008.

L. Arnold. *Stochastic differential equations, theory and applications*. Wiley, New York, 1973.

L. Bachelier. The theory of speculation. *Annales scientifiques de l'Ecole Normale Superieure*, 3(17):21–86, 1900.

Bank for International Settlements. Basel II: International Convergence of Capital Measurement and Capital Standards: A Revised Framework. Technical report, 2004.

O.E. Barndorff. Processes of normal inverse Gaussian type. *Finance and Stochastics*, 2:41–68, 1998.

O.E. Barndorff-Nielsen. Hyperbolic distributions and distributions on hyperbolae. *Scandinavian Journal of Statistics*, 5:151–157, 1978.

O.E. Barndorff-Nielsen. Normal inverse Gaussian distributions and stochastic volatility modelling. *Scandinavian Actuarial Journal*, 24:1–13, 1997.

D. Bates. Jumps and stochastic volatility: Exchange rate processes implicit in Deutsche mark options. *Review of Financial Studies*, 9(1):69–107, 1996.

T. Bayes. An essay towards solving a problem in the doctrine of chances. *Philosophical Transactions*, (53):370–418, 1763.

A. Bermúdez, M.R. Nogueiras, and C. Vázquez. Numerical solution of variational inequalities for pricing Asian options by higher order Lagrange–Galerkin methods. *Applied Numerical Mathematics*, 56(10-11):1256–1270, 2006.

J. Bertoin. *Lévy processes*, volume 121 of *Cambridge Tracts in Mathematics*. Cambridge University Press, Cambridge, 1996. ISBN 0-521-56243-0.

M. Bianchetti. Two curves, one price. *Risk*, 23:66, 2010.

F. Black and M. Scholes. The pricing of options and corporate liabilities. *Journal of Political Economy*, 81:637–654, 1973.

S. Borodin. *Handbook of Brownian motion*. Birkhäuser, second edition, 2002.

J.P. Boyd. *Chebyshev & Fourier spectral methods*. Springer Verlag, Berlin, 1989.

P. P. Boyle. Options: A Monte Carlo approach. *Journal of Financial Economics*, 4(3):323–338, 1977.

P.P. Boyle, M. Broadie, and P. Glasserman. Monte Carlo methods for security pricing. *Journal of Economic Dynamics and Control*, 21:1267–1321, 1997.

A. Brace, D. Gatarek, and M. Musiela. The market model of interest rate dynamics. *Mathematical Finance*, 7(2):127–155, 1997.

D. Breeden and R. Litzenberger. Prices of state-contingent claims implicid in option prices. *Journal of Business*, (51):621–651, 1978.

R.P. Brent. An algorithm with guaranteed convergence for finding a zero of a function. *The Computer Journal*, 14(4):422–425, 1971.

R.P Brent. *Algorithms for minimization without derivatives*. Courier Corporation, 2013.

D. Brigo and F. Mercurio. *Interest rate models – theory and practice: with smile, inflation and credit*. Springer Science & Business Media, 2007.

M. Broadie and Ö. Kaya. Exact simulation of stochastic volatility and other affine jump diffusion processes. *Operations Research*, 54(2):217–231, 2006.

M. Broadie and Y. Yamamoto. Application of the fast Gauss transform to option pricing. *Management Science*, 49:1071–1008, 2003.

P.P. Carr and D.B. Madan. Towards a theory of volatility trading, volatility: new estimation techniques for pricing derivatives, 417–427, 1998.

P.P. Carr and D.B. Madan. Option valuation using the Fast Fourier Transform. *Journal of Computational Finance*, 2:61–73, 1999.

P.P. Carr, D.B. Madan, and E.C. Chang. The variance gamma process and option pricing. *European Finance Review*, 2:79–105, 1998.

P.P. Carr, H. Geman, D.B. Madan, and M. Yor. The fine structure of asset returns: An empirical investigation. *Journal of Business*, 75:305–332, 2002.

C. Cattani. Shannon wavelets theory. *Mathematical problems in Engineering*, 2008:164808, 2008.

Chicago Board Option Exchange CBOE White Paper. The CBOE volatility index - VIX.

B. Chen, C.W. Oosterlee, and J.A.M. van der Weide. A low-bias simulation scheme for the SABR stochastic volatility model. *International journal Theoretical Applied Finance*, 15(2), 2012.

P. Cheng and O. Scaillet. Linear-quadratic jump-diffusion modelling. *Mathematical Finance*, 17(4):575–598, 2007.

K. Chourdakis. Option pricing using the fractional FFT. *Journal of Computational Finance*, 8(2):1–18, 2004.

I.J. Clark. *Foreign exchange option pricing: a practitioners guide*. Wiley, Chichester UK, 2011.

T.F. Coleman, Y. Li, and A. Verma. Reconstructing the unknown local volatility function. *Journal of Computational Finance*, 2(3):77–102, 1999.

R. Cont and P. Tankov. *Financial Modelling with Jump Processes*. Chapman and Hall, Boca Raton, FL, 2004.

J.C. Cox. Note on option pricing I: Constant elasticity of variance diffusions. *Stanford University Working Paper*, pages 229–263, 1975.

J.C. Cox, J.E. Ingersoll, and S.A. Ross. A theory of the term structure of interest rates. *Econometrica*, 53:385–407, 1985.

S.R. Das and S. Foresi. Exact solutions for bond and option prices with systematic jump risk. *Review of Derivatives Research*, 1:7–24, 1996.

C.S.L. de Graaf, Q. Feng, D.B. Kandhai, and C.W. Oosterlee. Efficient computation of exposure profiles for counterparty credit Risk. *International journal Theoretical Applied Finance*, 17(04):1450024, 2014. doi: 10.1142/S0219024914500241.

S. De Marco, P. Friz, and S. Gerhold. Rational shapes of local volatility. *Risk Magazine*, 2:82–87, 2013.

G. Deelstra and G. Rayée. Local volatility pricing models for long-dated FX derivatives. *Applied Mathematical Finance*, 1–23, 2012.

T.J. Dekker. Finding a zero by means of successive linear interpolation. *Constructive Aspects of the Fundamental Theorem of Algebra*, 37–51, 1969.

F. Delbaen and W. Schachermayer. A general version of the fundamental theorem of asset pricing. *Math. Ann.*, 300(3):463–520, 1994. ISSN 0025-5831.

E. Derman and I. Kani. Stochastic implied trees: Arbitrage pricing with stochastic term and strike structure of volatility. *International Journal Theoretical Applied Finance*, 1(1):61–110, 1998.

S. Desmettre. Change of measure in the Heston model given a violated Feller condition. arXiv Preprint in Quantitative Finance, https://arxiv.org/abs/1809.10955, 2018.

Y. d'Halluin, P.A. Forsyth, and G. Labahn. A penalty method for American options with jump diffusion processes. *Numerische Mathematik*, 97(2):321–352, 2004.

D. Duffie, J. Pan, and K. Singleton. Transform analysis and asset pricing for affine jump-diffusions. *Econometrica*, 68:1343–1376, 2000.

D. Duffie, D. Filipovic, and W. Schachermayer. Affine processes and applications in finance. *Annals of Applied Probability*, 13(3):984–1053, 2003.

D. Dufresne. The integrated square-root process. Working paper, University of Montreal, 2001.

B. Dupire. Pricing with a smile. *Risk*, 7:18–20, 1994.

E. Eberlein. Application of generalized hyperbolic Lévy motions to finance. In *Lévy processes*, pages 319–336. Birkhäuser Boston, Boston, MA, 2001.

B. Engelmann, F. Koster, and D. Oeltz. Calibration of the Heston stochastic local volatility model: a finite volume scheme. *Available at SSRN 1823769*, 2011.

F. Fang. *The COS method: An efficient Fourier method for pricing financial derivatives*. PhD Thesis, Delft University of Technology, Delft, the Netherlands., 2010.

F. Fang and C.W. Oosterle. Pricing early-exercise and discrete barrier options by fourier-cosine series expansions. *Numerische Mathematik*, 114:27–62, 2009.

F. Fang and C.W. Oosterlee. A novel option pricing method based on Fourier-cosine series expansions. *SIAM Journal on Scientific Computing*, 31(2): 826–848, 2008.

F. Fang and C.W. Oosterlee. A fourier-based valuation method for bermudan and barrier options under heston's model. *SIAM Journal on Financial Mathematics*, 2:439–463, 2011.

W. Feller. Two singular diffusion problems. *The Annals of Mathematics*, 54: 173–182, 1951.

Q. Feng. *Advanced estimation of credit valuation adjustment*. PhD Thesis, Delft University of Technology, Delft, the Netherlands., 2017.

Q. Feng and C.W. Oosterlee. Monte Carlo calculation of exposure profiles and greeks for Bermudan and barrier options under the Heston Hull-White model. In J. Belair R. Melnik, R. Makarov, (ed.), *Recent Progress and Modern Challenges in Applied Mathematics, Springer Fields Inst. Comm.*, pp. 265–301. Springer Verlag, 2017.

Q. Feng, S. Jain, P. Karlsson, D.B. Kandhai, and C.W. Oosterlee. Efficient computation of exposure profiles on real-world and risk-neutral scenarios for Bermudan swaptions. *Journal of Computational Finance*, 20(1):139–172, 2016.

P.A. Forsyth and K.R. Vetzal. Quadratic convergence for valuing American options using a penalty method. *SIAM Journal on Scientific Computing*, 23 (6):2095–2122, 2002.

J.-P. Fouque, G. Papanicolaou, R. Sircar, and K. Solna. Maturity cycles in implied volatility. *Finance and Stochastics*, 8:451–477, 2004. doi: https://doi.org/10.1007/s00780-004-0126-7.

H. Geman, N. El Karoui, and J.C. Rochet. Changes of numéraire, changes of probability measures and pricing of options. *Journal of Applied Probability*, 32:443–458, 1995.

H.U. Gerber and E.S.W. Shiu. Option pricing by Esscher transforms (Disc: p141-191). *Transactions of the Society of Actuaries*, 46:99–140, 1995.

M.B. German and S.W. Kohlhagen. Foreign currency option values. *Journal of International Money and Finance*, 3:231–237, 1983.

A. Giese. On the pricing of auto-callable equity securities in the presence of stochastic volatility and stochastic interest rates. Presentation at MathFinanceWorkshop: Derivatives and Risk Management in Theory and Practice, Frankfurt, 2006.

M.B. Giles. Improved multilevel Monte Carlo convergence using the milstein scheme. *Monte Carlo and Quasi-Monte Carlo Methods 2006*, page 343, 2007.

M.B. Giles. Multilevel Monte Carlo path simulation. *Operations Research*, 56(3):607–617, 2008.

M.B. Giles and C. Reisinger. Stochastic finite differences and multilevel Monte Carlo for a class of SPDEs in finance. *SIAM Journal on Financial Mathematics*, 3(1):572–592, 2012.

I.V. Girsanov. On transforming a certain class of stochastic processes by absolutely continuous substitution of measures. *Theory of Probability and its Applications*, 3(5):285–301, 1960.

P. Glasserman. *Monte Carlo methods in financial engineering*, volume 53. Springer Science & Business Media, 2003.

P. Glasserman and X. Zhao. Fast greeks by simulation in forward Libor models. *Journal of Computational Finance*, 3(1):5–39, 1999.

I.S. Gradshteyn and I.M. Ryzhik. *Table of integrals, series, and products*. Academic Press San Diego, 5th edition, 1996.

J. Gregory. *Counterparty credit risk: the new challenge for global financial markets*. Wiley, Chichester UK. ISBN 9780470685761.

L.A. Grzelak, J.A.S. Witteveen, M. Suárez-Taboada, and C.W. Oosterlee. The Stochastic Collocation Monte Carlo sampler: Highly efficient sampling from "expensive" distributions. *Quantitative Finance*, 0(0):1–18, 2018. doi: 10.1080/14697688.2018.1459807.

L.A. Grzelak and C.W. Oosterlee. On the Heston model with stochastic interest rates. *SIAM Journal on Financial Mathematics*, 1(2):255–286, 2011.

L.A. Grzelak and C.W. Oosterlee. From arbitrage to arbitrage-free implied volatilities. *Journal of Computational Finance*, 20(3):31–49, 2016.

L.A. Grzelak, C.W. Oosterlee, and S. van Weeren. Extension of stochastic volatility equity models with Hull-White interest rate process. *Quantitative Finance*, 12:89–105, 2012.

S. Guo, L.A. Grzelak, and C.W. Oosterlee. Analysis of an affine version of the Heston Hull-White option pricing partial differential equation. *Applied Numerical Mathematics*, 72:141–159, 2013.

J. Guyon and P. Henry-Labordère. Being particular about calibration. *Risk Magazine*, 25(1):88, 2012.

T. Haentjens and K.J. In't Hout. Alternating direction implicit finite difference schemes for the Heston-Hull-White partial differential equation. *Journal of Computational Finance*, 16(1):83, 2012.

P.S. Hagan and G. West. Methods for constructing a yield curve. *Wilmott Magazine*, 70–81, 2008.

P.S. Hagan, D. Kumar, A.S. Lesniewski, and D.E. Woodward. Managing smile risk. *Wilmott Magazine*, 3:84–108, 2002.

J.W. Harris and H. Stocker. *Maximum likelihood method*. New York-Verlag, 1998.

J.M. Harrison and D.M. Kreps. Martingales and arbitrage in multiperiod securities markets. *Journal of Economic Theory*, 20(3):381–408, 1979. ISSN 0022-0531.

J.M. Harrison and S.R. Pliska. Martingales and stochastic integrals in the theory of continuous trading. *Stochastic Processes and their Applications*, 11(3): 215–260, 1981. ISSN 0304-4149.

C. He, J.S. Kennedy, T. Coleman, P.A. Forsyth, Y. Li, and K. Vetzal. Calibration and hedging under jump diffusion. *Review of Derivatives Research*, 9:1–35, 2006.

D. Heath and M. Schweizer. Martingales versus PDEs in Finance: An equivalence results with examples. *Journal of Applied Probability*, 37:947–957, 2000.

D. Heath, R.A. Jarrow, and A. Morton. Bond pricing and the term structure of interest rates: a new methodology for contingent claims valuation. *Econometrica*, 1(60):77–105, 1992.

P. Henry-Labordère. Calibration of Local Stochastic Volatility Models to Market Smiles: a Monte-Carlo Approach. *Risk Magazine*, 112–117, 2009.

S.L. Heston. A closed-form solution for options with stochastic volatility with applications to bond and currency options. *Review of Financial Studies*, 6: 327–343, 1993.

D.J. Higham. *An introduction to financial option valuation: mathematics, stochastics and computation*, volume 13. Cambridge University Press, 2004.

A. Hirsa and D.B. Madan. Pricing American options under Variance Gamma. *Journal of Computational Finance*, 7(2):63–80, 2004.

J. Hull. *Options, futures and other derivatives. 8th edition.* Prentice Hall, 2012.

J. Hull and A. White. The pricing of options on assets with stochastic volatilities. *Journal of Finance*, 42(2):281–300, 1987.

J. Hull and A. White. Pricing interest-rate derivative securities. *Review of Financial Studies*, 3:573–592, 1990.

J. Hull and A. White. Forward rate volatilities, swap rate volatilities and the implementation of the Libor Market Model. *Journal of Fixed Income*, 10(2): 46–62, 2000.

J. Hull and A. White. The FVA debate continued. *Risk*, 10:52, 2012.

C. Hunter and G. Picot. Hybrid derivatives-financial engines of the future. The Euromoney-Derivatives and Risk Management Handbook, BNP Paribas, 2006.

K.J. In 't Hout. *Numerical partial differential equations in finance explained: an introduction to computational finance.* Palgrave McMillan, 2017.

K.J. In't Hout and S. Foulon. Adi finite difference schemes for option pricing in the Heston model with correlation. *International Journal of Numerical Analysis & Modeling*, 7(2):303–320, 2010.

P. Jäckel. *Monte Carlo methods in finance.* Wiley, Chichester UK, 2002.

P. Jäckel and R. Rebonato. Linking caplet and swaption volatilities in a BGM framework: Approximate solutions. *Journal of Computational Finance*, 6(4): 41–60, 2000.

S. Jain and C.W. Oosterlee. The Stochastic Grid Bundling Method: Efficient pricing of Bermudan options and their Greeks. *Applied Mathematics and Computation*, 269:412–431, 2015. doi: http://dx.doi.org/10.1016/j.amc.2015.07.085.

F. Jamshidian. Pricing of contingent claims in the one-factor term structure model. *Journal of Finance*, 111–122, 1987.

F. Jamshidian. An exact bond option formula. *Journal of Finance*, 44:205–209, 1989.

F. Jamshidian. Libor and swap market models and measures. *Financ. Stoch.*, 1 (4):293–330, 1997.

R. Jarrow and Y. Yildirim. Pricing treasury inflation protected securities and related derivatives using an HJM model. *Journal of Financial and Quantitative Analysis*, 38(2):409–430, 2003.

M. Jex, R. Henderson, and D. Wang. Pricing exotics under the smile. *Risk Magazine*, 12(11):72–75, 1999.

N.L. Johnson and S. Kotz. *Distributions in statistics: continuous univariate distributions 2*. Boston: Houghton Mifflin Company, first edition, 1970.

B. Jourdain and M. Sbai. Coupling index and stocks. *Quantitative Finance*, 12(5):805–818, 2012.

I. Karatzas and S.E. Shreve. *Brownian motion and stochastic calculus*. Springer Verlag, 1991.

I. Karatzas and S.E. Shreve. *Methods of mathematical finance*, volume 39 of *Applications of Mathematics*. Springer Verlag, New York, 1998. ISBN 0-387-94839-2.

A. Kawai and P. Jäckel. An asymptotic FX option formula in the cross currency Libor Market Model. *Wilmott Magazine*, 74–84, 2007.

J.S. Kennedy, P.A. Forsyth, and K.R Vetzal. Dynamic hedging under jump diffusion with transaction costs. *Operations Research*, 57(3):541–559, 2009.

Ch. Kenyon, A.D. Green, and M. Berrahoui. Which measure for PFE? the risk appetite measure A. *SSRN Electronic Journal*, 12, 2015. doi: 10.2139/ssrn.2703965.

J.L. Kirkby. Efficient option pricing by frame duality with the fast Fourier transform. *SIAM Journal on Financial Mathematics*, 6(1):713–747, 2016.

P.E. Kloeden and E. Platen. *Numerical solution of stochastic differential equations*. Springer Verlag, 1992, 1995, 1999.

W. Koepf. *Hypergeometric summation: an algorithmic approach to summation and special function identities*. Braunschweig, Germany: Vieweg, 1998.

S.G. Kou. A jump diffusion model for option pricing. *Management Science*, 48:1086–1101, 2002.

S.G. Kou and H. Wang. Option pricing under a double exponential jump-diffusion model. *Management Science*, 50:1178–1192, 2004.

E.E. Kummer. Über die hypergeometrische Reihe $F(a; b; x)$. *Journal für die Reine und Angewandte Mathematik*, 15:39–83, 1936.

Y.-K. Kwok. *Mathematical models of financial derivatives*. Springer Verlag, 2008.

B. Lauterbach and P. Schultz. Pricing warrants: An empirical study of the Black-Scholes model and its alternatives. *Journal of Finance*, 45(4):1181–1209, 1990.

R.W. Lee. Option pricing by transform methods: extensions, unification, and error control. *Journal of Computational Finance*, 7:51–86, 2004.

Á. Leitao, L.A. Grzelak, and C.W. Oosterlee. On an efficient multiple time step Monte Carlo simulation of the SABR model. *Quantitative Finance*, 17(10): 1549–1565, 2017a.

Á. Leitao, L.A. Grzelak, and C.W. Oosterlee. On a one time-step Monte Carlo simulation approach of the SABR model: Application to European options. *Applied Mathematics and Computation*, 293:461–479, 2017b.

A. Lesniewski. Notes on the CEV model. NYU working paper, 2009.

A.E. Lindsay and D.R. Brecher. Simulation of the CEV process and the local martingale property. *Mathematics and Computers in Simulation*, (82):868–878, 2012.

A. Lipton. The vol smile problem. *Risk Magazine*, 15(2):61–66, 2002.

A. Lipton and W. McGhee. Universal barriers. *Risk Magazine*, 15(5):81–85, 2002.

A. Lipton, A. Gal, and A. Lasis. Pricing of vanilla and first-generation exotic options in the local stochastic volatility framework: survey and new results. *Quantitative Finance*, 14(11):1899–1922, 2014.

F.A. Longstaff and E.S. Schwartz. Valuing American Options by Simulation: A Simple Least-squares Approach. 14(1):113–147, 2001. doi: 10.1093/rfs/14.1.113.

R. Lord and C. Kahl. Why the rotation count algorithm works. Tinbergen Institute Discussion Paper No. 2006-065/2, 2006.

R. Lord and C. Kahl. Complex logarithms in Heston-like models. *Mathematical Finance*, 20(4):671–694, 2010.

R. Lord, R. Koekkoek, and D. Dijk. A comparison of biased simulation schemes for stochastic volatility models. *Quantitative Finance*, 10(2):177–194, 2010.

M. Lorig, S. Pagliarani, and A. Pascucci. Explicit implied volatilities for multifactor local-stochastic volatility models. *Mathematical Finance*, 27: 926–960, 2015.

D.B Madan and E. Seneta. The variance gamma (VG) model for share market returns. *Journal of Business*, 63(4):511–524, 1990.

D.B. Madan, P.R. Carr, and E.C. Chang. The variance gamma process and option pricing. *European Finance Review*, 2:79–105, 1998.

W. Margrabe. The value of an option to exchange one asset for another. *The Journal of Finance*, 33(1):177–186, 1978.

D. Marris. Financial option pricing and skewed volatility. *Unpublished master's thesis, University of Cambridge*, 1999.

A.-M. Matache, T. Von Petersdorff, and C. Schwab. Fast deterministic pricing of options on Lévy driven assets. *ESAIM: Mathematical Modelling and Numerical Analysis*, 38(1):37–71, 2004.

F. Mercurio. Interest rates and the credit crunch: new formulas and market models. *SSRN,id 3225872*, 2010.

R.C. Merton. Option pricing when the underlying stocks are discontinuous. *Journal of Financial Economics*, 5:125–144, 1976.

P.A. Meyer. Un cours sur les integrales stochastiques. *Sem. Probab. 10, Lecture Notes in Math.*, 511:245–400, 1976.

P. Mikkelsen. Cross-currency Libor Market Models. Center for Analytica Finance Aarhus School of Business working paper no. 85, 2001.

G.N. Milstein, J.G.M. Schoenmakers, and V. Spokoiny. Transition density estimation for stochastic differential equations via forward-reverse representations. *Bernoulli*, 10:281–312, 2004.

K.R. Miltersen, K. Sandmann, and D. Sondermann. Closed form solutions for term structure derivatives with lognormal interest rates. *Journal of Finance*, 52(1):409–430, 1997.

M. Mori and M. Sugihara. The double-exponential transformation in numerical analysis. *Journal of Computational and Applied Mathematics*, 127:287–296, 2001.

S.M. Moser. Some expectations of a non-central chi-square distribution with an even number of degrees of freedom. TENCON 2007 - 2007 IEEE Region 10 Conference, Oct. 30–Nov. 2 2007.

M. Musiela and M. Rutkowski. *Martingale methods in financial modelling.* Springer Finance, 1997.

M. Muskulus, K. in't Hout, J. Bierkens, A.P.C. van der Ploeg, J. in't Panhuis, F. Fang, B. Janssens, and C.W. Oosterlee. The ING problem — a problem from financial industry; three papers on the Heston-Hull-White model. *Proc. Math. with Industry*, Utrecht, Netherlands, 2007.

V. Naik and M. Lee. General equilibrium pricing of options on the market portfolio with discontinuous returns. *Review of Financial Studies*, 3(4):493–521, 1990.

J. Jacod, M. Podolskij, N. Shephard, O. Barndorff-Nielsen, and S. Graversen. *A central limit theorem for realised power and bipower variations of continuous semimartingales.* Springer, Berlin, Heidelberg, 2006.

G.W. Oehlert. A note on the delta method. *American Statistician*, 46:27–29, 1992.

B. Øksendal. *Stochastic differential equations; An introduction with applications.* Springer Verlag, New York, 5th edition, 2000.

L. Ortiz-Gracia and C.W. Oosterlee. Robust pricing of European options with wavelets and the characteristic function. *SIAM Journal on Scientific Computing*, 35(5):B1055–B1084, 2013.

L. Ortiz-Gracia and C.W. Oosterlee. A highly efficient Shannon wavelet inverse Fourier technique for pricing European options. *SIAM Journal on Scientific Computing*, 38(1):B118–B143, 2016.

C. O'Sullivan. Path dependent option pricing under Lévy processes. EFA 2005 Moscow Meetings Paper, http://ssrn.com/abstract=673424, 2005.

M. Overhaus, A. Bermudez, H. Buehler, A. Ferraris, C. Jordinson, and A. Lamnouar. *Equity Hybrid Derivatives*. Wiley, Chichester UK, 2007.

A. Papapantoleon. An introduction to Lévy processes with applications to finance. *Lecture notes, University of Leipzig*, 2005.

A. Pascucci. *PDE and martingale methods in option pricing*. Springer Science & Business Media, 2011.

A. Pascucci and A. Mazzon. The forward smile in local-stochastic volatility models. *Journal of Computational Finance*, 20(3):1–29, 2017.

A. Pascucci, M. Suárez-Taboada, and C. Vazquez. Mathematical analysis and numerical methods for a PDE model of a stock loan pricing problem. *Journal of Mathematical Analysis and Applications*, 403(1):38–53, 2013.

P.B. Patnaik. The non-central χ^2 and F-distributions and their applications. *Biometrika*, 36:202–232, 1949.

A. Pelsser. A tractable interest rate model that guarantees positive interest rates. *Review of Derivatives Research*, 1:269–284, 1997.

A. Pelsser. *Efficient methods for valuing interest rate derivatives*. Springer Verlag London, 2000.

V.V. Piterbarg. Stochastic volatility model with time dependent skew. *Applied Mathematical Finance*, 12(2):147–185, 2005a.

V.V. Piterbarg. Time to smile. *Risk Magazine*, 18(5):71–75, 2005b.

V.V. Piterbarg. Smiling hybrids. *Risk Magazine*, 19:66–71, 2006.

V.V. Piterbarg. Markovian projection method for volatility calibration. *Risk Magazine*, April 2007.

J. Pitman and M. Yor. A Decomposition of Bessel Bridges. *Zeitschrift für Wahrscheinlichkeitstheorie und verwandte Gebiete*, 59:425–457, 1982.

J. Poirot and P. Tankov. Monte Carlo option pricing for tempered stable (CGMY) processes. *Asia-Pacific Financial Markets*, 13(4):327–344, 2006.

N. Privault. An extension of the quantum Itô table and its matrix representation. *Quantum Probability Communications*, 10:311–320, 1998.

P.E. Protter. *Stochastic integration and differential equations*. Springer Verlag, 2005.

M. Pykhtin and S. Zhu. A guide to modelling counterparty credit risk. *GARP Risk Review*, 16–22, July/August 2007.

S. Raible. *Lévy processes in finance: theory, numerics and empirical facts*. PhD Thesis, Inst. für Math. Stochastik, Albert-Ludwigs-Univ. Freiburg, 2000.

R. Rebonato. *Volatility and Correlation in the Pricing of Equity, FX, and Interest-rate Options*. Wiley, Chichester UK, 1999.

R. Rebonato. *Modern pricing of interest-rate derivatives: The LIBOR market model and beyond*. Princeton University Press, 2002.

R. Rebonato. *Volatility and correlation: the perfect hedger and the fox*. Wiley, Chichester UK, 2005.

R. Rebonato. *The SABR/LIBOR Market Model: Pricing, Calibration and Hedging for Complex Interest-Rate Derivatives*. Wiley, Chichester UK, first edition, 2009.

C. Reisinger. Analysis of linear difference schemes in the sparse grid combination technique. *IMA Journal of Numerical Analysis*, 33(2):544–581, 2012.

C. Reisinger and G. Wittum. On multigrid for anisotropic equations and variational inequalities "pricing multi-dimensional european and American options". *Computing and Visualization in Science*, 7(3-4):189–197, 2004.

C. Reisinger and G. Wittum. Efficient hierarchical approximation of high-dimensional option pricing problems. *SIAM Journal on Scientific Computing*, 29(1):440–458, 2007.

Y. Ren, D.B. Madan, and M.Q. Qian. Calibrating and Pricing with Embedded Local Volatility Models. *Risk Magazine*, 20(9):138–143, 2007.

L.C.G. Rogers. Which model for term-structure of interest rates should one use? *Mathematical Finance*, 65:93–115, 1995.

M. Rubinstein. Displaced diffusion option pricing. *Journal of Finance*, 38(1): 213–217, 1983.

M. Rubinstein. Implied binomial trees. *Journal of Finance*, 49:771–818, 1994.

M.J. Ruijter and C.W. Oosterlee. Two-dimensional Fourier cosine series expansion method for pricing financial options. *SIAM Journal on Scientific Computing*, 34(5):B642–B671, 2012.

P.A Samuelson. Rational theory of warrant pricing. *Industrial Management Review*, 6:13–31, 1965.

K. Sato. Basic results on Lévy processes. In *Lévy processes*, pp. 3–37. Birkhäuser Boston, Boston, MA, 2001.

E. Schlögl. *Advances in finance and stochastics: essays in honour of Dieter Sondermann*, chapter Arbitrage-free interpolation in models of market observable interest rates, pp. 197–218. Springer Verlag, Heidelberg, 2002a.

E. Schlögl. A multicurrency extension of the lognormal interest rate market models. *Finance and Stochastics*, 6(2):171–196, 2002b.

R. Schöbel and J. Zhu. Stochastic volatility with an Ornstein-Uhlenbeck process: An extension. *European Finance Review*, 3:23–46, 1999.

W. Schoutens. *Lévy processes in finance: pricing financial derivatives*. Wiley, Chichester UK, March 2003.

M. Schröder. Computing the constant elasticity of variance option pricing formula. *Journal of Finance*, 1(44):211–218, 1989.

K. Sennewald and K. Wälde. "Itô's lemma" and the Bellman equation for Poisson processes: An applied view. *Journal of Economics*, 89(1):1–36, 2006.

A. Sepp and I. Skachkov. Option pricing with jumps. *Wilmott Magazine*, 50–58, 2003.

R.U. Seydel. *Tools for computational finance. 4th edition.* Springer Verlag, 2017.

S.E. Shreve. *Stochastic calculus for finance II: continuous-time models*. New York: Springer, 2004.

J. Sippel and S. Ohkoshi. All power to PRDC notes. *Risk Magazine*, 15(11): 531–533, 2002.

J.C. Stein and E.M. Stein. Stock price distributions with stochastic volatility: An analytic approach. *Review of Financial Studies*, 4:727–752, 1991.

M. Suárez-Taboada and C. Vázquez. Numerical solution of a PDE model for a ratchet-cap pricing with BGM interest rate dynamics. *Applied Mathematics and Computation*, 218(9):5217–5230, 2012.

A. Takahashi and K. Takehara. Fourier transform method with an asymptotic expansion approach: an application to currency options. *International Journal Theoretical Applied Finance*, 11(4):381–401, 2008.

H. Tanaka. Note on continuous additive functionals of the 1-dimensional brownian path. *Zeitschrift für Wahrscheinlichkeitstheorie und verwandte Gebiete*, 1:251–257, 1963.

U. Ushakov. *Selected topics in characteristic functions*. De Gruyter, 1999.

A.W. van der Stoep, L.A. Grzelak, and C.W. Oosterlee. The Heston stochastic-local volatility model: Efficient Monte Carlo simulation. *International Journal Theoretical Applied Finance*, 17(7):1450045, 2014.

A. van Haastrecht and A. Pelsser. Generic pricing of FX, inflation and stock options under stochastic interest rates and stochastic volatility. *Quantitative Finance*, 655–691, 2010. doi: https://doi.org/10.1080/14697688.2010.504734.

A. van Haastrecht, R. Lord, A. Pelsser, and D. Schrager. Pricing long-maturity equity and FX derivatives with stochastic interest and stochastic volatility. *Insurance: Mathematics & Economics*, 45(3):436–448, 2009.

A. van Haastrecht, R. Lord, and A. Pelsser. Monte Carlo pricing in the Schöbel-Zhu model and its extensions. *Journal of Computational Finance*, 17(3):57, 2014.

O.A. Vašiček. An equilibrium characterization of the term structure. *Journal of Financial Economics*, 5:177–188, 1977.

C. Vázquez. An upwind numerical approach for an American and European option pricing model. *Applied Mathematics and Computation*, 97(2–3):273–286, 1998.

L. von Sydow, J. Höök, E. Larsson, E. Lindström, S. Milovanović, J. Persson, V. Shcherbakov, Y. Shpolyanskiy, S. Sirén, J. Toivanen, et al. BENCHOP — The BENCHmarking project in Option pricing. *International Journal of Computer Mathematics*, 92(12):2361–2379, 2015.

A.T. Wang. Generalized Itô's formula and additive functionals of Brownian motion. *Zeitschrift für Wahrscheinlichkeitstheorie und verwandte Gebiete*, 41:153–159, 1977.

I. Wang, J.W. Wan, and P.A. Forsyth. Robust numerical robust numerical valuation of European and American options under the CGMY process. *Journal of Computational Finance*, 10(4):31–70, 2007.

D. Williams. *Probability with martingales*. Cambridge University Press, Cambridge, 1991.

P. Wilmott. *Derivatives: The theory and practice of financial engineering*. Wiley Frontiers in Finance Series, 1998.

P. Wilmott, J. Dewynne, and S. Howison. *Option pricing: mathematical models and computation*. Oxford Financial Press, Oxford, 1995.

J.H. Witte and C. Reisinger. A penalty method for the numerical solution of Hamilton–Jacobi–Bellman (HJB) equations in finance. *SIAM Journal on Numerical Analysis*, 49(1):213–231, 2011.

B. Wong and C.C. Heyde. On changes of measure in stochastic volatility models. *Journal of Applied Mathematics and Stochastic Analysis*, 2006:1–13, 2006.

L. Wu and F. Zhang. Fast swaption pricing under the market model with a square-root volatility process. *Quantitative Finance*, 8(2):163–180, 2008.

Y. Yamamoto. Double-exponential fast Gauss transform algorithms for pricing discrete lookback options. *Publications of Research Institute of Mathematical Sciences*, 41:989–1006, 2005.

J. Zhu. *Modular pricing of options*. Springer Verlag, Berlin, 2000.

R. Zvan, P.A. Forsyth, and K.R. Vetzal. Penalty methods for American options with stochastic volatility. *Journal of Computational and Applied Mathematics*, 91(2):199–218, 1998.

R. Zvan, P.A. Forsyth, and K.R. Vetzal. A finite volume approach for contingent claims valuation. *IMA Journal of Numerical Analysis*, 21(3):703–731, 2001.

Index

3/2 model, 39
10y zero-coupon bond, 366
2D Itô's lemma, 200
2D Taylor series expansion, 30
2D characteristic function, 215, 216
2D experiment, 272
2D function, 115
2D geometric Brownian motion, 199, 215
2D structure, 112
2D system, 215
2D system of SDEs, 225

absorbing boundary, 463, 465
absorption condition, 467
accrual period, 447
adjusted log-asset price, 172
adjusted paths, 277
adjustment, 392
affine, 244, 245, 354
    process, 250
affine diffusion, 213, 215
    class, 193, 211, 213, 225, 405
    conditions, 212
    framework, 234
    model, 217
    process, 211, 215, 242
    with jumps, 216
affine form, 212, 217, 419, 514
affine Heston hybrid model, 432, 433
affine jump diffusion, AJD, process, 211, 217, 252
    class, 216, 217
    model, 217
affine short-rate, 362

AJD, 252
algebraic convergence, 175, 186
Almost Exact Simulation, AES, 288, 290, 292, 307, 330, 334
almost surely, 11, 16, 140
alternative model, 81, 86, 121, 146, 309, 310
alternative price model, 224
American option, 54, 55, 303
analytic expression, 183, 296
    expectation, 422
analytic solution, 128, 183, 251, 262, 426
    Black-Scholes equation, 63, 272
    option price, 130
annuity factor, 374
annuity function, 374
annuity measure, 384
antithetic sampling, 292
approximate Heston model, 427
arbitrage, 38, 40, 55, 100, 107, 109, 235, 384, 479, 491
    butterfly spread, 110, 114
    calendar spread, 109
    conditions, 81
    interpolation, 114
    opportunity, 110, 492
    spread, 110
arbitrage-free, 144, 145, 360, 493
    asset price model, 114
    conditions, 107, 115, 467
    constraint, 463
    HJM, 360, 496
    interpolation, 81, 107
    market, 448

    model, 111, 114
    short rate, 368
arbitrage-free conditions
    HJM, 346
arbitrage-free dynamics, 446
arbitrage-free model, 346
arbitrage-free option price, 48, 87, 103, 106, 242
arithmetic average, 25
Arithmetic Brownian Motion, ABM, 27, 44, 45, 78, 478
Arrow-Debreu, 69, 70, 89
Asian option, 160, 299, 300
asset, 51, 52, 56, 310, 319, 370, 410, 416
    jump, 159
asset class, 111, 406
asset density function, 68
asset dynamics, 39, 309, 437
asset models, 39
asset path, 9, 50, 115, 146, 257, 258, 274, 307
asset price, 9, 78
asset price process, 121
asset return, 50, 86
asset scenario, 48
asset variance, 223, 225
asset volatility, 438
asset-or-nothing option, 66, 208
asymmetry, 157
asymptotic expansion, 111, 187, 326
asymptotic limit, 31
asymptotic volatility formula, 114
at the money, ATM, 54, 89, 94, 239, 460

Bachelier, 44
back testing, 48
back-to-back transaction, 73
backward dynamic programming, 435
backward induction, 435
bank account, 55, 58, 139, 148, 406, 476
base curve, 484
basis function, 176, 304
basis point, 340, 368, 374, 418, 476
basis swap, 481
Bates model, 223, 251, 252, 506
    dynamics, 254
    jumps, 252
    ODEs, 252
    parameters, 253
Bayes theorem, 202

Bayes' formula, 287
BCVA, 397
Bermudan option, 54, 303, 434–438
Bessel function, 230
Bessel process, 229, 285, 467
BGM model, 446
big-O notation, 31
bilateral credit value adjustment, BCVA, 396
bilateral CVA, 397
bilateral CVA, BCVA, 396
bin method, 329
    Monte Carlo, 309
bivariate density, 8
bivariate samples, 272
Black volatility, 446
Black-76 valuation, 468
Black-Scholes, 56, 58, 81, 157, 178, 206, 267, 312, 446, 469
    delta, 138, 296, 303
    derivation, 60
    dividend, 59
    dynamics, 66, 86, 137, 187, 311, 336
    equation, 58
    formula, 65, 272, 446
    forward start, 311
    FX model, 493
    hedging, 74
    implied vol., 82, 111, 236, 337, 486
    inversion, 105
    log-transformed, 214
    operator, 69
    PDE, 214
    prices, 133
    risk-neutral, 208
    solution, 62, 63, 72, 82, 86, 91
    stoch. IR, 412
    theory, 86
    time-dependent, 239
    vega, 303
    volatility, 71
    world, 144
Black-Scholes Hull-White, 406, 412, 437, 439, 440, 523
Black-Scholes model, 56
BM distribution in time, 35
bond, 340, 343, 370, 416, 454
    discounted, 370
    fixed rate, 341, 342
    floating, 368

Index 543

foreign, 513
market, 340
mathematical price, 352
model, 366
portfolio, 416
price, 375, 379
products, 368
relative price, 455
risk-free, 484
yield, 375
zero-coupon, 359, 361, 362, 368
bounded Jacobi process, 276, 277
Breeden-Litzenberger method, 89, 95, 100, 102, 111, 118, 119, 460, 488
Brent's method, 85, 117
Brownian bridge, 292
Brownian Motion
    definition of, 211
    drifted, 146
Brownian motion
    drift, 137, 140
bullet payment, 491
bump and revalue method, 295
bundles, bins, 328
butterfly condition, 109
butterfly spread, 69–71
    arbitrage, 109, 114, 119

cádlág, 12
calendar spread, 109
calibration, 87, 187, 350, 375, 382, 406
    formulas, 495
    Heston, 236, 238, 319
    historical, 44
    hybrid model, 420
    market model, 460
    model, 499
    problem, 376
    procedure, 105, 320
    product, 376
    results, 433
    SDEs, 406
    vol. smile, 473
call option, 52, 102
    cash-or-nothing, 66, 173, 183
    CGMY model, 184
    COS method, 173
    DD model, 469
    delta, 297
    digital, 66

dynamics, 324
equivalence, 391
European, 55, 320, 334
experiments, 183
forward start, 310, 312
FX, 494, 520
hedging, 75, 138
Heston model, 249, 290
implied vol., 83
integral, 104
jumps, 131, 133
market, 105
Monte Carlo, 272
payoff, 52, 58
pricing, 138, 178, 468
solution, 63, 131
spread, 69
VG model, 186
ZCB, 363
cap, 378
cap rate, 378
caplet, 378, 379, 381, 446, 462, 478, 509
    pricing, 472
    solution, 462
    SV-LMM model, 474
caplet implied volatility, 382, 446
caplet valuation, 381, 477
caplet value, 382
cash, 55, 86, 144
cash amount, 73
cash flow, 56, 341, 368, 371, 372, 375, 481, 484
cash-or-nothing, 66, 173
Cauchy-Schwartz, 16
CDF, 1
    CIR, 226, 357, 421
    exposure, 389
    jumps, 134
    LMM CEV, 466
    lognormal, 210
    normal, 5, 73, 313, 413
    QE scheme, 282
    stock, 94, 110
CDS market, 434
central differences, 296, 297
central limit theorem, 258, 261
CGMY process, 151, 178
    characteristic function, 152
    experiments, 184
    model, 179

CGMYB model, 151
CGMYB price, 153
CGMYB process, 180
    characteristic function, 153
change of measure, 205, 207–209, 286, 344, 359, 361, 362, 371, 373, 381, 408, 409, 412, 452, 453, 471, 497, 500, 502
change of numéraire, 210
characteristic function, ChF, 3
    affine diffusion, 212
    BSHW, 407
    CGMY, 152
    CIR, 357
    definition, 3
    discounted, 212, 311
    forward, 311, 312
    forward BS, 311
    forward FX, 503
    forward FX process, 502
    forward Heston, 314
    FX-HHW1, 505
    Gamma, 147
    Heston, 242
    HHW, 420, 425
    HW, 353, 379
    jump diffusion, 138
    Kou, 135
    Lévy-Khinchine, 146
    lognormal, 169
    Merton, 133
    models, 179
    NIG, 156
    normal, 168
    Vašiček, 355
characteristic functions overview, 179
Chebyshev
    approximation, 176
    polynomial, 176
    series, 177
    series expansion, 175, 176
ChF and Feynman-Kac theorem, 66, 68
ChF and PDF, 165
Chicago Board of Exchange, CBOE, 102
Cholesky decomposition, 194, 199, 232, 429
Cholesky matrix, 205, 500
CIR process, 231, 236
clock, 231
combined root-finding method, 84, 85, 117

compensated Poisson process, 123
compensator, 335
complete market, 144, 149
complex plane, 4, 167
compound Poisson process, 133
compounded interest, 28
conditional characteristic function, 284
conditional density, 13, 465
conditional distribution, 28, 45
conditional expectation, 8, 12, 13, 44, 130, 145, 326, 328, 329, 435
conditional PDF, 8
conditional probability, 67
conditional random variable, 45
conditional SABR process, 111
conditional sampling, 283–285, 287, 292
confluent hyper-geometric function, 421
Constant Elasticity of Variance, CEV, 320, 322, 461, 462
    distribution, 464
    process, 111, 464
continuation value, 435
contract-level exposure, 390
control variate, 292
convergence
    strong sense, 266
    weak sense, 266
convex function, 54, 109
convexity, 457, 458
convexity correction, 448, 457, 458
convolution, 14, 68
corporate bonds, 340
correlated Brownian motion, 193–195, 232, 407, 410
correlated interest rates, 494, 496
correlated processes, 193
correlated SDE system, 194
correlated SDEs, 197
correlated stocks, 199
correlated Wiener processes, 415
correlation, 205, 225, 231, 453
    negative, 194, 292, 417
    nonzero, 419
    positive, 194, 417
    stochastic, 276
correlation coefficient, 9, 247
correlation matrix, 194, 205, 413, 497
correlation parameter, 236, 320, 330, 472
correlation structure, 194, 495, 505, 506, 511

correlations
    HHW, 419
COS FX result, 507, 519
COS method, 169
    2D, 329
    discounting, 411
    error analysis, 174
    Heston, 248
    integration range, 177
    results, 182, 249, 519
cosine series, 175
countable measure, 141
counterparties, 396
counterparty, 367, 368, 387, 434
    default, 368, 388, 397
    exposure, 367, 396
counterparty credit risk, CCR, 387, 434
counterparty risk, 392, 394
counterparty-level exposure, 390
coupon, 340
covariance matrix, 419
Cox-Ingersoll-Ross
    model, 225
Cox-Ingersoll-Ross, CIR, 225, 274, 314, 356, 420, 516
credit default swap, CDS, 388, 483
credit exposure, 433
credit value adjustment, CVA, 367, 386, 389, 392, 433, 434
creditworthiness counterparty, 387
cross-currency model, 489
cubature, 258
cumulant, 5, 177, 179
    overview, 181
cumulant characteristic function, 4
cumulant-generating function, 5, 6
cumulants, 4, 177, 190, 249
cumulative distribution function, CDF, 1
curse of dimensionality, 258
curvature, 137
    implied vol., 236, 254
    parameter, 239
    smile, 112, 511
CVA, 389, 394, 397, 434, 437
    approximation, 395
    charge, 395
    computation, 419, 434
    exposure, 433, 434
    FX, 508
    monitoring date, 434

percentage, 437
unilateral, 396

DD-SV LMM model, 472, 510, 512, 520
debt value adjustment, DVA, 396
decay, 166, 176, 191, 228
    density, 103, 411
decay rate, 151, 171, 175, 227
default, 340, 387
default probability, 434
default risk, 387, 482
default time, 395
default, counterparty, 481
defaultable counterparty, 387
defaulting counterparty, 367
degrees of freedom
    parameter, 331
degrees of freedom parameter, 226, 228, 290, 421, 462
delta hedge, 56, 74, 75
delta hedging, 73, 86, 137
delta method, 422, 467
density decay, 170
density recovery, 163, 168, 169, 189, 190, 440
dependent Brownian motion, 502
derivative pricing, 56, 69, 257
digital option, 66, 79, 118, 171, 173, 208, 264, 272, 297, 308
Dirac delta function, 7, 68, 71, 95, 102, 323
discounted asset, 493
discounted asset process, 41
discounted bond, 370
discounted cash flow, 373
discounted characteristic function, 67, 212, 218, 353, 407, 425
discounted expected payoff, 130, 234
discounted expected value, 72, 130
discounted option price, 60
discounted payments, 394
discounted payoff, 392
discounted short-rate, 346
discounted stock process, 42, 391
discounted tradable asset, 41
discounted volatility, 346
discrete asset path, 433
discrete average, 160
discrete cosine transform, DCT, 175
discrete dynamics, 270

discrete forward Libor rate, 446
discrete Fourier transform, DFT, 175, 191
discrete model, 297
discrete monitoring dates, 437
discrete paths, 43, 123, 263
discrete process, 86, 276
discrete random variable, 130
discrete sum, 49
discrete tenor structure, 454
discrete time model, 38
discretely re-balancing, 454
displaced diffusion, 472
displaced diffusion LMM, 445, 472
displaced diffusion stochastic volatility, DD-SV, 470
displaced diffusion, DD, 461, 467
    European option, 468
    model, 467, 470
    smile and skew, 470
displaced diffusion,DD
    model, 219, 468
displacement parameter, 468
dividend, 38, 56, 59
    payment, 38, 59
    proportional, 38
dividend paying stock, 55
dividend yield, 38, 59
domestic risk-neutral measure, 493, 496
dominated convergence, 15
double Heston model, 487
drift adjustment, 124
drift correction, 126, 150, 152, 156, 179, 181, 513
drift parameter, 28, 43, 58, 144, 147, 180
Duffie-Pan-Singleton, 213, 217
Dupire, 88
Dupire local vol. term, 105, 325, 332, 335
dynamic delta hedge, 56, 74, 75
dynamic hedging, 73
    jumps, 137
dynamic portfolio, 56

early exercise feature, 54
effective parameters, 512
effective skew, 473
effective volatility, 473
elementary process, 14, 15, 19
equality in distribution, 40, 71, 230

Equivalent Martingale Measure, 144, 145, 149, 234
error analysis, 170, 174, 335, 427
Euler approximation, 268
Euler discretization, 115, 266, 269, 270, 275, 327, 332, 333, 428, 430
    jumps, 274
Euler formula, 166
Euler scheme, 266, 271
    GBM process, 267
    reflecting, 277
    truncated, 277
Euler-Maruyama scheme, 266
Euribor basis swap, 481
European option, 52
European option on interest rate, 379
European swaption, 383, 384, 386, 400
exact simulation, 268, 278, 282
excess of kurtosis, 86, 147, 157
exotic option, 55, 71, 88, 309
expected (positive) exposure, 389, 395
expected exposure FX, 509
expected exposure, EE, 389, 391, 433, 509
expected payoff, 435
expected value, 2, 28, 258
    discounted, 130
    martingale, 60
    zero, 126
expected variance, 138
expiry date, 52
explosion of moments, 246
exponential convergence, 167, 169, 175
exponential decay, 151
exponential function, 134, 281, 362
exponential growth, 29, 258
exponential Lévy process, 139, 142, 146, 178, 179, 181, 183
exponential term, 315
exponential VG dynamics, 148
exposure, 367, 374, 388–390, 395, 396, 433
    closed form, 391
    negative, 397
    positive, 388
    reduction, 397, 398
    simulation, 388
exposure perspective, 391
exposure profile, 388, 389, 392
exposure reduction, 397

Index

Fast Fourier Transform, FFT, 188
fat tails, 147, 161, 178
fear index, 98
Feller condition, 226, 227, 275, 277, 279, 334, 357, 424
Feynman-Kac theorem, 59, 61, 63, 69, 79, 130, 243
    multi-D, 200
filtered probability space, 28, 464
filtration, 10, 11, 140, 148, 203, 311, 322, 414
financial derivative pricing, 56, 69, 257
financial terminology, 340, 379
finite activity jump models, 87
finite activity process, 140, 141, 143, 151
finite difference, 105, 295, 317, 332
finite measure, 142
finite moment, 153
finite number of jumps, 140, 145
finite variation, 142, 152
fixed rate, 371, 402, 484, 508
fixed-income security, 340
fixed-rate bond, 341, 342
flat extrapolation, 119
flat implied volatility, 413
flat volatility, 317, 335
flatten volatility, 317
float-rate note, 340
floating bonds, 368
floating coupon, 340
floating FX rate, 490, 508
floating rate, 378, 387
floating rate note, FRA, 368, 371
floating-floating IR swap, 481
floor, 318, 378
floorlet, 378, 379, 446, 460, 462, 474, 475
Fokker-Planck forward equation, 498
Fokker-Planck PDE, 102, 103, 524
foreign-exchange, FX, 489
forward, 343
forward ChF
    FX, 502
forward FX contract, 491
forward FX measure, 499
forward FX price, 498
forward FX rate, 499, 506
forward implied volatility, 313, 316
forward interest rate, 445
forward Libor rate, 369, 455

forward measure, 446
forward rate, 343, 359, 375, 447, 491, 499
    Libor, 369
    volatility, 448
forward rate agreement, 370
forward start option, 309
forward stock, 410
Fourier cosine coefficients, 171, 177
Fourier cosine expansion, 165, 166, 169, 412, 505
Fourier cosine inversion, 183
Fourier cosine series, 164
Fourier expansion, 165, 329
Fourier inversion, 407
Fourier methods
    class, 187
Fourier pair, 165
Fourier series, 165
    convergence, 175
Fourier technique, 68, 428
Fourier transform, 187, 411
    density, 4
    existence, 166
Fourier-based option pricing, 187, 504
Fourier-Stieltjes transform, 3
free-of-risk counterparty, 387
freezing Libor rate, 473, 514, 520
freezing technique, 472
frozen Libor rate, 518
full Fourier series, 175
fundamental solution, 68
fundamental theorem, 341
funding account, 73, 74
future FX rate, 492
FX asset class, 489
FX calibration, 495, 506
FX expected exposure, 509
FX foreign exchange, 490
FX forward ChF, 502
FX forward dynamics, 499
FX forward rate, 491, 502, 508
FX Heston model, 495
FX implied volatility, 520
FX introduction, 490
FX IR smile, 510
FX market, 490
FX model, 489
    advanced, 494
    forward measure, 499
    hybrid, 494

multi-currency, 495
SV, 494
FX multi-currency model, 510
FX option, 495, 505
    analytic, 494
    BS, 493
    pricing, 498
FX process, 493
FX products, 491
FX rate, 490
FX swap, 491, 508
FX world, 490
FX-HHW forward model, 500
FX-HHW model, 496
    experiment, 505
FX-HHW1 model, 504, 505
FX-HLMM model, 510, 517
FX-LMM model
    experiment, 517
FX-LMM1 model, 517
FX-SABR model, 522

Gamma function, 147, 152, 226, 230, 421, 467
Gamma process, 146, 150
gap option, 173
Gauss quadrature, 100
GBM geometric Brownian motion, 27
GBM distribution in time, 35
GBM drift term, 38
GBM Libor rate, 462
GBM model, 60, 135, 151, 179
GBM price, 39, 81
GBM process, 28, 67, 321
generalized Heston model, 250
geometric averaging, 486
Geometric Brownian Motion, GBM, 56
    dynamics under $\mathbb{Q}$, 28
geometric compounding, 486
Gibb's phenomenon, 175
Girsanov kernel, 361, 408, 501
Girsanov theorem, 201, 204, 206, 360, 452, 456
gold standard, 490
Greek delta, 59
Greek gamma, 59
Greek vega, 83
Greeks, 59, 64, 173
Greeks COS method, 173
Green's function, 66, 68, 164

Green's function theory, 467

H-SLV model, 309
H1-HW model, 425
hazard rate, 483
Heath-Jarrow-Morton, HJM, 343, 446, 496
hedge fund, 48
hedging approach, 234
hedging frequency, 76, 138
hedging FX, 491
hedging parameters, 59
hedging performance, 309
hedging portfolio, 59, 139
    Heston, 234
hedging strategy, 75
Heston dynamics
    AD class, 243
Heston FX model, 494
Heston Hull-White model, 415, 418, 430
Heston hybrid model
    approximation, 420
Heston SLV model, 309, 330
Heston SV model, 72, 223, 226, 231, 415
    calibration, 238, 241
    call option, 238
    characteristic function, 242, 244, 247, 316
    conditional density, 234
    COS formula, 248
    COS method, 248
    delta, 298
    dynamics, 236, 244
    experiment, 249
    forward implied vol., 316
    forward pricing, 314
    forward start, 314
    implied volatility, 238
    kappa, 237
    Monte Carlo, 283
    ODEs, 244, 245
    option price, 236
    parameters, 236, 240
    PDE, 233, 234
    PDE solution, 247
    piecewise parameters, 250
    process, 241
    real-world, 231
    risk-neutral, 232
    simulation, 278, 289

Index 549

skew smile, 236
standard, 474
variance process, 225, 279
volatility surface, 241
Heston type Libor model, 472
Heston-Stochastic Local Volatility, 309
Hilbert space, 13
histogram, 258, 263, 322
historical stock prices, 44, 47, 156
HJM condition, 346
HJM drift condition, 349
HJM framework, 344, 347, 349, 356, 447, 449, 451
HJM instantaneous forward rate, 348
HJM model, 359
HJM volatility, 351, 473
homogeneous trades, 391
Hull-White FX model, 489
short-rate process, 496
Hull-White, HW, 408
model, 351
characteristic function, 353
decomposition, 353, 355, 356
dynamics, 349, 360
interest rate process, 413
model, 339, 351, 352, 379, 381, 384, 477
SDE, 352
short-rate process, 351, 353
HW discounted characteristic function, 354
hybrid derivative product, 416
hybrid Heston model, 417
simulation, 428
hybrid model, 406

immediate exercise, 435
implied volatility
BSHW, 413
implied asset information, 48
implied Black volatility, 506
implied Black-Scholes volatility, 81
implied density function, 69, 81, 89–91
implied distribution, 100
implied forward rate, 343, 375
implied interest rate, 437
implied vol. skew
FX, 494
implied volatility, 81, 317, 463
ATM, 413

caplet, 382
CEV, 463
concept, 82
curve, 236
DD skew, 470
DD smile, 470
definition, 82
example, 83
forward, 313, 316
FX, 520
FX smile, 510
Heston, 236
jumps, 136
skew, 86, 90, 236, 463
slope, 137
smile, 86, 89, 90, 153, 157, 236, 463
stochastic IR, 413
structure, 406
surface, 86, 460
term-structure, 89, 413
importance sampling, 292, 293
in the money, ITM, 54, 81, 83, 85, 91, 111, 290
incomplete market, 144
independence, 395
independent Brownian motion, 151, 194–196, 199, 210, 232, 242, 256, 272, 289, 330, 408, 409, 429, 471, 473, 500, 501
independent increments, 11, 18, 20, 45, 140, 199
index option, 187
industrial standard, 494
infinite activity, 141, 308
infinite activity jump models, 87
infinite activity process, 140, 151, 161, 183, 186
infinite integration range, 170
infinite number of jumps, 140
infinite series, 128
infinite sum, 34, 145, 189
infinite variation, 142
initial yield curve, 455
instantaneous forward rate, 339, 343, 345, 346, 356, 449, 450
instantaneous short-rate, 343, 347
integrated CIR process, 284, 285
integration range, 166, 170, 174, 175, 249
integration range truncation, 174

intensity, 437
interest, 29
interest payment, 29
interest rate cap, 378
interest rate derivative, 368
interest rate floor, 378
interest rate models, 445
inverse Fourier transform, 7
inverse Gaussian, 155
inverse quadratic interpolation, 85
Itô calculus, 33
Itô differential, 124, 349
Itô integral, 14, 19, 261
    discrete version, 17
    existence, 19
    martingale, 20, 207
    properties, 24, 62, 201
    solution, 21, 33
Itô isometry, 17, 19, 24, 40
Itô process, 29, 30, 33, 268, 269, 321
Itô table, 97, 199
Itô's lemma, 29, 30, 33, 34, 57, 102, 127, 198, 233, 235, 320, 352, 357, 359, 408, 449, 474, 493
    Poisson, 124
Itô-Taylor expansion, 268
iterated expectations, 13, 71, 72

joint CDF, 7
joint density, 328
joint distribution, 8, 326, 395, 428
joint PDF, 7, 329
jump diffusion, 127
jump diffusion dynamics, 272
jump dynamics, 137, 156
jump intensity, 126, 151, 216, 217
jump magnitude, 122, 125, 128, 132, 135, 273
jump risk, 145

Kolmogorov equation, 102, 326
Kolmogorov PDE, 102, 326
Kou
    model, 87
Kou model, 128, 129, 135
    characteristic function, 135
    density, 129
Kou's model, 178, 179
kurtosis, 157

Lévy density, 150
    CGMY, 151
Lévy framework, 145
Lévy incomplete market, 144
Lévy measure, 140, 142, 151
    VG, 154
Lévy model, 121
    COS method, 181
    infinite activity, 146
    processes, 139
Lévy models, 87
Lévy price process, 145
Lévy process, 139, 141, 178
    characteristic function, 179
    computation, 184
    definition, 140
    finite activity, 143
    hyperbolic, 155
    infinite activity, 146
Lévy triplet, 143, 151
    CGMYB, 151
    NIG, 156
Lévy-Khinchine exponent, 143
Lévy-Khinchine representation, 153
Lévy-Khinchine theorem, 152
large jumps, 142
law of iterated expectations, 13
law of large numbers, 258, 259, 261
Lebesgue measure, 140, 141
left-side boundary, 266
left-side tail, 228
LGD, loss given default, 394, 434
Libor fixing, 368
Libor forward rate, 369
Libor market model, LMM, 359, 445, 446
Libor market SV model, 461
Libor model
    Heston, 472
    numéraire, 454
Libor rate, 368, 379, 446
    CEV, 462
    DD, 467
    definition, 368
    domestic, 511, 514
    dynamics, 448, 453
    European option, 468
    floating, 370, 371
    foreign, 511, 514
    forward, 369

Index 551

HJM framework, 447, 449
  variance, 511, 512
Libor rate dynamics, 453
Libor rates, FX, 510
Libor spot measure, 455
likelihood ratio, 293
likelihood ratio method, 295, 301, 303
likelihood ratio weight, 302
linear interpolation, 114
Lipschitz conditions, 30, 33, 204
liquidity, 91
little-o notation, 31
LMM
  definition, 446
  lognormal, 451
  measure change, 452
  terminal measure, 453
LMM CEV, 463
  CDF, 466
  density, 465
  implied vol., 463
  model, 468
local martingale, 321, 323
local volatility component, 320
local volatility FX model, 494
local volatility, LV, model, 81
localizing sequence, 323
log-dynamics, 213
log-likelihood, 45, 46
log-price, 63, 71, 179
log-process, 64, 131, 215, 246
log-return, 86
log-transformed GBM, 40
lognormal, 312
lognormal FX dynamics, 494
lognormal Libor rate, 451
lognormal LMM, 451
lognormal variance, 428
long-term mean parameter, 225
Longstaff-Schwartz method, 304
lower incomplete Gamma function, 226, 467

Maclaurin expansion, 5
Margrabe formula, 221
market completeness, 343
Markov process, 28, 229, 464
Markov projection, 495
Markov property, 28
Markovian affine process, 211

Markovian projection, 326
martingale, 9, 12
  asset price, 40
  definition, 12
  Itô integral, 20
  option, 60
martingale approach, 60
  Heston, 233
  jumps, 127
  SV, 233
martingale equality, 448
martingale measure, 143
martingale property, 9, 12, 24, 40, 60, 124, 323
  SV, 234
martingale representation theorem, 20, 201
maturity time, 52, 375
maximum likelihood estimation, MLE, 44
mean reversion parameter, 382, 419
mean reverting level, 353
mean reverting process, 224, 352, 353
mean reverting square-root process, 225
measurability, 20, 45, 203
measure transformation, 408
Merton jump diffusion, 135, 136
Merton jumps, 133
Merton model, 128, 129, 178, 179
  characteristic function, 133
Milstein scheme, 266, 268, 269
model-free, 100
modified Bessel function, 155, 227
moment explosion, 246
moment matching, 39, 40, 280
moment-generating function, MGF, 3–5, 314, 315
  chi-squared, 315
  CIR, 314
moments, 4, 5
money-savings account, 29
Monte Carlo AES scheme, 330
Monte Carlo algorithm, 259
Monte Carlo asset path, 259
Monte Carlo basics, 257
Monte Carlo convergence, 263, 292
Monte Carlo estimate, 259
Monte Carlo FX result, 507, 519
Monte Carlo Greeks, 294
  finite differences, 296

likelihood ratio, 301
  pathwise, 297
Monte Carlo integration, 260–262
Monte Carlo method, 187, 258
  bins, 326, 328
  exposure, 388
  SLV, 327
Monte Carlo path, 298
Monte Carlo simulation, 257
  bivariate, 272
  early-exercise, 303
  FX-HHW, 507
  Heston, 283
  HHW, 431
  HSLV, 333
  HW, 477
  hybrid, 428
  importance sampling, 293
  improvements, 292
  jumps, 272
  LV, 116
  variance reduction, 292
Monte Carlo standard error, 259
multi-currency model, 510
multi-D Itô's lemma, 198
multi-D process, 197
multi-D samples, 272
multi-dimensional Feynman-Kac, 200
multi-dimensional Itô, 199
multi-dimensional process, 193
multi-dimensional SDEs, 269, 359
multilevel Monte Carlo, 292
multiple asset classes, 405
multiple curves, 482, 484

near-singular behavior, 226, 227
negative correlation, 518
negative exposure, 397
negative interest rate, 475
netting, 397
netting exposure, 391
netting impact, 397
Newton iteration, 376
Newton-Raphson iteration, 82, 236
NIG model, 179
no-arbitrage, 60, 105, 343, 346, 482
  assumption, 102, 144, 145, 149, 157
  drift condition, 346
  yield curve model, 352
non-dividend paying stock, 83

non-parametric LV method, 102
non-unique measure, 145
non-unique solution, 229, 462
nonaffine square-root term, 503
noncentral distribution, 226, 227, 262, 278–280, 315, 331, 421, 422, 431, 462, 466
noncentrality parameter, 226, 279, 314, 421, 431, 432, 462, 466
nonexplosive solution, 229, 461
nonparametric form, 320
nonparametric method, 327
nonstandard European derivative, 111
nonsymmetric double exponential, 129
Normal Inverse Gaussian, 141
  process, 155
notional amount, 28, 340, 341, 370, 372, 378, 387, 401, 457, 492
notional foreign currency, 490
numerical integration, 164, 187, 188, 265, 266, 424
Nyquist relation, 188

ODEs, 212, 217, 357, 380
  analytic solution, 213
  FX, 516
  H1-HW, 425
  Heston, 244
optimal stopping time, 435
option, 51
option contract, 51
option definition, 51
option delta, 59
option expiry, 52
option gamma, 59
option maturity, 52
option payoff, 52
option vega, 83
option writer, 52, 53, 56, 59, 144, 157, 435
OU process, 353
OU, Ornstein-Uhlenbeck, 224, 352, 431, 487
out of the money, OTM, 54, 83, 85, 91, 92, 105, 111, 239, 290
over-the-counter, OTC, 55, 74, 372, 433
overnight index swap rate, 484
own default, 396

Index

ℙ-measure prices, 41
$P(t,T)$-dynamics, 348
parameter estimation, 44
parametric LV model, 461
part-dependent option, 224
partial differential equation, PDE, 56, 58, 61
partial integro-differential equation, PIDE, 127, 128, 130, 143
path-dependent option, 71, 90, 258, 259, 309
pathwise sensitivity, 295
pathwise sensitivity method, 297
pay delay, 447
payer swap, 372
payer swaption, 383
payoff, 52
    forward start, 310
PDE terminology, 68
PDF
    $S(t)$, 36
    $X(t)$, 36
    absorbed, 463
    bivariate, 8
    CEV, 465
    CGMY, 151
    conditional, 13
    definition, 1
    differentiation, 302
    example, 4
    fat tails, 147
    Gamma, 147
    Heston, 226, 234
    implied, 69, 90, 91
    Kou, 129
    Lévy, 143, 150
    lognormal, 169, 494
    LV, 116
    market implied, 320
    Merton, 129
    multi-variate, 411
    NIG, 155
    noncentral chi-squared, 227, 315
    normal, 2, 68, 168
    recovery, 168
    squared Bessel, 229, 465
    stock, 90, 103
    time evolution, 36, 102
    transitional, 71, 319
    variance, 227, 279, 421
    VG, 151
PDF and ChF, 165
peakedness, 147
percentage CVA, 437
percentage performance, 310
performance option, 309
performance period, 316
piecewise parameters, 250
plain vanilla FX option, 519
plain vanilla interest rate swap, 372
plain vanilla option, 56, 102, 143, 169, 171, 309
plain vanilla payoff, 90
PnL, 73
Pochhammer symbols, 422
Poisson distribution, 122, 133, 274, 315
Poisson jumps, 124
Poisson process, 121–123, 125, 216, 234, 251, 273
    Itô's lemma, 124, 127
Poisson random variable, 122
portfolio, 57
positive exposure, 388
potential future exposure, PFE, 389, 399, 433, 436
price formula, 61
probability density function, PDF, 1
probability measure, 12, 40, 41, 43, 145, 202, 203
probability of default, 387, 395
probability space, 12, 13, 202, 211, 321
profit, 52, 55, 69, 74, 78, 319
profit upon self-default, 397
projection on moments, 516
proportional dividend payment, 180
put option, 52, 53
    payoff, 53
put-call parity, 55, 79, 91, 93, 95, 184, 190, 255, 364

ℚ-martingale, 201
ℚ-measure prices, 42
quadratic convergence, 83
quadratic equation, 280
Quadratic Exponential scheme, QE, 279, 282, 330, 431, 437
quadratic expression, 150
quadratic model, 39
quadratic polynomial, 247

quadrature rule, 173, 258
quanto correction term, 497

Radon-Nikodym derivative, 202, 205, 286, 293, 359, 361, 383, 408, 452, 456
random seed, 333
random step function, 16
random time, 159
raw moments, 5
receiver swap, 372
receiver swaption, 383
recovery, 393
recovery rate, 394, 397, 398, 441
reflecting boundary, 230, 465–467
reflecting Euler scheme, 278
reflecting origin, 275
relative bond price, 455
replicating portfolio, 56, 224, 233, 234
replicating strategy, 491
replication, 144, 149
reset delay, 447
reset interval, 319
Riccati ODEs, 212, 217, 221, 244, 250, 357, 427, 442
risk management, 48, 187, 188, 295, 305, 367, 375, 389, 390, 405, 419, 445, 475, 481
risk underestimation, 395
risk-free bond, 484
risk-free interest rate, 41, 43, 235
risk-free money savings account, 58
risk-free portfolio, 235
risk-free ZCB dynamics, 348
risk-neutral dynamics, 149
risk-neutral measure, 43, 149
risky asset, 394
risky derivative, 393
root-finding algorithm, 82, 385
roots of polynomial, 247
Runge-Kutta method, 213, 250

SABR model, 111
    formula, 118
    implied volatility., 112
    parametrization, 114
    simulation, 292
safe investor, 392
Samuelson model, 28, 143, 146
scale parameter, 147, 156, 189
scaled Gamma process, 150

Schöbel-Zhu Hull-White model, 405, 415, 433
Schöbel-Zhu model, 72, 224, 320, 415, 432, 494
Schöbel-Zhu process, 225
SCMC, Stochastic Collocation Monte Carlo, 292
score credit quality, 388
score function, 293, 302
semi-analytic solution, 187
    Black-Scholes equation, 51
semi-infinite half space, 58
semi-martingale, 321
Shannon wavelet, 188
shape density, 186, 227
shape implied vol., 136, 236, 486
shape parameter, 147
shape smile, 319
shifted distribution, 262
shifted implied volatility, 487
shifted lognormal, 468
shock size, 296
shocked Monte Carlo paths, 297
short-rate, 343, 347
    CIR, 356
    forward measure, 361
    HW, 360
    hybrid model, 407
    multi-factor, 419
short-rate dynamics, 347
short-rate model, 349
    affine, 354
short-rate process, 344
    FX, 495
short-selling, 57, 59
sigma-algebra, 10, 233
sigma-field, 10, 12, 13, 202
single asset, 187
single curve, 482, 483
skew and smile, 461
skewness parameter, 156
slope IR vol. smile, 520
SLV expectation, 327
SLV framework, 320, 326
SLV local volatility component, 326
SLV model, 319
SLV model simulation, 330
smoothness, 263, 265
SPD, symmetric positive definite matrix, 194

speed of mean reversion, 225, 231, 236, 357, 407, 414, 495, 497, 511
spine points, 375
spline interpolation, 114
spot-Libor measure, 455, 457
spread, 368, 483
square-root model, 472
squared Bessel process, 228, 229, 284, 463, 464
standard deviation, 31, 107, 129, 159, 177, 259, 506, 520
standard discretization scheme, 274
standard error, 259, 260
standard interpolation, 114
standard normal CDF, 413
standard normal distribution, 261, 267
standard normal PDF, 83, 168
standard normal variable, 23
standard squared Bessel process, 229
static portfolio, 56
stationarity properties, 142
stationary increments, 122, 140
stationary, independent increments, 148
stochastic continuity, 140
stochastic Euler scheme, 266
Stochastic Grid Bundling Method, SGBM, 304
stochastic integration, 14
stochastic interest rate, 445
stochastic process, 9
stochastic volatility, 434
stochastic volatility LMM, 461
stochastic volatility model, 223
stochastic-local volatility model, 319, 320
stock split, 157
stopped process, 323
stopping time, 322
Stratonovich, 17
    calculus, 17
stress testing, 48
strong convergence, 266, 269
sub-martingale, 41, 43, 465
subordinated process, 147
survival probability, 2, 434, 437, 483
SV model, 223
    introduction, 224
SV-LMM, 461, 474
swap, 371
    definition, 371

swap measure, 384
swap rate, 374, 384, 441, 470
swaption, 362, 383, 384, 420, 446
    expiry, 511
    volatility, 470
SWIFT, 188
symmetric covariance matrix, 441
symmetric jumps, 129, 138
symmetric positive semi-definite, 194
symmetric random variable, 135
symmetry parameter, 156

T_i-forward measure, 448
T_{i-1}-forward measure, 452
Tanaka-Meyer formula, 321
Taylor expansion, 29, 101, 134, 199, 268, 296, 422, 467
Taylor series, 523
Taylor-based discretization, 292
Taylor-based simulation, 276
telescopic sum, 22, 373
tenor, 369, 370, 481
tenor structure, 453–455, 479, 484, 510
terminal measure, 453
time lag, 212
time value of money, 28
time-change, 229, 231
time-changed Bessel process, 464
time-changed Brownian motion, 230
time-changed function, 464
time-dependent parameters, 472
time-dependent volatility, 39
tower property, 13, 18, 71, 72, 130, 133, 134, 381, 393, 395, 436, 509
tradable asset, 211, 344, 374
tradable ZCBs, 383
traded asset, 124, 202
truncated Euler scheme, 277, 278
truncated integral, 166

uncorrelated Brownian motion, 461
unique discount curve, 482
unique factorization, 194
unique fair value, 81
unique martingale measure, 144
unique measurable, 234
unique role, 227
unique solution, 229, 461, 464
unique solution SDE, 228
unique strong solution, 465

uniqueness, 17
unsecured Libor, 483

Vašiček model, 353, 355
Variance Gamma VG process, 146
    asset price, 148
    drift adjustment, 149
    experiments, 184
variance swap, 96
    Heston, 240
    payoff, 96
    pricing, 98, 240
    VIX, 100
vector-valued process, 218
VG process, 148
    characteristic function, 148
VIX index, 98, 100, 102, 240
vol-vol parameter, 225, 236
volatility surface, 241
volatility term structure, 413
volatility-of-variance parameter, 497
volatility-of-volatility parameter, 225, 236

Wald's equation, 131
wavelet, 188
weak convergence, 266, 267, 269
    order, 272
Wiener increment, 29, 34, 225, 261
Wiener process, 11, 12, 14, 27, 32, 62, 147, 260, 415
    correlated, 200
worst-case exposure, 389
wrong way risk, WWR, 395

yield, 375
    expected, 476
yield curve, 345, 348, 352, 375, 379, 420, 476
    configuration, 376
    construction, 375, 377
    spine factors, 377

zero-coupon bond, ZCB, 344–346, 368, 374, 381, 386, 388, 408, 446, 448, 453
    alternative derivation, 356
    definition, 341
    differentiation, 356
    discounted, 346
    domestic, 492, 499, 501
    dynamics, 348, 358, 360, 408, 496, 513
    foreign, 492
    formula, 354
    forward rate, 345, 348
    FX world, 495
    Hull-White model, 351
    IR product, 364
    negative IR, 478
    numéraire, 473, 474, 489
    price, 341, 357, 380
    put option, 382
    risk-free, 369
    SDE, 357
    unsecured, 484

Milton Keynes UK
Ingram Content Group UK Ltd.
UKHW050000300824
447615UK00002B/4